This book comprehensively reviews incubation effects on the embryonic development of birds and reptiles and presents the first-ever synthesis of data from these two vertebrate classes. The book covers three major areas. The first deals with the structure, shape and function of eggs, whilst the second examines the effects of the four main parameters on the process of incubation: temperature, water relations, respiratory gas exchange, and turning. The third deals with early embryonic development and the methods used to investigate and manipulate the embryo. Further chapters deal with aestivation, megapodes and oviparity.

International experts in each field have contributed to this extensively referenced volume and it will be of great interest not only to research biologists, but also to bird and reptile breeders, whether in commercial organisations or in zoos.

T0171792

Egg incubation: its effects on embryonic
development in birds and reptiles

Key to photograph of participants of 'Physical Influences on Embryonic Development in Birds and Reptiles'

Front row (left to right)

Sarah Herrick, Mike J. Thompson, Andre Ancel

Seated

Charlie Deeming, Mike Ewert, Hiroshi Tazawa, Ralph Ackerman, Kathy Packard, Mark Ferguson, Ruth Bellairs, Hermann Rahn

First row standing

Vince DeMarco, Ray Noble, Carol Vleck, Nick French, Mark Richards, Cor Zonneweld, Gary Packard, Rick Shine, Alison Cree, Louis Guillette, Charles Paganelli, Andrew Coughlan, Rocky Tuan, Keenan Smart

Second row standing

Nancy Clark, Bruce Dunn, Henry Wilson, Cynthia Carey, Ian Smart, Mike B. Thompson, Jenny Harry, Bill Bardsley, David Booth, Dave Price, Louis Visschedijk, Rob Gerrits, Claudio Stern

Back row standing

Elsiddig Babiker, Gary Robbins, Dinah Burgess, Grenham Ireland, Nat Bumstead, Michael Kern, Harold White, Paul O'Shea, Henri Girard, Robert Fitzsimmons, Roger Cattermole, John Lovett, Scott Turner, Rob Harvey, Nigel Jarret, John Phillips, Brent Palmer, Nick Sparks, Cecilia Lo, Ken Simkiss, Amos Ar

Egg incubation:
its effects on embryonic development in birds and reptiles

Edited by

D. CHARLES DEEMING
and
MARK W. J. FERGUSON

Department of Cell and Structural Biology
University of Manchester, UK

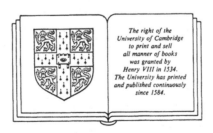

CAMBRIDGE UNIVERSITY PRESS

Cambridge
New York   Port Chester
Melbourne   Sydney

PUBLISHED BY THE PRESS SYNDICATE OF THE UNIVERSITY OF CAMBRIDGE
The Pitt Building, Trumpington Street, Cambridge, United Kingdom

CAMBRIDGE UNIVERSITY PRESS
The Edinburgh Building, Cambridge CB2 2RU, UK
40 West 20th Street, New York NY 10011–4211, USA
477 Williamstown Road, Port Melbourne, VIC 3207, Australia
Ruiz de Alarcón 13, 28014 Madrid, Spain
Dock House, The Waterfront, Cape Town 8001, South Africa

http://www.cambridge.org

First published 1991
First paperback edition 2004

A catalogue record for this book is available from the British Library

ISBN 0 521 39071 0 hardback
ISBN 0 521 61203 9 paperback

CONTENTS

CONTENTS

PREFACE

In September 1989, a meeting was held at Needham Hall, The University of Manchester, England, entitled 'Physical Influences on Embryonic Development in Birds and Reptiles'. The philosophy behind this meeting was to bring together scientists who specialised in the incubation and physiology of avian embryos with those workers interested in reptilian eggs and embryos. Leading authorities within many fields of incubation, oology and embryology were invited to give a presentation reviewing current knowledge within their sphere of research. The idea was to stimulate interest and collaboration between avian and reptilian researchers who may not have had any significant contact previously and to systematically compare and contrast events in birds and reptiles: a task never before attempted due in part to the paucity of reptilian data until the rapid expansion in knowledge of the past five years. Each speaker was also requested to produce a written review which forms the basis of this volume. This book is *not* a set of conference proceedings. Rather, it aims to be a comprehensive review of relevant reptilian and avian embryonic data: a text designed as a reference guide for the next few years.

The book aims to review incubation and embryology in birds and different kinds of reptiles and covers three major areas: the first deals with the chemical components and structure of eggs; the second examines the effects of the four main determinants of incubation, temperature, water relations, respiratory gas exchange and turning, upon embryonic development together with reviews of the evolutionary significance of incubation parameters. The final area deals with embryonic development and the possible methods used to investigate and manipulate the embryo. A review of the methods used in the generation of transgenic birds, originally presented at the meeting, has been withdrawn due to the ill-health of its author. Sadly, Professor Hermann Rahn, the leading international figure in avian egg physiology, died during the preparation of this book: we are privileged to publish one of his last seminal scientific contributions, authored before his death.

The chapters are extensively referenced either to the original literature or to other reviews so that this book might serve as a useful entrance into the field of avian and reptilian eggs and embryos. The book should be of value to a wide range of individuals: developmental biologists, zoologists, commercial incubation and breeding establishments, zoos, conservationists, wildlife managers, evolutionary biologists, herpetologists, physiologists, biochemists, the enlightened hobbyist and a variety of graduate students.

The meeting was supported by a number of organisations to whom we are very grateful: The Royal Society, London, The Nuffield Foundation, London, The University of Manchester Conference Support Fund, Brinsea Products Limited, A. B. Incubators Limited, Cherry Valley Farms Limited. We thank the staff of Needham Hall; Richard Neave for drawing the logo; and Tony Bentley for taking the group photograph.

D. Charles Deeming
Mark W. J. Ferguson

CONTRIBUTORS

RALPH A. ACKERMAN
Department of Zoology, Iowa State University, Ames, Iowa 50011, USA

ROBERT E. AKINS
Department of Orthopaedic Surgery, Thomas Jefferson University, Philadelphia, Pennsylvania 19107, USA

AMOS AR
Department of Zoology, Tel Aviv University, Ramat Aviv 69978, Israel

RUTH BELLAIRS
Department of Anatomy and Developmental Biology, University College London, Gower Street, London WC1E 6BT, UK

RONALD G. BOARD
School of Biological Sciences, University of Bath, Bath BA2 7AY, UK

DAVID T. BOOTH
Department of Organismal Biology and Anatomy, University of Chicago, 1025 East 57th Street, Chicago, Illinois 60637, USA

DENIS C. DEEMING
Department of Cell and Structural Biology, University of Manchester, Coupland III Building, Manchester M13 9PL, UK (Present address: Buckeye Company, Mill Lane, Lopen, South Petherton, Somerset TA13 5PL, UK)

VINCENT G. DeMARCO
Department of Zoology, University of Florida, Gainesville, Florida 32611, USA

CLAIRE J. DICKENS
The Department of Physiological Sciences, The Medical School, The University of Newcastle upon Tyne, Newcastle upon Tyne NE2 4HH, UK

BRUCE E. DUNN
Laboratory Service, John McClellan Memorial Veterans Hospital and Department of Pathology, University of Arkansas for Medical Sciences, Little Rock, Arkansas 72205, USA

MICHAEL A. EWERT
Department of Biology, Indiana University, Bloomington, Indiana 47405, USA

MARK W. J. FERGUSON
Department of Cell and Structural Biology, University of Manchester; Coupland III Building, Manchester M13 9PL, UK

JAMES I. GILLESPIE
The Department of Physiological Sciences, The Medical School, The University of Newcastle upon Tyne, Newcastle upon Tyne NE2 4HH, UK

JOHN R. GREENWELL
The Department of Physiological Sciences, The Medical School, The University of Newcastle upon Tyne, Newcastle upon Tyne NE2 4HH, UK

LOUIS J. GUILLETTE JR
Department of Zoology, University of Florida, Gainesville, Florida 32611, USA

DONALD F. HOYT
Department of Biological Sciences, California State University, Pomona, California 91768, USA

JOHN B. IVERSON
Department of Biology, Earlham College, Richmond, Indiana 47374, USA

MASAFUMI KOIDE
Department of Orthopaedic Surgery, Thomas Jefferson University, Philadelphia, Pennsylvania 19107, USA

CAROL MASTERS VLECK
Department of Ecology and Evolutionary Biology, University of Arizona, Tucson, Arizona 86721, USA

RAY C. NOBLE
Department of Nutrition and Microbiology, The West of Scotland College, Ayr KA6 5HW, Scotland, UK

TAMAO ONO
Department of Chemistry and Biological Chemistry, University of Essex, Wivenhoe Park, Colchester CO4 3SQ, UK

PAUL O'SHEA
Department of Chemistry and Biological Chemistry, University of Essex, Wivenhoe Park, Colchester CO4 3SQ, UK

GARY C. PACKARD
Department of Biology, Colorado State University, Fort Collins, Colorado 80523, USA

MARY J. PACKARD
Department of Biology, Colorado State University, Fort Collins, Colorado 80523, USA

CHARLES V. PAGANELLI
Department of Physiology, State University of New York, Buffalo, New York 14214, USA

BRENT D. PALMER
Department of Biological Sciences, Wichita State University, Wichita, Kansas 67208, USA

HERMANN RAHN (Deceased)
Department of Physiology, State University of New York, Buffalo, New York 14214, USA

ELSPETH RUSSELL
The Department of Physiological Sciences, The Medical School, The University of Newcastle upon Tyne, Newcastle upon Tyne NE2 4HH, UK

RICHARD SHINE
School of Biological Sciences, Zoology A08 The University of Sydney, New South Wales 2006, Australia

KENNETH SIMKISS
Department of Pure and Applied Zoology, University of Reading, PO Box 228, Reading RG6 2AJ, UK

IAN H. M. SMART
Department of Anatomy and Physiology, The University, Dundee DD1 4HN, Scotland, UK

NICK H. C. SPARKS
School of Chemistry, University of Bath, Bath BA2 7AY, UK
(Present address: Poultry Section, West of Scotland College of Agriculture, Auchincruive, Ayr KA6 5HW, Scotland, UK)

MICHAEL B. THOMPSON
School of Biological Sciences, Zoology A08, University of Sydney, New South Wales 2006, Australia

ROCKY S. TUAN
Department of Orthopaedic Surgery, Thomas Jefferson University, Philadelphia, Pennsylvania 19107, USA

J. SCOTT TURNER
Percy FitzPatrick Institute of African Ornithology, University of Cape Town, Rondebosch, South Africa
(Present address: Department of Environmental and Forest Biology, College of Environmental Science and Forestry, State University of New York, Syracuse, New York 13210, USA

DAVID VLECK
Department of Biology and Evolutionary Biology, University of Arizona, Tucson, Arizona 86721, USA

HAROLD B. WHITE III
Department of Chemistry and Biochemistry, University of Delaware, Newark, Delaware 19716, USA

Maternal diet, maternal proteins and egg quality

HAROLD B. WHITE, III

Introduction

> This egg type is, to us, a familiar and hence seemingly prosaic structure suitable for the breakfast table. It is, in reality, the most marvelous single 'invention' in the whole history of vertebrate life.
>
> Alfred S. Romer (1968, p. 183)

The relationship between the nutrient content of the maternal diet and egg production and hatchability has been studied for many nutrients in the domestic fowl (*Gallus gallus*) (Beer, 1969; Naber, 1979; Anon, 1979). The confounding influence of various strains, interacting nutrients, antibiotics, husbandry, and a host of other variables on this relationship continues to be the focus of research papers. There are thousands of birds and egg-laying reptiles, however, for which there is virtually no such information and on which there is little chance of doing the appropriate dietary studies. Some of these species are endangered, and our only hope for their continued survival may be through captive breeding programs where successful reproduction may depend upon manufactured and supplemented diets. To what extent can nutritional knowledge of a few domestic species be applied to the variety of species that exist? As yet, this question cannot be answered. However, this review will argue that a knowledge of the evolutionarily conserved processes that link maternal diet with egg quality, can provide a basis for evaluating the diverse nutritional needs of other species even when few individuals are available for study. In particular, there are a number of proteins whose function is to transport specific nutrients to the oocyte; these protein–nutrient complexes become yolk constituents for later use by the embryo. The amounts, and relative saturation, of these proteins may indicate the specific nutrient needs of the embryo and the nutrient status of the mother respectively. A review of these widely distributed yolk proteins will be presented in this chapter with particular focus on vitamin-binding proteins.

Nutritional perspective on the origin of the amniote egg

All terrestrial vertebrates are descended from a succession of organisms that were able, some 300 to 400 million years ago, to leave the aquatic environment that had bound their more distant ancestors for more than two billion years. Among the major anatomical and physiological changes permitting this lineage to succeed were the development of lungs to breathe air, transformation of appendages to support heavy bodies out of water, and adaptations to reduce water loss from the body surface. However, the evolution of an egg that could sustain an embryo away from aquatic habitats was the ultimate liberating adaptation (Baldwin, 1948; Romer, 1957). Both the terms *amniote* and *cleidoic* apply to the eggs of terrestrial vertebrates, but they are not synonyms. Many insect eggs, for example, are cleidoic but not amniote. *Amniote* refers to the membrane that envelops the embryo and contains the amniotic fluid that bathes the embryo within the egg. In effect, the amniotic membrane permits the embryo to develop in an aqueous environment protected from desiccation. *Cleidoic* refers to the shell that encloses the egg and isolates the embryo from the external environment.

In contrast to the shell-less eggs that preceded them, cleidoic eggs must contain all the nutrients required for embryonic growth and maintenance when they are laid. Therefore, in their evolution, mechanisms had to be developed to provision these eggs with nutrients that previously could have been scavenged from the aqueous environment during embryogenesis (Baldwin, 1948; Craik, 1986). In particular, trace nutrients such as vitamins and minerals must be deposited within the yolk during the vitellogenic stage of oogenesis. All oviparous reptiles, birds, and protherian mammals therefore produce shelled eggs containing large, nutrient-rich yolks. With the evolution of a placenta the necessity for a large yolky egg was lost. However, some of the systems used to deliver nutrients to the oocyte were retained and were adapted to deliver nutrients to the placenta (Adiga & Ramanamurty, 1982; White & Merrill, 1988).

While total provisioning of the egg represented a new dimension to maternal investment in evolution, it created the potential for eggs to be laid in other nutrient-poor environments, the ultimate of which is the terrestrial environment. Among terrestrial eggs the fowl egg has been extensively studied (Gilbert, 1971). It contains separate binding proteins for biotin (White, 1985; Bush & White, 1989), riboflavin (Rhodes, Bennett & Feeney, 1959; White & Merrill, 1988), cobalamin (Levine & Doscherholmen, 1983), thiamin (Muniyappa & Adiga, 1979), cholecalciferol (Fraser & Emtage, 1976), retinol (Heller, 1976), and iron (Williams *et al.*, 1982; Taborsky, 1980). These proteins are also present in the eggs of a turtle (*Chrysemys picta*), a snake (*Python molurus*) and an alligator (*Alligator mississippiensis*) (Abrams *et al.*, 1988, 1989) which suggests that all are present in the shelled eggs of terrestrial vertebrates (see later). These proteins have not been detected in the much smaller aquatic eggs of several species of bony fish (*Fundulus heteroclitis*, *Onchyonchus nerka* and *Cynscion regalis*) (H. B. White and associates, unpublished).

Vitellogenin, a protein satisfying many nutritional needs

In comparison with other metazoan cells, the mature oocyte is large, sometimes enormous. Its size is due to the accumulation of yolk, the mixture of maternal lipids and proteins which are used for growth and maintenance of the embryo to a stage where it can feed independently or be fed. None of the proteins in egg are derived directly from the maternal diet. Rather, dietary proteins are hydrolysed to free amino acids which are then precursors for the species-specific maternal proteins that become part of the yolk and albumen. Thus the amino acid resources available to the developing embryo are set by the female via the genetically determined amino acid composition and relative amounts of egg proteins. Therefore, within a species, eggs have a rather uniform protein content despite diverse protein sources in the maternal diet. Survival of the embryo depends upon a balance of essential amino acids and so natural selection should assure the production of high quality maternal proteins that satisfy these amino acid requirements for growth and maintenance of the embryo.

The bulk of yolk protein in both invertebrate and vertebrate egg yolk is derived from vitellogenins, a small family of large, extracellular, lipo-, glyco-, and sometimes phosphometalloproteins which are produced and secreted by the fat body in insects (Bownes, 1986), the intestine in nematodes (Spieth *et al.*, 1985), or the liver in vertebrates (Wahli *et al.*, 1981). A vitellogenin has not been reported from mammals (but see Baker, 1988). The protein is distributed by the blood or haemolymph to the ovaries where it is deposited via receptor-mediated endocytosis (Wallace, 1985; Griffin, Perry & Gilbert, 1984). In the oocytes of the frog, *Xenopus laevis*, more than 90% of the yolk proteins are products of the limited proteolysis of vitellogenin (Wallace *et al.*, 1983). At least one of the multiple genes for vitellogenin has been cloned and its nucleotide sequence determined from the nematode *Caenorhabditis elegans* (Spieth *et al.*, 1985), *X. laevis* (Wahli & Ryffel, 1985), and the fowl (van het Schip *et al.*, 1987). The genes are homologous to each other (Nardelli *et al.*, 1987a) and to the partial gene sequence from the sea urchin *Strongylocentrotus purpuratus* (Shyu, Blumenthal & Raff, 1987).

A comparison of the deduced amino acid

Table 1.1. *Amino acid composition of vitellogenins and apo-very low density lipoprotein (apo VLDL-II) from different species based on nucleotide and amino acid sequences.*

	Protein						
	Vitellogenin			apoVLDL-II			
	Caenorhabditis[a] *elegans*	*Xenopus*[b] *laevis*	*Gallus*[b] *gallus*	*Gallus*[c] *gallus*	*Meleagris*[d] *gallopavo*	*Dromaeus*[c] *novaehollandiae*	*Anas*[c] *platyrhynchos*
Amino acid							
Alanine	80	145	134	8	7	7	7
Arginine	98	90	109	6	7	5	6
Asparagine	73	86	76	4	3	4	2
Aspartate	58	69	81	4	4	4	4
Cysteine	20	31	36	1	0	0	0
Glutamine	100	105	76	3	4	3	3
Glutamate	172	115	102	5	5	5	5
Glycine	44	88	87	3	3	3	3
Histidine	33	53	47	0	0	0	0
Isoleucine	92	95	102	7	7	6	6
Leucine	141	139	145	9	7	7	9
Lysine	124	140	124	6	5	7	6
Methionine	35	47	46	1	2	3	2
Phenyl-alanine	70	65	51	2	3	4	3
Proline	63	75	88	2	2	3	3
Serine	118	200	258	4	4	3	3
Threonine	98	89	91	5	6	5	7
Tryptophan	14	13	18	1	2	2	2
Tyrosine	53	52	55	4	4	4	4
Valine	117	110	124	7	7	9	7
Total length	1603	1807	1850	82	82	84	82

[a]Spieth *et al.*, 1985. [b]Nardelli *et al.*, 1987*b*. (Corrected values provided by D. Nardelli and W. Wahli.) [c]Inglis & Burley, 1977. [d]Inglis *et al.*, 1979.

sequences of frog and fowl vitellogenins (Nardelli *et al.*, 1987*b*) reveals only a 40% identity. This considerable divergence among vertebrates implies there is the potential to evolve vitellogenins with distinctive amino acid compositions corresponding to distinctive nutritional needs of embryos of different species. A comparison of the deduced amino acid composition of three vitellogenins is presented in Table 1.1: vertebrate vitellogenins are over 200 amino acids longer than that of the nematode. The increase in size is accompanied by significant increases in the content of alanine, cysteine, glycine, histidine, and, in particular, serine. Only glutamic acid is significantly lower in fowl and frog vitellogenin than in nematode vitellogenin. While total amino acid composition of the yolk can be influenced by other yolk proteins, it is

nevertheless tempting to conclude, based on the similarity of two vertebrate vitellogenins, that the amino acid requirements of embryos of terrestrial vertebrates are quite similar. The maternal diet, provided it meets certain minimum standards for egg production, has little effect on the amino acid content of eggs.

In recent years there have been several studies on the yolk proteins of reptiles but few reports of amino acid composition of the proteins or yolk. Ho, L'Italien & Callard (1980) purified vitellogenin from the painted turtle (*C. picta*), and showed that its amino acid composition is quite similar to that of *X. laevis*. However, the amino acid composition of the total yolk, normalised to aspartic acid, was significantly different from vitellogenin, with methionine, isoleucine, lysine, histidine, and arginine being more than twofold higher in total yolk. This implies a significant contribution from other proteins with different compositions.

Not only does vitellogenin provide essential amino acids, it is also the primary source of phosphate for vertebrate embryos. Vertebrate vitellogenin has a disproportionately high serine content (Table 1.1). Virtually all of the excess serine is phosphorylated and occurs in a single segment of the protein coded by exon 23 (Byrne *et al.*, 1984). This region is highly variable among the different vitellogenin genes of a single species (Byrne *et al.*, 1989) and does not exist in the nematode gene (Spieth *et al.*, 1985). As a result, the limited proteolysis of vitellogenin that occurs after deposition in the oocyte produces several phosphoproteins, the phosvitins (Wallace & Morgan, 1986). These proteins can contain more than 100 phosphoryl groups, and phosphoserine can comprise more than 50% of the amino acid composition.

The large number of phosphoryl groups on vitellogenin provides yet another nutritional dimension to this protein. The highly charged surface provides binding sites for a variety of ions such as calcium (Guyer *et al.*, 1980) and iron (Taborsky, 1980). The specificity of these binding sites is not well characterised and so it is unclear whether the full spectrum of associated ions satisfies the trace mineral needs of the embryo. If not, there may be specific trace mineral-binding proteins in yolk. An uncharacterised protein fraction with bound zinc is observed in fowl egg yolk (C. Kittle & H. B. White, unpublished observations), but its possible identity with the lipovitellin–phosvitin is not excluded.

As it naturally occurs in maternal blood plasma, vitellogenin is composed of two homologous subunits, each more than 1600 amino acids long. The molecular weight of this dimer is close to 500,000 in vertebrates. In *X. laevis* a substantial portion of that mass (~12%) is not protein but lipid. There are approximately 100 molecules of associated phospholipid (70%) and neutral lipid (30%) per dimer (Ohlendorf *et al.*, 1977). The yolk lipovitellin-phosvitin complex, derived from vitellogenin by limited proteolysis, has been purified and crystallised. X-ray crystallography of the complex from yolk of the lamprey (*Ichthyomyzon unicuspis*) reveals two large discoidal cavities (68 nm^3) containing lipid. The cavities are well separated one per lipovitellin subunit (Raag *et al.*, 1988). ^{31}P Nuclear magnetic resonance analysis of the lipovitellin–phosvitin complex from *X. laevis* clearly distinguishes phosphoryl groups associated with phosphoserine and those associated with phospholipids (Banaszak & Seelig, 1982). The fact that there is a spectrum of lipids present, and that a variety of non-polar compounds such as biliverdin are present, suggests that vitellogenin is rather non-specific with respect to the lipids it escorts to the oocyte. Thus the lipid composition of egg yolk can vary and will be influenced by the maternal diet (Naber & Biggert, 1989). If essential fatty acids are missing from the diet, they will not be found in the egg. Since triglycerides and phospholipids are readily synthesised when the maternal diet is high in carbohydrate, the yolk lipid composition will normally be a mix of endogenous and exogenous lipids that have been processed in the liver.

Of all the maternal proteins to be discussed in this review, vitellogenin is clearly the most ancient and most important. It has evolved into a nearly complete, single-source food. It supplies lipid, phosphate, 20 amino acids, and various associated metal ions. Indirectly, by its sheer abundance, it serves as the primary precursor for other nutrients such as niacin (White, 1987), lipoic acid (Shih & Steinberger, 1981) and ascorbic acid (Romanoff & Romanoff, 1949). Because vitellogenin has such a central role in reproduction, the hormonal control of its synthesis (Tata *et al.*, 1987; Cochran & Deeley, 1988; Gordon *et al.*, 1988) and its receptor-mediated uptake by the oocyte (Woods & Roth, 1984; Stifani, George & Schneider, 1988) have been the objects of considerable study as models of gene control and protein targeting respectively. Considering that vitellogenin synthesis and metabolism form the core around which the yolk deposition and embryonic nutrition are built, it seems reasonable that vitellogenin synthesis might be directly sensitive to the availability of essential amino acids, phosphate, essential fatty acids, and certain vitamins. For example, fowl hens on riboflavin-deficient diets stop producing eggs (White & Whitehead, 1987). Such speculation needs to be explored as does the possibility that oestrogen production might be dependent upon these nutrients from the maternal diet. Either possibility could provide a mechanism that would link poor quality maternal diets with the lack of, or reduction in, egg production, a phenomenon well known in the fowl.

Very low density lipoproteins and yolk lipids

One-third of the weight of a yolk of the fowl egg is lipid (Romanoff & Romanoff, 1949; see Noble, Chapter 2), which provides for over 90% of the energy consumed by a fowl embryo (Romanoff, 1967; Visschedijk, 1968). A comparison of the energy density of yolk from a variety of bird eggs suggests that fowl eggs are about average for a property that shows little variation among birds (Ar et al., 1987). The use of lipid rather than carbohydrate or protein as an energy source is significant. Lipids generate about twice as much energy per gram as either carbohydrate or protein. Furthermore, lipids can be stored in a nearly anhydrous state, thereby conserving water and increasing the energy per unit weight. Oxidation of lipids also produces metabolic water: for every gram of lipid oxidised, about 1.07 g of water is produced (Baldwin, 1948). The nearly total dependence on lipid oxidation defines the β-oxidation of fatty acids as a major metabolic pathway in the embryo and, in turn, affects the trace nutrient requirements, e.g. riboflavin.

Among terrestrial vertebrates there are significant differences in the delivery of lipid to yolk. In fish and amphibians, oestrogen stimulates the hepatic synthesis of one major plasma lipoprotein, vitellogenin, which becomes the major source of both yolk lipid and protein (Wallace, 1985). In the fowl and other birds, two additional major plasma lipoproteins are produced by the liver for deposition in the oocyte (McIndoe, 1971). Both are associated with very low density lipoprotein (VLDL) complexes. Apolipoprotein B (apoB) is a normal plasma constituent whose synthesis in chicken is stimulated five- to sevenfold by oestrogen (Kirchgessner et al., 1987). It is an enormous protein with a molecular weight of over 500 KDa (Kirchgessner et al., 1987). The other much smaller protein, apoVLDL-II, is only 82 amino acids long, is very abundant, and is produced only during egg laying (Inglis et al., 1979). Together these two proteins make up 12.5% of the mass of VLDL. The remainder is lipid: triglycerides (66%), phospholipids (17%), and cholesterol and its esters (4.5%) (Nimpf, Radosavljevic & Schneider, 1989a).

VLDL contributes 60–70% of the solids of fowl egg yolk (MacKenzie & Martin, 1967; Burley et al., 1988). While VLDL is characteristic of bird eggs, it is not unique to birds as was once thought. In Crocodylus porosus crocodilian (the closest reptilian group to birds), 30% of the yolk is derived from VLDL (Burley et al., 1988). No mention of VLDL was made in reports of yolk proteins from a lizard (Anolis pulchellus) (DeMorales, Vallés & Baerga-Santini, 1987) and a turtle, C. picta (Ho et al., 1980). In contrast to the ancient and widely distributed vitellogenin, VLDL has been recruited recently for yolk deposition and its appearance indicates that it has a new role in avian and crocodilian eggs but as yet this is unclear.

ApoVLDL-II, because it is both small and abundant, may be an evolutionary target for adjusting the amino acid composition of the egg in birds. A single amino acid replacement in this protein would have an incremental effect on the overall amino acid composition of yolk 20 to 50 times that for vitellogenin or apoB. Comparison of the apoVLDL-II sequences from several birds (Table 1.1) shows the protein is evolving rapidly (Inglis et al., 1979); however, many of the replacements are compensatory.

The hormonal control of the synthesis of apoVLDL-II has been studied (Cochrane & Deeley, 1988; Bakker et al., 1988), but apparently regulation by amino acid or lipid has not been observed or considered. In the fowl it takes a little more than one hour after synthesis of the proteins, for the assembly of a VLDL particle in the endoplasmic reticulum and Golgi before secretion by a hepatocyte (Bamberger & Lane, 1988). ApoB in the plasma VLDL particle is recognised by an oocyte-specific receptor and deposited via receptor-mediated endocytosis (Hayashi, Nimpf & Schneider, 1989). Fowl hens with the sex-linked restricted ovulator mutation lack this oocyte receptor and as a consequence accumulate VLDL in their plasma rather than deposit it in yolk (Nimpf, Radosavljevic & Schneider, 1989b). Although apoB undergoes limited cathepsin-D-like proteolysis, the integrity of the VLDL is maintained and receptor recognition is retained (Nimpf et al., 1989a). It has been suggested that the fragmented apoB may continue to function by targeting VLDLs for uptake by embryonic cells. The function of apoVLDL-II is not known.

Assembly of VLDL is an intracellular process (Bamberger & Lane, 1988); therefore, the incorporated lipids are newly synthesised or newly reassembled lipids from the maternal diet. Except for insignificant lipid exchange, there should be little, if any, direct incorporation of dietary lipids into the VLDL complex. This is in contrast to the high-affinity binding proteins that probably scavenge their essential nutrients in the blood plasma rather than deplete intracellular supplies.

High-affinity binding proteins for vitamins

There seems to have been a trend toward more complex mixtures of proteins in the evolution of avian and reptilian eggs. For example, VLDL is present in avian eggs and some reptilian eggs, but not in fish or frog eggs. The phenomenon is also apparent in the proteins of albumen (see later). Proteins provide a means for an organism to gain genetic control over the nutrient content of its eggs. New proteins to transport vitamins may have presented the greatest evolutionary challenge to the total provisioning of a cleidoic egg. Unlike macronutrients over which the organism has control in the synthesis and deposition of vitellogenin, each vitamin represents a separate need that requires a distinct solution. In birds some, but not all, of these challenges have been accommodated with specific high-affinity binding proteins (White, 1987). Rather than duplicate that review, I focus here on some of the characteristics of maternal-egg vitamin-transport systems and discuss principles that might be used to assess the adequacy of maternal diets and the nutritional quality of eggs.

In order for a protein to bind a vitamin at the low concentrations normally found in the maternal diet and then to deliver it to the oocyte, several properties are essential. First, the protein must be specific for its vitamin and bind it with high affinity. Thus the vitamin will be scavenged effectively, and non-active compounds will not compete. Second, there needs to be a system to direct the protein–vitamin complex to the oocyte where it can be recognised and deposited. And third, the synthesis of the binding protein and its oocyte-transport system should be under oestrogen control. Such a combination of properties does not occur spontaneously. Thus the vitamin-binding proteins that are observed in the eggs of birds (White, 1985, 1987) and reptiles (Abrams *et al.*, 1988, 1989) must have been recruited, as the result of gene duplication and modification, from proteins that had some of these properties already. Most likely are membrane-associated vitamin-binding proteins that are responsible for the capture and cellular deposition of uncomplexed vitamins. Such systems must be associated with the oocytes of lower vertebrates and their embryos, although there is little direct evidence for their existence.

In contrast to the membrane-associated vitamin-binders, extracellular binding proteins deposited in yolk function as stoichiometric, rather than catalytic, carriers. Because vitamins are essential nutrients, an embryo's survival will depend upon having adequate resources in the egg (Fig. 1.1). In order to satisfy those needs, the maternal binding protein will be produced in

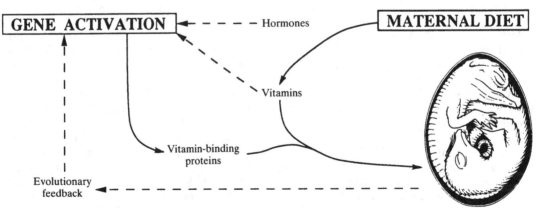

Fig. 1.1. Factors influencing the vitamin resources of an avian or reptilian embryo. Amounts and activity of maternal proteins responsible for the delivery of vitamins must satisfy the needs of the embryo in a cleidoic egg. The regulation of the amounts of these proteins is normally controlled by oestrogen but sometimes can be influenced by the availability of the vitamin. (The motif of an avian and a reptilian embryo in the same egg, by Richard Neave/D. C. Deeming, was used as the emblem for the Conference entitled 'Physical Influences on Embryonic Develoment in Birds and Reptiles' in Manchester, England.)

amounts that equal or exceed those needs without compromising maternal health and survival. This is the basis for equating stoichiometrically the amount of vitamin-binding protein deposited under optimal maternal diets with the optimal needs of the embryo and neonate. To the extent that the maternal diet does not satisfy the optimal needs, the amount of vitamin deposited in the egg will decrease. Depending on the regulatory processes involved the amount of binding protein may (White & Whitehead, 1987) or may not (White, Armstrong & Whitehead, 1986) decrease proportionately.

The best studied vitamin transport system in chickens is that for riboflavin (White & Merrill, 1988). A more detailed discussion of it will serve to illustrate the ideas and provide a framework for extending studies to other vitamins and other organisms. Riboflavin is the vitamin precursor of two coenzymes, flavin mononucleotide (FMN) and flavin adenine dinucleotide (FAD), that among other roles have important functions in the oxidation of fat. In particular, an examination of the β-oxidation pathway for fatty acids reveals the remarkable conjunction of three FAD-dependent enzymes (acyl CoA dehydrogenase, electron transferring flavoprotein [ETF], and ETF dehydrogenase) coupled in sequence to the first step of fatty acyl CoA oxidation. These reactions conduct electrons into the respiratory chain involved in mitochondrial oxidative phosphorylation. Although there are other important flavin-dependent enzymes in the complete oxidation of fats, it would seem that the initial step would be the most sensitive to riboflavin deficiency due to the multiplicative effects of FAD-depleted enzymes. In turn, the riboflavin requirements of the embryo would be determined in large measure by the high flux demands of the β-oxidation pathway.

The riboflavin content of a hen's diet influences the production, riboflavin content, and hatchability of eggs (Stamberg, Petersen & Lampman, 1946; Petersen, Lampman & Stamberg, 1947a,b). A high-affinity riboflavin-binding protein first discovered in egg albumen (Rhodes et al., 1959) and then in egg yolk (Ostrowski, Skarzynski & Zak, 1962) binds all the riboflavin in an egg (Blum, 1967). When the maternal diet was supplemented with excessive riboflavin, the excess was not deposited in the egg (Fig. 1.2; White et al., 1986). Rather, the amount deposited is limited by the amount of riboflavin-binding protein. This protein-dependent deposition of riboflavin is specifically demonstrated by a mutation in White Leghorn fowl that in a homozygous state results in embryonic death (Maw, 1954). This strain is unable to synthesise riboflavin-binding protein and consequently cannot deposit riboflavin in its eggs (Winter et al., 1967). The synthesis of the protein is under oestrogen control (DurgaKumari & Adiga, 1986). A specific oocyte receptor for riboflavin-binding protein has been sought but not detected (Benore-Parsons, 1986) causing alternative routes of entry to be considered. In particular, it has been proposed that the unusual phosphorylated region of this protein serves as a site of calcium-mediated association with vitellogenin (White & Merrill, 1988). Thus riboflavin-binding protein might gain entry to the oocyte by parasitising the pre-existing transport system for vitellogenin.

A graph of the riboflavin content of chicken egg yolk as a function of riboflavin in the maternal diet (Fig. 1.2) can be conceptually divided into a region where diet has an influence and a region where diet has little or no influence. In the latter region, riboflavin-binding protein is saturated with riboflavin. Thus the molar ratio of vitamin to its binding protein is 1 : 1, and it is the binding protein that limits the amount of vitamin that can be deposited. In the former region, there is insufficient riboflavin to saturate the binding protein, yet the protein continues to be produced and deposited. In this region the molar ratio of vitamin to protein falls below 1 : 1 in relation to the riboflavin content of the maternal diet.

In commercial fowl flocks where riboflavin requirement for egg production and hatchability are well known (Anon, 1979) and hatchlings are well fed, it is clear that the recommended requirements are below the transport capacity of the system and below the presumed evolutionary optimum. That even a 50% reduction in the riboflavin content of an egg would have only marginal effects on survival is not surprising because the sensitivity of flux in metabolic pathways to changes in enzyme activity is relatively small in most cases (Kacser & Burns, 1981). Eggs from hens heterozygous for the mutation in the gene for riboflavin-binding protein produce half as much protein as normal homozygotes, yet this is more than adequate for unimpaired hatching (E. G. Buss, personal communication). However, for species with unknown requirements, and where survival is clearly more

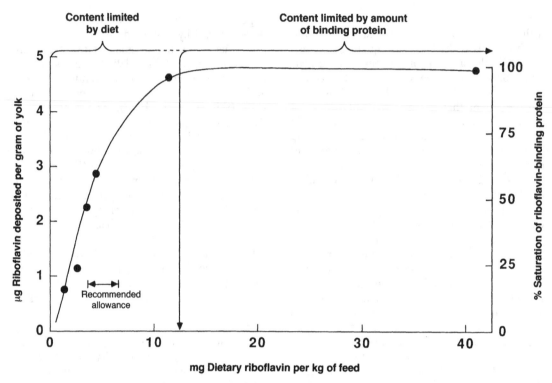

Fig. 1.2.   Riboflavin content of chicken egg yolk in response to dietary riboflavin (plotted from data in White *et al.*, 1987). Entry of riboflavin into the fowl hen oocyte is dependent on a specific maternal riboflavin-binding protein. The amount of this protein sets a stoichiometric limit on the amount of riboflavin that can be deposited in an oocyte. This amount is presumably set by natural selection and corresponds to the upper limit for the riboflavin needs of a fowl embryo and neonate. Below this limit, riboflavin deposition is related to the maternal diet. The recommended allowance for riboflavin does not saturate the transport system but supplies more than sufficient riboflavin for normal hatchability.

important than small differences in the economics of feed, it would seem prudent to supplement diets to saturate the transport system and maximise the resources of the embryo.

Riboflavin-binding protein has been purified from the yolk of alligator, python and turtle eggs (Abrams *et al.*, 1988, 1989); however, there have been no studies of the fractional saturation in relation to hatchability in any reptilian species. While this criterion may be useful in assessing maternal diets and embryonic requirements for riboflavin, other measures will be needed for other vitamins because the characteristics of each system vary (see later).

Biotin deposition in avian oocytes is mediated by a specific oestrogen-stimulated, biotin-binding protein (White, 1985; Bush & White, 1989). In both the fowl and the turkey (*Meleagris gallopavo*), a saturation phenomenon is observed in the egg, but at different levels related to the

amount of binding protein (White & Whitehead, 1987; White, Whitehead & Armstrong, 1987). The biotin-binding protein has been partially purified from alligator egg yolk (Abrams *et al.*, 1989), and it has been detected in turtle egg yolk (Korpela *et al.*, 1981). An analysis of eggs from a single clutch of the red-footed tortoise (*Geochelona carbonaria*) has revealed that there is little or no biotin and no biotin-binding proteins (H. B. White, unpublished observations). There are several possible explanations for this unexpected result. The most likely among them is that, with a few biotin-dependent enzymes and a long incubation period (~3 months), the tortoise embryo has very low requirements for biotin. A much less likely alternative is that the maternal diet of the captive animals was deficient in biotin. In poultry, the production of biotin-binding protein is dependent upon the availability of biotin. Thus a marginal deficiency in the maternal

Table 1.2 *Comparison of protein-mediated riboflavin and biotin deposition in the fowl egg.*

	Vitamin	
Characteristic	Riboflavin	Biotin
Distribution within the egg – albumen : yolk	60 : 40	10 : 90
Concentration ratio – plasma : yolk	1 : 6	1 : 25
Yolk binding protein		
a. product of same gene as albumen binding protein	yes	no
b. glycosylated	yes	no
c. phosphorylated	yes	no
d. acidic (pH $\leqslant$ 5)	yes	yes
Concentration of yolk binding protein dependent on		
a. estrogen	yes	yes
b. availability of vitamin	no	yes
Plasma and yolk binding protein always saturated	no	yes
Binding protein multimeric	no	yes
Deposition in yolk and albumen a simple function of total concentration of vitamin in plasma	yes	no
Egg production greatly reduced on deficient diet	yes	no

diet cannot be detected by the ratio of vitamin to binding protein like it can for riboflavin.

The differences in maternal response to dietary deficiency of different vitamins are well illustrated in the fowl for both biotin and riboflavin. Diets virtually devoid of available biotin have little or no effect on egg production (Cravens *et al.*, 1942). However, the eggs are deficient in biotin, and the embryos die in the first few days of incubation. By contrast, fowl hens fed riboflavin-deficient diets go out of production in about three weeks (White *et al.*, 1986). The mechanism by which riboflavin-deficiency terminates egg production is not known. Only limited information on vitamin metabolism in the fowl makes it evident that generalities about the details may be few. A comparison of riboflavin and biotin transport systems in the laying fowl hen are presented in Table 1.2 and serve to emphasise the many differences.

While specific vitamin-binding proteins in the maternal blood plasma mediate the deposition of some vitamins in yolk, there are other vitamins that must use other yet uncharacterised routes. For example, fowl and alligator egg yolk contain folate-binding activity (Abrams *et al.*, 1989). However, in the fowl at least, and probably for alligator as well, this protein accounts for less than 1% of the folate in the yolk. Plasma folate concentration in laying hens is much lower than the yolk concentrations and it is likely that a catalytic transporter for the uncomplexed folate exists in the oocyte membrane (T. A. Sherwood & H. B. White, unpublished observations). Despite these differences in the various vitamin-transport systems, saturation at high dietary levels is a general phenomenon (Whitehead, 1986). A knowledge of this maximal response can be used to assess problems with egg quality and maternal diet. Such an approach has been used for riboflavin (Squires, Naber & Waldman, 1988), vitamin B_{12} (Naber & Squires, 1989), and vitamin A (Squires & Naber, 1989) in the fowl.

Miscellaneous yolk proteins

The quality of an egg can be affected by maternal proteins that are not involved in nutrient delivery. Many normal plasma proteins are transferred to the oocyte (Williams, 1962). Among those proteins are immunoglobulins which confer temporary resistance in the neonate to infectious agents encountered by the mother (Brambell, 1970). This aspect of yolk deposition has been exploited in the vaccination of hens to protect offspring and as a method to conveniently produce antibodies. The fact that immunity is transferred implies that immunoglobulins are transferred without degradation to the embryonic blood (Kowalczyk *et al.*, 1985). This transfer of maternal proteins to the neonate also has been demonstrated for transferrin (Frelinger, 1971). It is not known if this type of transfer also occurs for the various vitamin-binding proteins in yolk and albumen, but it seems unlikely for yolk biotin-binding protein (White & Whitehead, 1987).

Maternal proteins and nutritional aspects of egg albumen

Egg albumen is secreted by the oestrogen-sensitised oviduct in response to progesterone (Edwards, Luttrell & Nir, 1976; Hyne *et al.*, 1979). It is a viscous, protein-rich layer that surrounds and protects the yolk and embryo (Li-Chan & Nakai, 1989). In contrast to yolk, which has a relatively uniform protein complement, the albumen is quantitatively, and qualitatively, variable between species (Wilson, Carlson & White, 1977). For example, the well characterised lysozyme that constitutes about 4% of the protein in albumen of the fowl egg is found in only two orders of birds while a different lysozyme is much more widespread (Prager, Wilson & Arnheim, 1974). Penalbumen, a major protein of albumen in penguin eggs, is not found in fowl eggs (Ho *et al.*, 1976).

While albumen proteins are consumed by the embryo (Romanoff, 1967; see Deeming, Chapter 19) and therefore contribute significantly to the amino acid resources of the embryo (Table 1.3), their primary function seems to be to protect the embryo from fungal and microbial attack (Silva & Buckley, 1962; Tranter & Board, 1982). In addition to lysozyme and a variety of protease inhibitors, there are several nutrient-binding proteins in albumen. The most prominent of these is the iron-binding protein ovotransferrin (Jeltsch & Chambon, 1982; Williams *et al.*, 1982) that constitutes more than 10% of the egg albumen of many birds (Feeney & Allison, 1969). Despite this large capacity to bind iron, there is little iron in albumen. By scavenging iron and being protected from proteolysis, transferrin prevents the growth of organisms, virtually all of which require iron.

The other nutrient-binding proteins in albumen are presumed to act in an analogous way; however, they do bind vitamins. Because many microorganisms do not require exogenous vitamins, the effectiveness of these vitamin-binding proteins as antimicrobial agents is less well documented than for ovotransferrin. They may have a more active role in embryonic nutrition by recovering for the embryo, vitamins that diffuse out of the yolk. Experiments on the eggs

from a single hen show that there is a progressive loss of biotin from the exterior of the yolk that is scavenged by avidin in the albumen adjacent to the yolk (Bush & White, 1989). This phenomenon has not been observed in a subsequent experiment (R. W. Schreiber & H. B. White, unpublished observations) but nevertheless, all of the vitamin-binding proteins in albumen are unsaturated and therefore have the potential to scavenge vitamins.

In cases where substantial amounts of a particular vitamin are protein-bound in the egg albumen when the egg is laid, it seems very likely that there is a nutritional role for the protein. For example, the fowl is unusual among birds in that more riboflavin is deposited in the albumen than in the yolk (Table 1.2). In the albumen of duck (*Anas platyrhynchos*) eggs, riboflavin-binding protein has no significant bound riboflavin but there is significantly more saturated binding protein in the yolk (Lotter *et al.*, 1982). Therefore, in the fowl, but not the Muscovy duck (*Carina moschata*), the adequacy of the maternal diet for riboflavin could be determined by analysis of the albumen.

A similar situation appears to occur with biotin in shag (*Stictocarbo p. punctatus*) eggs (White, 1985; C. R. Grau & H. B. White, unpublished observations). The avidin in these eggs is about 85% saturated with biotin. This observed variation among birds, and the known variability of albumen, make it impossible to generalise to reptiles where little work has been done. Albumen of alligator eggs, for example, appears to lack both avidin and riboflavin-binding protein (Abrams *et al.*, 1989), and thus the yolk must provide the total resources of these vitamins for the alligator embryo.

This review has attempted to combine evolutionary and biochemical perspectives in order to understand the relationship between maternal diet and egg quality in both birds and reptiles. The focus has been on maternal proteins that mediate the deposition of nutrients in the egg and are subject to natural selection through the survival of the embryo. While the mother has fairly good genetic control over the macronutrients she deposits in her eggs, she has less control over the micronutrients that must come from the diet. This is particularly true for captive animals whose diets are less varied than diets in the wild. An example may be found in captive alligators which grow faster and seem

Table 1.3. *Amino acid compositions of proteins of albumen of the fowl egg as determined from cloned genes and cDNA.*

	Albumen Protein						
	Ovalbumen[a]	Transferrin[b]	Ovomucoid[c]	Lysozyme[d]	Riboflavin binding protein[e]	Avidin[f]	Cystatin[g]
% of total protein	54	12	11	3.5	0.8	0.05	0.05
Amino acid							
Alanine	35	53	11	12	13	5	7
Arginine	15	29	6	11	5	8	8
Asparagine	17	32	14	13	8	10	4
Aspartate	14	46	16	8	11	5	6
Cysteine	6	30	18	8	18	2	4
Glutamine	15	22	1	3	11	3	7
Glutamate	33	45	13	2	24	7	9
Glycine	19	52	15	12	7	11	5
Histidine	7	12	4	1	8	1	1
Isoleucine	25	27	3	6	7	7	6
Leucine	32	50	11	8	13	7	10
Lysine	20	58	13	6	16	9	7
Methionine	16	11	2	2	7	2	2
Phenylalanine	20	26	5	3	6	7	3
Proline	14	28	7	2	8	2	5
Serine	38	47	12	10	30	9	12
Threonine	15	37	14	7	7	21	5
Tryptophan	3	11	0	6	6	4	1
Tyrosine	10	21	6	3	9	1	3
Valine	31	49	15	6	5	7	9
Total length	385	686	186	129	219	128	114

[a]McReynolds *et al.*, 1978. [b]Jeltsch & Chambon, 1982. [c]Catterall *et al.*, 1980. [d]Jung *et al.*, 1980. [e]Zheng *et al.*, 1988. [f]Gope *et al.*, 1987. [g]Schwabe *et al.*, 1984 (based on amino acid sequence).

healthier than their wild counterparts, yet have lower fertility and egg hatchability (Lance, Joanen & McNease, 1983; Lance, 1985). In this case, certain trace nutrients were considered to be limiting. In the absence of specific nutritional information, maternal diets of captive animals should be supplemented with trace nutrients to maximise the embryonic resources.

Acknowledgements

I thank the F. Hoffmann-LaRoche Co. for their support of recent unpublished work mentioned here. I also thank Drs E. G. Buss, C. R. Grau, W. J. Schneider, J. S. White and R. C. Noble for their comments on a draft of this manuscript.

References

Abrams, V. A. M., McGahan, T. J., Rohrer, J. S., Bero, A. S. & White, H. B., III (1988). Riboflavin-binding proteins from reptiles: A comparison with avian riboflavin-binding proteins. *Comparative Biochemistry and Physiology*, **90B**, 243–7.

Abrams, V. A. M., Bush, L., Kennedy, T., Sherwood, T. A. & White, H. B., III (1989). Vitamin transport proteins in alligator eggs. *Comparative Biochemistry and Physiology*, **93B**, 241–7.

Adiga, P. R. & Ramanamurty, C. V. (1982). Vitamin carrier proteins in birds and mammals. In *Molecular Biology of Egg Maturation*, ed. R. Porter & J. Whelan, pp. 111–31. London: Pitman.

Anonymous (1979). *Rationale for Roche Recommended Vitamin Fortification, Poultry.* Nutley: Hoffmann-La Roche Inc.

Ar, A., Arieli, B., Belinsky, A. & Yom-Tov, Y. (1987). Energy in avian eggs and hatchlings: Utilization and transfer. *Journal of Experimental Zoology*, Supplement, 1, 151–64.

Baker, M. E. (1988). Is vitellogenin an ancestor of apolipoprotein B-100 of human low density lipoprotein and human lipoprotein lipase? *Biochemical Journal*, **255**, 1057–60.

Bakker, O., Arnberg, A. C., Noteborn, M. H. M.,

Winter, A. J. & AB, G. (1988). Turnover products of the apo very low density lipoprotein II messenger RNA from chicken liver. *Nucleic Acids Research*, 16, 10109–18.

Baldwin, E. (1948). *An Introduction to Comparative Biochemistry*, 3rd edn., Cambridge: Cambridge University Press.

Bamberger, M. J. & Lane, M. D. (1988). Assembly of very low density lipoprotein in the hepatocyte-differential transport of apoproteins through the secretory pathway. *Journal of Biological Chemistry*, 263, 11868–78.

Banaszak, L. J. & Seelig, J. (1982). Lipid domains in the crystalline lipovitellin/phosvitin complex: A phosphorus-31 and deuterium nuclear magnetic resonance study. *Biochemistry*, 21, 2436–43.

Beer, A. E. (1969). A review of the effects of nutritional deficiencies on hatchability. In *The Fertility and Hatchability of the Hen's Egg*, ed. T. C. Carter & B. M. Freeman, pp. 93–108. Edinburgh: Oliver & Boyd.

Benore-Parsons, M. (1986). *The Transport of Riboflavin-binding Protein to the Hen Oocyte*. Unpublished Ph.D. Thesis, University of Delaware.

Blum, J.-C. (1967). *Le Métabolisme de la Riboflavine chez la Poule pondeuse*. Paris: F. Hoffmann-La Roche et Cie.

Bownes, M. (1986). Expression of the genes coding for vitellogenin (yolk protein). *Annual Review of Entomology*, 31, 507–31.

Brambell, F. W. R. (1970). *The Transmission of Passive Immunity from Mother to Young*. Amsterdam: North Holland Publishing Co.

Burley, R. W., Back, J. F., Wellington, J. E. & Grigg, G. C. (1988). Proteins and lipoproteins in yolk from eggs of the estuarine crocodile (*Crocodylus porosus*); a comparison with egg yolk of the hen (*Gallus domesticus*). *Comparative Biochemistry and Physiology*, 91B, 39–44.

Bush, L. & White, H. B., III (1989). Conversion of domains into subunits in the processing of egg yolk biotin-binding protein. *Journal of Biological Chemistry*, 264, 5741–5.

Byrne, B. M., van het Schip, A. D., van de Klundert, J. A. M., Arnberg, A. C., Gruber, M. & AB, G. (1984). Amino acid sequence of phosvitin derived from the nucleotide sequence of part of the chicken vitellogenin gene. *Biochemistry*, 23, 4275–9.

Byrne, B. M., de Jong, H., Fouchier, A. M., Williams, D. L., Gruber, M. & AB, G. (1989). Rudimentary phosvitin domain in a minor chicken vitellogenin gene. *Biochemistry*, 28, 2572–77.

Catterall, J. F., Stein, J. P., Kristo, P., Means, A. R. & O'Malley, B. W. (1980). Primary sequence of ovomucoid messenger RNA as determined from cloned complementary DNA. *Journal of Cell Biology*, 87, 480–7.

Cochrane, A. W. & Deeley, R. G. (1988). Estrogen-dependent activation of the avian very low density apolipoprotein II and the vitellogenin genes. *Journal of Molecular Biology*, 203, 555–67.

Craik, J. C. A. (1986). Levels of calcium and iron in the ovaries and eggs of cod, *Gadus morhua* L. and plaice, *Pleuronectes platessa* L. *Comparative Biochemistry and Physiology*, 83A, 515–17.

Cravens, W. W., Sebesta, E. E., Halpin, J. G. & Hart, E. B. (1942). Effect of biotin on reproduction in the domestic fowl. *Proceedings of the Society for Experimental Biology and Medicine*, 50, 101–4.

DeMorales, M. H., Vallés, A. M. & Baerga-Santini, C. (1987). Studies of the egg proteins of tropical lizards: Purification and partial characterization of yolk proteins of *Anolis pulchellus*. *Comparative Biochemistry and Physiology*, 87B, 125–36.

DurgaKumari, B. & Adiga, P. R. (1986). Hormonal induction of riboflavin carrier protein in the chicken oviduct and liver: A comparison of kinetics and modulation. *Molecular and Cellular Endocrinology*, 44, 285–92.

Edwards, N. A., Luttrell, V. & Nir, I. (1976). The secretion and synthesis of albumen by the magnum of the domestic fowl (*Gallus domesticus*). *Comparative Biochemistry and Physiology*, 53B, 183–6.

Feeney, R. E. & Allison, R. G. (1969). *Evolutionary Biochemistry of Proteins*. New York: Wiley-Interscience.

Fraser, D. R. & Emtage, J. S. (1976). Vitamin D in the avian egg – Its molecular identity and mechanism of incorporation into yolk. *Biochemical Journal*, 160, 671–82.

Frelinger, J. A. (1971). Maternally derived transferrin in pigeon squabs. *Science*, 171, 1260–1.

Gilbert, A. B. (1971). The female reproductive effort. In *Physiology and Biochemistry of the Domestic Fowl*, vol. 3, ed. D. J. Bell & B. M. Freeman, pp. 1153–62. London: Academic Press.

Gope, M. L., Keinänen, R. A., Kristo, P. A., Conneely, O. M., Beattie, W. G., Zarucki-Schulz, T., O'Malley, B. W. & Kulomaa, M. S. (1987). Molecular cloning of the chicken avidin cDNA. *Nucleic Acids Research*, 15, 3595–606.

Gordon, D. A., Shelness, G. S., Nicosia, M. & Williams, D. L. (1988). Estrogen-induced destabilization of yolk precursor protein mRNAs in avian liver. *Journal of Biological Chemistry*, 263, 2625–31.

Griffin, H. D., Perry, M. M., Gilbert, A. B. (1984). Yolk formation. In *Physiology and Biochemistry of the Domestic Fowl*, vol. 5, ed. D. J. Bell & B. M. Freeman, pp. 345–80. London: Academic Press.

Guyer, R. B., Grunder, A. A., Buss, E. G. & Clagett, C. O. (1980). Calcium-binding proteins in serum of chickens: Vitellogenin and albumin. *Poultry Science*, 59, 874–9.

Hayashi, K., Nimpf, J. & Schneider, W. J. (1989). Chicken oocytes and fibroblasts express different apolipoprotein-B-specific receptors. *Journal of Biological Chemistry*, **264**, 3131–9.

Heller, J. (1976). Purification and evidence for the identity of chicken plasma and egg retinol–retinol binding protein–prealbumen complex. *Developmental Biology*, **51**, 1–9.

Ho, C. Y.-K., Prager, E. M., Wilson, A. C., Osuga, D. T. & Feeney, R. E. (1976). Penguin evolution: Protein comparisons demonstrate relationship to flying aquatic birds. *Journal of Molecular Evolution*, **8**, 271–82.

Ho, S.-M., L'Italien, J. & Callard, I. P. (1980). Studies on reptilian yolk: *Chrysemys* vitellogenin and phosvitin. *Comparative Biochemistry and Physiology*, **65B**, 139–44.

Hyne, N. E., Groner, B., Sippel, A. E., Jeep, S., Wurtz, T., Nguyen-Huu, M. C., Giesecke, K. & Schütz, G. (1979). Control of cellular content of egg white protein specific RNA during estrogen administration and withdrawal. *Biochemistry*, **18**, 616–24.

Inglis, A. S. & Burley, R. W. (1977). Determination of the amino acid sequence of apovitellenin I from duck's egg yolk using an improved sequenator procedure: Comparison with other avian species. *FEBS Letters*, **73**, 33–7.

Inglis, A. S., Striki, P. M., Osborne, W. C. & Burley, R. W. (1979). Sequenator determination of the amino acid sequence of apovitellenin I from turkey's egg yolk. *FEBS Letters*, **97**, 179–82.

Jeltsch, J.-M. & Chambon, P. (1982). The complete nucleotide sequence of the chicken ovotransferrin mRNA. *European Journal of Biochemistry*, **122**, 291–5.

Jung, A., Sippel, A. E., Grez, M. & Schütz, G. (1980). Exons encode functional and structural units of chicken lysozyme. *Proceedings of the National Academy of Sciences USA*, **77**, 5759–63.

Kacser, H. & Burns, J. A. (1981). The molecular basis of dominance. *Genetics*, **97**, 639–66.

Kirchgessner, T. G., Heinzmann, C., Svenson, K. L., Gordon, D. A., Nicosia, M., Lebherz, H. G., Lusis, A. J. & Williams, D. L. (1987). Regulation of chicken apolipoprotein B: Cloning, tissue distribution, and estrogen induction of mRNA. *Gene*, **59**, 241–51.

Korpela, J. K., Kulomaa, M. S., Elo, H. A. & Tuohimaa, P. J. (1981). Biotin-binding proteins in eggs of oviparous vertebrates. *Experientia*, **37**, 1065–6.

Kowalczyk, K., Daiss, J., Halpern, J. & Roth, T. F. (1985). Quantitation of maternal–fetal IgG transport in the chicken. *Immunology*, **54**, 755–62.

Lance, V. A. (1985). Hormonal control of reproduction in alligators. In *Wildlife Management: Crocodiles and Alligators*, ed. G. J. W. Webb, S. C. Manolis & P. J. Whitehead. pp. 409–15. Sydney: Surrey Beatty and Sons.

Lance, V., Joanen, T. & McNease, L. (1983).

Selenium, vitamin E, and trace elements in the plasma of wild and farm-reared alligators during the reproductive cycle. *Canadian Journal of Zoology*, **61**, 1744–51.

Levine, A. S. & Doscherholmen, A. (1983). Vitamin B$_{12}$ bioavailability from egg yolk and egg white: Relationship to binding proteins. *American Journal of Clinical Nutrition*, **38**, 436–9.

Li-Chan, E. & Nakai, S. (1989). Biochemical basis for the properties of egg white. *Critical Reviews in Poultry Science*, **2**, 21–58.

Lotter, S. E., Miller, M. S., Bruch, R. C. & White, H. B., III (1982). Competitive binding assays for riboflavin and riboflavin-binding protein. *Analytical Biochemistry*, **125**, 110–17.

MacKenzie, S. L. & Martin, W. G. (1967). The macromolecular composition of hen's egg yolk at successive stages of maturation. *Canadian Journal of Biochemistry*, **45**, 591–601.

Maw, A. J. G. (1954). Inherited riboflavin deficiency in chicken eggs. *Poultry Science*, **33**, 216–17.

McIndoe, W. M. (1971). Yolk synthesis. In *Physiology and Biochemistry of the Domestic Fowl*. vol. 3, ed. D. J. Bell & B. H. Freeman, pp. 1208–23. London: Academic Press.

McReynolds, L., O'Malley, B. W., Nisbet, A. D., Fothergill, J. E., Givol, D., Fields, S., Robertson, M. & Brownlee, G. G. (1978). Sequence of chicken ovalbumin mRNA. *Nature, London*, **273**, 723–8.

Muniyappa, K. & Adiga, P. R. (1979). Isolation and characterization of thiamin-binding protein from chicken egg white. *Biochemical Journal*, **177**, 887–94.

Naber, E. C. (1979). The effect of nutrition on the composition of eggs. *Poultry Science*, **58**, 518–28.

Naber, E. C. & Biggert, M. D. (1989). Patterns of lipogenesis in laying hens fed a high fat diet containing safflower oil. *Journal of Nutrition*, **119**, 690–5.

Naber, E. C. & Squires, M. W. (1989). Vitamin profiles of eggs as indicators of nutritional status in the laying hen: III. Vitamin B$_{12}$ study. *Poultry Science*, **68**, 103 (Supplement 1).

Nardelli, D., Gerber-Huber, S., van het Schip, F. D., Gruber, M., AB, G. & Wahli, W. (1987a). Vertebrate and nematode genes coding for yolk proteins are derived from a common ancestor. *Biochemistry*, **26**, 6397–402.

Nardelli, D., van het Schip, F. D., Gerber-Huber, S., Haefliger, J.-A., Gruber, M., AB, G. & Wahli, W. (1987b). Comparison of the organization and fine structure of a chicken and *Xenopus laevis* vitellogenin gene. *Journal of Biological Chemistry*, **262**, 15377–85.

Nimpf, J., Radosavljevic, M. & Schneider, W. J. (1989a). Specific postendocytic proteolysis of apolipoprotein B in oocytes does not abolish receptor recognition. *Proceedings of the National Academy of Sciences, USA*, **86**, 906–10.

Nimpf, J., Radosavljevic, M. J. & Schneider, W. J. (1989*b*). Oocytes from the mutant restricted ovulator hen lack receptor for very low density lipoprotein. *Journal of Biological Chemistry*, **264**, 1393–8.

Ohlendorf, D. H., Barbarash, G. R., Trout, A., Kent, C. & Banaszak, L. J. (1977). Lipid and polypeptide components of the crystalline yolk system from *Xenopus laevis*. *Journal of Biological Chemistry*, **252**, 7992–8001.

Ostrowski, W., Skarzynski, B. & Zak, Z. (1962). Isolation and properties of flavoprotein from the egg yolk. *Biochimica et Biophysica Acta*, **59**, 515–19.

Petersen, C. F., Lampman, C. E. & Stamberg, O. E. (1947*a*). Effect of riboflavin intake on egg production and riboflavin content of eggs. *Poultry Science*, **26**, 180–6.

— (1947*b*). Effect of riboflavin intake on hatchability of eggs from battery confined hens. *Poultry Science*, **26**, 187–91.

Prager, E. M., Wilson, A. C. & Arnheim, N. (1974). Widespread distribution of lysozyme g in egg white of birds. *Journal of Biological Chemistry*, **249**, 7295–7.

Raag, R., Appelt, K., Xuong, N.-H. & Banaszak, L. (1988). Structure of the lamprey yolk lipid–protein complex lipovitellin–phosvitin at 2.8 Å resolution. *Journal of Molecular Biology*, **200**, 553–69.

Rhodes, M. B., Bennett, N. & Feeney, R. E. (1959). The flavoprotein–apoprotein system of egg white. *Journal of Biological Chemistry*, **234**, 2054–60.

Romanoff, A. L. (1967). *Biochemistry of the Avian Embryo*. New York: John Wiley.

Romanoff, A. L. & Romanoff, A. J. (1949). *The Avian Egg*. New York: Academic Press.

Romer, A. S. (1957). Origin of the amniote egg. *The Scientific Monthly*, **85**, 57–63.

— (1968). *The Procession of Life*. Cleveland and New York: The World Publishing Co.

Schwabe, C., Anastasi, A., Crow, H., McDonald, J. K. & Barrett, A. J. (1984). Cystatin, amino acid sequence and possible secondary structure. *Biochemical Journal*, **217**, 813–17.

Shih, J. C. H. & Steinberger, S. C. (1981). Determination of lipoic acid in chick livers and chicken eggs during incubation. *Analytical Biochemistry*, **116**, 65–8.

Shyu, A.-B., Blumenthal, T. & Raff, R. A. (1987). A single gene encoding vitellogenin in the sea urchin *Strongylocentrotus purpuratus*: Sequence at the 5′ end. *Nucleic Acids Research*, **15**, 10405–17.

Silva, M. & Buckley, H. R. (1962). Activity of egg white against fungi pathogenic to man. In *Fungi and Fungus Diseases*, ed. G. Dalldorf. pp. 277–91. Springfield: Charles C. Thomas, Publishers.

Spieth, J., Denison, K., Zucker, E. & Blumenthal, T. (1985). The nucleotide sequence of a nematode

vitellogenin gene. *Nucleic Acids Research*, **13**, 7129–38.

Squires, M. W. & Naber, E. C. (1989). Vitamin profiles as indicators of nutritional status in the laying hen: II. Vitamin A study. *Poultry Science*, **68**, 140 (Supplement 1).

Squires, M. W., Naber, E. C. & Waldman, R. (1988). Vitamin profiles in eggs as indicators of nutritional status in the laying hen. I. Riboflavin Study. *Poultry Science*, **67**, 160 (Supplement 1).

Stamberg, O. E., Peterson, C. F. & Lampman, C. E. (1946). Effect of riboflavin intake on the content of egg whites and yolks from individual hens. *Poultry Science*, **25**, 320–6.

Stifani, S., George, R. & Schneider, W. J. (1988). Solubilization and characterization of the chicken oocyte receptor. *Biochemical Journal*, **250**, 467–75.

Taborsky, G. (1980). Iron binding by phosvitin and its conformational consequences. *Journal of Biological Chemistry*, **255**, 2976–85.

Tata, J. R., Ng, W. C., Perlman, A. J. & Wolffe, A. P. (1987). Activation and regulation of the vitellogenin gene family. In *Gene Regulation by Steroid Hormones*, vol. 3, ed. A. K. Roy & J. H. Clark, pp. 205–33. New York: Springer-Verlag.

Tranter, H. S. & Board, R. G. (1982). The antimicrobial defense of avian eggs: Biological perspective and chemical basis. *Journal of Applied Biochemistry*, **4**, 295–338.

van het Schip, F. D., Samallo, J., Broos, J., Ophuis, J., Mojet, M., Gruber, M. & AB, G. (1987). Nucleotide sequence of a chicken vitellogenin gene and derived amino acid sequence of the encoded yolk precursor protein. *Journal of Molecular Biology*, **196**, 245–60.

Visschedijk, A. H. J. (1968). Air space and embryonic respiration. I. Pattern of gaseous exchange in fertile egg during closing stages of incubation. *British Poultry Science*, **9**, 173–84.

Wahli, W., Dawid, I. B., Ryffel, G. U. & Weber, R. (1981). Vitellogenesis and the vitellogenin gene family. *Science*, **212**, 298–304.

Wahli, W. & Ryffel, G. (1985). *Xenopus* vitellogenin genes. In *Oxford Surveys on Eukaryotic Genes*, ed. N. Maclean, pp. 96–120. Oxford: Oxford University Press.

Wallace, R. A. (1985). Vitellogenesis and oocyte growth in nonmammalian vertebrates. In *Developmental Biology*, Vol. 1, ed. L. W. Browder, pp. 127–77. New York: Plenum Publishing Corp.

Wallace, R. A. & Morgan, J. P. (1986). Chromatographic resolution of chicken phosvitin-multiple macromolecular species in a classic vitellogenin-derived phosphoprotein. *Biochemical Journal*, **240**, 871–8.

Wallace, R. A., Opresko, L., Wiley, H. S. & Selman, K. (1983). The oocyte as an endocytic cell. In *Molecular Biology of Egg Maturation*, ed. R. Porter &

J. Whelan, pp. 228–48. London: Pitman Publishers.

White, H. B., III (1985). Biotin-binding proteins and biotin transport to oocytes. *Proceedings of the New York Academy of Science*, **447**, 202–11.

— (1987). Vitamin-binding proteins in the nutrition of the avian embryo. *Journal of Experimental Zoology*, Supplement 1, 53–63.

White, H. B., III, Armstrong, J. & Whitehead, C. C. (1986). Riboflavin-binding protein-concentration and fractional saturation in chicken eggs as a function of dietary riboflavin. *Biochemical Journal*, **238**, 671–5.

White, H. B., III & Merrill, A. H., Jr. (1988). Riboflavin-binding proteins. *Annual Review of Nutrition*, **8**, 279–99.

White, H. B., III & Whitehead, C. C. (1987). Role of avidin and other biotin-binding proteins in the deposition and distribution of biotin in chicken eggs: Discovery of a new biotin-binding protein. *Biochemical Journal*, **241**, 677–84.

White, H. B., III, Whitehead, C. C. & Armstrong, J. (1987). Relationship of biotin deposition in turkey eggs to dietary biotin and biotin-binding proteins. *Poultry Science*, **66**, 1236–41.

Whitehead, C. C. (1986). Requirements for vitamins.

In *Nutrient Requirements of Poultry and Nutritional Research*, ed. C. Fisher & K. N. Boorman. pp. 173–89. Kent: Butterworth & Co.

Williams, J. (1962). Serum proteins and the livetins of hen's-egg yolk. *Biochemical Journal*, **83**, 346–55.

Williams, J., Elleman, T. C., Kingston, I. B., Wilkins, A. G. & Kuhn, K. A. (1982). The primary structure of hen ovotransferrin. *European Journal of Biochemistry*, **122**, 297–303.

Wilson, A. C., Carlson, S. S. & White, T. J. (1977). Biochemical evolution. *Annual Review of Biochemistry*, **46**, 573–639.

Winter, W. P., Buss, E. G., Clagett, C. O. & Boucher, R. V. (1967). The nature of the biochemical lesion in avian renal riboflavinuria-II. The inherited change of a riboflavin-binding protein from blood and eggs. *Comparative Biochemistry and Physiology*, **22**, 897–906.

Woods, J. W. & Roth, T. F. (1984). A specific subunit of vitellogenin that mediates receptor binding. *Biochemistry*, **23**, 5774–80.

Zheng, D. B., Lim, H. M. Pène, J. J. & White, H. B., III (1988). Chicken riboflavin-binding protein-cDNA sequence and homology with milk folate-binding protein. *Journal of Biological Chemistry*, **263**, 11126–9.

2

Comparative composition and utilisation of yolk lipid by embryonic birds and reptiles

RAY C. NOBLE

Introduction

The presence of an extensive yolk mass to supply the developing embryo with a large proportion of its nutrients is a feature common to all birds and reptiles. The yolk size, the proportional distribution of the major nutrients within the yolk and the rates of utilisation of the components during embryonic development differ widely between species. However, the major feature of all yolks is a high initial lipid content and rapid lipid utilisation during the later stages of embryo development when growth is maximal (Romanoff, 1960; Noble & Moore, 1964; Manolis, Webb & Dempsey, 1987). The yolk lipid performs a role both as the major energy source and as a supply of nutritionally essential tissue components (Romanoff, 1960). Utilisation of yolk lipid, therefore, constitutes a major part of the interrelated chain of events required for successful hatching and is characterised by extensive and, in many instances, distinctive metabolic features that are quite unique (Noble, 1987a; Noble et al., 1990a). Much is known concerning the utilisation of the yolk lipid during embryonic development in birds, especially the fowl (Gallus gallus) but similar precise and extensive data for lipids and lipid utilisation in reptiles are largely unavailable. However, the importance of understanding yolk lipid uptake in reptiles is of increasing importance, especially in view of the increasing need for captive breeding for conservation and commercial purposes (Ferguson, 1985). This chapter reviews some of the major features of yolk lipid uptake within the two classes, dwelling on some of the unique aspects of the metabolism that have evolved to accommodate to the lipid-rich environment.

The nature of lipids

The term lipid is used to cover a wide range of complex and heterogeneous substances that have an insolubility in water but solubility in organic solvents. It includes long chain hydrocarbons, fatty acids and their derivatives or metabolites, alcohols, aldehydes, sterols, terpenes, carotenoids and bile acids. The principal classes consist of fatty acid moieties linked by an ester bond to an alcohol, in the main, glycerol, or by amide bonds to long chain bases. They may also contain other moieties, e.g. phosphoric acid, organic bases or sugars. They can be divided into two major groups, simple lipids (or neutral lipids) containing one or two of the above hydrolysis products per molecule, e.g. tri- , di- and monoglycerides, free and esterified cholesterol, free fatty acids, and complex lipids (or polar lipids) containing three or more of the hydrolysis products, e.g. phosphoglycerides, sphingolipids. The apolar nature of lipids requires some form of soluble complex for their distribution and transport in the aqueous environment of the plasma and elsewhere. This is achieved through lipid–protein interactions to form complexes known as lipoproteins which, in turn, may be divided, by suitable analytical techniques, into distinct classes of families possessing metabolic and compositional distinctions. For further information on lipids and their divisions see Christie (1982) and Christie & Noble (1984).

Yolk lipid and the parent

Although the proportion of total egg weight that is associated with the yolk may differ considerably with the species, e.g. poultry 30%, crocodilians 40–46%, turtles up to 55% (Romanoff & Romanoff, 1949; Ewert, 1979; Manolis *et al.*, 1987), in all instances so far studied almost the entire lipid content of the egg is to be found within the yolk; negligible levels of lipid occur within the cuticle of the egg shell. Lipid deposition is, therefore, confined to ovum maturation with no contribution from subsequent elaborative processes (Gilbert, 1971*b*; Astheimer, Manolis & Grau, 1989). Maturation of the ovum occurs according to a species-specific sequence and is associated with a series of unique hepatic and circulatory lipid changes. These are most prominently displayed by the domestic fowl where the provision of lipid for an artificially extended period of egg laying is accommodated by the hepatic synthesis, turnover and subsequent plasma transfer of an enormous quantity of lipid (Gilbert, 1971*a*). As a result, extensive qualitative and quantitative lipid changes are discernible in the liver and the plasma with the onset of egg-laying (Husbands & Brown, 1965; Chung, Ning & Tsao, 1966; Yu, Campbell & Marquardt, 1976) to which the lipid composition of the egg yolk can be directly related. The development of the avian and reptilian ovum can be divided into several distinct phases, with different lipid transfer patterns being involved in each of the sequences (Bellairs, 1967; Astheimer *et al.*, 1989). In the fowl, extensive alterations to capillary and membrane structure of the ovarian follicle, in conjunction with specific recognition processes, allows a massive transfer of yolk lipid precursors from the plasma into the ovum (Griffin, Perry & Gilbert, 1984). Accumulation of lipid by the yolk proceeds by direct intact transfer of the plasma lipid components (Gornall & Kuksis, 1973; Griffin *et al.*, 1984).

At oviposition, the yolk is bounded by a vitelline membrane and divided into several distinct layers which have been laid down during follicular maturation. Development of the embryo sees the rapid dissolution of this structure through the development of the yolk sac membrane, with its elaborate folding microvillous structure and extensive vascularisation (Lambson, 1970) and formation of sub-embryonic fluid (Schlesinger, 1958).

Yolk lipid composition

The yolks of avian and crocodilian eggs are not homogeneous: they are divided into white and yellow components (Gilbert, 1971*b*; Astheimer *et al.*, 1989). The majority of the yolk mass is associated with the yellow yolk which, due to its deposition during the later stages of ovum formation, contains the major part of the yolk lipid in the form of large floating spheres (25–150 μm in diameter) emulsified within an aqueous–protein phase. Much smaller particles known as granules (up to 2 μm in diameter) exist within the spheres and the aqueous phase. The series of concentric layers of the yellow yolk is merely a manifestation of a cyclic pattern of lipid deposition; the layers represent circadian periodicity in lipid deposition. This process would appear to be very similar in crocodilians, with the probability that the period of yolk deposition is considerably longer (Astheimer *et al.*, 1989).

Almost all the lipid in the yolk is present as lipoprotein complexes with an overall lipid: protein ratio of about 2:1. Centrifugation of the yolk of the fowl egg at low g forces produces a sediment of the granular fraction which accounts for about 25% of the total yolk solid and 7% of the total yolk lipid (Cook, 1968; Gornall & Kuksis, 1973). Further separation of the granular fraction, by electrophoresis or ion exchange chromatography, yields two high density lipoprotein fractions (the α and β lipovitellins) and a small amount of a low density triacylglycerol-rich lipoprotein. Prolonged centrifugation at much higher g forces yields the vast bulk of the yolk lipid (> 90%) which is associated with a low density triacylglycerol-rich fraction found within the large spheres. This fraction can be further subdivided by centrifugation or gel filtration (Griffin *et al.*, 1984). A very small amount of residual lipid may also be recovered from the aqueous infranatant. Ultracentrifugal separation of crocodilian yolk yields notable differences from the fowl in the fractions obtained. Thus, in *Crocodylus porosus* the proportion of the very low density fraction is only about a half that displayed by the yolk of the fowl egg with a compensatory increase in the proportion of the granules (Burley *et al.*, 1988). In common with the fowl, the very low density fraction of *C. porosus* eggs has a very high lipid content which is mainly triacylglyceride. By contrast, the granular layer contains very little low density lipoprotein. Therefore, whereas in the fowl, most of the yolk lipid is present as very low density lipoprotein, in *C. porosus* the proportion of very low density material is considerably less. Corresponding differences in the levels of the

Table 2.1. *Total weights of eggs, yolks and yolk lipid and lipid compositions (major components, per cent of total) of the yolk of the domestic fowl, turkey, alligator and a lizard.*

	Domestic fowl[a] (*Gallus gallus*)	Domestic turkey[b] (*Meleagris gallopavo*)	Alligator[c] (*Alligator mississippiensis*)	Lizard[d] (*Sceloporus*) *jarrovi*)
Egg weight, g	60.3	84.2	69.6	2.4
Yolk weight, g	19.2	25.9	31.4	NA
Wt. lipid per yolk, g	6.0	8.2	6.4	NA
Lipid composition % Total lipid:				
Cholesterol ester	0.8	0.5	1.5	NA[e]
Triglyceride	71.4	66.8	69.5	86.9
Free fatty acid	0.9	0.4	1.6	NA[e]
Free cholesterol	5.6	8.6	7.7	NA[e]
Phosphoglyceride	20.7	23.7	19.8	9.0
Phosphoglyceride:				
Phosphatidyl ethanolamine	23.9	NA	18.0	NA
Phosphatidyl serine	2.7	NA	3.0	NA
Phosphatidyl choline	69.1	NA	71.5	NA
Sphingomyelin	1.0	NA	1.3	NA
Others	3.2	NA	6.4	NA

NA, not available.
[a]Noble & Moore (1967*b, c*); [b]Noble (unpublished observations); [c]Noble *et al.*, (1990,1991); [d]Hadley & Christie (1974); [e]combined levels 6–7%.

plasma precursors between these species are also indicated (Chapman, 1980).

By the very nature of its mode of synthesis there can be no 'standard' lipid composition for yolks of any avian or reptilian species. Clearly, modern methods and the controls exerted over poultry egg production considerably reduce the variation that may exist but, even under these circumstances, the composition is open to the influences of ever changing production criteria and consumer requirements (Noble, 1987*b*). However, in the case of the domesticated bird, a high degree of uniformity of egg lipid composition for any given species may be obtained. This may readily be achieved by providing the bird with a stable environment, assuring that it is of an age for high reproductive capacity and receives a diet that is adequate in all the necessary nutrients. Emphasis on hepatic synthesis for yolk lipid formation provides an enhanced assurance of uniform lipid supply but the lipids of the diet can and do influence yolk composition, particularly their fatty acids (Noble, 1987*b*). Clearly, extreme variation may exist in wild species (Romanoff & Romanoff, 1949) whilst in reptiles the influence of ambient temperature on the lipid metabolism of the female may affect lipid composition of the yolk. Indeed, in the

lizard *Sceloporus jarrovi*, some notable seasonal effects on fatty acid composition of egg yolk have been noted (Hadley & Christie, 1974). The short discussion on the content and composition of the lipid found in eggs of birds and reptiles that follows is, therefore, made with the above provisions in mind. The general absence of detailed information for reptiles makes it even more difficult to make definitive statements.

Compared with most avian species (Romanoff & Romanoff, 1949; Noble, 1987*b*), there is a higher proportion by weight of the yolk in the eggs of reptiles but the level of lipid in the yolk is much lower (Table 2.1) (Stewart & Castillo, 1984; Manolis *et al.*, 1987; Noble *et al.*, 1990). Generally, there appears to be a marked similarity between avian and reptilian species with respect to the distribution of the major lipid components (Table 2.1) (Noble & Moore, 1964; Noble, 1987*b*; Hadley & Christie, 1974; Noble *et al.*, 1990). By far the principal lipid fraction of the yolk, as would be expected from the plasma precursors, is triacylglyceride (70–80% of total lipid) which is accompanied by a substantial level of phosphoglyceride (20–25% of total lipid). Phosphatidyl choline and phosphatidyl ethanolamine are the major phosphoglyceride components, together accounting for about 90%

Table 2.2. *Fatty acid compositions, per cent of total, of the major lipid fractions of the yolk. Data sources as for Table 2.1.*

	Palmitic	Palmitoleic	Stearic	Oleic	Linoleic	Linolenic	Arachidonic	Docosa-hexaenoic
Cholesterol esters:								
Fowl	29.1	1.0	9.5	40.1	18.0	<0.5	0.9	0.5
Turkey	35.0	<0.5	6.0	41.2	23.7	<0.5	<0.5	<0.5
Alligator	54.4	8.4	5.9	13.3	3.1	0.9	8.9	1.3
Triacylglyceride:								
Fowl	24.5	6.6	6.4	46.2	14.7	1.1	<0.5	<0.5
Turkey	28.1	6.8	5.7	43.9	14.1	1.1	<0.5	<0.5
Alligator	28.9	18.6	6.3	32.8	6.4	4.4	1.0	1.0
Lizard	18.6	2.8	2.8	54.6	15.3	1.2	1.6	<0.5
Phosphoglyceride:								
Fowl	28.4	1.9	14.9	29.5	13.8	<0.5	6.2	4.1
Turkey	27.4	1.9	17.5	27.8	17.3	<0.5	4.5	2.8
Alligator	32.7	8.0	7.7	21.1	4.3	2.9	11.0	10.3
Phosphatidyl ethanol-amine:								
Fowl	21.7	1.1	30.1	15.3	9.2	<0.5	13.2	8.4
Alligator	16.7	7.7	13.1	18.0	3.9	3.2	15.4	16.7
Phosphatidyl serine:								
Fowl	33.6	5.4	27.3	15.9	7.3	<0.5	8.5	1.2
Alligator	15.2	<0.5	33.3	10.4	2.3	1.1	31.5	1.7
Phosphatidyl choline:								
Fowl	33.7	1.0	15.8	27.7	14.1	<0.5	4.4	1.8
Alligator	37.4	7.2	7.8	18.0	5.9	2.4	12.6	6.3
Lizard	28.7	<0.5	3.6	22.2	32.4	2.5	4.7	2.0
Sphingomyelin:								
Fowl	41.7	6.5	17.6	23.7	9.1	<0.5	<0.5	<0.5
Alligator	54.9	<0.5	22.7	15.5	2.9	<0.5	<0.5	<0.5

of the total present (Noble & Moore, 1965b, 1967c; Noble et al., 1991). The only other major component is free cholesterol (5–8%). Cholesterol ester and free fatty acid which constitute a significant proportion of the lipid content of most animal tissues, are minor components only. Overall proportions of other extractable lipid substances, e.g. pigments, carotenoids, are negligible.

In all cases, the major fatty acids that have been identified are palmitic, palmitoleic, stearic, oleic, linoleic, linolenic, arachidonic and docosahexaenoic (Table 2.2). In the yolk of poultry eggs, oleic is the major acid (40–45% of the total), palmitic and stearic acids together account for more than a third of the total and there is also present a substantial proportion of linoleic acid (Table 2.2). The phosphoglyceride shows a high level of C_{20} and C_{22} polyunsaturated fatty acids. In the phosphatidyl ethanolamine, phosphatidyl serine and phos-

phatidyl choline fractions palmitic and stearic acids together account for about 50% of the total fatty acids (Table 2.2). This is a feature that is common to similar phosphoglycerides from other animal tissues and is consistent with the structure of these moieties isolated from most natural sources (Strickland, 1973). There are also marked differences in the distribution of the saturated and polyunsaturated fatty acids between the major individual phosphoglycerides (Table 2.2), again a feature that is common to many animal tissues. The yolks of the two reptilian eggs appear to show notable differences in their distributions of the major fatty acids when compared to each other and to the yolk of the fowl. Thus, in the alligator (*Alligator mississippiensis*) by far the major fatty acid in the cholesterol ester is palmitic (Table 2.2). The triacylglyceride has very high levels of palmitoleic but low levels of oleic and linoleic acids compared with the fowl. By contrast, in the lizard (*S. jarrovi*), the triacylglyceride has a high level of oleic and a level of linoleic acid similar to that for the fowl (Table 2.2). In the alligator yolk, the phosphoglycerides show low levels of stearic and, in particular, linoleic acid but very high levels of C20 and C22 polyunsaturates; again there is a high level of palmitoleic acid (Table 2.2). The yolk of the lizard displays an entirely different distribution of polyunsaturates in its phosphoglyceride (Hadley & Christie, 1974). As in the case of poultry species, the yolk of the alligator displays a specific intramolecular association with respect to the distribution of the major fatty acids between the individual phosphoglyceride fractions (Noble *et al.*, 1991).

In the fowl, extensive analyses have been performed on the stereospecific positional distributions of the fatty acids in the triacylglyceride and phosphoglyceride of the yolk. In both phosphatidyl ethanolamine and phosphatidyl choline the selective distribution of fatty acids is similar to that widely displayed by most animal tissues: saturated fatty acids predominate in position 1 and unsaturated fatty acids in position 2 (Holub & Kuksis, 1969; Christie & Moore, 1972; Strickland, 1973). 1-palmitoyl-2-oleoyl (37%) and 1-palmitoyl-2-linoleoyl (20%) are the main species of phosphatidyl choline. Phosphatidyl ethanolamine shows a high proportion of stearic acid in position 1 in association with oleic and linoleic acid in position 2. The C20 and C22 fatty acid containing species in phosphatidyl ethanolamine display a preferential association with stearic acid in position 1. Both phosphatidyl ethanolamine and phosphatidyl choline have negligible proportions of disaturated species. Triacylglyceride shows a high degree of asymmetry between the fatty acids of positions 1 and 3 (Christie & Moore, 1970, 1972), the differences being much greater than is normally found in the triacylglyceride of most animal tissues and even those of the adult fowl. Although there are some similarities between the arrangement of the fatty acids in the triacylglyceride and phosphoglyceride, there are also many notable differences. Structural analyses have not been able to provide any substantive evidence of a biosynthetic relationship between the origins of the yolk triacylglyceride and phosphoglyceride. Comparative structural data for the yolk lipids of reptiles is extremely limited. However, results for *S. jarrovi* (Hadley & Christie, 1974) indicate that the lipid structures differ considerably from those of the fowl.

Extensive investigations have been made of the distribution and fatty acid profiles of the lipids associated with the lipoproteins in the yolk of the fowl egg (Cook, 1968; Cook & Martin, 1969; Gornall & Kuksis, 1973; Griffin *et al.*, 1984). The triacylglyceride-rich low density fraction consists of a non-polar core of virtually pure triacylglyceride surrounded by a mixture of apoproteins, phosphoglyceride and cholesterol. The lipid compositions of fractions 1 and 2 are similar, high levels of triacylglyceride being associated with 25–30% phosphoglyceride in which phosphatidyl ethanolamine and phosphatidyl choline predominate. The two high density fractions both have phosphatidyl ethanolamine and phosphatidyl choline as their major lipids in proportions similar to that displayed by the low density fraction. The lipids of the triacylglyceride-rich low density fraction in the granules do not appear to have been characterised but from the limited evidence available, they appear to be of similar composition to that of the spheres (Gornall & Kuksis, 1973). The lipid compositions and many physical characteristics of the triacylglyceride-rich low density fractions change considerably with differing physiological and dietary states (Evans, Flegal & Bauer, 1975). Stereochemical structural analyses of the lipid moieties within the major lipoproteins (Gornall & Kuksis, 1973) have shown close similarities for each lipid class; as a result it was suggested that the yolk lipid components are derived from a single liver lipid pool. Although a recent investigation of the lipid composition of the two main lipoprotein fractions of the yolk of *C. porosus* (Burley *et al.*, 1988) indicates very high levels of triacylglyceride in the low density fraction, a very much lower level of phosphoglyceride in the high density fraction, compared to the fowl, is indicated.

Yolk lipid uptake

General

Yolk lipid utilisation during incubation has been extensively studied in the fowl (Romanoff, 1960; Noble & Moore, 1964, 1967*b,c*; Noble, 1986, 1987*a*) and will therefore be used to describe the major events of yolk uptake. Investigations with other avian species indicate an overall similarity in the sequence of events and metabolic features to that demonstrated by the fowl embryo. Nett lipid loss from the yolk of a

Table 2.3. *Weights of lipid associated with the yolk contents, yolk sac membrane and embryo during incubation of the fowl egg (data from Noble & Moore, 1964, 1967b; Noble et al., 1986).*

Incubation day	0	13	15	17	19	21
Yolk contents, g	6.0	5.3	4.5	2.3	1.8	0.6
Yolk sac membrane, g	0	0.5	0.9	2.2	1.7	0.6
Embryo, g	0	0.2	0.5	1.2	1.6	2.1

Table 2.4. *Weights of eggs, yolks and yolk lipid during incubation of the alligator* (Alligator mississippiensis) *egg at 30 °C (data from Noble et al., 1990).*

Incubation day	8	32	40	48	55	64	75
Egg wt, g	69.6	70.2	71.1	69.8	72.7	75.1	64.7
Yolk wt, g	31.4	29.6	26.9	24.3	23.1	15.6	3.1
Wt. lipid per yolk, g	6.4	6.4	5.8	5.2	5.2	4.3	0.9

60 g fowl egg during the first 13 days of incubation only amounts to about 2–300 mg (Table 2.3). Between days 13 and 15 there is a further loss of about 200 mg. However, between days 15 and 21 the loss of lipid increases substantially such that during the last two days of incubation the yolk lipid is removed at a rate of nearly 1 g per day (Table 2.3). The amount of lipid within the yolk contents decreases from over 5 g at day 13 to less than 1 g by day 21, an overall loss of over 500 mg per day (Table 2.3). Preferential absorption of lipid, relative to other components from the yolk contents, is indicated by a decrease in the percentage of total lipid in the dry matter, 65% on day 13 to 44% on day 21. Between days 13 and 17 of incubation, loss of lipid from the yolk contents is accompanied by its rapid accumulation in the yolk sac membrane (Table 2.3) such that by day 17 the membrane contains as much lipid as the contents (Noble & Moore, 1967b; Noble et al., 1986). Just prior to hatching there is a rapid loss of lipid from the membrane and at the same time a marked drop in the percentage of lipid in the dry matter, from 82% at day 17 to 68% at day 21.

Similar detailed observations (Table 2.4) have been made on the changes for the weights and lipid contents of the yolk of the alligator egg over the period of incubation at 30 °C (Noble *et al.*, 1990). Over the first 56 days the weight of the yolk shows only a slight change, thereafter there is a rapid decline such that by day 75 only 10% of the initial yolk weight remains (Table 2.4). The majority of the loss of yolk weight and dry matter content is due to the removal of lipid.

The proportion of lipid within the yolk steadily increases as incubation proceeds; by the end of the incubation period, lipid accounts for some 31% of the remaining yolk weight compared to 21% of total yolk weight at the beginning of incubation. As in the fowl, this is probably due to re-location of lipid into the yolk sac membrane.

The amounts of lipid used during incubation differ markedly between, and within, avian and reptilian species. Thus, whereas a large proportion of the yolk lipid is used by embryos of the fowl, crocodilians and snakes, in turtles the amount used is very much lower (Congdon, Tinkle & Rosen, 1983; Stewart & Castillo, 1984; Wilhoft, 1986; Manolis *et al.*, 1987; Noble *et al.*, 1990). In reptiles, incubation temperature has a notable effect on the extent of yolk lipid uptake (Deeming & Ferguson, Chapter 10). For instance, in *C. johnstoni* (Manolis *et al.*, 1987) the period of incubation is extended at 28 °C resulting in a far greater absorption of yolk lipid (87% of that present at oviposition) compared to incubation at 34 °C (64% of that initially present). For both birds and reptiles, a wide difference exists between the extent of recoverable lipid from the combined remnant yolk and embryonic tissues at hatching compared to that initially present in the yolk.

Lipid compositional changes

In the fowl (Noble & Moore, 1964, 1965a) the proportions of the triacylglyceride and phosphoglyceride in the yolk remain largely unchanged during the last week of incubation

Table 2.5. *Major changes in the lipid composition of the yolk of the fowl between days 0 and 21 of incubation and of the alligator between days 8 and 75 of incubation.*

Fowl[a]	Cholesterol ester, %	Triacyl-glyceride, %	Free fatty acid, %	Free cholesterol, %	Phospho-glyceride, %
Whole yolk	0.8→4.0	71.4→72.3	<1.0	5.6→4.0	20.7→18.0
Yolk contents	0.8→0.5	71.4→75.9	<1.0	5.6→5.0	20.7→18.2
Yolk sac membrane	– →7.0	– →69.0	<1.0	– →2.3	– →17.5
Alligator[b], whole yolk	1.5→20.8	69.5→59.4	1.6→6.7	7.7→5.2	19.8→7.9

	Phosphatidyl ethanolamine, %	Phosphatidyl choline, %
Fowl:		
Whole yolk	23.9→9.1	69.1→75.2
Yolk contents	23.9→8.3	69.1→71.7
Yolk sac membrane	– →9.8	– →76.8
Alligator, whole yolk	18.0→24.8	71.5→62.6

[a]Noble & Moore (1967*b, c*); [b]Noble *et al.* (1990,1991).

(Table 2.5). In general, only low levels of partial glycerides, free fatty acids and lysophosphoglycerides are detected. However, analyses of the yolk, complete or separated into its contents and membrane, show that there is a preferential removal of phosphatidyl ethanolamine amongst the major phosphoglyceride fractions (Noble & Moore, 1965*b*, 1967*b,c*). Although the proportion of total cholesterol in the yolk does not change, the ratio of esterified to free cholesterol increases considerably (Noble & Moore, 1964); the accumulation of cholesterol ester is associated in particular with the yolk sac membrane (Noble & Moore, 1967*b*). The extent of yolk lipid changes observed during incubation in *A. mississippiensis* eggs are far more extensive than for the fowl (Noble *et al.*, 1990, 1991). Thus the absolute amount of cholesterol ester increases such that from being only a minor component present initially, it accounts for more than 20% of the total lipid just prior to hatching (Table 2.5). In contrast to the fowl, yolk lipid absorption is associated with a preferential removal of the phosphoglyceride such that 94% of the total has been absorbed by day 75 of incubation (Table 2.5). By comparison, in the fowl just prior to hatching, some 79% of the yolk phosphoglyceride has been absorbed. In the alligator, the level of free fatty acid in the yolk lipid increases markedly during incubation whilst within the phosphoglyceride fraction there is a reduction in the level of phosphatidyl choline and increases in that of phosphatidyl ethanolamine and lysophosphatidyl choline. Wide differences have

been noted in the time scale and extent of the yolk lipid compositional changes at different incubation temperatures (Noble, Deeming & Ferguson, unpublished observations).

In the fowl, several changes have been observed to occur in the proportions of the lipoproteins of the yolk during incubation (Saito, Martin & Cook, 1965).

Fatty acid compositional changes

Although in the fowl the fatty acid composition of triacylglyceride remains constant during incubation (Noble & Moore, 1964, 1965*a,b*), notable changes occur in the cholesterol ester and phosphoglyceride fractions. The cholesterol ester, in particular that associated with the yolk sac membrane, shows a progressive increase in oleic acid content (Noble & Moore, 1967*b*). The phosphoglyceride fraction is characterised by two major features, both involving the C20 and C22 polyunsaturated fatty acids. During incubation there is a marked decrease in phosphatidyl ethanolamine species containing high levels of docosahexaenoic acid whilst in comparison with the yolk contents, the yolk sac membrane shows a marked increase in its level of arichidonic acid (Noble & Moore, 1965*b*, 1967*b,c*). The suggestion that there is a similar preferential uptake of yolk triacylglycerides containing high levels of polyunsaturated fatty acids (Isaaks *et al.*, 1964) has not been substantiated by subsequent investigations. Apart from small changes in the levels of palmitic and stearic acids, the composition of the triacylglyceride fraction of the alligator yolk also remains constant during incubation (Noble *et al.*, 1990). As in the fowl, there is a large increase in the level of oleic acid in the cholesterol ester. In contrast to the fowl, the overall fatty acid composition of the phosphoglyceride in alligator yolk does not change to any appreciable extent in spite of marked changes in the fatty acid compositions of certain individual fractions, especially in their contents of C20 and C22 polyunsaturates (Noble *et al.*, 1990, 1991).

Process of yolk lipid uptake

Morphological and biochemical evidence has established that, in the fowl embryo, yolk lipid uptake by the yolk sac membrane occurs through non-specific phagocytosis. Electron microscopic studies (Lambson, 1970) have demonstrated the sequential engulfment of intact lipid droplets and their subsequent appearance at the apical surface of the endodermal cells. Although the presence of various lipolytic enzymes has been reported to be present within the yolk (Zacks, 1954), there is no real evidence for extracellular digestion playing a part in the uptake of the major lipids into the membrane. Almost all the comparative analytical data concerned with the yolk lipid contents over the incubation period are reconcilable with there being no lipid breakdown before removal (Noble & Moore, 1964, 1967*b,c*). The relative proportions of the major lipids remain unchanged, low levels only of lipid breakdown products, e.g. partial glycerides, free fatty acids, lysophosphoglycerides, are detected during the most intense period of yolk lipid uptake, whilst the fatty acid compositions and structures of the lipid show very little change. The similarities observed between the relative contents, fatty acid compositions and structures of the triacylglyceride and phosphoglyceride fractions of the yolk contents and membrane during incubation (Noble & Moore, 1967*b*) provide further evidence that there is no extensive breakdown and resynthesis during uptake into the membrane.

Once within the yolk sac membrane there is extensive evidence for the hydrolysis and re-esterification of the major lipids, together with re-assembly into newly synthesised lipoproteins, before their passage into the embryo. Lipase and other enzyme activity within the membrane and a sequence of lipid droplet elaborations within the individual cells, in a manner resembling synthesis known to involve hydrolysis and re-esterification, have both been observed (Zacks, 1954; Lambson, 1970). Distinct differences exist between the lipid content and the fatty acid composition of the yolk contents and embryonic plasma (Schjeide, 1963; Yafei & Noble, 1990); there are also differences between the structures and lipid contents of the major lipoprotein classes (Schjeide, 1963). Although comparable data are not available for reptiles, recent investigations on yolk lipid changes during incubation in the alligator (Noble *et al.*, 1990, 1991), in which marked increases in the levels of free fatty acids and lysophosphatidyl choline were observed just prior to hatching, may implicate an increased role for lipolytic breakdown in the assimilation of lipid.

Yolk lipid uptake and embryo accumulation

As well as a source of a large amount of lipid, the yolk is also involved in specific lipid changes before assimilation that have far-reaching and unique consequences on embryonic and neonatal metabolism. Most prominent are those involving the metabolism of cholesterol and the polyunsaturated fatty acids.

Cholesterol metabolism

A most unusual feature of the tissue lipid changes of the fowl embryo is that displayed by the liver. A very large accumulation of lipid is accounted for almost wholly by cholesterol ester, in particular cholesterol oleate (Moore & Doran, 1962; Noble & Moore, 1964). In the fowl, comparative data for the embryonic tissues does not indicate that the assimilation of the yolk lipid is directly responsible for the accumulation. The increasing presence in the yolk sac membrane, during yolk lipid assimilation, of cholesterol ester displaying levels of oleic acid similar to that of the liver suggested the yolk sac membrane as the major hepatic source (Noble & Moore, 1964, 1967*b*). Investigations have not only shown the presence within the yolk sac membrane of an active cholesterol esterifying system, but also that its pattern of fatty acid specificity is identical to that of the fatty acids present in choleterol ester which accumulates in the liver (Noble, Connor & Smith, 1984). This evidence

Table 2.6. *Polyunsaturated fatty acid levels (per cent of total fatty acids) in phosphoglycerides of the yolk and tissues of the fowl and alligator just prior to hatching.*

	Yolk contents	Yolk sac membrane	Plasma	Liver	Extra-hepatic tissues
Fowl[a]					
Linoleic	13.8	15.2	15.4	11.6	12.5
Arachidonic	3.2	7.6	8.4	22.7	7.7
Linolenic	1.4	1.3	1.1	1.5	1.3
Docosahexaenoic	2.1	4.3		4.9	9.8	5.6
Alligator[b]		Total yolk		Liver	
Linoleic		4.3		3.4	
Arachidonic		11.5		17.0	
Linolenic		1.9		1.9	
Docosahexaenoic		11.5		14.2	

[a]Noble & Cocchi (1989); [b]Noble *et al.* (1991)

substantiates the idea that the hepatic accumulation of cholesterol ester is intimately associated with, or derived from, its synthesis in the yolk sac membrane. The fact that this membrane is the major site of lipoprotein synthesis and assembly prior to passage into the embryo, together with the high concentration of cholesterol oleate in the lipids of the plasma of the embryo (Schjeide, 1963), suggests that cholesterol ester synthesis is important to the formation of the lipoprotein complexes required for lipid transport. Indeed, a known function for cholesterol ester is the maintenance of particle stability for chylomicron and large lipoprotein complexes (Vandenheuvel, 1962). The accumulation of cholesterol ester by the liver would therefore represent accumulation of remnant lipoprotein following hydrolysis of the mature portomicrons and very low density lipoprotein fractions arising from the yolk sac membrane. Although there have been no comparable investigations with reptiles, cholesterol ester containing a high level of oleic acid accumulates within the yolk of the alligator during incubation to a far greater extent than even in the fowl (Noble *et al.*, 1990) and may therefore assume an even greater role in yolk lipid assimilation.

Polyunsaturated fatty acid metabolism

A further feature of the lipids of the embryonic liver, and to a lesser extent other tissues, is the much higher proportion of C20 and C22 polyunsaturated fatty acids than exist in the yolk (Noble & Moore, 1964, 1965*a*, 1967*a,b,c*). The proportions of both arachidonic and docosahex-aenoic acids increase substantially within the phosphoglyceride between the yolk contents, yolk sac membrane, plasma and, in particular, the embryonic tissues (Table 2.6). By contrast the proportions of linoleic and linolenic acids fowl embryo liver the arachidonic acid, in specific association with the phosphatidyl ethanolamine and phosphatidyl choline fractions, accounts for more than 20% of the total fatty acids. By contrast, arachidonic acid comprises only 3 and 7% respectively of the fatty acids of these two fractions in the yolk contents. The relative concentration of docosahexaenoic acid is also some two-fold higher in the phosphoglycerides of the embryo tissues than in the yolk. In addition, there are surprisingly high levels of docosahexaenoic acid in the triacylglyceride of the liver and other tissues. The phosphoglyceride in the liver of the alligator also contains high levels of arachidonic and docosahexaenoic acids (Noble *et al.*, 1991). Durying embryonic development in the fowl there is a preferential absorption by the yolk sac membrane of yolk phosphatidyl ethanolamine species particularly rich in docosahexaenoic acid (Noble & Moore, 1965*b*, 1967*b,c*). No similar preferential absorption of arachidonic acid containing phosphoglyceride species is observed. However, the presence of appreciable $\triangle$6-desaturase activity (responsible for the first step in the conversion of linoleic to arachidonic acid) has been identified in the yolk sac membrane (Noble & Shand, 1986). During yolk lipid absorption, therefore, there is an opportunity for the enhancement of the arachidonic acid level of the absorbed lipid. The synthesis of arachidonic acid and the preferential absorption of docosa-hexaenoic acid are both presumably geared to a requirement by the rapidly developing embryonic tissues for a range of fatty acids that cannot be adequately satisfied by the more saturated fatty acid spectrum of the yolk. There is much similarity between this role of the yolk sac membrane to that of the placenta in various mammals (Noble & Cocchi, 1989). In the case of docosahexaenoic acid it is strange that following its preferential absorption, it should be incorporated to such an extent in triacylglyceride in the liver.

Data on the role of the yolk in the supply of polyunsaturated fatty acids in reptiles are confined to the recent observations on the utilisation of phosphoglycerides by the alligator embryo (Noble *et al.*, 1991). Here, the existence of a specific role for phosphatidyl ethanolamine in the provision of C20 and C22 polyunsaturates is also indicated. However, environmental temperature may have a considerable effect on the polyunsaturated fatty acid content of the alligator yolk through their rate of deposition (Noble, Deeming & Ferguson, unpublished data) or the influence on *de novo* desaturation (Hadley & Christie, 1974). In the event, the effect on the supply of polyunsaturates to the embryo may be quite extreme.

Post-hatching and yolk lipid

Following hatching, the presence of residual yolk material in both avian and reptilian species provides a continuation in the supply of nutrients, including lipids, during the critical change from embryonic to free living development (Cagle, 1950; Romanoff, 1960; Wilhoft, 1986). The time-scale of the existence of the residual yolk within the body varies considerably with species, from 5–6 days in the case of the fowl to 2–3 weeks in the turtle. Although the fate of the residual yolk lipid in reptiles is largely unknown, a recent study with the hatched fowl (Noble & Ogunyemi, 1989) has investigated its lipid and fatty acid compositional changes and compared them with the lipid changes in the liver occurring at the same time. Although in several respects the changes observed are accelerated progressions of those seen during embryonic development, there are also several unique features. However, the yolk and liver comparisons show a rapid alteration in their lipid metabolic relationship away from that which existed during the embryonic phase.

References

Astheimer, L. B., Manolis, S. C. & Grau, C. R. (1989). Egg formation in crocodiles: avian affinities in yolk deposition. *Copeia*, **1989**, 221–4.

Bellairs, R. (1967). Aspects of the development of the yolk spheres in the hen's oocyte studied by electron microscopy. *Journal of Embryology and Experimental Morphology*, **17**, 267–81.

Burley, R. W., Back, J. F., Wellington, J. E. & Grigg, G. C. (1988). Proteins and lipoproteins in yolk from eggs of the estuarine crocodile (*Crocodylus porosus*); a comparison with the egg of the hen (*Gallus domesticus*). *Comparative Biochemistry and Physiology*, **91B**, 39–44.

Cagle, F. R. (1950). The life history of the slider turtle, *Pseudemys scripta troostii* (Holbrook). *Ecology Monographs*, **20**, 31–54.

Chapman, M. J. (1980). Animal lipoproteins: chemistry, structure and comparative aspects. *Journal of Lipid Research*, **21**, 789–853.

Christie, W. W. (1982). *Lipid Analysis*. Oxford: Pergamon Press.

Christie, W. W. & Moore, J. H. (1970). The structure of egg yolk triglycerides. *Biochimica et Biophysica Acta*, **218**, 83–8.

— (1972). The lipid components of the plasma, liver and ovarian follicles in the domestic chicken (*Gallus gallus*). *Comparative Biochemistry and Physiology*, **41B**, 287–95.

Christie, W. W. & Noble, R. C. (1984). Recent developments in lipid analysis. In *Food Constituents and Food Residues*, ed. J. F. Lawrence, pp. 1–50. New York: Marcel Dekker.

Chung, R. A., Ning, J. M. J. & Tsao, Y. C. (1966). Effect of diethyl stilbesterol and cholesterol on the fatty acid metabolism of cockerels. *Poultry Science*, **45**, 661–7.

Congdon, J. D., Tinkle, D. W. & Rosen, P. C. (1983). Egg components and utilisation during development in aquatic turtles. *Copeia*, **1983**, 264–8.

Cook, W. H. (1968). Macromolecular components of egg yolk. In *Egg Quality: A study of the Hen's Egg*, ed. T. C. Carter, pp. 109–32. Edinburgh: Oliver & Boyd.

Cook, W. H. & Martin, W. G. (1969). Egg lipoproteins. In *Structural and Functional Aspects of Lipoproteins in Living Systems*, ed. E. Tria & A. M. Scanu, pp. 579–615. London: Academic Press.

Evans, R. J., Flegal, C. J. & Bauer, D. H. (1975). Molecular sizes of egg yolk very low density lipoproteins fractionated by ultracentrifugation. *Poultry Science*, **54**, 889–95.

Ewert, M. A. (1979). The embryo and its egg; development and natural history. In *Turtles,*

Perspectives and Research, ed. M. Harless & H. Morlock, pp. 333–413. New York: John Wiley & Sons.

Ferguson, M. W. J. (1985). Reproductive biology and embryology of the crocodilians. In *Biology of the Reptilia*, vol. 14, Development A, ed. C. Gans, F. Billet & P. F. A. Maderson, pp. 329–491. New York: John Wiley & Sons.

Gilbert, A. B. (1971a). The female reproductive effort. In *Physiology and Biochemistry of the Domestic Fowl*, vol. 3, ed. D. J. Bell & B. M. Freeman, pp. 1153–62. London: Academic Press.

— (1971b). The ovary. In *Physiology and Biochemistry of the Domestic Fowl*, vol. 3, ed. D. J. Bell & B. M. Freeman, pp. 1163–208. London: Academic Press.

Gornall, D. A. & Kuksis, A. (1973). Alterations in lipid composition of plasma lipoproteins during deposition of egg yolk. *Journal of Lipid Research*, **14**, 197–205.

Griffin, H. D., Perry, M. M. & Gilbert, A. B. (1984). Yolk formation. In *Physiology and Biochemistry of the Domestic Fowl*, vol. 5, ed. B. M. Freeman, pp. 345–80. London: Academic Press.

Hadley, N. F. & Christie, W. W. (1974). The lipid composition and triglyceride structure of eggs and fat bodies of the lizard *Sceloporus jarrovi*. *Comparative Biochemistry and Physiology*, **48B**, 275–84.

Holub, B. J. & Kuksis, A. (1969). Molecular species of phosphatidyl ethanolamine from egg yolk. *Lipids*, **4**, 466–72.

Husbands, D. H. R. & Brown, W. O. (1965). Sex differences in the composition and acetate incorporation into the liver lipids of the adult fowl. *Comparative Biochemistry and Physiology*, **14**, 445–51.

Isaaks, R. E., Davies, R. E., Ferguson, T. M., Reiser, R. & Couch, J. R. (1964). The selective utilisation of fatty acids by the chick embryo. *Poultry Science*, **43**, 113–20.

Lambson, R. O. (1970). An electron microscopic study of the endodermal cells of the yolk sac of the chick during incubation and after hatching. *American Journal of Anatomy*, **129**, 1–20.

Manolis, S. C., Webb, G. J. W. & Dempsey, K. E. (1987). Crocodile egg chemistry. In *Wildlife Management: Crocodiles and Alligators*, ed. G. J. W. Webb, S. C. Manolis & P. J. Whitehead, pp. 445–72. Sydney: Surrey Beatty Pty Ltd.

Moore, J. H. & Doran, B. M. (1962). Lipid metabolism in the normal and vitamin B_{12}-deficient chick embryo. *Biochemical Journal*, **84**, 506–13.

Noble, R. C. (1986). Lipid metabolism in the chick embryo. *Proceedings of the Nutrition Society*, **45**, 17–25.

— (1987a). Lipid metabolism in the chick embryo: some recent ideas. *Journal of Experimental Zoology*, Supplement 1, 65–73.

— (1987b). Egg lipids. In *Egg Quality: Current Patterns and Recent Advances*, ed. R. G. Wells & C. G. Belyavin, pp. 159–77. London: Butterworths.

Noble, R. C. & Cocchi, M. (1989). The relationship between the supply and demand for essential polyunsaturated fatty acids during mammalian and avian embryonic development. *Research and Development in Agriculture*, **6**, 65–90.

Noble, R. C., Connor, K. & Smith, W. K. (1984). The synthesis and accumulation of cholesteryl esters by the developing embryo of the domestic fowl. *Poultry Science*, **63**, 558–64.

Noble, R. C., Deeming, D. C., Ferguson, M. W. J. & McCartney, R. (1990a). Changes in the lipid and fatty acid composition of the yolk during embryonic development of the alligator (*Alligator mississippiensis*). *Comparative Biochemistry and Physiology*, **96B**, 183–7.

— (1990b). The utilisation of yolk phosphoglycerides by the alligator during embryonic development. *Journal of Comparative Physiology, B*, (in press).

Noble, R. C., Lonsdale, F., Connor, K. & Brown, D. (1986). Changes in the lipid metabolism of the chick embryo with parental age. *Poultry Science*, **65**, 409–16.

Noble, R. C. & Moore, J. H. (1964). Studies on the lipid metabolism of the chick embryo. *Canadian Journal of Biochemistry*, **42**, 1729–41.

— (1965a). Further studies on the lipid metabolism of the normal and vitamin B_{12}-deficient chick embryo. *Biochemical Journal*, **95**, 144–9.

— (1965b). Metabolism of the yolk phospholipids by the developing chick embryo. *Canadian Journal of Biochemistry*, **43**, 1677–86.

— (1967a). The liver phospholipids of the developing chick embryo. *Canadian Journal of Biochemistry*, **45**, 627–39.

— (1967b). The partition of lipids between the yolk and yolk sac membrane during the development of the chick embryo. *Canadian Journal of Biochemistry*, **45**, 949–58.

— (1967c). The transport of phospholipids from the yolk to the yolk-sac membrane during the development of the chick embryo. *Canadian Journal of Biochemistry*, **45**, 1125–33.

Noble, R. C. & Ogunyemi, D. (1989). Lipid changes in the residual yolk and the liver of the chick immediately after hatching. *Biology of the Neonate*, **56**, 228–36.

Noble, R. C. & Shand, J. H. (1985). Unsaturated fatty acid compositional changes and desaturation during the embryonic development of the chicken (*Gallus domesticus*). *Lipids*, **20**, 278–82.

Romanoff, A. L. & Romanoff, A. J. (1949). *The Avian Egg*. New York: John Wiley & Sons.

Romanoff, A. L. (1960). *The Avian Embryo*. New York: MacMillan.

Saito, Z., Martin, W. G. & Cook, W. H. (1965). Changes in the major macromolecular fractions of egg yolk during embryogenesis. *Canadian Journal of Biochemistry*, **43**, 1755–70.

Schjeide, O. A. (1963). Lipoproteins of fowl-serum, egg yolk and intracellular. In *Progress in the Chemistry of Fats and Other Lipids*, Vol. 5, ed. R. T. Holman, W. O. Lundberg & T. Malkin, pp. 253–90. Oxford: Pergamon Press.

Schlesinger, A. B. (1958). The structural significance of the avian yolk in embryogenesis. *The Journal of Experimental Zoology*, **138**, 223–58.

Stewart, J. R. & Castillo, R. E. (1984). Nutritional provision of the yolk of two species of viviparous reptiles. *Physiological Zoology*, **57**, 377–83.

Strickland, K. P. (1973). The chemistry of phospholipids. In *Form and Function of Phospholipids*, ed. G. B. Ansell, R. M. C. Dawson & J. N. Hawthorne, pp. 9–42. Amsterdam: Elsevier.

Vandenheuvel, F. A. (1962). The origin, metabolism and structure of normal human serum lipoproteins. *Canadian Journal of Biochemistry and Physiology*, **40**, 1299–326.

Wilhoft, D. C. (1986). Eggs and hatching components of the snapping turtle (*Chelydra serpentina*). *Comparative Biochemistry and Physiology*, **84**, 483–6.

Yafei, N. & Noble, R. C. (1990). Further observations on the association between lipid metabolism and low embryo hatchability in eggs from young broiler birds. *Journal of Experimental Zoology*, **253**, 325–9.

Yu, J. Y. L., Campbell, L. D. & Marquardt, R. R. (1976). Immunological and compositional patterns of lipoproteins in chickens (*Gallus domesticus*) plasma. *Poultry Science*, **55**, 1626–31.

Zacks, S. I. (1954). Esterases in the early chick embryo. *Anatomical Record*, **118**, 509–37.

3

Oviductal proteins and their influence on embryonic development in birds and reptiles

BRENT D. PALMER AND LOUIS J. GUILLETTE JR

Introduction

The amniotic egg provides an environment suitable for the development of the embryo. This environment, produced by the mother in the form of yolk and albumen, is enclosed by the eggshell. The eggshell consists of an underlying layer of proteinaceous fibres and a surface layer predominantly of calcium carbonate, whereas albumen consists primarily of proteins, although carbohydrates and lipids are found in trace quantities. Proteins produced by the oviduct of the mother, therefore, are largely responsible for the environment in which the embryo will develop.

The albumen and eggshell create a relatively homeostatic environment for embryonic development, the eggshell reducing the influence of the external environment but allowing gases and water to be exchanged. The composition of albumen is not static, but changes during embryonic development by interactions between the external environment and developing embryo. The composition and function of albumen alters during the course of embryonic development. In later development, the embryo is able to interact more directly with its environment through the apposition of extra-embryonic membranes to the eggshell and, therefore, plays a greater role in its own homeostasis.

This chapter will review the variety of proteins produced by the maternal oviduct and examine their influence on embryonic development. By far, the most studied is the albumen and eggshell membrane proteins of avian eggs, particularly those of the fowl (*Gallus gallus*). Those of reptilian eggs are only recently coming under investigation.

Biological properties of albumen proteins

A functional approach will be used to examine albumen proteins, rather than a discussion of each protein individually. A summary of characteristics of albumen proteins of the fowl egg is given in Table 3.1. Most have been shown to be glycoproteins, with carbohydrate content of each ranging between 2 and 22%. The functions described for albumen proteins have been largely examined *in vitro*, with virtually nothing known concerning *in vivo* functions. Further, these functions are not mutually exclusive, as some proteins may have multiple functions. Although albumen proteins have received intense investigation, specific biological functions have not been found for all, including the most abundant avian albumen protein, ovalbumin (Baker, 1968; Nisbet *et al.*, 1981; Li-Chan & Nakai, 1989).

Antimicrobial proteins

Albumen provides for microbial defense of the yolk and the developing embryo in two ways, mechanically and chemically (Board & Tranter, 1986). Mechanically, albumen supports and surrounds the yolk, keeping it from contact with the eggshell membranes. Additionally, albumen forms a colloid which, due to its viscous and fibrous nature, acts as a barrier to any bacteria which breach the eggshell. Ovomucin is largely responsible for this high viscosity (Robinson, 1972, 1987), which is further augmented by lysozyme complexes (Kato *et al.*, 1981; Miller, Kato & Nakai, 1982; Hayakawa *et al.*, 1983).

Chemically, albumen may prevent microbial infection by directly killing bacteria or by creating an environment unfavorable for their

Table 3.1. *Composition and physicochemical characteristics of major albumen proteins of the fowl*[a].

	% of albumen	Molecular weight (kD)	Isoelectric point	% Carbohydrate
Avidin	0.05	68.3	10.0	8
Cystatin	0.05	12.7	5.1	0
Lysozyme	3.4–3.5	14.3	10.7	0
Ovalbumin	54.0	45	4.5	3
Ovoglobulin G_2	4.0(?)	49	5.5	~6
Ovoglobulin G_3	4.0(?)	49	5.8	~6
Ovoglycoprotein	0.5–1.0	24.4	3.9	16
Ovoinhibitor	0.1–1.5	49	5.1	6
Ovomacroglobulin	0.5	760–900	4.5–4.7	9
Ovomucin	1.5–3.5	230–8300	4.5–5.0	19
Ovomucoid	11.0	28	4.1	22
Ovotransferrin	12–13	77.7	6.0	2
Riboflavin-binding protein	0.8	32	4.0	14

[a]Compiled from Gilbert, 1971; Osuga & Feeney, 1977; Powrie & Nakai, 1986; and Li-Chan & Nakai, 1989.

growth. Most studies on the antimicrobial properties of albumen proteins have been conducted *in vitro* and it is unclear if these properties are applicable *in vivo*. Lysozyme, the only albumen protein known to directly effect bacteria, does so by hydrolysing β(1–4) glycosidic bonds (Geoffroy & Bailey, 1975), a component of the cell wall of certain bacteria, thereby lethally disrupting cell wall integrity.

There are several other ways in which albumen can create an unfavourable environment for bacterial growth (Tranter & Board, 1982a; Board & Tranter, 1986). Albumen is alkaline, having a pH of about 9.5 (Heath, 1977), which is generally unsuitable for growth of many microorganisms (Board & Tranter, 1986). Additionally, some proteins bind nutrients, such as minerals or vitamins, making them unavailable for bacterial use. Iron is found in very low concentrations in egg albumen. Ovotransferrin strongly (dissociation constant $(K_D) \sim 10^{-29}$M) binds iron (Chasteen, 1977, 1983; Aisen & Listowsky, 1980; Brock, 1985), especially at the alkaline pH of albumen, and is effective in inhibiting bacterial growth by creating an essentially iron-free environment (Weinberg, 1977; Tranter & Board, 1982a). Additionally, ovotransferrin also binds copper, which may maintain the bactericidal properties of lysozyme, which are inhibited by this metal (Gilbert, 1971). Ovotransferrin has been identi-

fied as the major antibacterial component of fowl egg albumen (Tranter & Board, 1982a).

Other proteins bind vitamins, making them inaccessible to microorganisms. Avidin, well known for its ability to strongly bind $(K_D \sim 10^{-15}$M) the vitamin biotin (Green, 1963, 1975; Elo & Korpela, 1984), is strongly antibacterial and occurs largely in the unbound (apoprotein) form (Tranter & Board, 1982a,b; Banks, Board & Sparks, 1986). Avidin also binds to the surface of some bacteria (Korpela, 1984; Korpela *et al.*, 1984). A thiamin-binding protein (Muniyappa & Adiga, 1979; Adiga & Murty, 1983) occurs largely in the apoprotein form in albumen, and may inhibit bacterial growth, although its affinity for thiamin is low $(K_D \sim 3 \times 10^{-7}$M). Riboflavin-binding protein also occurs in the albumen and although it has been suggested that this apoprotein binds riboflavin too weakly $(K_D \sim 10^{-9}$M) to inhibit bacterial growth (Rhodes, Bennett & Feeney, 1959), *in vitro* studies have demonstrated inhibition (Li-Chan & Nakai, 1989).

Another group of antimicrobial proteins are protease inhibitors. Albumen has a very low concentration of free nitrogen, which is required by bacteria for protein synthesis. Substantial nitrogen for bacterial growth is found in albumen proteins, and many bacteria have proteases which are used to digest albumen proteins to free nitrogen. However, several of the albumen proteins have been demonstrated to be protease inhibitors (Li-Chan & Nakai, 1989), thereby preventing degradation and release of nitrogen for bacterial use. Ovomucoid, the third most common protein of fowl albumen, inhibits trypsin (Lineweaver & Murray, 1947; Feeney & Allison, 1969; Feeney, 1971). Ovoinhibitor, although a minor component of avian eggs, simultaneously binds two molecules of trypsin or chymotrypsin (Rhodes, Bennett & Feeney, 1960; Liu, Means & Feeney, 1971; Zahnley, 1980), as well as several types of bacterial and fungal proteases (Matsushima, 1958; Feeney, Stevens & Osuga, 1963; Tomimatsu, Clary & Bartulovich, 1966; Feeney, 1971). Both ovomucoid and ovoinhibitor belong to a family of serine proteinase inhibitors. Based on its primary structure, ovalbumin has also been placed into a superfamily of serine proteinase inhibitors (Breathnach et al., 1978; Hunt & Dayhoff, 1980; Woo et al., 1981; Carrell & Boswell, 1986; Ye et al., 1989), although no biological function has been found for this protein (Li-Chan & Nakai, 1989). Ovomacroglobulin also has serine protease inhibitory activity (Kitamoto, Nakashima & Ikai, 1982), and is believed to be evolutionarily related to human serum α_2-macroglobulin, which is inhibitory to trypsin and certain other proteases (Li-Chan & Nakai, 1989). Avian ovomacroglobulin also may be related to the crocodilian counterpart, which inhibits trypsin, subtisin and papain (Ikai, Kitamoto & Nishigai, 1983). Thiol proteases, which include ficin and papain, are inhibited by the minor egg protein cystatin (Fossum & Whitaker, 1968; Sen & Whitaker, 1973; Keilová & Tomášek 1974, 1975; Barrett, 1981; Anastasi et al., 1983; Barrett et al., 1986).

Although most defences are directed against bacterial or fungal invasion, two proteins have been suggested to possess antiviral properties. Ovomucin inhibits viral hemagglutination (Lanni & Beard, 1948; Lanni et al., 1949; Gottschalk & Lind, 1949a,b) and cystatin also may prevent viral infection (Barrett et al., 1986). These properties may be important as albumen proteins have been found in embryonic blood (Marshall & Deutsch, 1951; Wise, Ketterer & Hansen, 1964), where they may be involved in prevention of viral haemagglutination. The combined antimicrobial effect of different albumen proteins may be synergistic in preventing microbial attack (Banks et al., 1986).

Nutritive proteins

Albumen proteins represent a substantial supply of nutrients to the embryo (White, Chapter 1). By binding vitamins and minerals, albumen proteins not only act against microbial attack, but also supply nutrients to the developing embryo. Albumen proteins may be ingested by the embryo directly or selectively taken up phagocytotically by extraembryonic membranes (Marshall & Deutsch, 1951; Wise et al., 1964). This uptake may supply the embryo with either the protein itself, its amino acid constituents, or other molecules which are bound to the protein.

Several proteins may be involved in selective transport of vitamins to the embryo (White, Chapter 1). Avidin is well known to bind the vitamin biotin (Green, 1975), and may be used to transport biotin from mother to embryo (Adiga & Murty, 1983; White, 1987; White, Whitehead & Armstrong, 1987; Bush & White, 1989), although albumen in the fowl contains only 10% of egg biotin (Romanoff & Romanoff, 1949), with yolk supplying the remaining 90%. Riboflavin-binding protein is present in the albumen in approximately equal proportions of bound and unbound forms (Rhodes, Azari & Feeney, 1958; Rhodes et al., 1959). The albumen of fowl eggs contains about 50–70% of the riboflavin in the egg (Stamberg, Petersen & Lampman, 1946), indicating that it is a major source of this nutrient (White & Merrill, 1988). Riboflavin-binding protein is synthesised in the oviduct (Mandeles & Ducay, 1962) and riboflavin is transferred to it from the blood. Thiamin-binding protein in albumen may exist largely in the apoprotein form, since the majority of thiamin in fowl eggs is concentrated in yolk (White, 1987). This may limit the value of albumen thiamin-binding protein in embryonic nutrition.

Ovotransferrin is one protein which may be used to transport specific minerals to the embryo. Ovotransferrin is an iron-binding protein and is identical in amino acid structure to mammalian transferrin (Williams, 1962; Williams et al., 1982), differing only in carbohydrate moieties (Williams, 1968) and absence of sialic acid in ovotransferrin (Osuga & Feeney, 1968). Ovotransferrin may be involved in the transport of iron to the embryo as occurs with a related mammalian blood protein, transferrin (Faulk & Galbraith, 1979). Uteroferrin, an iron-containing protein from the pig, is secreted by the uterus and is taken up by the fetus as a source of iron (Roberts, Raub & Bazer, 1986; Roberts & Bazer, 1988). However, since ovotransferrin in albumen is predominantly in the apoprotein form, it is unlikely to supply much iron to the embryo, although it can release iron to embryonic red blood cells (Li-Chan & Nakai, 1989).

Support and cushioning proteins

The albumen supports the yolk and embryo within the shell, cushioning them from mechanical injury. In birds, the chalazae are fibrous ligaments which suspend the yolk within the centre of the albumen. The chalazae are formed from the twisting of ovomucin fibres as the egg rotates while descending through the oviduct (Conrad & Phillips, 1938; Scott & Huang, 1941). The remaining albumen aids in cushioning the yolk due to its viscosity. The role of albumen in support and cushioning of reptilian eggs is variable. Chalazae have not been reported in reptilian eggs (Ferguson, 1982), and the quantity and viscosity of albumen in reptilian eggs is extremely variable at oviposition. In species which have a thick albumen layer, it may help cushion the yolk when the egg drops into the nest cavity.

Water binding proteins

Most avian albumen proteins are water soluble and at least partially hydrophilic. Most also are glycoproteins, which are known to tightly bind water to the hydrophilic side groups of the peptide chain, as well as the carbohydrate residues (Fennema, 1977). In addition, because of the physical structure of the proteinaceous mass, some water is trapped by surrounding protein molecules. There also is free water in the albumen due simply to osmotic principles. It is, therefore, the combined qualities of albumen as a whole which enables it to act in water storage, creating an osmotic impediment between the embryo and surrounding environment. In the fowl egg, 88% of the weight of albumen is water (Shenstone, 1968; Osuga & Feeney, 1977), three times the amount of water present in yolk, thereby constituting the major source of water for development of the avian embryo. In reptiles, there is no single scheme of storage of water in albumen. Crocodilians, some chelonians and some squamates lay rigid-shelled eggs (Packard

& DeMarco, Chapter 5). In these, albumen is relatively hydrated in the uterus before oviposition (Tracy & Snell, 1985; Palmer & Guillette, 1988), although additional water may be absorbed from the substrate (Packard, Tracy & Roth, 1977; Packard, Chapter 13). In those reptiles with parchment-shelled (most squamates) or pliable-shelled (many turtles) eggs, the egg may absorb substantial quantities of water during development (Ackerman, Chapter 12; Packard, Chapter 13). In fact, eggs of many reptilian species must absorb water for successful embryonic development which is at least partially taken up by albumen in some species.

In birds, albumen is initially secreted as a viscous, concentrated protein mixture in the magnum of the oviduct, which later is saturated with 'plumping' water by the shell gland (Breen & de Bruyn, 1969; Wyburn et al., 1973; Solomon, 1983). The swelling of albumen of in utero eggs has also been observed in turtles (Agassiz, 1857). In squamates, albumen may be relatively unhydrated at oviposition, but obtain water from the substrate (Packard et al., 1977; Tracy, 1980; Tracy & Snell, 1985).

Reptilian albumen proteins

In birds, albumen proteins exhibit a wide variety of functional properties and any alterations in the composition of these proteins will change the functional propertes of the albumen as a whole. This may have important consequences on the embryo's ability to survive under different nest and incubation conditions (Muth, 1980). It is well known that there is variability in composition of albumen proteins from eggs of different avian species (Sibley, 1970; Sibley & Ahlquist, 1972). Embryos whose eggs have a protein composition with properties suitable to the prevailing nest and incubation conditions will have a selective advantage.

When compared to birds, reptiles have more ancient origins, are found in extremely diverse habitats, and exhibit both oviparity and viviparity, suggesting a less conservative pattern of reproductive anatomy and physiology than is observed in birds. This is easily seen in the reproductive anatomy of reptiles (Fox 1977), where studies on oviductal anatomy in lizards (Guillette & Jones, 1985; Adams & Cooper, 1988; Palmer & Guillette, 1988; Uribe et al., 1988; Guillette, Fox & Palmer, 1989), snakes

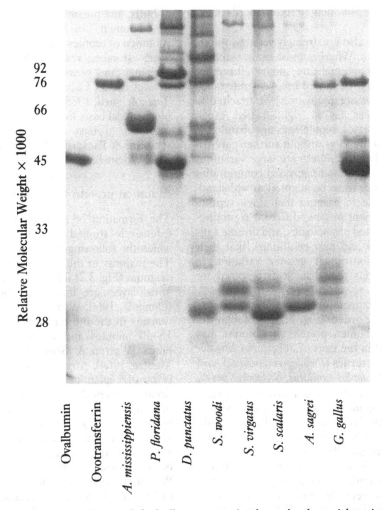

Fig. 3.1. Separation of reptilian and fowl albumen proteins by molecular weight using one-dimensional polyacrylamide gel electrophoresis. Two major avian proteins, ovalbumin and ovotransferrin are included for reference. Eggs of the fowl (*Gallus gallus*), alligator (*Alligator mississippiensis*) and turtle (*Pseudemys floridana*) were infertile, whereas snake (*Diadophis punctatus*) and lizard (*Sceloporus woodi, S. virgatus, S. scalaris,* and *Anolis sagrei*) eggs were fertile with developing embryos analysed after temporary incubation (embryonic stages 35–40; Dufaure & Hubert, 1961; Hubert & Dufaure, 1968).

(Mead, Eroschenko & Highfill, 1981), turtles (Aitken & Solomon, 1976; Palmer, 1987; Palmer & Guillette, 1988, 1990a) and crocodilians (Palmer & Guillette, unpublished data) have demonstrated substantial differences in functional morphology among the reptilian orders and the avian condition. In addition, there also must be differences in physiological control of egg production among these reptilian groups (Trauth & Fagerberg, 1984; Palmer & Guillette, 1988). These phylogenetic, anatom-ical, and physiological differences may effect the types and ratios of extraembryonic proteins.

The eggshell is extremely variable among reptilian groups (Packard & DeMarco, Chapter 5) and is an important mediator of environmental effects on the embryo. Differences in egg-shell structure may greatly affect the role of extra-embryonic proteins due to differences in water conductance (Deeming & Thompson, Chapter 17) and need for antimicrobial agents, thus selecting for differences in protein composition of the albumen (Packard & Packard, 1980).

The nest site also is extremely variable among reptilian species. Whereas most birds exhibit a large degree of parental care, almost all reptiles do not. Eggs may be buried in a substrate (such as sand, soil, or organic matter) or attached to exposed surfaces (as in rigid-shelled gecko eggs). In most cases, reptile eggs are abandoned and allowed to develop without further parental assistance. Therefore, there are large variations in moisture content and microbial communities to which the egg must be adapted to withstand. It is reasonable to assume that since reptiles have more ancient origins, different reproductive anatomies and physiologies, and diverse eggshell structures and nest conditions, that their eggs will exhibit much greater variation in albumen proteins than has been reported for avian eggs.

Reptilian albumen exhibits substantial differences among orders (Fig. 3.1). Although the major avian albumen protein, ovalbumin, has been detected in the eggs of alligators (*Alligator mississippiensis*), turtles (*Pseudemys floridana*), and squamates (*Diadophis punctatus, Sceloporus woodi, S. virgatus, S. scalaris, Anolis sagrei*), it comprises a relatively minor fraction of the albumen (Palmer, 1990). An ovotransferrin-like molecule has been detected in turtle (*P. floridana*) albumen (Palmer, 1988), based on molecular weight data, although more rigorous identification is required. Neither ovalbumin nor ovotransferrin were detected in eggs of *Crocodylus porosus* (Burley *et al.*, 1987), although a major protein of 59,000 molecular weight was identified, and has been subsequently detected in the eggs of both alligators and turtles (Palmer, 1990: Fig. 3.1). An α_2-macroglobulin-like protein has been found in the eggs of *Crocodylus rhombifer* (Ikai *et al.*, 1983), *C. porosus* (Burley *et al.*, 1987) and other reptiles (Palmer, 1990). Although avidin synthesis was observed in oviducts of the lizard *Lacerta s. sicula* (Botte, Segal & Koide, 1974; Botte & Granata, 1977), neither avidin, riboflavin-binding protein, nor thiamin-binding protein have been detected in the albumen of *A. mississippiensis* (Abrams *et al.*, 1988, 1989).

Clearly, the study of albumen proteins of reptiles is in its infancy. There appears to be a greater diversity in composition of albumen proteins of reptiles (Burley *et al.*, 1987; Palmer, 1988, 1989, 1990; Abrams *et al.*, 1989) than that

of birds, and possibly also in the functional role of albumen in embryonic development. Albumen of reptiles has been suggested to act in water storage, resist lethal rates of water exchange and possess antimicrobial properties (Movchan & Gabaeva, 1967; Ewert, 1979; Tracy & Snell, 1985). These properties, and the biochemical basis for them, may have evolved in response to nest and incubation conditions (Packard & Packard, 1980; Tracy, 1980, 1982; Tracy & Snell, 1985).

Albumen protein formation

The formation of albumen proteins has been extensively studied in the domestic fowl, on which the following discussion is largely based. The majority of the albumen is secreted by the magnum (Fig. 3.2) following ovulation, although initial layers are formed in the infundibulum (Dominic, 1960; Wyburn *et al.*, 1970). The egg remains in the infundibulum for approximately 15–30 minutes and in the magnum for 2–3 hours (Warren & Scott, 1935; Woodward & Mather, 1964). Anatomically, the infundibulum is mostly aglandular, but 'glandular grooves' occur posteriorly (Giersberg, 1923). True tubular glands occur near the infundibulum–magnum junction which resemble 'glandular grooves' in general cell structure (Dominic, 1960), although characteristic differences have been reported (Aitken & Johnston, 1963). Within the magnum, there are tubular endometrial glands in addition to a secretory luminal epithelium (Richardson, 1935; Wyburn *et al.*, 1970). The endometrial glands are composed of three types of cells: A cells, filled with electron dense granules; B cells, filled with low electron density secretory material; and C cells, with prominent Golgi and extensive rough endoplasmic reticulum; cell types A and C may, however, simply reflect different phases of secretory activity within the same cell type (Wyburn *et al.*, 1970).

Albumen proteins are secreted by specific cells within the magnum (Gilbert, 1979). Most albumen proteins are thought to be secreted by endometrial glands. Ovalbumin is secreted by A-cells whereas lysozyme is released by B-cells (Kohler, Grimley & O'Malley, 1968; Oka & Schimke, 1969; Wyburn *et al.*, 1970). The cells of the tubular glands also are known to secrete ovotransferrin and ovomucoid (Schimke *et al.*, 1977), whereas the luminal epithelium secretes

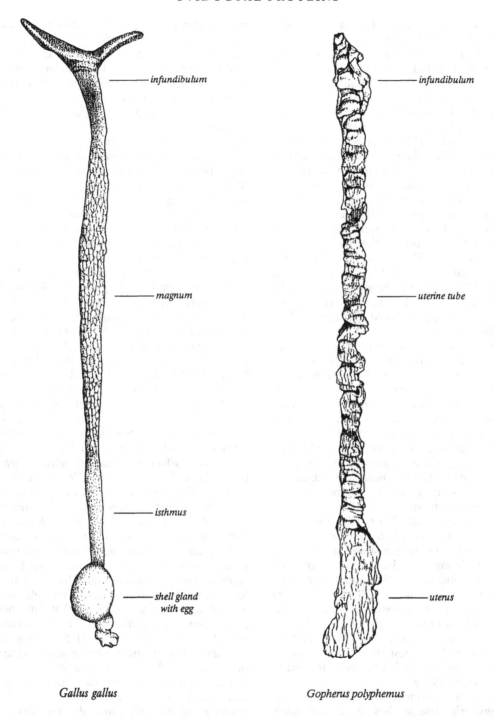

Gallus gallus   *Gopherus polyphemus*

Fig. 3.2.   External morphology of the avian (*Gallus gallus*) and reptilian (*Gopherus polyphemus*) oviduct (adapted from Palmer & Guillette, 1988).

both avidin and ovomucin (Kohler *et al.*, 1968; Wyburn *et al.*, 1970; Tuohimaa, 1975).

The magnum synthesises and secretes the albumen protein, unlike yolk proteins which are manufactured in the liver and transported to the ovary via the circulatory system (White, Chapter 1). This has been shown by *in vitro* incorporation of radiolabeled amino acids into various albumen proteins (Mandeles & Ducay, 1962; O'Malley, 1967). Additional work has shown *in vivo* incorporation of radioactive lysine and glycine into ovalbumin, ovotransferrin and lysozyme (Mandeles & Ducay, 1962). Some albumen proteins secreted by the oviduct are either identical (avidin) or similar (ovotransferrin) to proteins found in other body tissues or fluids (Green, 1975; Chasteen, 1977, 1983; Aisen & Listowsky, 1980; Elo & Korpela, 1984; Brock, 1985), however, these are known to be synthesised directly by the oviduct and not transported there by blood (Williams, 1962; O'Malley, 1967).

The enormous amount of protein synthesised for albumen has given molecular biologists an excellent model to study the regulation of genetic mechanisms involved in protein formation. Under stimulation from the adenohypophysis, ovarian follicles synthesise and release oestrogen (Norris, 1985). This is transported to the oviduct via the blood and readily passes through the plasmalemma of cells in the endometrial glands. Once inside the cell, oestrogen is bound by a nuclear receptor protein, which binds to chromatin, stimulating transcription (O'Malley *et al.*, 1979; Chambon *et al.*, 1984; Gorski, Hansen & Welshons, 1987; Leavitt, 1989). Oestrogen stimulates the gland cells of immature chicks to differentiate (Brant & Nalbandov, 1956) – initiates transcription of the ovalbumin gene (Roop *et al.*, 1978), and synthesis of ovalbumin, ovotransferrin, ovomucoid, and lysozyme (Palmiter & Gutman, 1972; Palmiter & Schimke, 1973). Further, treatment with progesterone and oestrogen causes the luminal epithelium to secrete avidin (Kohler *et al.*, 1968).

The cellular mechanism of ovalbumin and ovotransferrin release has been elucidated. Inhibition of glycosylation does not block the release of ovalbumin or ovotransferrin (Kato, Noguchi & Naito, 1987), indicating that these proteins are free within the endoplasmic reticulum and Golgi complex, and not bound to the organelle membranes. Further, inhibition of protein transport from the endoplasmic reticulum to the Golgi complex blocks release of these proteins, as does disrupting microtubules which have been implicated in secretory granule transport (Kato *et al.*, 1987). Thus, these data suggest that ovalbumin and ovotransferrin are enclosed within secretory granules which are transported from the Golgi complex to the plasmalemma by microtubules.

The signal for secretion of albumen proteins is still poorly understood, although the pressure of the descending yolk on the oviductal wall is generally thought to stimulate albumen secretion (Sturkie & Mueller, 1976; Laugier & Brard, 1980). Artificial mechanical stimulation, created by introducing objects into the oviduct, does induce the formation of albumen and eggshell, although these are grossly abnormal (Wentworth, 1960; Pratt, 1960). However, albumen is also secreted by isolated loops of the oviduct while an egg passes through the intact portion (Burmester & Card, 1939, 1941), indicating that neuronal or endocrine stimulation may play a role in secretion. Although the oviduct is known to be innervated by the autonomic nervous system, administration of acetylcholine (parasympathomimetic) and ephedrine sulfate (sympathomimetic) had negligible effects on albumen secretion (Sturkie & Weiss, 1950; Sturkie, Weiss & Ringer, 1954). Diffusible yolk proteins can affect oviductal metabolism *in vitro* (Eiler, González & Horvath, 1970), but it is unknown if they can alter albumen synthesis *in vivo*. Steroid hormones, which are involved in the ovulatory process, have been shown to induce albumen synthesis but not their release.

It remains to be determined how other hormones involved during ovulation, such as arginine vasotocin and prostaglandins (PG), may be involved in albumen release. Prostaglandins are potent paracrine hormones that influence blood flow and contraction of the female reproductive tract (Poyser, 1981). Stretching the reproductive tract causes prostaglandin release that can influence contractile patterns throughout the oviduct. Prostaglandin synthesis immediately after ovulation may be essential for transport of the yolk down the reproductive tract, release of oviductal proteins and egg rotation, which is important for normal shell formation. Recent work in our laboratory indicates that the reproductive tract of lizards is capable of synthesis and secretion of large amounts of PG-F and PG-E_2 during the first 24 hours after

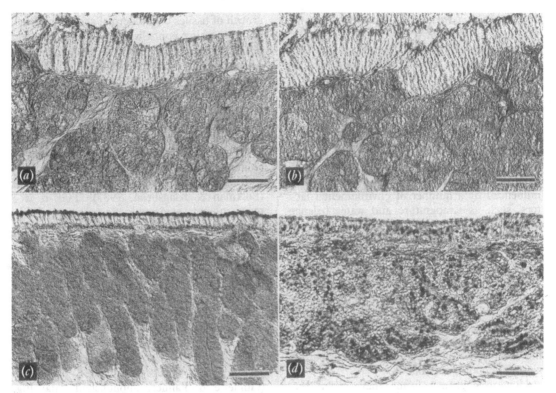

Fig. 3.3. Immunocytochemical localisation of albumen proteins and growth factors in reptilian oviducts. (*a*) Localisation of albumen secreting cells in the endometrial glands of the tortoise (*Gopherus polyphemus*) tube using whole fowl egg albumen antibodies. (*b*) Insulin-like growth factor-I (IGF-I) localisation in the tube of the tortoise (*G. polyphemus*). (*c*) Localisation of IGF-I in the uterus of the tortoise (*G. polyphemus*). (*d*) Epidermal growth factor (EGF) localisation in the posterior uterus of the alligator (*Alligator mississippiensis*). Bars = 50 μm.

ovulation (Guillette, unpublished data). In birds, a similar pattern is observed but on a more rapid time scale (Hertelendy *et al.*, 1984). Clearly, Mechanical stimulation of the oviduct may trigger a complex series of paracrine events are responsible for albumen and eggshell protein release. Further research into other mechanisms, such as the hypothesis regarding the role of prostaglandins, needs to be conducted.

Albumen of avian eggs is composed of four distinct portions: a chalaziferous layer, which is attached to the yolk and suspends it, an inner liquid layer around the chalazae, a thick layer, and an outer thin layer adjacent to the eggshell. These different layers are not obvious during their formation in the magnum but are the result of secretion of different components by consecutive regions of the oviduct (Gilbert, 1971), by the mechanical action of the egg twisting as it descends through the oviduct (Romanoff & Romanoff, 1949) and chemical changes during

plumping in the shell gland (Sturkie & Mueller, 1976). The chalazae are formed by the mechanical twisting of ovomucin fibres as the egg spirals down the magnum (Conrad & Phillips, 1938; Scott & Huang, 1941). The thin albumen layers are produced by addition of plumping water after the egg enters the shell gland (Sturkie & Polin, 1954). Like the fibrous chalazae, the gelatinous nature of the thick albumen is due to its high concentration of ovomucin (Robinson, 1987).

In contrast to birds, the formation of albumen proteins in reptiles is poorly understood. Avidin was the first albumen protein detected in oviductal secretions of a reptile (Botte *et al.*, 1974), and is induced by combined effects of oestrogen and progesterone, as in the fowl (Botte *et al.*, 1974; Botte & Granata, 1977). Recent work using tritiated leucine and explant cultures in the turtle, *Pseudemys s. scripta*, has shown that albumen proteins are synthesised and secreted *in vitro* by the uterine tube of the oviduct (Palmer *et al.*, 1990: Fig. 3.2). Further, antibodies to whole fowl albumen and purified ovalbumin show cross-reactivity to reptilian albumen proteins and bind to endometrial glands of the tube in the alligator, *A. mississippiensis* and the tortoise *Gopherus polyphemus* (Palmer, 1990: Fig. 3.3).

Growth factors

Control of embryonic growth is known to be influenced by a number of environmental factors, such as temperature and water balance (Hubert, 1985; Deeming & Ferguson, Chapter 10; Packard, Chapter 13). However, recent evidence suggests that the maternal system may also influence these phenomena by secreting growth factors (low molecular weight proteins) from the reproductive tract that are taken up by embryo or extraembryonic tissues (Guillette, 1989). This is not surprising in embryonic development of viviparous mammalian species, but the role of these compounds in oviparous species is poorly understood.

Numerous growth factors have been described but their specific functions are only now being elucidated. Somatomedins, polypeptide growth factors secreted by the liver and other tissues, are important during embryonic development (Gilbert, 1988). Specifically, insulin-like growth factor I (IGF-I or somatomedin C), a potent mitotic agent (Leof *et al.*, 1982; Froesch *et al.*, 1985; Baxter, 1986), is a polypeptide composed of 70 amino acids, having a proinsulin-like structure (Humbel & Rinderknecht, 1979). IGF-I inhibits adenylate cyclase and stimulates guanylate cyclase activity, thereby lowering cyclic AMP concentrations and elevating cyclic GMP levels in a manner similar to the action of insulin (Hadley, 1988). Cell differentiation is stimulated by somatotropin (STH, growth hormone), which also initiates IGF-I production (Isaksson, Edén & Jansson, 1985). IGF-I causes the newly differentiated cells to undergo mitosis, creating a clonal expansion of that cell type (Zezulak & Green, 1986). Somatomedins also have insulin-like actions, including accelerated amino acid and glucose uptake, and glucose oxidation (Martin, 1985).

Epidermal growth factor (EGF), a polypeptide composed of 53 amino acids, is synthesised by many tissues in the body (Carpenter & Cohen, 1979; Hadley, 1988). It modulates growth of tissues of ectodermal and endodermal origin, causing epithelial cell proliferation and differentiation (Carpenter & Cohen, 1979; Martin, 1985; Hadley, 1988). EGF activates protein kinases that catalyse phosphorylations of tyrosyl moieties of specific proteins (Cohen, Carpenter & King, 1980; Soderquist & Carpenter, 1983). Another growth factor of possible importance is transforming growth factor-α (TGF-α). This is a polypeptide composed of 50 amino acids having sequence homology (35%) with EGF, and binds to the EGF receptor (Folkman & Klagsbrun, 1987). TGF-α is a more potent angiogenic factor than EGF, requiring ten times less hormone *in vivo* to initiate new microvascular endothelial cell proliferation (Schreiber, Winkler & Derynck, 1986).

Recent work in our laboratory has demonstrated that the reptilian reproductive tract may synthesise and release both EGF and IGF-I. Using immunocytochemistry and specific antibodies, we have observed that IGF-I is localised in the endometrial glands of the tube and uterus, in the uterine luminal epithelium, and their secretions in *A. mississippiensis* and *G. polyphemus* (Guillette & Palmer, unpublished data: Fig. 3.3), whereas EGF is localised in uterine endometrial glands, luminal epithelium, and secretions of the alligator reproductive tract (Guillette & Palmer, unpublished data: Fig. 3.3). Although some growth factors have been localised in alligator embryos, *in situ* hybridisation for EGF and IGF-I in the embryo has revealed no RNA synthesis (Ferguson, personal communication). Growth factors may, therefore, be synthesised by the oviduct, secreted into the lumen, and incorporated into the egg, to aid in embryonic development.

Eggshell membranes

The proteinaceous fibres of the eggshell membranes play an important role in the maintenance of a suitable environment for the developing embryo. As with albumen, the eggshell membranes exhibit several functions, which are due to their mechanical structure as a whole, rather than from individual chemical properties. The protein of the eggshell membrane has proven difficult to investigate as it is highly insoluble (DeSalle, Veith & Sexton, 1984) but despite this the amino acid composition of eggshell membrane protein is known for a variety of species, including fowl (Candlish &

Scougall, 1969; Balch & Cooke, 1970; Britton & Hale, 1977; Starcher & King, 1980; Crombie *et al.*, 1981; Leach, Rucker & van Dyke, 1981), green sea turtle, *Chelonia mydas* (Solomon & Baird, 1977), and lizards, *Iguana iguana, Eumeces fasciatus* and *Opheodrys vernalis* (Cox, Mecham & Sexton, 1982; Cox *et al.*, 1984). The protein has, however, been assigned to a variety of classes over the years, including keratin (Balch & Cooke, 1970; Britton & Hale, 1977), collagen (Candlish & Scougall, 1969; Wong *et al.*, 1984) and elastin (Starcher & King, 1980; Leach *et al.*, 1981; Crombie *et al.*, 1981) leading to the conclusion that eggshell membranes are composed of unique proteins (Leach, 1978; Tullett, 1987).

Eggshell membranes have a variety of functions, including water balance, mechanical support and microbial defence (Packard & Packard, 1980; Tracy & Snell, 1985). As part of the eggshell, the fibrous membrane acts in water balance by forming a barrier to gas exchange and increasing the diffusion distance, thereby slowing down water movement (Ackerman, Seagrave, Dmi'el & Ar, 1985; Ackerman, Dmi'el & Ar, 1985). The membrane, however, has no effect on direction of water movement, which is determined by relative osmotic potential across the eggshell (Vleck, Chapter 15). The fibrous membrane also serves to maintain egg shape, support egg contents, and protect the embryo from mechanical injury (Packard & Packard, 1980; Tracy, 1982). Since fibres of the membrane form a dense crisscrossing mat, they act as a barrier to invading bacteria and fungus (Board & Tranter, 1986). Functions of eggshell membranes may be enhanced by the organic matrix of protein and carbohydrates surrounding the fibres. It is likely that albumen and eggshell coevolved in response to the environmental conditions to which eggs are subjected (Packard & Packard, 1980; Tracy, 1980, 1982; Tracy & Snell, 1985).

Fibres are produced by endometrial glands of the isthmus in fowl (Misugi & Katsumata, 1963; Simkiss, 1968; Simkiss & Taylor, 1971; Solomon, 1975; Aitken, 1971; Draper *et al.*, 1972; Solomon, 1983: Fig. 3.2) or uterus in the sea turtle, *C. mydas* (Aitken & Solomon, 1976), lizards, *Crotaphytus collaris, Eumeces obsoletus*, and *S. woodi* (Guillette *et al.*, 1989; Palmer, 1990; Palmer & DeMarco, unpublished data), and tortoise, *G. polyphemus* (Palmer & Guillette, 1988: Fig. 3.2). In *A. mississippiensis*, fibres are produced only in the anterior uterus whereas

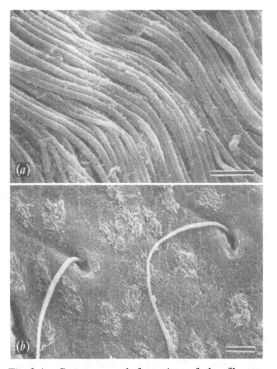

Fig. 3.4. Structure and formation of the fibrous eggshell membrane in reptiles. (*a*) Proteinaceous fibres of the eggshell membrane of the lizard *Crotaphytus collaris*. (*b*) Secretion of eggshell fibres in the uterus shortly after ovulation in the lizard *Sceloporus woodi*. Bars = 5 μm.

calcium is secreted by the posterior uterus (Palmer & Guillette, 1990*b*). The alligator oviduct is, therefore, morphologically and histologically similar to that of the avian oviduct, which also has separate regions for the formation of the eggshell membrane (isthmus) and calcareous layer (shell gland). Secretory granules are released into the lumen of the endometrial glands and coalesce as they are extruded from the glandular duct into the oviductal lumen forming fibres (Fig. 3.4; Solomon, 1983; Palmer, 1989, 1990; Palmer & DeMarco, unpublished data). These fibres have been described as possessing a protein core with a mucopolysaccharide mantle (Simons & Wiertz, 1963; Candlish, 1972; Solomon & Baird, 1977) but in some reptilian eggs, the fibres lack the mantle and exhibit a pitted appearance (Solomon & Baird, 1977; Andrews & Sexton, 1981; Trauth & Fagerberg, 1984; Sexton, Veith & Phillips, 1979; DeSalle *et al.*, 1984).

Future research directions

How oviductal proteins influence embryonic development is still a growing field. This is especially true in reptiles, which are only now coming under intense research. In particular, the composition of reptilian egg albumen, including protein characterisation, isolation, and determination of biological properties, requires extensive investigation. This may not only shed light on the physiological development of the embryo, but also on the evolution of proteins and the ecology of eggs in general. Although the genetics of albumen transcription has been studied intensely in the fowl, it remains to be examined in reptiles. The mechanism of protein release in birds and reptiles is still unclear. Whereas several growth factors are known to be synthesised and secreted by the oviduct, others remain to be investigated and may play an important role in embryonic development. Further, the anatomy and physiology of reptilian oviducts and production of albumen and eggshell components are areas requiring indepth investigation. The biochemical nature of eggshell membrane proteins continues to be debated, and may prove to be a unique class of proteins, as has been previously suggested. The comparative study of oviductal egg proteins, including their secretion, function in embryonic development, and role in ecological interactions of eggs with their nest environment, may shed new light on the evolution of amniotic eggs.

Acknowledgements

We wish to thank M. W. J. Ferguson and D. C. Deeming for discussions and assistance in producing the manuscript; Drs William Buhi and Harold White III for comments on the manuscript; Dr Henry Wilson for his discussions and assistance; Dr Henry Aldrich and the staff of the E. M. Core Facility of the Interdisciplinary Center for Biotechnology Research for assistance with electron micrographs; Daryl Harrison for line drawings; the Department of Zoology for financial support; Louis Somma, John Matter and Vincent DeMarco for assistance collecting specimens; and Sylvia Palmer for assistance and patience in preparing the manuscript.

References

Abrams, V. A. M., Bush, L., Kennedy, T., Schreiber, R. W. Jr, Sherwood, T. A. & White, H. B., III (1989). Vitamin-transport proteins in alligator eggs. *Comparative Biochemistry and Physiology*, **93B**, 291–7.

Abrams, V. A. M., McGahan, T. J., Rohrer, J. S., Bero, A. S. & White, H. B., III (1988). Riboflavin-binding protein from reptiles: A comparison with avian riboflavin-binding proteins. *Comparative Biochemistry and Physiology*, **90B**, 243–7.

Ackerman, R. A., Dmi'el, R. & Ar, A. (1985). Energy and water vapor exchange by parchment-shelled reptile eggs. *Physiological Zoology*, **58**, 129–37.

Ackerman, R. A., Seagrave, R. C., Dmi'el, R. & Ar, A. (1985). Water and heat exchange between parchment-shelled reptile eggs and their surroundings. *Copeia*, **1985**, 703–11.

Adams, C. S. & Cooper, W. E. Jr (1988). Oviductal morphology and sperm storage in the keeled earless lizard, *Holbrookia propinqua*. *Herpetologica*, **44**, 190–7.

Adiga, P. R. & Murty, C. V. R. (1983). Vitamin carrier proteins during embryonic development in birds and mammals. In *Molecular Biology of Egg Maturation, Ciba Foundation Symposium 98*, ed. R. Porter & J. Whelan, pp. 111–36. London: Pitman Books.

Agassiz, L. (1857). *Contributions to the Natural History of the United States of America*, Vol. I, II. Boston: Little, Brown and Co.

Aisen, P. & Listowsky, I. (1980). Iron transport and storage proteins. *Annual Review of Biochemistry*, **49**, 357–93.

Aitken, R. N. C. (1971). The oviduct. In *Physiology and Biochemistry of the Domestic Fowl*, vol. III, ed. D. J. Bell & B. M. Freeman, pp. 1237–89. New York: Academic Press.

Aitken, R. N. C. & Johnston, H. S. (1963). Observations on the fine structure of the infundibulum of the avian oviduct. *Journal of Anatomy*, **97**, 87–99.

Aitken, R. N. C. & Solomon, S. E. (1976). Observations on the ultrastructure of the oviduct of the Costa Rican green turtle (*Chelonia mydas* L.). *Journal of Experimental Marine Biology and Ecology*, **21**, 75–90.

Anastasi, A., Brown, M. A., Kembhavi, A. A., Nicklin, M. J. H., Sayers, C. A., Sunter, D. C. & Barrett, A. J. (1983). Cystatin, a protein inhibitor of cysteine proteinases. Improved purification from egg white, characterization, and detection in chicken serum. *Biochemical Journal*, **211**, 129–38.

Andrews, R. M. & Sexton, O. J. (1981). Water relations of the eggs of *Anolis auratus* and *Anolis limifrons*. *Ecology*, **62**, 556–62.

Baker, C. M. A. (1968). The proteins of egg white. In *Egg Quality: A Study of the Hen's Egg*, ed. T. C. Carter, pp. 67–108. Edinburgh: Oliver & Boyd.

Balch, D. A. & Cooke, R. A. (1970). A study of the composition of the hen's egg-shell membranes. *Annales de Biologie Animale, Biochime et Biophysique*, **10**, 13–25.

Banks, J. G., Board, R. G. & Sparks, N. H. C. (1986). Natural antimicrobial systems and their potential in food preservation of the future. *Biotechnology and Applied Biochemistry*, **8**, 103–47.

Barrett, A. J. (1981). Cystatin, the egg white inhibitor of cysteine proteinases. *Methods in Enzymology*, **80**, 771–8.

Barrett, A. J., Rawlings, N. D., Davies, M. E., Machleidt, W., Salvesen, G. & Turk, V. (1986). Cysteine proteinase inhibitors of the cystatin superfamily. In *Proteinase Inhibitors*, ed. A. J. Barrett & G. Salvesen, pp. 515–69. Amsterdam: Elsevier Scientific Publishing Co.

Baxter, R. C. (1986). The somatomedins: insulin-like growth factors. *Advances in Clinical Chemistry*, **25**, 49–115.

Board, R. G. & Tranter, H. S. (1986). The microbiology of eggs. In *Egg Science and Technology*, 3rd edn, ed. W. J. Stadelman & O. J. Cotterill, pp. 75–96. Connecticut: Avi Publishing Company, Inc.

Botte, V. & Granata, G. (1977). Induction of avidin synthesis by RNA obtained from lizard oviducts. *Journal of Endocrinology*, **73**, 535–6.

Botte, V., Segal, S. J. & Koide, S. S. (1974). Induction of avidin synthesis in the oviduct of the lizard *Lacerta sicula*, by sex hormones. *General and Comparative Endocrinology*, **23**, 357–9.

Brant, J. W. A. & Nalbandov, A. V. (1956). Role of sex hormones in albumen secretion by the oviduct of chickens. *Poultry Science*, **35**, 692–700.

Breathnach, R., Benoist, C., O'Hare, K., Gannon, F. & Chambon, P. (1978). Ovalbumin gene: evidence for a leader sequence in mRNA and DNA sequences at the exon-intron boundaries. *Proceedings of the National Academy of Science, USA*, **75**, 4853–7.

Breen, P. C. & de Bruyn, P. P. H. (1969). The fine structure of the secretory cells of the uterus (shell gland) of the chicken. *Journal of Morphology*, **128**, 35–66.

Britton, W. M. & Hale, K. K. Jr (1977). Amino acid analysis of shell membranes of eggs from young and old hens varying in shell quality. *Poultry Science*, **56**, 865–71.

Brock, J. H. (1985). Transferrins. In *Metalloproteins, Part 2: Metal proteins with Non-redox Roles*, ed. P. M. Harrison, pp. 183–262. Deerfield Beach, Florida: Verlag Chemie.

Burley, R. W., Back, J. F., Wellington, J. E. & Grigg, G. C. (1987). Proteins of the albumen and vitelline membrane of eggs of the estuarine crocodile, *Crocodylus porosus*. *Comparative Biochemistry and Physiology*, **88B**, 863–7.

Burmester, B. R. & Card, L. E. (1939). The effect of resecting the so-called 'chalaziferous region' of the hen's oviduct on the formation of subsequent eggs. *Poultry Science*, **18**, 138–45.

— (1941). Experiments on the physiology of egg white secretion. *Poultry Science*, **20**, 224–6.

Bush, L. & White, H. B., III (1989). Avidin traps biotin diffusing out of chicken egg yolk. *Comparative Biochemistry and Physiology*, **93B**, 543–7.

Candlish, J. K. (1972). The role of the shell membranes in the functional integrity of the egg. In *Egg Formation and Production*, ed. B. M. Freeman & P. E. Lake, pp. 87–105. Edinburgh: British Poultry Science.

Candlish, J. K. & Scougall, R. K. (1969). L-5-hydroxylysine as a constituent of the shell membranes of the hen's egg. *International Journal of Protein Research*, **1**, 299–302.

Carpenter, G. & Cohen, S. (1979). Epidermal growth factor. *Annual Review of Biochemistry*, **48**, 193–216.

Carrell, R. W. & Boswell, D. R. (1986). Serpins: The superfamily of plasma serine proteinase inhibitors. In *Proteinase Inhibitors*, ed. A. J. Barrett & G. Salvesen, pp. 403–20. Amsterdam: Elsevier Scientific Publishing Co.

Chambon, P., Dierich, A., Gaub, M., Jokowler, S., Jongstra, J., Krust, A., LePennec, J., Oudet, P. & Reudelhuber, T. (1984). Promoter elements of genes coding for proteins and modulation of transcription by estrogen and progesterone. *Recent Progress in Hormone Research*, **40**, 1–42.

Chasteen, N. D. (1977). Human serotransferrin: structure and function. *Coordination Chemistry Reviews*, **22**, 1–36.

— (1983). Transferrin: a perspective. *Advances in Inorganic Biochemistry*, **5**, 201–33.

Cohen, S., Carpenter, G. & King, L. Jr (1980). Epidermal growth factor (EGF)-receptor-protein kinase interactions: co-purification of receptor and EGF-enhanced phosphorylation activity. *Journal of Biological Chemistry*, **255**, 4834–42.

Conrad, R. M. & Phillips, R. E. (1938). The formation of the chalazae and inner thin white in the hen's egg. *Poultry Science*, **17**, 143–6.

Cox, D. L., Koob, T. J., Mecham, R. P. & Sexton, O. J. (1984). External incubation alters the composition of squamate eggshells. *Comparative Biochemistry and Physiology*, **79B**, 481–7.

Cox, D. L., Mecham, R. P. & Sexton, O. J. (1982). Lysine derived cross-links are present in a non-elastin, proline-rich protein fraction of *Iguana iguana* eggshell. *Comparative Biochemistry and Physiology*, **72B**, 619–23.

Crombie, G., Snider, R., Faris, B. & Franzblau, C. (1981). Lysine-derived cross-links in the egg shell membrane. *Biochimica et Biophysica Acta*, **640**, 365–7.

DeSalle, R., Veith, G. M. & Sexton, O. J. (1984). An enzymatic and histochemical analysis of eggshells of anoline lizards. *Comparative Biochemistry and Physiology*, **78A**, 237–42.

Dominic, C. J. (1960). On the secretory activity of the funnel of the avian oviduct. *Current Science*, **29**, 274–5.

Draper, M. H., Davidson, M. F., Wyburn, G. M. & Johnston, H. S. (1972). The fine structure of the fibrous membrane forming region of the isthmus of the oviduct of *Gallus domesticus*. *Quarterly Journal of Experimental Physiology*, **57**, 297–310.

Dufaure, J. P. & Hubert, J. (1961). Table de développement du lézard vivipare: *Lacerta (Zootoca) vivipara* Jacquin. *Archives d'anatomie microscopique et de morphologie experimentale*, **50**, 309–28.

Eiler, H., González, E. & Horvath, A. (1970). Effect of egg yolk substances on the metabolism of the magnum *in vitro*. *Biology of Reproduction*, **2**, 172–7.

Elo, H. A. & Korpela, J. (1984). The occurrence and production of avidin: a new conception of the high-affinity biotin-binding protein. *Comparative Biochemistry and Physiology*, **78B**, 15–20.

Ewert, M. A. (1979). The embryo and its egg: development and natural history. In *Turtles Perspectives and Research*, ed. M. Harless & H. Morlock, pp. 333–413. New York: John Wiley & Sons, Inc.

Faulk, W. P. & Galbraith, G. M. P. (1979). Trophoblast transferrin and transferrin receptors in the host-parasite relationship of human pregnancy. *Proceedings of the Royal Society of London, B*, **204**, 83–97.

Feeney, R. E. (1971). Comparative biochemistry of avian egg white ovomucoids and ovoinhibitors. In *Proceedings of the International Research Conference on Proteinase Inhibitors*, ed. H. Fritz & H. Tschesche, pp. 162–7. Berlin: Walter de Gruyter.

Feeney, R. E. & Allison, R. G. (1969). *Evolutionary Biochemistry of Proteins. Homologous and Analogous Proteins from Avian Egg Whites, Blood Sera, Milk, and Other Substances*. New York: Wiley-Interscience.

Feeney, R. E., Stevens, F. C. & Osuga, D. T. (1963). The specificities of chicken ovomucoid and ovoinhibitor. *Journal of Biological Chemistry*, **238**, 1415–18.

Fennema, O. (1977). Water and protein hydration. In *Food Proteins*, ed. J. R. Whitaker & S. R. Tannenbaum, pp. 50–90. Westport, Connecticut: Avi Publishing Co., Inc.

Ferguson, M. W. J. (1982). The structure and composition of the eggshell and embryonic membranes of *Alligator mississippiensis*. *Transactions of the Zoological Society of London*, **36**, 99–152.

Folkman, J. & Klagsbrun, M. (1987). Angiogenic factors. *Science*, **235**, 442–7.

Fossum, K. & Whitaker, J. R. (1968). Ficin and papain inhibitor from chicken egg white. *Archives of Biochemistry and Biophysics*, **125**, 367–75.

Fox, H. (1977). The urinogenital system of reptiles. In *Biology of the Reptilia*, vol. 6, ed. C. Gans, pp. 1–157. New York: Academic Press.

Froesch, E. R., Schmid, C., Schwander, J. & Zapf, J. (1985). Actions of insulin-like growth factors. *Annual Review of Physiology*, **47**, 443–67.

Geoffroy, P. & Bailey, C. J. (1975). The action of hen and goose lysozyme on the cell-wall peptidoglycan of *Micrococcus lysodeikticus*. *Biochemical Society Transactions*, **3**, 1212–14.

Giersberg, H. (1923). Untersuchungen über physiologie und histologie des eileiters der reptilien un vögel; nebst einem beitrag zur fasergenese. *Zeitschrift für Wissenschaftliche Zoologie*, **120**, 1–97.

Gilbert, A. B. (1971). Egg albumen and its formation. In *Physiology and Biochemistry of the Domestic Fowl*, vol. 3, ed. D. J. Bell & B. M. Freeman, pp. 1291–329. New York: Academic Press.

— (1979). Female genital organs. In *Form and Function in Birds*, vol. 1, ed. A. S. King & J. McLelland, pp. 237–360. London: Academic Press.

Gilbert, S. F. (1988). *Developmental Biology*, 2nd edn. Sunderland, Massachusetts: Sinauer Associates, Inc.

Gorski, J., Hansen, J. C. & Welshons, W. V. (1987). Estrogen receptors as nuclear proteins. In *Cell and Molecular Biology of the Uterus*, ed. W. W. Leavitt, pp. 13–29. New York: Plenum Press.

Gottschalk, A. & Lind, P. E. (1949*a*). Ovomucin, a substrate for the enzyme of influenza virus. I. Ovomucin as an inhibitor of haemagglutination by heated Lee virus. *British Journal of Experimental Pathology*, **30**, 85–92.

— (1949*b*). Product of interaction between influenza virus and ovomucin. *Nature*, London, **164**, 232–3.

Green, N. M. (1963). Avidin. I. The use of [^{14}C]biotin for kinetic studies and for assay. *Biochemical Journal*, **89**, 585–91.

— (1975). Avidin. *Advances in Protein Chemistry*, **29**, 85–133.

Guillette, L. J. Jr (1989). The evolution of vertebrate viviparity: morphological modifications and endocrine control. In *Complex Organismal Functions: Integration and Evolution in Vertebrates*, ed. D. B. Wake & G. Roth, pp. 219–33. New York: John Wiley & Sons.

Guillette, L. J. Jr, Fox, S. L. & Palmer, B. D. (1989). Oviductal morphology and egg shelling in the oviparous lizards *Crotaphytus collaris* and *Eumeces obsoletus*. *Journal of Morphology*, **201**, 145–59.

Guillette, L. J. Jr & Jones, R. E. (1985). Ovarian, oviductal and placental morphology of the reproductively bimodal lizard species, *Sceloporus aeneus*. *Journal of Morphology*, **184**, 85–98.

Hadley, M. E. (1988). *Endocrinology*, 2nd edn. Englewood Cliffs, New Jersey: Prentice Hall, Inc.

Hayakawa, S., Kondo, H., Nakamura, R. & Sato, Y. (1983). Effect of β-ovomucin on the solubility of α-ovomucin and further inspection of the structure of ovomucin complex in thick egg white. *Agricultural and Biological Chemistry*, **47**, 815–20.

Heath, J. L. (1977). Chemical and related osmotic changes in egg albumen during storage. *Poultry Science*, **56**, 822–8.

Hertelendy, F., Olson, D. M., Todd, H., Hammond, R. W., Toth, M. & Asboth, G. (1984). Role of prostaglandins in oviposition and ovulation. In *Reproductive Biology of Poultry*, ed. F. J. Cunningham, P. E. Lake & D. Hewitt, pp. 89–102. Harlow: Longman Group.

Hubert, J. (1985). Embryology of the squamata. In *Biology of the Reptilia*, vol. 15, ed. C. Gans & F. Billett, pp. 1–34. New York: John Wiley & Sons.

Hubert, J. & Dufaure, J. P. (1968). Table de développement de la vipère aspic, *Vipera aspis* L. *Bulletin de la Société Zoologique de France*, **91**, 135–48.

Humbel, R. E. & Rinderknecht, E. (1979). From NSILA to IGF (1963–1977). In *Somatomedins and Growth*, ed. G. Giordano, J. J. Van Wyk & F. Minuto, pp. 61–5. London: Academic Press, Inc.

Hunt, L. T. & Dayhoff, M. O. (1980). A surprising new protein superfamily containing ovalbumin, antithrombin-III, and alpha$_1$-proteinase inhibitor. *Biochemical and Biophysical Research Communications*, **95**, 864–71.

Ikai, A., Kitamoto, T. & Nishigai, M. (1983). Alpha-2-macroglobulin-like protease inhibitor from the egg white of cuban crocodile (*Crocodylus rhombifer*). *Journal of Biochemistry*, **93**, 121–7.

Isaksson, O. G. P., Edén, S. & Jansson, J.-O. (1985). Mode of action of pituitary growth hormone on target cells. *Annual Review of Physiology*, **47**, 483–99.

Kato, S., Noguchi, T. & Naito, H. (1987). Secretion of egg white proteins in primary cultured oviduct cells of laying Japanese quail (*Coturnix coturnix japonica*). *Poultry Science*, **66**, 1208–16.

Kato, A., Ogata, S., Matsudomi, N. & Kobayashi, K. (1981). A comparative study of aggregated and disaggregated ovomucin during egg white thinning. *Journal of Agricultural and Food Chemistry*, **29**, 821–3.

Keilová, H. & Tomášek, V. (1974). Effect of papain inhibitor from chicken egg white on cathepsin B$_1$. *Biochimica et Biophysica Acta*, **334**, 179–86.

— (1975). Inhibition of cathepsin C by papain inhibitor from chicken egg white and by complex of this inhibitor with cathepsin B$_1$. *Collection of Czechoslovak Chemical Communications*, **40**, 218–24.

Kitamoto, T., Nakashima, M. & Ikai, A. (1982). Hen egg white ovomacroglobulin has a protease inhibitory activity. *Journal of Biochemistry*, **92**, 1679–82.

Kohler, P. O., Grimley, P. M. & O'Malley, B. W. (1968). Protein synthesis: differential stimulation of cell-specific proteins in epithelial cells of chick oviduct. *Science*, **160**, 86–7.

Korpela, J. (1984). Avidin, a high affinity biotin-binding protein, as a tool and subject of biological research. *Medical Biology*, **62**, 5–26.

Korpela, J., Salonen, E.-M., Kuusela, P., Sarvas, M. & Vaheri, A. (1984). Binding of avidin to bacteria and the outer membrane porin of *Escherichia coli*. *FEMS Microbiology Letters*, **22**, 3–10.

Lanni, F. & Beard, J. W. (1948). Inhibition by egg-white of hemagglutination by swine influenza virus. *Proceedings of the Society for Experimental Biology and Medicine*, **68**, 312–13.

Lanni, F., Sharp, D. G., Eckert, E. A., Dillon, E. S., Beard, D. & Beard, J. W. (1949). The egg white inhibitor of influenza virus hemagglutination. I. Preparation and properties of semipurified inhibitor. *Journal of Biological Chemistry*, **179**, 1275–87.

Laugier, C. & Brard, E. (1980). Effects of estradiol benzoate and progesterone on egg white proteins secretion. *Poultry Science*, **59**, 643–6.

Leach, R. M. Jr (1978). Studies on the major protein component of eggshell membranes. *Poultry Science*, **57**, 1151.

Leach, R. M. Jr, Rucker, R. B. & van Dyke, G. P. (1981). Egg shell membrane protein: a nonelastin desmosine/isodesmosine-containing protein. *Archives of Biochemistry and Biophysics*, **207**, 353–9.

Leavitt, W. W. (1989). Cell biology of the endometrium. In *Biology of the Uterus*, 2nd edn, ed. R. M. Wynn & W. P. Jollie, pp. 131–73. New York: Plenum Medical Book Company.

Leof, E. B., Wharton, W., Van Wyk, J. J. & Pledger, W. (1982). Epidermal growth factor (EGF) and somatomedin C regulate G1 progression in competent BALB/c-3T3 cells. *Experimental Cell Research*, **141**, 107–15.

Li-Chan, E. & Nakai, S. (1989). Biochemical basis for the properties of egg white. *Critical Reviews in Poultry Biology*, **2**, 21–58.

Lineweaver, H. & Murray, C. W. (1947). Identification of the trypsin inhibitor of egg white with ovomucoid. *Journal of Biological Chemistry*, **171**, 565–81.

Liu, W.-H., Means, G. E. & Feeney, R. E. (1971). The inhibitory properties of avian ovoinhibitors against proteolytic enzymes. *Biochimica et Biophysica Acta*, **229**, 176–85.

Mandeles, S. & Ducay, E. D. (1962). Site of egg white protein formation. *Journal of Biological Chemistry*, **237**, 3196–9.

Marshall, M. E. & Deutsch, H. F. (1951). Distribution of egg white proteins in chicken blood serum and egg yolk. *Journal of Biological Chemistry*, **189**, 1–9.

Martin, C. R. (1985). *Endocrine Physiology*. Oxford: Oxford University Press.

Matsushima, K. (1958). An undescribed trypsin inhibitor in egg white. *Science*, **127**, 1178–9.

Mead, R. A., Eroschenko, V. P. & Highfill, D. R.

(1981). Effects of progesterone and estrogen on the histology of the oviduct of the garter snake, *Thamnophis elegans*. *General and Comparative Endocrinology*, **45**, 345–54.

Miller, S. M., Kato, A. & Nakai, S. (1982). Sedimentation equilibrium study of the interaction between egg white lysozyme and ovomucin. *Journal of Agricultural and Food Chemistry*, **30**, 1127–32.

Misugi, K. & Katsumata, T. (1963). Cytological architecture of the oviduct of the domestic hen with special reference to the formation of the egg. *Yokohama Medical Bulletin*, **14**, 259–67.

Movchan, N. A. & Gabaeva, N. S. (1967). On the antibiotic properties of the egg envelopes of grass frogs (*Rana temporaria*) and steppe turtle (*Testudo horsfieldi*). *Herpetological Review*, **1**, 6.

Muniyappa, K. & Adiga, P. R. (1979). Isolation and characterization of thiamin-binding protein from chicken egg white. *Biochemical Journal*, **177**, 887–94.

Muth, A. (1980). Physiological ecology of desert iguana (*Dipsosaurus dorsalis*) eggs: temperature and water relations. *Ecology*, **46**, 1335–43.

Nisbet, A. D., Saundry, R. H., Moir, A. J. G., Fothergill, L. A. & Fothergill, J. E. (1981). The complete amino-acid sequence of hen ovalbumin. *European Journal of Biochemistry*, **115**, 335–45.

Norris, D. O. (1985). *Vertebrate Endocrinology*, 2nd edn. Philadelphia: Lea & Febiger.

Oka, T. & Schimke, R. T. (1969). Progesterone antagonism of estrogen-induced cytodifferentiation in chick oviduct. *Science*, **163**, 83–5.

O'Malley, B. W. (1967). *In vitro* hormonal induction of a specific protein (avidin) in chick oviduct. *Biochemistry*, **6**, 2546–51.

O'Malley, B. W., Roop, D. R., Lai, E. D., Nordstrom, J. L., Catterall, J. F., Swaneck, G. E., Colbert, D. A., Tsai, M. J., Dugaiczyk, A. & Woo, S. L. C. (1979). The ovalbumin gene: organization, structure, transcription and regulation. *Recent Progress in Hormone Research*, **35**, 1–146.

Osuga, D. T. & Feeney, R. E. (1968). Biochemistry of the egg-white proteins of the ratite group. *Archives of Biochemistry and Biophysics*, **124**, 560–74.

— (1977). Egg proteins. In *Food Proteins*, ed. J. R. Whitaker & S. R. Tannenbaum, pp. 209–66. Westport, Connecticut: Avi Publishing Co.

Packard, G. C. & Packard, M. J. (1980). Evolution of the cleidoic egg among reptilian antecedents of birds. *American Zoologist*, **20**, 351–62.

Packard, G. C., Tracy, C. R. & Roth, J. J. (1977). The physiological ecology of reptilian eggs and embryos, and the evolution of viviparity within the class reptilia. *Biological Reviews*, **52**, 71–105.

Palmer, B. D. (1987). *Histology and Functional Morphology of the Female Reproductive Tracts of the Tortoise*, Gopherus polyphemus. Gainesville, FL: Univ. of Florida, thesis.

— (1988). Eggshell and albumen formation in turtles. *American Zoologist*, **28**, 49A.

— (1989). Albumen and eggshell formation in reptiles: physiological and morphological adaptations to ecological constraints. *Biology of Reproduction [Suppl. 1]*, **40**, 94.

— (1990). Functional Morphology and Biochemistry of Reptilian Oviducts and Eggs: Implications for the Evolution of Reproductive Modes in Tetrapod Vertebrates. Gainesville, FL: University of Florida, dissertation.

Palmer, B. D. & Guillette, L. J. Jr (1988). Histology and functional morphology of the female reproductive tract of the tortoise *Gopherus polyphemus*. *American Journal of Anatomy*, **183**, 200–11.

— (1990a). Morphological changes in the oviductal endometrium during the reproductive cycle of the tortoise, *Gorpherus polyphemus*. *Journal of Morphology*, **204**, 323–33.

— (1990b). Eggshell formation in alligators: implications for the evolution of mammals and birds. *Biological Reproduction [Suppl. 1]* **42**, 51.

Palmiter, R. D. & Gutman, G. A. (1972). Fluorescent antibody localization of ovalbumin, conalbumin, ovomucoid and lysozyme in chick oviduct magnum. *Journal of Biological Chemistry*, **247**, 6459–61.

Palmiter, R. D. & Schimke, R. T. (1973). Regulation of protein synthesis in chick oviduct. III. Mechanism of ovalbumin 'superinduction' by actinomycin D. *Journal of Biological Chemistry*, **248**, 1502–12.

Powrie, W. D. & Nakai, S. (1986). The chemistry of eggs and egg products. In *Egg Science and Technology*, 3rd edn, ed. W. J. Stadelman & O. J. Cotterill, pp. 97–139. Connecticut: Avi Publishing Co.

Poyser, N. L. (1981). *Prostaglandins in Reproduction*. New York: Research Studies Press, John Wiley & Sons.

Pratt, A. (1960). Introduction of foreign objects into the oviducts of laying hens. *Proceedings of the South Dakota Academy of Science*, **39**, 178.

Rhodes, M. B., Azari, P. R. & Feeney, R. E. (1958). Analysis, fractionation and purification of egg white proteins with cellulose-cation exchanger. *Journal of Biological Chemistry*, **230**, 399–408.

Rhodes, M. B., Bennett, N. & Feeney, R. E. (1959). The flavoprotein-apoprotein system of egg white. *Journal of Biological Chemistry*, **234**, 2054–60.

— (1960). The trypsin and chymotrypsin inhibitors from avian egg whites. *Journal of Biological Chemistry*, **235**, 1686–93.

Richardson, K. C. (1935). The secretory phenomena in the oviduct of the fowl, including the process of shell formation examined by the microincineration technique. *Philosophical Transactions of the Royal Society of London*, **B225**, 149–95.

Roberts, R. M. & Bazer, F. W. (1988). The functions

of uterine secretions. *Journal of Reproduction and Fertility*, **82**, 875–92.

Roberts, R. M., Raub, T. J. & Bazer, F. W. (1986). Role of uteroferrin in transplacental iron transport in the pig. *Federation Proceedings*, **45**, 2513–18.

Robinson, D. S. (1972). Egg white glycoproteins and the physical properties of egg white. In *Egg Formation and Production*, ed. B. M. Freeman & P. E. Lake, pp. 65–86. Edinburgh: British Poultry Science Ltd.

— (1987). The chemical basis of albumen quality. In *Egg Quality – Current Problems and Recent Advances*, ed. R. G. Wells & C. G. Belyavin, pp. 179–91. London: Butterworths.

Romanoff, A. L. & Romanoff, A. J. (1949). *The Avian Egg*. New York: John Wiley & Sons.

Roop, D. R., Nordstrom, J. L., Tsai, S. Y., Tsai, M.-J. & O'Malley, B. W. (1978). Transcription of structural and intervening sequences in the ovalbumin gene and identification of potential ovalbumin mRNA precursors. *Cell*, **15**, 671–85.

Schimke, R. T., Pennequin, P., Robins, D. & McKnight, G. S. (1977). Hormonal regulation of egg white protein synthesis in chick oviduct. In *Hormones and Cell Regulation*, vol. 1, ed. J. Dumont & J. Nunez, pp. 209–21. Amsterdam: Elsevier/North-Holland Biomedical Press.

Schreiber, A. B., Winkler, M. E. & Derynck, R. (1986). Transforming growth factor-α: a more potent angiogenic mediator than epidermal growth factor. *Science*, **232**, 1250–3.

Scott, H. M. & Huang, W.-L. (1941). Histological observations on the formation of the chalaza in the hen's egg. *Poultry Science*, **20**, 402–5.

Sen, L. C. & Whitaker, J. R. (1973). Some properties of a ficin-papain inhibitor from avian egg white. *Archives of Biochemistry and Biophysics*, **158**, 623–32.

Sexton, O. J., Veith, G. M. & Phillips, D. M. (1979). Ultrastructure of the eggshell of two species of anoline lizards. *Journal of Experimental Zoology*, **207**, 227–36.

Shenstone, F. S. (1968). The gross composition, chemistry and physico-chemical basis of organization of the yolk and white. In *Egg Quality: A Study of the Hen's Egg*, ed. T. C. Carter, pp. 26–58. Edinburgh: Oliver & Boyd.

Sibley, C. G. (1970). *A Comparative Study of the Egg-White Proteins of Passerine Birds*. New Haven: Peabody Museum of Natural History (Yale Univ.), Bulletin 32.

Sibley, C. G. & Ahlquist, J. E. (1972). *A Comparative Study of the Egg White Proteins of Non-Passerine Birds*. New Haven: Peabody Museum of Natural History (Yale Univ.), Bulletin 39.

Simkiss, K. (1968). The structure and formation of the shell and shell membranes. In *Egg Quality: A Study of the Hen's Egg*, ed. T. C. Carter, pp. 3–25. Edinburgh: Oliver & Boyd.

Simkiss, K. & Taylor, T. G. (1971). Shell formation. In *Physiology and Biochemistry of the Domestic Fowl*, vol. III, ed. D. J. Bell & B. M. Freeman, pp. 1331–43. New York: Academic Press.

Simons, P. C. M. & Wiertz, G. (1963). Notes on the structure of membranes and shell in the hen's egg: an electron microscopical study. *Zeitschrift für Zellforschung und mikroskopische Anatomie*, **59**, 555–67.

Soderquist, A. M. & Carpenter, G. (1983). Developments in the mechanisms of growth factor action: activation of protein kinase by epidermal growth factor. *Federation Proceedings*, **42**, 2615–20.

Solomon, S. E. (1975). Studies on the isthmus region of the domestic fowl. *British Poultry Science*, **16**, 255–8.

— (1983). Oviduct. In *The Physiology and Biochemistry of the Domestic Fowl*, vol. 4, ed. B. M. Freeman. pp. 379–419. London: Academic Press.

Solomon, S. E. & Baird, T. (1977). Studies on the soft shell membranes of the egg shell of *Chelonia mydas* L. *Journal of Experimental Marine Biology and Ecology*, **27**, 83–92.

Stamberg, O. E., Petersen, C. F. & Lampman, C. E. (1946). Ratio of riboflavin in the yolks and whites of eggs. *Poultry Science*, **25**, 327–9.

Starcher, B. C. & King, G. S. (1980). The presence of desmosine and isodesmosine in eggshell membrane protein. *Connective Tissue Research*, **8**, 53–5.

Sturkie, P. D. & Mueller, W. J. (1976). Reproduction in the female and egg production. In *Avian Physiology*, 3rd edn, ed. P. D. Sturkie, pp. 302–30. New York: Springer-Verlag.

Sturkie, P. D. & Polin, D. (1954). Role of magnum and uterus in the determination of albumen quality of laid eggs. *Poultry Science*, **33**, 9–17.

Sturkie, P. D. & Weiss, H. S. (1950). The effects of sympathomimetic and parasympathomimetic drugs upon egg formation. *Poultry Science*, **29**, 781.

Sturkie, P. D., Weiss, H. S. & Ringer, R. K. (1954). Effects of injections of acetylcholine and ephedrine upon components of the hen's egg. *Poultry Science*, **33**, 18–24.

Tomimatsu, Y., Clary, J. J. & Bartulovich, J. J. (1966). Physical characterization of ovoinhibitor, a trypsin and chymotrypsin inhibitor from chicken egg white. *Archives of Biochemistry and Biophysics*, **115**, 536–44.

Tracy, C. R. (1980). On the water relations of parchment-shelled lizard (*Sceloporus undulatus*) eggs. *Copeia*, **1980**, 478–82.

— (1982). Biophysical modeling in reptilian physiology and ecology. In *Biology of the Reptilia*, vol. 12, ed. C. Gans & F. H. Pough, pp. 275–321. London: Academic Press.

Tracy, C. R. & Snell, H. L. (1985). Interrelations among water and energy relations of reptilian eggs, embryos, and hatchlings. *American Zoologist*, **25**, 999–1008.

Tranter, H. S. & Board, R. G. (1982a). The antimicrobial defense of avian eggs: biological perspectives and chemical basis. *Journal of Applied Biochemistry*, **4**, 295–338.

— (1982b). The inhibition of vegetative cell outgrowth and division from spore of *Bacillus cereus* T by hen egg albumen. *Journal of Applied Bacteriology*, **52**, 67–74.

Trauth, S. E. & Fagerberg, W. R. (1984). Ultrastructure and stereology of the eggshell in *Cnemidophorus sexlineatus* (Lacertilia: Teiidae). *Copeia*, **1984**, 826–32.

Tullett, S. G. (1987). Egg shell formation and quality. In *Egg Quality – Current Problems and Recent Advances*, ed. R. G. Wells & C. G. Belyavin, pp. 123–46. London: Butterworths.

Tuohimaa, P. (1975). Immunofluorescence demonstration of avidin in the immature chick oviduct epithelium after progesterone. *Histochemie*, **44**, 95–101.

Uribe, M. C. A., Velasco, S. R., Guillette, L. J. Jr & Estrada, E. F. (1988). Oviductal histology of the lizard, *Ctenosaura pectinata*. *Copeia*, **1988**, 1035–42.

Warren, D. C. & Scott, H. M. (1935). The time factor in egg production. *Poultry Science*, **14**, 195–207.

Weinberg, E. D. (1977). Infection and iron metabolism. *American Journal of Clinical Nutrition*, **30**, 1485–90.

Wentworth, B. C. (1960). Fistulation of the hen's oviduct. *Poultry Science*, **39**, 782–4.

White, H. B., III (1987). Vitamin-binding proteins in the nutrition of the avian embryo. *Journal of Experimental Zoology*, Supplement, **1**, 53–63.

White, H. B., III & Merrill, A. H. Jr (1988). Riboflavin-binding proteins. *Annual Review of Nutrition*, **8**, 279–99.

White, H. B., III, Whitehead, C. C. & Armstrong, J. (1987). Relationship of biotin deposition in turkey eggs to dietary biotin and biotin-binding protein. *Poultry Science*, **66**, 1236–41.

Williams, J. (1962). A comparison of conalbumin and transferrin in the domestic fowl. *Biochemical Journal*, **83**, 355–64.

— (1968). A comparison of glycopeptides from ovotransferrin and serum transferrin. *Biochemical Journal*, **108**, 57–67.

Williams, J., Elleman, T. C., Kingston, I. B., Wilkins, A. G. & Kuhn, K. A. (1982). The primary structure of hen ovotransferrin. *European Journal of Biochemistry*, **122**, 297–303.

Wise, R. W., Ketterer, B. & Hansen, I. A. (1964). Pre-albumins of embryonic chick plasma. *Comparative Biochemistry and Physiology*, **12**, 439–43.

Wong, M., Hendrix, M. J. C., von der Mark, K., Little, C. & Stern, R. (1984). Collagen in the eggshell membranes of the hen. *Developmental Biology*, **104**, 28–36.

Woo, S. L. C., Beattie, W. G., Catterall, J. F., Dugaiczyk, A., Staden R., Brownlee, G. G. & O'Malley, B. W. (1981). Complete nucleotide sequence of the chicken chromosomal ovalbumin gene and its biological significance. *Biochemistry*, **20**, 6437–46.

Woodward, A. E. & Mather, F. B. (1964). The timing of ovulation, movement of the ovum through the oviduct, pigmentation and shell deposition in Japanese quail (*Coturnix coturnix japonica*). *Poultry Science*, **43**, 1427–32.

Wyburn, G. M., Johnston, H. S., Draper, M. H. & Davidson, M. F. (1970). The fine structure of the infundibulum and magnum of the oviduct of *Gallus domesticus*. *Quarterly Journal of Experimental Physiology*, **55**, 213–32.

— (1973). The ultrastructure of the shell forming region of the oviduct and the development of the shell of *Gallus domesticus*. *Quarterly Journal of Experimental Physiology*, **58**, 143–51.

Ye, R. D., Ahern, S. M., Le Beau, M. M., Lebo, R. V. & Sadler, J. E. (1989). Structure of the gene for human plasminogen activator inhibitor-2: The nearest mammalian homologue of chicken ovalbumin. *Journal of Biological Chemistry*, **264**, 5495–502.

Zahnley, J. C. (1980). Independent heat stabilization of proteases associated with multiheaded inhibitors. Complexes of chymotrypsin, subtilisin and trypsin with chicken ovoinhibitor and with lima bean protease inhibitor. *Biochimica et Biophysica Acta*, **613**, 178–90.

Zezulak, K. M. & Green, H. (1986). The generation of insulin-like growth factor-I-sensitive cells by growth hormone action. *Science*, **233**, 551–3.

4

Fluxes during embryogenesis

KENNETH SIMKISS

Introduction

A central problem in understanding the evolution of cellular homeostasis is the recognition that it often involves three systems with no clear inter-relationship in time and space. At some stage, cells produce specific hormone receptors that are linked to particular metabolic pathways and respond to distinctive extracellular molecules. How such functional relationships evolved to cope with a novel situation is a recurrent theme in evolutionary studies and to some extent the same difficulties arise in understanding the development of homeostatic systems in the embryo (Csaba, 1986). In this latter case it is possible to undertake experiments that provide some indication as to how the systems function. This is, however, a relatively recent area of study and the results are frequently unexpected.

Ten years ago, together with Nancy Clark, I attempted to pose some of these problems by studying the onset of calcium homeostasis in the avian embryo. We considered the problem in relation to four general systems: ion diffusion, active transport, passive storage and active resorption and posed a number of fundamental problems (Clark & Simkiss, 1980). In this chapter I will refer to some of the more recent work in these areas taking as examples the problems as they arise chronologically in a developing embryo.

Ion fluxes

Albumen formation

The egg is fertilised at the top of the oviduct and passes down the magnum region where the albumen is secreted. By the time it reaches the isthmus region, where the shell membranes are formed, it has, in the case of the fowl (*Gallus gallus*), reached approximately 16–32 cells in about 5 hours. It then enters the shell gland where additional albumen is added (plumping) and the shell is secreted (Solomon, 1983). This takes another 18–20 hours and the embryo contains about 60,000 cells by the time the egg is laid.

Albumen samples have been collected at various stages during the formation of the egg and dissolved in acid prior to analysing them for sodium and potassium ions (Simkiss & Luke, unpublished observations). The albumen that is originally secreted by the magnum region of the oviduct has a Na^+/K^+ ratio of about 10:1. By the time the egg is laid it is approximately 2:1. This is a large change and it will come as no surprise that the developing embryo does not seem to be able to tolerate these extremes for any period of time. In fact, it appears that the albumen that is initially deposited around the blastoderm is incompatible with the embryo's survival. Such a radical conclusion raises many questions. Why should the bird secrete albumen with such a composition? How is it changed to one compatible with embryonic life? What is the significance of these changes both physiologically and in evolutionary terms? The answers to some of these questions probably lie with the concepts of diffusion, pumps and storage.

The change in albumen composition that normally occurs in the oviduct is brought about by the addition of plumping fluid during the time that the egg is in the shell gland. The composition of this liquid changes progressively with time, initially being rich in sodium ions and then becoming sodium depleted and rich in potassium ions (Fig. 4.1). These changes drive the changes in albumen composition. The indications are, therefore, that sodium ions are removed from the egg during its time in the shell gland. This would be in keeping with the observations of Hurwitz, Cohen & Bar, (1970) who recorded a 10 mV potential difference (mucosa negative) for the shell gland of several species during shell formation. They suggest that a 'mucosa to serosa' directed sodium ion pump is at least partly responsible since the potential is both oxygen and ouabain sensitive. A potential of this size and orientation would be sufficient to drive calcium into the shell gland and facilitate shell formation. If this interpretation is correct, we have a maternal pump (Na^+/K^+) driving a diffusion process in the albumen related to the deposition of an embryonic store, the shell. Unfortunately, there seem to be few analyses of reptile albumen, particularly during its secretion (Palmer & Guillette, Chapter 3). It would be interesting to see first if there is any correlation between the degree of eggshell calcification and albumen composition, and second, whether the extent of

shell formation was related to the fluid fluxes that occur in the egg during the formation of the albumen.

Sub-embryonic fluid

Early in development, the blastoderm begins to transport ions from the albumen into a space that comes to separate the germinal disc from the yolk sac. This sub-embryonic fluid forms a compartment of up to 13 cm^3 in the fowl, i.e. approximately 25% of the total egg volume (Simkiss, 1980). Its formation and function are discussed elsewhere (Deeming, Chapter 19) but two aspects are pertinent to this analysis. First, Grau et al., (1962) show that this fluid has a Na^+/K^+ ratio of roughly 2:1 and that this is essential for the survival of the embryo. Second, Deeming, Rowlett & Simkiss, (1987) suggest that turning the egg facilitates this process of fluid formation by causing fresh albumen to move over these ionic pumps (see Deeming, Chapter 19). In this case, we have a diffusion-limited process limiting the activity of an ionic pump in an embryonic epithelium.

In neither the secretion of the albumen electrolytes by the hen nor in the resorption of these ions by the embryo is there any clear evidence for a regulatory system.

Calcium regulation in the embryo

The avian embryo has two major sources of calcium: the yolk and the eggshell. These supplies could be absorbed and regulated by two epithelia namely, the yolk sac membrane and the chorio-allantois. Both of these are extra-embryonic membranes which implies that any system of calcium regulation must be unique to the embryo and radically different from the adult, which depends on regulatory processes in the intestinal, renal and bone cells. Neither the ultimobranchial glands (which secrete calcitonin) nor the parathyroid glands (which secrete parathyroid hormone) appear to be functional during the first half of embryonic development to drive the homeostasis system (Fig. 4.2).

During this initial stage of development, all uptake across the yolk sac membrane is thought to occur by endocytosis. The endodermal cells of the yolk sac have a well-developed microvillus border oriented into the yolk material and showing well developed endocytotic activity (Mobbs & McMillan, 1979). There appears to be little

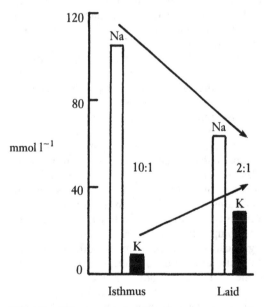

Fig. 4.1. Histograms showing the changes in the composition of albumen from eggs obtained as they enter (i.e. isthmus) or leave (i.e. laid) the shell gland. Note how the Na^+/K^+ ratio changes from 10:1 to 2:1. Values from Mongin & Sauveur (1970) show the change in composition of shell gland fluid (arrows) for comparison.

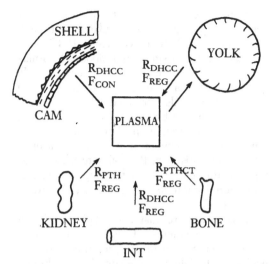

Fig. 4.2. Calcium regulatory systems for the embryo (*top*) and adult (*bottom*). During most of the incubation period, the renal, osseous and intestinal systems appear to play a relatively minor role and the parathyroid (PTH) and calcitonin hormones are not effective. Receptors (R) for these hormones and for 1-25 $(OH)_2D_3$ (DHCC) are found on a variety of membranes. At the present time, the evidence for constant (F_{CON}) and regulated calcium fluxes (F_{REG}) do not correspond well with the presence of receptors.

evidence of specific transport systems and, during this period, the avian embryo accumulates only small amounts of calcium (Johnston & Comar, 1955; Crooks & Simkiss, 1974).

When the chorio-allantoic membrane develops and fuses onto the shell membranes at about 10–11 days of incubation, it signals a major influx of calcium into the embryo. The cellular transport systems that are involved have been extensively studied and reviewed (Crooks & Simkiss, 1975; Tuan, 1987; Tuan et al., Chapter 27). It is generally agreed that the process is active, can occur against a concentration gradient and is normally specific for calcium. What is not clear is whether it is under hormonal control and, if it is, how the process is regulated.

One possible embryonic system for controlling the uptake of calcium from the shell would be to vary the rate at which it is dissolved. The calcareous eggshell is separated from the chorio-allantois by the shell membranes. It appears likely that the villus cavity cells of the chorio-allantois secrete an acid that passes outwards along these membrane fibres and dissolves the shell mineral. Calcium would then

pass back down these shell membranes to be absorbed by the chorio-allantois. In such a situation, the shell membranes might act as a diffusion barrier between the calcareous shell and the absorptive cells of the embryo. Rowlett & Simkiss (1990) have grown embryos of the fowl inside eggshells of the turkey (*Meleagris gallopavo*) which have considerably thicker membranes. They have calculated, for their system, that if the shell membranes are a diffusion barrier then calcium uptake would be reduced by up to 50%. In fact, the embryos develop with a perfectly normal calcium content suggesting that the shell membranes do not act as an effective diffusion barrier.

An alternative approach to this question has been to inject the vitamin D metabolite 1,-25 dihydroxycholecalciferol (1-25 $(OH)_2D_3$) and parathyroid hormones into the embryo. Both these treatments induce hypercalcaemia at least in the final few days of incubation. There is also evidence for hormone receptor molecules in bone, the chorio-allantoic membrane and kidney cells of the embryo (Scanes et al., 1987). There is, however, no evidence for any direct increase in the calcium transporting role of the chorio-allantois when stimulated by these hormones (Clark, Murphy & Lee, 1989). In the absence of that data one must consider the possibility that 1-25 $(OH)_2D_3$ may facilitate calcium uptake via carbonic anhydrase activity in the membrane resorbing system (Tuan & Ono, 1986). It remains a mute point, therefore, as to whether there is any regulation of calcium transport by the chorio-allantois or whether it is simply an input site with a constant calcium flux into the embryo.

If such was the function of the chorio-allantois there would, of course, have to be some other site of regulation if calcium homeostasis was to be maintained. Evidence is accumulating that the yolk sac membrane may perform such a function since 1-25 $(OH)_2D_3$ induces the mobilisation of calcium from the yolk into the circulation of the embryo (Tuan & Ono, 1986; Tuan et al., Chapter 27). If this is the case, however, the system must be effectively reversible since both birds and crocodilians (e.g. *Alligator mississippiensis*) increase the yolk sac calcium store in the last few days of incubation (Packard & Packard, 1989). The evidence on studies of embryonic calcium regulation suggest, therefore, that there are two large stores available to the embryo, that there are no obvious diffusion barriers to acquiring these ions and that there are active transport systems available at both these sites. These may be activated and possibly regulated during development, but this is poorly documented and raises the question as to whether this is a specific process or simply an increase in general resorption.

The yolk of birds and reptiles is not only a major store of calcium, it also acts as the major depository for a wide range of other ions and proteins that are transferred from the maternal circulation to the embryo. Thus 99% of the embryo's supply of zinc, 96% of its iron and 80% of its copper are also stored in the yolk. Unlike the 'class a' ionic metals such as Na^+, K^+ and Ca^{2+}, which are largely controlled by ion pumps, the 'class b-like' metals such as zinc and copper have a more covalent-bond chemistry and are usually regulated by protein binding (Simkiss, 1981). The metabolism of these metals is, therefore, an interesting comparison in terms of embryonic adaptations.

Zinc and copper metabolism

The major yolk proteins phosvitin and lipovitellin are synthesised as vitellogenin by the maternal liver and transferred to the developing ovary by the blood (White, Chapter 1). These proteins complex considerable amounts of calcium during this transfer thereby inducing the high levels of non-diffusible calcium found in most oviparous vertebrates (Simkiss, 1961). It is suggested by Panic et al. (1974) that the transfer of many trace elements to the oocyte occurs on these same phosphoproteins which, therefore, became packaged as a calcium complex in the yolk granules of the oocyte (Grau, Roudybush & McGibbon, 1979). This, therefore, provides an interesting dilemma for the embryo in that the stores of ions in the egg, for which there may be specific needs, are tied into the general metabolism of the yolk. The problem appears to be resolved by two mechanisms. First, the metal-rich yolk granules are rapidly metabolised during the first half of incubation (Saito, Martin & Cook, 1965). Second, an inducible and more easily regulated metal store is established in the embryonic liver.

The absorption of yolk by the endodermal cells of the yolk sac membranes is largely controlled by endocytosis and intracellular digestion (Mobbs & McMillan, 1981). Not all the components of the yolk sac granules are released at the same time, however, and zinc in particular appears to be retained and regulated by the yolk sac endoderm (Richards & Steele, 1987). The yolk sac is a major site of red blood cell formation throughout development and considerable quantities of iron are used directly to form haemoglobin. It also produces plasma proteins that are capable of binding metals and releasing them into the circulation. There is, therefore, during development, a general but controlled release of metals from the yolk into the circulation of the embryo where they are largely sequestered and utilised by the liver (Richards & Steele, 1987).

Hepatocytes are major regulating cells for many trace elements. They develop specific membrane receptors for such ion transport molecules as transferrin. Hepatocyte membranes possess copper and zinc uptake systems and these cells store these metals on ferritin and metallothionein molecules (Cousins, 1986). These molecules have been extensively studied in recent years (Theil, 1987; Kagi & Kojima, 1987) and are rapidly turned over within the cytoplasm. They have been studied in the mammal and are important in the fetus where they act as a short-term store of essential trace elements that carry the offspring through into neonatal life (Brady, Webb & Mason, 1982).

A similar situation probably exists in the avian embryo. Metallothioneins are found in both the yolk sac (Richards, 1984) and embryonic hepatocytes (Klasing et al., 1987) where they not only bind copper and zinc but can regulate these ions independently of each other (Darcey et al., 1986).

In conclusion

The regulation of ions by the developing avian embryo depends upon a variety of mechanisms. Some of these involve stores such as albumen, yolk and eggshell which have a composition that is initially established by the hen and transmitted to the presumptive embryo. In utilising these sources the embryo is faced with some diffusion limited supplies (e.g. albumen Na^+ and K^+) and others such as eggshell and yolk granules that are specifically mobilised (Fig. 4.3). The regulation of these processes has been poorly investigated but they may involve relatively uncontrolled fluxes (calcium from chorio-allantois, zinc and copper from yolk) upon which some mechanism for local controlled storage

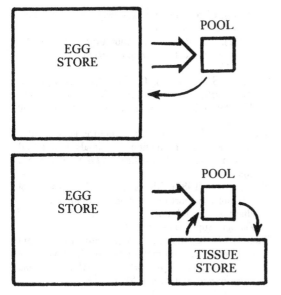

Fig. 4.3. Two possible systems that could be used to establish ion regulation during embryogenesis. In the first system (top) the ion pool is regulated through feedback to the original store (e.g. yolk calcium). While in the second scheme (bottom) regulation occurs by the control of a secondary tissue store (e.g. hepatic zinc).

may have been superimposed (yolk sac calcium, hepatic and yolk sac metallothionein). Clearly, this is an interesting area for further study.

References

Brady, F. O., Webb, M. & Mason, R. (1982). Zinc and copper metabolism in neonates. Role of metallothionein in growth and development. In *Biological Roles of Metallothionein*, ed. E. C. Foulkes. Elsevier.

Clark, N. B. & Simkiss, K. (1980). Time, targets and triggers: a study of calcium regulation in the bird. In *Avian Endocrinology*, pp. 191–208. New York: Academic Press.

Clark, N. B., Murphy, M. J. & Lee, S. K. (1989). Ontogeny of vitamin D action on the morphology and calcium transport properties of the chick embryo yolk sac. *Journal of Developmental Physiology*, 11, 243–51.

Cousins, R. J. (1986). Trace element transport in isolated hepatocytes. *Federal Proceedings*, 45, 2798–816.

Crooks, R. J. & Simkiss, K. (1974). Respiratory acidosis and eggshell resorption by the chick embryo. *Journal of Experimental Biology*, 61, 197–202.

— (1975). Calcium transport by the chick chorioal-lantois *in vivo*. *Quarterly Journal of Experimental Physiology*, 60, 55–63.

Csaba, G. (1986). Why do membrane receptors arise? *Experientia*, 42, 715–18.

Darcey, S. E., Richards, M. P., Klasing, K. C. & Steele, N. C. (1986). Zinc and copper metabolism in embryonic turkey hepatocytes. *Federal Proceedings*, 45, 1084a.

Deeming, D. C., Rowlett, K. & Simkiss, K. (1987). Physical influences on embryo development. *Journal of Experimental Zoology*, Supplement 1, 341–6.

Grau, C. R., Fritz, H. I., Walker, N. E. & Klein, N. W. (1962). Nutrition studies with chick embryo deprived of yolk. *Journal of Experimental Zoology*, 150, 185–95.

Grau, C. R., Roudybush, T. E. & McGibbon, W. H. (1979). Mineral composition of yolk fractions and whole yolk from eggs of restricted ovulatory hens. *Poultry Science*, 58, 1143–8.

Hurwitz, S., Cohen, I. & Bar, A. (1970). The transmembrane electrical potential difference in the uterus (shell gland) of birds. *Comparative Biochemistry and Physiology*, 35, 873–8.

Johnston, P. M. & Comar, C. L. (1955). Distribution and contribution of calcium from the albumen, yolk and shell to the developing chick embryo. *American Journal of Physiology*, 183, 365–70.

Kagi, J. H. R. & Kojima, Y. (1987). Metallothionein II. *Experientia*, Supplement 52, 755 pp.

Klasing, K. C., Richards, M. P., Darcey, S. E. & Laurin, D. E. (1987). Presence of acute phase changes in zinc, iron, and copper metabolism in turkey embryos. *Proceedings of the Society for Experimental Biology and Medicine*, 184, 7–13.

Mobbs, I. G. & McMillan, D. B. (1981). Transport across endodermal cells of the yolk sac during early stages of development. *American Journal of Anatomy*, 160, 285–308.

Mongin, P. & Sauveur, B. (1970). Composition du fluide utérin et de l'albumen durant le séjour de l'oeuf dans l'utérus chez la poule domestique. *Comptes Rendes du Academie de Science, Paris*, 270D, 1715–18.

Packard, M. J. & Packard, G. C. (1989). Mobilization of calcium, phosphorous and magnesium by embryonic alligators (*Alligator mississippiensis*). *American Journal of Physiology*, 257, R1541–7.

Panic, B., Bezbradica, L., Nedeljkov, N. & Istwari, A. G. (1974). Some aspects of trace element metabolism in poultry. In *Trace element metabolism in animals*. pp. 635–7. Hoekstra, W. G., Suttie, J. W., Ganther, H. E. & Mertz, W. Baltimore: University Park Press.

Richards, M. P. (1984). Synthesis of a metallothioein-like protein by developing turkey embryos maintained in long-term shell-less culture. *Journal of Pediatrics, Gastroenterology and Nutrition*, 3, 128–36.

Richards, M. P., Rosebrough, R. W. & Steele, N. C. (1984). Hepatic zinc, copper and iron of embryo turkeys (*Meleagris gallopavo*) maintained in long term shell-less culture. *Comparative Biochemistry and Physiology*, **78A**, 525–31.

Richards, M. P. & Steele, N. C. (1987). Trace element metabolism in the developing avian embryo: a review. *Journal of Experimental Zoology*, Supplement 1, 39–51.

Rowlett, K. & Simkiss, K. (1990). Differentiation and function of chorioallantoic cells in avian surrogate eggs. *Tissue and Cell*, **22**, 65–70.

Saito, Z., Martin, W. G. & Cook, W. H. (1965). Changes in the major macromolecular fractions of egg yolk during embryogenesis. *Canadian Journal of Biochemistry*, **43**, 1755–70.

Scanes, C. G., Hart, L. E., Deauypere, E. & Kuhn, E. R. (1987). Endocrinology of the avian embryo: an overview. *Journal of Experimental Zoology*, Supplement 1, 253–64.

Simkiss, K. (1961). Calcium metabolism and avian reproduction. *Biological Reviews*, **36**, 321–67.

— (1980). Water and ion fluxes inside the egg. *American Zoologist*, **20**, 385–93.

— (1981). Cellular discrimination processes in metal accumulating cells. *Journal of Experimental Biology*, **94**, 317–27.

Solomon, S. E. (1983). Oviduct. In *Physiology and Biochemistry of the Domestic Fowl*. Vol. 4, pp. 379–419, ed. B. M. Freeman. London: Academic Press.

Theil, E. C. (1987). Ferritin: Structure, gene regulation and cellular function in animals, plants and microorganisms. *Annual Review of Biochemistry*, **56**, 289–315.

Tuan, R. S. (1987). Mechanism and regulation of calcium transport by the chick embryonic chorioallantoic membrane. *Journal of Experimental Zoology*, Supplement 1, 1–13.

Tuan, R. S. & Ono, T. (1986). Regulation of extra embryonic calcium mobilization by the developing chick embryo. *Journal of Embryology and Experimental Morphology*, **97**, 63–74.

5

Eggshell structure and formation in eggs of oviparous reptiles

MARY J. PACKARD AND VINCENT G. DeMARCO

Introduction

The vast majority of contemporary reptiles are oviparous and deposit eggs with some sort of enveloping shell (Packard & Packard, 1988). The eggshell separates the developing embryo from its environment, modulates the movement of water and gases into and out of the egg, and serves as a source of calcium during embryogenesis (Packard & Packard, 1984, 1988, 1989a,b). Structure of eggshells is quite diverse, ranging from the small, flexible, parchment-shelled eggs of many squamates to the large, rigid-shelled eggs of crocodilians. None the less, certain common themes unite structure of eggs of all oviparous reptiles.

Eggshell structure

Scanning electron microscopy has been used to examine structure of eggshells from a large number of species (Table 5.1). Many studies of eggshells have been quite superficial, but a number of generalisations about structure of reptilian eggshells can still be made. Shells of reptilian eggs are composed of an outer, inorganic layer underlain by an organic (shell) membrane comprised of multiple layers of fibres (Fig. 5.1). Fibres within a layer of the membrane may form a random felt-work or may be highly parallel, and may be closely apposed to one another or widely spaced.

The number of fibrous layers in the shell membrane varies among species, and the different layers comprising the shell membrane vary morphologically. This variability, coupled with the occasional occurrence of avian-like air cells in reptilian eggs has led to the interpretation that some reptilian eggs exhibit both an inner and an outer shell membrane (Packard & Packard, 1979). However, air cells may also form between the shell membrane and the calcareous layer or between the shell membrane and the chorio-allantois (Manolis, Webb & Dempsey, 1987; Whitehead, 1987). Moreover, many reptilian eggs do not form air cells. Thus, our interpretation is that most reptilian eggs have a single, multi-layered shell membrane. Unless air cells are a regular and consistent aspect of shell structure, designation of inner and outer shell membranes is probably artificial.

The inner aspect of the fibrous shell membrane (i.e. that surface adjacent to the albumen and to which the extra-embryonic membranes will attach) has been given a number of names in studies of avian and reptilian eggshells (inner boundary = Tung, Garland & Gill, 1979; Packard et al., 1982a; Guillette, Fox & Palmer, 1989; limiting membrane = Bellairs & Boyde, 1969; Tranter, Sparks & Board, 1983; boundary layer = Erben, 1970; Schlëich & Kastle, 1988; inner boundary layer = Trauth & Fagerberg, 1984). In one case, the inner boundary was identified with a variety of different names in the same report (Sexton, Veith & Phillips, 1979; see also Packard et al., 1982a). We prefer the term inner boundary for the innermost layer of the shell membrane, primarily because this term avoids the word 'membrane'.

Some of the confusion concerning what name to apply to the inner boundary stems from the fact that this structure may appear to be relatively amorphous and unstructured when viewed by scanning electron microscopy (SEM). When examined with transmission electron microscopy (TEM), on the other hand, this part

Table 5.1. *Studies that have used scanning electron microscopy to elucidate structure of reptilian eggshells. [ ] = name under which the description was published.*

Crocodylia

Alligator mississippiensis	Ferguson (1982, 1985)
	Hirsch (1983, 1985)
	Silyn-Roberts & Sharp (1986)
	Hirsch & Packard (1987)
Alligator sinensis	Zikui (1986)
Caiman crocodilus	Kriesten (1975)
Caiman latirostris	Schlëich & Kastle (1988)
Crocodylus acutus	Hirsch (1983)
Crocodylus cataphractus	Erben (1970)
Crocodylus niloticus	Solomon & Tippett (1982)
	Solomon & Reid (1983)
	Hirsch (1983, 1985)
	Hirsch & Packard (1987)
	Grine & Kitching (1987)
	Schlëich & Kastle (1988)
	Deeming & Ferguson (1991)
Crocodylus porosus	Silyn-Roberts & Sharp (1986)
Melanosuchus niger	Erben (1970)
	Krampitz, Erben & Kriesten (1972)

Testudines

Caretta caretta	Packard, Packard & Boardman (1982*b*)
	Packard & Hirsch (1986)
	Schlëich & Kastle (1988)
Carettochelys insculpta	Erben (1970)
Chelodina expansa	Woodall (1984)
Chelonia mydas	Aitken & Solomon (1976)
	Solomon & Baird (1976)
	Baird & Solomon (1979)
Chelus fimbriatus	Schlëich & Kastle (1988)
Chelydra serpentina	Packard *et al.* (1979)
	Packard & Packard (1980)
	Packard (1980)
	Hirsch (1983)
Chrysemys picta	Packard *et al.* (1982*b*)
	Packard & Hirsch (1986)
Cuora amboinensis	Schlëich & Kastle (1988)
Emydoidea blandingii	Packard & Hirsch (1986)
Emys orbicularis	Erben (1970)
Geochelone carbonaria	Schlëich & Kastle (1988)
[*Chelonoidis carbonaria*]	
[*Testudo carbonaria*]	Erben & Newesely (1972)
Geochelone denticulata	Schlëich & Kastle (1988)
[*Chelonoidis denticulata*]	
Geochelone elephantopus	Hirsch (1983)
	Hirsch & Lopez-Jurado (1987)
[*Testudo elephantopus*]	Erben (1970)
	Kriesten (1975)
[*Chelonoidis elephantopus*]	Schlëich & Kastle (1988)
Geochelone elephantopus ephippium	Erben & Newesely (1972)
Geochelone paradalis	Schlëich & Kastle (1988)
[*Testudo paradalis*]	Grine & Kitching (1987)
Kinosternon alamosae	Packard *et al.* (1984*a*)
Kinosternon baurii	Packard *et al.* (1982*b*, 1984*a*)

Table 5.1. (*Contd.*)

Kinosternon flavescens	Packard *et al.* (1984*a*, *b*)
	Packard & Hirsch (1986)
Kinosternon hirtipes	Packard *et al.* (1984*a*)
	Hirsch & Packard (1987)
Kinosternon subrubrum	Schlëich & Kastle (1988)
Lepidochelys kempi	Packard & Packard (1980)
	Packard *et al.* (1982*b*)
	Hirsch (1983)
	Hirsch & Packard (1987)
Lepidochelys olivacea	Acuna M. (1984)
Malacochersus tornieri	Ewert, 1985
Mauremys caspica	Schlëich & Kastle (1988)
Mauremys nigricans	Ewert (1985)
[*M. mutica*]	
Melanochelys trijuga	Ewert *et al.* (1984)
	Ewert (1985)
Phrynops hilarii	Silyn-Roberts & Sharp (1985)
Podocnemis erythrocephala	Schlëich & Kastle (1988)
Trachyemys scripta elegans	Solomon & Tippett (1982)
[*Pseudemys scripta elegans*]	Solomon & Reid (1983)
Rhinoclemmys areolata	Ewert *et al.* (1984)
	Ewert (1985)
Siebenrockiella crassicollis	Silyn-Roberts & Sharp (1985, 1986)
Staurotypus triporcatus	Ewert (1985)
Sternotherus minor	Packard *et al.* (1982*b*, 1984*a*)
	Packard & Hirsch (1986)
Terrapene carolina	Erben (1970)
Testudo graeca	Schlëich & Kastle (1988)
Testudo hermanni	Schlëich & Kastle (1988)
Testudo marginata	Schlëich & Kastle (1988)
Trionyx spiniferus	Packard & Packard (1979)
	Packard *et al.* (1979)
	Packard & Packard (1980)
	Packard & Hirsch (1986)
Trionyx triunguis	Schlëich & Kastle (1988)
Squamata: Amphisbaenia	
Blanus cinereus	Schlëich & Kastle (1988)
Squamata: Sauria	
Agama agama lionotus	Schlëich & Kastle (1988)
Agama mutabilis	Schlëich & Kastle (1988)
Agama planiceps	Schlëich & Kastle (1988)
Agama tuberculata	Schlëich & Kastle (1988)
Amphibolurus barbatus	Schlëich & Kastle (1988)
Anolis carolinensis	Packard *et al.* (1982*b*)
	Schlëich & Kastle (1988)
Anolis limifrons	Sexton, Veith & Phillips (1979)
Anolis equestris	Schlëich & Kastle (1988)
Anolis sagrei	Sexton *et al.* (1979)
Basiliscus plumifrons	Schlëich & Kastle (1988)
Callisaurus draconoides	Packard & Packard (1980)
	Packard *et al.* (1982*a*)
Chamaeleo fischeri tavetanus	Schlëich & Kastle (1988)
Chondrodactylus angulier	Deeming (1988)
Cnemidophorus sexlineatus	Trauth and Fagerberg (1984)
Cnemidophorus uniparens	Cuellar (1979)
Crotaphytus collaris	Guillette *et al.* (1989)
Cyrtodactylus (Tenuidactylus) kotschyi	Schlëich & Kastle (1988)

Table 5.1. (*Contd.*)

Ctenosaura pectinata	Schlëich & Kastle (1988)
Dipsosaurus dorsalis	Packard *et al.* (1982*b*)
Eublepharis macularis	Packard & Hirsch (1986)
	Deeming (1988)
	Schlëich & Kastle (1988)
Eumeces obsoletus	Guillette *et al.* (1989)
Gehyra mutilata	Deeming (1988)
Gekko gecko	Schlëich & Kastle (1988)
	Packard & Hirsch (1989)
Gerrhosaurus flavigularis	Schlëich & Kastle (1988)
Hemidactylus bouvieri boavistensis	Schlëich & Kastle (1988)
Hemidactylus mabouia	Grine & Kitching (1987)
Hemidactylus turcicus	Packard *et al.* (1982*b*)
	Packard & Hirsch (1986)
	Schlëich & Kastle (1988)
Heloderma horridum	Solomon & Tippett (1982)
Hemitheconyx caudicinctus	Deeming (1988)
Iguana iguana	Kriesten (1975)
Lacerta agilis	Schlëich & Kastle (1988)
	Greven (1988)
Lacerta (Podarcis) dugesii	Schlëich & Kastle (1988)
Lacerta lepida nevadensis	Schlëich & Kastle (1988)
Lacerta (Scelarcis) perspicillata	Schlëich & Kastle (1988)
Lacerta (Podarcis) sicula campestris	Schlëich & Kastle (1988)
Lacerta (Gallotia) simonyi stehlini	Schlëich & Kastle (1988)
Lacerta viridis	Schlëich & Kastle (1988)
Lepidodactylus lugubris	Packard & Hirsch (1986)
Papuascincus stanleyana	Allison & Greer (1986)
[Lobulia stanleyana]	
Phelsuma agalegae	Deeming (1988)
Phelsuma guentheri	Deeming (1988)
	Deeming & Ferguson (1991)
Phelsuma madagascarensis	Hirsch & Packard (1987)
	Schlëich & Kastle (1988)
	Packard & Hirsch (1989)
Ptyodactylus hasselquistii	Schlëich & Kastle (1988)
Sceloporus aeneus	Guillette & Jones (1985)
Tarentola delalandii	Hirsch, Krishtalka & Stucky (1987)
Tarentola delalandii delalandii	Schlëich & Kastle (1988)
Tarentola mauritanica	Erben & Newesely (1972)
	Krampitz, Erben & Kriesten (1972)
	Hirsch & Packard (1987)
	Schlëich & Kastle (1988)
Varanus gouldii	Schlëich & Kastle (1988)
Varanus indicus	Schlëich & Kastle (1988)
Varanus salvator	Schlëich & Kastle (1988)
Zonosaurus madagascariensis	Schlëich & Kastle (1988)
Squamata: Serpentes	
Calloselasma rhodostoma	Schlëich & Kastle (1988)
[Agkistrodon (Daboia)	
rhodostoma]	
Chondropython viridis	Schlëich & Kastle (1988)
Coluber constrictor	Packard *et al.* (1982*b*)
	Packard & Hirsch (1986)
Elaphe guttata	Schlëich & Kastle (1988)
Elaphe longissima	Schlëich & Kastle (1988)
Elaphe obsoleta	Lillywhite & Ackerman (1984)

Table 5.1. (*Contd.*)

Elaphe quadrivirgata [*Elaphe obsoleta quadrivatta*]	Hirsch & Packard (1987)
Elaphe taeniura	Schlëich & Kastle (1988)
Hydrodynastes gigas	Packard & Hirsch (1986)
Lampropeltis getulus	Schlëich & Kastle (1988)
Liophis poecilogyrus [*Philodryas schottii*]	Schlëich & Kastle (1988)
Natrix maura	Schlëich & Kastle (1988)
Natrix natrix	Krampitz, Kriesten & Bohme (1973) Schlëich & Kastle (1988)
Pituophis melanoleucus	Deeming & Ferguson (1991)
Python molurus	Schlëich & Kastle (1988)
Python molurus bioittatus	Solomon & Tippett (1982) Solomon & Reid (1983)
Python regis	Solomon & Reid (1983) Packard & Hirsch (1986)
Python sebae	Schlëich & Kastle (1988)
Vipera (Gloydius) palaestinae	Schlëich & Kastle (1988)
Sphenodontia	
Sphenodon punctatus	Packard *et al.* (1982*c*, 1988) Packard & Hirsch (1986)

of the shell membrane is seen to consist of several layers of variable morphology (see Sexton *et al.*, 1979; Trauth & Fagerberg, 1984). Unfortunately, some authors seem not to have realised that the amorphous layer seen with SEM and the multilayered structure observed with TEM were simply different views of the inner boundary. No detailed study of the function of the inner boundary has been made in reptilian eggs. However, this layer apparently represents an important resistance to diffusion of oxygen in avian eggs during the first few days of incubation (Tranter *et al.*, 1983).

The inorganic layer of the eggshell consists of calcium carbonate in the form of calcite in crocodilians and oviparous squamates, but is aragonite in eggshells of testudinians (Erben, 1970; Solomon & Baird, 1976; Ferguson, 1982, 1985; Hirsch, 1983; Woodall, 1984; Packard & Hirsch, 1986; Deeming, 1988). Diet and/or other environmental factors may affect the morph of calcium carbonate. For example, captive sea turtles (*Chelonia mydas*) may lay eggs containing both calcite and aragonite (Baird & Solomon, 1979), and eggshells of some captive pythons (*Python molurus bioittatus*) may be aragonitic rather than calcitic (Solomon & Reid, 1983). Factors associated with captivity may also affect the chemical composition of the inorganic layer, in that shells of captive bearded dragons

(*Amphibolurus barbatus*) have been reported to consist of calcium phosphate (Schlëich & Kastle, 1988). However, shells of feral bearded dragons are calcitic (Packard & Hirsch, unpublished observations).

The calcareous layer of eggshells of crocodilians, turtles, and some geckos consists of individual columns or shell units attached to the underlying shell membrane (Fig. 5.1). Shell units in eggs of crocodilians and turtles have a knob-like or conical tip at one end that provides for the attachment between the calcareous layer and the membrane (Fig. 5.1; Packard & Packard, 1979; Packard, 1980; Ferguson, 1982, 1985; Ewert, Firth & Nelson, 1984; Packard, Hirsch & Iverson, 1984*a*; Woodall, 1984). The mode of attachment between shell units of rigid-shelled gecko eggs and fibres of the shell membrane has not been determined (Packard & Hirsch, 1989). However, an intimate intermingling of the small crystals at the tips of the columns and fibres of the shell membrane may link the crystalline layer and the shell membrane (Deeming & Ferguson, 1991; Deeming, personal communication; Packard & Hirsch, unpublished observations).

We assume that the shell membrane exhibits some flexibility in all eggshells and that flexibility or rigidity of an eggshell depends in large measure on the organisation of the calcareous

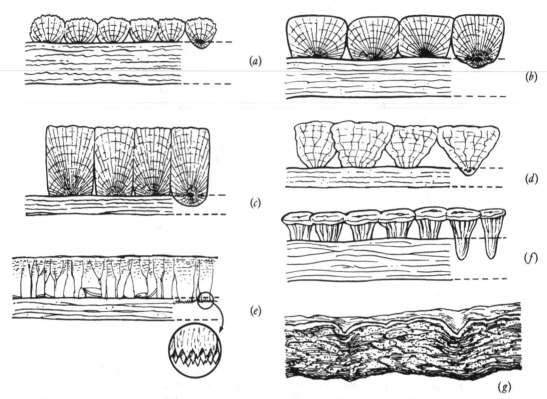

Fig. 5.1.   Highly schematic diagrams of cross-sections of reptilian eggshells. The area to the right in (*a*)–(*f*) has been cut away to show the inner aspect of shell units. (*a*) Flexible-shelled egg of a sea turtle (e.g. *Lepidochelys olivacea*). Shell units are small and irregular in shape, and the shell membrane is roughly the same thickness as the calcareous layer. (*b*) Flexible-shelled egg of turtles other than sea turtles (e.g. *Chelydra serpentina*). Shell units are more nodular, more regularly shaped, and the calcareous layer and shell membrane are about equal in thickness. (*c*) Rigid-shelled turtle egg (e.g. *Trionyx spiniferus*). Shell units are tall, columnar and abut tightly. The shell membrane is thin relative to the thickness of the calcareous layer. In all cases, shell units extend a short distance into the shell membrane and enclose fibres of the membrane, thereby providing for attachment between the calcareous layer and the membrane. (*d*) Rigid-shelled egg of a crocodilian (e.g. *Alligator mississippiensis*). Shell units are roughly wedge-shaped and the shell membrane is thin relative to the calcareous layer. The conical tip of a shell unit encloses fibres of the shell membrane and thereby anchors the calcareous layer to the membrane. (*e*) Schematic cross-section of a rigid-shelled gecko egg (e.g. *Gekko gecko*). Shell units are less well defined and are more difficult to trace through the thickness of the shell. The shell membrane is thin relative to the calcareous layer. The enlarged area in the circle shows the tip or inner aspect of shell units. The tip is comprised of fine, needle-like crystallites. The mode of attachment between the calcareous layer and the shell membrane has not been determined. (*f*) Schematic cross-section of a tuatara (*Sphenodon punctatus*) eggshell. The calcareous layer consists of irregularly-shaped shell units that extend deep into the fibrous shell membrane. (*g*) Schematic cross-section of the flexible eggshell laid by most oviparous squamates (e.g. *Callisaurus draconoides*). The shell consists of a fibrous membrane organised into a series of crests and troughs with the calcareous layer filling in the troughs and obscuring the crests to varying degrees. The calcareous layer seems simply to rest on the outer surface of the underlying shell membrane and does not enclose fibres of the membrane.

layer. When shell units abut closely and form a dense, highly organised, compact layer, the eggshell is rigid (Fig. 5.1). By contrast, if shell units are not closely apposed or the calcareous layer is not organised into discrete shell units, the eggshell is flexible (Fig. 5.1). In general, the calcareous layer is thick relative to the shell membrane in rigid-shelled eggs but is equal in thickness to the shell membrane or is thin relative to the shell membrane in flexible-shelled eggs (Fig. 5.1).

Flexible-shelled eggs, laid by most squamates, present a somewhat different situation from crocodilians and turtles in that the calcareous layer often occurs as a thin crust or as isolated deposits overlying a shell membrane that is itself organised into a series of crests and troughs (Fig. 5.1; Packard et al., 1982a; Packard & Hirsch, 1986). This topography may not be obvious with SEM because the overlying calcareous layer tends to fill in the troughs and obscure the crests. Also, swelling of eggs during incubation may cause a 'relaxation' of this morphology. Thus, radial (= cross sectional) and surface views may make it appear that the shell membrane has less relief than is actually the case.

The crystalline layer in squamate eggshells may be nodular or bulbous; may consist of plaques or rosettes or both; may have a bubbly, frothy, or fenestrated appearance; or may be relatively amorphous (Packard et al., 1982a; Packard, Packard & Boardman, 1982b; Packard & Hirsch, 1986; Deeming, 1988; Schlëich & Kastle, 1988). Although discrete shell units like those in eggs of turtles and crocodilians are absent, the crystalline layer is often organised into irregular blocks which follow closely the crest and trough topography of the shell membrane. Fissures or cracks generally occur between these blocks. No obvious attachment is evident between the calcareous layer and the fibrous shell membrane. However, an organic layer occurs just beneath the crystalline layer on the outer surface of the underlying shell membrane. This layer may be relatively amorphous or may consist of a felt-work of fine fibres embedded in an organic matrix (Packard et al., 1982a,b). The organic layer could be involved in anchoring the calcareous material to the underlying shell membrane and/or it may also have a role in the initiation of calcification (see later).

Some authors have reported that squamate eggshells lack a calcareous layer, have not addressed the question at all, or have provided no documentation for assertions regarding the presence or absence of crystalline material (e.g. Andrews & Sexton, 1981; Packard et al., 1982b; Schlëich & Kastle, 1988). Unfortunately, the surface of squamate eggshells can appear to lack a calcareous layer simply because this layer does not always appear in the SEM to be obviously crystalline. Moreover, squamates are prone to oviposit eggs prematurely in captivity (e.g. Plummer & Snell, 1988), and some descriptions of shell morphology therefore may have been based on eggs that had not completed shell formation. We are inclined to believe that all 'normal', fully-shelled squamate eggs have a crystalline component, and suggest that reports of the absence of calcareous material in these eggshells need to be confirmed.

Flexible-shelled eggs laid by the tuatara (Sphenodon punctatus) represent an exception to the general pattern manifested by flexible-shelled eggs of other squamates (Packard, Hirsch & Meyer-Rochow, 1982c; Packard & Hirsch, 1986; Packard et al., 1988). Tuatara eggshells consist of loosely organised shell units or columns as do the flexible-shelled eggs laid by some turtles. However, the columns extend deep into the shell membrane and exhibit a more extensive association with fibres of the membrane than occurs in flexible-shelled eggs of turtles (Fig. 5.1). This pattern of organisation occurs also in eggs of the lizard Amphibolurus barbatus (Packard & Hirsch, unpublished observations) and may characterise eggs of the lizard Varanus gouldii (Schlëich & Kastle, 1988).

Removal of calcium from the shell during embryogenesis may result in changes in morphology of the eggshell. For example, mobilisation of calcium from the shell by embryonic turtles (e.g. Chelydra serpentina) causes the shell membrane to detach from the crystalline layer. Remnants of the calcareous layer obscure the outer surface of the shell membrane of hatched eggs, and the inner aspect of the calcareous layer is flat rather than rounded (Packard, 1980). Similar observations have been reported for rigid-shelled eggs of the turtle Trionyx spiniferus (Packard & Packard, 1979). Changes in morphology stemming from mobilisation of calcium may occur in eggshells of all oviparous reptiles. These changes may be difficult to document for the flexible-shelled eggs of most oviparous squamates because the shell membrane and calcareus layer of hatched eggs normally do not separate (Packard, unpublished observations). Thus, an examination of the inner aspect of the crystalline layer is not possible. However, mobilisation of calcium may contribute to flaking and defoliation of the crystalline crust in hatched eggs (Packard, unpublished observations).

Changes in morphology of the outer surface of the eggshell owing to microbial degradation have been reported for eggs of *Alligator mississippiensis* (Ferguson, 1981*a,b*, 1982, 1985). These changes are presumed to be of critical importance to development and successful hatching of crocodilian embryos (Ferguson, 1985). However, these observations, as well as this interpretation itself, have been questioned recently (Silyn-Roberts & Sharp, 1985; Whitehead, 1987), and additional research is required to resolve the differences among these studies.

Exchange of water and gases between an egg and its environment may occur through recognisable channels or pores in the calcareous layer or through less well-defined gaps, cracks, or discontinuities. When used in reference to avian eggs, the term pore generally describes a well-defined channel arising at the intersection of several shell units and traversing the thickness of the calcareous layer (Board, 1982; Board & Sparks, Chapter 6). This description applies also to pores in eggshells of crocodilians, many turtles, and a few geckos (Ferguson, 1982, 1985; Packard & Hirsch, 1986, 1989). Use of the term 'pore' becomes more problematical when considering eggs of sea turtles and those of most squamates because of the loosely organised nature of the mineral layer. Unfortunately, some authors use this term to refer to any discontinuity in the outer surface of an eggshell even if it is not clear that the apparent channel actually penetrates through the thickness of the calcareous layer (Lillywhite & Ackerman, 1984; Schleich & Kastle, 1988). We prefer to limit use of 'pore' to describe well-defined channels occurring at the junction of recognisable shell units, and to use the term 'pore-like' to describe the gaps and discontinuities in the calcareous layer of flexible-shelled eggs that lack shell units altogether or have very ill-defined shell units.

Eggshell formation

Shell formation in reptiles bears a similarity to shell formation in birds (Board & Sparks, Chapter 6), but important differences also exist. For example, the time-scale of shell formation in reptiles is prolonged compared to birds. Birds exhibit sequential ovulation in which a single mature follicle is ovulated, shelled, and oviposited independent of other eggs in the clutch (Simkiss, 1968; Tyler, 1969; Simkiss & Taylor, 1971). The entire process from ovulation to oviposition requires about 24 hours in the fowl (*Gallus gallus*). By contrast, many reptiles exhibit simultaneous ovulation in which all eggs in a clutch are ovulated within a short period and enter the oviduct one after the other. Shelling of eggs often takes considerably longer than 24 hours.

The eggs of birds and oviparous reptiles are encased in several egg membranes at oviposition (Fig. 5.2), including the vitelline membrane (adjacent to the cell or plasma membrane of the oocyte), the albumen, the shell membrane(s), and the calcareous layer (Balinsky, 1975; Packard *et al.*, 1982*a*). The site of formation of these various egg membranes has been studied extensively in the fowl but comparable information is largely lacking for reptiles. Numerous studies of oviductal morphology in reptiles have been published (see Palmer & Guillette, 1988, Chapter 3; Guillette *et al.*, 1989), but most studies provide no accompanying information on the process of shell formation.

In birds, secretion of the vitelline membrane begins in the ovary and continues in the oviduct (Bellairs, Harkness & Harkness, 1963; Bellairs, 1971), but the site of formation of this membrane has not been determined with certainty in reptiles. The other egg membranes (albumen, shell membrane, and calcareous layer) are formed in the oviduct in both birds and reptiles. The inner boundary of the shell membrane may be formed in the infundibular region of the oviduct in squamates (Guillette *et al.*, 1989), and albumen probably is formed and applied to eggs in the uterine tube in turtles (Palmer & Guillette, 1988, Chapter 3). Formation of the shell membrane and calcareous layer occurs in the

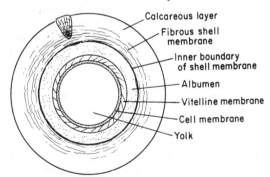

Fig. 5.2. Schematic cross-section through a generalised egg of an oviparous, amniotic vertebrate showing the yolk and the egg membranes. Drawing is not to scale and size relationships have been exaggerated for clarity.

uterus (Aitken & Solomon, 1976; Palmer & Guillette, 1988 Chapter 3 Guillette *et al.*, 1989).

We assume that formation of the mineral layer of any eggshell requires that two conditions be met, provision of: a supersaturated solution of ions from which growth of crystals can occur, and a site of nucleation which will overcome the free energy barrier to crystal growth (Simkiss, 1968; Garside, 1982). The nucleation site may be homogeneous (i.e. the same chemical composition as the crystal that is nucleated) or heterogeneous (i.e. different in chemical composition) (Garside, 1982). Chemical analyses have not been done on the sites of nucleation in reptilian eggs, but morphological evidence indicates that nucleation is heterogeneous (Packard & Packard, 1979; Packard, 1980; Packard, Iverson & Packard, 1984*b*; Woodall, 1984).

Probable nucleators on the surface of the shell membrane include small, roughly globular protrusions known as organic cores, membranous, radially symmetrical structures known as central plaques, and microscopic pits or saucer-shaped depressions in the shell membrane (Agassiz, 1857; Solomon & Baird, 1976; Packard & Packard, 1979; Packard, 1980; Packard *et al.*, 1984*a,b*; Woodall, 1984). Cores have been observed in rigid-shelled eggs of several turtles (Packard & Packard, 1979; Packard *et al.*, 1984*a,b*; Woodall, 1984), and can be inferred from studies of rigid-shelled eggs of other turtles and crocodilians (Ferguson, 1982, 1985; Ewert *et al.*, 1984). Central plaques have been identified only in the flexible-shelled eggs of snapping turtles (*C. serpentina*) and painted turtles (*Chrysemys picta*) (Agassiz, 1857; Packard, 1980, unpublished), and pits have been described in sea turtle (*C. mydas*) eggshells (Solomon & Baird, 1976). We suggest that central plaques will be found in flexible-shelled eggs laid by all turtles except sea turtles, but this prediction requires confirmation.

The organic cores in rigid eggshells are similar in many respects to those in avian eggshells (Simkiss, 1968; Tyler, 1969; Fujii & Tamura, 1970; Robinson & King, 1970; Simkiss & Taylor, 1971; Fujii, 1974; Creger, Phillips & Scott, 1976; Stemberger, Mueller & Leach, 1977; Leach, 1982). Cores in avian eggs apparently are secreted onto the shell membrane following its formation (Tyler, 1969), and the process is relatively short-lived (Creger *et al.*, 1976). We do not know if cores in reptilian eggs form in conjunction with the shell membrane or

subsequent to its formation. However, central plaques and pits seemingly are modifications of the shell membrane itself and are unlikely to be secreted *de novo* onto the membrane (Solomon & Baird, 1976; Packard, 1980).

Irrespective of its morphology, a nucleator on the surface of the shell membrane provides a site for initiation and deposition of the calcareous layer, with each nucleator presumably being responsible for initiation of a single column or shell unit (Fig. 5.3; Packard & Packard, 1979; Packard *et al.*, 1984*a,b*; Silyn-Roberts & Sharp, 1985, 1986). Shell units grow a short distance into the shell membrane and enclose both the organic core and fibres of the membrane (Fig. 5.3; Packard & Packard, 1979; Packard *et al.*, 1984*a,b*; Woodall, 1984; Silyn-Roberts & Sharp, 1986). However, the primary directions for growth are lateral and away from the shell membrane. As additional mineral is deposited, shell units may come to abut closely and tightly, thus forming the dense, highly organised calcareous layer of rigid-shelled eggs (Fig. 5.1; Packard & Hirsch, 1986; Hirsch & Packard, 1987). Alternatively, shell units may grow upward without close apposition, and large numbers of spaces may remain between shell units. In this case, the resulting eggshell is flexible rather than rigid (Fig. 5.1; Packard & Hirsch, 1986).

Interestingly, rigid-shelled eggs of the turtle *Kinosternon flavescens* pass through morphological stages during shell formation that resemble the fully-formed, flexible eggshells of other turtles (Packard *et al.*, 1984*b*). Partially shelled eggs of *K. flavescens* and fully formed eggs of many turtles that produce flexible-shelled eggs are characterised by rounded, nodular shell units with numerous spaces between shell units (Fig. 5.1). However, fully-formed shells from *K. flavescens* have a dense, relatively non-porous calcareous layer with tightly apposed columnar shell units (Fig. 5.1; Packard *et al.*, 1984*a,b*; Packard & Hirsch, 1986). These results indicate that formation of a rigid or flexible shell in turtles may depend in part on the duration of shell formation (Packard *et al.*, 1984*b*). Thus, very similar processes may operate during formation of both kinds of turtle eggshells, and rigidity or flexibility may simply represent different end points along the continuum of shell growth.

The presence of organic cores in several species of turtles that produce hard-shelled eggs composed of aragonite (Packard & Packard,

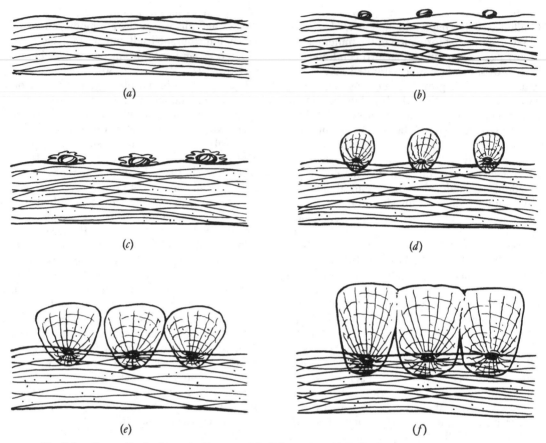

Fig. 5.3. Generalised schematic diagram of shell formation. (a) Fully-formed shell membrane. (b) Shell membrane with small, organic nucleators (cores) on the outer surface. (c) First deposition of crystal around the nucleators. (d)–(f) Growth of shell units. Shell units grow in all directions initially, but the main directions of growth are lateral and upward. Shell units grow a short distance into the shell membrane and enclose fibres of the shell membrane. We assume that this general scheme applies to shells of all turtles, crocodilians, those squamates that lay rigid-shelled eggs, and the tuatara. See text for exceptions exhibited by the tuatara.

1979; Packard *et al.*, 1984*a,b*; Woodall, 1984) obviates assertions that formation of aragonitic shells is associated with saucer-shaped nucleation sites on the shell membrane (Solomon & Tippett, 1982; Solomon & Reid, 1983). The contention that aragonitic shells invariably are soft, pliable, and lack discrete pores (Solomon & Tippett, 1982; Solomon & Reid, 1983) is also untenable.

The organic matrix present within the crystalline material of many eggshells generally is assumed to play an important role in control of crystal growth (Krampitz, 1982). However, recent crystallographic analyses of avian, crocodilian, and rigid testudinian eggshells indicate that the organic matrix may hinder (rather than facilitate) deposition of crystal (Silyn-Roberts & Sharp, 1986). The organic matrix may act as a network that helps to reinforce the crystalline matrix, but may not be a central determinant of crystal growth (Silyn-Roberts & Sharp, 1986).

Silyn-Roberts & Sharp (1986) propose that crystals comprising the shell units of avian, crocodilian, and rigid testudinian eggshells grow outward from a common centre and grow in the same crystallographic direction throughout the length of the shell unit. Differences in the nature of the organic matrix, and in its distribution, may obscure the direction of crystal growth and may influence the apparent texture of the fracture surface revealed when radial preparations are made (Silyn-Roberts & Sharp, 1986). Organic matter is evenly distributed in eggs of crocodilians and virtually absent from rigid testudinian eggs (Silyn-Roberts & Sharp, 1986). The almost total absence of organic matter in rigid eggshells of turtles results in a radial fracture of uniform morphology throughout the length of individual shell units (Silyn-Roberts & Sharp, 1986).

We would like to believe that similar principles underlie shell formation in all oviparous, amniotic vertebrates, but evidence pertaining to shell formation in squamates is scarce. Indeed, shell formation in gekkonids that lay rigid-shelled eggs has not been examined at all. Moreover, morphology of shell units offers no insight into how the calcareous layer might be initiated (Packard & Hirsch, 1989). Thus, for the present, we can make no inference concerning shell formation in these rigid-shelled eggs.

Shell formation has been examined in several squamates, the tuatara (*S. punctatus*; Packard *et al.*, 1988), four iguanids (Guillette *et al.*, 1989; DeMarco, 1988, unpublished observations; Palmer & DeMarco, unpublished observations), and a skink (Guillette *et al.*, 1989). We shall consider shell formation in the tuatara first because the process in this species differs from that in other squamates. The very earliest stages of shell formation in the tuatara have not been examined (Packard *et al.*, 1988). However, we assume that shell formation begins with nucleation of columns on the surface of a partially-formed shell membrane (Packard *et al.*, 1988). Columns and membrane grow more-or-less in concert, but growth of the membrane slightly precedes that of columns initially. Growth of the membrane eventually comes to dominate at what will be the inner aspect of the shell and growth of columns comes to dominate at the future outer surface of the shell (Packard *et al.*, 1988). The result is an eggshell in which the columns enclose fibres of the membrane throughout much of their length, not just at the tip as is the case in other eggshells comprised of calcareous columns or shell units (Packard *et al.*, 1982b, 1988; Packard & Hirsch, 1986).

Studies of shell formation in *Crotaphytus collaris*, *Eumeces obsoletus* (Guillette *et al.*, 1989), *Sceloporus clarki*, *Sceloporus woodi*, *Sceloporus virgatus*, and *Sceloporus scalaris* (DeMarco, 1988 and unpublished observations; Palmer & DeMarco, unpublished observations) include information on morphology of the oviduct and identification of those regions of the oviduct responsible for formation of various egg membranes. We have tried to incorporate generalisations concerning shell formation that are common to the several species studied and are likely to apply to formation of flexible-shelled eggs in most squamates (Fig. 5.4). However, we recognise that species-specific differences exist in detail.

When an oocyte is released by the ovary but prior to its entry into the oviduct, the surface of the oocyte appears highly convoluted (Fig. 5.5). These convolutions may be the surface of the vitelline membrane, but studies with TEM are required to confirm this prediction.

The inner boundary of the shell membrane is formed shortly after the oocyte enters the oviduct. The developing shell is flat at this time and follows the contours of the underlying egg (Fig. 5.4; Guillette *et al.*, 1989). Fibres of the shell membrane are secreted onto the inner boundary in a process that may require several days for completion. None the less, the majority of fibres are probably laid down within the first 24 hours. Fibres are secreted into the lumen of the oviduct from glands in the oviductal epithelium (Palmer & DeMarco, unpublished observations). We assume that the orientation of fibres arises, at least in part, from rotation of eggs as they pass down the oviduct. The last layer of the shell membrane to be formed consists of a tightly woven mat of fine fibres embedded in an organic matrix (Figs 5.4 and 5.6). This layer covers the outer aspect of the membrane and obscures the underlying fibres of the shell membrane (Figs 5.4 and 5.6).

Several days post-ovulation, the shell membrane begins to take on the crest and trough morphology characteristic of fully formed eggshells (Figs 5.4, 5.6 and 5.7). The crests become more pronounced and come to resemble blocky, fibrous polygons separated by fissures that reveal the underlying fibres of the shell membrane (Figs 5.4, 5.6 and 5.7). Formation of this morphology may involve reorganisation of the shell membrane and/or may result from secretion of additional fibres and matrix material onto the crests of the shell membrane.

As these changes in morphology of the shell membrane occur, small deposits of crystalline material begin to appear on the outer surface of the membrane in association with crests (Figs 5.4, 5.6 and 5.7). The deposits may occur as small crystalline blocks or spheres (Figs 5.6 and

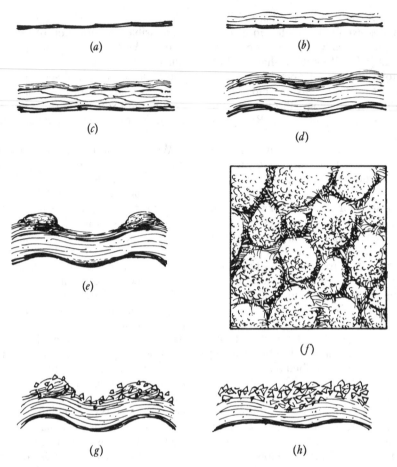

Fig. 5.4.   Highly schematic diagram of shell formation in squamates. All views except (*f*) are cross-sections; (*f*) is a surface view. (*a*) Inner boundary of the shell membrane forms after the oocyte enters the oviduct. (*b*) Fibres of the shell membrane are secreted onto the inner boundary by glands in the oviductal epithelium. (*c*) A layer of fine fibres in an organic matrix is laid down as a surface film on the outer aspect of the shell membrane. (*d*) The shell membrane begins to exhibit the crest and trough morphology characteristic of squamate eggshells. (*e*) The crest and trough morphology becomes more distinct, and the crests come to resemble fibrous, blocky polygons. (*f*) Fissures develop between the polygons, and the continuity of the surface film of small fibres and organic matrix may be disrupted so that fibres of the shell membrane can be seen at the bottom of the fissures. (*g*)–(*h*) Small deposits of crystal form in association with the polygons and eventually cover the entire surface of the shell membrane.

5.7). Additional crystal is added to these initial deposits so that the outer surface of the shell membrane becomes covered by irregular blocks or by elliptical nodules, depending on the species (Figs 5.7 and 5.8). Oviposited eggs are covered by a dense layer of closely spaced blocks or nodules (Fig. 5.8). Organic material may be secreted onto the outer surface of the calcareous layer in some species.

We do not know what nucleates deposition of the initial crystalline structures that begin to appear on the shell membrane 3–4 days post-ovulation. Cores similar to those of other reptilian eggshells have not been observed on the shell membrane prior to appearance of crystals nor are such cores revealed when eggshells are decalcified. Calcification may be initiated by reorganisation of the shell membrane into blocky polygons and/or secretion of fibres and matrix onto the membrane. The relief provided by the crests themselves may play a role in this process or subtle chemical changes independent of changes in morphology may be responsible for the initial deposition of crystal.

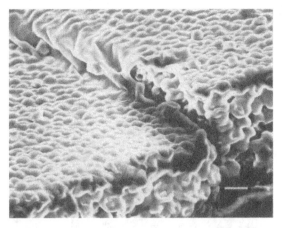

Fig. 5.5. Scanning electron micrograph of the outer surface of a freshly ovulated oocyte of *Sceloporus virgatus* prior to entry into the ostium. The surface shown here may be the vitelline membrane but additional work is required to confirm this possibility. Scale bar = 17μm.

Any (or all) of these factors may contribute to the initiation of calcification in flexible-shelled eggs of oviparous squamates.

Implications for future research

Clearly, many questions concerning shell formation in oviparous reptiles remain to be answered. The endocrine control of shell formation has not been addressed, and we have little or no information concerning the chemical milieu within the oviduct during shell formation. Sources of calcium for formation of yolk and eggshell have not been determined, and genetic and environmental factors affecting shell formation have not been examined.

Additional studies of shell formation in oviparous squamates undoubtedly will contribute importantly to our understanding of this process in reptiles generally. Squamates offer some advantages for study in that many of them are small and can be maintained relatively easily in captivity. Squamates are particularly interesting in light of insights to be gained concerning the relationship between shell formation, structure of eggshells, retention of eggs, and the evolution of viviparity. Structure of eggshells in this group is more variable than in any other assemblage of oviparous reptiles, thus raising the potential for similar variability in structure of the oviduct and the mechanism of shell formation. Observations of detailed structural changes

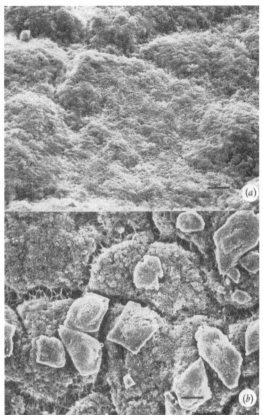

Fig. 5.6. (*a*) Scanning electron micrograph of the outer surface of the shell membrane of *Sceloporus woodi* approximately three days post-ovulation. The surface of the shell membrane has been covered by a layer of fine fibres in an organic matrix. Organisation of shell membrane into crests and troughs has begun. Scale bar = 3 μm. (*b*) Scanning electron micrograph of the outer surface of the shell membrane of *S. woodi* approximately 5 days post-ovulation. The surface of the membrane consists of crests and troughs. Crests have a blocky, irregular shape and are separated by fissures through which the underlying fibres of the shell membrane can be seen. Small blocks of crystal have been deposited onto the crests in some areas. Scale bar = 17 μm.

in the shell during its formation, coordinated studies with SEM and TEM, and correlations between shell structure and morphology of the oviduct will provide much greater understanding of this process in the future.

Acknowledgements

We thank D. C. Deeming and M. W. J. Ferguson for inviting our participation in the

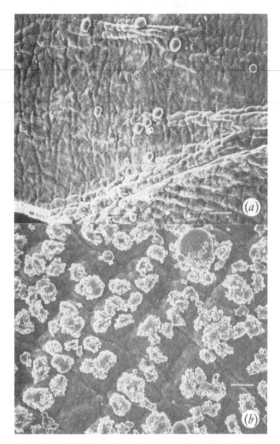

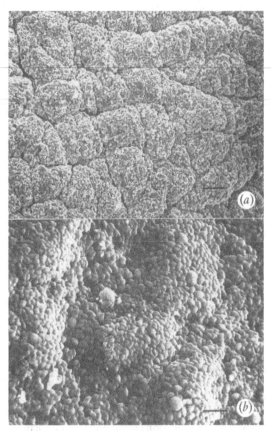

Fig. 5.7. (a) Scanning electron micrograph of the outer surface of the shell membrane of *S. virgatus* approximately 4 days post-ovulation. Small spheres of crystal have been deposited on the surface of the membrane. Scale bar – 85 μm. (b) Scanning electron micrograph of the outer surface of the shell membrane of *S. virgatus* showing a more advanced stage of crystal deposition. Scale bar = 36 μm.

Fig. 5.8. Scanning electron micrographs of the outer surface of shells of oviposited eggs of *Sceloporus clarki* (a) and *Sceloporus scalaris* (b). Outer surface of both shell membranes has been covered with a crust of crystalline material, but the crest and trough morphology still is apparent. Scale bar = 61 μm, (a); 43 μm, (b).

international conference on Physical Influences on Embryonic Development in Birds and Reptiles. G. C. Packard, K. F. Hirsch, and L. J. Guillette, Jr. critically reviewed drafts of the manuscript. Our thanks to G. Zug for verifying the scientific names in Table 5.1. Our work was supported, in part, by grants from the U.S. National Science Foundation (DCB 87–18191 to MJP and DCB 84–16707 to L. J. Guillette, Jr), from the Society of the Sigma Xi (to VGD), and from the Electron Microscopy Center at Colorado State University (to MJP). Travel by MJP was funded by monies from the Vice President for Research, the Dean of the College of Natural Sciences, and the Chair of the Department of Biology at CSU. Travel by VGD was funded by the Department of Zoology and the Dean of the Graduate School at UF. Our thanks to G. Erdos and H. Aldrich of the Electron Microscope Core Laboratory at UF for advice and help with SEM. D. Carlson drew the schematic diagrams.

References

Acuna M., R. A. (1984). La ultraestructura superficial de la cascara del huevo de la tortuga marina *Lepidochelys olivacea* Eschscholtz. *Brenesia*, **22**, 299–308.

Agassiz, L. (1857). *Contributions to the Natural History of the United States of America*, First Monograph, vol. II, pp. 1–643. Boston: Little, Brown & Co.

Aitken, R. N. C. & Solomon, S. E. (1976). Observations on the ultrastructure of the oviduct of the

Costa Rican green turtle (*Chelonia mydas* L.). *Journal of Experimental Marine Biology and Ecology*, **21**, 75–90.

Allison, A. & Greer, A. E. (1986). Egg shells with pustulate surface structures: basis for a new genus of New Guinea skinks. *Journal of Herpetology*, **20**, 116–19.

Andrews, R. M. & Sexton, O. J. (1981). Water relations of the eggs of *Anolis auratus* and *Anolis limifrons*. *Ecology*, **62**, 556–62.

Baird, T. & Solomon, S. E. (1979). Calcite and aragonite in the egg shell of *Chelonia mydas* L. *Journal of Experimental Marine Biology and Ecology*, **36**, 295–303.

Balinsky, B. I. (1975). *An Introduction to Embryology*, 4th edition. Philadelphia: W. B. Saunders Co.

Bellairs, R. (1971). *Developmental Processes in Higher Vertebrates*. Coral Gables, Florida: University of Miami Press.

Bellairs, R. & Boyde, A. (1969). Scanning electron microscopy of the shell of the hen's egg. *Zeitschrift für Zellforschung und Mikroskopische Anatomie*, **96**, 237–49.

Bellairs, R., Harkness, M. & Harkness, R. D. (1963). The vitelline membrane of the hen's egg: a chemical and electron microscopical study. *Journal of Ultrastructure Research*, **8**, 339–59.

Board, R. G. (1982). Properties of avian egg shells and their adaptive value. *Biological Reviews of the Cambridge Philosophical Society*, **57**, 1–28.

Creger, C. R., Phillips, H. & Scott, J. T. (1976). Formation of an eggshell. *Poultry Science*, **55**, 1717–23.

Cuellar, H. S. (1979). Disruption of gestation and egg shelling in deluteinized oviparous whiptail lizards *Cnemidophorus uniparens* (Reptilia: Teiidae). *General and Comparative Endocrinology*, **39**, 150–7.

Deeming, D. C. (1988). Eggshell structure of lizards of two sub-families of the Gekkonidae. *Herpetological Journal*, **1**, 230–4.

Deeming, D. C. & Ferguson, M. W. J. (1991). Incubation and embryonic development in reptiles and birds. In *Avian Incubation*, ed. S. G. Tullett, 3–37. London: Butterworths.

DeMarco, V. G. (1988). The timing of eggshell deposition in the lizard genus *Sceloporus* and the evolution of viviparity in reptiles. *American Zoologist*, **28**, 131A.

Erben, H. K. (1970). Ultrastrukturen und Mineralisation rezenter und fossiler Eischalen bei Vogeln und Reptilien. *Biomineralisation*, **1**, 1–66.

Erben, H. K. & Newesely, H. (1972). Kristalline Bausteine und Mineralbestand von Kalkigen Eischalen. *Biomineralisation*, **6**, 32–48.

Ewert, M. A. (1985). Embryology of turtles. In *Biology of the Reptilia*, vol. 14, ed. C. Gans, F. Billett & P. F. A. Maderson, pp. 75–267. New York: John Wiley & Sons.

Ewert, M. A., Firth, D. J. & Nelson, C. E. (1984).

Normal and multiple eggshells in batagurine turtles and their implications for dinosaurs and other reptiles. *Canadian Journal of Zoology*, **62**, 1834–41.

Ferguson, M. W. J. (1981*a*). Increasing porosity of the hatchling alligator eggshell caused by extrinsic microbial degradation. *Experientia*, **37**, 252–4.

— (1981*b*). Extrinsic microbial degradation of the alligator eggshell. *Science*, **214**, 1135–7.

— (1982). The structure and composition of the eggshell and embryonic membranes of *Alligator mississippiensis*. *Transactions of the Zoological Society of London*, **36**, 99–152.

— (1985). Reproductive biology and embryology of the crocodilians. In *Biology of the Reptilia*, vol. 14, ed. C. Gans, F. Billett & P. F. A. Maderson, pp. 329–491. New York: John Wiley & Sons.

Fujii, S. (1974). Further morphological studies on the formation and structure of hen's eggshell by scanning electron microscopy. *Journal of the Faculty of Fisheries and Animal Husbandry, Hiroshima University*, **13**, 29–56.

Fujii, S. & Tamura, T. (1970). Scanning electron microscopy of shell formation in hen's eggs. *Journal of the Faculty of Fisheries and Animal Husbandry, Hiroshima University*, **9**, 65–81.

Garside, J. (1982). Nucleation. In *Biological Mineralization and Demineralization*, Dahlem Konferenzen 1982, ed. G. H. Nancollas, pp. 23–35. Berlin: Springer-Verlag

Greven, H. (1988). Zür Feinstruktur der Eischale von *Lacerta agilis* Linnaeus, 1758. *Mertensiella*, **1**, 66–74.

Grine, F. E. & Kitching, J. W. (1987). Scanning electron microscopy of early dinosaur egg shell structure: a comparison with other rigid sauropsid eggs. *Scanning Microscopy*, **1**, 615–30.

Guillette, L. J., Jr, Fox, S. L. & Palmer, B. D. (1989). Oviductal morphology and egg shelling in the oviparous lizards *Crotaphytus collaris* and *Eumeces obsoletus*. *Journal of Morphology*, **198**, 1–16.

Guillette, L. J., Jr & Jones, R. E. (1985). Ovarian, oviductal and placental morphology of the reproductively bimodal lizard, *Sceloporus aeneus*. *Journal of Morphology*, **184**, 85–98.

Hirsch, K. F. (1983). Contemporary and fossil chelonian eggshells. *Copeia*, **1983**, 382–97.

— (1985). Fossil crocodilian eggs from the Eocene of Colorado. *Journal of Paleontology*, **59**, 531–42.

Hirsch, K. F., Krishtalka, L. & Stucky, R. K. (1987). Revision of the Wind River fauna, early Eocene of central Wyoming. Part 8. First fossil lizard egg (?Gekkonidae) and list of associated lizards. *Annals of the Carnegie Museum*, **56**, 223–30.

Hirsch, K. F. & Lopez-Jurado, L. F. (1987). Pliocene chelonian fossil eggs from Gran Canaria, Canary Islands. *Journal of Vertebrate Paleontology*, **7**, 96–9.

Hirsch, K. F. & Packard, M. J. (1987). Review of fossil eggs and their shell structure. *Scanning Microscopy*, **1**, 383–400.

Krampitz, G. P. (1982). Structure of the organic matrix in mollusc shells and avian eggshells. In *Biological Mineralization and Demineralization, Dahlem Konferenzen 1982*, ed. G. H. Nancollas, pp. 219–32. Berlin: Springer-Verlag.

Krampitz, G. P., Erben, H. K. & Kriesten, K. (1972). Über Aminosaurenzusammensetzung und Struktur von Eischalen. *Biomineralisation*, 4, 87–99.

Krampitz, G. P., Kriesten, K. & Bohme, W. (1973). Untersuchungen uber Ultrastruktur und Aminosaurenzusammensetzung der Eischalen von *Natrix natrix*. *Experientia*, 29, 416–18.

Kriesten, K. (1975). Untersuchungen uber Ultrastruktur, Proteinmuster und Aminosaurenzusammensetzung der Eischalen von *Testudo elephantopus, Caiman crocodilus* und *Iguana iguana. Zoologische Jahrbucher, Abteilung fur Anatomie und Ontogenie der Tiere*, 94, 101–22.

Leach, R. M., Jr (1982). Biochemistry of the organic matrix of the eggshell. *Poultry Science*, 61, 2020–47.

Lillywhite, H. B. & Ackerman, R. A. (1984). Hydrostatic pressure, shell compliance and permeability to water vapor in flexible-shelled eggs of the colubrid snake *Elaphe obsoleta*. In *Respiration and Metabolism of Embryonic Vertebrates*, ed. R. S. Seymour, pp. 121–35. Dordrecht: Dr W. Junk Publishers.

Manolis, S. C., Webb, G. J. W. & Dempsey, K. E. (1987). Crocodile egg chemistry. In *Wildlife Management: Crocodiles and Alligators*, eds. G. J. W. Webb, S. C. Manolis & P. J. Whitehead, pp. 445–72. Chipping Norton, New South Wales: Surrey Beatty and Sons Pty Limited.

Packard, G. C. & Packard, M. J. (1980). Evolution of the cleidoic egg among reptilian antecedents of birds. *American Zoologist*, 20, 351–62.

— (1988). The physiological ecology of reptilian eggs and embryos. In *Biology of the Reptilia*, vol. 16, ed. C. Gans & R. B. Huey, pp. 523–605. New York: Alan R. Liss.

Packard, G. C., Taigen, T. L., Packard, M. J. & Shuman, R. D. (1979). Water-vapor conductance of testudinian and crocodilian eggs (class: Reptilia). *Respiration Physiology*, 38, 1–10.

Packard, M. J. (1980). Ultrastructural morphology of the shell and shell membrane of eggs of common snapping turtles (*Chelydra serpentina*). *Journal of Morphology*, 165, 187–204.

Packard, M. J., Burns, L. K., Hirsch, K. F. & Packard, G. C. (1982a). Structure of shells of eggs of *Callisaurus draconoides* (Reptilia, Squamata, Iguanidae). *Zoological Journal of the Linnean Society*, 75, 297–316.

Packard, M. J. & Hirsch, K. F. (1986). Scanning electron microscopy of eggshells of contemporary reptiles. *Scanning Electron Microscopy*, 1986, 1581–90.

— (1989). Structure of shells from rigid-shelled eggs of the geckos *Gekko gecko* and *Phelsuma madagascarensis. Canadian Journal of Zoology*, 67, 746–58.

Packard, M. J., Hirsch, K. F. & Iverson, J. B. (1984a). Structure of shells from eggs of kinosternid turtles. *Journal of Morphology*, 181, 9–20.

Packard, M. J., Hirsch, K. F. & Meyer-Rochow, V. B. (1982c). Structure of the shell from eggs of the tuatara, *Sphenodon punctatus. Journal of Morphology*, 174, 197–205.

Packard, M. J., Iverson, J. B. & Packard, G. C. (1984b). Morphology of shell formation in eggs of the turtle *Kinosternon flavescens. Journal of Morphology*, 181, 21–8.

Packard, M. J. & Packard, G. C. (1979). Structure of the shell and tertiary membranes of eggs of softshell turtles (*Trionyx spiniferus*). *Journal of Morphology*, 159, 131–43.

— (1984). Comparative aspects of calcium metabolism in embryonic reptiles and birds. In *Respiration and Metabolism of Embryonic Vertebrates*, ed. R. S. Seymour, pp. 155–79. Dordrecht: Dr W. Junk Publishers.

— (1989a). Environmental modulation of calcium and phosphorus metabolism in embryonic snapping turtles (*Chelydra serpentina*). *Journal of Comparative Physiology B*, 159, 501–8.

— (1989b). Mobilization of calcium, phosphorus, and magnesium by embryonic alligators *Alligator mississippiensis. American Journal of Physiology*, 257, R1541–7.

Packard, M. J., Packard, G. C. & Boardman, T. J. (1982b). Structure of eggshells and water relations of reptilian eggs. *Herpetologica*, 38, 136–55.

Packard, M. J., Thompson, M. B., Goldie, N. K. & Vos, M. (1988). Aspects of shell formation in eggs of the tuatara (*Sphenodon punctatus*). *Journal of Morphology*, 197, 147–57.

Palmer, B. D. & Guillette, L. J., Jr (1988). Histology and functional morphology of the female reproductive tract of the tortoise *Gopherus polyphemus. American Journal of Anatomy*, 183, 200–11.

Plummer, M. V. & Snell, H. L. (1988). Nest site selection and water relations in the snake, *Opheodrys aestivus. Copeia*, 1988, 58–64.

Robinson, D. S. & King, N. R. (1970). The structure of the organic mammillary cores in some weak egg shells. *British Poultry Science*, 11, 39–44.

Schlëich, H. H. & Kastle, W. (1988). *Reptile Egg-Shells SEM Atlas*. Stuttgart: Gustav Fischer.

Sexton, O. J., Veith, G. M. & Phillips, D. M. (1979). Ultrastructure of the eggshell of two species of anoline lizards. *Journal of Experimental Zoology*, 207, 227–36.

Silyn-Roberts, H. & Sharp, R. M. (1985). Preferred orientation of calcite and aragonite in the reptilian eggshells. *Proceedings of the Royal Society of London B*, 225, 445–55.

— (1986). Crystal growth and the role of the organic network in eggshell biomineralization. *Proceedings of the Royal Society of London B*, **227**, 303–24.

Simkiss, K. (1968). The structure and formation of the shell and shell membranes. In *Egg Quality A Study of the Hen's Egg*, ed. T. C. Carter, pp. 3–25. Edinburgh: Oliver & Boyd.

Simkiss, K. & Taylor, T. G. (1971). Shell formation. In *Physiology and Biochemistry of the Domestic Fowl*, vol. 3, ed. D. J. Bell & B. M. Freeman, pp. 1311–43. London: Academic Press.

Solomon, S. E. & Baird, T. (1976). Studies on the egg shell (oviducal and oviposited) of *Chelonia mydas* L. *Journal of Experimental Marine Biology and Ecology*, **22**, 145–60.

Solomon, S. E. & Reid, J. (1983). The effect of the mammillary layer on eggshell formation in reptiles. *Animal Technology*, **34**, 1–10.

Solomon, S. E. & Tippett, R. (1982). The diversity of crystal structure within the eggshells of the class Reptilia. *Journal of Anatomy*, **136**, 605–6.

Stemberger, B. H., Mueller, W. J. & Leach, R. M., Jr (1977). Microscopic study of the initial stages of egg shell calcification. *Poultry Science*, **56**, 537–43.

Tranter, H. S., Sparks, N. H. C. & Board, R. G. (1983). Changes in structure of the limiting membrane and in oxygen permeability of the chicken egg integument during incubation. *British Poultry Science*, **24**, 537–47.

Trauth, S. E. & Fagerberg, W. R. (1984). Ultrastructure and stereology of the eggshell in *Cnemidophorus sexlineatus* (Lacertilia: Teiidae). *Copeia*, 1984, 826–32.

Tung, M. A., Garland, M. R. & Gill, P. K. (1979). A scanning electron microscope study of bacterial invasion in hen's egg shell. *Journal de l'Institut Canadien de Science et Technologie Alimentaires*, **12**, 16–22.

Tyler, C. (1969). Avian egg shells: their structure and characteristics. *International Review of General and Experimental Zoology*, **4**, 81–130.

Whitehead, P. J. (1987). Respiration of *Crocodylus johnstoni* embryos. In *Wildlife Management: Crocodiles and Alligators*, eds. G. J. W. Webb, S. C. Manolis & P. J. Whitehead, pp. 473–97. Chipping Norton, New South Wales: Surrey Beatty and Sons Pty Limited.

Woodall, P. F. (1984). The structure and some functional aspects of the eggshell of the broad-shelled river tortoise *Chelodina expansa* (Testudinata: Chelidae). *Australian Journal of Zoology*, **32**, 7–14.

Zikui, Z. (1986). The ultrastructure of the eggshell of Chinese alligator. *Acta Herpetologica Sinica*, **5**, 129–33.

6

Shell structure and formation in avian eggs

RONALD G. BOARD AND NICK H. C. SPARKS

Introduction

Four names dominated the literature on eggshell composition and structure in the era when studies of the former were based on 'wet' chemical methods and those of the latter on light microscopy: von Nathusius, Romanoff, Schmidt and Tyler. Access to these early studies is easy. The pioneering studies of von Nathusius (1821–1899), who trained as a chemist, managed the family estate and studied eggshells as a hobby, were translated into English and published in a single volume by Tyler (1964). Romanoff collaborated with his wife, Anastasia, in the publication (1949) of the classic monograph, *The Avian Egg*. In 1969 Tyler reviewed the literature on many aspects of eggshell structure and composition and highlighted Schmidt's contribution to early studies of the crystallography of the mineral part of the shell. These four workers laid down a substantial staddle upon which later workers, blessed as they were with developments in electron optics, built up knowledge of eggshell structure and composition. There has been a tendency for such studies to be done in isolation. Little if any attempt has been made, however, to consider structure in the broad context of the breeding biology of birds.

Such isolation is no doubt due in part to the apparent independence of the avian embryo for much in the way of communication with the nest-bulk environment. Needham (1950) stresses the extent to which the avian embryo is closed off from the environment, the term 'cleidoic' (closed box) is used to designate this state. Tyler (1969) discusses pore configuration and numbers in the eggshells of many species of birds and stresses that porosity not only permits the exchange of respiratory gases but also loss of water from the reservoir in an egg at oviposition. These conflicting attributes of eggshells have been studied extensively by many (Rahn & Paganelli, 1981; Paganelli, Chapter 16; Rahn, Chapter 21) and it is now apparent that shell porosity has been influenced by natural selection in eggs brooded in harsh environments. As the pores are of micron dimensions in cross section, they are vulnerable to occlusion by nest debris and infiltration by microbes unless adapted to counter one or both of these threats to the embryo's well-being (Lack, 1968; Board, 1982).

Both Lack and Board contend that avian eggshells are subjected to selection pressures such that each egg is adapted to its particular nest environment. The shell, acting as a mediating boundary, contributes in many ways to successful embryogenesis (Table 6.1). It is highly probable that a nest environment, and maybe even the parents' life style, impose conflicting demands on a shell such that its structure and chemical composition have been selected to achieve the best compromise between two or more demands (Board, 1982).

Although it may appear naive to invoke the concept of natural selection leading to amendment of shell structure, we have stated it boldly simply because inimical factors leading to selection, and the means whereby shell structure has or may be changed, have not been appreciated to any worthwhile extent. Consequently this chapter has three objectives: 1) To discuss the variations in eggshell structure and composition so that morphologists are encouraged to pose two key questions. (a) Is a particular shell structure

Table 6.1. *Contributions of eggshell to the well-being of the developing avian embryo.*

Chemical communication between embryo and nest environment
Conservation of the water contained in albumen and yolk at oviposition
Reservoir of calcium and magnesium
Protection of embryo against harm due to:
  1. Insolation
  2. Predation (cryptic markings)
  3. Physical damage resulting from quasi-static, or shock loadings
  4. Microorganisms

Table 6.2. *Nomenclature applied to eggshell structures by various authors. As one descends the table the structures get closer to the albumen.*

Romanoff & Romanoff (1949)	Schmidt (1962)	Tyler (1969)	This chapter
—	Cover (pure organic) or (calcified)	Cover	Shell accessory materials[a]
Cuticle	—	True cuticle	
—	—	Surface (vertical) crystal layer	Surface crystal layer
Spongy layer	Palisade layer (column layer)	Column layer	Palisade layer
Mammillary layer	Cone layer	Cone layer	Cone layer
	Basal caps	Basal caps	Basal caps
Membranes	Membranes	Membranes	Shell-membranes
—	—	—	Limiting membrane[b]

[a]This term was introduced into the literature by Board & Scott (1980) and [b] by Bellairs & Boyde (1969).

correlated with a biological function? (b) What are the control mechanism(s) within the shell gland that amend shell structure? 2) To consider recent developments in our understanding of biomineralisation such that the physiology of the shell gland is viewed in a new light. 3) To draw attention to possible inimical features of the nest environment that may threaten embryogenesis through negating one or more of the essential attributes of an eggshell. Hopefully, discussion of the third topic will encourage field workers to enlarge the repertoire of methods they use to study the breeding biology of birds thereby providing challenging questions to morphologists and avian physiologists about structure, functional relationships and biomineralisation.

Eggshell structure

Various names have been used to describe the radial structure of the mineral portion (the true shell of Tyler (1969)) of unincubated eggshells examined with the light microscope (Table 6.2). Four of these are in current use: basal caps, cone layer, palisade layer and surface crystal layer. A study of the radial face of the eggshells of upwards of 200 bird species with electron optics has led Board (1982) to propose an arbitrary classification based on the latter three layers (Fig. 6.1). The inner surface of all eggshells is formed by a densely crystalline cone layer. The palisade layer of the eggshells of different species of birds, on the other hand, vary in respect of their content of vesicles, spherical, gas filled voids. In general, the thick, densely crystalline shells of the ostrich (*Struthio camelus*), for example, contain relatively few vesicles whereas these occur in very large numbers in the palisade layer of the eggshells of the parrot *Agapornis roseicollis*. There is marked variation

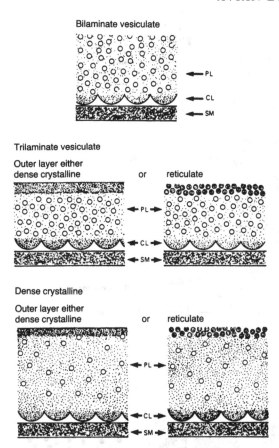

Fig. 6.1. An arbitrary classification of eggshells based on the density of crystals and vesicles (gas filled spherical voids) as seen in radial sections of shells examined with scanning electron-microscopy. Reproduced with permission from Board (1982). PL, palisade layer; CL, cone layer; SM, shell membranes.

also in the outer portion of the shell. In some eggshells the palisade layer forms the outer edge of the shell: the vesicles are an obvious feature of the surface of the bilaminate vesiculate eggshells of *A. roseicollis*. In other eggshells, the palisade layer is capped with the surface crystal layer which is either complete, except where pores vent, or reticulate (Fig. 6.1).

Yet another arbitrary classification is based on the radial configuration of pores observed with scanning electron microscopy of eggshells of nearly 200 species of bird (Board, Tullett & Perrott, 1977; Board, 1982). In the simplest case the pore, posthorn in shape, vents immediately at the shell surface (Fig. 6.2). In eggshells of species of tinamou (e.g. *Endromia elegans*) and

the jaçana (*Micropara capensis*) (Board & Perrott, 1979a), the pores vent through the surface crystalline layer capping the palisade layer but their outer orifices are plugged (Fig. 6.2). Board (1982) states that the outer orifice of the pores in kiwi (*Apteryx* sp.) eggshells is plugged with spheres but pores in unincubated eggshells of the North Island brown kiwi (*Apteryx australis mantelli*) are unplugged (Silyn-Roberts, 1983). Pores in incubated eggs are plugged with debris, presumably from the nest, in which fissures presumably assure the diffusion of respiratory gases (Silyn-Roberts, 1983). The validity of the plugged-pore category of the classification for the eggshells (Fig. 6.2) is, therefore, questionable. The plugs in eggshells of tinamous consist of spheres and fibres and electron probe analysis shows that these plugs are rich in sulphur but poor in calcium and phosphates (Board & Perrott, 1979a). Additionally, many of the peripheral spheres in the plugs are partially cupped in shallow depressions in the surface crystalline layer. These observations are taken as evidence of the concomitant deposition of the latter layer and the plugs. Although the plugs in tinamou eggshells are amorphous, their high content of iron (protoporphyrin-IX?) was taken by Board & Perrott (1979b) as evidence of oviductal rather than nest-debris origin.

Assuming that plugs are oviductal in origin, they cannot be assigned easily to the categories of 'cover' and/or 'true cuticle' of the earlier classification of shell structure (Table 6.2). Indeed Board & Scott (1980) have encountered problems when attempting to apply these terms in studies of shells with electron microscopy. Consequently these authors propose that the term 'shell accessory materials' ought to be used for structures, whether predominantly organic or inorganic in nature, occurring on the outer surface of the surface crystalline layer. Shell accessory material normally envelopes the entire outer surface of the shell such that the outer orifice of the pore canals are capped and, in many instances, roughly plugged (Fig. 6.2). Eggshells of megapodes have the shell accessory materials which bridge the outer orifice of the pore canals which, in the case of Malee fowl (*Leipoa ocellata*), is crudely plugged with a calcium-rich material. In the reticulate pore system (Fig. 6.2) of the eggshells of storks (Ciconiidae), cassowaries (Casuariidae), emus (Dromaiidae) (Board & Tullett, 1975; Board et al., 1977; Mikhailov, 1987) and some, but not all, of the

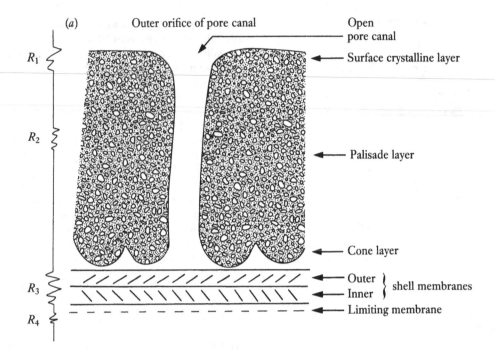

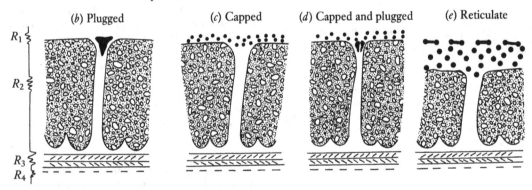

Fig. 6.2. A classification of pore systems in avian eggshells. It is based on the publication by Board *et al.* (1977). A straight tube with both orifices open and traversing a shell radially (*a*) is taken to be the simple system. Types (*b*) to (*d*) arise from deposition of shell accessory materials at or towards the end of the formation of the calcitic shell. Type (*e*) arises from major but poorly understood amendment of calcite deposition.

Falconiformes (Tyler, 1966), the pore canals cross only a part of the shell. The pore lumens vent into a honeycomb of chambers in the outer portion of the shell which vent to the outside of the shell via orifices unrelated to a pore.

So far in this discussion, attention has been given to pores that originate in single orifices in clefts in the cone layer and remain as a single tube until venting to the outer surface of the surface crystalline layer or into the reticulate system. This is a characteristic feature of thin eggshells. In thicker eggshells, however, it is common for some of the pores to fork (Tyler, 1969). The ostrich eggshell has extensive branching of the original pore canal, such that a whorl of canals having a common origin in the cone layer vents in a saucer-shaped depression in the outer surface of the shell (Tyler & Simkiss, 1959; Sauer, 1972; Mikhailov, 1986).

The inner surface of the shell is lined with the

inner and outer shell membranes (Fig. 6.2). The inner surface of the latter is covered with amorphous material, the limiting membrane (Bellairs & Boyde, 1969) or inner boundary (Packard & DeMarco, Chapter 5). Both of the shell membranes are formed from anastomosing fibres having a proteinaceous core and a polysaccharide mantle (Tranter, Sparks & Board, 1983). In mature shell membranes, the core and mantle are separated by a cleft containing lipid (Candlish, 1972). This description relates to the shell structure immediately following oviposition. During incubation alterations in structure can occur. The physical characteristics of a nest, as well as the life style of the brooding parent, may amend the structure of the outer surface of the shell of incubating eggs. As all embryos are dependent upon the shell for calcium and magnesium ions, resorption of these elements changes the morphology of the inner surface of eggshells, particularly in the later stages of incubation and especially in those areas in juxtaposition to the chorio-allantois (Doskocil *et al.*, 1985). Observations of the fine structure of avian eggshells during incubation have shown that changes are first detectable at 65% of the total incubation time of the eggs of the domestic fowl (*Gallus gallus*) and Japanese quail (*Coturnix c. japonica*). By 80% of the incubation period, large central cavities had developed within the cones (Bond, Board & Scott, 1988*a*). The changes occurring in the eggshells of altricial species of birds were less extensive than those in precocial ones presumably due to the former's restricted needs for Ca^{2+} and Mg^{2+}. Bond *et al.* (1988*a*) surmise that the resorption of these two elements is confined to the cone tips so that the porosity of the shell is not increased, nor its strength diminished, during incubation (see later).

Eggshell composition

Shell membranes

The shell membranes are 95% protein, 2% carbohydrate and 3% fat (Tullett, 1987; Palmer & Guillette, Chapter 3), an elastin-like protein being the main component (Leach, Rucker & Van Dyke, 1981). The inner surface of the inner shell membrane is covered with a thin (2.7 μm) limiting membrane: an amorphous layer of unknown composition (Tullett, 1987).

The true shell

This part of the shell, which consists of the cone and palisade layers as well as the basal caps (Table 6.2), is composed mainly of calcite permeated by an organic network (Silyn-Roberts & Sharp, 1986) or a matrix, described as a protein–acid mucopolysaccharide complex (Simkiss & Tyler, 1957). This matrix, which is variously described as a net formed from cross-linked fibrous-sheets (Silyn-Roberts & Sharp, 1986) or fine fibrils (e.g. Tullett, 1987), accounts for *circa* 2% of the weight of the true shell.

The following organic components had been found in the true shell: unspecified but non-collagenous proteins, carbohydrates consisting of a range of hexosamines, neutral sugars, hexuronic acid and mucopolysaccharides, as well as unspecified fats (Tullett, 1987). Neutral polysaccharides and sialomucins occur within the individual cones (Tullett, 1987). According to Weiser & Krampitz (1987), the organic matrix of the eggshell of the fowl is made up of highly sulphated glycoproteins, the soluble precursors of which are synthesised in the liver under oestrogen control. Both a soluble and insoluble organic matrix have been distinguished and both parts of this structure are formed from a relatively small number of polypeptides synthesised in the liver (Schade, 1987). The peptides become attached to a great range of carbohydrates, including sulphated sugars, in the shell gland.

Apart from calcium, magnesium ions also occur in eggshells but in trace amounts only. Using 'wet' chemical methods, Brooks & Hale (1955) and Itoh & Hatano (1964) have found that the concentration of Mg^{2+} rises from a very low level towards the inside of the shell to a peak concentration in the surface crystal layer in eggs of the fowl. A similar profile is noted in the eggshells of Japanese quail using electron probe analysis (Quintana & Sandoz, 1978). Using this technique to survey the eggshells of nearly 200 species of bird, Board & Love (1980, 1983) have found that the radial distribution of Mg^{2+} in the eggshells of these two species (pattern (*a*) in Fig. 6.3) is common to many but not all members of the Galliformes and Anseriformes. Another pattern ((*b*) in Fig. 6.3) is common to the eggshells of the majority of bird orders included in these studies.

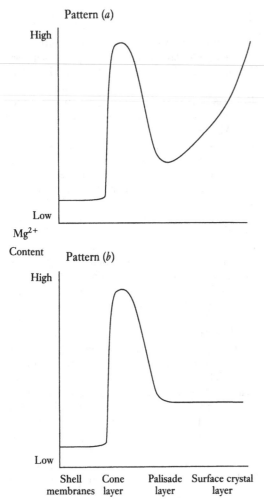

Fig. 6.3. Magnesium distribution across the radial plane of the eggshells of Galliformes/Anseriformes (Pattern (*a*)) and the majority of other orders of birds (Pattern (*b*)). Electron probe microanalysis was used to survey the eggshells of a large number of bird species in the studies conducted by Board & Love (1980, 1983).

Shell accessory materials

'Up to the point when the main shell crystals have ceased to grow, most eggshells have a similar structure; but various additions may be made afterwards' (Tyler, 1969). Diverse terms have been applied to such additions (Table 6.2). Tyler prefers to retain the term 'cuticle' for a purely organic layer immediately above the true shell and for anything above this to be referred to as 'cover'. If such a distinction cannot be made, then 'covering' is an appropriate term. All

too often, however, the nomenclature applied in anatomical studies directs attention to a structure without at the same time exciting curiosity about its biological origin or purpose. It is for these reasons that the term 'shell accessory materials' is used to describe structures present on the outer surface of the true shell (Board & Scott (1980)). These authors hope that an interest in the overall breeding biology of birds will be encouraged in order to identify the role of accessory materials and to explain their synthesis and deposition. Shell accessory materials range from mainly organic, for example the glycoproteins on the outer surface of the eggshells of the fowl or mixtures of organic and inorganic materials (Table 6.3).

Of inorganic materials, vaterite and amorphous calcium phosphate are the two major contributors. It needs to be stressed that shell accessory materials normally occur as small spheres (Tyler, 1969; Board *et al.*, 1984). Even with scanning electron optics and electron probe analysis, it has not proved possible as yet to interpret completely their fine structure or chemical structure. Indeed such interpretations may have to await studies of the physiology of the shell gland. The accessory materials occur predominantly as spheres on grebe (Podicipitiformes) eggshells but the very outside of the shell is enveloped in a thin, amorphous layer which is extensively fissured (Board *et al.*, 1984). Infra red analysis indicates that calcium phosphate is present in the shell accessory materials of these eggshells. X-ray diffraction studies have failed to demonstrate the occurrence of crystalline material, and electron probe microanalysis shows that the sulphur content of the accessory material is greater than that of the true shell (Board *et al.*, 1984). These authors have found similar patterns of sulphur concentration in other eggs in which the shell accessory materials are dominated by spheres of amorphous calcium phosphate or vaterite.

Shell pigments

The following pigments have been shown to occur in some but not all avian eggshells (Kennedy & Vevers, 1975): protoporphyrin-IX, biliverdin-IX and its zinc chelate and occasionally traces of coproporphyrin-III. Although pigment deposition occurs throughout the formation of the true shell (Baird, Solomon & Tedstone, 1975), the rate tends to accelerate

Table 6.3. *Accessory materials on the surface of avian eggshell.*

Material	Where found
	Covering entire surface of shells of:
Glycoprotein (Cuticle)	Domestic hen (*Gallus gallus*)[a]
Vaterite	Gannet (*Sula bassana*)[b]
	Snake bird (*Anhinga anhinga*)[c]
Amorphous calcium phosphate	Grebes (e.g. *Podiceps cristatus*)[d]
	Megapodes (e.g. *Leipoa ocellata*)[e]
	Patches on shell surface
Vaterite	Nonparasitic cuckoo (*Guira guira*)[f]
	Confined to outer orifice of pore canal
Uncharacterised	Tinamous, e.g. Crested tinamou (*Eudromia elegans*)[g]

[a]Wedral, Vadehra & Baker (1974); [b]Tullett *et al.* (1976); [c]Colacino *et al.* (1985); [d]Board *et al.* (1984); [e]Board *et al.* (1982); [f]Board & Perrott (1979*b*); [g]Board & Perrott (1979*a*).

towards the end of shell formation and, in the case of the fowl, to peak during the formation of the cuticle (Lang & Wells, 1987).

Formation of the eggshell

Within 3 hours of being shed from the ovary of the fowl, an ovum is encapsulated in albumen, which in turn is enveloped in the limiting membrane (Tullett, 1987). The formation of two shell membranes in the isthmus of the oviduct can be taken to be the first step in shell formation because 'preferential sites' (Solomon, 1987), the mammillary cores of Tullett (1987), on the outer surface of the outer shell membrane are the sites at which mineralisation of the true shell is initiated. The spacing of the nucleation sites may influence shell porosity (Tullett, 1975) and shell strength (Solomon, 1987). In the 4.5–24 hour period post ovulation, bulk mineralisation occurs in the shell gland pouch. This overview of shell formation reflects the accounts given in the literature over many years. It needs to be stressed that such accounts invariably deal with shell formation in the egg of the fowl and leave the impression of a step-by-step process in which each step is a self-contained event. It is difficult to comprehend the diversity of morphologies and pore systems among bird eggs should all shells be fashioned in this manner. In the following discussion, we emphasise the dynamic nature of eggshell formation before considering shell mineralisation

in terms of modern concepts relating to biomineralisation.

The deposition of a major component of the shell can occur concomitantly with some other event in egg formation, e.g. deposition is followed by a maturation phase. The shell membranes, for example, are immature when laid down and they mature (Fig. 6.4) whilst plumping is occurring. Indeed, plumping continues during the early phase of mineralisation of the true shell. We contend (Fig. 6.5) that the eventual shape, and mass, of an egg is determined in large part by an interaction between the expansive forces generated by plumping and the gradual development of a resistance to such forces arising from maturation of the shell membranes and the deposition of the mineral phase and organic matrix of the true shell. This is not discussed further but see Smart (Chapter 8).

The termination of the mineralisation phase can be abrupt such that there is a clear-cut demarcation between the true shell and shell accessory materials. In the fowl, for example, a demarcation results from the ending of Ca^{2+} and HCO_3^- secretion and a brief phase during which glycoprotein is secreted by the oviduct (Tullett, 1987). This material matures post-oviposition to form the cuticle (Sparks & Board, 1985). A clear-cut demarcation between the outer surface of the true shell and shell accessory materials is also evident in the eggshells of the flamingo (*Phoenicopterus ruber roseus*) and gannet (*Sula bassana*). In these species a change in the ionic,

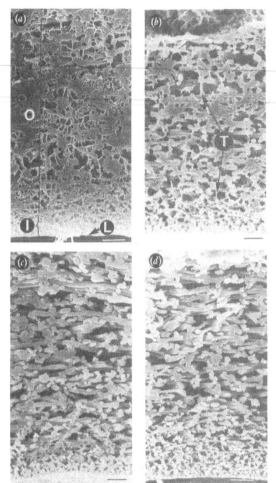

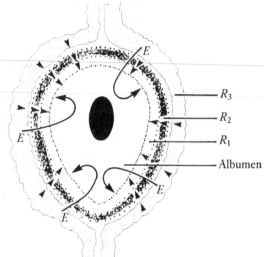

Fig. 6.5. Forces in the oviduct contributing to the ultimate shape and size of an avian egg. Expansive forces (E) are the product of water adsorption (plumping) by the albumen. Constraints are imposed by the maturing shell membranes (R_1), the developing shell (R_2) and the wall of the shell gland (R_3).

Fig. 6.4. A sequence of scanning electron micrographs showing radial sections through the shell membranes of a duck (*Anas platyrhynchos*) egg during formation. Bar marker is 10 μm in all cases. (*a*) 2 hours after entering isthmus. The immature membrane has a trilaminate appearance – the inner membrane (I) consisting of fine fibres and the outer membrane (O) of a layer of thick and a layer of thin fibres. The fibres are linked by fine threads of an amorphous material. The limiting membrane (L) is clearly visible at this stage. (*b*) 2 hours after entering isthmus. Note the loss of the trilaminate appearance and the diminution of the amorphous material (T) linking the fibres (*a*). (*c*) 4 hours after entering isthmus. There is no evidence by this stage of the amorphous material that previously linked the fibres. (*d*) immediately prior to oviposition. A radial section through a mature shell membrane.

with vaterite (Tullett *et al.*, 1976). With the grebe (*Podiceps cristatus*) eggshell, on the other hand, the outer part of the true (calcitic) shell is laid down concomitantly with amorphous calcium phosphate, the latter occurring in globules having a nut-like morphology (Board *et al.*, 1984). As ultimately only the latter are formed, there is a transition zone between the main part of the shell and its covering of shell accessory materials. The discussion so far assumes that the majority of pore systems depicted in Fig. 6.2 can be fashioned by changes in the composition of the uterine secretions during the time that an egg is resident in the shell gland. Mikhailov (1987) suggests another contributor to eggshell structure and hypothesises that in the cassowary (*Casuarius casuarius*) the shell membranes and the bulk of the true shell of eggs are laid down as the egg passes through the isthmus and shell gland. Such passage is not followed by oviposition but rather retroperistalsis takes the egg back to the isthmus region of the oviduct. Seeding sites are formed and mineralisation results in irregularly shaped blocks of calcite forming a reticulate layer. An abbreviated residence in the shell gland results in a dense crystalline cover being laid down on the reticulate layer (Mikhailov, 1987).

Using cryogenic methods, Sparks (1985) has followed the development of the shell mem-

rather than the bulk composition of uterine secretions presumably results in the true shell of the former being coated with amorphous calcium phosphate (Board, 1982) and the latter branes of the domestic turkey (*Meleagris gallopavro*) and duck (*Anas platyrhynchos*) as the egg passes along the oviduct (Fig. 6.4). Less than two hours after entering the isthmus region of the oviduct the membrane fibres appeared as three distinct bands, one thick one sandwiched between thin ones, the inner surface of the inner membrane being bordered by a fully developed limiting membrane and on the outer surface of the outer membrane by a smooth amophorus layer (Fig. 6.6). The inter-fibre space has a

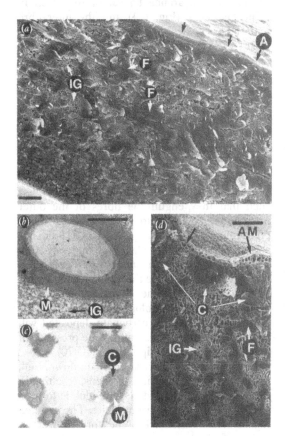

honey-combed appearance. With the onset of mineralisation (Fig. 6.7) (2 hours after entering the isthmus) there is a marked decrease in the interstitial 'honey-combing' coinciding with a change in the gross membrane morphology such that the trilaminate appearance gave way to a bilaminate structure.

The mammillary cores, the site of shell crystal nucleation, are formed by secretions of the tubular shell gland coming into contact with the fibres of the outer shell membrane. The distribution of the cores appears to be independent of the local membrane topography (Fujii, Tamura & Okamoto, 1970). Hence the spatial distribution of the cone tips is determined by the organisation of the secretory cells of the tubular shell gland (Tullett *et al.*, 1976). According to Baumgartner *et al.* (1978) maturation of the shell membranes (Fig. 6.4) is crucial to normal shell development. This process appears to be dependent upon the production of lysine-derived cross-linking within the membrane fibres. These cross-links can be disrupted in copper-depleted hens or in hens that have been exposed to the lathyrogen-like compound, *beta* aminopropionitrile (BAPN), resulting in abnormal shell formation. The implications of these conditions to shell formation will be considered later. The precise mechanism of mammillary core development, or the way in which it triggers crystal nucleation is not clear. The mammillary cores have a high sialic acid content (Cooke & Balch, 1970) which readily forms complexes with calcium. Its presence in the core may cause a local increase in calcium concentration sufficient to exceed the solubility product of calcium carbonate thus initiating crystal nucleation and subsequent growth (Cooke & Balch, 1970). Carbonic anhydrase has also been observed in

Fig. 6.6. A series of scanning electron micrographs of the shell membranes of turkey (*Meleagris gallopavo*) egg during formation. (*a*) 2 hours after entering the isthmus. This sample was prepared and viewed under cryogenic conditions – note the retention of fine structure compared with a similar sample (Fig. 6.4(*a*)) prepared under standard (freeze dried) conditions. The membrane has a trilaminate appearance – a layer of thick fibres (F) sandwiched between layers of thinner fibres. The outer surface of the outer membrane (arrowed) is covered by an amorphous layer. Bar marker 10 μm. (*b*) 2 hours after entering isthmus. Radial thin section through an outer membrane fibre. The amorphous material (G) appears to be condensing onto the mantle (M). The membrane

is embedded in resin during the preparation of the sample, resulting in the preservation of fine structures such as the amorphous material, which is shown to be homogenous in appearance – cf. Fig. 6.4(*a*) and 6.6(*a*). Bar marker 1 μm. (*c*) At oviposition. Radial thin section showing the inner membrane fibres and the limiting membrane. The core of the fibre is separated from the mantle by an electron-translucent halo (arrowed). Bar marker 2 μm. (*d*) 2 hours after entering isthmus. This sample was prepared and viewed under cryogenic conditions. The mineralisation of the developing cone tips (C) takes place under the layer of amorphous material (arrowed) that is visible in Fig. 6.6(*a*). The role of this material in controlling or influencing mineralisation has yet to be elucidated. Bar marker 20 μm.

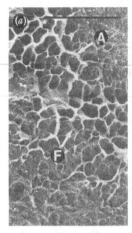

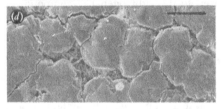

Fig. 6.7. A series of scanning electron micrographs of duck eggs during formation. (*a*) 2 hours after entering isthmus. The outer surface of the outer shell membrane. Peaks (an artefact of preparation – see Fig. 6.6(*a*)) of amorphous material (A) overlying fibres (F). Bar marker 100 µm. (*b*) 2 hours after entering isthmus. Mineralisation of the protein core leading to the formation of the cone layer. Fibres of the outer shell membrane (F) can be seen underlying the cones (C). Bar marker 100 µm. (*c*) 4 hours after entering isthmus. Radial section through the cone layer showing the intimate association between the fibres (F) of the outer shell membrane. Bar marker 100 µm. (*d*) View of the developing, surface face of the cone layer, showing the individual cones prior to fusion. Bar marker 100 µm.

mammillary core preparations (Robinson & King, 1963) and it too may be involved in the initiation or maintenance of shell crystal growth.

Whatever the method of initiation it is known that the first crystals to develop are needle-like, radiating from the centre of the core, to form the primary spherulite (Tyler, 1969). Crystalisation continues, giving rise to the secondary spherulites, until the inward growth to give the basal caps is restricted by the shell membranes and sideways growth by neighbouring spherulites – this process leading to the formation of the cone or mammillary layer (Table 6.2).

The membranes continue to mature as shell mineralisation proceeds so that five hours after entering the isthmus the membranes have a bilaminate appearance typical of that seen in oviposited eggs (Fig. 6.4) and the inter-fibre material has coalesced onto the mantle (Fig. 6.6; Sparks, 1985). Individual fibres also mature, the halo associated with the core (Fig. 6.6) being absent, or very indistinct, in the early stages of maturation. The limiting membrane however remains unchanged with time, apparently being deposited either in, or very close to, its mature form.

The mechanisms that control the development of the cone tips or mammillary layer and the reason why the tips retain a rounded appearance and do not continue to mineralise such that they engulf the fibres of the outer membrane, have/yet to be elucidated. Using cryogenic SEM techniques Sparks (1985) has shown that the initial stages of mineralisation take place beneath an amorphous cover (Fig. 6.6). This will affect the flux of ions to the developing cone tip and hence has the ability to influence the morphology. In addition, the polymorph of calcium carbonate deposited will be affected and may account for the presence of aragonite in the cone tips (Erben, 1970).

Once the mammillary layer is laid down, bulk mineralisation of the shell proceeds, resulting in formation of the palisade layer which is formed almost entirely (98%) from calcite. The palisade region was considered to be formed from a series of single crystals originating from the mammillae (Schmidt, 1962; Terepka, 1963). On the basis of polarised light studies Terepka (1963) proposed that the c-axes of the calcite crystals were normal to the shell surface. More recent studies (Perrott, Scott & Board, 1980; Silyn-Roberts & Sharp, 1986) have shown, however, that the columns are polycrystalline. X-ray diffraction has been used to examine the orientation of these crystals but with conflicting results. Cain & Heyn (1964) report that the c-axis is inclined at angles of between 12 and 44° while Perrott *et al.* (1980) have found little or no orientation with respect to the shell surface. Silyn-Roberts & Sharp (1986) contend that while crystal deposition is initiated from a single location, a mammilla, crystals grow out in all directions but there is an increase in orientation (c-axis normal to the surface) as the distance from the nucleation site increases. This accounts for earlier reports (Faverjee *et al.*, 1965; Perrott *et al.*, 1980; Sharp & Silyn-Roberts, 1984) which indicate that there is limited orientation within the surface layer of the shell.

The inorganic shell is permeated by an organic mesh which, while it only accounts for 2–4% of the weight, has a profound influence on shell morphology. Masshoff & Stolpmann (1961) contend that this organic 'micronet' is continuous with the outer shell membrane. An EM study by Simons & Wiertz (1963) has shown, however, that there is clear demarcation between the organic material of the cone tip and the palisade layer, the intervening region having only a sparse organic content. The organic fibrils have been variously described as being *circa* 10 µm long and 0.01 µm wide (Simons & Wiertz, 1963) and of between 1 to 2 µm wide (Fujii & Tamura, 1969). The distribution of organic material across the fowl eggshell is highly organised and several studies report a relationship between the organic content and the crystal size (Terepka, 1963; Quintana & Sandoz, 1978). Indeed, historically (Simkiss & Tyler, 1957, 1958) the organic matrix has been assigned an epitaxial role. The ability of matrix preparations to act as chelating agents (Simkiss & Tyler, 1958) suggests a mechanism whereby spatial ordering of constituent ions necessary for shell development can be achieved. The ability of organic matrices to organise inorganic phases may, however, have been over emphasised in the past (Mann, 1983; Silyn-Roberts & Sharp, 1986). Silyn-Roberts & Sharp (1986) contend rather that the development of well-ordered calcite crystals within the shell is hindered by the deposition of the organic phase. Petersen & Tyler (1967) demonstrate that the incorporation of an organic phase confers strength upon the shell as shells with a higher organic content tend to be stronger. This could be due to the mechanical properties of the organic material or the reduction in shell crystal size (Terepka, 1963; Williams, 1984). These findings support the contention that the organic component does not control crystal growth, but instead acts as a reinforcing network (Silyn-Roberts & Sharp, 1986).

The calcitic shell is made porous to respiratory gases and water vapour by the numerous pores that traverse it radially. Although the pore is often displayed schematically as the simple posthorn shape, there are a range of pore morphologies of increasing complexity (Board *et al.*, 1977). The mechanisms responsible for keeping these sometimes complicated pores open during mineralisation is not clear. Stewart (1935) surmises that protein material within the crystallising mass forms threads which, after shell formation, dry and move away from the pore sides thereby allowing gaseous diffusion. Tullett (1975) contends that in the fowl, pore formation is the result of a combination of the irregular packing of the calcite columns and the flux of plumping fluid, a suggestion made earlier by Tyler & Simkiss (1959). This contention gains support from the findings of Tyler (1969) who reports that the pigmented layer on the surface of shells of birds of the Phaethontidae lined the pore canals for up to 25% of the shell thickness, indicating a dynamic state within the pore lumen at the time of deposition. More recently, Mikhailov (1986) has examined pore formation in ratite eggs and proposes a mechanism for pore formation which relies upon the change of angle between the epithelial cells and the mineralising face. The invaginated regions of the epithelial layer, Mikhailov contends, correspond to areas of minimal mineralisation and consequently pore formation.

The organic cuticle of the hen egg is probably secreted in the shell gland as a granular substance (MacCallion, 1953; Cooke & Balch, 1970) although the secretory cells have yet to be identified. At oviposition the cuticle is a slippery lustrous layer which dries rapidly on exposure to the atmosphere. As with the other organic components of the avian egg (i.e. the shell membranes, and the shell matrix), the cuticle undergoes a maturation process (Sparks & Board, 1985). In its 'wet', immature, state the cuticle has a frothy, open appearance (Fig. 6.8) which changes with time, as the cuticle dries out, into the flakey, fissured layer that is typical of a mature cuticle (Fig. 6.8). The implications of this maturation process for bacterial penetration of the shell integument have been discussed elsewhere (Sparks & Board, 1985).

The eggs of the gannet, shag (*Phalacrocorax*

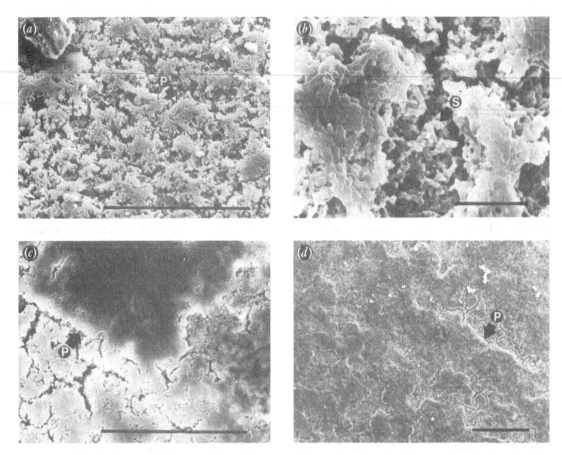

Fig. 6.8(*a*) Scanning electron micrograph of the cuticle on a fowl egg which has been fixed immediately following oviposition. Compared with a mature (*d*) cuticle the structure is very open and porous with minimal plaque (P) formation. Bar marker 100 μm. (*b*) Enlargement of (*a*) showing the development of spheres within the cuticle. Bar marker 10 μm. (*c*) Scanning electron micrograph of the cuticle on a fowl egg that has been fixed 30 s after oviposition. Plaques (P) are forming over the underlying open structure cf. (*a*). Bar marker 100 μm. (*d*) Scanning electron micrograph of a mature cuticle, the plaques (P) covering the underlying porous structure and consequently making the cuticle more resistant to bacterial penetration cf. (*a*). Bar marker 100 μm.

aristotelis) and cormorant (*Phalacrocorax carbo*) (Pelicaniformes) and the nonparasitic cuckoos are covered with an inorganic layer of vaterite (Table 6.3) (Tullett *et al.*, 1976; Board & Perrott, 1979*b*). In all species where it has been found, vaterite is present as spheres 0.2–6.3 μm in diameter. The surface morphology of the spheres varies, the larger ones being dimpled whereas the smaller spheres appear smooth. The change from calcitic shell to vaterite cover, the so called transition zone, is quite distinct in the gannet egg whereas in grebe eggs, the transition zone is ill-defined (Board *et al.*, 1984).

It is interesting to note that, whatever the shell accessory material, the unit particle often has a spherical appearance. This is true even of the organic cuticle on the fowl egg (Simons, 1971). In order to bring about region-specific, physicochemical changes biological systems often revert to compartmentalisation. Could it be that in the avian egg the various shell accessory materials are a function of the control exerted over the chemical environment through the use of vesicles? For example, the cover of the gannet egg is rich in organic compounds and phosphates. Both of these can be used to affect the kinetics of calcium carbonate precipitation, resulting in the formation of vaterite, which compared with calcite is the less thermodynamically stable form of calcium carbonate. Vaterite crystals are stabilised so that even at 80 °C they do not transform to calcite. This could be brought about by the presence of phosphorus or by the organic phase which may provide a hydrophobic layer, preventing water from coming into contact with the vaterite which could then recrystallise to calcite.

Role of the eggshell

Two broad and interacting groupings are evident in the list of contributions of eggshells to the well-being of the developing embryo (Table 6.1): 1) chemical communication between the inside and outside of an egg, and 2) mechanical protection of an egg's contents. The former has been intensively studied by Rahn, Paganelli and collaborators (Paganelli, Chapter 16). The conductance of the shell to respiratory gases and water vapour is closely correlated to egg mass, which in turn is allometrically related to such eggshell properties as surface area, thickness and especially porosity (Paganelli, Chapter 16). As yet, however, such studies have failed to even hint at the range of pore types shown in Fig. 6.2. Why should this be? If the pore systems are considered to be resistances in series, then R_2 and, to a lesser extent, R_3 in Fig. 6.2, both of which have their conductance potential 'fixed' in the oviduct, are the principal controllers of the diffusion rates of O_2, CO_2 and water vapour. Resistance R_4 is negligible and transient, associated with the limiting membrane, a structure that breaks down in the first few days of incubation (Tranter et al., 1983). Shell accessory materials (resistance R_1) also impose a minor barrier to the movement of water vapour across the shells of Anhinga eggs (outer surface covered with vaterite spheres) (Colacino et al., 1985) or those (covered with cuticle) of poultry (Deeming, 1987). In practice R_1 ought not to be regarded as merely an impediment to gaseous diffusion but rather as an adaptation that ensures that such diffusion continues unabated throughout incubation even with eggs in hostile nest environments (Board, 1982).

Shell accessory materials perform one or more of the following functions. 1) Protection of the pore canal in the true shell from blockage with preening oil and/or nest debris. Such an event would raise the value of R_2 (Fig. 6.2) to unacceptable levels. Evidence of this role of shell accessory materials in the well-being of embryos was presented by Board & Perrott

(1981) and Board et al. (1984). 2) Waterproofing of eggshells such that bacterial penetration of pores and subsequent infection of the egg contents is unlikely to occur. This role of the cuticle on the eggshells of domestic hens has been investigated (Sparks & Board, 1984). 3) The defence of eggs in humid environments against microbial colonisation of the shell surface, infiltration of the pore canals and eventual infection of the egg contents. The vulnerability of the glycoprotein cuticle on the eggshell of the fowl to fungal colonisation and digestion has been demonstrated (Board et al., 1979). If shell accessory materials are considered in the context of the nest environment (Table 6.4), then there is circumstantial evidence to support these notions. Thus open-pored eggshells are associated with dry nests and parents with little if any preening oil on their feathers, as is the case with pigeons. At the opposite extreme, a megapode egg in a mound of fermenting vegetation (Booth & Thompson, Chapter 20), the outer orifice of the pore canal is plugged and the outer surface of the shell covered with amorphous calcium phosphate. It would seem reasonable to assume that this adaptation precludes the colonisation of the shell by fungi. The outside of the true shell of grebe eggshells is covered with a layer of spheres of amorphous calcium phosphate which in its turn is enveloped in a fissured skin of amorphous material. Board et al. (1984) conclude that this arrangement impedes microbial colonisation of the egg contents and prevents blockage of the pore canals in the true shell with mud from the sodden platform of soil and vegetation that passes as a nest.

It has been shown that the efficacy of the various roles ascribed to pore systems depicted in Fig. 6.2 can be seriously impaired by shell fracture (Board, 1982). Commonly, the eggshell is considered to be an external skeletal support for the egg such that the developing embryo is not crushed by the incubating parent (Ar, Rahn & Paganelli, 1979). These authors conclude that the mechanical strength of an eggshell must not be such that it imposes a serious impediment to the hatching process. There is a correlation between hatching strategies and the mechanical properties of eggshells (Bond, Scott & Board, 1986; Bond, Board & Scott, 1988b). Thus embryos in hard brittle shells make relatively few holes in the shell before pushing away a cap from the broad pole. Those in tough, more flexible shells perforate extensively the shell along a

Table 6.4. *Pores systems and nest types.*

Pore system	Nest type
Outer orifice open	A bundle of twigs : pigeons (*Columba* spp)[a]
Outer orifice contains a plug	A platform of vegetation on surface of water: the lily trotter (*Micropara capensis*)[b]
Outer orifice covered and partially plugged with spheres of organic material	Scrape in ground: guinea fowl (*Numida meleagris*)[c]
Vaterite	A platform of rotting vegetation: blue-eyed shag (*Phalacrocorax atriceps*)[d]
Amorphous calcium phosphate	A platform of sodden vegetation floating on water: great crested grebe (*Podiceps cristatus*)[e]
	A cone-shaped mound of mud: greater flamingo (*Phoenicopterus ruber roseus*)[f]
Outer orifice plugged and overlaid with stratum of spheres	A mound of soil and rotting vegetation: Mallee fowl (*Leipoa ocellata*)[g]

[a]Board (1974); [b]Board & Perrott (1979*b*); [c]Board & Perrott (1981); [d]Board & Perrott (1979*a*); [e]Board *et al.* (1984); [f]Board (1981); Board *et al.*, 1982.

latitude before attempting to push away a portion of the shell. Indeed in some instances, Bond *et al.* (1988*b*) note that perforations extend for more than a complete revolution of the shell before a hatchling attempts to escape. As far as we are aware the literature on shell strength has not distinguished between tough-flexible and hard-brittle shells. Consequently, little if any attention has been paid to the contribution of the organic matrix to the mechanical properties of shells post oviposition (Silyn-Roberts & Sharp, 1986).

The probable mechanical properties of the eggshell types shown in Fig. 6.1 and their possible correlation with egg mass and/or the nest environment have not been studied in detail. Board (1982) postulates that the theoretical order of resistance of the various types of shell depicted in Fig. 6.1 to damage by impact would be vesiculate > trilaminate vesiculate > dense crystalline. With erosion resistance, the order would be dense crystalline and trilaminate > bilaminate vesiculate. As shell integrity is of such obvious importance to shell function, and thus the successful development of the embryo, future studies ought to explore the mechanical properties of eggshells, the physical insults to which they are exposed in the nest and the physiological basis of amendments that 'fit' a shell to a particular set of insults.

References

Ar, A., Rahn, H. & Paganelli, C. V. (1979). The avian egg: mass and strength. *Condor*, 81, 331–7.

Baird, T., Solomon, S. E. & Tedstone, D. R. (1975). Localization and characterization of egg shell porphyrins in several avian species. *British Poultry Science*, 16, 201–8.

Baumgartner, S., Brown, D. J., Salevsky Jr, E. & Leach Jr, R. M. (1978). Copper deficiency in the laying hen. *Journal of Nutrition*, 108, 804–11.

Bellairs, R. & Boyde, A. (1969). Scanning electron microscopy of the shell membranes of the hen's egg. *Zeitschrift für Zellforschung und Mikroskopische Anatomie*, 96, 237–49.

Board, R. G. (1974). Microstructure, water resistance and water repellency of the pigeon egg shell. *British Poultry Science*, 15, 415–19.

— (1981). The microstructure of avian eggshells, adaptive significance and practical implications in aviculture. *Wildfowl*, 32, 132–6.

— (1982). Properties of avian eggshells and their adaptive value. *Biological Reviews*, 57, 1–28.

Board, R. G., Loseby, S. & Miles, V. R. (1979). A note on the microbial growth on the avian eggshell. *British Poultry Science*, 20, 413–20.

Board, R. G. & Love, G. (1980). Magnesium distribution in avian eggshells. *Comparative Biochemistry and Physiology*, 66A, 667–72.

— (1983). Magnesium distribution in avian eggshells with particular reference to those of wildfowl (Anatidae). *Comparative Biochemistry and Physiology*, 75A, 111–16.

Board R. G. & Perrott, H. R. (1979*a*). The plugged pores of the tinamou (Tinamidae) and jaçana (Jacanidae) eggshells. *Ibis*, 121, 469–74.

— (1979*b*). Vaterite, a constituent of the eggshells of the non-parasitic cuckoos, *Guira guira* and *Crotophagi ani*. *Calcified Tissue International*, 29, 63–9.

— (1981). The fine structure of incubated eggs of the

Helmeted guinea fowl (*Numida meleagris*). *Journal of Zoology, London*, **96**, 445–51.

Board R. G., Perrott, H. R., Love, G. & Scott, V. D. (1984). The phosphate-rich cover of the eggshells of Grebes. *Journal of Zoology, London*, **203**, 329–43.

Board R. G., Perrott, H. R., Love, G. & Seymour, R. S. (1982). A novel pore system in the eggshells of the Mallee fowl, *Leipoa ocellata*. *Journal of Experimental Zoology*, **220**, 131–6.

Board R. G. & Scott, V. D. (1980). Porosity of avian eggshells. *American Zoologist*, **20**, 339–49.

Board, R. G. & Tullett, S. G. (1975). The pore arrangement in the emu (*Dromaius novaehollandiae*) eggshell as shown by plastic models. *Journal of Microscopy*, **103**, 281–4.

Board R. G., Tullett, S. G. & Perrott, H. R. (1977). An arbitrary classification of the pore system in avian eggshells. *Journal of Zoology, London*, **182**, 251–65.

Bond, G. M., Board, R. G. & Scott, V. D. (1988*a*). An account of the hatching strategies of birds. *Biological Reviews*, **63**, 395–415.

— (1988*b*). A comparative study of changes in the structure of avian eggshells during incubation. *Zoological Journal of the Linnean Society*, **92**, 105–13.

Bond, G. M., Scott, V. D. & Board, R. G. (1986). Correlation of mechanical properties of avian eggshells with hatching strategies. *Journal of Zoology, London*, **209**, 225–37.

Brooks, J. & Hale, H. P. (1955). Strength of the shell of hen's egg. *Nature, London*, **175**, 848–9.

Cain, C. J. & Heyn, A. N. (1964). X-ray diffraction studies of the crystalline structure of the avian egg shell. *Biophysical Journal*, **4**, 23–9.

Candlish, J. K. (1972). The role of the shell membranes in the functional integrity of the egg. In *Egg Formation and Production*. B. M. Freeman, & P. E. Lake, (eds). pp. 87–105. Edinburgh: British Poultry Science Ltd.

Colacino, J. M., Hamel, P. B., Rahn, H., Board, R. G. & Sparks, N. H. C. (1985). The vaterite cover on the eggs of *Anhinga anhinga* and its effect on gas conductance. *Journal of Zoology, London*, **205**, 425–33.

Cooke, A. S. & Balch, D. A. (1970). Studies of membrane, mammillary cores and cuticle of the hens egg shell. *British Poultry Science*, **11**, 353–65.

Deeming, D. C. (1987). Effect of cuticle removal on the water vapour conductance of egg shells of several species of domestic bird. *British Poultry Science*, **28**, 231–7.

Doskocil, V. M., Blazek, J., Nemcova, P. & Starkova, B. (1985). The structure of the hen's egg shell and its changes during incubation. A SEM study. *Anat Az Jena*, **159**, 117–26.

Erben, H. K. (1970). Ultrastrukturen und Mineralisation rezenter und fossiler Eischalen bei Vogeln und Reptilien. *Biomineralisation*, **1**, 1–66.

Faverjee, J. Ch. L., Van der Plas, R., Schoorl, R. & Floor, P. (1965). X-ray diffraction of the crystalline structure of the avian egg shell: some critical remarks. *Biophysical Journal*, **5**, 359–61.

Fujii, S. & Tamura, T. (1969). Scanning electron-microscopy of the hen's egg shell. *Journal of the Faculty of Fisheries and Animal Husbandry, Hiroshima University*, **8**, 85–98.

Fujii, S., Tamura, T. & Okamoto, T. (1970). Scanning electron microscopy of shell membrane formation in hen's eggs. *Journal of the Faculty of Fisheries and Animal Husbandry Hiroshima University*, **9**, 139–50.

Itoh, H. & Hatono, T. (1964). Variation of magnesium and phosphorus deposition rates during egg shell formation. *Poultry Science*, **43**, 77–80.

Kennedy, G. Y. & Vevers, H. G. (1975). A survey of avian egg shell pigments. *Comparative Biochemistry and Physiology*, **55B**, 117–23.

Lack, D. (1968). *Ecological adaptations for breeding in birds*. London: Chapman & Hall.

Lang, M. R. & Wells, J. W. (1987). A review of egg shell pigmentation. *World's Poultry Science Journal*, **43**, 238–46.

Leach Jr, R. M., Rucker, R. B. & Van Dyke, G. P. (1981). Eggshell membrane protein: A nonelastin desmosine-isodesmosine – containing protein. *Archives of Biochemistry and Biophysics*, **207**, 353–9.

Mann, S. (1983). Mineralization in biological systems. *Structural Bonding*, **54**, 125–74.

Masshoff, W. & Stolpmann, H. J. (1961). Licht-undelektronenmikroskopische Untersuchungen an der Schalenhaut und Kalkschale des Huhnereies. *Zeitschrift für Zellforschung und Mikroskopische Anatomie*, **55**, 818–32.

MacCallion, D. J. (1953). A cytological and cytochemical study of the shell gland of the domestic hen. *Canadian Journal of Zoology*, **21**, 577–89.

Mikhailov, K. E. (1986). Pore complexes of ratite eggshells and mode of pore formation. *Palaeontological Journal, Academy of Sciences of the USSR*, **3**, 84–93.

— (1987). New data on the structure of the Emu egg shell. *Zoology Journal*, **66**, 1349–53.

Needham, J. (1950). The evolution of the cleidoic egg. In *Biochemistry and Morphogenesis Part 1. The Morphogenetic sub-stratum*, section 1.62 pp. 33–8. Cambridge: Cambridge University Press.

Perrott, H. R., Scott, V. D. & Board, R. G. (1980). Crystal orientation in the shell of the domestic fowl: an electron diffraction study. *Calcified Tissue International*, **33**, 119–24.

Petersen, J. & Tyler, C. (1967). The strength of guinea fowl (*Numida meleagris*) eggshells. *British Poultry Science*, **7**, 291–6.

Quintana, C. & Sandoz, D. (1978). Coquille de l'oeuf de caille: etude ultrastructurale et crystallographique. *Calcified Tissue Research*, **25**, 145–59.

Rahn, H. & Paganelli, C. V. (1981). *Gas exchange in*

avian eggs. State University of New York at Buffalo: New York.

Robinson, D. S. & King, N. R. (1963). Carbonic anhydrase and formation of the hen's egg shell. *Nature, London*, **199**, 497.

Romanoff, A. L. & Romanoff, A. J. (1949). *The Avian Egg*. Wiley: New York; Hall: London.

Sauer, E. G. F. (1972). Ratite eggshells and phylogenetic questions. *Zoologisches Forschungsintitut und Museum Koening, Bonn*.

Schade, R. (1987). Biochemishche und Molekulargenetishche Aspekte der Eischalenbildung. *Archiv für Gelflugelkunde*, **51**, 81–7.

Schmidt, W. J. (1962). Liegt der Eischalenkalk der Vogel als sumickroskopiche Kristallite vor? *Zeitschrift für Zellforschung und Mikroskopische Anatomie*, **57**, 848–80.

Sharp, R. M. & Silyn-Roberts, H. (1984). The development of preferred orientation in the egg shell of the domestic fowl. *Biophysical Journal*, **46**, 175–80.

Silyn-Roberts, H. (1983). The pore geometry and structure of the egg shell of the North Island Brown Kiwi. *Journal of Microscopy*, **130**, 23–36.

Silyn-Roberts, H. & Sharp, R. M. (1986). Crystal growth and the role of the organic network in eggshell biomineralization. *Proceedings of the Royal Society, London*, B, **227**, 303–24.

Simkiss, K. & Tyler, C. (1957). A histochemical study of the organic matrix of the hen egg shells. *Quarterly Journal of Microscopical Science*, **98**, 19–28.

— (1958). Reactions between egg shell matrix and metallic cations. *Quarterly Journal of Microscopical Science*, **99**, 5–13.

Simons, P. C. M. (1971). Ultrastructure of the hen eggshell and its physiological interpretation. *Commun. No. 175. Centr. Inst. Poultry Res.* Beekbergen, The Netherlands.

Simons, P. C. M. & Wiertz, G. (1963). Notes on the structure of membranes and shell in the hen's egg. An electron microscopical study. *Zeitschrift für Zellforschung und Mikroskopische Anatomie*, **59**, 555–67.

Solomon, S. E. (1987). Egg shell pigmentation. In *Egg Quality – Current Problems and Recent Advances*. R. G. Wells, & C. G. Belyavin, eds. pp. 147–57 London: Butterworths.

Sparks, N. H. C. (1985). The hen's eggshell – a resistance network. Unpublished Ph.D. thesis, University of Bath.

Sparks, N. H. C. & Board, R. G. (1984). Cuticle, shell porosity and water uptake through hens' egg shells. *British Poultry Science*, **25**, 267–76.

— (1985). Bacterial penetration of the recently oviposited shell of hen's egg. *Australian Veterinary Journal*, **62**, 169–70.

Stewart, G. F. (1935). The structure of the hen's eggshell. *Poultry Science*, **14**, 24–32.

Terepka, A. R. (1963). Structure and calcification in the avian egg shell. *Experimental Cell Research*, **30**, 171–82.

Tranter, H. S., Sparks, N. H. C. & Board, R. G. (1983). Changes in the limiting membrane and oxygen permeability of chicken eggshells during incubation. *British Poultry Science*, **24**, 537–47.

Tullett, S. G. (1975). Regulation of avian eggshell porosity. *Journal of Zoology, London*, **177**, 339–48.

— (1987). Eggshell formation and quality. In *Egg Quality – Current Problems and Recent Advances*. R. G. Wells, & C. G. Belyavin, eds. pp. 123–46 London: Butterworths.

Tullett, S. G., Board, R. G., Love, G., Perrott, H. R. & Scott, V. D. (1976). Vaterite deposition during eggshell formation in the cormorant, gannet and shag, and in the 'shell-less' eggs of the domestic fowl. *Acta Zoologica, Stockholm*, **57**, 79–87.

Tyler, C. (1964). *Wilhelm von Nathusius on avian eggshells*. Reading: The University.

— (1966). A study of the eggshells of the Falconiformes. *Journal of Zoology, London*, **150**, 413–25.

— (1969). Avian eggs: their structure and characteristics. *International Review of General and Experimental Zoology*, **4**, 82–127.

Tyler, C. & Simkiss, K. (1959). A study of the eggshells of ratite birds. *Proceedings of the Zoological Society, London*, **133**, 201–43.

Wedral, E. M., Vadehra, D. U. & Baker, R. C. (1974). Chemical composition of the cuticle and inner and outer membranes from eggs of *Gallus gallus*. *Comparative Biochemistry & Physiology*, **47B**, 631–40.

Weiser, D. & Krampitz, G. (1987). Oestrogens activate egg shell protein-genes in the liver of laying hens. *Zuchthygience, Berlin*, **22**, 267–74.

Williams, R. J. P. (1984). An introduction to biominerals and the role of organic molecules in their formation. *Philosophical Transactions of the Royal Society, London*, **304**, 411–24.

7

Physical characteristics of reptilian eggs and a comparison with avian eggs

JOHN B. IVERSON AND MICHAEL A. EWERT

Introduction

Most female oviparous reptiles ovulate and shell a few to several eggs at the same time to produce a clutch that is normally laid as a unit. This differs significantly from the normal procedure in birds, in which individual eggs in a clutch are each ovulated, shelled, and oviposited separately (usually on a 24-hour cycle in the fowl (*Gallus gallus*); Gilbert, 1971). Egg size and number within a clutch are clearly important aspects of an organism's life history 'strategy' (Smith & Fretwell, 1974; Stearns, 1976). Egg size, shape and number also may be related to the anatomy of the female (e.g. ovarian follicle size, oviductal length and diameter, abdominal volume, and size of the pelvic canal), as well as to the physiological ecology of the clutch in the nest. For reptiles, evidence is available that suggests that life history strategies (Moll, 1979; Ferguson, Brown & DeMarco, 1982; Dunham, Miles & Reznick, 1988; Seigel & Ford, 1987; Ford & Seigel, 1989), pelvic canal size (Congdon & Gibbons, 1987; Long & Rose, 1989), and nest physiology (Packard & Packard, 1988) do place important constraints on egg size, shape, and number. However, the influence of the anatomy and physiology of the female reproductive tract on egg parameters has not been explored in reptiles.

Although numerous studies report clutch sizes, lengths and widths of eggs in reptiles, and some even report data on mass, volume, or other aspects of egg-shape, no synthesis of egg size and shape has been attempted for reptiles at the level done for birds, which includes aspects of size, volume, density, surface area, and shape (Preston, 1953, 1969, 1974; Paganelli,

Olszowka & Ar, 1974; Tatum, 1975; Rahn, Paganelli & Ar, 1975; Hoyt, 1976; Smart, Chapter 8). As a beginning to that synthesis we have focused on the eggs of turtles, many of which have as rigid shells as those of birds (Ewert, 1979), and hence do not change significantly in size or shape during development. We have only recently begun a review of the pliable-shelled eggs of squamates, because the substantial increases in egg size that normally (though not always) occur after oviposition in those reptiles (Packard & Packard, 1988) alter the significance of shape relative to these variables in turtles and birds.

The typical avian egg is asymmetrical and is tapered at one end (Preston, 1969, 1974). This shape is generally considered (Welty, 1962; Preston, 1969) to be an adaptation to 1) maximise the number of eggs that can be packed under a brooding female (i.e. a thermoregulatory hypothesis) and/or 2) reduce the likelihood of an egg rolling out of a nest (the rolling characteristic hypothesis). However, the typical reptile egg is symmetrical, i.e. spherical or ellipsoid and is not tapered. The reasons why this is the case have not been discussed. By focusing on turtle eggs we hope to be able to make direct comparisons regarding the adaptive significance of egg size and shape in at least turtles and birds. Any extrapolation of conclusions based on turtle eggs to reptiles in general must await the collection of additional data from, in particular, squamate reptiles.

Basic measurements of reptilian eggs

Reptilian eggs exhibit extraordinary variation in size and shape (Table 7.1) with a nearly 4900–

Table 7.1. *The range in variation in egg size and shape in reptiles, birds and mammals. Length (L) and width (W) in mm; mass (M), in grams.*

Group	Smallest egg L×W; M (species; source)	Largest egg L×W; M (species; source)	Least elongate L×W; L/W (species; source)	Most elongate L×W; L/W (species; source)
Turtles	23.6×13.5; 2.57 (*Sternotherus odoratus*; Gross, 1982)	59.7×55.2; 106.9 (*Geochelone elephantopus*; Shaw, 1966)	26.5×25.3; 1.05 (*Chelydra serpentina*; this paper)	71.5×31.6; 2.26 (*Rhinoclemmys annulata*; this paper)
Lizards and Amphisbaenians	6.57×4.48; 0.072 (*Lygodactylus klugei*; Vitt, 1986)*	86.5×55.5; 124.7 (*Varanus komodoensis*; Auffenberg, 1981)	14.02×13.1; 1.07 (*Ptyodactylus hasselquistii*; Werner, 1989)	31×2; 15.5 (*Chirindia ewerbecki*; Loveridge, 1941)
Snakes	25×2.5; ? (*Leptotyphlops blanfordi*; Minton, 1966)	?×?; 226–303 (*Python molurus*; Pope, 1961)	26.9×19.2; 1.4 (*Coluber constrictor*; Fitch, 1963)	25×2.5; 10.0 (*Leptotyphlops blanfordi*; Minton, 1966)
Tuatara	ca.30×20; 4–6 (*Sphenodon punctatus*; Moffat, 1985)	–	30×20; ca. 1.5 (same)	–
Crocodylia	68×34; 51.7 (*Alligator sinensis*; Ferguson, 1985)	76×51; 113 (*Crocodylus siamensis*; Ferguson, 1985)	86×67; 1.28 (*Gavialis gangeticus*; Ferguson, 1985)	68×34; 2.0 (*Alligator sinensis*; Ferguson, 1985)
Birds	?×?; 0.3 (*Chlorostilbon canivetii*; Schönwetter, in Paganelli *et al.*, 1974)	170.7×?; 1480 (*Struthio camelus*; Paganelli *et al.*, 1974)	34.3×30.5; 1.12 (*Otus asio*; Reed, 1965)	?×?; 1.87 (*Megapodius pritchardi*; Preston, 1969)
Monotremes	16.5×13.0; ? (*Tachyglossus*; Griffiths, 1978)	20×?; ? (*Ornithorhynchus*; Cockrum, 1962)	17.25×14.0; 1.23 (*Ornithorhynchus*; Griffiths, 1978)	?

*Eggs of *Gymnophthalmus multiscutatus* have a mass of 0.062 g but other dimensions were not reported (Vitt, 1982)

fold range in mass and from nearly spherical to over 15 times as long as wide. The range in egg mass found in reptiles is the same as that found among living birds (4930-fold) although avian eggs reach larger absolute sizes. However, the range of elongation of reptile eggs far exceeds that of birds, which only range from off-round to not even twice as long as wide (Table 7.1). Reptiles also exhibit a greater range in eggshell type than birds, from pliable to brittle in the former and all brittle in the latter (Packard & DeMarco, Chapter 5; Board & Sparks, Chapter 6). Avian eggs do, however, vary significantly more than in reptiles in symmetry, with the typical bird egg being asymmetrical and the typical reptile egg being symmetrical. However, irregularly shaped eggs do occur commonly (but not exclusively) in the turtle genera *Platemys*, *Rhinoclemmys*, *Heosemys* and *Melanochelys* and pillow-shaped (i.e. elliptical in cross-section) eggs are the rule in *Kinixys* turtles (Ewert, 1979, unpublished).

Within the reptiles, lizards and snakes have the most variable eggs, with an approximate 2000-fold range in egg mass in each of the two groups, and with 14.5- and 7.1-fold ranges in elongation (length/width), respectively (Table 7.1). A lack of round eggs in snakes is conspicuous, but not unexpected given the constraints of their elongate body morphology.

Comparisons among formulae to describe shape

Three models for estimating egg volume from length (L) and width (W) data for reptile eggs seem most realistic: 1) a cylinder formula, 2) an ellipsoid formula, and 3) a bicone formula. A fourth model, two cones base-to-base, can be used, but since no reptile eggs even approach this shape, it is considered unrealistic and therefore, is not considered further.

The cylinder formula assumes that an egg is shaped like a cylinder of L-W height and W diameter with a hemisphere of diameter W on each end:

$$V = (\pi/1000)[(W^3/6) + (W^2/4)(L-W)] \tag{7.1}$$

Where V is in cm^3 and W and L are in mm.

The ellipsoid formula assumes an egg is shaped like an ellipsoid with length L and diameter W:

$$V = (4\pi/3000)(L/2)[(W/2)^2]$$
$$\text{or}$$
$$V = (\pi/6000)LW^2 \tag{7.2}$$

Based on bird egg shape data from Preston (1974), Tatum (1975) showed that the volume of symmetrical eggs (as in nearly all reptiles, including turtles) can be estimated by the bicone formula:

$$V = (\pi LW^2/6000)[1 + 2/5c_2 + 3/35c_2^2] \tag{7.3}$$

where c_2 = bicone which is negative for eggs with pointed ends and positive for those with blunt ends relative to a basic ellipsoid, when $c_2 = 0$ (Preston, 1969).

This formula can be rearranged and solved for c_2:

$$c_2 = ([(6.3 \times 10^5 V/\pi LW^2) - 56]^{1/2} - 7)/3 \tag{7.4}$$

c_2 can then be calculated and used to compare egg shape, but can also be used to estimate egg volume in those eggs for which volume measurements are lacking (equation 7.3). As a means of comparing the utility of these three formulae in estimating egg volume, and to permit comparisons with egg data in birds, we have measured length (L, mm) and width (= diameter; W, mm), mass (g), and volume (cm^3; by water displacement) for 236 fully shelled oviductal or freshly oviposited eggs representing 32 turtle species (Table 7.2). An effort was made to include a representative survey of egg types (i.e. rigid vs pliable; Ewert, 1979, 1985), elongations (E; round, to more than twice as long as wide), and higher taxa (i.e. family and subfamily). Unfortunately, we lack data from sea turtles (families Cheloniidae and Dermochelyidae) or tortoises (Testudinidae).

For our formulae comparisons, we employ the bicone method in two ways. First, c_2 is calculated for all 236 eggs, but the average c_2 value is used in equation 7.3 to estimate volume. Second, c_2 is calculated for all eggs, but a mean c_2 by species is used to estimate volume as before. Thus, we are able to evaluate four methods of estimating the volume of turtle eggs by comparing estimated values with measured volumes.

The cylinder method provides the poorest estimate of turtle egg volume, tending to overestimate volume (231 of 236 eggs) by an average error of 14.5% of actual volume (Table 7.3). It does not provide the best average estimate of egg volume for the eggs of any species examined. The ellipsoid method is considerably more accurate, but tends to underestimate values (198 of 236 eggs) by an average error of 4.8% of actual volume (Table 7.3). This method provides the best average estimate of egg volume in *Chelydra serpentina* (8 of 14 eggs) and *Chrysemys picta* (5 of 7 eggs), and as good an estimate as either of the bicone methods in *Sternotherus carinatus* (5 vs 4 eggs), *Trionyx spiniferus* (4 vs 4 eggs), *Kinosternon flavescens* (5 vs 4 eggs), and *Graptemys geographica* (3 vs 2 eggs). Using a single mean c_2 across all eggs (mean of all 236 eggs = 0.111) tends slightly to underestimate volume (121 of 236 eggs) by an average error of 3.4% of actual volume (Table 7.3). Using separate mean c_2 values by species (Table 7.2) provides the best estimates of egg volume, though this technique still tends slightly to underestimate volume (117 of 236 eggs) by an average error of 2.2% of actual volume (Table 7.3). In summary, across all 236 eggs, the cylinder method provides the closest estimate of actual egg volume in only 10, the ellipsoid method in 60, the bicone method with universal c_2 value in 80, and the bicone method with individual species c_2 values in 95 (the total exceeds 236 because of equally good estimates by two methods).

Our results indicate, not surprisingly, that using species-specific c_2 values in equation 7.3 is the most accurate means of estimating egg

Table 7.2. *Mean values for egg measurements and constants from 32 species of turtles. Taxonomy follows Iverson (1986).*

Species	N	Length (mm)	Width (mm)	Elongation	Mass (g)	Volume (cm^3)	Density (g cm^{-3})	c_2* (range)
Rigid-shelled eggs								
Kinosternon flavescens	9	32.16	17.88	1.79	6.12	5.53	1.107	0.020 (−.14/.11)
Kinosternon leucostomum	7	37.17	19.03	1.95	8.60	7.59	1.132	0.184 (.12/.38)
Kinosternon scorpioides	12	35.72	18.28	1.95	7.44	6.44	1.156	0.070 (−.01/.21)
Sternotherus carinatus	9	30.80	17.55	1.75	6.11	5.10	1.199	0.042 (−.05/.18)
Sternotherus minor	7	29.04	17.19	1.69	5.32	4.63	1.149	0.070 (−.05/.17)
Sternotherus odoratus	27	26.17	15.24	1.71	3.80	3.35	1.134	0.108 (−.04/.38)
Staurotypus triporcatus	17	44.69	24.92	1.79	17.14	15.37	1.116	0.121 (.04/.21)
Chinemys reevesii	6	42.88	19.47	2.20	10.17	9.27	1.097	0.210 (.11/.26)
Cistoclemmys flavomarginata	1	45.60	23.40	1.95	15.87	14.10	1.126	0.180
Heosemys spinosa	1	64.70	31.30	2.07	40.98	36.50	1.123	0.230
Mauremys mutica	1	47.00	22.70	2.07	15.70	14.40	1.090	0.310
Melanochelys trijuga	12	49.13	26.18	1.88	20.35	18.28	1.114	0.093 (−.05/.30)
Rhinoclemmys annulata	2	71.45	31.55	2.26	46.77	41.55	1.125	0.270 (.26/.28)
Rhinoclemmys areolata	6	63.05	29.57	2.14	35.30	31.55	1.120	0.205 (.12/.32)
Rhinoclemmys pulcherrima	8	53.18	29.93	1.78	28.68	25.34	1.132	0.026 (−.11/.13)
Siebenrockiella crassicollis	1	49.80	26.90	1.85	22.68	20.50	1.106	0.200
Trionyx spiniferus	10	26.95	25.08	1.07	9.91	9.10	1.090	0.045 (−.03/.13)
Phrynops gibbus	1	55.00	28.90	1.90	27.46	24.40	1.125	0.030
Platemys platycephala	5	51.76	27.62	1.87	25.74	22.44	1.148	0.190 (.14/.24)
Pliable-shelled eggs								
Chelydra serpentina	15	26.49	25.25	1.05	9.84	8.87	1.109	−0.005 (−.12/.17)
Chrysemys picta	8	31.56	19.10	1.65	6.67	6.19	1.077	0.059 (−.01/.18)
Clemmys guttata	7	31.91	17.13	1.86	5.82	5.32	1.095	0.204 (.14/.35)
Clemmys insculpta	12	35.08	23.62	1.48	11.19	10.45	1.070	0.031 (−.08/.17)
Graptemys geographica	7	37.20	22.04	1.69	10.57	9.89	1.069	0.097 (−.01/.30)
Graptemys nigrinoda	2	47.65	23.00	2.07	14.82	13.75	1.077	0.095 (−.05/.24)
Graptemys ouachitensis	8	36.18	22.08	1.64	10.06	9.56	1.052	0.081 (−.03/.21)
Trachemys scripta	3	35.63	21.83	1.63	9.74	9.23	1.054	0.090 (.06/.14)

Table 7.2. *cont.*

Species	N	Length (mm)	Width (mm)	Elongation	Mass (g)	Volume (cm^3)	Density (g cm^{-3})	c_2* (range)
Pseudemys concinna	12	42.03	24.92	1.68	15.44	14.77	1.045	0.189 (.11/.28)
Ocadia sinensis	3	39.50	19.77	2.00	9.46	8.75	1.081	0.200 (.15/.25)
Elseya latisternum	4	35.13	19.80	1.77	8.19	7.70	1.063	0.165 (.12/.22)
Pelomedusa subrufa	6	31.20	18.52	1.68	6.69	6.32	1.058	0.292 (.25/.40)
Pelusios castaneus	7	37.69	19.69	1.92	9.06	8.55	1.059	0.280 (.15/.47)

*c_2 calculated using equation 7.4.

Table 7.3. *Comparison of methods (see text for descriptions) used to estimate egg volume. Errors are per cent of actual volume. Positive values indicate overestimates; negative, underestimates.*

Method	Actual error		Absolute value of error
	Mean±S.D.	Range	Mean±S.D.
Cylinder	+14.5±6.3	−2.6 to +28.9	14.5±6.3
Elliptical	−4.4±4.5	−17.1 to +4.9	4.8±3.2
Bicone (mean c_2)	−0.1±4.5	−13.4 to +9.7	3.4±3.2
Bicone (actual species-specific c_2)	−0.1±3.2	−10.5 to +6.0	2.3±1.8
Bicone (c_2 estimated from equation 7.5 and mean species E)	+0.2±3.2	−11.7 to +10.2	3.0±3.2
Bicone (c_2 estimated from equation 7.5 and individual egg E)	+0.2±3.2	−11.1 to +11.0	2.9±3.2

volume in turtles. However, given that most published data on turtle eggs include only measurements of length and width (and sometimes mass), the relationship of c_2 to these variables (in particular its predictability from them) has been examined. Species sample sizes are unequal and so correlates with bicone value are examined using species means, rather than data from individual eggs. In any case, the conclusions drawn here are supported by analysis from both approaches.

Bicone values (c_2) are positively related to egg elongation (Fig. 7.1) and length, but not width (Table 7.4), suggesting that more elongate eggs have blunter ends than rounder eggs. These relationships generally hold for rigid- and pliable-shelled eggs analysed separately (Table 7.4), although the relationship between length and bicone for pliable-shelled eggs is not significant. In addition, in those relationships that *are* significant, those for the sample of pliable-shelled eggs are consistently more variable (as measured by r^2) than those for the rigid-shelled egg or overall egg samples. Average c_2 values in those species with rigid-shelled eggs are not significantly different from those for pliable-shelled eggs ($r = 0.015$; $p = 0.99$). Eggshell type has no significant effect on the relationships of c_2 with E, L, or W (ANCOVA, $F = 2.06$, 1.08, and 0.02, respectively; $p = 0.16$, 0.31, and 0.88, respectively).

The relationship of egg size to bicone value was also investigated via multiple regression of L and W on c_2 (Table 7.4). Regression equations

Table 7.4. *Regression equations relating bicone value* (c_2) *to the independent variables elongation* (E), *egg length* (L), *and egg width* (W). *No simple regression involving width (or mass) was significant at less than p=0.15; their equations are therefore omitted. Asterisk indicates data set without outlying data from* Graptemys nigrinoda.

Independent variable(s)	Subgroup	N	Regression equation	r	p value
E	All eggs	32	$c_2=0.196E-0.216$	0.59	0.001
	All rigid	19	$c_2=0.230E-0.293$	0.65	0.003
	All pliable	13	$c_2=0.206E-0.213$	0.58	0.04
	All oblong	30	$c_2=0.250E-0.317$	0.54	0.003
	Obl, rigid	18	$c_2=0.394E-0.613$	0.74	0.001
	Obl, pliable	12	$c_2=0.213E-0.225$	0.43	0.16
	Obl, pliable	11*	$c_2=0.399E-0.536$	0.67	0.02
L	All eggs	32	$c_2=0.003L-0.011$	0.40	0.02
	All rigid	19	$c_2=0.004L-0.031$	0.55	0.01
	All pliable	13	$c_2=0.003L+0.042$	0.16	0.62
	All oblong	30	$c_2=0.002L+0.040$	0.32	0.08
	Obl, rigid	18	$c_2=0.004L-0.022$	0.51	0.033
	Obl, pliable	12	$c_2=0.002L+0.217$	0.10	0.75
	Obl, pliable	11*	$c_2=0.002L+0.095$	0.06	0.85
L and W	All eggs	32	$c_2=0.008L-0.015W+0.151$	0.59	0.003
	All rigid	19	$c_2=0.008L-0.013W+0.071$	0.66	0.01
	All pliable	13	$c_2=0.005L-0.024W+0.469$	0.67	0.049
	All oblong	30	$c_2=0.011L-0.022W+0.182$	0.55	0.008
	Obl, rigid	18	$c_2=0.015L-0.029W+0.120$	0.073	0.003
	Obl, pliable	12	$c_2=0.007L-0.027W+0.475$	0.58	0.16
	Obl, pliable	11*	$c_2=0.018L-0.035W+0.246$	0.70	0.07

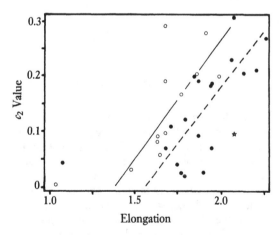

Fig. 7.1. The relationship between mean egg elongation ($E = L/W$) and mean bicone value (c_2) for 32 species of turtles. Solid circles are species with rigid-shelled eggs; open circles and star, those with pliable-shelled eggs. Regression equations (see text and Table 7.2) are plotted separately for rigid-shelled egg species (dashed line) and pliable-shelled egg species (minus the outlier *Graptemys nigrinoda*: open star).

using L and W were better predictors of c_2 than those using either independent variable alone, and about as accurate as those based on E alone. Thus the best equations predicting c_2 are those simple regressions using elongation as the independent variable, specifically (for data from all 32 turtle species):

$$c_2 = 0.196 E - 0.216 \qquad (7.5)$$

However, in all of the above regressions, species with nearly spherical eggs (i.e. where $E < 1.3$) have the greatest residuals (i.e. deviate most from the regression line; Fig. 7.1). This suggests that c_2 values for turtle eggs might be more accurately estimated from separate regressions based on round versus oblong eggs. Unfortunately, our sample includes only two species with round eggs (one rigid and one pliable), precluding the possibility of evaluating the relationships of E, L, and/or W to c_2 in these species. However, adequate samples of species with oblong eggs of both pliable and rigid shell types are available for regression analysis (Table 7.4). These separate analyses reveal that the significant relationships of E and L to c_2 hold for species with oblong, rigid-shelled eggs, but not for those with oblong, pliable-shelled eggs (unless data for the obvious outlier *Graptemys nigrinoda* are removed; see Fig. 7.1 and Table 7.4). However, for eggs of any given elongation, pliable-shelled eggs tend to be more bicone (more bluntly ended) than rigid-shelled eggs (ANCOVA, $F = 4.19$; $p = 0.048$), although the former may become more ellipsoid as they swell during development (Quay, 1976; Packard, Tracy & Roth, 1977; Packard & Packard, 1988). There is no significant difference in the relationship of bicone to either length or width between egg types (ANCOVA, $F = 1.10$ and 0.18, respectively; $p = 0.30$ and 0.68, respectively).

As a final test of the utility of these equations in estimating c_2, and therefore estimating egg volume from length and width data, volumes of every egg have been estimated in two ways based on equation 7.5, and compared to measured volumes of the individual eggs of each species. The first estimate employs separate species-specific c_2 values estimated from the regression equation relating E to c_2 for all 32 species (equation 7.5) using the mean E values in Table 7.2. The second estimate uses the same equation, but instead uses individual egg values for E. Volume estimates by the first method tend very slightly to overestimate actual volume (123 of 236 eggs) with an average of the absolute value of the error of 3.0% (Table 7.3), whereas those by the second method tend very slightly to overestimate actual volume (119 of 236 eggs) by an average of the absolute value of the error of 2.9% (Table 7.3). This suggests that both methods of estimating egg volume from length and width data are equally good, although using the separate equations for the subsets of oblong eggs would presumably increase that accuracy even more.

Although elongation explains a great deal of the variation in bicone value (and thus volume) in turtle eggs, we can only speculate on a causal relationship between the two. More elongate eggs might be more likely to be crowded along the oviduct and so it seems logical that eggs that are more blunt might well be a result either of natural selection to maximise oviductal space for eggs or from simple space restrictions during the shelling process. If the latter were true, it would be predicted that c_2 values would vary in the same female from clutch to clutch, and from oviduct to oviduct depending on the number of

eggs present, but we have yet to test this prediction. Other potential correlates of c_2 have been examined (mean female size and body mass, mean species clutch size and clutch mass), but none demonstrate a significant relationship.

Shape comparisons

Turtle eggs are typically positive bicone in shape (i.e. more blunt ended than an ellipsoid; mean species bicone = 0.137), even those that are nearly spherical (as in *Chelydra serpentina*). Calculation of the bicone value from the mean egg length, width, and volume data presented by Deeming & Ferguson (1990) for *Alligator mississippiensis* ($c_2 = 0.095$) suggests that crocodilian eggs are also positive bicones. Unfortunately, volume and shape data from other reptilian groups are unavailable for comparison (Fitch, 1970), however, our preliminary studies (and those of Preston, 1969 and Seymour, 1979) indicate that reptile eggs (including those of dinosaurs) are typically symmetrical, positive bicones. Bird eggs are usually very different, being asymmetrical and negative bicone (mean c_2 = −0.06; range, −0.17 to + 0.25; Preston, 1969). In addition, the range of elongations of turtle eggs (*circa* 1.0 to *circa* 2.2) far exceeds that of birds (1.19 to 1.64; mean = 1.39; Table 7.2; Preston, 1969, although the addendum to that paper extends the range to 1.87).

Among the birds, the most reptile-like eggs are found in the hummingbirds ($E = 1.48$; c_2 = 0.25; asymmetry = 0.019; Preston, 1969) and the megapodes (mean $E = 1.73$; Preston, 1969). Given that egg bicone and elongation are positively correlated with relative surface area (Hoyt, 1976), the adaptive advantage of an oblong, positive bicone reptilian egg may be related to maximising the surface for physiological exchanges (especially of gases) with the environment. Perhaps hummingbirds, because of their tiny eggs, and megapodes, because their eggs are buried in soil, have responded to similar selective pressures (Seymour & Ackerman, 1980; Booth & Seymour, 1987; Booth & Thompson, Chapter 20).

In birds in general, surface area for exchanges seems to be less important to the developing egg, as evidenced by the lack of correlation between elongation (hence, surface area) and pore area, pore density, and shell thickness (Ar *et al.*, 1974; Hoyt, 1976). In fact, Hoyt (1976) concludes that the functional significance of surface area variation in avian eggs is unrelated to exchanges with the environment. However, reptile eggs exhibit much greater variation in at least egg shell type, shell thickness, and elongation, so exchange of gases and water with the environment is therefore probably much more critical to their embryonic development (Packard & Packard, 1988). Variation in shape and shell type, as well as density and functional pore area, in reptile eggs may have important physio-ecological correlates (Seymour, 1979). However, elucidating these correlates may be difficult given that egg shape seems in part to be related to clutch size and body size, which are themselves correlated in turtles (Moll, 1979). Species with relatively large clutches (e.g. Cheloniids (sea turtles), Chelydrids (snapping turtles), Trionychids (softshell turtles), and some large Testudinids (e.g. the giant tortoises) and Chelids (e.g., the matamata)) tend to have round eggs, suggesting that oviducal/abdominal space limitation may be important (Ewert, 1979). It has also been suggested that diameter of the pelvic canal may affect egg shape by constraining the diameter of eggs (Congdon & Gibbons, 1987). This constraint is most evident in turtles with small body size (Congdon & Gibbons, 1987) and, perhaps more significantly, small clutch size (Iverson, 1991). However, that constraint probably represents an ultimate, rather than proximate one.

Smart (Chapter 8) discusses the importance of oviductal anatomy in determining egg-shape in birds. In the avian oviduct, albumen is secreted by the magnum region and the shell membrane is secreted by the lining of the more distally located isthmus, after which the egg passes to the shell gland region (i.e. uterus) for deposition of the eggshell (Simkiss & Taylor, 1971; Gilbert, 1979; Board & Sparks, Chapter 6). Smart hypothesises that in birds, egg-shape may, therefore, be primarily determined in the isthmus. In reptiles, the uterine tube (structurally and functionally similar to and probably homologous to the magnum of birds) secretes the albumen, but the more distal isthmus region is short, non-glandular, and appears to contribute little to albumen, membrane or shell formation (Palmer & Guillette, 1988, Chapter 3). Ova with albumen move instead to the uterus, where water is added ('plumping') and where the fibrous shell membrane and calcareous eggshell are both secreted (Aitken & Solomon, 1976; Guillette, Fox & Palmer, 1989; Palmer &

Guillette, 1988; Chapter 3). Thus, the contribution of oviductal anatomy to egg-shape is fundamentally different in reptiles and birds.

This difference probably explains the main difference in egg shape in birds and reptiles. Upon receipt of a number of ova surrounded by albumen, the reptilian uterus must exhibit differential constriction along its length (similar to the action Smart attributes to the isthmus in the avian oviduct). This differential constriction must operate on the differential resistance against the uterine wall provided by each ovum versus the region between each one. The result is the formation of a series of bola of albumen-surrounded ova around which the fibrous and calcareous eggshells are secreted. The source of the information the uterine wall depends upon for the determination of egg width and length is unknown in reptiles (as in birds), but it must be fairly tightly programmed. Of particular interest is the extent of dilation of the uterine wall (i.e. egg width) that is determined by programming of the uterus itself, as opposed to dilation that is simply a direct reflection of the diameter of the ovum at the time of ovulation (i.e., programmed via the ovary itself).

The rare production of 'miniature' eggs with tiny yolks (Caldwell, 1959; Miller, 1982) confirms that ovum size at ovulation has at least some effect on egg size. Perhaps egg width in reptiles is primarily related to ovum size at ovulation, whereas egg length and symmetry might logically be more under uterine control. Observations of uterine mistakes support this suggestion. For example, turtles occasionally produce eggs with two yolks (Caldwell, 1959; N.B. normal twinning (Yntema, 1970; Lehmann, 1984) apparently results from the development of two blastodiscs from a single yolked ovum (Ewert, unpublished)); small eggs with albumen and/or small yolks only (so-called 'yolkless' eggs, although they don't always lack yolk; Caldwell, 1959; Miller, 1982; Limpus, 1985); 'double' eggs, where two adjacent eggs are fused end to end (Archer, 1948; Ferguson & Joanen, 1983); and eggs missing part of the shell on one end, where double eggs apparently broke apart during oviposition (Ewert, unpublished). All of these oviducal errors are apparently more common among species that lay huge clutches of relatively small eggs, as one might logically predict. In addition, the eggs of *Rhinoclemmys* species, which often lay only a single, huge egg at a time (Moll, 1979), are sometimes slightly

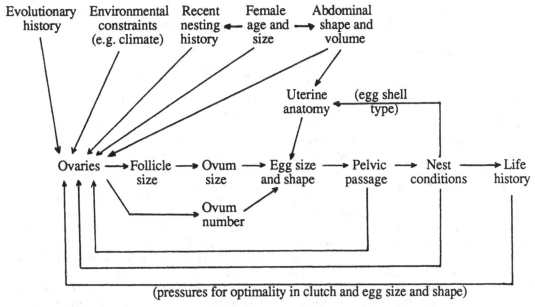

Fig. 7.2. An interactive model for the determination of egg size and shape in reptiles.

tapered (i.e., asymmetrical, as in birds), and the more pointed end passes out the cloaca first (Ewert, 1985, unpublished observations). This suggests that it forms near the terminus of the uterus and perhaps is constrained by the shape of the posterior abdominal cavity.

These observations suggest that a proximate, dual-control mechanism of egg-shape via both ovary and uterus (with any pelvic, optimal off-spring size, or nest physiology constraints presumably representing ultimate effects operating indirectly via follicle size in the ovary; Fig. 7.2), would explain the general pattern among turtles of having more variability among egg lengths than among egg widths (as measured by the coefficient of variation) within a single clutch (Mitchell, 1985a,b; Iverson, unpublished observations). It may also explain reports of variability in egg size and shape related to order of oviposition (presumably a reflection of location along the uterus), with eggs laid later tending to be smaller (Caldwell, 1959; Limpus, 1985) and less elongate (Limpus, 1985; Ewert, unpublished data for *Chelydra* and *Trionyx*). This mechanism also logically explains why reptile eggs are typically symmetrical. Asymmetry would require differential programming of the constriction forces along different sections of the uterus, which would be complicated by changes in the number of ova present in the uterus at any one time during the life of the individual. One simple, resistance-based uterine mechanism would reduce such problems, but would result in symmetrical eggs by default. Thus reptiles, which typically carry more than one egg in their oviducts, normally have symmetrical eggs, whereas birds, which shell only a single egg at a time and have a greatly reduced abdominal cavity volume, can have a more complicated series of resistances which results in asymmetry of the egg.

Finally, as mentioned above, egg size and shape may to some degree be determined proximately, as well as ultimately (Smith & Fretwell, 1974), by clutch size. A better understanding of the proximate determinants of clutch size, e.g. resource constraints, climatic conditions, time of season, age and size of female, etc., thus becomes very important to our understanding of egg size and shape variability in reptiles. In addition, the importance of the uterus as a determinant of egg size and shape can be tested by more detailed study of eggs from individual females, especially from those species that produce multiple, large clutches within the same year (e.g. marine turtles). To date, studies of seasonal changes in clutch size in individual females have yielded inconsistent results (Congdon & Tinkle, 1982; Frazer & Richardson, 1985; Mortimer & Carr, 1987; Bjorndal & Carr, 1989) and this clearly deserves more attention. Similarly, the possible correlations between egg length or width, and clutch sizes or oviposition order within an individual female should also be investigated further.

All of these considerations lead us to propose an interactive model for the determination of egg size and shape in reptiles (Fig. 7.2). This takes into account not only the direct effects of anatomy and physiology of the reproductive tract, but also the more indirect effects of pelvis size, gaseous conditions in the nest, life history strategy (including optimality considerations), local environmental conditions, and female age, size, and reproductive history. The relative importance of each of these various factors is as yet unclear, but all deserve more consideration than they have previously received.

As mentioned earlier, the typically asymmetrical bird egg is generally considered to relate to a thermoregulatory hypothesis and/or a rolling characteristic hypothesis. However, given that in birds, egg-shape has been so well studied quantitatively, we are surprised that there has been no broad quantitative study of ecological correlates (especially nest types) of egg size, shape and volume in bird eggs (but see Packard, Sotherland & Packard, 1977; Sotherland et al., 1980). In fact, we predict that such an analysis will reveal that asymmetry in bird eggs is primarily a function of abdominal shape as well as oviductal anatomy in the group (Smart, Chapter 8), and only secondarily responding to nest characteristics (e.g. rounder eggs in tree hole nesters). Support for our hypothesis comes from the observation that about 90% of eggs form in the oviduct with the pointed end toward the cloaca (Romanoff & Romanoff, 1949). The eggs that are laid blunt end first have turned in the oviduct (Romanoff & Romanoff, 1949). The reduced abdominal space typical of birds (presumably an adaptation for flight, as is the habit of carrying only a single shelled egg at a time) may therefore be the most important determinant of egg-shape in birds. This does not deny any adaptive advantage to bird egg shape in the nest, but does suggest that those advantages may be secondary to maximising egg volume in an organism with a reduced body cavity. A comparison of egg-shape and size versus abdominal size and shape would test this hypothesis, as would the absence of any relationship between egg-shape and such other factors as clutch size or nest type. The fact that a few birds deviate from the normal asymmetrical bicone egg suggests that there may be intense selection pressures on egg-shape (or at least different ones) in those species (Seymour & Ackerman, 1980).

Egg density comparisons

In the turtle eggs described earlier, mean values of density are significantly greater ($r = 6.77$; $p = 0.0001$) for species with rigid-shelled eggs (1.126 ± 0.026; $N = 19$; range of means 1.090 to 1.199) than for those with pliable-shelled eggs (1.070 ± 0.018; $N = 13$; range of means 1.045 to 1.109). Values of density for individual eggs also show that rigid-shelled eggs (1.129 ± 0.045; $N = 137$; range 1.011 to 1.261) are significantly denser ($t = 11.15$; $p<0.0001$) than pliable-shelled eggs (1.076 ± 0.032; $N = 99$; range 1.023 to 1.183). This difference is not surprising given that the relative dry mass of the eggshell in the former eggs is twice that of the latter (Congdon & Gibbons, 1985). This same positive relationship between eggshell thickness and density was demonstrated for avian eggs by Romanoff & Romanoff (1949), even though the variation in bird eggshell thickness is a fraction of that in turtle eggs. However, densities in bird eggs generally range from only about 1.03 to 1.09 (Romanoff & Romanoff, 1949; Barth, 1953; Paganelli et al., 1974), despite their heavy, brittle eggshells. The density of the rigid-shelled eggs of the crocodilian Alligator mississippiensis using an estimate of volume is 1.125 (Deeming & Ferguson, 1990), but using measured volumes the mean value for density is 1.089 (Deeming & Ferguson, unpublished). This is in line with those of turtle eggs, but higher than those of typical bird eggs.

Despite our prediction that more elongate eggs should have greater relative surface area and thus be composed of a greater proportion of eggshell and therefore have a higher overall density, in turtles no relationship is found between egg density and elongation ($r = 0.15$; $p = 0.56$).

Phylogenetic considerations of egg-shape

In an attempt to determine the ancestral egg condition for turtles, as well as the primary directions of evolutionary change in turtle egg shape and shell type, the distribution of eggshell and shape character states has been examined in light of the current best phylogenetic hypothesis of the relationships of the extant higher taxa

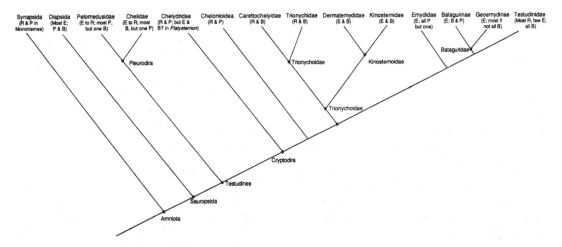

Fig. 7.3. Phylogenetic distribution of turtle egg character states. Phylogeny redrawn from Gaffney & Meylan (1988). Egg-shape is coded as R (round or nearly so) or E (elongate); eggshell type, as B (rigid) or P (pliable). Higher taxon names from Gaffney & Meylan (1988) apply to nearest node (indicated by solid dot).

(Gaffney & Meylan, 1988). If that phylogeny (Fig. 7.3) is correct, and we assume no character reversals, the most parsimonious conclusion is that the primitive turtle egg was round and pliable. However, this would require a minimum of seven independent origins of oblong eggs and five independent origins of rigid-shelled eggs. If the ancestral condition was an oblong and pliable-shelled egg, round eggs would have had to have evolved independently at least eight times and pliable-shelled eggs, at least six times. Thus, the support for any state being ancestral is not particularly strong, especially when one outgroup (the Synapsida: Monotremata) has round, pliable-shelled eggs (Cockrum, 1962; Griffiths, 1978), and the other (the Diapsida) has mostly elongate eggs, but mostly pliable-shelled eggs in some groups (Sphenodontia and Squamata) and all rigid-shelled eggs in others (Crocodylia and Aves).

Clearly, both egg elongation and eggshell type are highly subject to convergence, with spherical eggs favoured in species with larger clutches in order to permit maximal oviductal packing of eggs and/or to minimise surface area for water loss in the nest. Elongate eggs will be favoured in species which need to maximise offspring size within the constraints of pelvic canal size (Hailey & Loumbourdis, 1988) and/or to increase relative surface area for gas exchange in humid nest substrates (Seymour & Ackerman, 1980). Pliable-shelled eggs should be favoured in species in which desiccation during development is not a danger (and high rates of gas and water exchange are advantageous; e.g. within sandbars and beaches), and rigid-shelled eggs favoured in species in which risk of desiccation, predation or infection during development is high. We are continuing to test these hypotheses by examining additional possible correlates of egg-shape and shell type, including variation in adult habitat, nest site, relative clutch size, body size, developmental strategy, incubation time, eggshell pore density, and egg size relative to pelvic canal size.

This chapter reveals the general lack of attention paid to reptilian egg-shape as compared to bird egg-shape. Unfortunately, simple egg measurements are not reported for many reptiles even when the data were likely recorded by the investigators (as is often true in husbandry and egg and developmental physiology studies). Particularly lacking are data for volume (and to a lesser extent mass) for squamate eggs. We therefore urge investigators of all disciplines involving reptile eggs to report regularly egg size and shape (and female size and clutch size) data. Only when more such data are available will a real synthetic analysis of egg-shape in reptiles be possible. Toward that goal our own research is continuing with an analysis of egg size and shape and their correlates in a turtle egg data set that now includes our own measurements of over 12,000 eggs of 90 species, as well as those of many additional species reported in the literature (Ewert, Iverson & Moll, unpublished observations).

Acknowledgements

Earlham College and Indiana University provided space and support during our studies. The National Science Foundation (BSR-8404462) provided microcomputer hardware. Additional support to JBI was provided by the Hughes Medical Institute. H. B. Hanes scrutinized our algebra and discussions with W. H. Buskirk stimulated the development of our ideas. The critical comments of G. C. Packard and K. A. Bjorndal on early drafts of the manuscript are greatly appreciated.

References

Aitken, R. N. C. & Solomon, S. E. (1976). Observations on the ultrastructure of the oviduct of the Costa Rican green turtle (*Chelonia mydas* L.). *Journal of Experimental Marine Biology and Ecology*, **21**, 75–90.

Ar, A., Paganelli, C. V., Reeves, R. B., Greene, D. G. & Rahn, H. (1974). The avian egg: Water vapor conductance, shell thickness, and functional pore area. *Condor*, **76**, 153–8.

Archer, W. H. (1948). The mountain tortoise (*Giochelone pandalis*). *African Wildlife*, **3**, 75–9.

Auffenberg, W. (1981). *The Behavioral Ecology of the Komodo Monitor*. Gainesville: University Presses Florida.

Barth, E. K. (1953). Calculations of egg volume based on loss of weight during incubation. *Auk*, **70**, 151–9.

Bjorndal, K. A. & Carr, A. F. (1989). Variation in clutch size and egg size in the green turtle nesting poulation at Tortuguero, Costa Rica. *Herpetologica*, **45**, 181–9.

Booth, D. T. & Seymour, R. S. (1987). Effect of eggshell thinning on water vapor conductance of Malleefowl eggs. *Condor*, **89**, 453–9.

Caldwell, D. K. (1959). The loggerhead turtles of Cape Romain, South Carolina. *Bulletin of the Florida State Museum, Biological Sciences*, **4**, 319–48.

Cockrum, E. L. (1962). *Introduction to Mammalogy*. New York: Ronald Press.

Congdon, J. D. & Gibbons, J. W. (1985). Egg components and reproductive characteristics of turtles: Relationships to body size. *Herpetologica*, **41**, 194–205.

— (1987). Morphological constraints on egg size: A challenge to optimal egg size theory. *Proceedings of the National Academy of Science*, **64**, 4145–7.

Congdon, J. D. & Tinkle, D. W. (1982). Reproductive energetics of the painted turtle (*Chrysemys picta*). *Herpetologica*, **38**, 228–36.

Deeming, D. C. & Ferguson, M. W. J. (1990). Methods for the determination of the physical characteristics of eggs of *Alligator mississippiensis*: A comparison with other crocodilian and avian eggs. *Herpetological Journal*, **1**, 456–62.

Dunham, A. E., Miles, D. B. & Reznick, D. N. (1988). Life history patterns in squamate reptiles. In *Biology of the Reptilia*, vol. 16, Ecology B, ed. C. Gans & R. B. Huey, pp. 441–522. New York: A. R. Liss, Inc.

Ewert, M. A. (1979). The embryo and its egg: Development and natural history. In *Turtles: Perspectives and Research*, ed. M. Harless & H. Morlock, pp. 333–413. New York: Wiley & Sons.

— (1985). Embryology of Turtles. In *Biology of the Reptilia*, vol. 14, Development A, ed. C. Gans, F. Billett & P. Maderson, pp. 75–267. New York: Wiley & Sons.

Ferguson, M. W. J. (1985). Reproductive biology and embryology of the crocodilians. In *Biology of the Reptilia*, vol. 14, Development A, ed. C. Gans, F. Billett & P. Maderson, pp. 329–491. New York: Wiley & Sons.

Ferguson, M. W. J. & Joanen, T. (1983). Temperature-dependent sex determination in *Alligator mississippiensis*. *Journal of Zoology*, **200**, 143–77.

Ferguson, G. W., Brown, K. L. & DeMarco, V. G. (1982). Selective basis for the evolution of variable egg and hatchling size in some iguanid lizards. *Herpetologica*, **38**, 178–88.

Fitch, H. S. (1963). Natural history of the racer *Coluber constrictor*. *University of Kansas Publications, Museum of Natural History*, **15**, 351–468.

— (1970). Reproductive cycles in lizards and snakes. *University of Kansas Museum of Natural History, Miscellaneous Publications*, **52**, 1–247.

Ford, N. B. & Seigel, R. A. (1989). Relationships among body size, clutch size, and egg size in three species of oviparous snakes. *Herpetologica*, **45**, 75–83.

Frazer, N. B. & Richardson, J. I. (1985). Seasonal variation in clutch size for loggerhead sea turtles, *Caretta caretta*, nesting on Little Cumberland Island, Georgia. *Copeia*, **1985**, 1083–5.

Gaffney, E. S. & Meylan, P. A. (1988). A phylogeny of turtles. In *The Phylogeny and Classification of Tetrapods*, vol 1, ed. M. Benton, pp. 157–219. London: Oxford University Press.

Gilbert, A. B. (1971). Transport of the egg through the oviduct and oviposition. In *Physiology and Biochemistry of the Domestic Fowl*, vol. 3, ed. D. J. Bell & B. M. Freeman, pp. 1345–53. London and New York: Academic Press.

— (1979). Female genital organs. In *Form and Function in Birds*, Vol. 1, ed. A. S. King & J. McClelland, pp. 237–360. London: Academic Press.

Griffiths, M. (1978). *The Biology of the Monotremes.* New York: Academic Press.

Gross, D. T. (1982). Reproductive biology of the stinkpot, *Sternotherus odoratus*, in a central Florida lake system. Unpublished M.A. thesis. Tampa: University of South Florida.

Guillette, L. J., Fox, S. L. & Palmer, B. D. (1989). Oviductal morphology and egg shelling in the oviparous lizards, *Crotaphytus collaris* and *Eumeces obsoletus. Journal of Morphology*, **201**, 145–59.

Hailey, A. & Loumbourdis, N. S. (1988). Egg size and shape, clutch dynamics, and reproductive effort in European tortoises. *Canadian Journal of Zoology*, **66**, 1527–36.

Hoyt, D. F. (1976). The effect of shape on the surface-volume relationships of birds' eggs. *Condor*, **78**, 343–9.

Iverson, J. B. (1986). *A Checklist with Distribution Maps of the Turtles of the World.* Richmond, Indiana: Iverson.

— (1991). Population dynamics of a long-lived vertebrate: The yellow mud turtle. *Herpetological Monographs* (in press).

Lehmann, H. (1984). Ein Zwillingsschlupf bei *Sternotherus minor minor. Salamandra*, **20**, 192–6.

Limpus, C. J. (1985). A study of the loggerhead sea turtle, *Caretta caretta*, in eastern Australia. Unpubl. Ph.D. Dissertation. St Lucia, Australia: University of Queensland.

Long, D. R. & Rose, F. L. (1989). Pelvic girdle size relationships in three turtle species. *Journal of Herpetology*, **23**, 315–18.

Loveridge, A. (1941). Revision of the African lizards of the family Amphisbaenidae. *Bulletin of the Museum of Comparative Zoology, Harvard*, **88**, 467–524.

Miller, J. D. (1982). Development of marine turtles. Unpublished Ph.D. Dissertation. Armidale, NSW, Australia: University of New England.

Minton, S. A. (1966). A contribution to the herpetology of West Pakistan. *Bulletin of the American Museum of Natural History*, **134**, 27–184.

Mitchell, J. C. (1985*a*). Female reproductive cycle and life history attributes in a Virginia population of painted turtles, *Chrysemys picta. Journal of Herpetology*, **19**, 218–26.

— (1985*b*). Female reproductive cycle and life history attributes in a Virginia population of Stinkpot turtles, *Sternotherus odoratus. Copeia*, **1985**, 941–9.

Moffat, L. A. (1985). Embryonic development and aspects of reproductive biology in the tuatara, *Sphenodon punctatus*. In *Biology of the Reptilia*, vol. 14, Development A, ed. C. Gans, F. Billett & P. Maderson, pp. 493–521. New York: Wiley & Sons.

Moll, E. O. (1979). Reproductive cycles and adaptations. In *Turtles: Perspectives and Research*, ed. M.

Harless & H. Morlock, pp. 305–31. New York: Wiley & Sons.

Mortimer, J. A. & Carr, A. F. (1987). Reproduction and migrations of the Ascension Island green turtle (*Chelonia mydas*). *Copeia*, **1987**, 103–13.

Packard, G. C. & Packard, M. J. (1988). The physiological ecology of reptilian eggs and embryos. In *Biology of the Reptilia*, vol. 16, Ecology B, ed. C. Gans & R. B. Huey, pp. 523–605. New York: A. R. Liss, Inc.

Packard, G. C., Sotherland, P. R. & Packard, M. J. (1977). Adaptive reduction in permeability of avian eggshells in water vapour at high altitude. *Nature, London*, **266**, 255–6.

Packard, G. C., Tracy, C. R. & Roth, J. J. (1977). The physiological ecology of reptilian eggs and embryos, and the evolution of viviparity within the class Reptilia. *Biological Reviews*, **52**, 71–105.

Paganelli, C. V., Olszowka, A. & Ar, A. (1974). The avian egg: Surface area, volume, and density. *Condor*, **76**, 319–25.

Palmer, B. D. & Guillette, L. J. (1988). Histology and functional morphology of the female reproductive tract of the tortoise *Gopherus polyphemus. American Journal of Anatomy*, **183**, 200–11.

Pope, C. H. (1961). *The Giant Snakes.* New York: A. A. Knopf.

Preston, F. W. (1953). The shapes of birds' eggs. *Auk*, **70**, 160–82.

— (1969). Shapes of birds' eggs: Extant North American families. *Auk*, **86**, 246–64.

— (1974). The volume of an egg. *Auk*, **91**, 132–8.

Pritchard, P. C. H. (1979). *Encyclopedia of Turtles.* Neptune, New Jersey: T.F.H. Publ.

Quay, W. B. (1976). Impression tonometry of chelonian eggs (Reptilia, Testudines), a guide to internal pressure and membrane phenomena. *Journal of Herpetology*, **10**, 55–62.

Rahn, H., Paganelli, C. V. & Ar, A. (1975). Relation of avian egg weight to body weight. *Auk*, **92**, 750–65.

Reed, C. A. (1965). *North American Bird Eggs.* New York: Dryer Publications, Inc.

Romanoff, A. L. & Romanoff, A. J. (1949). *The Avian Egg.* New York: Wiley & Sons.

Seigel, R. A. & Ford, N. B. (1987). Reproductive ecology. In *Snakes: Ecology and Evolutionary Biology*, ed. R. A. Seigel, J. T. Collins & S. S. Novak, pp. 210–52. New York: MacMillan.

Seymour, R. S. (1979). Dinosaur eggs: Gas conductance through the shell, water loss during incubation and clutch size. *Paleobiology*, **5**, 1–11.

Seymour, R. S. & Ackerman, R. A. (1980). Adaptations to underground nesting in birds and reptiles. *American Zoologist*, **20**, 437–47.

Shaw, C. E. (1966). Breeding the Galapagos Tortoise – Success Story. *Oryx*, **9**, 119–26.

Simkiss, K. & Taylor, T. G. (1971). Shell formation. In *Physiology and Biochemistry of the Domestic Fowl*,

Vol. 3, ed. D. J. Bell & B. M. Freeman, pp. 1331–43. London & New York: Academic Press.

Smith, C. C. & Fretwell, S. D. (1974). The optimal balance between size and number of offspring. *American Naturalist*, **108**, 499–506.

Sotherland, P. R., Packard, G. C., Taigen, T. L. & Boardman, T. J. (1980). An altitudinal cline in conductance of cliff swallow (*Petrichelidon pyrrhonota*) eggs to water vapor. *Auk*, **97**, 177–82.

Stearns, S. D. (1976). Life history tactics: A review of ideas. *Quarterly Review of Biology*, **51**, 3–47.

Tatum, J. B. (1975). Egg volume. *Auk*, **92**, 576–80.

Vitt, L. J. (1982). Sexual dimorphism and reproduction in the microteiid lizard, *Gymnophthalmus multiscutatus*. *Journal of Herpetology*, **16**, 325–9.

— (1986). Reproductive tactics of sympatric gekkonid lizards with a comment on the evolutionary and ecological consequences of invariant clutch size. *Copeia*, **1986**, 773–86.

Welty, J. C. (1962). *The Life of Birds*. Philadelphia: W. B. Saunders.

Werner, Y. L. (1989). Egg size and egg shape in near-Eastern Gekkonid lizards. *Israel Journal of Zoology*, **35**, 199–213.

Yntema, C. L. (1970). Twinning in the common snapping turtle, *Chelydra serpentina*. *Anatomical Record*, **166**, 491–8.

8

Egg-shape in birds

IAIN H. M. SMART

Introduction

The egg of a bird can be envisaged as a naturally occurring organ culture which, given sufficient heat and oxygen, allows the original single cell to transform the acellular substrate into a complex multicellular animal. The feature of the system which concerns us in this chapter is the calcareous shell which imparts a fixed shape to this system. A typical avian egg presents to the eye a regular profile, symmetrical about its long axis with one end more pointed than the other yet this shape has been strangely difficult to characterise mathematically. An equation describing egg curvature is useful as a shape descriptor and for understanding the forces which give rise to the ovoid form. Also, since an egg is axisymmetric, an equation that can be solved from simple linear measurements allows the volume and surface area of the solid of revolution generated by its profile to be calculated by integration (Paganelli, Olszowka & Ar, 1974).

Egg volume and surface area are rather special parameters as they are constants of the system. It is remarkable that a given volume of organic molecules, constituting a new-laid egg, is constrained within an envelope of constant area through which all respiratory exchanges must be conducted. This predetermined surface area, established when only one diploid nucleus is present, must eventually be compatible with mediating the gaseous transfer required by the many millions of aerobically respiring cells of the embryo (Paganelli, Chapter 16).

The curvature of an egg also contains information about the forces that were in equilibrium when the egg received its final shape. D'Arcy Thompson (1942) points out that the form of an

object is a diagram, the algebraic sum as it were, of the forces that have acted on it in the past. The form of an egg may, thus, be read as a diagram of the action of the surrounding oviduct wall at the time the egg is receiving its permanent shape. This concept of form, thus leads us to make enquiries about the structure and mechanical properties of the wall of the oviduct imposing this form.

Mathematical analysis of egg-shape

An equation for egg curvature based on transformed coordinates

We begin by deriving empirically an equation which describes the curvature of a 'typical' bird egg that is easy to visualise and to demonstrate how this may be solved for a particular egg by making four linear measurements (Smart, 1969). Only three are required to solve the equation: the fourth is used as a check on the accuracy of the predicted curvature. We show also that, if the equation does not, in fact, describe the curvature of a particular egg, it can be adapted to show how the egg's curvature differs from the standard ovoid. The initial equation is derived from an application of a method of transformed coordinates (D'Arcy Thomson, 1942).

We start with a spherical egg and set its circular profile in a cartesian grid with its centre at the origin (Fig. 8.1). If the grid is made rectangular without affecting its area, the inscribed

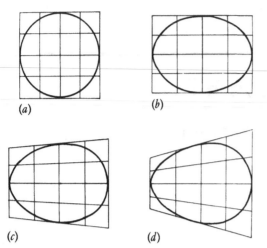

(a) (b)

(c) (d)

Fig. 8.1. (a) A circle plotted on Cartesian coordinates: The same curve transferred point by point to (b) a rectangular and (c) and (d) increasingly trapezoidal systems. (Smart, 1969).

circle becomes an ellipse of the same area as the original circle (Fig. 8.1). If the rectangular grid is transformed into a trapezoid, the quondam ellipse takes up something of an egg-shape and, if the grid is made even more trapezoidal, the contained egg-shape becomes more pointed (Fig. 8.1).

The equation of these curvatures is easily extracted. The equation of the initial ellipse is

$$\frac{x^2}{a^2} + \frac{y^2}{b^2} = 1 \qquad (8.1)$$

where a is the semi-major axis (half the egg's length along its long axis) and b is the semi-major axis (half the egg's breadth at the mid-point of its long axis).

Rotating the line tangent to the mid-point is an empirical procedure. The effect of this is to magnify the y coordinates of those points on an ellipse with positive x coordinates by an amount proportional to the increase in height of the system, and to diminish those with negative x coordinates by an amount proportional to the equivalent decrease in height to the left of the origin, i.e. according to the gradient of the tangent line (see Smart, 1969 for further explanation). The equation of a straight line is $y = b + x \tan \theta$; where b is the intercept with the y axis, in this case the semi-diameter of the egg at its mid-point, and $\tan \theta$ is the slope of the mid-point tangent. This expression will modify the y coordinate appropriately if made the denomina-

tor of the y term in equation 8.1, so that equation 8.1 becomes

$$\frac{x^2}{a^2} + \frac{y^2}{b + (x \tan \theta)^2} = 1 \qquad (8.2)$$

In order to draw the profile of an actual egg, equation 8.2 has to be solved but the length ($2a$), the diameter at the mid-point ($2b$), and the angle θ must be known. In practice, it is difficult to determine, on the surface of the egg, the plane passing at right angles through the mid-point of the long axis where $2b$ must be measured, and the value of θ must be inferred somehow from the asymmetry of the egg's profile. Both may be calculated in the following way. First, the length of the egg ($2a$) is measured with a Vernier caliper. Then using a second Vernier caliper, whose jaws have been suitably shortened, a depth of about 1 cm is about right for eggs of the dimensions of the domestic fowl (*Gallus gallus*), the diameter of the egg at a distance k from each end equal to the depth of the abbreviated jaws is measured (Fig. 8.2). The semi-diameter R, a distance k from the blunt end, and the semi-diameter r, a distance k from the pointed end, correspond to the y coordinates of two points which have x coordinates of the same numerical value but with different signs, i.e. $+x$ is given by $a–k$, and $–x$ also by $a–k$ (see Fig. 8.2). Equation 8.2 may now be written:

$$b + x \tan \theta = \frac{y}{\sqrt{1 - (x^2/a^2)}} \qquad (8.3)$$

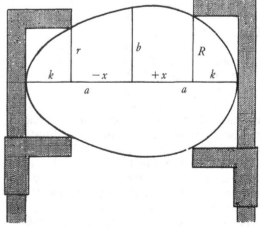

Fig. 8.2. Diagram showing use of Vernier caliper for measuring the diameter of an egg a constant distance from each end (Smart, 1969).

By substituting in the above equation, R for y when x is positive and r for y when x is negative, we get a pair of simultaneous equations:

$$b + x \tan \theta = \frac{R}{\sqrt{1 - (x^2/a^2)}} \qquad (8.4)$$

and

$$b + x \tan \theta = \frac{r}{\sqrt{1 - (x^2/a^2)}} \qquad (8.5)$$

By addition of equations 8.4 and 8.5 we get:

$$b = \frac{R + r}{2\sqrt{1 - (x^2/a^2)}} \qquad (8.6)$$

and by subtraction of equation 8.5 from 8.4:

$$\tan \theta = \frac{R - r}{2x\sqrt{1 - (x^2/a^2)}} \qquad (8.7)$$

Since the values of a, R and r are known from direct measurement and the x coordinates of R and r are the same and can be calculated from $a = k$, equations 8.6 and 8.7 can be solved and the value of b and θ obtained. Thus, if equation 8.2 is valid, the curvature of an egg can be derived from three simply made linear measurements.

The validity of the equation may be tested in three ways: 1) Using the method just described, the predicted curve of a particular egg may be drawn directly on graph paper, the area within the curve cut out, then the paper template fitted round the egg and the goodness of fit observed. In practice, a good fit is obtained by this method for typical, 'egg-shaped' eggs. 2) By taking a fourth measurement, the maximum breadth of the egg ($2B$), the predicted and actual values of $2B$ may be compared. 3) A useful alternative is to reverse the procedure, and by making serial measurements of breadth along the length of the egg to reconstruct the coordinate tangential to the egg at its mid-point; if equation 8.3 is valid, then the points should lie on a straight line. In practice, this can be done by photographing the egg and projecting its outline on to graph paper, so that its total length is magnified to 20 cm. The long axis of the egg is arranged to lie along the x axis of the graph paper, with its mid-point at the origin. The outline of the egg is then drawn onto the graph paper. Magnification helps to reduce the error in measurement making it possible to analyse the curve of even very small eggs.

The mid-point tangent is reconstructed from the following equation

$$Y = \frac{y}{\sqrt{1 - (x^2/a^2)}} \qquad (8.8)$$

where Y is the expected y coordinate of the point on the tangent related to a point x, y on the egg's profile. As a is constant at 10 cm, a table of values for the divisor may be drawn up for 1 cm changes in the value of x which can be used for all eggs. Thus, by measuring the height of y at 1 cm intervals of x, the values of Y for each value of x may be quickly calculated and plotted (Fig. 8.3).

These tests have confirmed that equation 8.2 gives a remarkably good fit for the individual eggs of many species of bird, and may be taken as describing the curvature of the 'typical' egg. Fig. 8.3 depicts the reconstructed mid-point

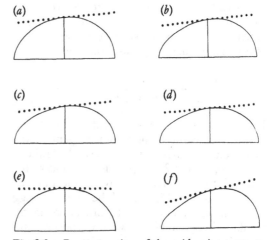

Fig. 8.3. Reconstruction of the mid-point tangents of eggs of a variety of species of birds photographically enlarged to the same length. In (a)–(d) the slope of the tangent is in each case about 6° but the proportion of length to breadth is different: (a) thrush (*Turdus ericetorum*) (b) common tern (*Sterna macrura*) (c) herring gull (*Larus argentatus*) (d) sparrow (*Passer domesticus*) (e) The practically horizontal mid-point tangent of the egg of the ostrich (*Struthio camelis*) denoting a form approximating a true ellipsoid, (f) at the opposite extreme the mid-point tangent of the egg of the curlew (*Numenius arquata*) which makes an angle of some 16° with the x axis (Smart, 1969).

tangents of eggs of a variety of different species and show the straight lines obtained. The value of θ varies from practically zero in the case of such near ellipsoids as the egg of the ostrich (*Struthio camelus*; Fig. 8.3) to 15° or so in the case of the pointed eggs of wading birds such as the curlew (*Numenius arquata*; Fig. 8.3).

The form of 'atypical' eggs

Certain eggs are atypical in that reconstruction of their mid-point tangents does not produce a straight, or even approximately straight line. In the egg of the kiwi (*Apteryx* spp.) each half of the tangent slopes away from the x axis denoting it has both ends more bulbous than would be expected of a linearly transformed ellipsoid (Fig. 8.4), and in the emu (*Dromaius novaehollandiae*) egg the opposite situation is observed, where each half of the tangent slopes towards the x axis (Fig. 8.4). In the case of the guillemot (*Uria aalge*) egg, the tangent has an approximately linear slope in its middle reaches but curves in opposite directions at its ends (Fig. 8.4). Similar non-linear mid-point tangents are properties of hyper- and hypo-ellipses. These are produced by increasing or decreasing, respectively, the value of the exponent in the standard ellipse equation 8.1, or its transformed version 8.2, so that a more general equation would be

$$\frac{x^n}{a^n} + \frac{y^n}{(b + x \tan \theta)^n} = 1 \qquad (8.9)$$

If n is >2, a hyper-elliptical profile is produced and its reconstructed mid-point tangent curves away from the x axis as in the case of the kiwi egg (Fig. 8.4); if n is <2, a hypo-ellipse results with a mid-point tangent curving towards the x axis as in the emu egg (Fig. 8.4). The guillemot egg always much admired for its difficult geometry would require n to be <2 when x is positive and >2 when x is negative in order to produce appropriately curved tangents. For example, an exact fit for a particular guillemot egg of elongation 1.56 is obtained from equation 8.9 when $\theta = 15°$ and $n = 1.8$ (x, positive) and $\theta = 17°$ and $n = 2.2$ (x, negative).

An equation based on approximation by polynomial

Preston (1953) presents complementary equations which, although they do not give such an immediate visual impression of the deformations producing a particular egg-shape, they are mathematically more convenient, particularly when dealing with eggs of atypical shape. Preston also starts with the circle as the base form defined by the parametric form of the circle equation

$$x = r \cos \theta \qquad (8.10)$$

$$y = r \sin \theta \qquad (8.11)$$

where r is the radius vector and θ the angle it makes with the x axis.

If the circle is rotated about the x axis, so that is seen more edge on, its fore-shortened profile is an ellipse. Only the values of y change during this operation and equations 8.10 and 8.11 become

$$x = a \cos \theta \qquad (8.12)$$

$$y = b \sin \theta \qquad (8.13)$$

Equations 8.10 and 8.12 are identical for the change from radius r to semi-major axis a and the semi-minor diameter b is the apparent length of the fore-shortened radius at the point where $x = 0$.

Further transformations of this ellipse to make it egg-shaped (Fig. 8.1) involve changes in the value of the y coordinates only and the change will be a function of the variable in the x equation 8.10, namely $\cos \theta$. Preston (1953) uses a polynomial expression incorporating $\cos\theta$ to modify equation 8.13 to give values of y which will approximate the curvature of any given egg:

$$x = a \cos \theta \qquad (8.12)$$

$$y = b \sin \theta \, (1 + c_1 \cos \theta + c_2 \cos^2 \theta + c_3 \cos^3 \theta) \qquad (8.14)$$

in which the c terms are dimensionless constants, empirically determined for an individual egg. If the c_1 term is made positive, it imposes a blunt end, when x is positive and a pointed end when negative (Fig. 8.1). The c_2 term affects both poles equally making them either more bulbous (hyper-elliptic) when positive or more pointed (hypo-elliptic) when negative. The c_3 term, like the c_1 term, produces opposite effects on each pole also according to its sign.

Most eggs can be described by c_1 and c_2 terms (Preston, 1953). The c_2 term acting alone or as the major term produces egg profiles like the kiwi or emu (Fig. 8.4). A c_3 term is only required to describe the most egregious of egg shapes

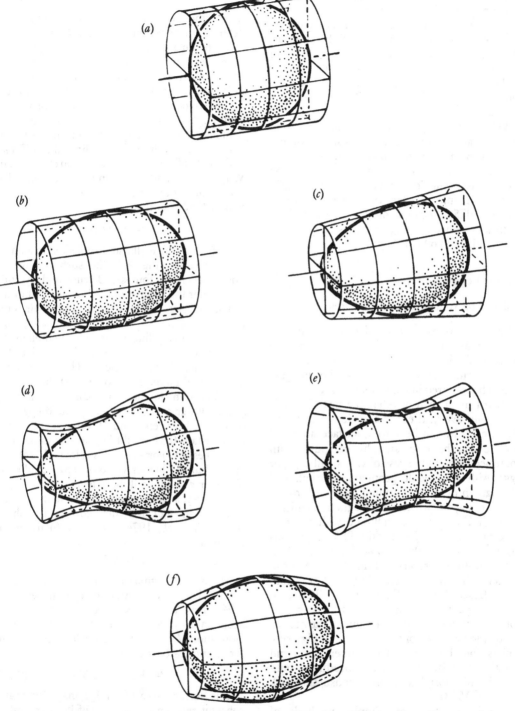

Fig. 8.4.   Diagram showing the effect of different transformations of the coordinate system on the contained solids of revolution. (*a*) the base forms a spheroid. (*b*) the same stretched into an ellipsoid; (*c*) the trapezoidal coordinates of the 'typical egg'; (*d*) the complex curvature of the guillemot egg; (*e*) the flaring coordinates of the kiwi egg; (*f*) the crimped barrel-shaped coordinates of the emu egg (Todd & Smart, 1984).

such as that of the guillemot family. A c_4 term is never required. The c terms, therefore, exert an effect analogous to the mid-point tangents. As an example of the magnitude of the c terms, Preston (1953) gives values for c_1 and c_2 of $+0.264$ and -0.241 respectively for the herring gull (*Larus argentatus*); and for the common guillemot: $c_1 = +0.374$, $c_2 = -0.0415$ and $c_3 = -0.1291$.

Values for the c terms can be calculated from four linear measurements (length, maximum breadth, breadth at mid-point and distance of maximum breadth from blunt end) by using intermediary equations and referring to specially prepared graphs (Carter, 1968; Preston, 1974; Tatum, 1975). The procedure is tedious and has been superceded by computer assisted curve fitting routines (Todd, 1986).

Applied egg geometry

Calculating volume and surface area

Since an egg is axisymmetrical, or very nearly so, the area of the surface of revolution and the volume of the solid of revolution generated by the egg's curvature can be calculated by integration. The required integrals may be found in any standard mathematical reference text (Daintith & Nelson, 1989). If either equations 8.2, 8.9 or 8.12 and 8.14 can be solved, surface area and volume can be calculated by computer with some accuracy. In the case of very small eggs, or eggs with a difficult geometry, an enlarged photograph of the egg printed or traced from a negative may be placed on a magnetic digitising tablet and its perimeter followed with a cursor (Paganelli et al., 1974; Todd, 1986). Ingenious machines for directly copying the profile of an egg to avoid distortions that may be inherent in photography have been designed (Preston, 1953; Besch, Sluka & Smith, 1967). The data, however gathered, can be used to compute volume, surface area and the value of the c terms in equation 8.14. Prior to the advent of digitising tablets and the power of computers for rapid iterative calculation, such a procedure would have been a formidable, time consuming task (Preston, 1953).

There are, however, simpler methods of estimating the volume, V, and surface area, A, of an egg using measurements of length, L and maximum diameter, D. The equation of Hoyt (1976) for volume is

$$V = 0.51LD^2 \qquad (8.15)$$

where the coefficient 0.51 is an empirically determined constant which gives a good estimate, on average, for a wide spectrum of egg shapes and sizes. LD^2 is, in effect, the volume of the rectangular box into which the egg can just be fitted. Equation 8.15 states that an egg occupies about 51% of the volume of its surrounding rectangular box. That this is so can be demonstrated non-empirically in the case of an ellipsoid. The volume of an ellipsoid is $4/3 \, \pi \, ab^2$ and that of its surrounding box $2a \times (2b)^2$. Where $2a = L$ and $2b = d$ (in the case of an ellipsoid d and D are the same). This simplifies to $\pi/6$ or 0.524, so that

$$V = 0.524Ld^2 \qquad (8.16)$$

In the case of an ellipsoid subjected to a linear transformation (Fig. 8.1) the surrounding box would become a four-sided truncated pyramid but neither the volume of the box nor that of the contained egg would change and their ratio would remain constant. In calculating the volume of such a linearly transformed ellipsoid, the diameter at its midpoint, d is used rather than the maximum diameter D. Hoyt's use of the latter would increase the volume of the box and would account for his empirical constant being slightly lower at 0.51. In very pointed eggs, the difference between d and D would increase and the constant 0.51 would be too low to produce the necessary correction as Hoyt (1976) has noted, but the higher value of 0.524 applied to diameter d should still be effective.

Similarly, a close approximation to the surface area A of an egg can be derived from its volume (Paganelli et al., 1974; Hoyt, 1976) using the equation:

$$A = K V^{2/3} \qquad (8.17)$$

where K is an empirically determined constant averaged out for a variety of egg shapes. $K = 4.94$ gives a fair estimate, but for greater accuracy Hoyt recommends taking into account elongation, $L/D = E$, when calculating a value for K and his best estimate was

$$K = 4.393 + 0.394E \qquad (8.18)$$

The surface area of an egg may also bear a constant relation to the area of its surrounding box. In the special case of a sphere with a surface area of $4\pi r^2$ surrounded by a cube of area $6 \times (2r)^2$ the ratio is again $\pi/6$ or 0.524. There is no simple expression for the area of an ellipsoid that allows a similarly neat expression of this ratio. However, the ratio can be estimated using volume calculated from equations 8.17 and 8.18. If the surface area, A, of an ellipsoid is 52.4% of that of its surrounding box, its equation will be

$$A = 0.524 \, [4(2a \times 2b) + 2(2b)^2] \quad (8.19)$$

Applying the last three equations to the case of an ellipsoid whose dimensions are, say, $L = 2a = 10$, $d = 2b = 8$ and $V = 0.524 \, Ld^2$, equation 8.19 gives an estimate of 234.8 for surface area; and equations 8.17 and 8.18 236.3 which suggests that the box method also may be valid for estimating surface area.

The box concept is also useful in visualising the relationship of atypical egg shapes to that of the standard ellipsoid. For example, hyper-ellipsoids will occupy more than 52.4% of their box and hypo-ellipsoids less. The coefficients in equations 8.16 and 8.19 will have to be determined for such eggs.

Practical egg measurement

In view of the range of procedures available, it may be useful to suggest a scheme for the measurement of eggs. The choice of measurements depends on the task in hand. The two easiest measurements to make are length (L) and maximum diameter (D). In the case of eggs that are linearly transformed ellipsoids (in which θ in equation 8.2 is of the order of 5°) equations 8.15 and 8.17 (Hoyt, 1976) give good estimates for volume and surface area and L/D or elongation is a useful shape descriptor. For many purposes, this simple procedure may suffice. In dealing with species with more pointed eggs equations 8.16 and 8.19 are more accurate. In this case, d must be calculated from equation 8.6 and this requires two extra measurements taken with abbreviated calipers (Fig. 8.2). The extra trouble is worth it as an estimate for θ is provided and matching predicted D and calculated D is a check on linearity of the mid-point tangent. If dealing in species with atypically shaped eggs (non-linearly transformed ellipsoids or hyper- and hypo-ellipsoids) it will be easier to revert to using D as d is difficult to measure directly. It is also necessary to calculate an appropriate value for the exponent in equations 8.15 and 8.16 and check volume predictions against volume measured by displacement (Hoyt, 1976).

The transformation an egg has undergone may be identified using equation 8.8 to reconstruct its mid-point tangent or by

determining the values of the c terms in equation 8.14. These, and other calculations, are best done by digitising the profile of the egg from a photograph and allowing a computer to perform the arithmetical drudgery. In any case, if working with one particular species, it is advisable to perform the latter procedures on a sample of the population in order to confirm its status and to have background knowledge of size and shape variation. Similarly, if examining the physiological properties of a particular single egg, it is advisable to define its properties individually rather than from coefficients averaged out from large populations. A photograph of its profile would also be a useful permanent record which could be used for subsequent detailed calculations if required.

Much basic work remains to be done in the field of egg measurement. No species has yet been subjected to full shape analysis using the above battery of mensuration procedures. They have also yet to be applied to a single large clutch of eggs to discover the variations in the sculptural capabilities of one oviduct.

Egg-shape and the mechanics of egg formation

Origin of egg form

Egg form does not result from cohesive forces within the albumen or yolk since the contents of a broken egg flatten out. The initial, possibly final, shape is received in the isthmus segment of the oviduct prior to entry into the shell gland (Fig. 8.5). The shape imparted by the action of the isthmus wall is, therefore, retained by the shell membranes secreted by the cells lining the isthmus lumen where the egg remains stationary for a time, pointed end first, while these membranes are laid down (Gilbert, 1979).

It is convenient to consider first the symmetry of an egg about its long axis. Even an egg as massive as that of an ostrich (1400 g) or of the extinct *Aepyornis* (4000 g) is circular in cross section (or very nearly so). To maintain such axisymmetry the pull of gravity must be precisely counteracted. If this is done by the oviduct, its wall would have to be correspondingly endowed with muscle and connective tissue. However, the oviduct wall looks inappropriately thin and flaccid (Fig. 8.5) and it seems unlikely that the precision of an egg's axisymmetry is imposed by the action of the oviduct alone. A simpler and less expensive way in terms of energy expenditure to acquire axisymmetry could result from egg and oviduct floating among the fluid viscera and fat filling the abdominal cavity. Just as the yolk floating in the fluid albumen tends towards a spheroid so would the whole egg tend to float among the viscera and take up a spheroidal form. If, as seems to be the case in most eggs, the volume of the egg is greater than the oviduct diameter can accommodate as a spheroid, the surplus volume extends along the lumen to produce a spheroid elongated along one axis which is the definition of an ellipsoid. Neutral buoyancy would still operate to maintain symmetry about the extending axis. It is a reasonable assumption that the ellipsoid form is the most economical to maintain after a spheroid when subjected to this type of constraint.

The egg occupies only a part of the length of the isthmus (Fig. 8.5) so that isthmus length is not a determinant of egg length. The egg at this time must, therefore, be prevented from spreading along the isthmus lumen by a sphincteric action, which is not evident in Fig. 8.5. An observation, possibly relevant here, is that in reconstructing the mid-point tangents it is often found that the extreme ends of the tangent deviate from the linear and became irregular. This suggests that they are picking up perturbations produced by the required sphincteric action of the living oviduct.

The asymmetry of the classic egg-shape must result from differential action of the walls of the isthmus. However, the muscle and connective tissue coats have not been described in enough detail to be helpful in working out their action on the contained egg. In the absence of such basic data the various possibilities may be considered: 1) The isthmus itself may be egg-shaped and the entering egg may adapt to the space it is offered as if entering a mould. However, the isthmus does not look particularly egg-shaped (Fig. 8.5). Thus, preshaping, at least in the case of the domestic fowl, is not a factor in determining egg form. 2) The isthmus wall may be differentially elastic (Mallock, 1925). By inflating elastic capsules with different thicknesses of elastic along their long axis the shape of a variety of eggs can be simulated (Mallock, 1925). The same would happen in the isthmus if its walls were differentially extensible, either by variations in the amount of frank elastin or in the arrangement of collagen meshworks. Again the isthmus looks disappointingly flaccid (Fig. 8.5) which suggests that it is, at best, weakly elastic. 3) Muscle is,

thus, the most obvious candidate for imposing form on the egg. A muscle coat has been noted in the wall of the isthmus with its component fibres orientated longitudinally (Gilbert, 1979) but no further data seems to be available. Mural muscle could act either by variation in the number of fibres present, and therefore, of the power to exert differential force or, if the muscle coat is uniform, by an appropriate variation in the length and tone of individual fibres. The latter arrangement need not waste energy by exerting pressure on the egg. Muscle fibres could adjust the shape of the isthmus lumen to provide a low pressure mould with a volume appropriate to that of the entering yolk and albumen. Variations in the volume of material entering the isthmus will evoke a corresponding adjustment in geometry of the lumen. Analysis of egg form of large clutches of eggs would reveal something of how an individual oviduct responds to variation in egg volume. A shape arising from differential action of a uniform muscle coat would imply that some memory of the required shape exists, either in the neural plexus of the isthmus wall or, perhaps, centrally in the hypothalamus.

Histological and physiological analysis of the isthmus with these questions in mind will indicate what blend of the above morphogenetic agents impresses shape on the egg mass while it is held steady during the time it is receiving its definitive form. Pressure readings from within the egg while it lies within the isthmus could reveal if the egg is subject to a compressive force from elastic or muscular action by the surrounding isthmus wall. It is surprising that such basic data either does not exist or is not commented on in major reviews. However, Preston (1953) refers en passant to a paper by Okabe (1952) in which egg-shape was treated as a problem in the forces that mould that shape, 'exactly as D'Arcy Thompson would have wished'.

Finally, it is worth noting that in all birds, with the exception of the domestic fowl, the oviduct functions only for a brief period once a year. During its long resting period it is shorter and more thin-walled than during the egg producing period, as the body minimises the energy it expends maintaining temporarily unrequired tissue (Gilbert, 1979). The structural agents of morphogenesis, that is the muscle, collagen, elastic and neural components of the oviduct wall must also participate in these cycles of regression and reconstruction. A knowledge of how the shape of the eggs of a single bird varies

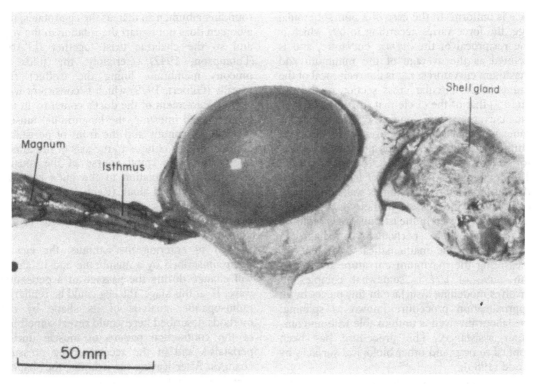

Fig. 8.5. Photograph of domestic hen oviduct, showing shape established in the isthmus is retained by the cell membranes (Gilbert, 1979).

from year to year would give some insight into the variability occurring in the annual reconstitution of the oviduct.

Forces acting on the egg

Since the egg is assumed to be held motionless in the isthmus, at least during the time the egg membranes are being deposited, we need not consider propulsive forces, only those maintaining the shape of the static egg constant. If the egg happens to be subject directly to gravity at this time, the force trying to make the egg flatten out would have to be counteracted by complex differential action of the oviduct which would have to vary as the bird changes posture. We are, however, assuming that the egg is neutrally buoyant, or very nearly so, so that there is no need for the isthmus to counteract differentially the force of gravity. Let us also assume, for the sake of initial simplicity, that the lumen of the isthmus is pre-shaped by muscular adjustment, and that the volume of the egg is just sufficient to fill the proferred mould like a shell in the breach of a gun. Under these conditions, the minimum force to be exerted will be that preventing the egg from becoming a spheroid, the form it would naturally seek under conditions of neutral buoyancy. In order to produce, and maintain, a non-spheroidal egg-shape, the isthmus, even if acting as a passive mould, must exert differential pressure along the long axis of the egg and suffer a corresponding differential tension in its walls. The normal force exerted by an elastic surface under tension is proportional to the average curvature of its surface. Thus, under conditions of equilibrium the local curvature at any point on an egg's surface is proportional to the normal force exerted by the isthmus wall at that point. A series of measurements of local curvature along the long axis of the egg will give a description of the differential action of the isthmus wall required to maintain the shape in equilibrium.

The curvature of a line on a plane is defined as the reciprocal of the radius of the circle that 'just fits' the curvature at a particular point. In the case of a spheroid of radius r, the curvature $1/r$ is constant in all directions on the surface and the normal force acting at all points on the surface is uniform. In the case of a non-spheroidal egg, the force varies according to $1/H$ which is the reciprocal of the *average* curvature, and is defined as the average of the minimum and maximum curvatures: r_{min} is the reciprocal of the radius of the circular cross-section of the egg and r_{max} that of the circle that 'just fits' the surface curvature of the long axis of the egg at the same point, i.e. the curvatures of the lines of latitude and longitude at the point with respect to the long axis of the egg (D'Arcy Thompson, 1942; Todd, 1986). Thus $1/H$ is a measure of the local tension in the isthmus wall and a graph of $1/H$ along the axis of symmetry is a graph of the relative tension in the isthmus walls required to maintain that particular egg-shape in equilibrium. The mathematical treatment for measuring the maximum curvature and fitting the curve of $1/H$ is somewhat complex. It involves modelling a surface in tiny pieces by an approximation procedure known as splining: the labour involved is unthinkable without computer assistance. The procedure has been applied to eggs and other biological surfaces by Todd (1986).

The graphs of the differential tension along the axis of symmetry for different egg shapes obtained by Todd's procedure are shown in Fig. 8.6. The curves of relative tension show a basic similarity. A greater force is impressed on the pointed end in order to decrease its curvature. There is a corresponding decrease in the force exerted on the blunt end; it should be noted, however, that, towards the blunt pole, the tension in the isthmus wall tends to show a slight increase (Fig. 8.6), perhaps, associated with sphincteric muscle contraction.

Movements in the oviduct

Prior to entering the isthmus, the egg is assumed to be propelled along the oviduct by peristalsis (conventionally described as a wave of muscular contraction preceded by a zone of relaxation which is propagated along the viscus for a short distance before dying away to be followed by another wave). It is possible that the blunt end of the egg lies in the preceding zone of relaxation and expands correspondingly.

An egg also rotates about its long axis as it passes along the oviduct as is indicated by the spiralling of the chalazae (Palmer & Guillette, Chapter 3). There is little friction between the membrane circumscribing the yolk and the sur-

rounding albumen so that, as the egg rotates, the albumen does not impart the rotation to the yolk and so the chalazae twist together (D'Arcy Thompson, 1942). Certainly, the folds of mucous membrane lining the oviduct run spirally (Gilbert, 1979) which is consistent with a spiral movement of the duct's contents. In the human small intestine, the longitudinal muscle is arranged spirally and the front of peristaltic contraction lies oblique to the axis of the lumen and follows the spiral course of the muscle so imparting a rotation to the gut's contents as they are moved along (Reid, Ivy & Quigley, 1934). Whether a similar spiral configuration of oviducal muscle also exists in birds is unknown.

Prior to entering the isthmus, the egg is uncircumscribed by a membrane and its shape will change during the passage of a peristaltic wave. If, at this stage, the egg could be rendered radio-opaque, analysis of its shape by the methods described here would reveal something of the contraction pattern of muscle during peristalsis and of the accompanying pressure changes. After leaving the isthmus, the shaped, membrane-bound egg enters the shell gland where the inflexible calcareous shell is deposited. It is possible that the shape of the egg is modified here prior to mineralisation by the action of the uterine wall. In this case, it would be necessary to re-apply the analytic procedures described for the isthmus to the profile of the egg in the shell gland and then compare the two sets of data. It is interesting to note that the wall of the shell gland is capable of rotating the egg about its short axis so that it can leave for the cloaca, and eventually be laid, blunt end first (Gilbert, 1979). For this movement to occur, the shell gland must be able to relax sufficiently to allow the long axis of the egg to rotate through the transverse diameter of the lumen suggesting that the egg is under no great pressure at this time. This rotation may result from the position of the shell gland lumen which does not lie in line with the rest of the oviduct; it is more a pocket projecting from the main lumen of the oviduct. Chance may determine which end of the calcified egg re-enters the lumen first. If the blunt end happens to catch a peristaltic wave while the pointed end remains caught in the egg gland wall peristaltic action could lever the blunt end across the transverse diameter of the lumen and so initiate a rotation (Pitman, 1964). The variability of this curious manoeuvre accounts

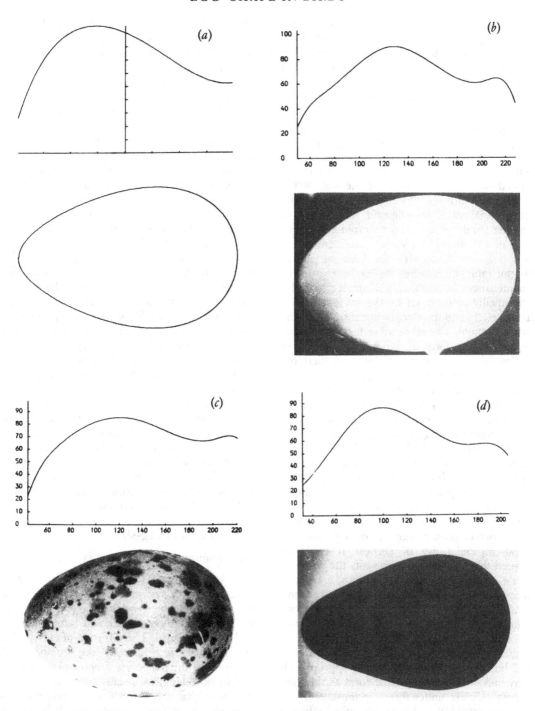

Fig. 8.6. Graphs of $1/H$ showing the relative tension in the walls of the oviduct associated with generating the various egg shapes: (*a*) standard egg shape; (*b*) fowl (*Gallus gallus*) (*c*) herring gull (*Larus argentatus*) and (*d*) guillemot (*Uria aalge*) (Todd & Smart, 1984).

for the controversy about which end of an egg is laid first yet both may be correct at least in the domestic fowl. Whether similar occasional rotation about the egg's short axis also occurs in other species is unknown but if present, it is more likely to occur in less elongated eggs.

Relation of the equations describing egg shape to the functional histology of the oviduct

The various equations presented earlier indicate something about the action of the wall of the oviduct and indicate why egg-shape tends to be some variation of an ellipsoid rather than of another family of curves. A spheroid is a special case of an ellipsoid (x, y and z axes are equal) and it is the form with the least surface to volume ratio and requires the least energy for its maintenance. In most eggs, diameters y and z are equally constrained by the oviduct to give axisymmetry and the third x, increases along the oviduct lumen. The ellipsoid so formed possibly represents the form of minimum surface area under the prevailing constraints and therefore the most economical to maintain in terms of energy expenditure by the oviduct wall. This may account for the absence of curvatures belonging to other mathematical families. A perfectly ellipsoidal egg would tell us that its surrounding isthmus had adjusted to give the minimum energy expenditure for that length and breadth of egg. It is possible that this could be the typical response of an isthmus of uniform histology to a contained body whose volume cannot be accommodated as a sphere. What of the hyper-ellipsoidal egg of the kiwi and the hypo-ellipsoidal egg of grebes? It could be argued that the former represents the response of an overfilled segment of isthmus, a segment that for some reason cannot lengthen to accommodate its allotted egg volume as an ellipsoid. The hyper-ellipsoid form may be the next most energy efficient configuration to maintain. Similarly, a segment of isthmus that is underfilled but cannot adjust its length appropriately may impose a hypoellipsoidal form as the most energy efficient under these circumstances.

Asymmetry is possibly an adventitial effect of peristalsis. The zone of relaxation said to precede the zone of contraction may come to rest on meeting the proximal pole of the egg in which case this end of the egg may reside in a zone where intermittently arriving relaxation waves are stored. Such a process, initially an incidental happening, may have been retained and developed because of some adaptive value. That peristaltic relaxation should modify an ellipsoid by what can be expressed mathematically as a linear transformation may reflect the fact that the extra volume that enters the proximal expanded end is equal to that transferred from the distal end. The straight line mid-point tangent merely indicates that what is lost by one end is gained by the other. Symmetrical eggs may retain their symmetry from lying in oviducts which are unsubjected to peristaltic disturbance during the time of form imposition. These are perhaps eggs that spend a relatively long time at rest during this stage of their development.

The graphs of $1/H$ in Fig. 8.6, on the other hand, indicate the differential energy expenditure along the egg's long axis required of the isthmus wall in order to maintain the related shape in equilibrium. In the case of an ellipsoid, the graph of $1/H$ would, by hypothesis, represent the minimum total energy required and would be consistent with the passive response of a segment of a uniformly elastic tube to a volume greater than its diameter can accommodate. In the case of asymmetric forms, the energy expenditure is asymmetrically distributed and presumably greater than that required to maintain an ellipsoid; the source of this additional energy and its non-uniform distribution must reside in differential development of passively acting tissue in the isthmus wall, or an active differential response from a tissue component, the only candidate for the latter is muscle.

Implications of egg-shape

The significance of surface area

The surface of a rigid-shelled avian egg is penetrated by a system of pores through which all respiratory changes take place, namely the diffusion of carbon dioxide and water outward and oxygen inward. The diffusion of these gases is controlled *inter alia* by the number and diameter of the pores (the functional pore area) and by pore length – usually equivalent to shell thickness (Rahn & Ar, 1974). During incubation, an egg suffers a weight loss. It has been observed (Rahn & Ar, 1974) that, on average, an egg loses 18% of its initial laying weight from water loss during incubation whatever the length of the incubation period may be. This is a remarkable finding as the relative volume and chemical content of yolk and albumen can vary quite widely from species to species. The loss is principally from water vapour and is linear, i.e. an equal weight of water is lost each day ·(Romanoff & Romanoff, 1949; Ar, Chapter 14).

Clearly, the pore system of a particular egg must be constructed to allow a certain weight loss in a certain time. For example, a spheroidal egg which has the minimum S/V ratio for a given volume would be expected to have a greater pore area and/or lesser shell thickness and/or longer incubation period than an ellipsoidal egg of the same volume. Similarly, two eggs of the same volume and S/V ratio but with different laying weights would be expected to have different permutations of the key parameters if each was to achieve the same weight loss. Thus, absolute surface area of an egg represents only a potential upper limit to the functional pore area. This limit may, in fact, never be reached as an egg already has a considerable respiratory capacity, in spite of a less than maximum density of pores. For example, the domestic fowl produces a hatchling of about 40 g in 21 days of incubation; a rat, by contrast, generates a neonate of about 4 g in a gestation period of the same length. Clearly, the rigid-shelled egg permits the contained system to develop at an impressive rate and, in this respect, cannot be an impediment to avian development. The egg surface may, indeed, offer to the avian embryo a relatively greater area for respiratory exchange than the placenta offers to the embryo of mammals. An egg also has the additional advantage of its surface interfacing directly with atmospheric gases thus obtaining the maximum pressure gradient to drive gas exchange with the embryonic circulation, whereas the gas tensions of the placental circulation are nearer to those of the maternal bloodstream. Moreover, each egg of a clutch is exposed to the same abundant atmosphere whereas in a multiparous mammal several fetuses must be irrigated by a single cardiovascular system and share the limited metabolic resources of one mother. Indeed, the avian embryonic circulation may be concerned with regulating exchange with an over-endowed respiratory interface.

Variation in an egg's surface and surface to volume ratio resulting from differences in shape are, therefore, probably of minor functional importance. As Hoyt (1976) summarises, 'Since pore area, pore density and shell thickness, the factors that determine the respiratory parameters of the shell are independent of surface area, there would appear to be no effect of shape on respiratory physiology. The effect of a small variability in surface area on heat exchange would seem to be insignificant when compared with the effect of nest insulation, microhabitat selection and incubation patterns. Therefore, it seems likely that the functional significance of the variation in the shapes of bird eggs is not related to physiological exchanges with the environment.'

Inter-relation of egg parameters

In the foregoing section, weight loss is taken as the independent variable and the other dimensions are considered to vary in permutations capable of permitting the required weight loss. Other studies have used the initial laying weight as the independent variable. Equations relating egg weight to water vapour conductance, daily weight loss and incubation time have been produced by Rahn & Ar (1974), to eggshell thickness, total pore area and water vapour conductance by Ar et al. (1974) and to surface area, volume, shell weight and shell density by Paganelli et al. (1974). Each of these studies provide compatible alignment graphs so that the value of the dependent variables related to a particular egg weight can be read off. Ar, Rahn & Paganelli (1979) also find that the force required to dent an eggshell is directly proportional to the square of shell thickness. Equations also exist relating egg weight to the basal metabolic rate (BMR) and body weight of the female and to the BMR of the egg and through the latter to hatchling brain weight (Bennet & Harvey, 1985). It must be remembered, however, that these equations have been reached by averaging out data gathered from large populations of eggs of different shapes and sizes and from different species with shell surfaces varying from enamelled to matt or chalky. They may not give accurate predictions for a single egg or for species with an atypical egg shape or special surface characteristics.

That so many variables can be cross-related by simple equations probably stems from the fact that the egg makes relatively simple demands on its environment and does so through an inert interface with fixed properties. For example, pore area during incubation is constant and the respiratory gases move passively through an inert shell according to the physical laws of gas diffusion, i.e. there is no biochemically mediated active transport through a varying interface. The eggshell is, therefore, a mask through which the embryo breathes and it seems unlikely that the mask is too finely adjusted to the minimum needs of the embryo as this would render the egg vulnerable to accidental reduction of its functional pore area by dirt or faeces.

The size range over which these interrelations can operate is extensive. At the small end of the range is the egg of the unidentified species of hummingbird referred to by Pitman (1964) which has reported minimum dimensions of length, 1.3 cm, breadth, 0.8 cm, and weight, 0.5 g. This gives an elongation of 1.63 and a surface to volume ratio of about 6.5 both of which are high. Thus, there is apparently no need, at the small end of the size-scale, to minimise surface to volume ratio by becoming spheroidal which again suggests that functional pore area is more important than shape in incurring weight loss. Also, at this small size, the shell is fragile enough to allow the hatchling to break out yet is strong enough to survive incidental trauma and retains a conventional egg-shape. This suggests that there may be some intrinsic property of the standard egg-shape which maximises the strength of a minimal amount of material. The upper end of the known size range is reached by the extinct *Aepyornis* with a 4 kg egg of unremarkable ostrich-type shape. The factors limiting maximum egg size are unknown but do not appear to involve specialisation of egg-shape. The shell thickness required beyond this size may present too strong a barrier for the hatchling to break out or some limitations in pore geometry may be encountered.

An egg must be strong enough to resist incidental traumas received in the nest. This ability is bestowed principally by the hard calcified shell. However, as noted above, it is possible that intrinsic egg-shape and structure may also contribute to eggshell strength. This topic has received little attention (Sluka, Besch & Smith, 1966). Modern computers are used to model the mechanical properties of much more complicated systems than an egg and it should be relatvely easy to model the reaction of a whole egg to various stresses and strains and to predict the effects of different shapes. It would be interesting, for example, to find out why an egg has such great resistance to a force applied along its axis of symmetry. More prosaically it might be possible to design the perfect egg-box.

The evolution of egg-shape

Birds, like mammals evolved from reptiles. The shape of reptile eggs varies from species to species; some lack axisymmetry and at least one (*Draco volans*), lays a pointed egg of typical avian shape (Bellairs, 1969). That this shape should occur in one of the few airborne reptiles must surely be no more than an amusing coincidence. However, reptile eggs are, for the most part, spheroids or hyper-ellipsoids (Iverson & Ewert, Chapter 7). Birds evolved from Archosaurs, the same stem as dinosaurs which appear to have laid hyper-ellipsoidal eggs. Crocodiles, the only other extant descendants of the Archosaurs, also lay an egg of this type and it is, therefore, possible that the first avian eggs were of this shape, a feature retained, perhaps, by the kiwi (Fig. 4). Since egg-shape is less important than pore configuration (Hoyt, 1976), and as reptile eggs lack an architecturally complex pore structure (Packard & DeMarco, Chapter 5), the change to a particular egg-shape is probably secondary to the evolution of a sophisticated pore structure (Board & Sparks, Chapter 6).

The observed diversity of egg-shapes must have other origins than maintaining a certain surface to volume ratio. Obviously a spheroidal egg, with an inbuilt tendency to roll, would be a lethal shape for a guillemot laying on a vertiginous ledge but no disadvantage to an owl nesting in a deeply cupped recess. The guillemot's virtuoso performance on the ellipsoid theme can thus be plausibly attributed to reducing the probability of falling off its ledge as experiments have indicated (Drent, 1975). The ability to roll in a narrow arc is produced by the almost straight section of the egg's surface between its terminal curvatures. This feature is unlikely to have any other role.

Most birds, however, lay in less exposed sites than sloping cliff ledges and yet do not lay spheroidal eggs. The practical factor tending to produce non-spheroidal eggs may be the mundane need to pass a large volume of egg through a bony pelvis of finite cross-sectional area. A relative increase in egg volume is presumably a desirable adaptation and an elongated egg will have a smaller diameter per unit volume than a spheroid and will be correspondingly easier to push through a pelvis, the geometry of which is subject to more important functional priorities than preserving the mathematical simplicities of a spheroidal egg. Nevertheless, in addition to being elongated, the majority of bird eggs are modestly pointed, linearly transformed ellipsoids with a value for θ in equation 8.2 of around 5° which suggests that this configuration has properties which are mechanically favoured by the oviduct or physiologically advantageous to the contained embryo. Some possible general benefits of an asymmetrically pointed egg can be suggested.

The blunt end accommodates the air cell and the shell surface overlying the cell has a greater pore density than the remainder of the shell (Rokitka & Rahn, 1987). In the fowl egg, gas conductance is nearly 60% higher over the air cell than over the remaining chorio-allantoic part of the egg surface. There is also a difference in the gas tensions in the air cell and the chorio-allantoic circulation maintained, at least in part, by a venous shunt (Seymour & Visschedijk, 1988). Blunt-endedness may therefore be a means of bringing about an increase in the area of this specialised respiratory site relative to the rest of the egg. If so, more pointed eggs may have a relatively large proportion of porous blunt end. Very pointed eggs are characteristic of precocial wading birds, and the presence of this respiratory specialisation may be a factor in permitting the more complete maturation of the embryo prior to hatching particularly of metabolically expensive neural tissue. A direct relationship between egg respiration and hatchling brain weight has been reported by Bennet & Harvey (1985). In the case of precocial birds, any relative increase in pre-hatching brain size would be an adaptive advantage.

In symmetrical eggs, the point furthest from the surface, i.e. the point with the longest pathway to be traversed by a heat gradient (the thermal centre of the egg) lies at the intersection of its axes. Increasing the pointedness of an egg results in this point moving towards the blunt end which may have some advantage in placing the thermal centre closer to the core of the developing chick.

Finally, egg-shape may have some role in determining the pattern of turning during incubation (Deeming, Chapter 19). The optimum turning regime is unexpectedly complicated. For example, the egg should lie with its long axis at 20° to the horizontal, blunt end highest and undergo multiplane rotation (Drent, 1975). A pointed egg may produce these positions and movements by virtue of its shape when turned by the parent or otherwise jostled in the nest. It is, however, by no means certain that these precise movements are intrinsically beneficial to development and that egg-shape has adapted to obtain them. It may also be that the behaviour of the embryo has adapted to make the most of movements imposed by pointedness evolved for some other reason.

References

Ar, A., Paganelli, C. V., Reeves, R. B., Greene, D. G. & Rahn, H. (1974). The avian egg: water vapour conductance, shell thickness and functional pore area. *Condor*, **76**, 153–8.

Ar, A., Rahn, H. & Paganelli (1979). The avian egg: mass and strength. *Condor*, **81**, 331–7.

Bellairs, A. (1969). *The Life of Reptiles* vol. 2. London: Weidenfeld & Nicolson.

Bennet, P. M. & Harvey, P. H. (1985). Brain size, development and metabolism in birds and mammals. *Journal of Zoology, London* (A) **207**, 491–509.

Besch, E. L., Sluka, S. J. & Smith, A. H. (1967). Determination of surface area using profile recordings. *Poultry Science*, **47**, 82–5.

Carter, T. C. (1968). The hen's egg: a mathematical model with three parameters. *British Poultry Science*, **9**, 165–71.

Daintith, R. D. & Nelson, R. D. (1989). *Dictionary of Mathematics*. London: Penguin Books.

Drent, R. (1975). Incubation. In *Avian Biology*, vol. 5, Farnar, D. S. & J. R. King, eds. pp. 333–420 New York: Academic Press.

Gilbert, A. B. (1979). Female genital organs. In: *Form and Function in Birds*, vol. 1. A. S. King, & J. McLelland, eds. pp. 237–360. London: Academic Press.

Hoyt, D. F. (1976). The effect of shape on the surface–volume relationship of birds' eggs. *Auk*, **96**, 73–7.

Mallock, A. (1925). The shapes of birds' eggs. *Nature, London*, **116**, 311–12.

Okabe, J. (1952). On the form of hen's eggs. *Reports of Research Institute for Applied Mechanics, Kyushu University*. Quoted by Preston, 1953.

Paganelli, C. V., Olszowka, A. & Ar, A. (1974). The avian egg: surface area, volume and density. *Condor*, **76**, 319–25.

Pitman, C. R. S. (1964). Natural History of Eggs. In *A New Dictionary of Birds*. A. L. Thomson (ed.), Edinburgh: Nelson.

Preston, F. W. (1953). The shapes of birds' eggs. *Auk*, **70**, 160–82.

— (1974). The volume of an egg. *Auk*, **91**, 132–8.

Rahn, H. & Ar, A. (1974). The avian egg and water loss. *Condor*, **76**, 147–52.

Reid, P. E., Ivy, A. C. & Quigley, J. P. (1934). Spiral propulsion of a bolus in the intestine. *American Journal of Physiology*, **109**, 483–7.

Rokitka, M. A. & Rahn, H. (1987). Regional differences in shell conductance and pore density of avian eggs. *Respiration Physiology*, **68**, 371–6.

Romanoff, A. L. & Romanoff, A. J. (1949). *The Avian Egg*. New York: Wiley.

Seymour, R. S. & Visschedijk, A. H. J. (1988). Effects of variation in total and regional shell conductance on air cell gas tension and regional gas exchange in chicken eggs. *Journal of Comparative Physiology*, **B158**, 229–36.

Sluka, S. J., Besch, E. L. & Smith, E. H. (1966). Calculation and analysis of stresses in eggshells.

Paper No. 60–808. Presented at the Winter Meeting, American Society of Agricultural Engineers.

Smart, I. H. M. (1969). The method of transformed co-ordinates applied to the deformations produced by the walls of a tubular viscus on a contained body: the avian egg as a model system. *Journal of Anatomy*, **104**, 507–18.

Tatum, J. B. (1975). Egg volume. *Auk*, **92**, 576–80.

Thompson, D'Arcy W. (1942). *On Growth and Form*. Cambridge: Cambridge University Press.

Todd, P. H. (1986). Intrinsic geometry of biological surface growth. *Lecture Notes in Biomathematics*, **67**, 1–128.

Todd, P. H. & Smart, I. H. M. (1984). The shape of birds' eggs. *Journal of Theoretical Biology*, **106**, 239–43.

9

The thermal energetics of incubated bird eggs

J. SCOTT TURNER

Introduction

A bird embryo can generate only a limited amount of heat (Romijn & Lokhorst, 1960; Tazawa *et al.*, 1988). The ability of the egg to dissipate heat is also limited (Spotila, Weinheimer & Paganelli, 1981; Turner, 1985; Sotherland, Spotila & Paganelli, 1987). Consequently, the temperature of an egg can be controlled only slightly by the embryo and deviate from ambient temperature by roughly -0.5 °C to $+3$ °C at most (Turner, 1985; Sotherland *et al.*, 1987). Most bird embryos must develop within fairly narrow temperature limits (Barott, 1937; Drent, 1970, 1973, 1975; Deeming & Ferguson, Chapter 10), and environments that provide these temperatures are rare. It follows that an egg will rarely be at an ideal temperature for development unless fluxes of heat into and out of it are somehow supplemented and controlled. The management of heat flux into and out of the egg, effected by the brooding parent, *is* incubation: all other attributes of incubation (gas and water exchange and turning) are, at root, secondary to this one function.

In this chapter, I wish to address the fundamental problem of how incubation 'works', i.e. how heat flow between an egg and incubating parent is managed. Although the thermal interaction of the embryo and the incubating parent has been extensively studied, it remains poorly understood. The physics and physiology of temperature of incubated bird eggs are examined, and the energy cost to the incubating parent of keeping the egg at that temperature, is determined. The various ways this problem has been studied in the past, and what these approaches have implied and predicted about

the way incubation works will be discussed. A more informative approach to the problem will be introduced which is able to predict many important features of incubation energetics that previous approaches have missed. My goal is to put the thermal energetics of avian incubation onto a sound physiological and physical foundation; a foundation that, in my opinion, has heretofore been missing.

Approaches to incubation energetics

Some definitions

Bird eggs may be incubated in one of two ways. One way is to place the egg in air that is warm enough to bring it to a suitable temperature. This type of incubation, which is typical of a commercial incubator, is termed *convective incubation*. The other way is to warm, and regulate the temperature of, part of the egg's surface. Here, the air surrounding the rest of the egg can be any temperature. Any variation of heat loss to the air is compensated by a similar variation of heat flow into the egg across the warmed surface: egg temperature will then stay more or less constant. This type of incubation, which is what an incubating bird does with its brood patch, is termed *contact incubation*.

Contact incubation is a thermal interaction between the parent and the egg and although it has been studied in many ways (Grant, 1984) conceptually, these approaches can be reduced to just two. The first method, the 'analogical approach', regards the thermal behavior of convectively incubated eggs and contact incubated eggs to be sufficiently similar that an analogy may be drawn between them: what is known about one type of incubation can inform the other. The second method, the 'lumped conductance approach', regards the thermal behavior of a contact incubated egg to be sufficiently unique to make an analogical approach inappropriate. Further, it presumes that the analysis of the flows of heat in a contact incubated egg can be greatly simplified yet still yield robust answers.

The analogical approach

The most influential analogy between convectively incubated and contact incubated eggs is an equation, derived by Kendeigh (1963), which predicts the heat input required to keep a clutch of eggs at an appropriate temperature for development. Kendeigh's formula (slightly rewritten from the original) is:

$$Q_i = nMc_p (T_e - T_a) (1-c)/\tau \qquad (9.1)$$

where: Q_i = instantaneous power requirement for incubation (W), n = number of eggs in a clutch, M = egg mass (kg), cp = egg specific heat ($J\,kg^{-1}\,°C^{-1}$), τ = time constant for cooling (s), T_e = egg temperature (°C), T_a = ambient temperature (°C), and c = fraction of the egg surface covered by the brood patch. The quantity ($M\,c_p/\tau$) is equivalent to the egg's thermal conductance, K ($W\,°C^{-1}$).

Equation 9.1 is an analogy due to its reliance on the time constant for cooling, τ, which Kendeigh measured for eggs cooling in still air. Such eggs, of course, are completely surrounded by air, which is characteristic of convective incubation (Fig. 9.1). Therefore, the heat an incubating parent must apply to a contact incubated egg is simply: the heat normally lost from a convectively incubated egg, minus whatever heat would be lost from an exposed surface equivalent in area to that covered by the brood patch (Fig. 9.1).

The lumped conductance approach

This approach starts with the presumption that an analogical approach to incubation cannot work. Put simply, the analogical approach cannot predict a very fundamental attribute, namely the temperature distribution inside a contact incubated egg. A convectively incubated egg is a well-mixed body (Turner, 1987): interior temperatures differ only slightly from surface temperatures (Sotherland et al., 1987). The analogical approach suggests the temperature distribution in contact incubated eggs should be similar. Yet, contact incubated eggs have marked temperature gradients between the top (near the brood patch) and bottom (Rahn, Krog & Mehlum, 1983; Vleck et al., 1983; Turner, 1987). The energy flows within an egg and cannot reliably be assessed without realistic temperature distributions (nor can temperatures be predicted from energy flows). The analogical approach cannot provide this and therefore must fail.

Analysing the actual temperature distribution (and hence energy flow) in a contact incubated egg is a difficult problem. The lumped conductance approach attempts to simplify the problem by lumping the flows of heat into one or a few discrete elements, or *nodes*. The aim in doing this is to make the problem tractable without simplifying away what is interesting.

There have been three principal applications of the lumped conductance approach: 1) the clutch-mass method regards each egg as an extension of the parent's body: the heat required to incubate the eggs is simply that required to keep an equal mass of parent warm (West, 1960). 2) The egg is treated as a single node interposed between the brood patch and nest environment (Walsberg & King, 1978a,b; Webb & King, 1983). 3) The most important model, developed by Ackerman & Seagrave (1984), is the one I shall consider in detail through this paper. Ackerman & Seagrave (1984) distribute the conductance of a 60 g egg into three adjacent nodes (Fig. 9.1). One node is that portion of the egg which contacts the brood patch, a second node is that portion of the egg contacting the substratum and sandwiched between the two, and comprising the third node, is the remainder of the egg. The three-node egg lets heat flow in two dimensions (Fig. 9.1): the middle node can exchange heat with both of the adjacent egg nodes (one dimension) and with the nest air (the other dimension). Additionally, the egg has physiology: the 'embryo' produces heat and circulates blood (Ackerman & Seagrave, 1984).

A third way: the numerical model approach

In a contact incubated egg, heat flow vectors emanate from the egg surface warmed by the brood patch, and spread symmetrically about the egg's vertical axis (Fig. 9.1; Turner, 1987). Both the analogical approach and the lumped con-

Three approaches to the energetics of
contact incubated eggs

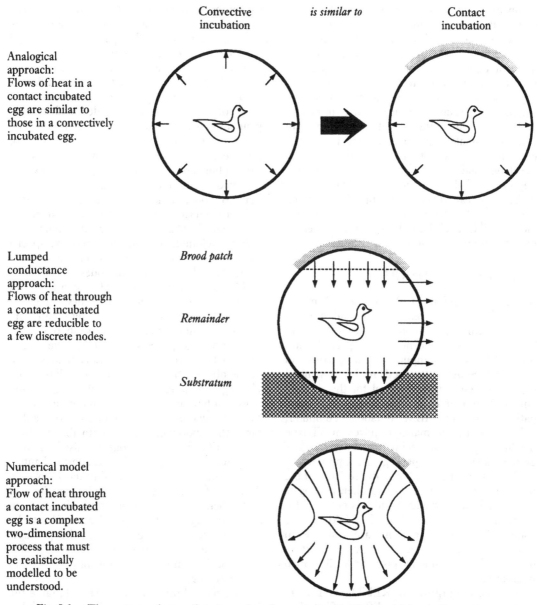

Convective
incubation

is similar to

Contact
incubation

**Analogical
approach:**
Flows of heat in a
contact incubated
egg are similar to
those in a convectively
incubated egg.

**Lumped
conductance
approach:**
Flows of heat through
a contact incubated
egg are reducible to
a few discrete nodes.

Brood patch

Remainder

Substratum

**Numerical model
approach:**
Flow of heat through
a contact incubated
egg is a complex
two-dimensional
process that must
be realistically
modelled to be
understood.

Fig. 9.1.   Three approaches to the energetics of contact incubated eggs. Light shading indicates
the area of egg covered by the brood patch. Arrows indicate the flows of heat. *Top panel:* the alleged
similarity between radial heat flow in convectively incubated eggs and contact incubated eggs.
Middle panel: heat flows in the lumped conductance approach, where the egg is divided into three
nodes. *Bottom panel:* detailed heat flows in the numerical model approach.

ductance approach attempt to simplify this
temperature distribution, albeit in different
ways. The analogical approach assumes it does
not exist and the lumped conductance approach

assumes it can be reduced to three discrete
temperatures. A third approach is to model the
temperatures and flows of heat in the egg in as
much detail as practicable.

This is the approach introduced here: a numerical finite-difference model of temperatures and flows of heat through a contact incubated egg has been developed, which divides the egg into not a few, but many hundreds of nodes. This makes it possible to estimate the temperature distribution in the egg in great detail, which in turn makes possible realistic estimates of energy flows through the egg. This model is designed to be flexible and general. For example, it allows an examination of how the temperatures and flows of heat in contact incubated eggs are affected by various environmental conditions, incubation stages, developmental types, egg size and embryonic physiology. To honour the man who first tried to make physical sense out of the thermal energetics of incubation, I have named this model KENDEIGH and hereafter it is referred to as the numerical model.

Description of KENDEIGH

A brief description of the numerical model KENDEIGH is presented here and a fuller mathematical description is the subject of another paper (Turner, in preparation). KENDEIGH is a computer package, written in Turbo PASCAL, designed to be a general exploratory and predictive tool for the thermal energetics of contact incubated eggs. It is compatible with any IBM personal computer, although best performance requires an AT type machine, preferably with an EGA colour graphics adapter. The package is the property of the University of Cape Town and free use of the program is granted to any researcher, as long as its source is acknowledged and it is not used for monetary gain. Copies of the PASCAL source code and compiled programs are available on request to the author, and upon provision either of a 5.25 inch floppy disc or a 3.5 inch microdiscette.

The basic purpose of KENDEIGH is to estimate the temperatures and flows of heat in a contact incubated egg. It is a numerical finite-difference model. The model egg is divided into 401 interior nodes and 20 surface nodes which are laid out according to a system of spherical coordinates. The model is two-dimensional and temperature is presumed to vary only in the r and ς dimensions. Consequently, with the exceptions of the centre node and two of the surface nodes, the nodes are toroidal in shape.

Temperatures are calculated by performing a simultaneous heat balance on all nodes. Through various iterative numerical methods, the temperatures required to balance the energy budgets of all nodes simultaneously can be calculated. Heat flows in the egg may be estimated from the solved temperature distribution.

KENDEIGH allows heat to flow into, and out of, each node by conduction from adjacent nodes, circulation of blood between certain adjacent and non-adjacent nodes, and by heat generation within a node. Each node is differentiated according to whether it is embryo, albumen, yolk, chorio-allantois or shell. Depending upon its classification, the node is assigned a certain thermal conductivity. If it is an embryo node, a heat generation term and a circulation term is included in its heat balance. If it is a chorio-allantoic node, a circulation term only is included. Additionally, there are special terms for the circulation between the chorio-allantoic nodes and embryo nodes.

The package KENDEIGH is comprised of three separate programs: The first program, EGGINIT, 'builds' an egg for later simulation. EGGINIT allows the user to specify one of 21 incubation ages, egg size, whether the egg is precocial or altricial, and it allows one to 'switch' on or off metabolism and/or circulation within the egg. The latter feature enables one to build eggs that have either metabolism but no circulation, circulation but no metabolism, or neither circulation nor metabolism. With the specified inputs, EGGINIT maps out the distribution of the embryo, albumen, yolk and chorio-allantoic nodes and assigns the heat flux terms to each node. EGGINIT then writes an information text file, which contains basic descriptive information about the egg, and an initialisation file, to be used by the second program in the package, which identifies the type of node and contains the variable heat balance terms for each node in the model egg.

The second program, EGGHEAT, takes the information about the egg 'built' by EGGINIT, and performs the actual energy balance calculations. EGGHEAT further allows the user to specify how much of the egg is covered by the brood patch, what the temperature is of the surroundings, and allows one to set a 'gain' for the circulation and metabolism terms, so that metabolism can be varied continuously and independently of circulation and vice versa. The program solves the energy balance equations using standard iterative methods, and solves them to between-iteration error of 0.00001. Upon completion of the simulation, the program writes a data file containing the temperatures of all 421 nodes. It also writes a simulation report, which may be printed out, and which contains descriptive information on the simulation, the energy balance for the incubated egg, and weighted average temperatures for the entire interior of the egg, the surface of the egg not covered by the brood patch and for the nodes identified as embryo.

The third program, EGGDRAW, requires an EGA colour monitor. EGGDRAW reads the temperature data file created by EGGHEAT and draws a colour map of the temperature distribution inside the egg and on the surface of the egg.

Evaluating the three approaches: an artificial brood patch system

To evaluate the three ways of approaching incubation energetics, five 'case studies', which consider different aspects of the problem of incubation energetics, are investigated and the respective predictions of the analogical, lumped conductance and numerical approaches are compared. Experimental data, obtained from an artificial brood patch system which measures the incubation energy budget for contact incubated eggs are presented. These data will allow us to evaluate the predictions of the three approaches.

The artificial brood patch system is based upon a device originally described by Drent (1970) and is fully described in Turner (1990). It consists of two parts. First, the artificial brood patch itself, a custom-built Dewar flask that sits over the upper surface of the egg (Fig. 9.2). The flask contains a heating coil in its cup, and the heating coil and egg are thermally coupled by dental alginate which also serves to secure the artificial brood patch to the egg. This arrangement ensures that the major share of the heat dissipated by the heating coil must exit through the egg although there are minor extraneous fluxes resulting from conduction through the glass walls of the flask. If the temperature of the brood patch thermocouple is regulated, the heat dissipated by the heating coil is a close approximation to the heat an incubating parent must pass into the egg *via* its brood patch.

The second part of the system is the machinery, which has three major elements, that

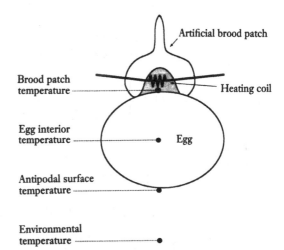

Fig. 9.2. Detail of an egg incubated with an artificial brood patch, showing location of various thermocouples.

estimates the incubation energy budget for the egg under the artificial brood patch (Fig. 9.3). This is comprised of three major elements. First, there is machinery to regulate the temperature of the brood patch thermocouple. The output of the thermocouple is fed to a time-proportional on-off temperature controller, which operates a relay that feeds current through the heating coil (Fig. 9.3). Second, there is a system for estimating the heat production of the embryo based on standard methods of flow-through respirometry. Third, a micro-computer controls the operation of the machinery, senses the signals for the relevant variables, and when the experiment is complete, writes a report detailing the results of the experiment.

Some case studies in the energetics of contact incubated eggs

Case 1: The energy cost of incubating an unembryonated egg

The most obvious test of any model is to ask whether it gives the correct answer to what it is meant to predict. For example, in the case of the analogical approach, does equation 9.1 correctly predict the energy cost of keeping a contact incubated egg at a certain temperature? As an initial test, let us consider the simplest case: a single unembryonated egg, comprised of albumen and yolk, with neither internal generation of heat nor circulation of blood. This is

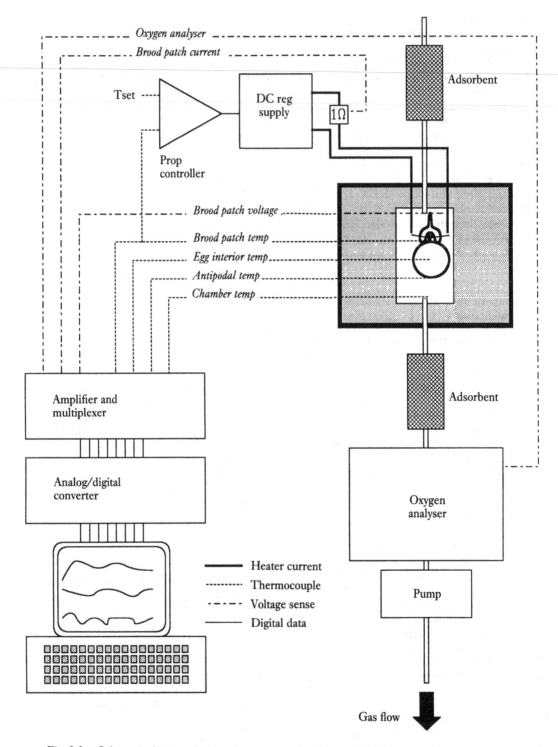

Fig. 9.3. Schematic diagram showing the components of the artificial brood patch system.

characteristic of an egg at the start of incubation.

Application of equation 9.1 is straightforward: A 55 g fowl (*Gallus gallus*) egg cooling in still air has a time constant of 2492 s, which gives a thermal conductance (K) of 72.9 mW °C^{-1} (Turner, 1985). For egg temperature in a single egg the most reasonable temperature is that actually regulated by an incubating parent, namely brood patch temperature (White & Kinney, 1974; Tøien, Aulie & Steen, 1986; Swart, Rahn & de Kock, 1987). Let us assume the bird keeps the brood patch at 40 °C in an environment of 30 °C, and that the brood patch covers 20% of the egg's surface. Equation 9.1 estimates that the parent must pass heat into the egg through its brood patch at a rate of 583 mW. By comparison, the numerical model predicts that for the same egg under the same conditions, the parental heat input is 244 mW. How much power actually is required? The actual heat requirement determined for unembryonated fowl eggs using the artificial brood patch system is 253 mW (Turner, 1990).

The reason the analogical approach overestimates energy costs is simple: it implies that the surface temperatures of a contact incubated egg are uniform (Fig. 9.4). However, the surface temperature of a contact incubated egg declines as one goes from the brood patch to the surface antipodal to the brood patch (Fig. 9.4). Thus, the analogical approach predicts that the average surface temperature of a contact incubated egg is higher than it really is. Consequently, it will always overestimate the heat loss from the egg surface, and by inference, the parent's energy requirements for keeping the egg warm.

What about eggs that are larger or smaller than a 55 g fowl egg? The analogical approach suggests that the incubation conductance for contact incubation has the same scaling exponent as the thermal conductance of eggs cooling in air (Kendeigh, 1963), which is roughly 0.60 (Turner, 1985). The surface temperatures of eggs cooling in air are nearly independent of egg size (Turner, 1987), and the analogical approach predicts the same for contact incubated eggs. On the other hand, the numerical model predicts that larger eggs will have cooler surfaces than smaller eggs. The consequence is a smaller scaling exponent for the incubation conductance of contact incubated eggs (Fig. 9.5). This prediction applies only to unembryonated eggs: the problem of egg size and incubation energetics for eggs containing living embryos is discussed later.

Analogical approach

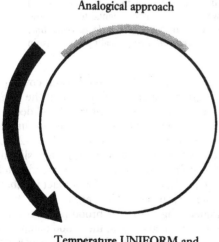

Temperature UNIFORM and
HIGH over entire surface
of egg

Numerical model approach

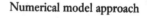
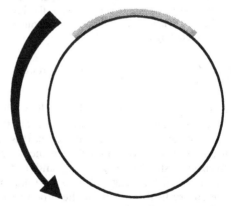

Surface temperature HIGH
near brood patch and
DECLINES with distance
from brood patch

Fig. 9.4. How surface temperature is distributed in a contact incubated egg, as predicted by the analogical approach and by the numerical model approach.

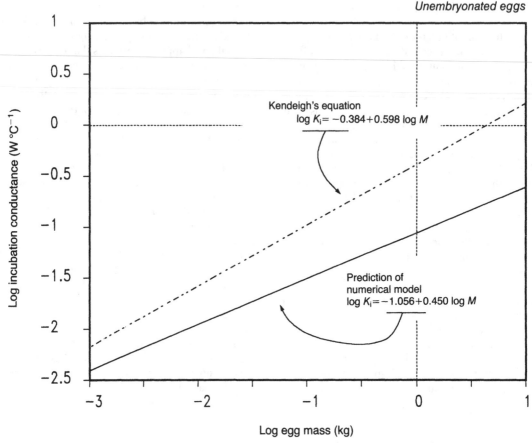

Unembryonated eggs

Fig. 9.5. The scaling of incubation conductance, K_i, defined as the heat input from the brood patch (W) required to keep brood patch temperature 1 °C warmer than environmental temperature. Both scaling relationships assume the brood patch covers 15% of the egg's surface. *Solid line:* Scaling relationship predicted by the numerical model. *Dashed and dotted line:* Scaling relationship predicted by equation 9.1.

Case 2: Coverage of the egg by the brood patch

The factor $(1-c)$ in equation 9.1 predicts what effect brood patch coverage will have on the energy cost for contact incubating eggs. It is an intuitively sensible prediction. When the brood patch covers a small area of the egg, and there is much surface for the egg to lose heat from, the compensatory heat input from the brood patch must be large (Fig. 9.6). Conversely, at large coverages, when there is little surface area for the egg to lose heat from, the compensatory heat input from the brood patch is smaller (Fig. 9.6).

The numerical model makes a very different prediction (Fig. 9.6). At very small brood patch coverage, the heat input to the egg should be small and it should increase as brood patch coverage increases. Once brood patch coverage passes a critical value (roughly 18% of the egg's surface area), energy expenditure should decline slowly as brood patch coverage increases further and for brood patch coverages of 10–25%, energy cost of incubation varies only slightly (Fig. 9.6). The effects of differing brood patch coverage on the energy cost for contact incubating unembryonated fowl eggs have been determined (Fig. 9.6). For brood patch coverages of 9%, 19% and 29%, the brood patch heat inputs are indistinguishable. This is consistent with the prediction of the numerical model.

The analogical approach has failed again because (as in case 1) it imposes inappropriate temperatures on the egg. The analogical approach presumes the egg surface to be of high and uniform temperature, and to be unaffected

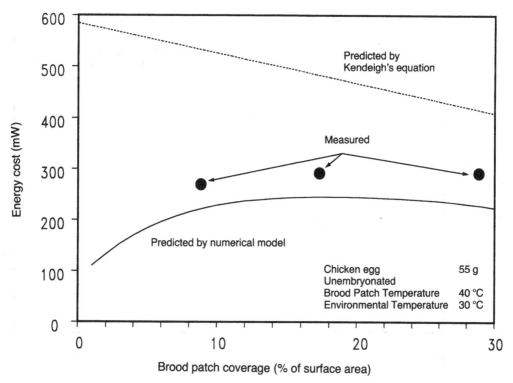

Fig. 9.6. The effect of brood patch coverage on the brood patch heat input for an infertile fowl egg. The brood patch temperature is kept 10 °C warmer than the surroundings. *Solid curve:* Relationship predicted by the numerical model. *Broken line:* Relationship predicted by equation 9.1. *Solid circles:* Mean measured values for 6 infertile fowl eggs at 9%, 19% and 29% coverage by an artificial brood patch. There is no significant difference between the three groups at $p = 0.05$.

by brood patch coverage. Heat loss from the egg (and hence parental energy cost) is determined solely by the reduction of exposed surface area that results from greater brood patch coverage. On the other hand, the numerical model shows that increased brood patch coverage warms the egg and its surface (Fig. 9.7). If, as brood patch coverage increases, surface temperature increases faster than surface area is reduced, the total loss of heat from the egg will increase. This appears to be the case for brood patch coverages smaller than 18% of the egg's surface. At brood patch coverages larger than 18%, temperature does not increase rapidly enough to offset the effect of the reduced surface area. Consequently, heat loss from the egg (and hence heat requirement from the parent) declines.

Case 3: The thermal interaction between incubating parent and living egg

A living embryo generates heat and circulates blood and these have the potential to influence

both the temperature of the egg and the energy costs to the incubating parent. The temperature of a contact incubated egg should be elevated by both the embryo's metabolism and its circulation (Fig. 9.8). Metabolism raises egg temperature by direct generation of heat. Circulation raises egg temperature by distributing heat from the brood patch more readily through the egg. With respect to the energy cost of incubation, metabolism opposes the flow of heat from the brood patch into the egg (Fig. 9.8). As metabolism of the embryo increases, this should drive the energy cost of incubation down. Circulation, on the other hand, facilitates the flow of heat from the brood patch into the egg (Fig. 9.8). As circulation increases, this should drive the energy cost of incubation up.

The net effect of metabolism and circulation will depend upon their relative importances to the egg's thermal energetics. For example, metabolism's effect may so dominate the egg's thermal energetics that circulation's effects are negligible. The analogical approach implicitly

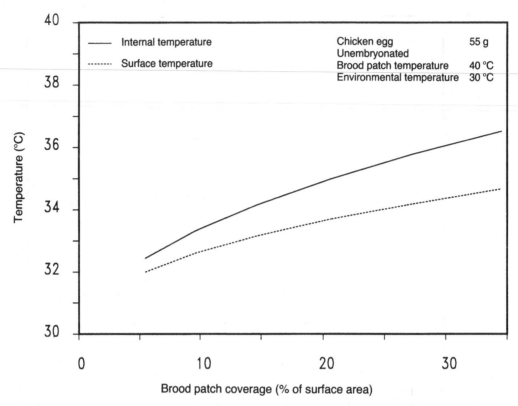

Fig. 9.7. The effect of brood patch coverage on internal temperature and surface temperature as predicted by the numerical model for a 55 g infertile fowl egg. Brood patch temperature is 40 °C. Environmental temperature is 30 °C. Temperatures are weighted averages for all interior nodes (weighted by node volume) and all surface nodes (weighted by surface area) not covered by the brood patch.

denies circulation any role in the energetics of contact incubated eggs. The embryo's circulation has little influence on the cooling rates, and hence the thermal conductances, of eggs cooling in air (Turner, 1987). By extension of the analogy implied in equation 9.1, circulation should then have little influence on the energetics of contact incubated eggs. The lumped conductance approach deals with embryonic metabolism and circulation more explicitly. Nevertheless, Ackerman & Seagrave (1984) conclude that temperature of a contact incubated egg is determined mainly by the embryo's metabolism, and that circulation has only a small effect. By contrast, the numerical model predicts that the influence of the embryo's circulation, not its metabolism, is the most important factor in the thermal energetics of contact incubated eggs.

It is worthwhile outlining how this conclusion is reached. The numerical model allows us to

'construct' eggs that have either circulation only, or metabolism only, or both metabolism and circulation, or neither. This enables one to evaluate how much influence either circulation or metabolism has. If embryonic metabolism dominates the egg's thermal energetics, and circulation is negligible, then a simulated egg with both metabolism and circulation, or a living egg (which also has both), should behave similarly to a simulated egg with *metabolism only* and *no circulation*. Conversely, if circulation is most important and metabolism is negligible, then a simulated egg with both metabolism and circulation, or a living egg, should behave similarly to an egg with *circulation only* and *no metabolism*.

Let us first consider temperatures of a simulated fowl egg, with the embryo just prior to internal pipping. The average internal temperature of the simulated egg with both metabolism and circulation should be roughly

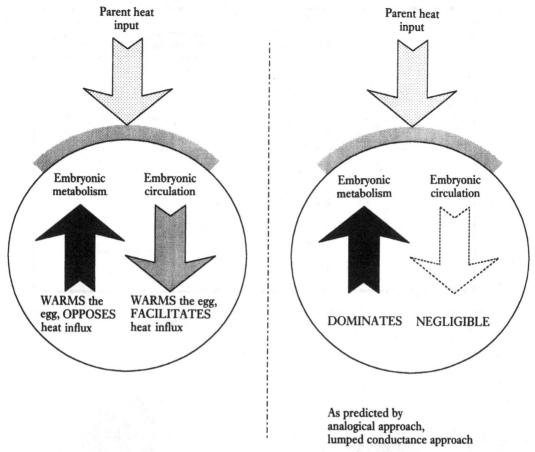

Fig. 9.8. How embryonic metabolism and circulation interact to influence egg temperature and brood patch heat inputs. Arrows represent heat flow vectors. *Left panel:* the effect of embryonic metabolism and circulation on temperature and parental heat input. *Right panel:* The analogical and lumped conductance approaches predict the effect of circulation to be negligible.

37 °C (Fig. 9.9). The temperature of the egg with circulation alone is similar to this, while the temperature of the egg with metabolism alone is not at all similar (Fig. 9.9). By the logic outlined above, the embryo's circulation, not its metabolism, appears to be the most important influence on the temperature of the contact incubated egg.

Turning now to the energy cost of incubation, a simulated egg with both metabolism and circulation should require heat from the parent at a rate of roughly 400 mW (Fig. 9.9). The parental contribution for a simulated egg with metabolism alone and no circulation is roughly 180 mW (Fig. 9.9). If the egg has only circulation and no metabolism, the parent's contribution is roughly 440 mW (Fig. 9.9). Again, the egg with circulation and metabolism is similar to

the egg with circulation alone, and is quite different from the egg with metabolism alone (Fig. 9.9). Again, we may infer that circulation's influence is stronger than that of metabolism.

The predictions just outlined may be tested experimentally by comparing living eggs (which have both metabolism and circulation) with killed eggs (which have neither). When a contact incubated egg is killed, it should cool, irrespective of whether circulation or metabolism dominates (Fig. 9.9). However, if circulation dominates, killing the egg should *reduce* the energy cost to the parent and if metabolism dominates, killing the egg should *increase* the energy cost (Fig. 9.9).

Using the artificial brood patch system, the brood patch heat inputs for chicken eggs containing living embryos with those containing kil-

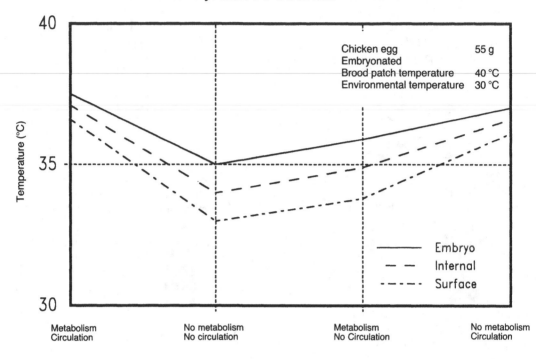

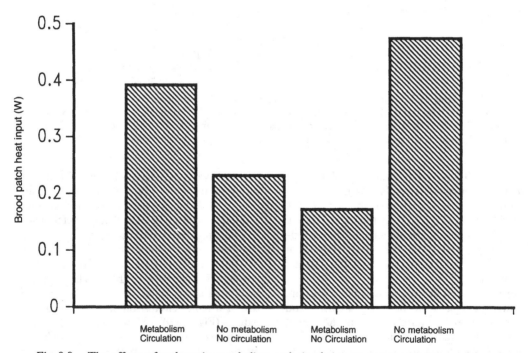

Fig. 9.9. The effects of embryonic metabolism and circulation on egg temperature and brood patch heat input as predicted by the numerical model. The egg is a 55 g fowl egg. The incubation age is just prior to internal pipping. The brood patch covers 15% of the egg's surface area. The brood patch temperature is 40 °C. The environmental temperature is 30 °C. The model considers four conditions: (1) Simulated egg, with metabolism and circulation; (2) Simulated egg, with metabolism and no circulation; (3) Simulated egg, with circulation and no metabolism; (4)

Table 9.1. *Measured brood patch heat inputs, Q_i required to contact incubate embryonated fowl eggs, embryo heat inputs, Q_e and egg temperatures. Embryos are 1–4 days pre-pipping. Data from Turner (1990). Numbers are mean values for at least ten replicates.*

Type	Heat inputs		Temperatures (°C)		
	Q_e (mW)	Q_i (mW)	Brood patch	Embryo	Ambient
Living	138	403	39.3	35.8	29.9
Killed	0	322	39.6	33.8	30.1

led embryos have been compared (Turner, 1990; Table 9.1). The results are clear: killing the egg reduces the brood patch heat input (Table 9.1). A living egg requires roughly 400 mW from the parent to keep warm, almost exactly as predicted (Fig. 9.9). This is about 25% more than the brood patch heat required to keep a killed egg warm (Table 9.1). Compared to an unembryonated egg, a living egg requires about 40% more energy from the parent (Table 9.1). Clearly, it is circulation, not metabolism that is most influential in the thermal energetics of a contact incubated egg. This important finding cannot have been predicted by either the analogical approach or the lumped conductance approach.

One additional matter concerns the consequences of egg size. In case 1, equation 9.1 predicts a scaling exponent of 0.60 for incubation conductance (Fig. 9.5), while the numerical model predicts a smaller scaling exponent (0.44). The reason for the discrepancy lies in different predictions about egg size and surface temperature (Figs 9.4 & 9.10). Circulation, which distributes heat effectively through the egg, has the interesting effect of making temperatures of contact incubated eggs roughly uniform, irrespective of egg size (Fig. 9.10). Thus, the assumption that proved fatal for applying equation 9.1 to unembryonated eggs, i.e. size independent egg temperatures, is vindicated for mature embryonated eggs! The result is a scaling exponent for incubation conductance that is very close to that predicted by Kendeigh's analogical equation (equation 9.1; Fig. 9.11).

The intercepts of the respective equations still differ considerably, however (Fig. 9.11).

It is important to emphasise that this similarity of scaling exponents in no way vindicates Kendeigh's analogical equation. The similar scaling exponent applies only to eggs at the end of incubation (Fig. 9.11), and does not apply to earlier stages (Fig. 9.5). Second, the reason for the coincidence, high net transport of heat through the egg by circulation, is denied by the analogical approach. Third, equation 9.1 still overestimates the actual costs of incubation by nearly two-fold (Fig. 9.11).

Case 4: Incubation stage

During development of an avian embryo, both circulation and metabolic heat production inside the egg increase. The yolk sac membrane and the chorio-allantois have substantial amounts of blood circulating through them after the first third and half of incubation, respectively (Romanoff, 1960; Tazawa, 1980; Ar, Girard & DeJours, 1987). The embryo's major period of growth occurs in the second half of the incubation period (Hoyt, 1987), and during this time there are concomitant increases in heat production and circulation (Tazawa, 1980; Hoyt, 1987; Vleck & Vleck, 1987). Circulation, and secondarily, metabolism both influence the thermal energetics of mature contact incubated eggs (Case 3; Fig. 9.9; Table 9.1). Is this true throughout the embryo's incubation period, with its complicated variations of both metabolism and circulation?

Fig. 9.9 *contd*

Simulated egg, with neither circulation nor metabolism. *Top panel:* Average egg temperatures. *Solid line:* average embryo temperature. *Dashed line:* internal temperature. *Dashed and dotted line:* exposed surface temperature. *Bottom panel:* Brood patch heat inputs.

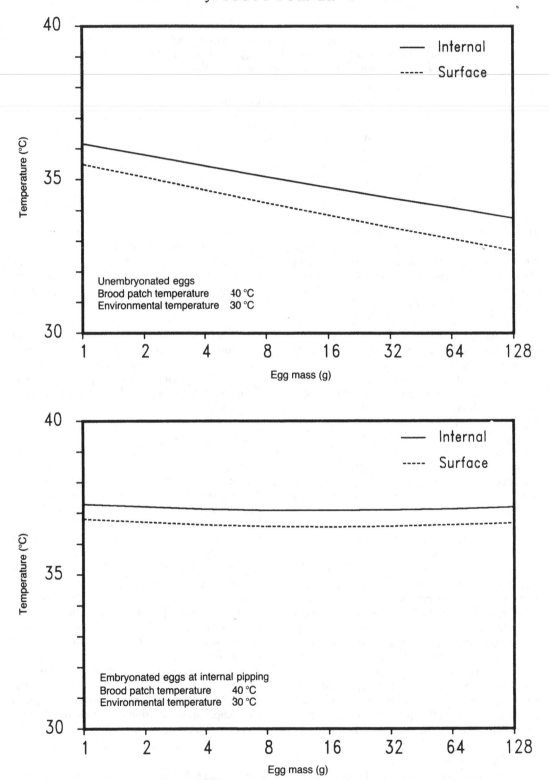

Fig. 9.10. The effect of egg size on temperatures of eggs as predicted by the numerical model. *Solid lines:* average internal temperature. *Broken lines:* average exposed surface temperature. The

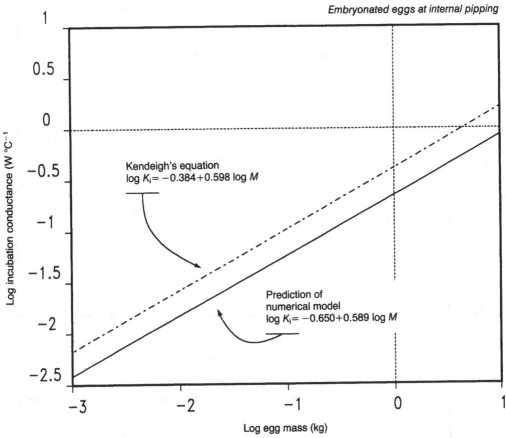

Embryonated eggs at internal pipping

Kendeigh's equation
$\log K_i = -0.384 + 0.598 \log M$

Prediction of numerical model
$\log K_i = -0.650 + 0.589 \log M$

Fig. 9.11. The scaling of incubation conductance, K_i, as defined for Fig. 9.5, for fertile eggs just prior to internal pipping. The brood patch covers 15% of the egg's surface area. *Solid line:* Scaling relationship predicted by the numerical model. *Dashed and dotted line:* Scaling relationship predicted by equation 9.1.

The analogical approach makes no prediction of the course of temperature of a contact incubated egg during incubation. Equation 9.1 contains no term for incubation age and it implicitly assumes that the energy cost for contact incubation is constant through the incubation period. As it develops, the embryo generates more and more heat and presumably this will be heat the parent does not have to contribute. Consequently, the analogical approach implicitly predicts that energy demand on the parent should decline as incubation proceeds. The lumped conductance approach predicts that the egg warms during the incubation period (Ackerman & Seagrave, 1984). Furthermore, it

attributes this warming mostly to the embryo's increasing production of heat and has little explicit to say about fluxes of heat into the egg through the incubation period (Ackerman & Seagrave, 1984).

The numerical model predicts internal and surface temperatures of the egg (Fig. 9.12) that are similar to those predicted by Ackerman & Seagrave (1984). Average egg temperature is roughly steady for the first third of incubation, and increases sigmoidally thereafter (Fig. 9.12). However, the numerical model predicts an interesting course of embryo temperature during incubation. At the start of incubation, embryo temperature is very close to brood patch

Fig. 9.10 *contd*
brood patch covers 15% of the egg's surface area. The brood patch temperature is 40 °C. The environmental temperature is 30 °C. *Top panel:* temperatures of infertile eggs. *Bottom panel:* temperatures of fertile eggs just prior to internal pipping.

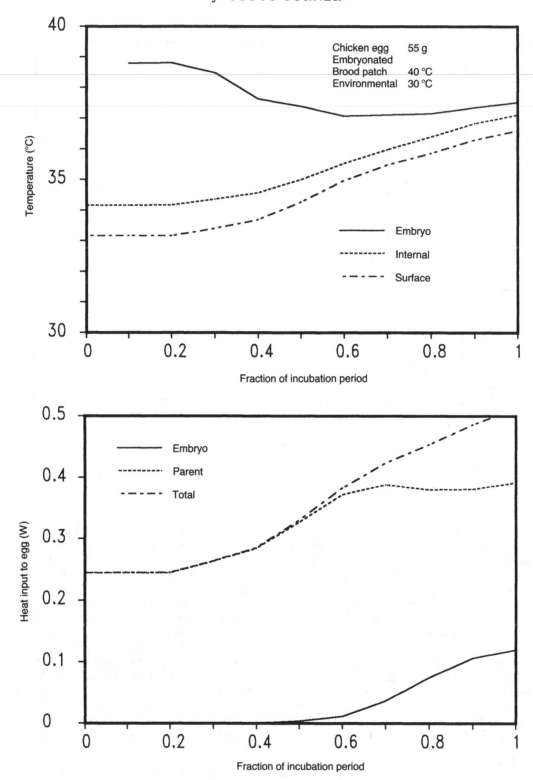

Fig. 9.12. The effect of incubation stage (defined as a fraction of incubation period ending with internal pipping) on egg temperature and heat input from brood patch, as predicted by the numeri-

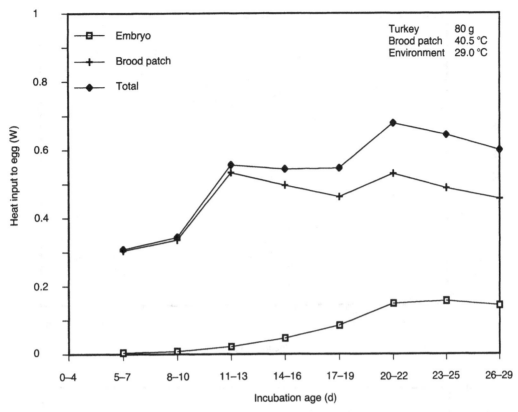

Fig. 9.13.   Observed incubation energy budget observed for living turkey (*Meleagris gallopavo*) eggs as measured by the artificial brood patch system. The brood patch covers roughly 20% of the egg's surface area. The brood patch temperature averaged roughly 40.5 °C. The environmental temperature averaged roughly 29 °C. Symbols are means of 10–12 replicates in each age group. *Open squares:* embryo heat production. *Crosses:* brood patch heat production. *Open diamonds:* Total heat input to egg.

temperature, and as incubation proceeds, it declines.

The numerical model predicts that energy demand on the parent will increase through incubation, not decrease (Fig. 9.12). During the first 20% of incubation, there should be no change in parental energy cost, during roughly the middle third of incubation, parental energy cost should increase and during the last third of incubation, when the embryo's rate of heat production becomes significant, the parent's energy cost should be roughly constant (Fig. 9.12). Note especially that total heat input to the egg

(embryo's plus parent's) *increases in parallel* with increasing embryo metabolism (Fig. 9.12).

I measured incubation energy budgets for turkey (*Meleagris gallopavo*) eggs through the incubation period (Figs 9.13 & 9.14), and the results agree with the numerical model. During the initial third of incubation, heat input from the artificial brood patch is low (Figs 9.13 & 9.14). As incubation proceeds, heat input from the brood patch first increases, and steadies during the latter half of the incubation period (Figs 9.13 & 9.14). These changes cannot be attributed to some structural change in the egg,

Fig. 9.12 *contd*
cal model. The egg is a 55 g fowl egg. The brood patch covers 15% of the egg's surface area. The brood patch temperature is 40 °C. The environmental temperature is 30 °C. *Top panel:* Average egg temperatures. *Solid line:* embryo temperature. *Broken line:* internal temperature. *Dashed and dotted line:* Exposed surface temperature. *Bottom panel:* Heat input to the egg. *Solid line:* heat production by the embryo. *Broken line;* heat input from the brood patch. *Dashed and dotted line:* Total heat input to the egg (embryo plus brood patch).

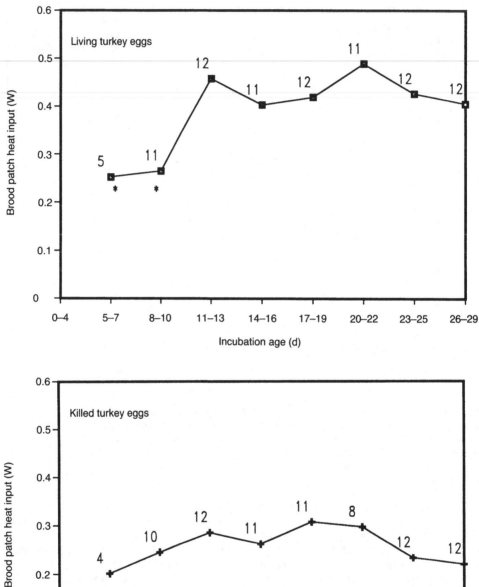

Fig. 9.14. Brood patch heat inputs for turkey eggs corrected to a brood patch temperature 10 °C warmer than environmental temperature. Symbols are the average for each age group. Number of replicates are indicated next to each point. *Top panel:* Brood patch heat inputs for living eggs. Asterisks indicate age groups that are significantly different ($p = 0.05$) from those without asterisks. *Bottom panel:* Brood patch heat inputs for the eggs once the embryos are killed.

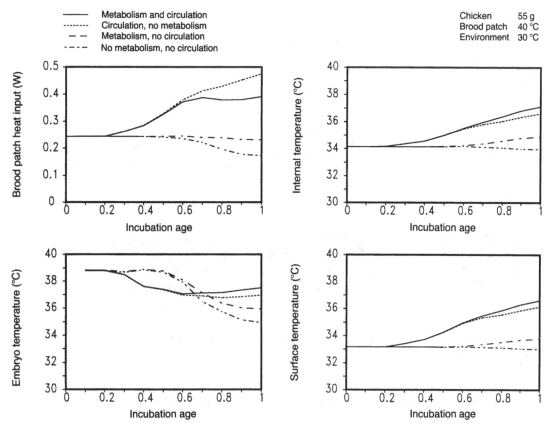

Fig. 9.15. The interaction of embryonic metabolism and circulation during the incubation period of a 55 g fowl egg, as predicted by the numerical model. The model considers four conditions: (1) Simulated egg, with metabolism and circulation (*solid line*); (2) Simulated egg, with metabolism and no circulation (*dashed line*); (3) Simulated egg, with circulation and no metabolism (*broken line*); (4) Simulated egg, with neither circulation nor metabolism (*dashed and dotted line*). The brood patch covers 15% of the egg's surface area. The brood patch temperature is 40 °C. The environmental temperature is 30 °C. *Top left panel:* The course of brood patch heat input. *Top right panel:* The course of average internal temperature. *Bottom left panel:* The course of average embryo temperature. *Bottom right panel:* The course of average internal egg temperature.

such as the development of the air cell, because the brood patch heat input for killed eggs is steady over the entire incubation period (Fig. 9.14). It is clear that the analogical approach cannot correctly predict how the energy costs for contact incubation change as the embryo develops. It is unclear whether the lumped conductance approach can do any better but the numerical model approach does clearly predict the pattern correctly.

The embryo's heat production and circulation influence the thermal energetics of the contact incubated egg in a complex way through incubation. Are there times when metabolism is more important than circulation or vice versa? By comparing simulated eggs with both metabolism and circulation with simulated eggs that have either, or neither, we may sort out these complexities.

Circulation in the egg appears to be an important influence on the energy cost for incubation of the parent, irrespective of incubation stage (Fig. 9.15). For the first 60% of the incubation period, the energy cost for incubating an egg with circulation but no metabolism is nearly identical to the cost for incubating an egg with both whereas the cost for the egg with metabolism alone is dissimilar (Fig. 9.15). During the last third of incubation, the influence of embryonic metabolism is greater but nevertheless it is secondary to that of the circulation. If the simulated egg has only circulation, the

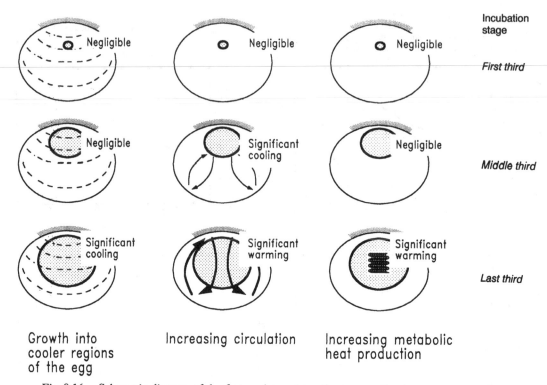

Fig. 9.16. Schematic diagram of the factors determining the course of embryo temperature in a fertile fowl egg as predicted by the numerical model. Full explanation is in the text.

parent's energy cost continues to rise through the entire incubation period but when the simulated egg has both metabolism and circulation, the parent's energy cost steadies through the last third of incubation (Fig. 9.15). Thus, metabolism appears to just compensate for the increased heat loss caused by the embryo's increasing circulation.

Circulation also appears to be most important in determining the egg's internal and surface temperatures (Fig. 9.15). The surface and interior temperatures of a simulated egg with circulation alone is similar to the simulated egg with both metabolism and circulation but when the simulated egg has metabolism alone, the temperatures are dissimilar (Fig. 9.15). While both the lumped conductance approach and the numerical model approach predict surface and interior egg temperatures that increase through incubation, it should be emphasised that the reasons for the similarity are very different: metabolism for the former, and circulation for the latter.

The predicted decline of average embryo temperature through incubation has a complex

explanation, involving the interplay of three factors: the position of the embryo in the egg, circulation, and metabolism (Fig. 9.16). The embryo usually occupies the upper portion of the egg, which happens to be warmer because it is closer to the brood patch. As the embryo grows, it cannot help but grow into cooler regions of the egg (Fig. 9.16). The consequence, even for an egg with neither circulation nor metabolism, is an average embryo temperature that cools as the embryo grows (Fig. 9.15). The major influence of embryo position, as shown by temperatures of the egg with neither circulation nor metabolism, occurs during the last third of the incubation period (Figs 9.15 & 9.16).

Circulation makes its mark on the course of embryo temperature from the time it begins to increase substantially, after the first third of incubation (Tazawa, 1980; Figs 9.15 & 9.16). Its influence on embryo temperature is in two phases. In the middle third of incubation, circulation cools the embryo, mostly because blood is flowing between the embryo and the membranes at the cooler surface of the egg (Fig. 9.15). In the last third of incubation, circulation appears to warm the embryo. This is mainly a secondary effect of the overall warming of the egg elicited by the circulation. The influence of metabolism on embryo temperature is secondary to, and moderates, the effects of circulation. With metabolism, the embryo is slightly warmer than it would be without but this is only significant in the last third of the incubation period (Figs 9.15 & 9.16).

Case 5: The effect of ambient temperature

Clearly, the lower the ambient temperature, the greater will be the rate of heat loss from the exposed surface of the egg, and the greater must the rate of heat input be from the brood patch to balance this loss. According to the analogical approach, the energy cost to the parent for incubation is a linear function of the difference between egg temperature and environmental temperature (equation 9.1). Where environmental temperature and egg temperature are equivalent, the parent's energy cost should be nil.

The numerical model also predicts that the energy cost for incubation is a linear function of environmental temperature (Fig. 9.17). However, the environmental temperature at which parental energy cost is nil is lower than brood patch temperature. Furthermore, the relative thermal importance of circulation and metabolism vary with ambient temperature in some complex ways.

Circulation and metabolism affect energy cost in two ways. If an egg has neither heat production nor circulation, parental energy costs are nil when environmental and brood patch temperatures are equal (Fig. 9.17), as predicted by equation 9.1. When circulation alone is present, this still is true, but now the egg's incubation conductance is greater, i.e. circulation has increased the energy *gain* for incubation (Fig. 9.17). Energy cost at any temperature (except where $T_e = T_b$) is, therefore, greater than it would be if circulation were absent. When metabolism alone is present, the environmental temperature at which parental energy cost is nil is now less than brood patch temperature (Fig. 9.17). Also, energy cost at any temperature is less (except where the egg must be cooled, i.e. $Q_i < 0$) than it would be if metabolism were absent (Fig. 9.17). Metabolism has changed the egg's energy *offset* for incubation.

When both circulation and metabolism are present, they independently affect both gain and offset of the energy balance of the egg during incubation: the consequence is a complex change in the thermal physiology of the egg at environmental temperatures close to brood patch temperature (Fig. 9.17). At an environmental temperature of 36.7 °C the opposing effects of metabolism and circulation exactly balance. At this temperature, removing both circulation and metabolism (equivalent to killing a living egg) should have no effect whatsoever on the parent's energy cost for incubation (Figs 9.17 & 9.18). When environmental temperature is less than 36.7 °C, killing the egg reduces the energy cost of incubating it, i.e. circulation dominates over metabolism and the lower the environmental temperature, the more circulation will dominate (Fig. 9.18).

At environmental temperatures exceeding 39.0 °C the parent must now dissipate heat to incubate the egg. More heat must be dissipated for an egg with circulation and metabolism than is required for an egg with metabolism only (Fig. 9.17). Thus, at high temperatures, circulation again dominates the egg's thermal energetics. Only from 36.7–39.0 °C, a range of 2.3 °C, can the metabolic rate of the embryo be said to dominate the egg's thermal energetics. Within this range, the closer environmental temperature approaches brood patch temperature, the more metabolism will dominate (Fig. 9.18).

Experimental measurements show results virtually identical to the predictions of the numerical model. As ambient temperature declines, the parental energy cost increases linearly (Fig. 9.19). The incubation conductance for the living egg is higher than it is for the killed egg. The energy offset for incubating a living fowl egg is negative with respect to that of a killed egg (Fig. 9.19). When environmental temperature is 35.5 °C, the contrary effects of metabolism and circulation balance (Figs 9.19 & 9.20). At cooler environmental temperatures, circulation dominates the incubated egg's thermal energetics (Figs 9.18 & 9.20). Once environmental temperature exceeds 38.5 °C, the extrapolated results show the parent must *dissipate* heat to incubate the egg and warmer environments should again make circulation important (Fig. 9.17).

Environmental temperature dramatically affects the thermal physiology of the contact incubated egg. At most environmental temperatures, embryonic circulation is the most

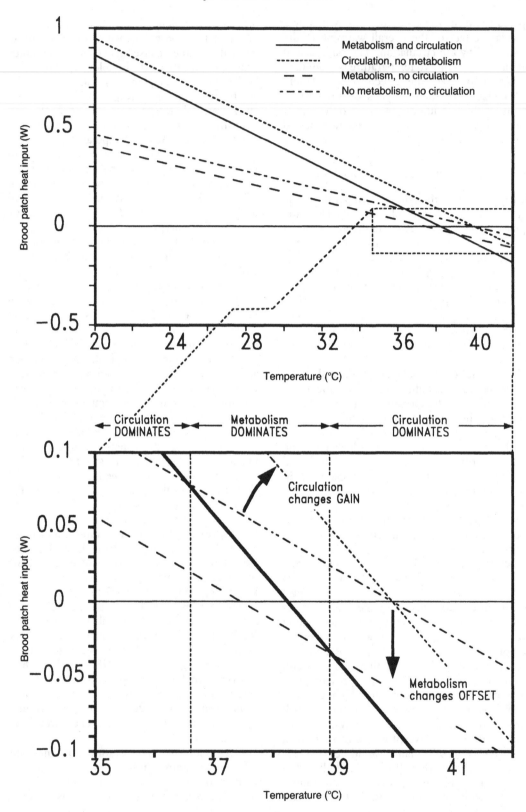

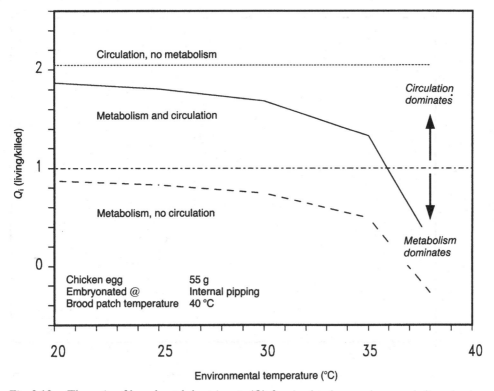

Fig. 9.18. The ratio of brood patch heat inputs (Q_i) for simulated eggs when metabolism dominates (metabolism, no circulation), when circulation dominates (circulation, no metabolism) and when both are present (metabolism and circulation), and its variation with ambient temperature. When the ratio is unity, the effects of circulation and metabolism offset one another.

important influence. But in a narrow range of environmental temperatures close to brood patch temperature, embryonic metabolism comes to the fore. No such thing is seen in convectively incubated eggs. These effects of ambient temperature have a straightforward explanation: as environmental temperature increases, egg temperature increases, and the egg becomes more uniform in temperature (Fig. 9.21). Circulation requires temperature gradients to transport heat. As increased environmental temperature makes temperature

gradients inside the egg disappear, then circulation will become less and less capable of altering flows of heat in the egg. This will leave metabolism as the only factor capable of influencing the egg's thermal energetics.

These results make it noteworthy that the simulations by Ackerman & Seagrave (1984) typically assume an air temperature in the nest only a few degrees less than brood patch temperature. This may partly explain why the lumped conductance approach and the numerical model approach come to such radically different con-

Fig. 9.17. Effect of environmental temperature on the brood patch heat input for incubating a 55 g fowl egg just prior to internal pipping. The model considers four conditions: (1) Simulated egg, with metabolism and circulation (*solid line*); (2) Simulated egg, with metabolism and no circulation (*dashed line*); (3) Simulated egg, with circulation and no metabolism (*broken line*); (4) Simulated egg, with neither circulation nor metabolism (*dashed and dotted line*). The brood patch covers 15% of the egg's surface area. The brood patch temperature is 10 °C. *Top panel:* Effect over the environmental temperature range 20–42 °C. *Bottom panel:* magnified picture of top panel from 35–42 °C, showing in more detail the complex changes in brood patch heat input at environmental temperatures near the brood patch temperature. Also, the independent effects of embryonic metabolism and circulation on the brood patch heat input are shown.

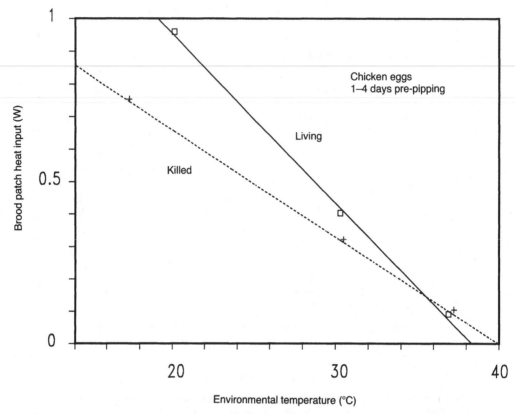

Fig. 9.19. Brood patch heat inputs experimentally measured for incubating fowl eggs at various environmental temperatures. Brood patch temperature is maintained at 40 °C. Symbols indicate means for groups of eggs. *Open squares:* Living fowl eggs, 1–4 days prior to external pipping. *Crosses:* Eggs after they have been killed. Lines indicate least-squares best fit to the means.

clusions regarding circulation. It cannot explain all the discrepancy, however, for even at environmental temperatures as low as 35 °C, the numerical model and experimental results show circulation to be at least as important as metabolism (Figs 9.19 & 9.20). It is likely that the three–node model of the thermal energetics of the egg (Ackerman & Seagrave, 1984) has somehow simplified away one of the most important thermal properties of a contact incubated egg: the influence of the embryo's circulation.

Discussion

The problem of how to estimate the energy cost of incubating eggs has been approached in many imaginative ways (Grant, 1984). Most of these approaches have involved simplified analyses of heat exchange between an egg and its incubating parent. It now appears that these approaches have not been adequate for the problem. For

example, the analogical approach of Kendeigh (1963), however appealingly simple or broadly applied, fails to predict accurately virtually any attribute of the energetics of contact incubation. The lumped conductance approach, even though it is more physically realistic, appears to simplify away some interesting, even fundamental, attributes of contact incubation. It is also clear that the energetics of contact incubation need to be analysed as the complex problem it really is. The numerical model approach accounts for this complexity and is able to successfully predict many attributes of contact incubation.

The importance of the embryo's circulation is one striking, and heretofore unappreciated, feature of contact incubation. Almost universally, the embryo's metabolism has been regarded as the most important influence on the thermal energetics of contact incubated eggs (Drent, 1970; Ackerman & Seagrave, 1984;

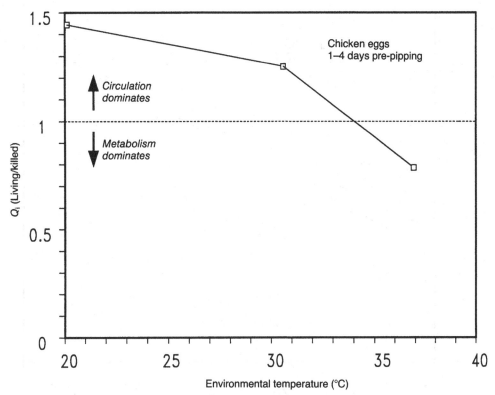

Fig. 9.20.   Ratio of brood patch heat inputs experimentally measured for living fowl eggs, 1–4 days prior to internal pipping, and for eggs after they have been killed. Brood patch temperature is maintained at 40 °C. When the ratio is unity, the effects of circulation and metabolism offset one another.

Grant, 1984). This is not generally true: except for a narrow range of environmental temperatures, embryonic circulation, not metabolism, plays the dominant role in the egg's temperatures and in the energy cost to the incubating parent for incubation. The failure to appreciate this has fostered many notions about incubation energetics that are manifestly false, e.g. that costs to the parent should decrease as the embryo grows and produces more heat. When modelled in appropriate detail, and when measured in living eggs, the opposite appears to be true.

Both the analogical approach and the lumped conductance approach have overlooked the important role of embryonic circulation for the same reason, although their respective paths to that reason differ. It is informative to explore what this reason is. The analogical approach has presumed contact incubated eggs to behave essentially as convectively incubated eggs. The developing circulation of the embryo has little influence on the thermal conductance of a con-

vectively incubated egg (Turner, 1987) because convectively incubated eggs are nearly isothermal (Sotherland *et al.*, 1987; Turner, 1987): the internal temperature gradients needed for circulation to transport heat simply are not there. The lumped conductance approach has missed the importance of circulation for the same reason. In varieties of this approach which presume the egg to be a single node (Walsberg & King, 1978*a,b*), an isothermal egg is axiomatic: nodes are, by definition, isothermal. Where the egg is distributed into a few nodes (Ackerman & Seagrave, 1984) temperature variation within the simulated egg conceivably could arise. Nevertheless, Ackerman & Seagrave's (1984) typical simulation assumes nest temperatures high enough to possibly render the egg isothermal. Again, because the temperature gradients necessary for circulation to transport heat are not there, circulation will not emerge as an important factor in these simulated eggs.

Nevertheless, naturally incubated eggs sustain

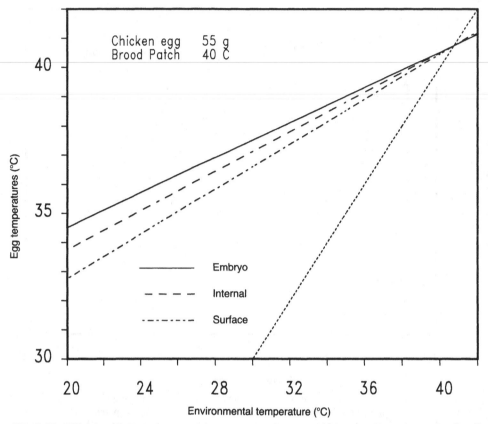

Fig. 9.21. Weighted average egg temperatures as environmental temperature varies as predicted by the numerical model. *Dotted line:* indicates equality between egg and environmental temperature. *Solid line:* indicates embryo temperature. *Dashed line:* indicates average internal temperature. *Dashed and dotted line:* indicates temperature of exposed surface.

large internal temperature gradients, even in very well insulated nests, where one might reasonably expect warm nest air. The egg of the eider (*Somateria mollissima*), for example, sustains a temperature difference of nearly 9 °C between the top and bottom (Rahn *et al.*, 1983). Large temperature gradients in contact incubated eggs are evident for a variety of other eggs and nest types (Grant *et al.*, 1982; Vleck *et al.*, 1983; Swart *et al.*, 1987). To the extent that heat flow in embryonic blood depends upon these temperature differences, so circulation will play an important (and heretofore unappreciated) role in the egg's thermal energetics.

The predicted course of embryo temperature during incubation is another radical difference between contact incubated and convectively incubated eggs. During incubation, the embryo's metabolic heat production warms a convectively incubated egg, and the embryo along with it (Tazawa *et al.*, 1988). The numerical model predicts that the combination of a number of factors, including temperature gradients within the egg, the position in the embryo within the egg, and the metabolism and circulation of the embryo, should result in the embryo actually *cooling* during incubation (Fig. 9.15).

Fig. 9.22. How visitation times and absence times for intermittent incubators might be affected by embryonic development. *Dashed lines:* represent expected rates of temperature change at the start of incubation. *Solid lines:* represent expected rates of temperature change at the end of incubation. *Top panel:* Egg temperature vs time absent from the nest. *Bottom panel:* Egg temperature vs time visiting the nest.

Start of incubation ----------
End of incubation ——————

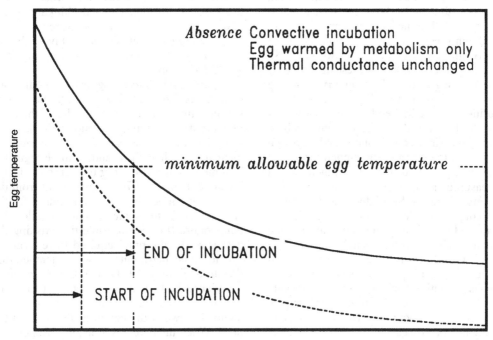

Absence Convective incubation
Egg warmed by metabolism only
Thermal conductance unchanged

Egg temperature

minimum allowable egg temperature ----

END OF INCUBATION

START OF INCUBATION

Time absent from nest

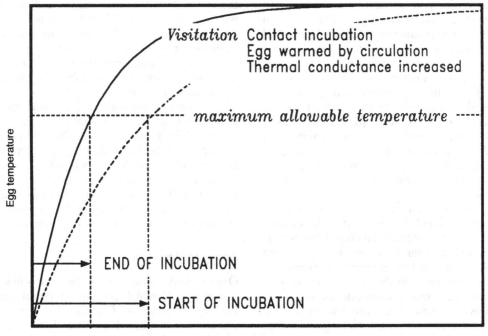

Visitation Contact incubation
Egg warmed by circulation
Thermal conductance increased

Egg temperature

maximum allowable temperature ----

END OF INCUBATION

START OF INCUBATION

Time visiting nest

Egg temperatures of naturally incubated eggs are usually assessed by implanting one, or at most a few, temperature sensors in an egg, which a bird is then allowed to incubate. Often, the egg is a 'dummy egg' of some sort, which is usually filled with some inert substance although some dummy eggs are of exceptionally clever design (Buttemer, Astheimer & Dawson, 1988). The course of temperature of these dummy eggs is often thought to reflect embryo or egg temperature in some way. The validity of this supposition may be limited by three factors now known to be crucial in determining embryo temperature. First, contact incubated eggs will have no such thing as an 'egg temperature', only an egg temperature distribution, which cannot be measured adequately by one or a few temperature sensors. Secondly, a static array of thermocouples inside a dummy egg will not reflect the changing distribution of embryo mass within the egg. Thirdly, a single dummy egg does not reflect the complex changes in heat flow elicited by a living embryo's circulation and metabolism. Without taking these factors into account, temperatures of dummy eggs can do little to inform us about embryo temperature in naturally incubated eggs.

Dummy eggs also are sometimes used as experimental tools to measure energy costs of incubation. For example, a dummy egg might be perfused with water of different temperatures (Toien et al., 1986; Vleck, 1981). The change in metabolic heat production of the incubating bird is then measured to assess the energy cost of incubating the egg. In a living egg, circulation is very important in determining this cost. Consequently, if the water-perfused dummy is vigorously circulated with water, which is common practice, the cost to the parent is likely to be greatly exaggerated. To the extent that experimenters err on the side of vigorous perfusion of a dummy egg used in this way, so will they err in overestimating the energy cost of incubation.

In nature, most avian eggs are contact incubated. Overall, it appears that the thermal physiology of the contact incubated egg is radically different from the convectively incubated egg. To understand the energetics of incubation, the fundamental attributes of the contact incubated egg must be taken into account. Most previous approaches to this problem seem not to have done this, and current thinking on incubation energetics may require some revision. For example, incubating eggs appears to be energetically much cheaper than we have previously thought. Additionally, energy cost to the parent should increase as incubation proceeds, not decrease as the conventional thinking would have it. Finally, embryo temperature in contact incubated eggs actually should decline slightly as the embryo matures: conventional thinking would have it increase.

A particularly interesting case in point concerns birds that intermittently vacate their nests for short time periods (on the order of the egg's time constant) to forage, defend territories, and so forth. The visitation schedules of intermittent incubators such as these appear to be regulated in some way by rates of egg temperature change (White & Kinney, 1974). Time absent from the nest is determined by the time required for the egg to cool to some minimum allowable temperature (Fig. 9.22), while time visiting the nest is set by the time required to rewarm the egg to some maximum allowable temperature (White & Kinney, 1974). These eggs routinely experience both contact incubation (when the parent is sitting on the nest) and convective incubation (when the parent is absent from the nest). When the parent is away, the temperature of the egg as it cools should therefore be influenced only by the embryo's metabolic rate. As the embryo grows and produces more heat, the egg will stay warmer longer, and absence times should increase. When the parent is on the egg, circulation will be the main influence determining the egg's rate of rewarming. As the embryo grows and circulates more blood, the egg should warm more quickly and visitation times should decrease.

To conclude, the thermal physiology of the contact incubated egg is radically different from the convectively incubated egg. Trying to understand how contact incubation works, either as an analogy with convectively incubated eggs, or by simplifying the problem excessively, has not proved fruitful. The process of contact incubation is complex, and understanding it requires an approach that recognises this complexity.

Acknowledgements

Original work reported in this paper was funded in part by a National Institutes of Health grant PO1-HL-28542 to Charles V. Paganelli, Department of Physiology, State University of New York at Buffalo, and in part by a Foundation for Research Development Comprehensive Research Support Grant to W. Roy Siegfried, University of Cape Town.

References

Ackerman, R. A. & Seagrave, R. C. (1984). Parent-egg interactions. Egg temperature and water loss. In *Seabird Energetics*, eds G. C. Whittow & H. Rahn, pp. 73–88. New York: Plenum Press.

Ar, A., Girard, H. & DeJours, P. (1987). Oxygen consumption of the chick embryo's respiratory organ, the chorioallantoic membrane. *Respiration Physiology*, **68**, 377–88.

Barott, H. G. (1937). *Effect of Temperature, Humidity and Other Factors on Hatch of Hen's Eggs and on Energy Metabolism of Chick Embryos.* United States Department of Agriculture Technical Bulletin 553. Washington DC.

Buttemer, W. A., Astheimer, L. B. & Dawson, T. J. (1988). Thermal and water relations of emu eggs during natural incubation. *Physiological Zoology*, **61**, 483–94.

Drent, R. (1970). Functional aspects of incubation in the herring gull. *Behaviour* Supplement, **17**, 1–132.

— (1973). The natural history of incubation. In *The Breeding Biology of Birds*, ed. D. S. Farner, pp. 262–320. Washington DC: National Academy of Sciences, USA.

— (1975). Incubation. In *Avian Biology*, vol. 5, ed. D. S. Farner & J. R. King, pp. 333–420. London: Academic Press.

Grant, G. S. (1984). Energy cost of incubation to the parent bird. In *Seabird Energetics*, eds G. C. Whittow & H. Rahn, pp. 59–71. New York: Plenum Press.

Grant, G. S., Pettit, T. N., Rahn, H., Whittow, G. C. & Paganelli, C. V. (1982). Water loss from Laysan and Black-footed albatross eggs. *Physiological Zoology*, **55**, 405–14.

Hoyt, D. F. (1987). A new model of avian embryonic metabolism. *Journal of Experimental Zoology*, Supplement 1, 127–38.

Kendeigh, S. C. (1963). Thermodynamics of incubation in the House wren, *Troglodytes aedon. Proceedings of the XIIIth International Ornithological Congress*, 884–904.

Rahn, H., Krog, J. & Mehlum, F. (1983). Microclimate of the nest and egg water loss of the Eider, *Somateria mollissima* and other waterfowl in Spitsbergen. *Polar Research*, **1**, 171–83.

Romanoff, A. L. (1960). *The Avian Embryo. Structural and Functional Development.* New York: Macmillan Press.

Romijn, C. & Lokhorst, W. (1960). Foetal heat production in the fowl. *Journal of Physiology (London)*, **150**, 239–49.

Sotherland, P. R., Spotila, J. R. & Paganelli, C. V. (1987). Avian eggs: Barriers to the exchange of heat and mass. *Journal of Experimental Zoology* Supplement 1, 81–6.

Spotila, J. R., Weinheimer, C. J. & Paganelli, C. V. (1981). Shell resistance and evaporative water loss from bird eggs: Effects of wind speed and egg size. *Physiological Zoology*, **54**, 195–202.

Swart, D., Rahn, H. & de Kock, J. (1987). Nest microclimate and incubation water loss of eggs of the African ostrich (*Struthio camelus* var. *domesticus*). *Journal of Experimental Zoology* Supplement, **1**, 239–46.

Tazawa, H. (1980). Oxygen and CO_2 exchange and acid-base regulation in the avian embryo. *American Zoologist*, **20**, 395–404.

Tazawa, H., Wakayama, H., Turner, J. S. & Paganelli, C. V. (1988). Metabolic compensation for gradual cooling in developing chick embryos. *Comparative Biochemistry and Physiology*, **89A**, 125–9.

Toien, O., Aulie, A. & Steen, J. B. (1986). Thermoregulatory responses to egg cooling in incubating bantam hens. *Journal of Comparative Physiology*, **156B**, 303–7.

Turner, J. S. (1985). Cooling rate and size of birds' eggs. A natural isomorphic body. *Journal of Thermal Biology*, **10**, 101–4.

— (1987). Blood circulation and the flows of heat in an incubated egg. *Journal of Experimental Zoology* Supplement, **1**, 99–104.

— (1990). The thermal energetics of an incubated chicken egg. *Journal of Thermal Biology* (in press).

Vleck, C. M. (1981). Energetic cost of incubation in the zebra finch. *Condor*, **83**, 229–37.

Vleck, C. M. & Vleck, D. (1987). Metabolism and energetics of avian embryos. *Journal of Experimental Zoology* Supplement, **1**, 111–25.

Vleck, C. M., Vleck, D., Rahn, H. & Paganelli, C. V. (1983). Nest microclimate, water vapor conductance, and water loss in heron and tern eggs. *Auk*, **100**, 76–83.

Walsberg, G. E. & King, J. R. (1978a). The heat budget of incubating mountain white-crowned sparrows (*Zonotrichia leucophrys oriantha*) in Oregon. *Physiological Zoology*, **51**, 92–103.

— (1978b). The energetic consequences of incubation for two passerine species. *Auk*, **95**, 644–55.

Webb, D. R. & King, J. R. (1983). An analysis of the heat budgets of the eggs and nest of the white-crowned sparrow *Zonotrichia leucophrys* in relation to parental attentiveness. *Physiological Zoology*, **56**, 493–505.

West, G. C. (1960). Seasonal variation in the energy balance of the tree sparrow in relation to migration. *Auk*, **77**, 306.

White, F. N. & Kinney, J. L. (1974). Avian incubation. *Science*, **186**, 107–14.

10

Physiological effects of incubation temperature on embryonic development in reptiles and birds

DENIS C. DEEMING AND MARK W. J. FERGUSON

Introduction

Generally, the incubation temperature of bird eggs is conservative: within a species there is little variation in incubation temperature at which normal development can proceed. By contrast, the incubation temperature of oviparous reptiles is relatively labile; normal patterns of development in individual embryos can ensue at a wide range of temperatures. In addition, the incubation temperature of avian eggs is usually higher than for reptile eggs which, as a group, have a much wider range of viable incubation temperatures. Average incubation temperatures for birds are tabulated by Rahn (Chapter 21) and comprehensive reviews of the thermal tolerances of avian embryos have been prepared by Drent (1975) and Webb (1987). Comparable data for reptiles are available for turtles (Ewert, 1979, 1985; Miller, 1985a), crocodilians (Ferguson, 1985) and squamates (Hubert, 1985).

Incubation temperature is very important in determining rates of embryonic growth and development and to a large extent the length of the incubation period. It also has other effects, as yet predominantly observed in reptiles. Incubation temperature determines sex in many species of reptile and also affects the pigmentation pattern of hatchlings, post-hatching growth rates and moulting cycles as well as thermoregulatory and sexual behaviour patterns. These topics are reviewed for reptiles, using primarily the American alligator (*Alligator mississippiensis*) as the example, but the possible effects of temperature on avian development are also examined. In addition, other aspects of development, such as heart rate, studied in bird embryos

but not yet in reptiles, are described. We suggest that all of these diverse processes may be linked by a common regulatory gene cascade and the development, and embryonic setting, of the hypothalamus.

Effects of incubation temperature on embryonic growth and development

Incubation period

In reptiles, a continuous increase in incubation temperature does not produce a simple linear pattern of decrease in the duration of incubation: as temperature increases the length of incubation shortens but at higher temperatures there is much less of an effect. This is illustrated well by eggs of the lizard *Dipsosaurus dorsalis* (Fig. 10.1). An increase in incubation temperature from 28 °C to 32 °C shortens the incubation period by 27 days but an increase from 32 °C to 36 °C only reduces the incubation period by 11 days (Muth, 1980), whilst an increase from 36 °C to 40 °C increases the incubation period by one day (Fig. 10.1). This pattern is seen in both oviparous lizards (*Sceloporus undulatus*) and snakes (e.g. *Pituophis melanoleucus*), viviparous lizards (*Sceloporus jarrovi*) and many turtles (including *Chelonia mydas, Caretta caretta, Chelydra serpentina, Chrysemys picta* and *Trionyx sinensis*) (Sexton & Marion, 1974; Yntema, 1978; Miller & Limpus, 1981; Miller, 1985a; Choo & Chou, 1987; Gutzke & Packard, 1987; Gutzke *et al.*, 1987; Packard *et al.*, 1987; Beuchat, 1988; Mrosovsky, 1988). Compared with many reptiles, crocodilians have a relatively restricted range of incubation temperatures, 28–34 °C (Ferguson, 1985), but

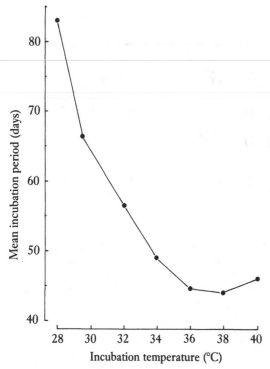

Fig. 10.1. The effect of incubation temperature upon the length of the incubation period for eggs of the lizard *Dipsosaurus dorsalis* (Data from Muth, 1980).

incubation temperature has similar effects on the length of incubation as in other reptiles (Hutton, 1987; Joanen, McNease & Ferguson, 1987; Webb *et al.*, 1987; Lang, Andrews & Whitaker, 1989; Webb & Cooper-Preston, 1989).

There is, however, a strong genetic component to the length of the incubation period in reptiles. Eggs of *A. mississippiensis* have a short, relatively invariable (at a fixed incubation temperature) incubation period relative to other crocodilians (Ferguson, 1985). This is a characteristic of the species and is considered to reflect its northerly range (Deeming & Ferguson, 1990a). In the turtle *C. serpentina*, the length of incubation at constant incubation temperature, correlates with geographical latitude (Ewert, 1985). Eggs obtained from northerly populations (high latitudes between 40–45° N) take significantly less time to develop than those at more southerly latitudes (25–30° N). In addition, the difference in incubation period between constant temperature incubation at 25 °C and 30 °C decreases with increasing latitude:

20 days at 30° N and 13 days at 40° N (Ewert, 1985). Australian chelid turtles from cooler climates also have shorter incubation periods under constant conditions than those from tropical and sub-tropical latitudes (Legler, 1985).

The effect of transient exposure of developing eggs of the lizard *S. undulatus* to cold temperatures is to prolong the incubation period, normally 40 days at 30 °C (Christian, Tracy & Porter, 1986). Exposing eggs after 15 days of incubation at 30 °C, to various lengths of time at 15 °C before returning them to the normal incubation temperature, prolongs the incubation period by the number of days of exposure. Up to 4 days at 15 °C does not affect embryonic mortality, exposure to the low temperature for nine days results in 50% mortality whereas exposure for 40 days produces 100% mortality. Embryos become more tolerant to cold exposure as they develop: up to day 32 of incubation exposure to 15 °C for 5 days increases the length of incubation by the same period but thereafter, low temperature has less effect and transfer of eggs after 38 days, for the rest of development, only increases the incubation period by 1–2 days (Christian *et al.*, 1986).

In birds, incubation period is more vigorously fixed by both genetic mechanisms and the uniform incubation conditions of temperature and humidity (Rahn, Chapter 21). In the fowl (*Gallus gallus*), a reduction of 1–2 °C can increase the length of incubation by 1–2 days but any greater reduction, or an increase of 1–2 °C, in temperature causes mortality of embryos before they hatch (Romanoff, Smith & Sullivan, 1938). Incubation period appears to be labile, however, in several species of seabird, including the fork-tailed storm petrel (*Oceanodroma furcata*), diving petrel (*Pelecanoides georgicus*) and Xantus' murrelet (*Synthliboramphus hypoleucus*) (Boersma & Wheelwright, 1979; Wheelwright & Boersma, 1979; Boersma *et al.*, 1980; Vleck & Kenagy, 1980; Murray *et al.*, 1983; Roby & Ricklefs, 1984). In these species, which are typically pelagic feeders, incubation is often interrupted when the adult leaves to forage and the eggs are left to cool to ambient temperatures (Boersma, 1982). These periods of neglect extend the length of incubation by almost 1.5 times the number of days of chilling (Boersma & Wheelwright, 1979). This increase in incubation period is presumably due to a reduction in development rate: an egg of *O. furcata* set in an incubator at 37 °C hatches in half the incubation period of eggs under natural incubation conditions with an incubation temperature of 34 °C (Vleck & Kenagy, 1980). Unlike the fowl embryo, which is more tolerant to low temperatures early in incubation (Tazawa & Rahn, 1986), embryos of *P. georgicus* are tolerant of long periods of low temperatures at the end of incubation (Roby & Ricklefs, 1984) but the thermal tolerance of embryos of this species throughout development is unknown.

Assessment of growth patterns

One assessment of how different incubation conditions affect embryonic development is that of growth, either in mass or in length. Growth of embryonic reptiles is commonly considered to be accelerated at high incubation temperatures, but, in squamates, only limited data are available for embryonic mass for the lizard *Lacerta vivipara* (Holder & Bellairs, 1962; Maderson & Bellairs, 1962) and the snake *Python molurus* (Vinegar, 1973). Body length and carapace length have been used to show the rate of development of *C. serpentina* at 20 °C and 30 °C (Yntema, 1968). The best data set for growth of a reptilian embryo at different incubation temperatures is for *A. mississippiensis* (Deeming & Ferguson, 1988, 1989a; Joss, 1989). At 33 °C, the rate of increase in mass and total body length is clearly greater than at 30 °C (Fig. 10.2).

In the fowl, departure from the normal incubation temperature affects embryonic growth rate (Romanoff *et al.*, 1938). Suboptimal incubation temperatures retard embryonic growth from the start of incubation (Fig. 10.3) and temperatures below 34.5 °C cannot support normal growth so that the embryos die early in incubation (Romanoff *et al.*, 1938; Romanoff, 1960). High incubation temperatures are also deleterious to embryonic development and retard growth (Fig. 10.3; Romanoff *et al.*, 1938).

The pattern of embryonic development of birds and reptiles is often described as a series of morphological stages. Tables of stages are available for the fowl (Hamburger & Hamilton, 1951), crocodilians (Ferguson, 1985), turtles (Yntema, 1968; Miller, 1985a) and squamates (Hubert, 1985). Advantages of assessing stage of development include standardisation of embryos in experiments, prediction of incubation time elapsed in the field and comparison of embryos from different incubation conditions, or from

different but closely related species (Deeming & Ferguson, 1989a, 1990a). As incubation temperature increases, the rate of embryonic development also increases; hence, the rate of development at different temperatures can be assessed by determining the time it takes for each embryo to reach a particular stage at each temperature (Yntema, 1978; Pieau & Dorizzi, 1981; Webb *et al.*, 1987). However, this use of stage of development as an assessment of the rate of embryonic development has its pitfalls.

Staging relies upon the presence, or absence, of a suite of characters, of which one or two are often critical (Ferguson, 1985; Miller, 1985b). However, if alterations in incubation temperature result in asynchronous development (Deeming & Ferguson, 1988, 1989a), then stage assignment becomes more difficult as some of the characters within a stage suite may conflict. Moreover, staging is not a uniform scale. At the onset of development, when there are often rapid changes in external embryonic morphology, the real time within and between stages is short. However, during the second half of incubation there are fewer changes in external morphology and the period for each stage is longer in real time. The embryo is also growing at these stages and differences in incubation temperature may have a major effect (Deeming & Ferguson, 1989a), which would be obscured by the relative insensitivity of the later staging criteria. Moreover, stage based comparisons between embryos from even closely related species (Deeming & Ferguson, 1990a), which usually differ significantly in the later stages of development, may tell us more about the arbitrary construction of the staging scale than real differences in development between these species. It is for these reasons that staging scales in mammals, e.g. man, are only constructed and described, for the embryonic period, when there is rapid external development; assessment of development and growth in the later fetal period is by a morphometric assay, e.g. crown–rump, or crown–heel length (O'Rahilly & Muller, 1987). Indeed, in mammals the different terms 'embryonic' and 'fetal' are used to distinguish these periods. Following the redefinition of these terms, and their loss of association with anything related to placentation, there is no reason why these terms should not be applied to birds and reptiles. The end of the embryonic period proper is defined in man when 90% of the named adult structures (4500 in man) have

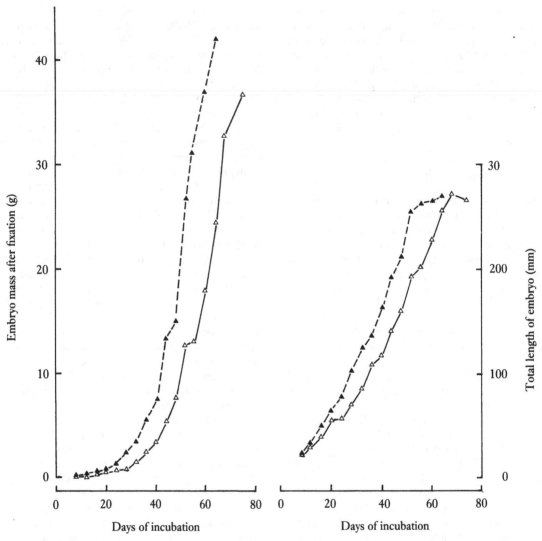

Fig. 10.2. The effects of incubation temperature on growth of embryos of *Alligator mississippiensis* incubated at 30 °C at 33 °C (open and closed symbols respectively). The graphs show the increase in total mass (*left*) and total length of the embryos (*right*) against days of incubation (Data from Deeming & Ferguson, 1989*a*).

appeared (O'Rahilly & Muller, 1987): a similar definition could be applied to birds and reptiles.

Methods of estimating relative growth rates

Development rate coefficient (DRC) has been used as a method of assessing relative rates of development at different incubation temperatures (Yntema, 1978; Webb *et al.*, 1987). The principle of a DRC can be illustrated using embryos of *A. mississippiensis* incubated at 33 °C. Development through time

of embryos incubated at 30 °C and 33 °C is assessed using stage of development. The time ('morphological age') at which each stage of development occurs for embryos incubated at 33 °C is expressed as the number of days that embryos at 30 °C take to reach the same stage. The relationship between 'morphological age' and real age (in days) for embryos at 33 °C is then analysed using a linear regression method which forces the line through the origin. The development rate coefficient at 33 °C, DRC_{33}, is equal to the slope of the regression line (Webb *et*

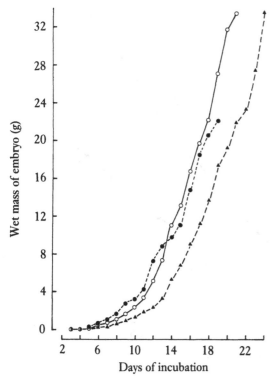

Fig. 10.3. The effects of incubation temperature on the increase in wet mass through time for embryos of the domestic fowl (*Gallus gallus*). Values are means for embryos incubated at 35.5 °C (triangles), 37.5 °C (open circles) and 39.5 °C (solid circles). Embryos at 39.5 °C failed to develop to hatching (Data from Romanoff, Smith & Sullivan, 1938).

al., 1987); by definition DRC_{30} is unity. Using this analysis, development of alligator embryos is accelerated at an incubation temperature of 33 °C by a factor (i.e. DRC_{33}) of 1.209 (± S.D. of 0.019) relative to incubation at 30 °C (Deeming & Ferguson, 1989*a*). This method has been used to assess relative rates of growth at various viable incubation temperatures for embryos of several crocodilians (*Crocodylus johnstoni*, *Crocodylus porosus* and *Crocodylus palustris*) and confirms the obvious relationship between incubation temperature and rate of development (Webb *et al.*, 1987; Lang *et al.*, 1989).

The use of the Development Rate Coefficient is, however, a poor assessment of embryonic development for three major reasons. First, the method relies upon the subjective assessment of embryonic stage and the arbitrary staging scale (see earlier). Secondly, temperature significantly affects the size of alligator embryos at equivalent morphological stages (Deeming & Ferguson,

1989*a*) and the effect is not always uniform, i.e. increased temperatures do not always mean accelerated growth. For example, embryos of *C. porosus* incubated at 33 °C apparently develop faster than at 34 °C (Webb *et al.*, 1987). In *A. mississippiensis*, embryos incubated at 33 °C are smaller between stages 18 and 22 than embryos incubated at 30 °C but they show rapid growth between stages 22 and 25 (Deeming & Ferguson, 1989*a*). During incubation at 30 °C growth is more sustained, with maximal growth occurring during stages 23 to 27 (Deeming & Ferguson, 1989*a*). Not only does the DRC obscure these differences but it also over-estimates the size of embryos at 33 °C between stages 18 to 22. Thirdly, the DRC at any temperature is an average of the rate of development through time. In *A. mississippiensis*, the relationship between time and rate of differentiation is clearly not linear (Deeming & Ferguson, 1989*a*). The DRC gives no indication of any changes in the rates of growth through incubation and forces everything into a linear relationship.

Much better assessments of embryonic growth rates are based on mathematical analyses of changes in embryonic dimensions through time. This approach involves modelling of growth patterns; a computer program, which fits seven non-linear models commonly used to describe growth patterns (W. G. Bardsley, unpublished observations), has been used to provide the best mathematical model for describing the pattern of growth of alligator embryos, in terms of mass and various length measurements (Deeming, Bardsley & Ferguson, unpublished data). The program reveals that the Gompertz growth model best fits the data for embryonic mass during incubation, whereas the Logistic model best fits the data for total embryonic length.

The Gompertz model is described by the equation:

$$\frac{dS}{dt} = k \cdot S \cdot (\log A - \log S) \qquad (10.1);$$

whereas the Logistic model has the equation:

$$\frac{dS}{dt} = (k/A) \cdot S \cdot (A - S) \qquad (10.2).$$

In both cases: dS/dt is the rate of growth of size S in time t; k is a constant; and A is the maximum size of the animal. The computer program

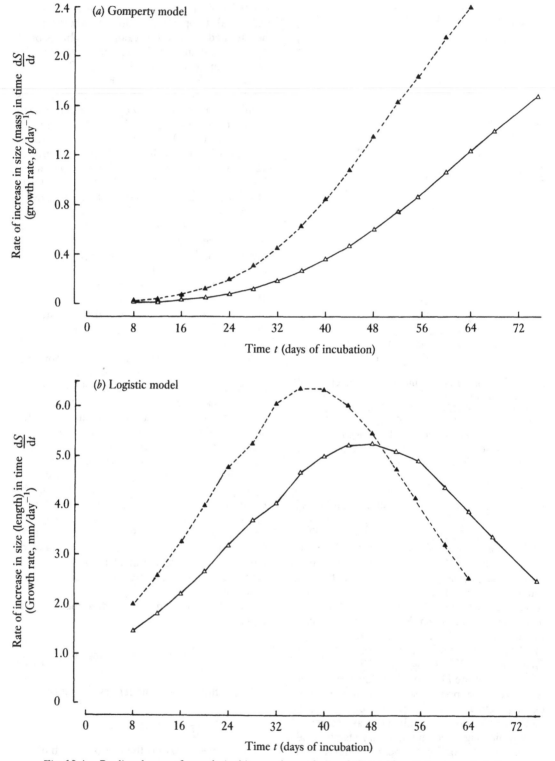

Fig. 10.4. Predicted rates of growth, in (*a*) mass (g per day) and (*b*) total length (mm per day), for embryos of *Alligator mississippiensis* incubated at 30 °C and 33 °C (open and solid symbols respect-

generates predicted values for A, k and the theoretical size S' at each time point for each model. Using this information an estimate of the rate of growth $(\frac{dS}{dt})$ can be calculated for each time point at both temperatures and plotted (Fig. 10.4). The changes in growth rate in embryonic mass and total length can be assessed quickly: the rapid rate of growth in mass of alligator embryos incubated at 33 °C compared with 30 °C, and how this increases through incubation, are clear (Fig. 10.4). The rate of growth in embryonic length at 33 °C, compared with 30 °C, is greater earlier in incubation and the maximum rate of growth is higher (Fig. 10.4).

Utilisation of egg components and formation of extra-embryonic fluids

Data on the effects of incubation temperature on the utilisation of the albumen and yolk by most reptile embryos are lacking. The pattern of albumen utilisation by crocodilian embryos is influenced by incubation temperature (Manolis, Webb & Dempsey, 1987; Deeming & Ferguson, 1989a): in A. mississippiensis, albumen is utilised at a constant rate at 33 °C but at 30 °C it is utilised at a slower rate during the middle third of incubation (Fig. 10.5). The amount of residual yolk in hatchlings is affected by incubation temperature in both turtles (C. serpentina and C. picta) and crocodilians (C. johnstoni and A. mississippiensis). The longer incubation periods at low temperatures mean that less residual yolk is present at hatching than at higher incubation temperatures (Fig. 10.5) (Gutzke et al., 1987; Packard et al., 1987; Manolis et al., 1987; Deeming & Ferguson, 1989a). In alligators, the pattern of yolk lipid utilisation is also affected by incubation temperature (Noble et al., 1991).

Data for the effects of incubation temperature on formation of extra-embryonic fluids in reptile eggs are almost limited to A. mississippiensis (Fig. 10.5). At 33 °C, the amount of sub-embryonic fluid both peaks and decreases earlier than in eggs at 30 °C (Deeming & Ferguson, 1989a). This pattern also occurs in C. johnstoni (Manolis

et al., 1987) and is repeated for allantoic fluid and amniotic fluid in A. mississippiensis (Fig. 10.5).

In the fowl, incubation temperature affects the utilisation of albumen and yolk, and formation of the extra-embryonic fluids (Romanoff et al., 1938; Romanoff, 1943; Romanoff & Hayward, 1943; Deeming, 1989). An increase in incubation temperature initially accelerates utilisation of yolk and albumen (and a decrease in temperature slows utilisation), but the adverse effects of temperature on embryonic development disrupt the normal pattern of utilisation later in incubation (Romanoff et al., 1938; Romanoff, 1943). Formation of sub-embryonic fluid is affected in the same way but at both low and high temperatures this fluid persists into the later stages of incubation (Romanoff, 1943). An incubation temperature of 39.5 °C induces rapid formation of allantoic fluid, at 34.5 °C formation is retarded but, in both cases, the volume of fluid is greatly reduced compared with eggs at 37.5 °C (Romanoff et al., 1938; Romanoff & Hayward, 1943). Formation of amniotic fluid in eggs incubated at 34.5 °C is retarded compared to eggs at 37.5 °C but the amount of fluid is similar. At 39.5 °C two peaks, before and after the normal, are observed for amniotic fluid (Romanoff & Hayward, 1943).

Effects of temperature on the metabolism, heart rate and pulmonary rate of avian embryos

The effects of incubation temperature on embryonic metabolism, cardio-vascular and pulmonary function are unknown for almost all reptiles. Oxygen consumption ($\dot{V}O_2$) has been recorded for embryos of C. johnstoni during incubation at 29 °C and 31 °C (Whitehead, 1987). At 31 °C, $\dot{V}O_2$ increases more rapidly and reaches a higher peak earlier in incubation (Whitehead, 1987). There is much more information on the effects of temperature on the physiology of avian embryos which is due, in part, to an interest in the pattern of development of endothermy during incubation. Work by Tazawa and his colleagues has shown that there are metabolic restrictions on the avian embryo during late incubation which limit its ability to

Fig. 10.4 contd
ively) using data from Deeming & Ferguson (1989a). Plots were calculated by a computer which fitted the Gompertz (a) and Logistic (b) models of growth to the data (Deeming, Bardsley & Ferguson, unpublished observations).

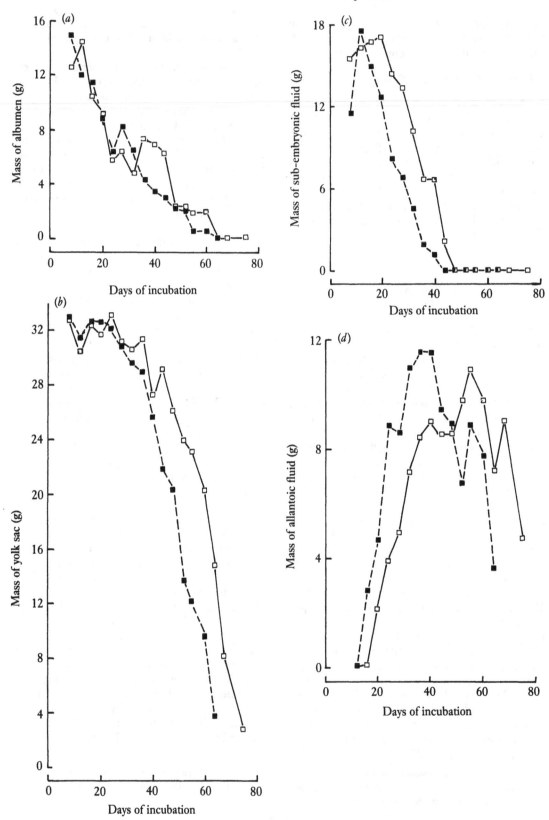

become endothermic, and that there are differences between altricial and precocial species. A hypothesis to explain these results relies on the metabolic limitations of embryos by the eggshell but space does not allow any major discussion of this hypothesis; readers are directed to Tazawa *et al.* (1988) for further details.

Measurement of $\dot{V}O_2$ of avian embryos during imposed hypothermia indicates their metabolic responses to cooling of the egg (Freeman, 1967; Tazawa, 1973; Nair, Baggott & Dawes, 1983; Booth, 1987; Tazawa & Rahn, 1987; Tazawa *et al.*, 1988, 1989*a,b*; Matsunaga *et al.*, 1989). Under normal incubation conditions (38 °C) 11 day-old fowl embryos can raise egg temperature by 0.2 °C but by internal pipping (20 days) they can increase egg temperature by 2 °C. However, when exposed to 28 °C 11 day-old embryos can only raise the egg temperature by 0.1 °C and at 20 days by 1.2 °C. After pipping and during hatching egg temperatures can be raised by 6–10 °C above ambient by increasing $\dot{V}O_2$ (Tazawa & Rahn, 1987). Prior to day 17 of incubation fowl embryos are poikilothermic; $\dot{V}O_2$ decreases as incubation temperature falls (Tazawa *et al.*,

1988) and at 24 °C is maintained at 40% of the normal resting metabolism of embryos at 38 °C (Tazawa & Rahn, 1987). However, from day 17 to the time of internal pipping there is some metabolic compensation: $\dot{V}O_2$ is unaffected by decreases in temperature to about 35 °C, below which there is no compensation (Tazawa *et al.*, 1988). After external pipping $\dot{V}O_2$ increases in response to cooling. This metabolic compensation is abolished by injection of thiourea, which inhibits thyroid function, and by low oxygen tensions in the incubator, but can be enhanced by incubation in 100% oxygen (Tazawa *et al.*, 1989*a*).

Values for Q_{10} are used to assess metabolic compensation in response to temperature. For temperature changes from 38 °C to 36 °C, Q_{10} for oxygen consumption of 16 day-old fowl embryos is about 1.25 but this increases to 1.5 at 32 °C and reaches a value of 2 at 28 °C (Nair *et al.*, 1983; Tazawa *et al.*, 1989*b*). As the embryo develops the physiological effects of temperature change and at external pipping Q_{10} at 36 °C is 0.8–1.0, increasing to 1.75–1.0 at 28 °C (Tazawa *et al.*, 1989*b*). In the emu (*Dromaius novaehollandiae*), Q_{10} is lower at temperature changes from 36–38 °C compared with 30–34 °C but Q_{10} decreases as incubation proceeds, nearing a value of 1 at hatching (Buttemer, Astheimer & Dawson, 1988).

Acute hypothermia, exposure to 30 °C for 2 hours after 16 days of incubation, affects the blood chemistry of the fowl embryo. Decreases in temperature reduce oxygen consumption and carbon dioxide production which lowers the carbon dioxide tension (pCO_2), but increases the oxygen tension (pO_2) of the blood (Freeman, 1967; Tazawa, 1973; Tazawa & Mochizuki, 1978; Nair, Baggott & Dawes, 1984). These changes make the blood more alkaline (Tazawa, 1973; Nair *et al.*, 1984) but the effects on plasma bicarbonate are equivocal. Tazawa (1973) reports that HCO_3 concentration increases during acute hypothermia but Nair *et al.* (1984) report that it is decreased; more data are required to resolve this issue. Acute hypothermia also causes a drop in blood sugar and hepatic stores of glycogen but the free fatty acid titre of the blood of fowl embryos increases (Freeman, 1967). Brown adipose tissue, a significant source of heat for the neonatal mammal, is absent in the fowl hatchling (Freeman, 1967).

The heart rate of avian embryos is highly sensitive to incubation temperature (Romanoff & Sochen 1936; Oppenheim & Levin, 1975; Ben-

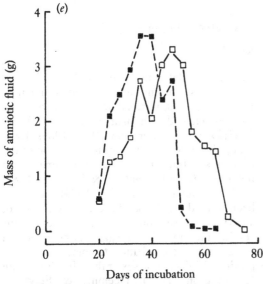

Fig. 10.5. The effect of incubation temperature upon the amount of the extra-embryonic components of eggs of *A. mississippiensis* during development at 30 °C and 33 °C (open and closed symbols respectively). Values are means, normalised to a mean egg mass of 72.19 g, and plotted against days of incubation: (*a*) albumen, (*b*) yolk sac, (*c*) sub-embryonic fluid, (*d*) allantoic fluid and (*e*) amniotic fluid (Data from Deeming & Ferguson, 1989*a*).

nett & Dawson, 1979; Deeming & Tullett, unpublished observations). Within a few minutes of a decrease in temperature there is a rapid fall in heart rate (Fig. 10.6) which exhibits an exponential pattern (Romanoff & Sochen, 1936; Tazawa & Nakagawa, 1985; Tazawa & Rahn, 1986). When the temperature is restored to normal, the cardiac response is immediate (Fig. 10.6; Tazawa & Rahn, 1986). The thermal tolerance of the heart changes with development

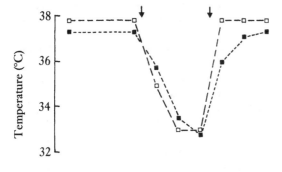

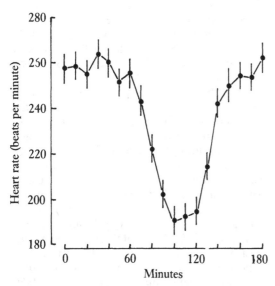

Fig. 10.6. The response of embryonic heart rate to lowered incubator temperature in pheasant (*Phasianus colchicus*) embryos after 85% of the incubation period. The upper graph shows the air temperature read from the incubator dial (open squares) and the actual temperature recorded inside an infertile egg (solid squares). The arrows indicate the point at which the set temperature of the incubator was lowered by 5 °C and then returned to normal (37.8 °C). The lower graph shows the heart rate with standard error bars (Deeming & Tullett, unpublished observations.)

(Romanoff & Sochen, 1936). Young embryos are more tolerant to thermal stress: 6 day-old fowl embryos can survive 26 hours at 8 °C but as the embryo ages this ability diminishes to the point that 20 day-old embryos can only survive 8 hours exposure (Tazawa & Rahn, 1986). A 3-hour exposure to 27.2 °C reduces both the systolic and diastolic blood pressure of 12 day fowl embryos (Tazawa & Nakagawa, 1985). As incubation proceeds, blood pressure increases and up to day 17, low temperature has adverse effects but on day 18 of incubation, a lowered temperature no longer affects blood pressure (Tazawa & Nakagawa, 1985).

Ambient temperature also affects the breathing rates of externally pipped embryos. In the fowl, domestic quail (*Coturnix coturnix japonica*) and domestic duck (*Anas platyrhynchos*) a lowering of ambient temperature decreases the frequency of breathing but increases the amplitude of each breath (Oppenheim & Levin, 1975; Nair & Dawes, 1980; Dawes, 1981; Nair *et al.*, 1983; Spear & Dawes, 1983). Warming of the egg, above the normal incubation temperature has the reverse effect: breaths become more frequent but shallower (Oppenheim & Levin, 1975; Dawes, 1979).

Sex determination

Reptiles

Sex can be determined in vertebrates by two mechanisms: chromosomal and environmental. In many reptiles, incubation temperature is the environmental determinant of sex (Bull, 1980, 1983; Deeming & Ferguson, 1988, 1989*b*; Lang *et al.*, 1989). Reports that water potential of the nesting substrate affects sex determination in the turtle *C. picta* (Gutzke & Paukstis, 1983; Paukstis, Gutzke & Packard, 1984) have now been discredited (Packard, Packard & Birchard, 1989) whilst restriction of gas and water vapour conductance in alligator eggs has no effect on the sex ratio at incubation temperatures which produce 100% of one sex or the other (Deeming & Ferguson, 1991). Deeming & Ferguson (1988) review the occurrence of temperature-dependent sex determination in reptiles, and Ewert & Nelson (1990) provide significantly more data for a large number of turtle species.

Three patterns of temperature dependent sex determination (TSD) can be recognised (Fig. 10.7). For type I, which is typified by many tur-

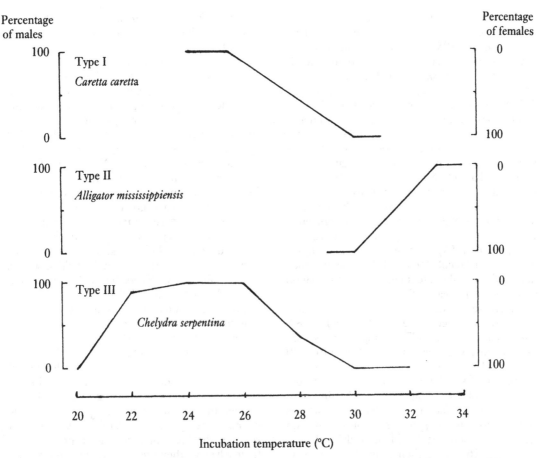

Fig. 10.7. Three patterns of temperature-dependent sex determination that can be recognised in reptiles. Each graph indicates the percentage of each sex against different incubation temperatures within the viable range (indicated by the limits of the line) for the species shown. (Data for *Caretta caretta* from Limpus, Reed & Miller, 1985; data for *Alligator mississippiensis* from Joanen, McNease & Ferguson, 1987; Data for *Chelydra serpentina* from Yntema, 1976.)

tles, females develop at higher temperatures than those which induce development of males (e.g. *C. caretta*) (Bull & Vogt, 1979, 1981; Deeming & Ferguson, 1988). For type II, typified by some crocodilians (e.g. *A. mississippiensis*) and a few lizards (e.g. *Eublepharis macularius*), male determining temperatures are higher than female determining temperatures (Warner, 1980; Ferguson & Joanen, 1982, 1983; Bull, 1987*a,b*). For type III, females are produced at low and high incubation temperatures with males at intermediate temperatures. This pattern has been observed in a lizard (*Gekko japonicus*), some crocodilians (e.g. *C. johnstoni*) and some turtles (e.g. *C. serpentina*) (Yntema, 1976, 1979, 1981; Webb & Smith, 1984; Tokunga, 1985; Webb *et al.*, 1987; Deeming & Ferguson,

1988). Type III is likely to be the ancestral pattern, the other two patterns being derived by selection (possibly through embryo survivorship) deleting either high or low temperature females (Deeming & Ferguson, 1988).

Many other lizards (e.g. *Lacerta viridis*) and turtles (e.g. *Emydura macquarii*) exhibit genetic chromosomal sex determination (GSD) (Raynaud & Pieau, 1972; Olmo, 1986; Deeming & Ferguson, 1988; Thompson, 1988; Ewert & Nelson, 1990). TSD has not been reported for any species of snake (Deeming & Ferguson, 1988, 1989*b*) but incubation temperature affects differential embryonic mortality in the snake *P. melanoleucus* (Burger & Zappalorti, 1988). Differentiated sex chromosomes are absent in crocodilians but present in various species of turtles, many lizards and all snakes as either ZW/ZZ or XY/XX systems (Cohen & Gans, 1970; Olmo, 1986). However, the apparent presence of differentiated sex chromosomes in the lizard *G. japonicus* (Yoshida & Msahiro, 1974) which also exhibits TSD (Tokunga, 1985) suggests that the presence of sex chromosomes does not necessarily mean the absence of TSD and provides further evidence for the multiple and recent evolution of sex chromosomes in reptiles, particularly squamates. It also suggests that TSD and GSD may not be so different in molecular mechanisms.

The phenomenon of TSD has been extensively studied in *A. mississippiensis* (Ferguson & Joanen, 1982, 1983; Deeming & Ferguson, 1988, 1989*b*, 1991) and will serve as an example. Alligator eggs incubated at 30 °C or below produce 100% female hatchlings, but as incubation temperature increases the proportion of males also increases so that, at 33 °C and above, 100% males are produced (Ferguson & Joanen, 1982, 1983; Joanen *et al.*, 1987). Temperature switch experiments, where eggs are transferred from 30 °C to 33 °C and vice versa, appear to show that the timing of sex determination depends on the original incubation temperature. The timing of sexual differentiation in alligator embryos appears to vary depending upon incubation temperature between 14–21 days of incubation (stages 14–16) for a switch from 30 °C to 34 °C and between 28–35 days (stages 21–23) for a switch from 33 °C to 30 °C (Ferguson & Joanen, 1983; Deeming & Ferguson, 1989*b*). It would seem at first sight, that sex is determined later at the higher incubation temperature despite the more rapid rate of development generally, and of the gonad in particular (Deeming & Ferguson, 1988, 1989*a*). Double switch experiments, where eggs from one incubation temperature (e.g. 33 °C) are exposed to the other (e.g. 30 °C) for only 7 days, reveal that it is more difficult to induce development of males than it is to induce development of females (Deeming & Ferguson, 1988, 1989*b*). Embryos do not respond in a switch-like fashion at a precise developmental time but rather the effects of incubation temperature appear more integrative and cumulative. These double shift experiments have led to a reinterpretation of the so-called 'temperature sensitive periods' derived from single switch experiments (Deeming & Ferguson, 1988, 1989*a*).

Similar experiments on the turtles *C. serpen-* *tina* and *Emys orbicularis* also suggest that it is easier to induce female development (Yntema, 1979; Pieau & Dorizzi, 1981). Moreover, if turtle eggs are cycled between 100% male and 100% female producing incubation temperatures, such that the net effect is that they are equivalent to developing at an intermediate constant incubation temperature, then the sex ratio of the offspring is identical between the cycled eggs and those incubated at the constant incubation temperature (Ewert & Nelson, 1990) further arguing for the integrative effects of incubation temperature on sex determination. Such findings make good sense in view of the varying daily patterns of temperature within wild nests of many reptiles (Webb & Smith, 1984; Packard *et al.*, 1985). Nest map experiments confirm that TSD operates in the field as well as in the laboratory (Ferguson & Joanen, 1982, 1983; Webb & Smith, 1984).

Adult sex ratios for crocodilians with TSD tend to be biased (often heavily) towards females (Ferguson & Joanen, 1982, 1983; Webb & Smith, 1984; Deeming & Ferguson, 1988). This is not difficult to explain from a developmental standpoint: females are produced in higher frequencies at a larger range of incubation temperatures than males. Moreover, in alligators, there is evidence that females select nest sites on the basis of temperature probably for themselves as they are ectothermic and remain around the nest site for most of the long incubation period (Ferguson 1985). As the temperature of egg incubation determines the preferred adult thermoregulatory temperature (see later), the small percentage of females which select warm nest sites, that subsequently produce males, probably come from eggs which were at the higher temperatures for female development (e.g. 32°C). As the percentage of females hatching from eggs incubated at higher incubation temperatures is low compared with those hatching at lower incubation temperatures (e.g. 30 °C) this serves to reinforce and stabilise the female bias in the adult sex ratio. Moreover, it will also be evident that eggs from higher temperature females subsequently produce more grandsons than their low temperature cohorts: such considerations may be important in an evolutionary context.

Our hypothesis to explain a molecular basis for TSD in *A. mississippiensis* is here summarised but discussed in depth in Deeming & Ferguson (1988, 1989*b*). Female is assumed to

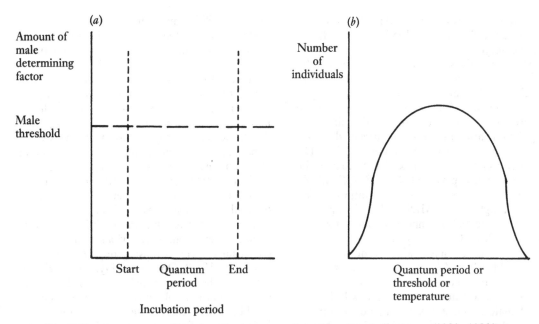

Fig. 10.8. A summary of the hypothesis proposed by Deeming & Ferguson (1988, 1989*b*) to explain temperature-dependent sex determination. (*a*) Male development is dependent upon the production of a male determining factor (MDF) above a threshold level during a quantrum of the incubation period. If these conditions are not met, then the embryo will default into a female. (*b*) Within a population of embryos there will be variation in the length of the quantum period, and/or the height of the male threshold and/or the optimal temperature for the production of sufficient MDF to cross the threshold.

be the default state in *A. mississippiensis* (and possibly in all species with TSD). Development of males is reliant upon important criteria which have to be met during incubation. Specifically, development of a testis depends upon the production of sufficient amounts of a male determining factor (MDF) during a critical quantum period of embryonic development (Fig. 10.8). Transcription, splicing and translation of the gene encoding for MDF and the activity of the gene product itself may be temperature sensitive (in some combination). Individual embryos vary: in the amount of MDF they require to cross the male determining threshold, in the length of time they require exposure to MDF, and in the sensitivity of transcription, splicing, translation and activity of MDF to temperature. However, at 33 °C sufficient active MDF is produced at the right time in all embryos, for each embryo to cross the male threshold, and therefore, all embryos develop testes and are male. As incubation temperature decreases, however, transcription and translation of the MDF gene, and/or activity of the gene product, are reduced. Therefore, in some embryos, the conditions for

male determination are not met and these embryos default into females. As the incubation temperature decreases further, more and more embryos fail to attain the male threshold and at 30 °C no embryo is able to produce sufficient active MDF at the right time to develop into a male.

The hypothesis predicts that the process of male determination is integrative over a prolonged quantum of developmental time (during the first 40 days of incubation in *A. mississippiensis*). Diurnal fluctuations in incubation temperature, or days when ambient temperature falls below the optimum, do not have a drastic effect on the process of sex determination. The hypothesis also places great importance on natural variation in the requirements for maleness (and the sensitivity to temperature) in a population of embryos. Each embryo has individual requirements, in the length of the quantum period of incubation and the threshold for MDF. In addition, the optimal temperature for maximal expression of the male determining factor will vary between embryos. Therefore, the 50:50 sex ratio which is produced at an intermediate temperature, around 31.5 °C in *A. mississippiensis* (Joanen *et al.*, 1987), may be due to some embryos having a low threshold for MDF, and/or a low exposure time requirement, and/or a low optimal temperature for transcription, splicing and translation of the MDF gene and/or activity of its product. Single switch experiments merely determine the limits of population variation in the quantum of developmental time and/or the amount of MDF required to develop maleness. In the case of the double switch experiment, from 33–30–33 °C, embryos which need a long quantum period and/or a high dose of MDF will be disrupted by a 7-day reduction in temperature and so default into females. Conversely in 30–33–30 °C switches some embryos, which require a small dose of MDF and/or a short quantum of time, will attain these conditions and so develop into males: most will not, however, and will develop into females. Hence it is easier to default into a female than to meet the specifications for male development (Deeming & Ferguson, 1988, 1989*b*).

More complex models of sex determination, e.g. optimal synthesis and activity of MDF at one temperature and either an agonist or an antagonist at another incubation, can also explain the current experimental data.

The identity, and function, of MDF has yet to be ascertained but genes similar to the presumptive testis determining factor, *ZFY* (Page *et al.*, 1987) have been found in the genome of both sexes in alligators and other reptiles, with either TSD or GSD (Deeming & Ferguson, 1988; Bull, Hillis & O'Steen, 1988). We have further characterised alligator *ZFY* by gene sequencing (it is 91% homologous at the amino acid level to human *ZFY*) and by studying its expression during embryonic development (Sharpe, Ferguson & Valleley, unpublished data). *ZFY* appears to be on the sex determining gene cascade but may not be the master gene. Instead, it may be involved in regulating growth, setting of the hypothalamus (see later) and spermatogenesis (Deeming & Ferguson, 1988, 1989*b*; Palmer *et al.*, 1989; Koopman *et al.*, 1989; Burgoyne, 1989).

Birds

Birds exhibit genetic sex determination with female sex chromosome heterogamety (ZW/ZZ) in over 100 species (Bloom, 1974;

Shields, 1987). Interestingly, classical genetic analysis suggests that avian sex determination is caused by the number of Z chromosomes (i.e. a dose) and not the presence of a W chromosome (Sittman, 1984). However, in species of ratite (e.g. the ostrich, *Struthio camelus*), no heteromorphic sex chromosomes can be identified (Bloom, 1974; Shields, 1987) suggesting that these species may represent an early evolutionary step in the morphological differentiation of avian sex chromosomes. The ratites are phylogenetically primitive birds, more closely related to reptiles in a number of characteristics than other birds such as the fowl.

The uniform incubation conditions of birds (brooding) and mammals (viviparity) likely selected against TSD in these groups (Deeming & Ferguson, 1988). One possibility, however, is to search for the remnants of TSD in birds, but the narrow thermal tolerances of avian embryos (Webb, 1987) complicate investigations of a wide range of temperatures. Moreover, the effects of incubation temperature, other than the optimal, on sexual differentiation of avian embryos have not been extensively studied. A reduction of incubation temperature to 22 °C after 72 hours of incubation affects the protein and nucleic acid metabolism of embryos of the fowl on day 20 of incubation (Vedeneeva, 1970; Shubina, Zhmurin & Vedeneeva, 1972). This is associated with an increase in the proportion of male hatchlings: 150 male embryos are produced for every 100 females (Shubina *et al.*, 1972). This experiment has yet to be repeated so its full significance cannot be assessed. An increase in incubation temperature, to 40.5 °C from day 6 of incubation, leads to retention of the Mullerian ducts by male fowl and quail embryos but has no effect on the gonads (Stoll, 1944, 1948; Lutz-Ostertag, 1966; Weniger & Thiebold, 1981). Mullerian duct retention is reversible and is unrelated to production of anti-Mullerian hormone by the embryonic testis (Weniger & Thiebold, 1981). Retention of the duct is considered to be associated with a reduced sensitivity of the duct to anti-Mullerian hormone (Weniger & Thiebold, 1981) or an inactivation of proteolytic enzymes responsible for the destruction of the duct (Salzgeber, 1959).

Like reptiles, birds also possess *ZFY* genes and type H–Y positive in the female (Page *et al.*, 1987; Nakumara *et al.*, 1987). The gonadal/extra-gonadal H–Y status of reptiles with TSD, e.g. *E. orbicularis*, other turtles and *A. mississip-*

piensis often conflict (Engel, Klemme & Schmid, 1981; Zaborski, Dorizzi & Pieau, 1982, 1988; Wellins, 1987) and is used as evidence for a theory that TSD in reptiles represents an environmental overriding (sex reversal) of a weak genetic ZW/ZZ sex determining mechanism (Engel *et al.*, 1981; Zaborski *et al.*, 1982, 1988).

Incubation temperature, developmental abnormalities and pigmentation patterns in reptiles

Developmental abnormalities associated with incubation temperature are common in birds and oviparous reptiles and can occur spontaneously or be induced (Romanoff *et al.*, 1938; Landauer, 1967; Romanoff, 1972; Ewert, 1979; Bellairs, 1981; Ferguson, 1985; Miller, 1985*a*). Incubation temperatures at the extremes of the viable range for development commonly induce a high proportion of malformations in both birds and reptiles (Romanoff *et al.*, 1938; Romanoff, 1972; Ewert, 1979; Ferguson, 1985; Gutzke *et al.*, 1987; Gutzke & Packard, 1987; Burger, Zappalorti & Gochfeld, 1987). For example, high incubation temperatures induce herniation of the viscera in fowl embryos (Ancel & Lallemand, 1941) whereas, in the snake *P. melanoleucus*, low temperatures induce a high incidence of ventral hernia and many hatchlings leave unabsorbed yolk in the eggshell (Burger *et al.*, 1987). Further information on the effects of incubation temperature on avian development can be obtained from Romanoff (1972). Typical malformations of reptilian embryos include those of the central nervous system and eyes (Yntema, 1960; Ferguson, 1985; Burger *et al.*, 1987), a loss of the lower jaw (Webb & Messel, 1977; Webb *et al.*, 1983), facial clefting (Ferguson, 1985) and limb anomalies (Yntema, 1960; Ferguson, 1985).

High incubation temperatures disrupt normal development of the vertebral column inducing loss of the tail or malformations of the spine (Yntema, 1960; Bustard, 1969, 1971; Vinegar, 1973, 1974; Webb *et al.*, 1983; Ferguson, 1985; Burger *et al.*, 1987). In the viviparous snake *Natrix fasciata* developmental temperature affects the number of vertebrae: both low and high temperatures induce increases in the number of vertebrae in the trunk region (Osgood, 1978). In both *P. molorus* and *N. fasciata*, incubation temperature also affects head scalation

(Vinegar, 1974; Osgood, 1978) but in *P. melanoleucus* temperature has no effect (Burger *et al.*, 1987). Similar effects occur in fish, where the number of vertebrae and dorsal, anal and pectoral fins are determined by larval temperature during a particular period of development (Taning, 1952).

Incubation temperature has been reported to have adverse effects on pigmentation of hatchlings. In the turtle *Graptemys pseudogeographica*, postocular striping patterns are influenced by incubation temperature. At 25 °C a *Graptemys kohni*-type pattern is found in 59 specimens out of a total of 181, but at 30 °C eggs from the same clutches only yield 5 out of 165 *G. kohni*-type specimens (Ewert, 1979). The intensity of the dark stripes in *G. pseudogeographica* is also affected by incubation temperature: paler stripes occur at 25 °C. In addition, the integrity of the dark plastral figure is reduced at 25 °C compared with 30 °C due to the lack of invasion of dark colour (Ewert, 1979). In *P. molorus* an incubation temperature of 27.5 °C can not support normal development through to hatching but the low temperature affects the pigmentation pattern of near-term embryos compared with hatchlings from the incubation temperature of 30 °C (Vinegar, 1973, 1974). Dark stripes on these embryos are paler at the lower temperature and stripes are more prevalent than the more normal blotches observed in these snakes (Vinegar, 1974).

These effects of incubation temperature on pigmentation patterns may reflect, however, the relative rates of growth of reptilian embryos and how these effect the formation of pigmentation patterns. In *A. mississippiensis*, the number of white stripes on the dorsal surface of hatchlings is greater in those individuals incubated at 33 °C (20 stripes) than those incubated at 30 °C (18 stripes), despite being the same length (Deeming & Ferguson, 1989*b*). This is not a sexual effect, however, for both male and female hatchlings derived from eggs incubated at the same temperature have the same number of white stripes (Murray, Deeming & Ferguson, 1990). Rather, the number of stripes may be determined by the length of the individual embryo at the time that the pigmentation pattern is established. It is the geometry of the embryo which determines the skin patterning (Murray *et al.*, 1990). Whilst the mechanism which produces these patterns is unknown, the pattern can be predicted mathematically by a mechanochemical model which describes the interactions between cells and the extra-cellular matrix in which they move. This model has been successfully tested using measurements of alligator embryos at different stages of development at 30 °C and 33 °C (Murray *et al.*, 1990).

Temperature during development also affects pigmentation of amphibians. In the frog *Pseudacris ornata*, there is a higher incidence of brown morphs with increasing temperature during larval development (Harkey & Semlitsch, 1988). In the leopard frog, *Rana pipiens*, the number of spots on an animal is related to temperature during larval development (Davison, 1964; Browder & Davison, 1964). The coloration of atlantic eels (*Anguilla rostrata* and *A. anguilla*) also correlates with larval temperature (Tucker, 1959). Insect larvae incubated at different temperatures also yield different colour morphs and patterns (Hudson, 1966; Kettlewell, 1973; Nijhout, 1980). Thus variation in reptilian colour patterns may not simply reflect genetic traits but rather the incubation history of the individual animal. Such effects have yet to be reported in birds.

Post-hatching effects of incubation temperature

Physiological effects

The post-hatching growth rates of crocodilians are affected by incubation temperature. In *A. mississippiensis*, hatchling mass and length are unaffected by incubation temperature. However, post-hatching alligators, grown in constant environmental chambers set at 30 °C with a controlled nutritional regimen, exhibit different growth rates according to their incubation history (Joanen *et al.*, 1987). Animals from the extreme incubation temperatures (29.4 °C and 32.8 °C) did not grow as fast as those animals from intermediate temperatures (30.6 °C and 31.7 °C). At these intermediate incubation temperatures, both males and females develop and the two sexes have different growth patterns. After 18 months post-hatching growth, female hatchlings from an incubation temperature of 30.6 °C are heavier than males from this incubation temperature but are of similar length (Fig. 10.9). By contrast, for animals from 31.7 °C incubation conditions males are both heavier and longer by 12 months of age (Fig. 10.9). Animals from either sex seem

to grow better when they are incubated at intermediate temperatures within the range at which they can develop: 30.6 °C for females and 31.7 °C for males (Joanen *et al.*, 1987; Webb & Cooper-Preston, 1989).

Hatchlings of *Crocodylus niloticus* from 34 °C incubation conditions are shorter in length than those at 28 °C and 31 °C although there are no differences in hatchling mass (Hutton, 1987). However, by 3 months of age under (variable) farming conditions those animals from 34 °C have outgrown their counterparts from 31 °C and 28 °C; animals from 28 °C are 20 mm shorter than those from 34 °C (Hutton, 1987). In *C. porosus*, hatchling mass and length are unaffected by incubation temperature but the size (snout–vent length) of females at 2 years of age increases as incubation temperature increases; data for males are limited but they are larger than females (Webb & Cooper-Preston, 1989).

Skin-shedding in *Pituophis melanoleucus* occurs within 10–14 days after hatching and the time to first shedding is related to both ambient temperature (Semlitsch, 1979) and incubation temperature (Burger *et al.*, 1987; Burger, 1989). Those snakes from low incubation temperatures (21 °C or 23 °C) shed their skin earlier than snakes from intermediate (26 °C or 28 °C) or high incubation temperatures (32 °C). If the post-hatching maintenance temperature is increased this pattern is maintained although the time to first shedding is uniformly shortened (Burger *et al.*, 1987).

Behavioural effects

The thermal preferences of juvenile crocodiles are influenced by their incubation temperature. Hatchlings from eggs from a single clutch of *Crocodylus siamensis*, split between a male determining (32.5–33 °C) and a female determining incubation temperature (27.5–28 °C), show markedly different post-hatching thermoregulatory behaviours (Lang, 1987). In a thermal gradient, the males, from the higher incubation temperature, select a higher preferred body temperature than females from the lower incubation temperature and this effect persists for at least 60 days post-hatching (Lang, 1987). A sexual basis for this observation cannot be ruled out but it may indicate that differences in preferred body temperatures between individuals, and between species (Huey, 1982), are not solely genetic traits (Huey & Webster, 1976) but may be physiologically acquired traits established during incubation. These experiments need repeating on a larger scale with a full range of incubation temperatures including those that produce both males and females. These results have profound implications for understanding the ecology and evolution of TSD: perhaps it is only the small percentage of females hatched from higher incubation temperatures which subsequently select warmer environments in which to nest thereby producing predominately male offspring and maintaining the female biased sex ratio.

Egg incubation temperature also affects the behaviour of *P. melanoleucus* hatchlings. In a series of laboratory tests under controlled conditions, Burger (1989) has shown that the locomotor ability of hatchling and juvenile snakes is affected by incubation temperature. Low incubation temperatures (21 °C or 23 °C) produce hatchlings which seem less able to cope with their environment. They are less active and drink water at a slower rate than hatchlings from intermediate incubation temperatures (26 °C or 28 °C). They take longer to move across a fixed distance and are poorer at maintaining balance. In addition, the predator response of hatchlings is different: low incubation temperature hatchlings are more likely to remain motionless when danger approaches whereas hatchlings from higher temperatures flee when the danger becomes acute. Snakes from high incubation temperatures (32 °C) are better than intermediate temperature animals in drinking rate but in many behavioural tests they are less adept and are comparable with snakes from low incubation temperatures (Burger, 1989).

The behaviour of adults of *E. macularius*, a lizard with TSD (Warner, 1980; Bull, 1987a,b; Gutzke & Crews, 1988), has been studied in relation to their incubation temperature. Females that develop at a high incubation temperature (32 °C) are more aggressive to both males and females than are females from lower incubation temperatures (26 °C and 29 °C). Similarly males produced at a higher temperature (32 °C) are more aggressive than are males from a lower incubation temperature (29 °C). High temperature females have not been observed to mate or produce eggs in 2 years (Gutzke & Crews, 1988) and their aggressive stance to males effectively makes them functionally sterile. The titre of oestrogens in both sexes is unaffected by incubation temperature but in males at 29 °C the titre of androgens is higher than in males at 32 °C. In females, as incubation temperature increases the titre of androgens is elevated although the levels do not approach those of adult males at either incubation temperature (Gutzke & Crews, 1988).

The possible effects of incubation temperature on hypothalamic control of physiological processes

In conclusion, it is interesting to speculate upon how incubation temperature affects these many different processes both during development and post-hatching. Temperature may have direct effects upon cells and tissues during incubation but the fact that incubation temperature can be correlated with aspects of an animal's physiology after hatching implies that incubation temperature is irreversibly affecting the structure and function of some organs in the embryo. A likely target organ to be 'set' is the hypothalamus: perhaps temperature determines the pulsatile release of hypothalamic releasing hormones (Deeming & Ferguson, 1988, 1989b).

The hypothalamus secretes a variety of peptide releasing hormones most of which act on the pituitary to cause the release of pituitary hormones which in turn feedback on the hypothalamus. In adults, these neuroendocrine feedback loops are important in determining: the structure and function of the reproductive organs, via follicle stimulating hormone and luteinising hormone; pigmentation, via melanocyte stimulating hormone; metabolic rate and thermal set point, via thyroid stimulating hormone; growth, via growth hormone; skin sloughing, via sex and thyroid hormones; and behaviour, particularly sexuality via the sex hormones. It is clear that all of the effects so far described for egg incubation temperature in reptiles: sex, embryonic and post-hatching growth, pigmentation patterns, thermal preference, skin sloughing, behaviour are also processes under the control of the hypothalamus. It is known that the embryonic hypothalamus becomes histologically distinct during the period when egg incubation temperature has these effects (up to 40 days in alligators) although functional and immunocytochemistry studies have yet to be conducted. It is, therefore, possible that incubation temperature determines the

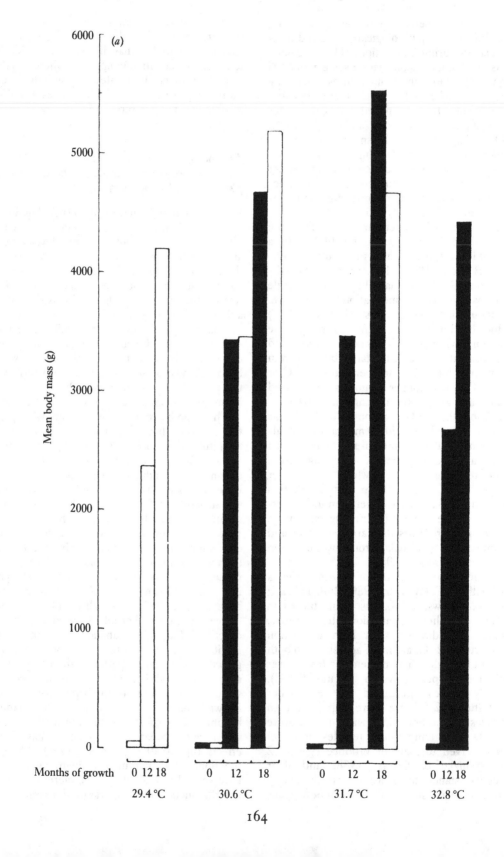

set points of the hypothalamus during development. This may be a direct effect, e.g. the hypothalamus matures, starts to monitor embryonic temperature, blood gases and osmomality, etc, and is thereby set. Alternatively, the same mechanisms that determine sex, e.g. dosage of a gene product, may also set the hypothalamus; thus the primary action of the male sex determining gene may not be on the gonad but on the hypothalamus. Whatever the cause and effect relationship, between sex determination and hypothalamic set points in embryonic reptiles, it is probable that all of these processes (sex, growth, behaviour, etc) are regulated by a common cascade of gene expression. This may have important implications for understanding and manipulating such processes in birds.

Importantly, these effects of incubation temperature on reptilian embryonic develop-

ment, produce a heterogenous population of animals. Within any species with TSD, there is a wide range of animals 'preadapted' to various environmental conditions, i.e. they vary in their optimal food demands, temperature requirements, potential growth rates, etc. It is this variance in a whole range of physiological parameters that may represent the selective advantage of temperature dependent sex (and size and growth, etc) determination. This effect endows the animal with incredible flexibility: to rapidly scale up or down its growth rate and final adult body size, and/or to colonise new habitats. Such changes would require long selection time-scales if these properties were regulated by classical genetic (chromosomal) mechanisms. Indeed, we have argued elsewhere that it was this incubation temperature induced plasticity that allowed crocodilians, turtles and squamates

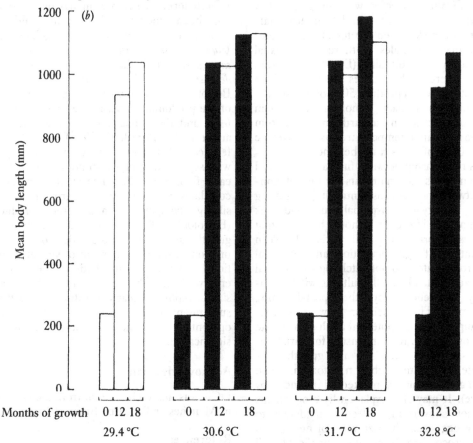

Fig. 10.9. The effects of incubation temperature upon the post-hatching growth rates in *Alligator mississippiensis*. Histograms show the size, in mass (*a*) and total length (*b*), of alligators (from four incubation temperatures) at hatching, after 12 months and 18 months under a controlled temperature and feeding regimen. Males are represented by black histograms, females by white histograms (Data from Joanen, McNease & Ferguson, 1987).

to survive virtually unchanged the global changes which led to the extinction of the dinosaurs (Deeming & Ferguson, 1989c): by our theory, dinosaurs had GSD and not TSD.

Interestingly, in alligators, when everything is optimal, i.e. egg incubation environment and post-hatching temperature, food availability, etc, then the largest animals are male (Deeming & Ferguson, 1988, 1989b). This is important as large males are reproductively superior: they control larger harems of females and produce sperm earlier and for longer during the breeding season. Thus in alligators, if you are a large animal there is a selective advantage in being male (Deeming & Ferguson, 1988, 1989b). In turtles, where warm incubation temperatures produce females, and cooler temperatures males, then the adult females are almost inevitably larger than the males (Ewert & Nelson, 1990). In those species with high and low temperature females and intermediate temperature males, then the females are either smaller than the males or there is no sexual dimorphism in adult size (Ewert & Nelson, 1990). Interestingly, turtles hatching from temperatures which produce 50% of one sex are not nearly so sexually dimorphic in certain aspects of their anatomy as turtles hatching from extreme incubation temperatures which produce nearly 100% of one sex or the other.

This relationship between final adult size and sex strengthens the arguments for the relationship between mechanisms controlling the setting of the embryonic hypothalamus and sex determination. However, it should be noted that the best correlation will exist between temperature of egg incubation and potential adult growth rate, and not hatchling size or final adult size. Certainly, final adult size will be profoundly influenced by the post-hatching environment. This association between incubation temperature and potential adult growth rate may allow reptiles to compensate for variations in egg size (final adult size and growth rates correlate better with incubation history than with hatchling size, which is markedly influenced by egg size). It may also explain why crocodilians have not shown any selective tendency to increase egg size. The size ratio of neonate to adult in crocodilians is one of the smallest for all vertebrates: one would therefore suppose that, with the anatomical constraints of the female, there would be selective pressures to increase egg size, but these appear to be absent. We argue that no such pressures exist because 1) alterations in egg size (and number) would alter the thermal profile of the nest and reduce the selective advantage of producing a number of hatchlings all preadapted to differing post-hatching environments (i.e. reduce the population variance); and 2) the association of incubation history with potential adult growth rate, allows animals which have experienced optimal incubation and post-hatching environments to grow fast, i.e. the largest animals in the population were not the largest hatchlings, are not always the oldest, but will always constitute a small percentage as the coincidence of prolonged optimal environments is rare.

Is there evidence of similar mechanisms in birds? To date, detailed studies are scarce. However, there are tantalising data that some of these effects may occur in birds, perhaps in an altered form. A sex bias in hatching sequence has been documented for the bald eagle (*Haliaeetus leucocephalus*), the golden eagle (*Aquila chrysaetos*), the lesser snow goose (*Chen caerulescens*) and the ring billed gull (*Larus delavarensis*) (Ankney, 1982; Ryder, 1983; Bortololti, 1986). In these species, the frequency of the predominant sex in the first egg is about 64% and the opposite sex in the second egg about 65% (Bortololti, 1986). This sex difference in hatching sequence occurs in species with sexual dimorphism in adult size (e.g. bald eagle females are 25% heavier than males) and correlates with post-hatching growth dynamics, sibling competition and brood reduction (Bortololti, 1986). How such hatching and growth asynchrony occur develomentally is unknown, but is reminiscent of similar associations between incubation temperature, embryonic growth rates, sex and potential adult size in reptiles. Clearly, careful investigation of such phenomena in birds is of profound developmental, evolutionary and commercial significance.

Acknowledgements

DCD was funded by a grant from the University of Manchester Research Support Fund.

References

Ancel, P. & Lallemand, S. (1941). Sur l'action teratogene des hautes temperatures d'incubation chez l'embryon de poulet. *Comptes Rendes de Societé de Biologie*, **135**, 221–3.

Ankney, C. D. (1982). Sex ratio and egg sequence in lesser snow goose. *Auk*, 99, 662–6.

Bellairs, A. d'A. (1981). Congenital and developmental diseases. In *Diseases of the Reptilia*, vol. 2, eds J. E. Cooper & O. F. Jackson, pp. 469–85. London: Academic Press.

Bennett, A. F. & Dawson, W. R. (1979). Physiological responses of embryonic Heermann's gulls to temperature. *Physiological Zoology*, 52, 413–21.

Beuchat, C. A. (1988). Temperature effects during gestation in a viviparous lizard. *Journal of Thermal Biology*, 13, 135–42.

Bloom, S. E. (1974). Current knowledge about the avian W chromosome. *Bioscience*, 24, 340–4.

Boersma, P. D. (1982). Why some birds take so long to hatch. *American Naturalist*, 120, 733–50.

Boersma, P. D. & Wheelwright, N. T. (1979). The costs of egg neglect in the Procellariiformes: reproductive adaptations in the fork-tailed stormpetrel. *Condor*, 81, 157–65.

Boersma, P. D., Wheelwright, N. T., Nerini, M. K. & Wheelwright, E. S. (1980). The breeding biology of the fork-tailed storm-petrel (*Oceanodroma furcata*). *Auk*, 97, 268–82.

Booth, D. T. (1987). Metabolic response of Mallee fowl *Leipoa ocellata* embryos to cooling and heating. *Physiological Zoology*, 60, 446–53.

Bortololti, G. R. (1986). Influence of sibling competition on nestling sex ratios of sexually dimorphic birds. *American Naturalist*, 127, 495–507.

Browder, L. W. & Davison, J. (1964). Spotting variations in the leopard frog. A test for the genetic basis in the *Rana burnsi* variant. *Journal of Heredity*, 55, 234–41.

Bull, J. J. (1980). Sex determination in reptiles. *Quarterly Review of Biology*, 55, 3–21.

— (1983). *Evolution of Sex Determining Mechanisms*. Menlo Park, California: Benjamin Cummings.

— (1987*a*). Temperature-sensitive periods of sex determination in a lizard: similarities with turtles and crocodilians. *Journal of Experimental Zoology*, 241, 143–8.

— (1987*b*). Temperature-dependent sex determination in reptiles: validity of sex diagnosis in hatchling lizards. *Canadian Journal of Zoology*, 65, 1421–4.

Bull, J. J., Hillis, D. M. & O'Steen, S. (1988). Mammalian *ZFY* sequences exist in reptiles regardless of sex-determining mechanism. *Science*, 242, 567–9.

Bull, J. J., & Vogt, R. C. (1979). Temperature-dependent sex determination in turtles. *Science*, 206, 1186–8.

— (1981). Temperature-sensitive periods of sex determination in emydid turtles. *Journal of Experimental Zoology*, 218, 435–40.

Burger, J. (1989). Incubation temperature has long term effects on behaviour of young pine snakes

(*Pituophis melanoleucus*). *Behavioural and Ecological Sociobiology*, 24, 201–7.

Burger, J. & Zappalorti, R. T. (1988). Effects of incubation temperature on sex ratios in pine snakes; differential vulnerability of males and females. *American Naturalist*, 132, 492–505.

Burger, J., Zappalorti, R. T. & Gochfeld, M. (1987). Developmental effects of incubation temperature on hatchling pine snakes *Pituophis melanoleucus*. *Comparative Biochemistry and Physiology*, 87A, 727–32.

Burgoyne, P. S. (1989). Mammalian sex determination: thumbs down for zinc finger? *Nature, London*, 342, 860–2.

Bustard, H. R. (1969). Tail abnormalities in reptiles resulting from high temperature incubation. *British Journal of Herpetology*, 4, 121–3.

— (1971). Temperature and water tolerances of incubating crocodile eggs. *British Journal of Herpetology*, 4, 198–200.

Buttemer, W. A., Astheimer, L. B. & Dawson, T. J. (1988). Thermal and water relations of emu eggs during natural incubation. *Physiological Zoology*, 61, 483–94.

Choo, B. L. & Chou, L. M. (1987). Effect of temperature on the incubation period and hatchability of *Trionyx sinensis* Wiegmann eggs. *Journal of Herpetology*, 21, 230–2.

Christian, K. A., Tracy, C. R. & Porter, W. P. (1986). The effect of cold exposure during incubation of *Sceloporus undulatus* eggs. *Copeia*, 1986, 1012–14.

Cohen, M. M. & Gans, C. (1970). The chromosomes of the order Crocodylia. *Cytogenetics*, 9, 81–105.

Davison, J. (1964). A study of spotting patterns in the frog (*Rana pipiens*). *Journal of Heredity*, 55, 47–56.

Dawes, C. M. (1979). The effects of heating the egg on the respiratory movements of the hatching chick. *Comparative Biochemistry and Physiology*, 64A, 405–10.

— (1981). The effects of cooling the egg on the respiratory movements of the hatching fowl, *Gallus g. domesticus*, with a note on vocalisation. *Comparative Biochemistry and Physiology*, 68A, 399–404.

Deeming, D. C. (1989). Failure to turn eggs during incubation: development of the area vasculosa and embryonic growth. *Journal of Morphology*, 201, 179–86.

Deeming, D. C. & Ferguson, M. W. J. (1988). Environmental regulation of sex determination in reptiles. *Philosophical Transactions of the Royal Society of London*, 322B, 19–39.

— (1989*a*). Effects of incubation temperature on the growth and development of embryos of *Alligator mississippiensis*. *Journal of Comparative Physiology*, 159B, 183–93.

— (1989*b*). The mechanism of temperature dependent sex determination in crocodilians: a hypothesis. *American Zoologist*, 29, 973–85.

— (1989c). In the heat of the nest. *New Scientist*, **121**, 33–8.

— (1990). Morphometric analysis of embryonic development in *Alligator mississippiensis, Crocodylus johnstoni* and *Crocodylus porosus. Journal of Zoology, London*, **221**, 419–29.

— (1991). Reduction in eggshell conductance to respiratory gases has no effect on sex determination in *Alligator mississippiensis. Copeia*, **228**, 228–31.

Drent, R. H. (1975). Incubation. In *Avian Biology*, vol. V, eds D. S. Farner & J. R. King, pp. 333–420. New York: Academic Press.

Engel, W., Klemme, B. & Schmid, M. (1981). H–Y antigen and sex determination in turtles. *Differentiation*, **20**, 152–6.

Ewert, M. A. (1979). The embryo and its egg: development and natural history. In *Turtles, Perspectives and Research*, eds M. Harless & H. Morlock, pp. 333–413. New York: John Wiley & Sons.

— (1985). Embryology of turtles. In *Biology of the Reptilia*, vol. 14. Development A, eds C. Gans, F. Billet & P. F. A. Maderson, pp. 75–267. New York: John Wiley & Sons.

Ewert, M. A. & Nelson, C. E. (1990). Sex determination in turtles: diverse patterns and some possible adaptive values. *Copeia*, (in press).

Ferguson, M. W. J. (1985). The reproductive biology and embryology of crocodilians. In *Biology of the Reptilia*, vol. 14. Development A, eds C. Gans, F. Billet & P. F. A. Maderson, pp. 329–491. New York: John Wiley & Sons.

Ferguson, M. W. J. & Joanen, T. (1982). Temperature of egg incubation determines sex in *Alligator mississippiensis. Nature, London*, **296**, 850–3.

— (1983). Temperature dependent sex determination in *Alligator mississippiensis. Journal of Zoology, London*, **200**, 143–77.

Freeman, B. M. (1967). Some effects of cold on the metabolism of the fowl during the perinatal period. *Comparative Biochemistry and Physiology*, **20**, 179–93.

Gutzke, W. H. N. & Crews, D. (1988). Embryonic temperature determines adult sexuality in a reptile. *Nature, London*, **332**, 832–4.

Gutzke, W. H. N. & Packard, G. C. (1987). Influence of the hydric and thermal environments on eggs and hatchlings of bull snakes *Pituophis melanoleucus. Physiological Zoology*, **60**, 9–17.

Gutzke, W. H. N., Packard, G. C., Packard, M. J. & Boardman, T. J. (1987). Influence of the hydric and thermal environments on eggs and hatchlings of painted turtles (*Chrysemys picta*). *Herpetologica*, **43**, 393–404.

Gutzke, W. H. N. & Paukstis, G. L. (1983). Influence of the hydric environment on sexual differentiation of turtles. *Journal of Experimental Zoology*, **226**, 467–70.

Hamburger, V. & Hamilton, H. L. (1951). A series of normal stages in the development of the chick embryo. *Journal of Morphology*, **88**, 49–92.

Harkey, G. A. & Semlitsch, R. D. (1988). Effects of temperature on growth, development and color polymorphism in the ornate chorus frog. *Copeia*, **1988**, 1001–7.

Holder, L. A. & Bellairs, A. d'A. (1962). The use of reptiles in experimental embryology. *British Journal of Herpetology*, **3**, 54–61.

Hubert, J. (1985). Embryology of the Squamata. In *Biology of the Reptilia*, vol. 15. Development B, eds C. Gans & F. Billet, pp. 3–34. New York: John Wiley & Sons.

Hudson, A. (1966). Proteins in the haemolymph and other tissues of the developing tomato hornworm *Protoparce quinquemaculata* Haworth. *Canadian Journal of Zoology*, **44**, 541–55.

Huey, R. B. (1982). Temperature, physiology and the ecology of reptiles. In *Biology of the Reptilia*, vol. 12, eds C. Gans & F. H. Pough, pp. 25–91. London: Academic Press.

Huey, R. B. & Webster, T. P. (1976). Thermal biology of *Anolis* lizards in a complex fauna: the *cristatellus* group of Puerto Rico. *Ecology*, **57**, 985–94.

Hutton, J. M. (1987). Incubation temperatures, sex ratios and sex determination in a population of Nile crocodiles (*Crocodylus niloticus*). *Journal of Zoology, London*, **211**, 143–55.

Joanen, T., McNease, L. & Ferguson, M. W. J. (1987). The effects of egg incubation temperature on post-hatching growth of American alligators. In *Wildlife Management: Crocodiles and Alligators*, eds G. J. W. Webb, S. C. Manolis & P. J. Whitehead, pp. 533–7. Sydney: Surrey Beatty Pty Ltd.

Joss, J. M. P. (1989). Gonadal development and differentiation in *Alligator mississippiensis* at male and female producing incubation temperatures. *Journal of Zoology, London*, **218**, 679–87.

Kettlewell, B. (1973). *The Evolution of Melanism*. Oxford: Clarendon Press.

Koopman, P., Gubbay, J., Collignon, J. & Lovell-Badge, R. (1989). *ZFY* gene expression patterns are not compatible with a primary role in mouse sex determination. *Nature, London*, **342**, 940–2.

Landauer, W. (1967). *The Hatchability of Chicken Eggs as Influenced by Environment and Heredity*, Monograph 1 (revised). Storrs, Connecticut: University of Connecticut Storrs Agricultural Experiment Station.

Lang, J. W. (1987). Crocodilian thermal selection. In *Wildlife Management: Crocodiles and Alligators*, eds G. J. W. Webb, S. C. Manolis & P. J. Whitehead, pp. 301–17. Sydney: Surrey Beatty Pty Ltd.

Lang, J. W., Andrews, H. & Whitaker, R. (1989). Sex determination and sex ratios in *Crocodylus palustris*. *American Zoologist*, **29**, 935–52.

Legler, J. M. (1985). Australian chelid turtles:

Reproductive patterns in wide-ranging taxa. In *Biology of Australasian Frogs and Reptiles*, eds G. Grigg, R. Shine & H. Ehmann, pp. 117–23. Sydney: Surrey Beatty Pty Ltd.

Limpus, C. J., Reed, P. C. & Miller, J. D. (1985). Temperature dependent sex determination in Queensland Sea turtles: intraspecific variation in *Caretta caretta*. In *Biology of Australasian Frogs and Reptiles*, eds G. Grigg, R. Shine & H. Ehmann, pp. 343–51. Sydney: Surrey Beatty Pty Ltd.

Lutz-Ostertag, Y. (1966). Action de la chaleur sur le développement de l'appareil génital de l'embryon de caille (*Coturnix coturnix japonica*). *Comptes Rendus des Acadamie des Sciences, Paris*, **262D**, 133–5.

Maderson, P. F. A. & Bellairs, A. d'A. (1962). Culture methods as an aid to experiment on reptile embryos. *Nature, London*, **195**, 401–2.

Manolis, S. C., Webb, G. J. W. & Dempsey, K. E. (1987). Crocodile egg chemistry. In *Wildlife Management: Crocodiles and Alligators*, eds G. J. W. Webb, S. C. Manolis & P. J. Whitehead, pp. 445–72. Sydney: Surrey Beatty Pty Ltd.

Matsunaga, C., Mathiu, P. M., Whittow, G. C. & Tazawa, H. (1989). Oxygen consumption of Brown Noddy (*Anous stolidus*) embryos in a quasiequilibrium state at lowered ambient temperatures. *Comparative Biochemistry and Physiology*, **93A**, 707–10.

Miller, J. D. (1985*a*). Embryology of marine turtles. In *Biology of the Reptilia*, vol. 14. Development A, eds C. Gans, F. Billet & P. F. A. Maderson, pp. 269–328. New York: John Wiley & Sons.

— (1985*b*). Criteria for staging reptilian embryos. In *Biology of Australasian Frogs and Reptiles*, eds G. Grigg, R. Shine & H. Ehmann, pp. 305–10. Sydney: Surrey Beatty Pty Ltd.

Miller, J. D. & Limpus, C. J. (1981). Incubation period and sexual differentiation in the green turtle *Chelonia mydas* L. *Proceedings of the Melbourne: Herpetological Symposium*, pp. 66–73. Melbourne Zoological Board of Victoria.

Mrosovsky, N. (1988). Pivotal temperatures for loggerhead turtles (*Caretta caretta*) from northern and southern nesting beaches. *Canadian Journal of Zoology*, **66**, 661–9.

Murray, K. G., Winnett-Murray, K., Eppley, Z. A., Hunt, G. L. & Schwartz, D. B. (1983). Breeding biology of Xantus' murrelet. *Condor*, **85**, 12–21.

Murray, J. D., Deeming, D. C. & Ferguson, M. W. J. (1990). Size dependent pigmentation pattern formation in embryos of *Alligator mississippiensis*: time of initiation of pattern generation mechanism. *Proceedings of the Royal Society, London*, **B**, **239**, 279–93.

Muth, A. (1980). Physiological ecology of desert iguana (*Dipsosaurus dorsalis*) eggs: temperature and water relations. *Ecology*, **61**, 1335–43.

Nair, G., Baggott, G. K. & Dawes, C. M. (1983). The

effects of a lowered ambient temperature on oxygen consumption and lung ventilation in the perinatal quail (*Coturnix c. japonica*). *Comparative Biochemistry and Physiology*, **76A**, 271–7.

— (1984). The effect of temperature on the acid-base status of the blood of the hatching quail (*Coturnix c. japonica*). *Comparative Biochemistry and Physiology*, **79A**, 81–6.

Nair, G. & Dawes, C. M. (1980). The effects of cooling the egg on the respiratory movements of the hatching quail (*Coturnix c. japonica*). *Comparative Biochemistry and Physiology*, **67A**, 587–92.

Nakumara, D., Wachtel, S. S., Lance, V. & Becak, W. (1987). On the evolution of sex determination. *Proceedings of the Royal Society, London*, **B232**, 159–80.

Nijhout, H. F. (1980). Ontogeny of the color pattern formation on the wings of *Precis coenia* (Lepidoptera: Nymphalidae). *Developmental Biology*, **80**, 275–88.

Noble, R. C., Deeming, D. C., Ferguson, M. W. J. & McCartney, R. (1991). The effects of incubation temperature on the lipid metabolism of embryos of *Alligator mississippiensis*. *Journal of Comparative Physiology*, **B** (in press).

Olmo, E. (1986). *Animal Cytogenetics 4, Chordata 3, A Reptilia*. Gebruder Borntraeger, Berlin.

Oppenheim, R. W. & Levin, H. L. (1975). Short-term changes in incubation temperature: behavioural and physiological effects in the chick embryo from 6 to 20 days. *Developmental Psychobiology*, **8**, 103–15.

O'Rahilly, R. & Muller, F. (1987). Developmental stages in human embryos. *Carnegie Institute of Washington Publication* **637**.

Osgood, D. W. (1978). Effects of temperature on the development of meristic characters in *Natrix fasciata*. *Copeia*, **1978**, 33–47.

Packard, G. C., Packard, M. J. & Birchard, G. F. (1989). Sexual differentiation and hatching success by painted turtles incubating in different thermal and hydric environments. *Herpetologica*, **45**, 385–92.

Packard, G. C., Packard, M. J., Miller, K. & Gutzke, W. H. N. (1987). Influence of moisture, temperature and substrate on snapping turtle eggs and embryos. *Ecology*, **68**, 983–93.

Packard, G. C., Paukstis, G. L., Boardman, T. J. & Gutzke, W. H. N. (1985). Daily and seasonal variation in hydric conditions and temperature inside nests of common snapping turtles (*Chelydra serpentina*). *Canadian Journal of Zoology*, **83**, 2422–9.

Page, D. C., Mosher, R., Simpson, E. M., Mardon, G., Pollack, J., McGillivary, B., de la Chapelle, A. & Brown, L. G. (1987). The sex determining region of the human Y chromosome encodes a finger protein. *Cell*, **51**, 1091–104.

Palmer, M. S., Sinclair, A. H., Berta, P., Ellis, N. A., Goodfellow, P. N., Abbas, N. E. & Fellows, M.

(1989). Genetic evidence that *ZFY* is not the testis determining factor. *Nature, London*, **342**, 937–9.

Paukstis, G. L., Gutzke, W. H. N. & Packard, G. C. (1984). Effects of substrate water potential and fluctuating temperatures on sex ratios of hatchling painted turtles (*Chrysemys picta*). *Canadian Journal of Zoology*, **62**, 1491–4.

Pieau, C. & Dorizzi, M. (1981). Determination of temperature sensitive stages for sexual differentiation of the gonads in embryos of the turtle, *Emys orbicularis*. *Journal of Morphology*, **170**, 373–82.

Raynaud, A. & Pieau, C. (1972). Effet de diverse températures d'incubation sur le développement somatique et sexuel des embryons de lizard vert (*Lacerta viridis* Laur.). *Comptes Rendes hebd. Seanc. Acad. Sci., Paris*, D **275**, 2259–62.

Roby, D. D. & Ricklefs, R. E. (1984). Observations on the cooling tolerance of embryos of the diving petrel *Pelecanoides georgicus*. *Auk*, **101**, 160–1.

Romanoff, A. L. (1943). Assimilation of avian yolk and albumen under normal and extreme incubating temperatures. *Anatomical Record*, **86**, 143–8.

— (1960). *The Avian Embryo*. New York: MacMillan.

— (1972). *Pathogenesis of the Avian Embryo*. New York: Wiley-Interscience.

Romanoff, A. L. & Hayward, F. W. (1943). Changes in volume and physical properties of allantoic and amniotic fluids under normal and extreme temperatures. *Biological Bulletin*, **84**, 141–7.

Romanoff, A. L., Smith, L. L. & Sullivan, R. A. (1938). Biochemistry and biophysics of the developing hen's egg. III. Influence of temperature. *Cornell University Agricultural Experimental Station Memoranda*, **216**, 1–42.

Romanoff, A. L. & Sochen, M. (1936). Thermal effect on the rate and duration of the embryonic heart beat of *Gallus domesticus*. *Anatomical Record*, **65**, 59–68.

Ryder, J. P. (1983). Sex ratio and egg sequence in ring billed gulls. *Auk*, **100**, 726–8.

Salzgeber, B. (1959). Contribution a l'étude de l'action d'une température élevée d'incubation sur les canaux de Muller cultives *in vitro*. *Comptes Rendes de Acadamie de Science, Paris*, **248**, 1707–9.

Semlitsch, R. D. (1979). The influence of temperature on ecdysis rates in snakes (Genus *Natrix*) (Reptilia, Serpentes, Colubridae). *Journal of Herpetology*, **13**, 212–14.

Sexton, O. J. & Marion, K. R. (1974). Duration of incubation of *Sceloporus undulatus* eggs at constant temperature. *Physiological Zoology*, **47**, 91–8.

Shields, G. F. (1987). Chromosomal variation. In *Avian Genetics: A Population and Ecological Approach*, eds F. Cooke & P. A. Buckley, pp. 95–104. London: Academic Press.

Sittman, K. (1984). Sex determination in birds: progeny of nondisjunction canaries of Durham (1926). *Genetic Research, Cambridge*, **43**, 173–80.

Shubina, G. N., Zhmurin, L. M. & Vedeneeva, V. A.

(1972). Effect of short-term temperature drop of egg incubation on sex determination of chicken embryos. *Tr. Vses. Nauch.-Issled. Inst. Fiziol., Biokhim. Pitan. Sel'skokhoz. Zhivotn.*, **11**, 260–8 (in Russian).

Spear, H. G. & Dawes, C. M. (1983). The effects of cooling the egg on the respiratory movements of the hatching duck (*Anas platyrhynchos*). *Comparative Biochemistry and Physiology*, **74A**, 861–7.

Stoll, R. (1944). Sur l'évolution normale des canaux de Muller de l'embryo de poulet dans la deuxième partie de l'incubation. *Comptes Rendes de Société de Biologie*, **138**, 7–8.

— (1948). Action de quelques agents physiques et chimiques sur les canaux de Muller de l'embryon de poulet. *Archive de Anatomie, Microscopie et Morphologie experimentale*, **37**, 333–46.

Taning, A. V. (1952). Experimental study of meristic characters in fishes. *Biological Reviews*, **27**, 169–93.

Tazawa, H. (1973). Hypothermal effect on the gas exchange in chicken embryos. *Respiration Physiology*, **17**, 21–31.

Tazawa, H. & Mochizuki, M. (1978). Oxygen transport in chicken embryos under hypothermal exposures. *Respiration Physiology*, **32**, 325–34.

Tazawa, H. & Nakagawa, S. (1985). Response of egg temperature, heart rate and blood pressure in the chick embryo to hypothermal stress. *Journal of Comparative Physiology*, **155B**, 195–200.

Tazawa, H., Okuda, A., Nakazawa, S. & Whittow, G. C. (1989b). Metabolic responses of chicken embryos to graded, prolonged alterations in ambient temperature. *Comparative Biochemistry and Physiology*, **92A**, 613–17.

Tazawa, H. & Rahn, H. (1986). Tolerance of chick embryos to low temperatures in reference to the heart rate. *Comparative Biochemistry and Physiology*, **85A**, 531–4.

— (1987). Temperature and metabolism of chick embryos and hatchlings after prolonged cooling. *Journal of Experimental Zoology*, Supplement 1, 105–9.

Tazawa, H., Wakayama, H., Turner, J. S. & Paganelli, C. V. (1988). Metabolic compensation for gradual cooling in developing chick embryos. *Comparative Biochemistry and Physiology*, **89A**, 125–9.

Tazawa, H., Whittow, G. C., Turner, J. S. & Paganelli, C. V. (1989a). Metabolic responses to gradual cooling in chicken eggs treated with thiourea and oxygen. *Comparative Biochemistry and Physiology*, **92A**, 619–22.

Thompson, M. B. (1988). Influence of incubation temperature and water potential on sex determination in *Emydura macquarii* (Testudines: Pleurodira). *Herpetologica*, **44**, 86–90.

Tokunga, S. (1985). Temperature-dependent sex determination in *Gekko japonicus* (Gekkonidae,

Reptilia). *Development, Growth and Differentiation*, 27, 117–20.

Tucker, D. W. (1959). A new solution to the atlantic eel problem. *Nature, London*, 183, 495–501.

Vedeneeva, V. A. (1970). Effect of a reduction in incubation temperature on the nucleic acid level in the tissues of developing chick embryos. *Byull. Vses. Nauch-Issled Inst. Fiziol. Biokhim. Sel'skokhoz Zhivotn.*, 3, 22–4 (in Russian).

Vinegar, A. (1973). The effects of temperature on the growth and development of embryos of the Indian python *Python molorus* (Reptilia: Serpentes: Boidae). *Copeia*, 1973, 171–3.

— (1974). Evolutionary implications of temperature induced anomalies of development in snake embryos. *Herpetologica*, 30, 72–4.

Vleck, C. M. & Kenagy, G. J. (1980). Embryonic metabolism of the fork-tailed storm petrel: physiological patterns during prolonged and interrupted incubation. *Physiological Zoology*, 53, 32–42.

Warner, E. (1980). Temperature-dependent sex determination in a gekko lizard. *Quarterly Review of Biology*, 55, 21.

Webb, D. R. (1987). Thermal tolerance of avian embryos. A review. *Condor*, 89, 874–98.

Webb, G. J. W., Beal, A. M., Manolis, S. C. & Dempsey, K. E. (1987). The effects of incubation temperature on sex determination and embryonic development rate in *Crocodylus johnstoni* and *C. porosus*. In *Wildlife Management: Crocodiles and Alligators*, eds G. J. W. Webb, S. C. Manolis & P. J. Whitehead, pp. 507–31. Sydney: Surrey Beatty Pty Ltd.

Webb, G. J. W. & Cooper-Preston, H. (1989). Effects of incubation temperature on crocodiles and the evolution of reptilian oviparity. *American Zoologist*, 29, 953–71.

Webb, G. J. W. & Messel, H. (1977). Abnormalities and injuries in the estuarine crocodile *Crocodylus porosus*. *Australian Wildlife Research*, 4, 311–19.

Webb, G. J. W., Sack, G. C., Buckworth, R. & Manolis, S. C. (1983). An examination of *Crocodylus porosus* nests in two northern Australian freshwater swamps, with an analysis of embryo mortality. *Australian Wildlife Research*, 10, 311–19.

Webb, G. J. W. & Smith, A. M. A. (1984). Sex ratio and survivorship in the Australian freshwater crocodile *Crocodylus johnstoni*. In *The Structure, Development and Evolution of Reptiles*, ed. M. W. J. Ferguson, pp. 319–55. London: Academic Press.

Wellins, D. J. (1987). Use of an H–Y antigen assay for sex determination in sea turtles. *Copeia*, 1987, 46–52.

Weniger, J.-P. & Thiebold, J.-J. (1981). On the mechanism of the retention of the Mullerian ducts in the chick embryo incubated at a temperature above normal. *Anatomy and Embryology*, 163, 345–50.

Wheelwright, N. T. & Boersma, P. D. (1979). Egg chilling and the thermal environment of the fork-tailed storm petrel (*Oceanodroma furcata*) nest. *Physiological Zoology*, 52, 321–9.

Whitehead, P. J. (1987). Respiration of *Crocodylus johnstoni* embryos. In *Wildlife Management: Crocodiles and Alligators*, eds G. J. W. Webb, S. C. Manolis & P. J. Whitehead, pp. 473–97. Sydney: Surrey Beatty Pty Ltd.

Yntema, C. L. (1960). Effects of various temperatures on the embryonic development of *Chelydra serpentina*. *Anatomical Record*, 136, 305–6.

— (1968). A series of stages in the embryonic development of *Chelydra serpentina*. *Journal of Morphology*, 125, 219–52.

— (1976). Effects of incubation temperature on sexual differentiation in the turtle *Chelydra serpentina*. *Journal of Morphology*, 150, 453–61.

— (1978). Incubation times for eggs of the turtle *Chelydra serpentina* (Testudines; Chelydridae) at various temperatures. *Herpetologica*, 34, 274–7.

— (1979). Temperature levels and periods of sex determination during incubation of *Chelydra serpentina*. *Journal of Morphology*, 159, 17–28.

— (1981). Characteristics of gonads and oviducts in hatchling and young of *Chelydra serpentina* resulting from three incubation temperatures. *Journal of Morphology*, 167, 297–304.

Yoshida, M. C. & Msahiro, I. (1974). Heteromorphic sex chromosomes in gecko lizards. *Chromosome Information Service*, 17, 29–31.

Zaborski, P., Dorizzi, M. & Pieau, C. (1982). H–Y antigen expression in temperature sex-reversed turtles (*Emys orbicularis*). *Differentiation*, 22, 73–8.

Zaborski, P., Dorizzi, M. & Pieau, C. (1988). Temperature-dependent gonadal differentiation in the turtle *Emys orbicularis*: concordance between sexual phenotype and serological H–Y antigen expression at threshold temperature. *Differentiation*, 38, 17–20.

Cold torpor, diapause, delayed hatching and aestivation in reptiles and birds

MICHAEL A. EWERT

Introduction

This chapter considers temporary suspension of active development within the oviposited eggs of reptiles and birds. These various processes appear to be 'natural' and adaptive; however, much of the study about them has occurred in the laboratory.

As presently known, all oviposited reptilian and avian embryos have advanced in development to blastulae or beyond (Bellairs, Chapter 23). In many, perhaps all, turtles embryonic development becomes arrested as a late gastrula within the oviducts and resumes normally only after oviposition (Ewert, 1985). In lizards, developmental arrest may occur in the oviducts in a few cases, e.g. *Sceloporus jarrovi*, which is ovoviviparous (Goldberg, 1971), but these must be very rare (Shine, 1983, 1985). The egg of the Tuatara (*Sphenodon punctatus*) is laid at an early embryonic stage, as in turtles, but whether this stage is also associated with pre-ovipositional arrest is unknown (Moffat, 1985). In snakes, crocodilians and birds, pre-ovipositional arrest is unknown and unlikely (Ferguson, 1985; Blackburn & Evans, 1986; Lance, 1987).

After oviposition, development within reptilian or avian eggs of many species can be divided into periods of active embryonic differentiation and growth and periods of relative inactivity, when as time passes there is little change. Such prolongations can be grouped as 1) cold torpor, 2) diapause, and 3) delayed hatching, which grades into aestivation in some reptiles. A previous survey of these topics (Ewert, 1985) focuses on turtles, but gives some reference to lizards and other vertebrates. Although turtles appear to possess the richest

variety of ways to prolong the egg stage, both data and inferential observations on lizards allow that this group may be as richly endowed. Birds exhibit much less diverse mechanisms of prolonging their egg stage; however, some parallels can be drawn to the cold torpor and delayed hatching of reptiles.

Cold torpor

In general, cold torpor is a suspension, or perhaps, near suspension of development that already has begun. In the present context, this suspension occurs only because temperatures of the egg and embryo are too low to meet thermal requirements for development. By contrast, development may not proceed because the embryo is in diapause at the time of chilling or because chilling has induced a diapause (see later). A question relevant to cold torpor is whether a potentially active embryo of a given species and stage of development can sustain a given degree of chilling without being damaged. According to one view, early embryos may endure very cool temperatures for several weeks whereas slightly warmer ones will cause teratogenesis and still warmer ones will permit slow but normal development for long periods, at least in reptiles (Yntema, 1960; Lundy, 1969). It is not known if more than just a few species give teratogenic responses to a band of cool temperatures or how such sensitivities change with embryonic stage. Still, a view encompassing both specific temperature and embryonic stage can provide perspective in examining the literature on various, often anecdotal tests of cold tolerance, some of which seem unrelated to potential adaptive value.

Reptiles

Eight-somite embryos of the turtle *Chelydra serpentina* can endure a couple of weeks at 10 °C and go on to develop normally at room temperature. Similar embryos at 15 °C for three weeks become terata, and embryos at 20 °C develop almost to hatchling (Yntema, 1960). At 21.5 °C, many embryos are able to complete normal hatching (Ewert, unpublished data). Although these observations are compatible with a scenario of a teratogenic range of cool temperatures, certainly, broader testing at 10 °C, and perhaps 5 °C would help produce a firm conclusion.

Two species of turtles, *Pseudemys floridana* and *Pseudemys nelsoni*, are broadly sympatric in northern Florida. *P. floridana* nests during the autumn, winter, and spring whereas *P. nelsoni* nests during the late spring and early summer. Freshly laid eggs of *P. floridana* can tolerate continuous exposure to 20 °C for longer than 30 days, whereas eggs of *P. nelsoni* cannot. Laboratory incubation at 22.5 °C allows eggs of *P. floridana* to hatch in 118–122 days, whereas viable eggs in natural nests can remain unhatched for at least to 150 days (Goff & Goff, 1932; Jackson, 1988). Hence, assuming normal diel and weather-related temperature fluctuations in shallow turtle nests (Packard & Packard, 1988) and expected accelerations of development during warm periods, eggs of *P. floridana* probably endure long periods of exposure to temperatures cooler than 20 °C. An alternative form of arrest, embryonic diapause, does not appear to occur in *P. floridana* in Florida. Oviductal eggs obtained in early February immediately commence development at 27 °C. Mid-spring eggs of third and partially sympatric species, *Pseudemys concinna suwanniensis* also begin immediate but slow development at constant 22.5 °C and closely approach term in 161–166 days. However, all of these embryos ($N = 9$) were deformed and died before pipping. The adaptive value of cold torpor in *P. floridana* appears to permit late autumn and winter nesting without risk of teratogenesis during periods of presumably sub-optimal temperatures. Another turtle, *Sternotherus minor*, nests in the autumn in northern Florida, where winters are cool (Cox & Marion, 1978). According to laboratory incubation, development nearly ceases during 30 days at 16–18 °C (29 eggs incubated) but very few eggs (2 of >40 incubated at 22.5–24 °C) show diapause.

Although the above instances given for cold torpor fit circumstantial evidence well, confirmation through examination of embryonic stages from the field is lacking. In other reptiles with eggs in nests during a cold season, confirmation of cold torpor is also lacking. In such species, eggs laid in the spring or early summer of temperate or montane climates do not hatch until the following spring or summer. The Tuatara is such a candidate for cold torpor (Moffat, 1985), the turtle *Geochelone pardalis* of southern Africa is another (Rose, 1962; Cairncross & Greig, 1977) but neither diapause nor aestivation in the unpipped egg are presently ruled out. In the montane, ovoviviparous lizard, *S. jarrovi*, uterine embryos overwinter for 3–4 months as blastoderms (Goldberg, 1971) but again the possibility of them being in diapause has not been ruled out. Eggs of montane populations of oviparous *Sceloporus undulatus* have been incubated at 15 °C following 15 days of incubation at 30 °C. Embryonic mortality increased from 0% following four days of chilling to 50% following 9 days and 100% following 40 days. In another test, embryos at nearly any stage of post-ovipositional development were able to tolerate five days of chilling. Such short cold spells are to be expected in the local habitat (Christian, Tracy & Porter, 1986).

A few anecdotal tests of embryonic chilling involve turtles. Fresh eggs of *Chrysemys picta* are able to follow a month at 'close to freezing' with normal development at room temperature. While cold, the embryos remain as late gastrulae (Cunningham, 1922). Fresh eggs of *Trachemys scripta* seem unable to survive two weeks at about 10 °C (Cagle, 1950). Such conditions are unnatural for early embryos of either species. Embryos of *Terrapene ornata* close to hatching can survive two days' exposure to approximately 4.5 °C, and may find this necessary in late season nests (Legler, 1960).

Birds

Cold torpor is common in taxonomically diverse groups of birds, particularly in unincubated eggs. It also functions in the later development of at least a few species but in no case does it appear obligatory for the individual embryo. Rather, it suits the needs of the parents. A review of temperature tolerance focusing on partially incubated embryos is provided by Webb (1987).

Many species begin brooding their eggs only after the last egg in a clutch has been laid. Cold torpor arrests development in the earlier eggs so that the clutch as a whole can follow the developmental schedule of the last egg (Welty, 1982). Hence, even in small species with only modest clutch sizes, the first egg may experience arrest for an appreciable fraction of its post-ovipositional state, e.g. about 20% in the warbler, *Dendroica discolor* (Nolan, 1978). Cold tolerance in unincubated eggs of domestic species is seen as a practical means of storing viable eggs (Rol'nik, 1970; Romanoff, 1972). For the fowl (*Gallus gallus*), storage at 13–14 °C gives the greatest short-term viability, about 80% of eggs stored for two weeks (Romanoff, 1972), and the least sublethal histological damage (Arora & Kosin, 1968) over an approximate test range of 0–28 °C. The holding of unincubated fowl eggs at warm (29–32 °C) but sub-optimal incubation temperatures, as opposed to cool ones (15 °C), for 4–6 days reduces embryonic viability (El Jack & Kaltofen, 1968). Hence, an absence of sufficiently cool temperatures during the building of large clutches in warm climates could be limiting on breeding strategies favouring use of cold torpor to synchronise development. Conversely, the fowl seems no less tolerant of partial freezing than the mallard, *Anas platyrhynchos*, even though freezing temperatures are a greater natural threat to the mallard (Greenwood, 1969; Rol'nik, 1970).

Once incubation begins, avian embryos of rather few species experience developmental arrest, although development may slow during brief periods when the eggs are not being brooded (Webb, 1987). Durations of tolerated chilling decline dramatically in the fowl. For instance, between four and five days of incubation, initiation of two days of chilling at 10–13 °C causes hatchability to decline from 70% to 17%, and at 12 days only about 50% hatch following 19 hours of chilling (Rol'nik, 1970). A similar relationship applies to pheasant (*Phasianus colchicus*) embryos (MacMullan & Eberhardt, 1953). However, this species seems to be less tolerant to given temperatures despite its overall adaptation to more chill-prone natural environments. Advanced embryos of the hummingbird, *Selasphorus platycercus* are able to

survive at least two hours at an estimated 6.5 °C, when the brooding adult enters hypothermia during stressful conditions (Calder & Booser, 1973).

Embryos of the fork-tailed storm petrel (*Oceanodroma furcata*) tolerate the longest known periods of chilling during incubation (Boersma & Wheelwright, 1979; Wheelwright & Boersma, 1979). Interruptions ('egg neglect') occur in early as well as late incubation and often last 1–3 days at 10–12 °C although one 7-day period of neglect has been survived. Eggs placed in a laboratory simulation of incubation with periodic chilling (cycles of 4 days at 34 °C plus 1 day at 10 °C for 32 days ending in hatching) reduce their rate of oxygen consumption during chilling to about 5% of that of warm incubation (Vleck & Kenagy, 1980). The peculiar pattern of incubation results from the brooders needing to make long journeys during uncertain weather in search of food (Boersma *et al.*, 1980).

Embryos of several other procellariiform birds survive chilling well (Deeming & Ferguson, Chapter 10). A fairly rough determination of natural survival to chilling (mostly near but slightly <17 °C) indicates that tolerance, as measured by duration vs. subsequent hatching, increases with developmental age (or stage) in shearwater (*Puffinus puffinus*) embryos (Matthews, 1954). This contrast to the pattern tolerance in galliform birds suggests that some avian embryos differ appreciably in developmental physiology in relation to chilling.

Embryonic diapause

Diapause is used here to refer to arrested development in healthy organisms that occurs when the immediately proximate environment would normally foster active development. The eggs reside in adequate temperature, water potential, and gaseous atmosphere and yet the embryos do not develop. Most diapausing organisms respond to external stimuli that cause release from the arrest and sometimes, though not necessarily, promote development. Common stimuli are temperature, water potential, and daylength. Experimental proof of the occurrence of diapause requires identification of the stimulus that terminates it. However, in absence of formal testing, careful comparison of developmental schedules among siblings, and even among moderately related species, can permit reasonably safe conclusions to be drawn. Embryonic diapause is widespread in turtles, is present in chamaeleonid lizards but is unknown in birds.

Reptiles

Reptilian diapause is known only in late gastrulae (Blanc, 1974; Ewert, 1985). A tentative diagnosis of diapause can result from direct observation that presomite development is appreciably prolonged relative to the remainder of development. In most turtles, two good reference stages that are easily observable through candling are Stage 8+ (circa 15–16 somites and a fully defined terminal sinus) and the end of Stage 25 (pre-hatching) of Yntema (1968). Eggs are best maintained at constant temperature, as temperature influences much or all of the rate of active development (Yntema, 1978; Pieau & Dorizzi, 1981; Ewert, 1985; Webb et al., 1987; Deeming & Ferguson, 1989, Chapter 10). Reports detailing instances of diapause tend to relate that development 'takes forever to get going', since the gastrula is difficult to see and observable changes, proliferations of pre-vascular cells and concentrations of haemoglobin, are quite small. For initial study, it is particularly helpful if a given clutch of eggs or group of clutches includes both diapausing and non-diapausing individuals, or alternatively, deeply and lightly diapausing individuals. The observer can then find an array of times between oviposition and Stage 8+, and can readily conclude which embryos have been arrested relative to others. Such asynchronous development can be demonstrated through measurement of metabolism (Fig. 11.1) as well as by keeping records on candled eggs. Later, mixed states of shallow and deep diapause, together with non-diapause, complicate formal testing for stimuli to terminate diapause.

As a caveat to equating arrested early development with diapause, development can be prevented because it has become abnormal though not immediately fatal. Diagnosis of diapause requires detection of healthy development following the period of arrest. Further, since death as a term or near-term embryo follows some abnormal instances of arrest, yield of a normal hatchling gives conservative support that the observed arrest was a normal feature of total development.

In some instances, embryonic diapause is so deep that release does not come without a stimulus, e.g. chilling in eggs of the turtle Deirochelys reticularia from northern Florida (Ewert, 1985; Jackson, 1988). Embryos trapped in diapause eventually lose potency and die. To an observer, however, this occurrence may appear like arrest caused by congenital defects. Hence, demonstrating the presence of diapause is best achieved through identifying the stimulus that terminates it. Temperature clearly has a strong influence on diapause in some species, and will be the principal subject of this section. However, moisture is implicated in one specialised case (see later).

As responses to temperature have been found to vary both within and among species, testing for stimuli has become difficult, especially when only small numbers of eggs are available. Four variables considered here are: 1) temperature ranges that give maximum expression of diapause; 2) temperature ranges that best terminate diapause; 3) durations of exposure to terminating temperatures; and 4) effects of pre-ovipositional conditions on subsequent diapause. Examples of these four variables are drawn from turtles.

Regarding the first variable, all freshly laid embryos of Staurotypus salvinii respond to warm temperatures (28–30 °C) with diapause whereas embryos of Kinosternon baurii respond to 30 °C and almost always to 28 °C with active development. Cool temperatures (22.5–24 °C) hold several embryos of K. baurii in diapause, whereas 22.5 °C terminates it in S. salvinii and then fosters slow but active development. Holding fresh embryos of S. salvinii at 25 °C leads to diapause in most, but this ends endogenously after a variable period (30–100 days among embryos) and completely normal development follows. Hence, these two species differ appreciably in temperature ranges which sustain or suppress diapause.

Is there any one temperature more effective than another in terminating diapause? Design for sound statistical testing becomes complicated by usually large variation among embryos. For instance, assume random assignment of freshly laid eggs to treatment (say, an initial month of chilling and then transfer to warmth) vs. control (unchilled, kept at the warm transfer temperature for the chilled group) and that the measured variable is days lapsed from oviposition to Stage 8+. A few embryos in the control group may lack diapause and advance to Stage 8+ before the period of chilling is over in the

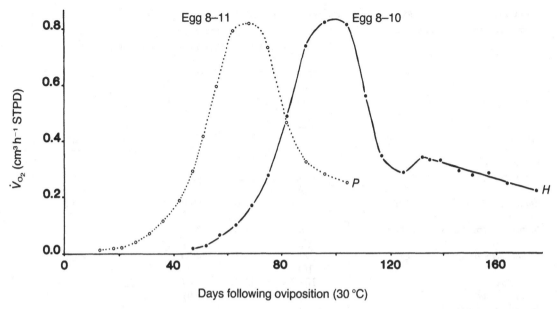

Fig. 11.1. Metabolic monitoring and developmental arrest. Patterns of oxygen consumption were followed in two individual eggs representing a single clutch from the turtle *Kinosternon scorpioides* (female #8, eggs laid on 23 September 1984) as measured by manometry ('Baralyme' as a CO_2 absorbant) in a Gilson Differential Respirometer. Measurements (at intervals of 3–8 days, represented as dots) were commenced at Stage 8+ (vitelline circulation) and terminated near pipping or hatching (as shown), or after hatching of egg #8–10. Differentiation and growth are complete 9–11 days after the peaking of respiration rates. Although the two eggs have similar patterns of oxygen consumption during differentiation and growth, egg #8–10 lags 30 days behind egg #8–11. Egg #8–10 had an initial month of diapause at 30 °C whereas egg #8–11 began active development just after oviposition. Egg #8–11 pipped naturally after 25 days of delayed hatching (the period of rapid decline in respiration rate). Egg #8–10 was broken open after its period of delayed hatching plus an additional 55 days of aestivation (the period of mainly gradual decline in respiration rate). V_{O_2} of confined hatchling #8–10 was 0.36 $cm^3 h^{-1}$ vs. 0.25 $cm^3 h^{-1}$ late in aestivation. As a consequence of arrests, egg 8–10 remained intact for 178 days whereas egg 8–11 lasted only 105 days.

treated group. Other embryos in the control group may persist in diapause for months. In the chilled group, a few (possibly deeply) diapausing embryos may not respond to the chilling; perhaps the duration of chilling is insufficient. However, in practice, nearly all embryos do respond by advancing to Stage 8+ during the two to three weeks that follow transfer from chilling. Hence, variance in the control group tends to be larger than in the treated group. Nonparametric statistical methods must, therefore, be applied to these data, which tend to require large samples to demonstrate significance. Despite this, the capacity of chilling to reduce the duration of early development can be demonstrated (Table 11.1).

When nearly all embryos appear to be indefinitely arrested (e.g. in *D. reticularia* and

Rhinoclemmys pulcherrima at 25 °C), congenital failure becomes a strong hypothetical alternative to diapause. Hence, all eggs should be treated. If chilling is suspected, its effect can be demonstrated by grouping eggs into immediately chilled vs. delayed chilled treatments (Table 11.1). As the immediately chilled embryos will receive their stimulus to terminate diapause sooner, they will advance to Stage 8+ sooner (Table 11.1). As all embryos have a chance to respond to chilling, variance within the group is uniformly low, although for the occasional embryo, the duration of chilling may be too short to bring arrest to a quick end.

While the possibility that warming, rather than chilling, terminates diapause has not undergone rigorous testing, it is nevertheless implicated. Embryos of *Kinosternon baurii* (at

Table 11.1. *Reduction in incubation of turtle embryos by a period of chilling.*

		Mean elapsed time (days) from oviposition to stage 8$^+$				
		Unchilled[c] (or Chill-delayed[d])		Chilled[e]		
Species	Unit of analysis	Mean (N)	S.D.	Mean (N)	S.D.	Significance[a, b]
Diapause group[f]						
Kinosternon scorpioides	clutch	76.5 (6)	26.0	42.3 (6)	1.3	<.02
Staurotypus salvinii	clutch	89.4 (13)	27.0	59.0 (13)	17.9	<.004
Staurotypus triporcatus	clutch	108.5 (6)	21.0	63.7 (6)	13.1	<.02
Melanochelys trijuga	egg	90.3 (16)	27.3	46.0 (22)	2.7	<0.001
Rhinoclemmys pulcherrima[d]	egg	104.3 (8)	11.9	55.7 (9)	3.3	<.003
Non-diapause group[g]						
Sternotherus minor	egg	11.6 (14)	3.7	47.9 (14)	3.2	<.001
Kinosternon subrubrum	egg	10.0 (12)	1.7	40.8 (9)	4.9	<.001

[a]Significance testing involved Wilcoxon's signed-ranks test (for paired means from split clutches) and Mann-Whitney U-tests or exact probabilities (for means derived from individual eggs).
[b]Within each species, each treatment includes two or more temporally separated runs — except for *Kinosternon subrubrum*, which was tested during just one season.
[c]Normal (unchilled) incubation was conducted at 25 °C.
[d]*Rhinoclemmys pulcherrima* is the only species subjected to a chill-delayed procedure. The eggs were held at 25 °C for 40 days and then chilled for 40 days.
[e]Chilling was conducted at 18 °C, 22.5 °C, or both. Durations of chilling were 30 days for all species other than *Rhinoclemmys pulcherrima*, which received 40 days. Listed means include the period of chilling.
[f]Mean values for unchilled groups are expected to exceed those for chilled groups because development remains arrested without chilling.
[g]Mean values for unchilled groups are expected to be smaller than those for chilled groups because chilling only slows development.

22.5–24 °C) and of *Phrynops geoffroanus* (at 25 °C) assume active development within 5–9 days of transfer to 30 °C even though they have appeared arrested for several weeks. When following Stage 8$^+$, when such embryos are returned to their former cooler temperatures, development continues through hatching. Hence, the cool temperatures are sufficient to sustain normal development in absence of diapause. The system can be viewed as reciprocal to one that responds to chilling. For instance, all early embryos of *S. salvinii* and *Staurotypus triporcatus* die if they are maintained at 30 °C without a period of chilling. However, following chilling and post-chilling development to Stage 8$^+$ (at <27 °C), nearly all embryos are able to complete development at 30 °C.

Chilling-sensitive diapause seems not to respond to single, brief chills of 10 days or less. Further, alternating brief chills (2–5 days) with similar, sometimes longer periods of warming does not appreciably hasten onset of active development in *Kinosternon scorpioides*, *D. reticularia*, or *Melanochelys trijuga*. It is when the sum of the days in the several brief periods of chilling approaches the duration of single continuous periods that active development follows. Still, it remains possible that there might be truly 'optimum' temperatures for chilling, and that these might greatly shorten the necessary exposure. In species that respond to warming, the necessary exposure appears to be up to five days.

The fourth variable, pre-ovipositional experience of the embryo, is mentioned here as a possible explanation for some seemingly wide variation in incidences of diapause. For example, consider a captive trio (two females, one male) of *S. triporcatus* that have reproduced nearly annually. In successive years, many of their eggs have been placed on substrates with a high water potential at 25 °C. In 1985, all 15 embryos (two clutches of eggs) showed diapause at 25 °C (required 74–187, av. = 147 ± 16 days to advance to Stage 8^+) whereas in 1981–1984, all 19 embryos (5 clutches) relatively quickly assumed active development at 25 °C (required 14–38, av. = 23 ± 6 days to advance to Stage 8^+) without a period of arrest or any post-ovipositional chilling (significance $p < 0.001$). Further, the frequency of diapause in autumn–early winter eggs of *K. baurii* seems a little greater than in later winter–spring eggs (80% vs. 20%, respectively with arrest, $N = 38$ eggs, $p = 0.037$).

The list of species with demonstrated or implicated diapause has grown in number and detail since the last review (Ewert, 1985). The first family groups described here are turtles, the last is lizards. Data not clearly documented with citation are my original data. Personally incubated eggs have been maintained a third to totally buried in damp vermiculite in aerated boxes in controlled temperature chambers.

Chelidae

In *Chelodina expansa*, development within nests seems to be prolonged in excess of a possible period of cold torpor (Goode & Russell, 1968). During laboratory incubation of eggs of this species and similar species (mostly at constant 25 °C or 28 °C), 'there seem to be pauses [in development] when nothing is "happening"' (Legler, 1985). Although early stages in the development of *C. expansa* are capable of moderately rapid development at 30 °C (Goode & Russell, 1968), evidence still allows a model in which embryos at high temperatures do not diapause, those at intermediate ones have prolonged diapause, and those kept cool undergo termination of diapause and will resume development when slightly warmed.

Moisture is implicated in influencing developmental arrest in *Chelodina rugosa*. If freshly laid (unchalked) eggs are submerged in water, the eggs remain viable but development remains arrested (in an essentially pre-ovipositional state?) for at least a few weeks of submergence (R. M. Kennett and A. Georges, personal communication).

An initial incubation of 41 days at 25 °C gave no sign of post-gastrular development in either of two eggs of *Chelodina parkeri*. Transfer of one of these eggs to 30 °C for ten days (days 42–52) induces active development during and after the transfer and advances development to Stage 8^+ by 58 days of incubation. The other egg, still at 25 °C, advanced to Stage 8^+ by 73 days. Incubation at warmer and perhaps fluctuating temperatures (27–33 °C) lacks evidence of diapause (Fritz & Jauch, 1989). Still, the data from cooler conditions indicate that early development is appreciably prolonged relative to data on other species that clearly lack diapause and yet develop slowly (e.g. *Platemys platycephala*; Ewert, 1985).

Unincubated embryos of *P. geoffroanus* placed at 25 °C usually remain arrested (10 of 13 embryos observed), whereas others placed at 30 °C usually become active (11 of 15 embryos observed; $p = 0.021$ for the comparison). Arrested embryos transferred from 25 °C to 30 °C become active, even after 90 days of arrest at 25 °C. Transferring nine seemingly arrested embryos to 30 °C induces obvious active development after 7–9 days of exposure, whereas only 4–5 days may be necessary. Embryos that have advanced to Stage 8^+ readily complete development at 25 °C. Among two apparent forms (subspecies?) of this species, arrest seems common only in one (the hatchling plastron shows areas of pale pink to white and lacks areas of red).

Embryos of *Platemys* (*Acanthochelys*) *spixii* show the first signs of vascularisation (*circa* Stage 8^+) only after 97–116 days of incubation at 29–30 °C (versus <14 days typical in species lacking developmental arrest). Two (of three) arrested embryos survived beyond hatching (Lehmann, 1988).

Kinosternidae

Early development in *Kinosternon acutum* (6 embryos observed) at 25 °C is appreciably prolonged relative to later development ($N = 4$) and by comparison with related species such as *Sternotherus carinatus* (Ewert, 1985).

Among 11 embryos of *K. baurii* arrested during initial incubation at 22.5–24 °C arrest persists up to 95 days in advance of heating. Six embryos have remained arrested until heated and then gone on to complete development. Once diapause is broken, viability through hatching at 22.5 °C is high (20 of 22 eggs hatched).

In *Kinosternon creaserii*, attempts to initiate active development at moderately warm temperatures (7 eggs at 25 °C, 4 at 27 °C, 1 at 28 °C) have failed except for two eggs at 27 °C. About 30 days of chilling at 19 °C or 22.5 °C allows development to proceed after a return to 25 °C.

Development of *Kinosternon alamosae* embryos remains suspended during an initial incubation period of two months at 25 °C or 30 °C, or 3–4 months at 22.5 °C (3 eggs, each temperature). Active development resumes following periods of chilling at 16.5 °C (2 months following initial 25 °C or 30 °C, 1 month following 22.5 °C) and a warming to 27 °C. With post-chilling warming sustained (at 27 °C), active development commences within 2–4 weeks without obvious variation among protocols.

Thirty-four eggs of *Kinosternon hirtipes* have been divided among the same protocols given to eggs of *K. alamosae* (except that initial 30 °C is replaced with 32 °C). Active development commenced in 28 eggs and 17 eggs have hatched at this writing. Although several of the eggs may have been defective from outset (in a premature condition when dissected from the oviducts), in general, the protocols have worked well and may represent more chilling than is necessary. Eight eggs of *Kinosternon integrum* have been divided among the same protocols given to eggs of *K. hirtipes* and active development has commenced in six eggs, with each protocol represented. During the warming that follows chilling, the lag in commencement of active development is similar to that of *K. alamosae*.

Evidence for diapause in *Kinosternon leucostomum* is based on the prolongation of presomite development at 25 °C or 30 °C (Ewert, 1985). In addition to direct observation (candling of eggs), oxygen consumption has been measured for two eggs from one clutch with different lengths of arrest following oviposition. The results are similar to those shown in Fig. 11.1.

Arrest in *K. scorpioides* embryos responds to chilling (Table 11.1). Viability through hatching at continuous 22.5 °C is high (36 of 45 eggs

hatched). At 30 °C, persistence of arrest tends to be weaker than at 25 °C (Ewert, 1985). However, in either case, few embryos linger at presomite stages and then die.

No embryos of *Kinosternon sonoriense* ($N = 28$) have advanced to somite stages within two months of oviposition, despite attempted incubation at 22.5 °C, 25 °C and 32 °C. The same general series of protocols applied to eggs of *K. alamosae* have been applied to 23 eggs of which 19 embryos have commenced active development and 1 remains arrested at this time of writing. So far, 16 eggs have hatched; in these, 4–12 months of arrest preceded active development. Initial incubation at 32 °C, followed by 25 °C, has given the best results. It appears that chilling at 22.5 °C does not help terminate arrest, that two months of chilling at 16.5 °C may be too brief for many embryos, or that 16.5 °C may be insufficiently cold. Embryos of *K. sonoriense* clearly have the deepest state of diapause within the Kinosternidae, and perhaps among turtles in general.

Diapause is mentioned in a report on *Claudius angustatus* by Vogt & Flores-Villela (1986) but without accompaniment of supporting data. Arrest in *S. salvinii* responds to chilling (Table 11.1) and arrest during early development has been described by Sachsse & Schmidt (1976). Arrest of embryos of *S. triporcatus* responds to chilling (Table 11.1) and a temperature of 22.5 °C not only terminates diapause but also allows development to proceed through hatching (43 eggs hatched).

Emydidae

Although periods of chilling of *D. reticularia* eggs terminate diapause (Ewert, 1985; Jackson, 1988), completely effective durations and optimum temperatures remain to be identified. Diapause is deep in individuals from northern Florida. Some of these embryos persist in diapause following 60 days of chilling, the longest uninterrupted period applied. However, pre-somite embryos are difficult to test because they seem to deteriorate fairly rapidly in absence of chilling. In unchilled eggs, proliferation of extra-embryonic tissue assumes large vacuoles and sheets that are abandoned if the embryo is able to resume development.

In embryos of *M. trijuga*, arrest responds to chilling (Table 11.1) (Ewert, 1985). In *M. trijuga coronata* nearly all unchilled eggs from two pairs of adults have shown arrest at 25 °C and 30 °C, and several at 30 °C have died. In *M. t. thermalis*, only one in over 100 embryos from four pairs of adults has shown arrest. Crosses between the two subspecies has given only embryos with short periods of arrest or with arrest lacking.

In *R. pulcherrima*, arrest responds to chilling (Table 11.1). Only one of five embryos maintained (at 25 °C or 30 °C) without any chilling has been able to resume active development. In this species (primarily *R. p. incisa*) as well as in *D. reticularia*, the optimum temperature for chilling appears to be too cool to promote active development once diapause is terminated. Hence, the post-diapause embryos must be warmed to become active.

Testudinidae
Presomite development at 28 °C, 30 °C, and 32 °C is much prolonged in *Malacochersus tornieri* (to 2 months at 30 °C) relative to that following Stage 8⁺ (just over 3 months at 30 °C). In species without arrest, proportionate times are 1:7 to 1:10 (Ewert, 1985) as opposed to 2:3 in *M. tornieri*. Still, following the inactive period, development resumes spontaneously at each of the three listed temperatures (11 embryos observed). A peculiarity in this species, as well as in *C. parkeri* and *R. pulcherrima*, is that some eggs delay adhesion of the vitelline membrane to the shell membrane for several weeks but eventually do achieve it prior to active development. In most turtles, adhesion follows oviposition by 1–4 days and closely precedes chalking or banding. However, eggs of *M. tornieri* do not chalk and eggs of the other two species chalk slowly and irregularly.

Given the great variation in durations of incubation reported for various tortoises (Ewert, 1979, 1985), the presence of diapause is a possible explanation although existing observations do not exclude cold torpor or aestivation.

Trionychidae
Preliminary evidence suggests the presence of diapause in *Trionyx (Aspideretes) gangeticus*. As determined through candling on 11 March 1987, thirty nest eggs collected from 21 August through 22 November 1986 included 28 eggs with seemingly healthy presomite embryos and just two with more advanced stages (both from 21 August, at Stage 8⁺ and at Stage 13–14). Six embryos in this group eventually advanced development and one hatched after 260 days of incubation. Although these eggs had been

allowed to incubate under ambient conditions, careful records indicate that temperatures during the month following 21 August held primarily around 29–30 °C. During the winter, 50 days passed at 25–26 °C, 29 at 24–25 °C and just two at 23–24 °C. Two weeks prior to candling, temperatures had warmed to 26–27 °C (J. W. Lang & H. Andrews, unpublished data). Hence, ambient conditions should have allowed appreciably more development in the absence of diapause. Also, given that the warm period initially available to the clutch laid in August had failed to promote development and yet the slightly cooler one of late winter gave some success, the data suggest that termination of diapause requires chilling. This species is native to northern India (R. Whitaker, personal communication) while the observations were made in southern India (Madras Crocodile Bank). Perhaps the persistence of many pre-somite embryos after late winter warming reflects a lack of sufficient natural chilling in the south to terminate diapause in many embryos.

Evidence in support of diapause in *Lissemys punctata* is similar to that for *T. gangeticus*, except that the eggs were laid between early November and February, the coolest part of the year. One nest egg found on 12 November 1986 still retained a pre-somite embryo on 11 March 1987 but later resumed development and hatched 270 days after having been found.

Chamaeleonidae
Chamaeleonid lizards are presently the only reptiles than turtles in which embryonic diapause is known or very likely. Although strong evidence is limited to a few species, diapause may be widespread among the oviparous forms because the egg stage typically is long (Bustard, 1989) and nesting often occurs in the late summer or autumn, e.g. *Chamaeleo zeylanicus* (Minton, 1966; Rao & Rao, 1988).

According to a detailed embryonic series of *Chamaeleo (Furcifer) lateralis*, incubation at 19–22 °C includes 18 weeks in gastrulation, almost all at stages 10–11, and 23.5 weeks in postgastrular active development, stages 11–44 (Blanc, 1974). Hatching can be enhanced by giving the eggs a two-month period of chilling (at 10–18 °C) in advance of incubation at ambient 25–30 °C (Schmidt, 1986). By withholding initiation of a 21-day period of chilling at 10 °C from day 30 to day 230 of incubation it has been possible to vary completion of development (at 20 °C) from 154 to 378 days (Schmidt, 1988).

Evidence for diapause in *Chamaeleo chamaeleon*, while old, is not as good as for *C. lateralis*. Following oviposition in the autumn, embryos develop through about two Blanc (1974) stages (from 8 to 10) and become arrested as advanced gastrulae that persist despite ambient temperatures warm enough to sustain development (Peter, 1935). Arrest persists in the presence of warmer temperatures (27–28 °C) and many embryos die while others that do live and resume development yield a high proportion of deformities (Bons & Bons, 1960). Eggs incubated at ambient 24–29 °C for five to six months have yielded embryos suitable for the study of limb development (Hurle *et al.*, 1987).

Chamaeleo (*Furcifer*) *polleni*: Eggs of this insular species that have been held at 28–31° throughout incubation have succeeded in developing to hatching (17 of 35 eggs) in 260–270 days (Leptien, 1988). Although no data are given on developmental arrest, this seems likely given the long incubation period and the data on other species.

Delayed hatching and embryonic aestivation

During delayed hatching and aestivation, the fully developed embryo prolongs its stay within the egg, usually remaining unpipped. In the present context, delayed hatching tends to last only a fairly brief period, a few days in birds to three weeks in reptiles. However, such instances of delay are clearly evident relative to a lack of delay in other individuals of the same and related species. The periods of delay are accompanied by changing, usually decreasing metabolism. Aestivation occurs when the embryo remains in the egg for very long periods. Metabolism has become very low and tends to remain steady or to decline slightly. Interest in delayed hatching extends to reptiles (Webb, Choquenot & Whitehead, 1986; Auffenberg, 1988) and birds (Vleck, Vleck & Hoyt, 1980; Vince, Ockleford & Reader, 1984; Cannon, Carpenter & Ackerman, 1986). Embryonic aestivation in turtles is treated briefly in Ewert (1985) but is unknown in birds. In both reptiles and birds, chronological changes in embryonic metabolism range from patterns that are incompatible with delayed hatching through those that indicate it and this allows comparisons of similarities and differences between reptiles and birds.

Reptiles

Delayed hatching can precede pipping in the crocodilians *Crocodylus johnstoni*, *Crocodylus porosus* and *Alligator mississippiensis* (Whitehead, 1987a,b; Thompson, 1989) and in the turtles *Emydura macquarii* and *Carettochelys insculpta* (Webb *et al.*, 1986; Thompson, 1989). It probably is widespread in turtles including Kinosternidae, Chelydridae, and Emydidae in addition to the above, but Ewert (1985) does not clearly differentiate it from aestivation. Delayed hatching, possibly in combination with aestivation, may occur in varanid lizards (Auffenberg, 1988).

Three types of evidence support a concept of delayed hatching. In the most superficial, a clutch of eggs hatches asynchronously over several days to weeks even though the eggs had been incubated together. This may, however, indicate temperature gradients within a nest, e.g. *A. mississippiensis* (Ferguson & Joanen, 1983; Packard & Packard, 1988). As candling early in development reveals that the embryos were at about the same stage and developing actively, diapause could not be responsible for any delay. In most species of turtles, the embryo remains enclosed in the chorio-allantois (from the perspective of candling) until shortly before hatching. The chorio-allantois parts following rupture of the amnion and gradually slips toward the posterior of the turtle leaving its head and forelimbs exposed to pip the eggshell. If the turtle does not pip within a typical few days, its hatching may be regarded as delayed. Hence, a record of elapsed time from the parting of the chorio-allantois provides a second line of evidence for delayed hatching. However, in the turtle *K. baurii*, considerable heterochrony may accompany the parting and this may well diminish its value in diagnosis (Ewert, 1985). Following the initial parting of the chorio-allantois, nearly the entire body of the turtle eventually becomes exposed within the eggshell and the turtle gradually enters aestivation.

Another indicator of delayed hatching is internalisation of yolk. Many turtles begin to hatch while their yolk sac lies external to the plastron (Fig. 11.2). With yolk internalisation, embryonic development is complete, except for final withdrawal of blood from any remaining chorio-allantois. This stage has been observed up to 25 days before pipping in *C. insculpta* (Webb *et al.*, 1986).

The third type of evidence comes from the measurement of metabolism. This work provides a means to distinguish species in which delayed hatching is unknown and unlikely from others that might have delayed hatching as well as those that do have it. Chronological patterns of oxygen consumption ($\dot{V}O_2$) vary from exponential through pipping to sigmoid and peaked, with pipping coming well after and below the peak (Whitehead, 1987*b*; Thompson, 1989; Vleck & Hoyt, Chapter 18). Exponential patterns typify many snakes (Dmi'el, 1970; Black *et al.*, 1984) and only slightly exponential (i.e. toward sigmoid) patterns which occur in some other snakes (Clark, 1953) and typify sea turtles (Ackerman, 1981). These species may not have the capacity for delayed hatching. Rather, tolerance for premature pipping and hatching may be more likely. At least some snakes compartmentalise their yolk (Stewart & Blackburn, 1988) and seem adept at abandoning portions of it under both adverse and apparently normal conditions of incubation (Burger, Zappalorti & Gochfeld, 1987; Deeming, 1989). Sea turtles tend to pip with large external yolks (Miller, 1985), which they rapidly internalise and use for posthatching dispersal (Kraemer & Bennett, 1981), rather than for basal metabolism within the egg.

Patterns of oxygen consumption in the turtle, *E. macquarii*, in *C. porosus* and *C. johnstoni* and in *A. mississippiensis* are peaked at about 90% of incubation, with pipping following a modest (20–25%) decline in oxygen consumption (Thompson, 1983, 1985, 1989; Whitehead, 1987*a,b*). Both workers consider that pipping can be delayed slightly. As patterns in the turtles, *Chelydra serpentina* (Gettinger, Paukstis & Gutzke, 1984; Ewert, unpublished data), *Trionyx spiniferus* (Gettinger *et al.*, 1984), and in eight of ten additional species (Fig. 11.1; Ewert, unpublished data) are also peaked, this pattern may typify most turtles, even without delayed hatching, and may pre-adapt them to delayed hatching. Turtles and crocodilians do not compartmentalise their yolk and (except for rare instances) must internalise all of it during hatching, even if subsequent health is compromised (Ewert, 1979, 1985; Ferguson, 1985). Hence, delayed hatching may be a natural derivation from this need.

Delayed hatching in the turtle *C. insculpta* is accompanied by a steady decline in oxygen consumption over a maximum observed period of 30 days and to a minimum of 33% of the peak rate at 30 °C (Webb *et al.*, 1986). Oxygen consumption in eggs (incubated and measured at 30 °C) of the turtles *M. trijuga* and *S. minor* declines to average minima of 50% ($N = 10$ eggs observed, extreme = 36%) and 56% ($N = 16$ eggs, extreme = 42%), respectively, over averages of 36 and 24 days, respectively, following peak oxygen consumption. Hence, these turtles probably delay hatching as in *C. insculpta*. However, in *M. trijuga* pipping occasionally lags up to 75 days after the peak and from a logical perspective, the decline in oxygen consumption during this period must cease at some low plateau, as in aestivation.

In a consideration of aestivation, details of certain temporal and anatomical details are useful in an understanding of the phenomenon. In personally observed turtles with peaked patterns of oxygen consumption, the parting of the chorio-allantois follows the peaking in $\dot{V}O_2$ by several days to weeks. A few turtles may remain in their unpipped eggs for an additional several weeks to months and eventually, may become weakened in absence of a stimulus to pip. Some example durations leading up to striking an egg, or otherwise causing it to break, include 40 days in *P. geoffroanus*, 41 days in *Phrynops gibbus*, 116 days in *K. scorpioides*, and 232 days in *Kinosternon flavescens*. Eggs opened within a month or two of the onset of retraction of the chorio-allantois show that the embryo retains a well-vascularised portion of it. It forms a ventral pad in *K. scorpioides* (Fig. 11.2; Ewert, 1985). In *P. gibbus*, the embryo comes to lie on its back (Lehmann, 1987) and the chorio-allantois covers the plastron, much of the margin of the turtle shell and the posterior carapace. It remains attached to the turtle by a narrow umbilical cord (Fig. 11.2).

Metabolism has been followed in *K. scorpioides* for up to 60 days (at 30 °C) following retraction of the chorio-allantois (Fig. 11.1). Oxygen consumption initially declines rapidly and then nearly levels off at an average of 22–23% ($N = 10$ monitored eggs) of the peak value and at 56% of the low average value for the same turtles measured in a resting state a few days after hatching.

Delayed hatching in *C. insculpta* has been terminated (i.e. hatching induced) by soaking

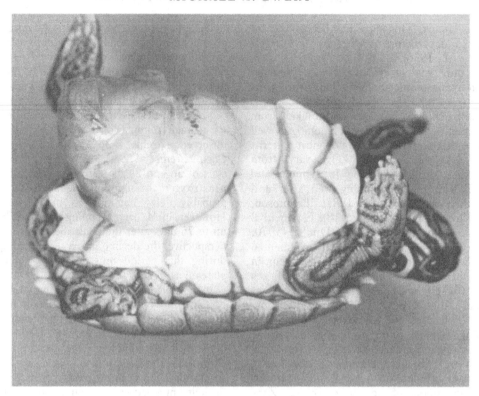

(a)

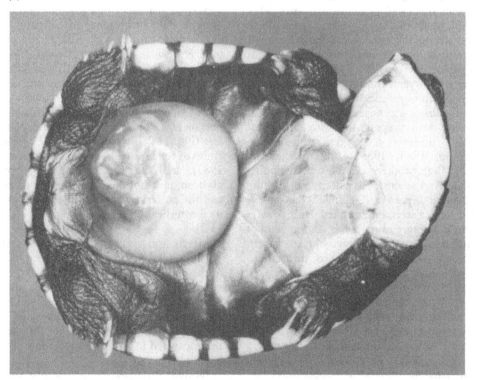

(b)

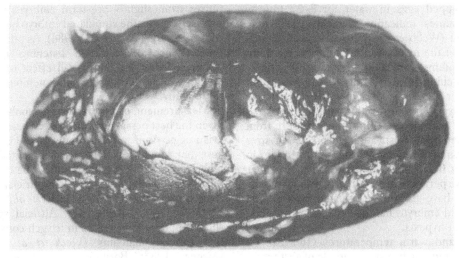

(c)

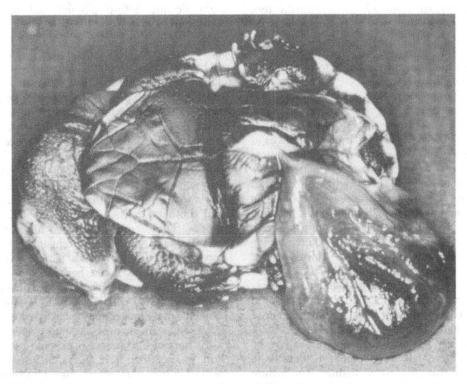

(d)

Fig. 11.2. Status of yolk internalisation in turtles that pip just after completion of differentiation and growth (a), (b) and others that have periods of delayed hatching and aestivation (c), (d). (a) *Graptemys nigrinoda*, 1 day after pipping. (b) *Pelomedusa subrufa*, 1 day after pipping – the chorio-allantois is extensively withered. (c) *Kinosternon scorpioides*, 1 hour after pipping – note the small, vascularised pad of chorio-allantois covering the umbilical region. (d) *Phrynops gibbus*, 1 day after pipping – although the umbilical seam is almost closed, a large chorio-allantoic sac remains intact and vascularised suggesting continued active circulation of blood.

the unpipped eggs in water for 3–27 (mostly 3–4) minutes indicating that hypoxia is the stimulant (Webb *et al.*, 1986). Very substantial wetting of the incubation medium (to the point of near submersion of the eggs in water) stimulates hatching in aestivating *K. scorpioides*, but only after an average period of 4 days (range 1–14 days). One aestivating turtle became so hydrated that the egg split under pressure from its swollen body. Aestivating eggs of *K. flavescens arizonense* have hatched after submersion in water for 0.4–3 hours (*N* = 6 eggs). Six other eggs have persisted for 4–8 hours without hatching and have had to be assisted because the unhatched embryos had become weak, presumably from hypoxia.

Cool and warm temperatures (16.5/32 °C) stimulate pipping in a small sample of aestivating eggs of *K. scorpioides* from 25 °C, whereas the same protocol initially lacked effect on eggs of *K. f. arizonense* from 27 °C. Eventually (two months after the initial testing), hatching accompanied the cooling phase. Hence, increased moisture may be the stronger natural stimulus but the most natural form of application remains unclear.

Birds

It is possible to view two separate phenomena as delayed hatching. In one, the advanced embryo of the wedge-tailed shearwater (*Puffinus pacificus*) pips its eggshell about eight days before it completes hatching. As this lag of hatching behind pipping is unusual among birds, hatching can be viewed as delayed, or prolonged. However, it is pipping time that is altered having heterochronously advanced to a less mature stage of development (Ackerman *et al.*, 1980). The second form of delayed hatching is a postponement of hatching beyond the normal, or at least necessary time or stage for completion of development (Vleck, Vleck & Hoyt, 1980). Alternatively, 'normal' hatching time may be optimally late but development is flexible to allow full hatching at a slightly less mature embryonic stage (Cannon *et al.*, 1986). This flexibility occurs in several species of precocial birds that continue to add eggs to a clutch after initiating incubation, and thus onset of development in the first eggs.

Eggs of bobwhite quail (*Colinus virginianus*) can be caused to delay hatching by applying a rigid protocol of 'clicks' (to simulate mechanical pulses from slightly pre-term embryos) rather precisely before the stimulated embryo begins its own breathing (Vince *et al.*, 1984). As other protocols tend to accelerate hatching in less advanced embryos, the overall effect of 'click' stimulation on synchrony in hatching awaits more study.

Measurement of embryonic metabolism has given the best perspective on delayed hatching in avian eggs. As in reptiles, patterns of $\dot{V}O_2$ suggest that some species have less potential than others for altering hatching times. The cleanest dichotomy distinguishes clearly altricial from clearly precocial species (Vleck *et al.*, 1979; Vleck & Hoyt, Chapter 18). Altricial species show exponential increases in oxygen consumption through hatching (Vleck *et al.*, 1979; Bucher, 1983; Bartholomew & Goldstein, 1984). Precocial species show sigmoid increases in oxygen consumption, often with peaking and a slight decline before pipping and hatching (Romanoff, 1967; Hoyt, Vleck & Vleck, 1978; Calder, 1979; Vleck *et al.*, 1979, 1980; Hoyt *et al.*, 1979; Vleck, Vleck & Seymour, 1984).

Oxygen consumption in older eggs (within a clutch) of the common rhea (*Rhea americana*) and Emu (*Dromicaius novaehollandiae*) peaks at about 75% of incubation and then declines by about 25% of the peak value. The postpeak period, which is long among birds, is viewed as a delaying of hatching (Vleck *et al.*, 1980). However, in experimentally staggered incubation schedules, eggs of the Darwin's rhea (*Pterocnemia pennata*) that started incubation behind others, hatched after shorter incubation periods. These embryos hatched just after the peak in $\dot{V}O_2$ as opposed to remaining in their eggs for the full post-peak period. Whereas these observations might beg a distinction of whether it is 'normal' or prolonged to have a post-peak period of incubation, young that hatched after longer post-peak periods had better long term survival, and thus may be more nearly normal (Cannon *et al.*, 1986).

In species that have peaks, followed by a decline, in oxygen consumption prior to pipping, there is a very rapid increase, an approximate doubling of the pre-pipping $\dot{V}O_2$ during pipping and hatching (Romanoff, 1967; Vleck *et al.*, 1980). This increase may be associated with initiation of homeothermy, although a functional homeothermic response is not always present at hatching (Booth, 1987a). The need for this great increase, however, may constrain delayed hatching. In megapodes, which hatch with greater precocity than other birds (Booth & Thompson, Chapter 20), achievement of this advanced state is not usually associated a peaked pattern of oxygen consumption; rather, it is sigmoid with an increase to pipping or with a brief plateau occurring just before (Vleck *et al.*, 1984; Booth, 1987*b*). The megapode *Leipoa ocellata* becomes moderately endothermic during the three days immediately preceeding hatching (Booth, 1987*a*). Just prior to hatching, metabolic rates in megapodes are high in contrast to other birds yet following hatching, $\dot{V}O_2$ doubles as in other pre-cocial species (Vleck *et al.*, 1984).

In conclusion

Mechanisms used by birds to prolong the egg stage are temporally less varied and show less physiological specialisation than those used by reptiles. Further, cold torpor, apparently the most common mechanism in birds, seems relatively rare in reptiles. It is replaced by embryonic diapause. This difference probably reflects different objectives. Birds appear to need mechanisms that offer short-term flexibility and responsiveness to nest attendance and active incubation. It is hard to imagine how embryonic diapause or aestivation can be adaptive in species with appreciable parental investment in tending or guarding their eggs. Such major prolongations of the egg stage would only increase the cost of this investment. Among reptiles, neither diapause nor aestivation are known in crocodilians or in the few lizards and snakes that show parental care of eggs (Deeming, Chapter 19).

Cold torpor in avian embryos represents a sacrifice by the embryo to meet ovipositional and incubational schedules of the attending adult. From the perspective of the individual embryo, cold torpor only increases risk of flawed development or exposure to predation. However, among various procellariiform birds or birds that produce large clutches of precocial young, probably few embryos are spared significant cold torpor. Alternatively, embryonic diapause and aestivation are clearly advantageous to the embryo or young because they prolong a presumably 'safe' egg stage through periods of inclement weather (e.g. dry seasons, chills) and allow synchronisation of hatching with optimal conditions for hatchling growth. Ultimately, the schedule of the nesting adult

may be involved in the timing of hatching, i.e. to produce eggs during and at the close of periods when food is abundant. However, given climatic variability and transient nesting conditions (e.g. dry season hardening of clay soils), it might be impossible for the adults of diapausing and aestivating species to schedule nesting so that active development just fills the time from laying to the best time for hatching.

An interpretation of delayed hatching in birds (Vleck *et al.*, 1980), while equivocal (Cannon *et al.*, 1986), has inspired parallel interpretation in reptiles (Thompson, 1989). However, pipping leads to a marked contrast among the two groups. Metabolism in reptiles tends to continue to decline at the same rate that it was declining a little before pipping (Thompson, 1989) whereas metabolism in birds increases enormously (Vleck *et al.*, 1980). The pre-pipping decline in metabolism seen in some birds is slight in contrast to reptiles and only lasts a few days. The decline in avian metabolism may be limited by the onset of homeothermy. Reptilian metabolism, by contrast, can continue a decline into aestivation. Further, at least one of the species (*K. flavescens*) that aestivates within the egg also aestivates as an adult (Seidel, 1978).

The geography of develomental arrest in reptiles promises to be an interesting area for future study. Presently, data allow only suggestions of trends. For instance, embryonic diapause appears to be most closely associated with strongly seasonal climates that have large fluctuations in rainfall combined with modest fluctuations in temperature that exclude appreciable freezing. Thus, diapause seems particularly well represented in peninsular India and in northwestern and southern Mexico, where perhaps half of the local species of turtles have it. In both areas, there is appreciable summer, especially late summer rain and a cooler dry season. In a contrast with Mexico, diapause is relatively rare in the turtle fauna of the less xeric and more freeze-prone southeastern United States and has yet to be established in any species from northern South America (five species examined so far and only one promising candidate among the remainder), which has little seasonal variation in temperature. However, seasonal subtropical climates are not limited to India and Mexico. In particular, a diapausing species has yet to be identified in southeastern Asia and southern China and good negative data are available for at least six species.

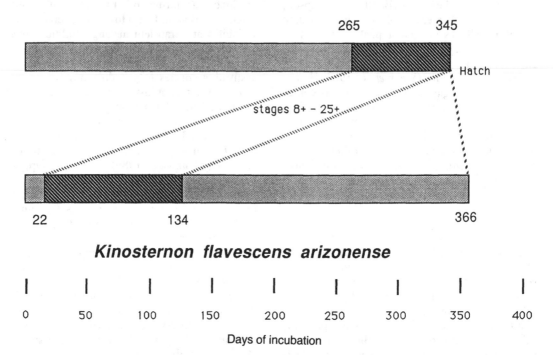

Fig. 11.3. Contrasting patterns of developmental arrest in two kinosternid turtles of the Sonoran Desert, Arizona and Sonora. The horizontal bars represent developmental chronologies of the two listed species. Dark shading denotes all but the first two to three weeks of the period of active differentiation and growth (Stages 8+ – 25+ of Yntema, 1968) at 27 °C (*K. sonoriense*) or at 25 °C (*K. flavescens*). *Upper bar:* the large light area represents a period of embryonic diapause in a pre-somite embryo. Incubation temperature was changed many times (from 20–32 °C and intermediate shifts) in an attempt to stimulate development before it actually commenced. *Lower bar:* light shading to the right represents a period of delayed hatching and embryonic aestivation at 25 °C. Stimulation of hatching required about an hour of submergence in water. Either strategy of developmental arrest prolongs the egg stage to nearly a year under laboratory conditions.

There is even less information available on the geography of delayed hatching and aestivation except that extreme cases appear to be related to drought conditions. On occasion, diapause and aestivation appear to be alternative adaptations toward the same goal, such as to synchronise nesting in the mid to late summer with hatching during rains of the following summer (Fig. 11.3).

Acknowledgements

For providing me with breeding turtles, gravid females, or fertile eggs, or for field assistance with the collection of this material, I thank staff of the Columbus Zoo (M. Goode, R. E. Hatcher, D. D. Badgley) and C. R. Etchberger, J. B. Iverson and L. A. Lantz. Constant temperature incubators were made available by M. A. Watson, R. C. Richmond and through a loan from the Carnegie Museum of Natural History (C. J. McCoy). The Gilson Differential Respirometer was made available by H. C. Prange, who, with P. J. Clark, provided instruction on its operation. C. E. Nelson provided space for some aspects of this project. I thank A. Georges for permission to mention the unpublished findings of R. M. Kennett and A. Georges.

References

Ackerman, R. A. (1981). Oxygen consumption by sea turtle (*Chelonia, Caretta*) eggs during development. *Physiological Zoology*, 54, 316–24.

Ackerman, R. A., Whittow, G. C., Paganelli, C. V. &

Pettit, T. N. (1980). Oxygen consumption, gas exchange, and growth of embryonic wedge-tailed shearwaters (*Puffinus pacificus chlororhynchus*). *Physiological Zoology*, 53, 210–21.

Arora, K. L. & Kosin, I. L. (1968). The response of the early chicken embryo to preincubation temperature as evidenced from its gross morphology and mitotic pattern. *Physiological Zoology*, 41, 104–12.

Auffenberg, W. (1988). *Gray's Monitor Lizard*. Gainesville: University of Florida Press.

Bartholomew, G. A. & Goldstein, D. L. (1984). The energetics of development in a very large altricial bird, the brown pelican. In *Respiration and Metabolism of Embryonic Vertebrates*, ed. R. S. Seymour, pp. 347–57. Dordrecht: Dr W. Junk Publishers.

Black, C. P., Birchard, G. F., Schuett, G. W. & Black, V. D. (1984). Influence of incubation water content on oxygen uptake in embryos of the Burmese python (*Python molurus bivittatus*). In *Respiration and Metabolism of Embryonic Vertebrates*, ed. R. S. Seymour, pp. 137–45. Dordrecht: Dr W. Junk Publishers.

Blackburn, D. G. & Evans, H. E. (1986). Why are there no viviparous birds? *American Naturalist*, 128, 165–90.

Blanc, F. (1974). Table de développement de *Chamaeleo lateralis* Gray, 1831. *Annales d'Embryologie et de Morphogénèse*, 7, 99–115.

Boersma, P. D. & Wheelwright, N. T. (1979). Egg neglect in the Procellariiformes: reproductive adaptations in the fork-tailed storm-petrel. *Condor*, 81, 157–65.

Boersma, P. D., Wheelwright, N. T., Nerini, M. K. & Wheelwright, E. S. (1980). The breeding biology of the fork-tailed storm-petrel (*Oceanodroma furcata*). *Auk*, 97, 268–82.

Bons, J. & Bons, N. (1960). Notes sur la reproduction et le développement de *Chamaeleo chamaeleon* (L.). *Bulletin Société des Sciences Naturelles et Physiques du Maroc*, 40, 323–35.

Booth, D. T. (1987a). Metabolic response of mallee fowl *Leipoa ocellata* embryos to cooling and heating. *Physiological Zoology*, 60, 446–53.

— (1987b). Effect of temperature on development of mallee fowl *Leipoa ocellata* eggs. *Physiological Zoology*, 60, 437–45.

Bucher, T. L. (1983). Parrot eggs, embryos, and nestlings: patterns and energetics of growth and development. *Physiological Zoology*, 56, 465–83.

Burger, J., Zappalorti, R. T. & Gochfeld, M. (1987). Developmental effects of incubation temperature on hatchling pine snakes *Pituophis melanoleucus*. *Comparative Biochemistry and Physiology*, 87A, 727–32.

Bustard, R. (1989). Keeping and breeding oviparous chameleons. *British Herpetological Society Bulletin*, 27, 18–33.

Cagle, F. R. (1950). The life history of the slider turtle, *Pseudemys scripta troostii* (Holbrook). *Ecological Monographs*, 20, 33–54.

Cairncross, B. L. & Greig, J. C. (1977). Note on variable incubation period within a clutch of eggs of the leopard tortoise (*Geochelonia ardalis*) (Chelonia: Cryptodia: Testudinidae). *Zoologica Africana*, 12, 255–6.

Calder, W. A. (1979). The kiwi and egg design: evolution as a package deal. *Bioscience*, 29, 461–7.

Calder, W. A. & Booser, J. (1973). Hypothermia of broadtailed hummingbirds during incubation in nature with ecological implications. *Science*, 180, 751–3.

Cannon, M. E., Carpenter, R. E. & Ackerman, R. A. (1986). Synchronous hatching and oxygen consumption of Darwin's rhea eggs (*Pterocnemia pennata*). *Physiological Zoology*, 59, 95–108.

Christian, K. A., Tracy, C. R. & Porter, W. P. (1986). The effect of cold exposure during incubation of *Sceloporus undulatus* eggs. *Copeia*, 1986, 1012–14.

Clark, H. (1953). Metabolism of the black snake embryo. II. Respiratory exchange. *Journal of Experimental Biology*, 30, 502–5.

Cox, W. A. & Marion, K. R. (1978). Observations on the female reproductive cycle and associated phenomena in spring-dwelling populations of *Sternotherus minor* in north Florida (Reptilia: Testudines). *Herpetologica*, 34, 20–33.

Cunningham, B. (1922). Some phases in the development of *Chrysemys cinerea*. *Journal of the Elisha Mitchell Scientific Society*, 38, 51–73.

Deeming, D. C. (1989). The residues in the eggs of squamate reptiles at hatching. *Herpetological Journal*, 1, 381–5.

Deeming, D. C. & Ferguson, M. W. J. (1989). Effects of incubation temperature on growth and development of embryos of *Alligator mississippiensis*. *Journal of Comparative Physiology*, 159B, 183–93.

Dmi'el, R. (1970). Growth and metabolism in snake embryos. *Journal of Embryology and Experimental Morphology*, 23, 761–72.

El Jack, M. H. & Kaltofen, R. S. (1968). The effect of high holding and housing temperatures on hatchability of chicken eggs. *Poultry Science*, 48, 1013–18.

Ewert, M. A. (1979). The embryo and its egg: development and natural history. In *Turtles: Perspectives and Research*, eds M. Harless & H. Morlock, pp. 333–413. New York: John Wiley & Sons.

— (1985). Embryology of turtles. In *Biology of the Reptilia*, vol. 14, eds C. Gans, F. Billett & P. F. A. Maderson, pp. 76–267. New York: John Wiley & Sons.

Ferguson, M. W. J. (1985). Reproductive biology and embryology of the crocodilians. In *Biology of the Reptilia*, vol. 14, eds C. Gans, F. Billett & P. F. A. Maderson, pp. 329–491. New York: John Wiley & Sons.

Ferguson, M. W. J. & Joanen, T. (1983). Temperature dependent sex determination in *Alligator mississippiensis*. *Journal of Zoology, London*, **200**, 143–77.

Fritz, U. & Jauch, D. (1989). Haltung, Balzverhalten und Nachzucht von Parkers Schlangenhalsschildkröte *Chelodina parkeri* Rhodin & Mittermeier, 1976 (Testudines: Chelidae). *Salamandra*, **25**, 1–13.

Gettinger, R. D., Paukstis, G. L. & Gutzke, W. H. N. (1984). Influence of hydric environment on oxygen consumption by embryonic turtles *Chelydra serpentina* and *Trionyx spiniferus*. *Physiological Zoology*, **57**, 468–73.

Goff, C. C. & Goff, D. S. (1932). Egg laying and incubation of *Pseudemys floridana*. *Copeia*, **1932**, 92–4.

Goldberg, S. R. (1971). Reproductive cycle of the ovoviviparous iguanid lizard *Sceloporus jarrovi* Cope. *Herpetologica*, **27**, 123–31.

Goode, J. & Russell, J. (1968). Incubation of eggs of three species of chelid tortoises, and notes on their embryological development. *Australian Journal of Zoology*, **16**, 749–61.

Greenwood, R. J. (1969). Mallard hatching from an egg cracked by freezing. *Auk*, **86**, 752–4.

Hoyt, D. F., Board, R. G., Rahn, H. & Paganelli, C. V. (1979). The eggs of Anatidae: conductance, pore structure, and metabolism. *Physiological Zoology*, **52**, 438–50.

Hoyt, D. F., Vleck, D. & Vleck, C. M. (1978). Metabolism of avian embryos: ontogeny and temperature effects in the ostrich. *Condor*, **80**, 265–71.

Hurle, J. M., Garcia-Martinez, V., Gañan, Y., Climent, V. & Blasco, M. (1987). Morphogenesis of the prehensile autopodium in the common chameleon (*Chamaeleo chamaeleo*). *Journal of Morphology*, **194**, 187–94.

Jackson, D. R. (1988). Reproductive strategies of sympatric freshwater emydid turtles in northern peninsular Florida. *Bulletin of the Florida State Museum, Biological Sciences*, **33**, 115–58.

Kraemer, J. E. & Bennett, S. H. (1981). Utilization of posthatching yolk in loggerhead sea turtles, *Caretta caretta*. *Copeia*, **1981**, 406–11.

Lance, V. A. (1987). Hormonal control of reproduction in crocodilians. In *Wildlife Management: Crocodiles and Alligators*, eds G. J. W. Webb, S. C. Manolis & P. J. Whitehead, pp. 409–15. Chipping Norton, New South Wales: Surrey Beatty & Sons.

Legler, J. M. (1960). Natural history of the ornate box turtle, *Terrapene ornata ornata* Agassiz. *University of Kansas Publications, Museum of Natural History*, **11**, 527–669.

— (1985). Australian chelid turtles: reproductive patterns in wide-ranging taxa. In *Biology of Australasian Frogs and Reptiles*, eds G. Grigg, R. Shine & H. Ehmann, pp. 117–23. Chipping Norton, New South Wales: Surrey Beatty & Sons.

Lehmann, H. (1987). Hypothetische Überlegungen zur Schlupfproblematik von künstlich inkubierten Gelegen südamerikanischer Schildkrötenarten der Familie Chelidae. *Salamandra*, **23**, 73–7.

— (1988). Beobachtungen bei einer ersten Nachzucht von *Platemys spixii* (Dumeril & Bibron, 1835) (Testudines, Chelidae). *Salamandra*, **24**, 1–6.

Leptien, R. (1988). Haltung und Nachzucht von *Furcifer polleni* (Peters, 1873) (Sauria: Chamaeleonidae). *Salamandra*, **24**, 81–6.

Lundy, H. (1969). A review of the effects of temperature, humidity, turning and gaseous environment on the hatchibility of the hen's egg. In *The Fertility and Hatchability of the Hen's Egg*, eds T. C. Carter & B. M. Freeman, pp. 143–76. Edinburgh: Oliver & Boyd.

MacMullan, R. A. & Eberhardt, L. L. (1953). Tolerance of pheasant eggs to exposure. *Journal of Wildlife Management*, **17**, 322–30.

Matthews, G. V. T. (1954). Some aspects of incubation in the manx shearwater, *Procellaria puffinus*, with particular reference to chilling resistance in the embryo. *Ibis*, **96**, 432–40.

Miller, J. D. (1985). Embryology of marine turtles. In *Biology of the Reptilia*, vol. 14, eds C. Gans, F. Billett & P. F. A. Maderson, pp. 270–328. New York: John Wiley & Sons.

Minton, S. A., Jr. (1966). A contribution to the herpetology of West Pakistan. *Bulletin of The American Museum of Natural History*, **134**, 29–184.

Moffat, L. A. (1985). Embryonic development and aspects of reproductive biology in the tuatara, *Sphenodon punctatus*. In *Biology of the Reptilia*, vol. 14, eds C. Gans, F. Billett & P. F. A. Maderson, pp. 494–521. New York: John Wiley & Sons.

Nolan, V. (1978). The ecology and behavior of the prairie warbler *Dendroica discolor*. *Ornithological Monographs*, **26**, 1–595.

Packard, G. C. & Packard, M. J. (1988). The physiological ecology of reptilian eggs and embryos. In *Biology of the Reptilia*, vol. 16, eds C. Gans & R. B. Huey, pp. 523–605. New York: Alan R. Liss.

Peter, K. (1935). Die innere Entwicklung des Chamäleonkeimes nach der Furchung bis sum Durchbruch des Urdarms. *Zeitschrift für Anatomie und Entwicklungsgeschichte*, **104**, 1–60.

Pieau, C. & Dorizzi, M. (1981). Determination of temperature-sensitive stages for sexual differentiation of the gonads in embryos of the turtle, *Emys orbicularis*. *Journal of Morphology*, **170**, 373–82.

Rao, M. V. S. & Rao, K. K. (1988). Eggs and hatching of the Indian chamaeleon, *Chamaeleo zeylanicus*, Laurenti. In *10th International Herpetological Symposium on Captive Propagation & Husbandry*, ed. K. H. Peterson, pp. 124–6. Thurmont, Maryland: Zoological Consortium, Inc.

Rol'nik, V. V. (1970). *Bird Embryology*. Jerusalem: Israel Program for Scientific Translations.

Romanoff, A. L. (1967). *Biochemistry of the Avian Embryo*. New York: John Wiley & Sons.
— (1972). *Pathogenesis of the Avian Embryo*. New York: John Wiley & Sons.
Rose, W. (1962). *The Reptiles and Amphibians of Southern Africa*, 2nd edn. Cape Town: Maskew Miller.
Sachsse, W. & Schmidt, A. A. (1976). Nachzucht in der zweiten Generation von *Staurotypus salvinii* mit weiteren Beobachtungen zum Fortpflanzungsverhalten (Testudines, Kinosternidae). *Salamandra*, **12**, 5–16.
Schmidt, W. (1986). Über die Haltung und Zucht von *Chamaeleo lateralis* (Gray, 1831) (Sauria: Chamaeleonidae). *Salamandra*, **22**, 105–12.
— (1988). Zeitigungsversuche mit Eiern des madagassischen Chamäleons *Furcifer lateralis* (Gray, 1831) (Sauria: Chamaeleonidae). *Salamandra*, **24**, 182–3.
Seidel, M. E. (1978). Terrestrial dormancy in the turtle *Kinosternon flavescens*: respiratory metabolism and dehydration. *Comparative Biochemistry and Physiology*, **61A**, 1–4.
Shine, R. (1983). Reptilian reproductive modes: the oviparity–viviparity continuum. *Herpetologica*, **39**, 1–8.
— (1985). The evolution of viviparity in reptiles: an ecological analysis. In *Biology of the Reptilia*, vol. 15, eds C. Gans & F. Billett, pp. 606–94. New York: John Wiley & Sons.
Stewart, J. R. & Blackburn, D. G. (1988). Reptilian placentation: structural diversity and terminology. *Copeia*, **1988**, 839–52.
Thompson, M. B. (1983). *The Physiology and Ecology of the Eggs of the Pleurodiran Tortoise* Emydura macquarii *(Gray), 1831*. PhD Thesis, Adelaide: University of Adelaide.
— (1985). Functional significance of the opaque white patch in eggs of *Emydura macquarii*. In *Biology of Australasian Frogs and Reptiles*, eds G. Grigg, R. Shine & H. Ehmann, pp. 387–95. Chipping Norton, New South Wales: Surrey Beatty & Sons.
— (1989). Patterns of metabolism in embryonic reptiles. *Respiration Physiology*, **76**, 243–56.
Vince, M. A., Ockleford, E. & Reader, M. (1984). The synchronization of hatching in quail embryos: aspects of development affected by a retarding stimulus. *Journal of Experimental Zoology*, **229**, 273–82.
Vleck, C. M., Hoyt, D. F. & Vleck, D. (1979). Metabolism of avian embryos: patterns in altricial and precocial birds. *Physiological Zoology*, **52**, 363–77.
Vleck, C. M. & Kenagy, G. J. (1980). Embryonic metabolism of the fork-tailed storm petrel: physio-

logical patterns during prolonged and interrupted incubation. *Physiological Zoology*, **53**, 32–42.
Vleck, D., Vleck, C. M. & Hoyt, D. F. (1980). Metabolism of avian embryos: ontogeny of oxygen consumption in the rhea and emu. *Physiological Zoology*, **53**, 125–35.
Vleck, D., Vleck, C. M. & Seymour, R. S. (1984). Energetics of embryonic development in the megapode birds, mallee fowl *Leipoa ocellata* and brush turkey *Alectura lathami*. *Physiological Zoology*, **57**, 444–56.
Vogt, R. C. & Flores-Villela, O. A. (1986). Determinación del sexo en tortugas por la temperatura de incubación de los huevos. *Ciencia* (México) **37**, 21–32.
Webb, D. R. (1987). Thermal tolerance of avian embryos: a review. *Condor*, **89**, 874–98.
Webb, G. J. W., Beal, A. M., Manolis, S. C. & Dempsey, K. E. (1987). The effects of incubation temperature on sex determination and embryonic development rate in *Crocodylus johnstoni* and *C. porosus*. In *Wildlife Management: Crocodiles and Alligators*, eds G. J. W. Webb, S. C. Manolis & P. J. Whitehead, pp. 507–37. Chipping Norton, New South Wales: Surrey Beatty & Sons.
Webb, G. J. W., Choquenot, D. & Whitehead, P. J. (1986). Nests, eggs, and embryonic development of *Carettochelys insculpta* (Chelonia: Carettochelidae) from northern Australia. *Journal of Zoology, London B*, **1**, 521–50.
Welty, J. C. (1982). *The Life of Birds*, 3rd edn. New York: Saunders College Publishing.
Wheelwright, N. T. & Boersma, P. D. (1979). Egg chilling and the thermal environment of the fork-tailed storm petrel (*Oceanodroma furcata*) nest. *Physiological Zoology*, **52**, 231–9.
Whitehead, P. J. (1987a). Respiration of *Crocodylus johnstoni* embryos. In *Wildlife Management: Crocodiles and Alligators*, eds G. J. W. Webb, S. C. Manolis & P. J. Whitehead, pp. 473–97. Chipping Norton, New South Wales: Surrey Beatty & Sons.
— (1987b). *Respiration and Energy Utilization in the Eggs of the Australian Freshwater Crocodile*, Crocodylus johnstoni *Krefft, 1873*. MSc Thesis, Adelaide: University of Adelaide.
Yntema, C. L. (1960). Effects of various temperatures on the embryonic development of *Chelydra serpentina*. *Anatomical Record*, **149**, 305–6.
— (1968). A series of stages in the embryonic development of *Chelydra serpentina*. *Journal of Morphology*, **125**, 219–52.
— (1978). Incubation times for eggs of the turtle *Chelydra serpentina* (Testudines: Chelydridae) at various temperatures. *Herpetologica*, **34**, 274–7.

12

Physical factors affecting the water exchange of buried reptile eggs

RALPH A. ACKERMAN

Introduction

The embryonic development of many reptiles takes place within an egg oviposited outside the body of the female. Energy stored in the egg is transformed into tissue as the embryo grows from a few cells to a hatchling capable of independent existence. The transformation requires oxygen while producing carbon dioxide, water and heat. All are exchanged between the embryo and the environment across the egg-shell. If development is to occur successfully, these exchanges must operate within appropriate limits which describe an environmental niche within which development is feasible. Even within these limits development may vary as a function of environmental gradients.

The reptile egg is typically exposed to an environment shaped by local edaphic, hydrologic, climatic and biotic variables. The female can influence the environment to which her eggs are exposed in several ways. This influence may occur through the timing of reproduction, through nest site selection, through nest construction, through variation in clutch size and in a few cases, through the exercise of parental care. The physical character of the incubation environment and of the eggs, as well as the physiological processes occurring inside the eggs, determine the microclimate in which embryonic development occurs (Fig. 12.1).

Our understanding of how reptilian embryonic development is adapted to the incubation environment is not extensive. There is little information available to describe how variation in soil characteristics, hydrology, topography or local climates influence the environment around eggs. Water exchange between the egg and the natural environment has not been extensively described. The mechanisms by which this exchange occurs remain unsettled (Packard & Packard, 1988). Reptile eggs have been studied intensively for many years (Packard & Packard, 1988; Packard, Chapter 13) yet there is remarkably little information on the hydric microclimate or water exchange of reptile eggs in natural nests (Ewert, 1985; Packard & Packard, 1988).

In this chapter I develop several hypotheses about the hydric environment and water exchange of eggs incubating in soil. The focus is on water because, in addition to influencing the hydric and hydraulic properties of soil, the water in soil also influences its thermal, respiratory and osmotic properties. Sufficient information is available in the soil physics literature (Campbell, 1985; Hillel, 1980a,b) to develop hypotheses about the thermal and respiratory microclimates and exchanges of buried eggs as well; however, limitations of space restrict my discussion to water.

The hydraulic properties of soils and the transport of water through soil have been extensively described (van Genuchten, 1980; Hillel, 1980a,b; Campbell, 1985; Koorevaar, Menelik & Dirksen, 1983; Mualem, 1986) and applied to the analysis of the exchange of water between plants and soil (Hillel, 1980b; Nobel, 1983; Campbell, 1985) but only rarely to analysis of the exchange of water between buried eggs and soil (Tracy, Packard & Packard, 1978; Tracy, 1982; Ackerman et al., 1985a; Ackerman & Seagrave, 1987). In this chapter I present data for three soils which are specific and homogeneous having been processed under laboratory conditions.

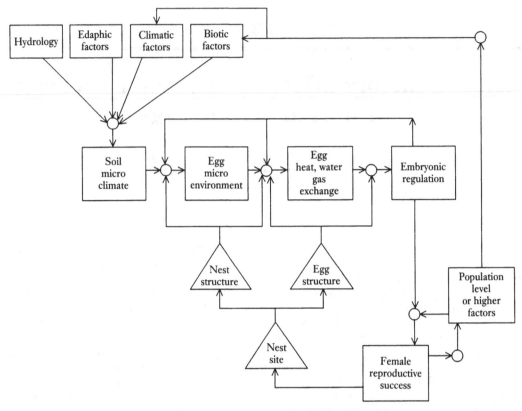

Fig. 12.1. Some of the factors describing the environmental, physiological and ecological relationships of buried vertebrate eggs.

Water retention relationship of soils

Dry soil comprises particulate or solid materials and gas. When water is present, gas is displaced until the water completely fills the gas space. The soil is then considered to be saturated with water. At saturation, the hydraulic pressure in the water is at zero with reference to atmospheric pressure. As the water content of the soil decreases from saturation, a negative pressure occurs in the soil water. This pressure is often described as water potential (the potential energy per unit volume, J m^{-3}; Corey & Klute, 1985; Campbell, 1985) whose dimensions are equivalent to those of pressure (kPa). The water potential is a matric potential if it is due to the capillary effects of surface tension in soil pores or to the binding of water to the surface of soil particles (Koorevaar et al., 1983). As the soil dries, the negative matric potential increases. The characteristic curve (Fig. 12.2) describes graphically the relationship between soil water content and the matric potential of water. The curves for drying soils and wetting soils are often different. A hysteresis may be present that produces, in a wetting soil, a less negative water potential at the same water content as in a drying soil and is produced when bulk water is present in the soil pores forming menisci (Koorevaar et al., 1983).

Negative water potentials may occur through forces other than matric forces. An osmotic potential may be present in the water if dissolved materials are present. If the liquid water in the soil is connected hydraulically to a saturated water source (phreatic water) in the soil column, then a gravitational potential may be present. Finally, there may be an overburden potential present due to the weight of overlying layers but since most vertebrate eggs are usually buried within the top 1 m of soil, or within a nest cavity, this potential is unlikely to be important. The relevant, total potential in soil water is thus the sum of the matric, osmotic and gravitational potentials. The vapour pressure of water in the soil gas phase is generally assumed to be in

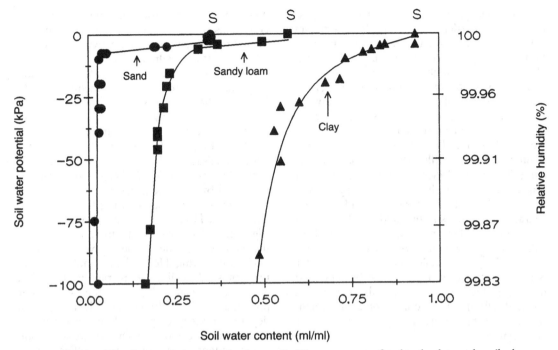

Fig. 12.2. The characteristic curves for three soils. The curves were fitted to the data as described in the text. 'S' represents the saturation point for each soil. The curves were fitted to data to around −1500 kPa but only the section of the curve from 0 to −100 kPa is illustrated so as to retain sufficient detail.

equilibrium with water potential. Relative humidity (H_r, f) can be related to water potential (ψ, 1 kPa = 1 J kg⁻¹) by:

$$\psi = (R{\cdot}T/M_w){\cdot}\ln(H_r) \qquad (12.1)$$

where the gas constant (R) is 8.31 J mol⁻¹ K⁻¹, the molar mass of water, (M_w) is 0.018 kg mol⁻¹ and T is temperature (K). Characteristic curves established for the relationship between water potential and water content for three soils, using data collected using the pressure plate technique (Klute, 1986) are shown in Fig. 12.2. Samples of soil are saturated with water and then dried. The curves represent the equation described by van Genuchten (1980) fitted using non-linear regression techniques. Only a small portion of the complete characteristic curve is shown in Fig. 12.2 since totally dry soil is assumed to have a water potential of −1 000 000 kPa (Koorevaar et al., 1983). The horizontal section of the characteristic curve represents the presence of bulk water in the sand. The vertical section of curve represents bound water remaining after the bulk water has drained from the soil (Koorevaar et al., 1983). Bound water covers the surface of soil particles and requires drastically

increasing pressure to remove it, which accounts for the steep slope. Hysteresis displaces wetting curves above drying curves and disappears as the curve turns downward (Hillel, 1980b).

When the external pressure exceeds the capillary pressure holding water in the soil, pores in the soil empty of water. When bulk water is present in the soil, water potential decreases slowly as water is lost. This is most noticeable for sand (Fig. 12.2), where water potential is still above −10 kPa even though most of the water has been lost. Data for five different sands from various reptile nesting locations, including marine and riverine beaches show that these soils all release their bulk liquid water at water potentials less negative (lower) than −10 to −15 kPa (Ackerman, unpublished data). In sands, the pores are relatively large and of similar size so that emptying (or filling) occurs abruptly as matric forces are exceeded by drying (or wetting) forces (Koorevaar et al., 1983). Clays, by contrast, are composed of much smaller pores with a greater variation in pore sizes than sand (or any other natural soil). This results in a more gradual emptying of the soil at greater negative pressures (Fig. 12.2). Clays retain more water bound to soil particles than other soils because of the small particle sizes which have a greater combined surface area to bind water. The position and shape of characteristic curves of the clays are more variable as a group. The two extremes of shape and position of characteristic curves among different soil types may be represented by sands and clays. This can be seen for some mixed soil data (Ruibal, Tevis & Roig, 1969). The bulk density of the soil and temperature also influence the characteristic curve (Campbell, 1985). Since bulk density and soil structure are likely to be altered by processing in the laboratory, extrapolation of laboratory characteristic curves to the field should be done with caution. Where possible, it is always advisable to make *in situ* measurements with particular attention to the relationship between water potential and H_r (Fig. 12.2). As has been pointed out, H_r is high even in dry soils (Packard & Packard, 1988). This requires some adjustment to the scale that we use to think about wetness and dryness.

Water transport through soils

A difference in soil water potential drives liquid water transport in the soil and determines the direction of water flow. The water potential difference gives no indication whether or not water will flow. The hydraulic conductivity ($K_1 = cm^2$ d^{-1} kPa^{-1}) is a measure of the tendency for water to flow. Steady state flow may be represented by the equation:

$$\dot{M}_{H_2O} = K_1 \cdot A \cdot (\Delta\psi / \Delta x) \qquad (12.2)$$

where $\dot{M}_{H_2O}$ = water exchange (cm^3 d^{-1}), A = area (cm^2) and x = distance (cm). The terms hydraulic conductivity (K_1), conductance (G = cm^3 d^{-1} kPa^{-1}) and permeability (L_p = cm^3 cm^{-2} d^{-1} kPa^{-1}) are easily confused and may be defined by the following equations:

$$L_p = K_1 / \Delta x \qquad (12.3)$$

$$G = L_p \cdot A \qquad (12.4)$$

Conductance may also be described on a mass basis (e.g. water vapour conductance, G_{H_2O} = mg d^{-1} kPa^{-1}) by converting from volume to mass using density. Bulk flow of water may be considered in two categories: 1) saturated flow (in saturated soil); and 2) unsaturated flow (in soil not saturated with water). Unsaturated flow is likely to be more relevant to the water exchange of buried reptile eggs during develop-

ment, because gas exchange under saturated conditions would be severely limited by the low diffusivity of gases, particularly oxygen, in water. The unsaturated hydraulic conductivity of soil is a function of soil water content (Fig. 12.3) and may be calculated (van Genuchten, 1980; Mualem, 1986) or measured (Klute & Dirksen, 1986). Results using the calculation of van Genuchten (1980) are shown in Fig. 3(*a*) (relative K) and Fig 3(*b*) (absolute hydraulic conductivity). The saturated hydraulic conductivity (K_s) of the three soils in Fig. 12.2 can be measured as described by Klute & Dirksen (1986) and data for vermiculite measured using the method of McBride & Horton (1985) are also included in Fig. 12.3. Hydraulic conductivity decreases as the largest (and most conductive) soil pores empty with drying. Hence the hydraulic conductivity of sand declines more rapidly than clay. The exact relationship between soil water potential and hydraulic conductivity also depends on the soil type. Hydraulic conductivity decreases more rapidly with a decrease in soil water content in sand than in clay reflecting the difference in the pore structure. There is little tendency for liquid water to move in sands drier than -50 to -100 kPa or clays drier than -500 to -1500 kPa. This is because little movable bulk water remains and bound water is held more tightly as water potential decreases. Rigid-shelled eggs may experience water potentials considerably drier ($<-10,000$ kPa) than those used in laboratory studies (Leshem & Dmi'el, 1986; Leshem *et al.*, 1986; Leshem, 1989). There is not likely to be bulk water present in these soils. The use of water potentials drier than -150 kPa in natural soils means that little bulk water is present in the soil. Other reports of water potentials in soils around nests of pliable-shelled turtle eggs during incubation, suggest that water potentials may often be wetter than -100 kPa (Packard *et al.*, 1985; Ratterman & Ackerman, 1989) and bulk water is likely to be present in these soils.

Water vapour transport represents a second and parallel pathway for water movement. Campbell (1985) calculates the water vapour hydraulic conductivity (K_v = cm^2 d^{-1} kPa^{-1}) of soil using an equation of the form:

$$K_v = D_v \cdot C_v' \cdot H_r \cdot M_w / RT \qquad (12.5)$$

where D_v = vapour diffusivity (cm^2 d^{-1}), C_v' = saturated vapour concentration (cm^3 cm^{-3}) and the unit for R has been converted to kJ. Note

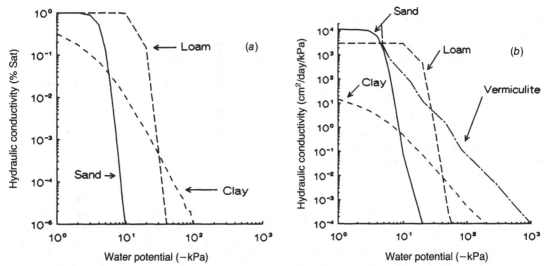

Fig. 12.3. The hydraulic conductivity (K_1) of the soils shown in Fig. 12.2. Relative hydraulic conductivity (a) is calculated as described by van Genuchten (1980). The absolute hydraulic conductivity (b) was calculated by multiplying the relative hydraulic conductivity by the saturated hydraulic conductivity. The line for vermiculite was constructed by drawing a line (by eye) through data that were directly measured (see text).

that the units of water vapour conductivity are the same as the units of hydraulic conductivity. The total hydraulic conductance (K_t) of the soil is the sum of the hydraulic conductivity and water vapour conductivity (Fig. 12.4). Water vapour conductivity is several orders of magnitude smaller than hydraulic conductivity in moist soil since liquid water fills or blocks many spaces but it approaches, and exceeds, hydraulic conductivity as the soils dry and gas spaces connect. This may occur even though a substantial quantity of water remains bound to the soil particles (e.g. clay). The relationships vary with soil type although the water vapour conductivity is similar for all the soils shown here. Liquid water transport is dominant at moist water potentials while water vapour transport is dominant at dry water potentials. The definition of soil wetness and dryness is relative and depends on the soil type and on hydraulic conductivity and not on water potential.

Reptile eggs are thought to exchange both water vapour and liquid water with the soil contacting them (Packard & Packard, 1988) but the type of exchange will depend strongly on the hydraulic properties of the soil that contacts the eggs. The hydraulic properties of soils vary as a function of water potential. Thus, extrapolating the water exchange of an egg in contact with one type of soil to another type must be done with caution. This may be especially true when

generalising the results obtained by incubating eggs on artificial soils, such as vermiculite, whose hydraulic properties have not been defined.

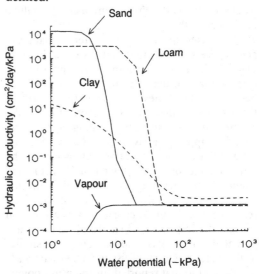

Fig. 12.4. The total hydraulic conductivity (K_t) of the soils shown in Fig. 12.2. The lines labelled as soils represent the sum of hydraulic conductivity and water vapour conductivity where water vapour conductivity is calculated as described in the text. The line labelled vapour represents the calculated water vapour conductivity for sand. For the soils shown, vapour transport becomes dominant when the curves for K_t flatten out.

Soil permeability and water exchange of buried eggs

Knowledge of the hydraulic conductivity of soil is essential if the transport of water in soil or between eggs and soil is to be assessed. However, the distance over which water is moving must be known or approximated. As described by equation 12.3, hydraulic conductivity may be transformed into a permeability. Steady-state egg water exchange with the soil may be described by:

$$\dot{M}_{H_2O} = L_p \cdot A \cdot (\psi_{EH_2O} - \psi_{sH_2O}) \qquad (12.6)$$

where ψ_{EH_2O} = egg ψ (kPa) and ψ_{sH_2O} = soil ψ (kPa). The permeability represents a total permeability per unit area. It is the sum, in series, of the eggshell permeability ($L_p{}^e$) and the soil permeability ($L_p{}^s$):

$$1/L_p = 1/L_p{}^e + 1/L_p{}^s \qquad (12.7)$$

A value for soil permeability may be approximated using the Nusselt (Sherwood) relationship (Bird, Stewart & Lightfoot, 1960; Seagrave, 1971; Ackerman et al., 1985a):

$$L_{sp} = 2.0 \cdot K_1/d \qquad (12.8)$$

where d (cm) is the linear dimension (usually a diameter) of the egg. When equations 12.7 and 12.8 are used to predict the water exchange of an egg, the soil is treated as if it were a thin film of soil with thickness d, around the egg. In these calculations the diameter of a 10 g sphere has been used as the linear dimension for describing the water exchange of a hypothetical egg. The permeability for soil surrounding a 1 g egg would be approximately twice as large while that for a 100 g egg would be one-half as large. The water uptake of a 10 g egg can be estimated when the permeability for the shell is infinite (i.e. when the soil functions as the only resistance) and can be considered as a function of the driving force (Fig. 12.5). The water potential of the egg is −800 kPa, which is similar to the water potential of tissue (Ackerman et al., 1985a) and to that reported for the water potential of the eggs of the turtle Trionyx triunguis (Leshem, 1989) and the lizard Dipsosaurus dorsalis (Muth, 1981). The choice of egg water potential is arbitrary. However, the result is insensitive to values for egg water potential within a physiologically realistic range.

Soil type influences liquid water exchange while water vapour exchange is relatively

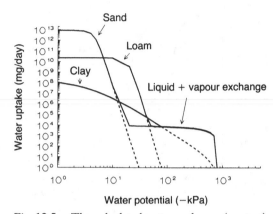

Fig. 12.5. The calculated water exchange (see text) of buried vertebrate eggs when only the soil surrounding the egg is controlling water transport. The effect of the eggshell is disregarded. The solid lines represent the calculation using K_t and the dashed lines represent the calculation using hydraulic conductivity.

independent of soil type. Liquid exchange in sand occurs only at low water potentials while liquid exchange in clay can occur at considerably drier water potentials. The development of a greater driving force as soils dry out, becomes counter productive (Hillel, 1980b) producing less and less flow. This is because hydraulic conductivity decreases faster than the driving force increases. Exchange of water vapour is a quantitatively important mechanism for exchange at low water potentials. It should be noted, however, that the curve for vermiculite would be shifted by an order of magnitude or more to the right of the clay curve (Fig. 12.5). Liquid water exchange in vermiculite is likely to be quantitatively important at water potentials drier than −1000 kPa, if liquid water is exchanged between the egg and the vermiculite. The report by Thompson (1987) of the existence of a liquid water connection between freshly laid, rigid-shelled turtle (Emydura macquarii) eggs contacting sand at −710 kPa is interesting in this context. Virtually no liquid water flow should have occurred in the sand at −710 kPa. On the other hand, Thompson (1987) reports using sand with a water content similar to that of the egg (around 72% water). This water content does not seem to be consistent with the measured water potential because sands are usually saturated with water at 50% of their volume or less (Fig. 12.2; Hillel, 1971; Koorevaar et al., 1983). Much experimental research describing the water exchange of reptile eggs is based on the use of water potentials around -150 to -1500 kPa in vermiculite (Tracy et al., 1978; Packard et al., 1979a; Packard et al., 1980; Tracy, 1980; Packard, Packard & Boardman, 1981a; Packard et al., 1981b; Packard et al., 1983; Morris et al., 1983; Gutzke & Packard, 1987; Gutzke et al., 1987; Miller, Packard & Packard, 1987). Liquid water exchange may be possible in vermiculite at these water potentials. However, it is considerably less likely in natural soils at the same water potentials. The assessment as to whether or not and to what extent liquid water exchange between the egg and the surrounding soil is even possible is dependent on the type of soil and on the range of water potentials considered.

The analysis of the water exchange of 10 g eggs using soil as their only shell (Fig. 12.5) indicates that water would be exchanged at high rates. For example, eggs in soils at -600 kPa would take up 4 g d^{-1} while eggs in soils at -1000 kPa it would lose 4 g d^{-1}. These rates seem excessively high. It is interesting to examine how the presence of an eggshell might influence egg water exchange. There are several estimates of the water vapour conductivity (G_{H_2O}) of eggshells of buried reptile and megapode eggs available (Harrison et al., 1978; Packard et al., 1979b; Tracy, 1982; Ackerman, Dmi'el & Ar, 1985b; Booth & Seymour, 1987; Seymour et al., 1987; Deeming & Thompson, Chapter 17) and a single estimate of the liquid conductivity (Tracy et al., 1978). The relative contribution of water vapour or liquid water exchange to the total water exchanged between the egg and the surroundings is unknown (Packard & Packard, 1988). The lack of quantitative estimates for liquid conductivity of eggshell makes it difficult to even guess at the contribution of liquid water to the total water exchange. The water vapour conductance data for some reptile eggs exchanging water vapour in air are summarised in Fig. 12.6. For parchment-shelled eggs, the external film of air around the egg contributes about half of the total water vapour conductance (Ackerman et al., 1985b). When the eggs are buried in soil the air film is lost and the egg water vapour conductance is doubled. However, because a significant fraction of the shell surface is occluded by the solid soil particles, only that part of the shell not occluded is available for water vapour exchange. A 10 g egg buried in sand (whose dry, gas filled volume, or porosity, is 0.50 ml gas ml^{-1} sand) would have

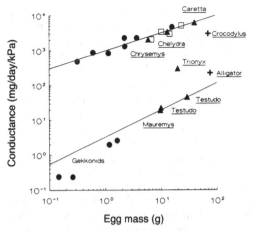

Fig. 12.6. Water vapour conductance (G_{H_2O}) of a variety of reptile eggs. The data were collected using the technique described by Ackerman et al. (1985b). The upper line represents the equation $G_{H_2O} = 992$ mass$^{0.51}$ from Ackerman et al. (1985b) and describes the influence of egg mass on water vapour conductance of parchment-shelled reptile eggs. The lower line represents the equation $G_{H_2O} = 3.3$ mass$^{0.78}$ and describes a relationship reported for avian eggs. The triangles represent turtle species, the crosses crocodilian species, the squares snake species and the solid circles represent lizard species.

a water vapour conductance of 3200 mg d^{-1} kPa^{-1}. In this case, $L_p = G_{H_2O} A^{-1}$ and is equivalent to 2.7×10^{-6} cm^3 cm^{-2} d^{-1} kPa^{-1} on a volumetric basis (where 1 kPa of liquid is equal to 2.3×10^{-5} kPa of vapour at 25 °C). A permeability for clay would be somewhat greater due to the greater porosity of clay. The resistance to water vapour exchange of the rigid-shelled eggs is located almost entirely in the eggshell so the air film need not be considered (Tracy & Sotherland, 1979; Spotila, Weinheimer & Paganelli, 1981). As a limiting approximation, a 10 g rigid-shelled egg might have a water vapour conductance of around 19 mg d^{-1} kPa^{-1}, similar to that of an avian egg of the same mass, or a permeability of 8.0×10^{-9} cm^3 cm^{-2} d^{-1} kPa^{-1}. In addition, a total permeability can be estimated that includes the contribution of both liquid and vapour exchange to the extent each is occurring. Many reptile eggs used in laboratory studies are around 10 g. The water exchange of these eggs falls within a range of about $\pm 50\%$ of initial mass for water potentials between -50 and -1500 kPa. If the water potential of the egg is -800 kPa, then a permeability for the shell of about 2.8×10^{-6} cm^3

cm^{-2} d^{-1} kPa^{-1} would produce a change in egg mass within the ±50% envelope over typical incubation periods (50–90 days). Data for eggs of the turtle *Chrysemys picta* in field nests yield a similar value, as do data for eggs of the turtle *Chelydra serpentina* incubating fully buried in sand at −7 kPa (Kam, 1988; Ratterman & Ackerman, 1989). The liquid permeability estimated by Tracy *et al.* (1978) is several orders of magnitude greater than this value and produces unrealistically large ranges of water exchange. A value of 2.8×10^{-6} cm^3 cm^{-2} d^{-1} kPa^{-1} can be used for the total permeability for parchment-shelled eggs but permeability for rigid-shelled eggs is 2–3 orders of magnitude smaller. No assumption is made about the relative contribution of liquid and vapour water exchange to permeability.

The water exchange of a 10 g egg can now be estimated (equation 12.6) as a function of soil water potential using the permeability estimated (from equation 12.7). The estimated daily water exchange of a 10 g egg with an internal water potential of −800 kPa is shown in Fig. 12.7. The result is not sensitive to egg water potential, because the change in driving force is offset by the change in egg permeability (calculated with the same water potential) and only the intercept with zero water exchange changes. The rate of water uptake depends strongly on the exchange mechanism. Water vapour exchange is independent of soil type. Liquid water exchange, on the other hand, is only possible over rather limited ranges of water potential, depending on soil type. Liquid water may not be available even at what appear to be moist water potentials, a situation similar to that for plants (Hillel, 1980*b*; Campbell, 1985). If the permeability for a rigid-shelled egg is used, water exchange (both uptake and loss) would be reduced proportionally to the reduction in permeability. A permeability of 8.0×10^{-9} cm^3 cm^{-2} d^{-1} kPa^{-1} for a 10 g egg would produce a water loss of 1.3 g over 50 days or 9% of initial mass at an ambient H_r of 50% (Fig. 12.7).

If the permeability of the eggshell is sufficiently smaller than the permeability of the surrounding soil, the egg shell controls egg water exchange. The shell permeability of parchment-shelled eggs is several orders of magnitude smaller than soil permeability down to water potentials of around −10,000 kPa. Even if shell permeability is increased by one order of magnitude (increasing the water exchange

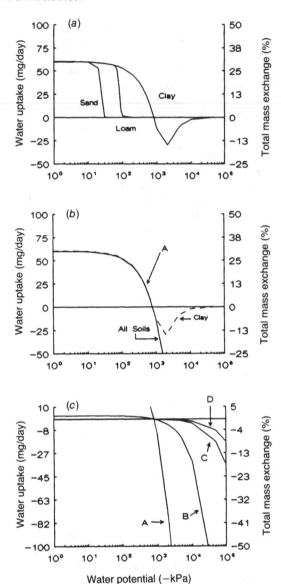

Fig. 12.7. The calculated water exchange (see text) of buried parchment-shelled eggs when water transport through the shell and the soil are considered together. The left axis is the water uptake at a given and constant water potential (bottom axis). The right axis represents the water exchange (as % of initial mass) that would occur if the water uptake were maintained constant throughout the 50 day incubation of a 10 g egg. The permeabilities of the eggshell and soil are summed in series. (*a*) shows the estimated exchange that would occur if only liquid water were moving in the sand, loam and clay used in the text. Water exchange in sand goes to zero at around −25 kPa, in loam at around −100 kPa and in clay at around −10,000 kPa because the permeability of the soil approaches zero. The clay line crosses over the

envelope to ±500%), it remains at least one order of magnitude smaller than soil permeability. Thus for water potentials from −1 to around −10,000 kPa, the soil surrounding eggs does not limit water exchange of parchment-shelled eggs. It probably never limits the water exchange of rigid-shelled vertebrate eggs at any water potential. This finding appears to justify the assumptions to this effect made by Tracy *et al.* (1978) and Ackerman *et al.* (1985*a*). However, soil may limit water exchange of parchment-shelled eggs at water potentials drier than −10,000 kPa. This is because the density of water vapour molecules is greatly reduced and the boundary layer/film effect described by Tracy & Sotherland (1979), Spotila *et al.* (1981) and Ackerman *et al.* (1985*b*) must be considered. The report by Gutzke (1984) on the effect of water loss by eggs of *C. serpentina* on dry vermiculite, presumably with water potentials much less than −100,000 kPa, may be related to the physical properties of vermiculite. Dry vermicu-

lite would act as a thick film of unstirred gas around the egg: vermiculite is about 90% air when dry (Ackerman, unpublished data). In addition, the very great capacitance of vermiculite for water, when compared to natural soils, would result in the storage of much more water than would occur in a natural soil. Eggs of *C. serpentina* do not incubate successfully in dry natural soils in my laboratory. Several workers conclude that soil resistance to water exchange limits water exchange by the egg (Tracy, 1982; Packard & Packard, 1988). This conclusion may be justified when applied to liquid water transport (depending on soil type and the range of water potentials considered) but is not justified when the additional contribution of water vapour transport is considered. If soil resistance to water exchange is not limiting then buried eggs are likely to have little influence on the water potential of the soil surrounding them. The sand in the centre of nests of *T. triunguis* has been found sometimes to be drier than the sand surrounding the nest (Leshem, 1989). However, this sand appears to have fallen into the nest from the drier sand above the nest because it appeared as the surface layer dried out (see later). This may occur wherever drying occurs in the layer of soil close to the top of the nest. The hydric microclimate of the eggs is probably largely determined by the soil rather than by the eggs. It becomes important, therefore, to know what water potential exists in the soil surrounding the eggs.

---

horizontal line representing zero water exchange when egg water potential equals soil water potential (−800 kPa). Water uptake in clay goes to zero only because the difference in water potential between the egg and the soil diminishes as soil water potential increases to −800 kPa. The eggshell limits water exchange when bulk water is present in the soil. The soil becomes limiting (when water exchange approaches zero) when bulk water is removed from the soil. The water potentials at which this occurs depends on soil type. (*b*) shows the estimated exchange when water vapour transport is added to liquid transport. All three soils are similar, and water uptake goes to zero only because the water potential difference between the egg and the soil goes to zero. Water loss increases only because the water potential difference between the egg and the soil increases. Total (liquid plus vapour) appears not to be a factor down to at least −100,000 kPa. Water transport is limited only by the eggshell. The broken line (in (*b*)) represents the line for clay shown in (*a*). The effect of decreasing eggshell water vapour permeability is shown in (*c*). Line A in (*b*) and (*c*) represents the calculated water exchange for parchment-shelled eggs. Line D represents the limiting lower eggshell water vapour permeability for rigid-shelled reptile and avian eggs as given in the text. Lines B and C are for water vapour permeabilities increased 10-fold and 100-fold over the lower limit and represent eggshells with intermediate permeabilities. Lines C and D coincide with the zero water uptake line at the scale shown on the figure. Egg water loss can be effectively limited by decreasing the water vapour permeability of the eggshell.

The distribution of water in soils

There are few data available with which to describe the hydric environment of soils around buried eggs but it is possible, using concepts in common use among soil physicists, to describe the distribution of water in a soil profile (Keulen & Hillel, 1974; Hillel & van Bavel, 1976; Hillel, 1980*a,b*; Koorevaar *et al.*, 1983; Campbell, 1985). The distribution of water is a function of the balance between water entering and leaving a layer of soil in the profile and if the distribution is stable it is because the same quantity of water enters a layer as leaves it. If this distribution is not stable, then the water in the profile varies in time and depth. The elements of such a balance for the two relevant situations are summarised in Fig. 12.8. In one case (Fig. 12.9) a source of liquid (phreatic) water is present in the soil not far from the surface (1–5 metres) and in the

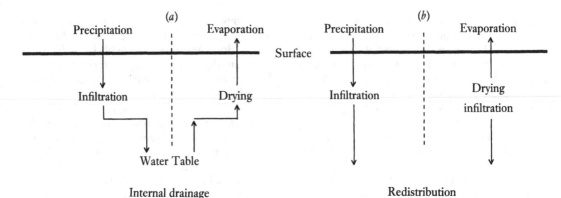

Fig. 12.8. A simplified moisture balance for internal drainage (*a*) and redistribution (*b*) of water in a soil profile. In case (*a*), water drains due to gravitational forces. This occurs until drying forces reduce water potential sufficiently to reverse the direction of flow and draw water from the subsurface water source to the surface. In case (*b*), water continues to be lost by evaporation through the surface and by infiltration downwards into the profile. However, after a short period of time, evaporation at the surface will approach zero as explained in the text.

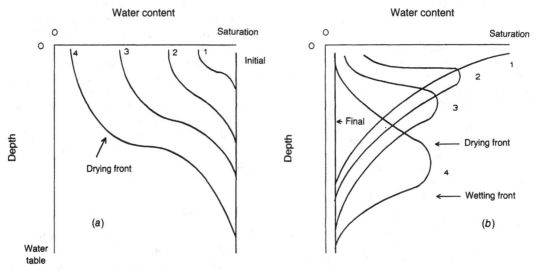

Fig. 12.9. Hypothetical water content profiles for internal drainage (*a*) in presence of a water table and redistribution (*b*) with no water table. The initial distribution in (*a*) is set arbitrarily after infiltration. The final distribution in (*b*) occurs after the profile dries out. The numbered curves represent sequential changes in the profile with the passage of time with no further surface infiltration of water. In (*a*), at equilibrium, water lost at the surface due to evaporation is balanced by water drawn up from the hydrostatic forces from the water table. In (*b*), water is lost at the surface due to evaporation and through the bottom of the profile due to hydrostatic forces. The time-scale is weeks to months.

other case (Fig. 12.9) no phreatic water is close to the surface. When a phreatic surface is present, water enters the soil and drains downward until it equilibrates with the gravitational field referenced to the water surface (Fig. 12.9). When no water surface is present, the water enters the soil and redistributes continuously downward through it (Fig. 12.9). If drying at the surface occurs, water is lost from the soil to the atmosphere (Hillel, 1980*b*; Campbell, 1985).

It is the balance of water exchange at the surface with water exchange deeper in the soil that produces characteristic distributions of soil water in depth and time. This process is strongly affected by the type of soil present and its structure. If phreatic water is present, water lost across the surface may be balanced by water drawn up from the liquid water source. This balance produces a characteristic profile in which, after a period of drying, a layer of dry soil comes to overlie a layer of wetter soil (Fig. 12.9). The top, dry soil layer limits water loss to the atmosphere from the wetter layer. Liquid water may then be drawn up at a rate sufficient to balance the water lost across the surface (Hillel, 1980a,b; Campbell, 1985). This occurs because of the relationship between soil water content and hydraulic conductivity (Fig. 12.3) (Hillel, 1980a,b; Koorevaar et al., 1983). Evaporation from the soil surface is self-limiting. This is because the rate of water vapour transport decreases as the length of the diffusion path increases. It is also because the density of vapour in the pathway decreases and approaches zero after short periods of time (Hillel, 1980b; Campbell, 1985). The width of the dry layer will depend, among other things, on the time since the infiltration of surface water, on the soil composition and structure and on the depth of phreatic water. Given sufficient time after surface infiltration, an equilibrium may be reached. This is likely to require weeks to months (Hillel, 1980b) however. The deeper the liquid water surface, the wider the dry layer is likely to be. It is also likely to be wider in better draining soils such as sands than in poorly draining soils such as clays. However, Hillel & van Bavel (1976) point out that better draining (coarser) soils will ultimately conserve the subsurface water to a greater extent than the more poorly draining (finer) soils. This is because the top dry layer forms more quickly in coarse soils, reducing water loss to the atmosphere. As time passes after infiltration, a discrete 'drying front' may develop (Hillel, 1980b). A large water potential difference develops across this layer whatever its width. Since most moist soils have water potentials of around −5 to −50 kPa (Hillel, 1980b), a difference across the dry layer of 50,000 to 100,000 kPa would not be unusual. Despite this enormous driving force, little water is lost across the surface. When redistribution occurs in the absence of phreatic water, water lost across the surface produces a dry layer similar to that found in internal drainage. However, water is lost continuously downward due to drainage, until eventually the profile dries out. Quantitative models of various aspects of these processes have been produced (Keulen & Hillel, 1974; Hillel & van Bavel, 1976; Campbell, 1985).

Soil profiles and the hydric microclimate of reptilian eggs

The ideas discussed here could easily be applied in the field. For example, soil cores can be taken periodically, sectioned, and the water content of the sample measured (Gardner, 1986) as a function of depth. If this is carried out together with the measurement of soil water potential (Cassell & Klute, 1986; Campbell & Gee, 1986; Rawlins & Campbell, 1986) close to buried eggs, and if the hydraulic properties of the soil can be measured or estimated (Klute & Dirksen, 1986; Mualem, 1986), then reasonable inference may be drawn about the hydric microclimate in the soil surrounding and contacting the eggs. The soil hydric microclimate around nests of the turtle C. picta in Iowa, where the water table was 1–3 m deep and rain occurred periodically, has been measured (Ratterman & Ackerman, 1989). Typical profiles for water content and water potential are shown in Fig. 12.10. A dry layer at the surface is evident with wetter soil underneath, within 48 h after rain. The water potential near the surface was at least −50,000 to −75,000 kPa. The water potential was around −30 kPa in the wetter soils below a depth of 5 cm. The water content profile shifted to the left as water drained through the profile or was lost to evaporation. However, soil water potential around the nest was not markedly affected because of the relationship between soil water content and water potential when bulk water is present in the soil. The eggs do not influence the hydric microclimate of the soil because the soil hydraulic conductivity is not limiting at the observed soil water potentials. The gradient between the dry soil and the wet underlayer was around 10,000 kPa cm^{-1}. Eggs protruding into the drier surface layer are likely to experience extremely desiccating conditions.

The soil water potentials around the nest are typical of the 'field capacity' for soils (Hillel, 1980b). Field capacity apparently remains useful (Hillel, 1980b; Cassell & Nielsen, 1986) and refers to the soil condition in which water drainage has been reduced to a minimum. Equilibrium distributions of water in the profile shown in Fig. 12.10 probably did not occur.

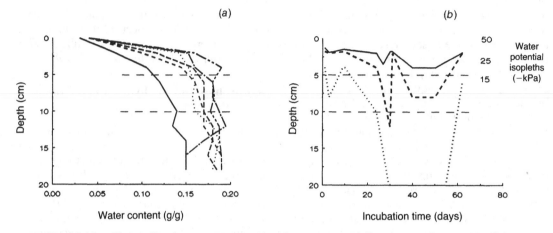

Fig. 12.10. The profiles for water content (*a*) and water potential (*b*) in loam soil near nests of the turtle *Chrysemys picta* in Iowa. The curves in (*a*) represent profile changes over the course of the incubation season and shift to the left as drying occurs after rains. This has little effect on soil water potential due to relationship between soil water content and soil water potential (see Fig. 12.2). The curves in (*b*) represent water potentials calculated from the water content using *in situ* characteristic curves. The dashed lines represent the average depths of the top and bottom of the nests. The average water potential between 4 and 10 cm as measured by tensiometry was −30 kPa. (The data are from Ratterman & Ackerman, 1989.)

Consistent with the low and constant water potential, the *C. picta* eggs incubating in field nests gained mass linearly and averaged a water uptake of about 35 mg d^{-1}. Eggs of *C. serpentina* are also deposited in the same area but deeper in the soil column (Ackerman, unpublished observations) and presumably would take up substantial quantities of water.

Data for an incubation site of *T. triunguis* eggs at the top of a 5 m high sand dune bordering the Alexander river in Israel are shown in Fig. 12.11 (Leshem, 1989; Ackerman, unpublished data). The characteristic curve of this sand is shown in Fig. 12.2. The hydrology of the dune is unknown but the water table was probably around 5 m deep. The last rain had fallen some 20 days prior to taking the initial core sample and did not re-occur for over 6 months. Water appears to be redistributing in this profile, suggesting that the water table is sufficiently deep it does not influence the top 60 cm of the profile. The drying front disappears after about 4 months. At another nesting site on a dune 500 m away, the drying front stabilises between 20 and 40 cm of depth and was similar to the 60-day profile (Fig. 12.11) (Leshem, 1989). This dune had a similar elevation and distance from the river. The dry sand at the surface of both sites was around −100,000 kPa (air dry). The water potentials in the moist sand below the drying

fronts were −7 to −10 kPa as measured by tensiometry. There is only 2–3% (by mass) water in this 'moist' sand even though the sand holds around 35% at saturation. The gradient for water potential in the drying front is around 2000 to 4000 kPa cm^{-1} and the drying front is relatively wide. In both cases, water loss to the atmosphere must be near zero. There are other reports of water held in dunes (Bagnold, 1941; Seely & Mitchell, 1987) and the water content profiles may be similar.

The change in mass of the eggs of *T. triunguis* in natural nests depends on the position of the egg in the nest (Leshem & Dmi'el, 1986; Leshem, 1989). Eggs at the top of the nest lose more mass than eggs at the bottom of the nest. However, the eggs do not survive incubation in the laboratory in sand dried to the same extent as the surface layer (Leshem, 1989). *T. triunguis* eggs appear to incubate in the drying front for all or a substantial fraction of their incubation period, 50–70 days (Leshem, 1989). Eggs of the turtle *Mauremys caspicus* incubate in the dry surface layer where the incubation period can be more than 100 days (Sidis, unpublished data) and in the laboratory, they incubate successfully in the air dry sand (water potential = −100,000) losing 10% to 30% of their initial mass (Ackerman, unpublished data).

An example of internal drainage with a stable

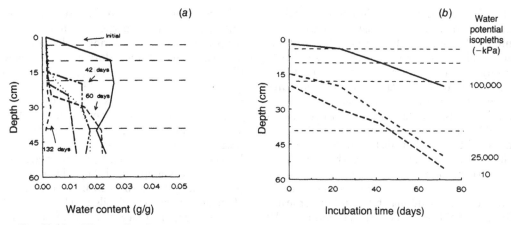

Fig. 12.11. The profiles for water content (*a*) and water potential (*b*) in a sand dune near nests of the turtles *Trionyx triunguis* and *Mauremys caspica* at the edge of the Alexander River in Israel. The times (*a*) represent the time since the initial sample. The initial sample was collected approximately 20 days after the last rain. The curves in (*b*) represent water potentials calculated from laboratory characteristic curves. The values around -10 kPa were confirmed directly by tensiometry in the field. The *T. triunguis* eggs in this location are deposited between 20 and 40 cm while the *M. caspica* eggs in this location are deposited between 4 and 10 cm.

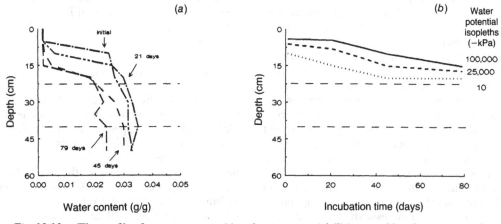

Fig. 12.12. The profiles for water content (*a*) and water potential (*b*) in a sand beach near nests of the sea turtle *Caretta caretta* located in Habonim Beach, Israel. The times (*a*) represent the time since the initial sample. The initial sample was collected approximately 20 days after the last rain. The curves in (*b*) represent water potentials calculated from laboratory characteristic curves. The values around -10 kPa were confirmed directly by tensiometry in the field. The top of the *C. caretta* nest is at a depth of around 20 cm.

drying front in a beach (Habonim, Israel) sand profile next to sea turtle (*Caretta caretta*) nests (Fig. 12.12). In this study rain occurred some 20 days before the initial sample and did not re-occur for over 5 months. The eggs were deposited 15–20 days after the initial sample and incubated for 50–60 days (Silberstein, 1988). A drying front appears to stabilise between 15 and

20 cm of depth. It may be narrower than this since samples were not collected between 15 and 20 cm. The dry sand at the surface was around $-100,000$ kPa. The water potential in the moist underlayer was measured at -7 kPa. The transition in the drying front from dry sand to moist sand occurs over 5–10 cm and the water potential gradient may be 10,000 to 20,000 kPa

cm^{-1}. The drying front stabilises by around 70 days after the last rain (49 days after the initial sample). The water table is located at 1 m at this site. Similar distributions occur in sea turtle nesting beaches in Costa Rica and Florida (Ackerman, unpublished data). However, the water content in the moist layer is substantially greater in Florida and Costa Rica (Ackerman, 1977) probably due to the greater frequency of rainfall on those beaches. This distribution (Fig. 12.12) is likely to be familiar to anyone who has walked on a marine or riverine beach where a water table is close to the surface. If the surface sand is removed, moist sand may be found underneath. This sand holds water at potentials around −10 kPa or wetter while the surface sand may be air dry or drier.

Snell & Tracy (1985) report the water potentials in soils at the end of the incubation of eggs of the lizard *Amblyrhynchus cristatus*. Packard & Packard (1988) point out some difficulties with this measurement. If soil drying occurs during incubation as appears likely, measurement only at the end of incubation yields little information on the hydric microclimate during incubation. Packard *et al.* (1985) report the water potentials inside several clutches of eggs of *C. serpentina* eggs, and in the soil (sand) immediately contacting these eggs, using thermocouple psychrometry. These data are difficult to analyse because sand water contents were not provided. Water potential excursions of the type observed by Packard *et al.* (1985) may represent temperature differences between the psychrometer and the soil rather than water potential differences. The eggs in the nest could be warmer than the surrounding soil due either to metabolic activity or to daily temperature cycling. This would increase the temperature of the thermocouple psychrometer contacting the eggs above that of the soil, introducing an error into the measurement: the soil would appear to be much drier than it is. A 0.1 °C difference between the psychrometer and the soil produces a 1000 kPa error (Rawlins & Campbell, 1986). This is a small error at −100,000 kPa but it is not at −100 kPa. Leshem & Dmi'el (1986) estimate the water potential profile around eggs of *T. triunguis* to be similar to those shown above. However, the characteristic curve for the sand measured with a thermocouple psychrometer could not be confirmed using the more standard pressure plate technique (Leshem, 1989).

Thermocouple psychrometry should be used with caution and with particular attention paid to temperature control during calibration and thermal equilibrium during measurement (Rawlins & Campbell, 1986). The psychrometers should not be placed in locations where metabolic activity is likely to change the temperature of the psychrometer relative to the soil. The introduced error can be estimated if measurement of the temperature differences between the soil and the hygrometer are made (Rawlins & Campbell, 1986). Thermocouple psychrometry has also been used to assess the characteristic curves of soils in which eggs have been incubated (Tracy *et al.*, 1978; Packard *et al.*, 1987; Packard & Packard, 1988). This technique is often considered useful only for water potentials drier than −100 kPa (Bruce & Luxmoore, 1986) for the reasons discussed above. This may be acceptable if medium water potentials drier than −100 kPa are used, as is commonly done in the laboratory studies of water exchange of eggs. There are few reports of temperature control during calibration or measurement in the zoological literature so it is difficult to assess the error involved in the measurement.

Hypotheses and conclusions

The characteristic patterns of distribution of water and water potential in natural soil profiles (Ackerman, 1987) appear to be predictable but more research is required to identify how soil composition and structure, hydrology, topography and climate influence the local hydric microclimates occurring in the soil. The local hydrology is likely to be a variable of particular interest (Chorley, 1978; Kirkby, 1978). However, it is possible at this stage to suggest several hypotheses (Fig. 12.13).

The profile of water in soils in which buried eggs incubate may be categorised into three or four zones depending on the water potential (or humidity) in each. A top, dry zone may be present containing soil with water potentials from around −20,000 to greater than −100,000 kPa. In some locations, this dry zone may be the predominant zone. Beneath the upper dry zone, a humid zone may be present with water potentials around −1 to −50 kPa depending on soil type. This humid zone may be stable in time, or it may move slowly or it may also disappear. Sandy soils will tend to have wetter water poten-

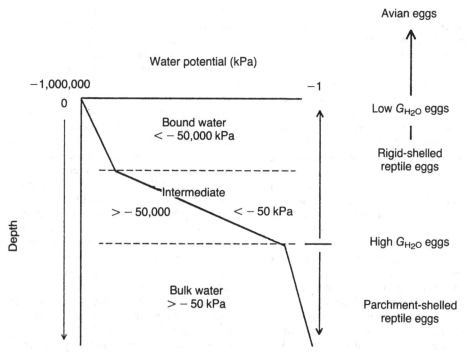

Fig. 12.13. A summary of the hypotheses about the hydric microclimate of buried vertebrate eggs as discussed in the text.

tials than clay soils. If both dry and humid zones are present, then a transitional zone will occur between them with water potentials around −50 kPa to −20,000 kPa or drier. The water potential gradient in this zone is likely to be steep (1000s of kPa cm⁻¹). The transitional zone also may be stable or moving slowly over time. The transitional zone is likely to be narrow where phreatic water is close to the surface and wider where phreatic water is farther from the surface. It will also be wide where no phreatic water is present. When phreatic water is present, a fourth zone will be present. This wet zone will include the flooded capillary fringe, just above the liquid water and will gradually decrease in wetness as it grades into the humid zone. The water logged soil in the wet zone will severely impede gas exchange.

It is also possible to hypothesise that the timing of egg deposition, the location of eggs buried in the soil, the type of shell characterising the egg and the water exchange of the egg are all directly related to the water profile in the soil in which the eggs are deposited. Muth (1980) hypothesised that distribution of the lizard, *D. dorsalis*, is correlated with the regional occurrence of hydric (and thermal) microclimates in

the soil suitable for the incubation of eggs. The distribution of animals whose eggs require specific hydric microclimates is likely to be associated with the differences in the regional and local distribution of water in soils. My hypotheses may be summarised as follows:

1) Pliable-, or parchment-shelled reptile eggs may be deposited in the humid zone beneath a drier surface layer. This zone is stable and close to the surface if phreatic water is close to the surface. This may be the case for many aquatic reptiles that deposit their eggs close to the water in which they live (Alho & Padua, 1982; Whitaker & Whitaker, 1984; Rao & Singh, 1985). Drying of the soil profile may continue during incubation with little effect on the water potential of the soil. This may be occurring in the soil surrounding iguana (*Iguana iguana*) eggs (Werner, 1988) where a water table is probably present close to the surface. However, the water table may be so close to the surface that the eggs can only be put into the wet zone. Successful incubation is unlikely in this zone and other accommodations must be made if hatchlings are to be produced. Eggs of various turtle genera (*Chrysemys*, *Trionyx* and *Kinosternon*) have been found incubating in the nest mounds of Ameri-

can alligators (*Alligator mississippiensis*) in swamps where liquid water is close to the surface (Dietz & Jackson, 1979).

Where phreatic water is not a consideration and where the soil has received sufficient rainfall, a humid soil zone will be present but its depth will vary over time. Parchment-shelled eggs may be incubated in humid zone, if it is present for an appropriate length of time and at an accessible depth. Otherwise, eggs may not be deposited or if deposited may not incubate successfully. It would be interesting to know how much water must be infiltrated into the soil to produce a humid zone that can support successfully the incubation of parchment-shelled eggs. Soil structure will be an important variable (Bagnold, 1941; Hillel, 1980; Campbell, 1985). In the humid microclimate in which they incubate, pliable- or parchment-shelled eggs will take up substantial quantities of water during incubation (Packard, Chapter 13). The quantity of water taken up will depend on other variables such as the thermal conductivity of the soil and the heat production of the eggs (Ackerman *et al.*, 1985a; Ackerman & Seagrave, 1987) because both the viscosity of liquid water and the saturation pressure of water vapour are dependent on temperature. Pliable- or parchment-shelled eggs will be in danger of lethal desiccation if they protrude into either the transitional or dry zones. This is because the permeability (or G_{H_2O}) of parchment-shelled eggs is very large (Fig. 12.6) and because the gradient between the wet and dry soil is also likely to be large (1000 to 10,000 kPa cm^{-1}). The eggs will lose too much water if exposed to these water potentials. Legler (1954) reports that some eggs of *C. picta* laid in clay in Minnesota are dehydrated at the top of the clutch while eggs at the bottom are not. The top eggs might well protrude into the dry surface layer. How typical this situation is remains to be determined. Neither eggs of *C. picta* nor *C. serpentina* survive incubation in conditions in natural soils in the laboratory drier than -3000 kPa (Ackerman, unpublished data). Under these conditions, water losses approaching 50% of initial mass or more occur.

2) Rigid-shelled eggs with low permeabilities or water vapour conductances, including megapode eggs, are likely to incubate in the dry zone near the surface of the soil profile where the water potential is likely to be around $-10,000$ to $-100,000$ kPa or drier and reasonably constant during incubation. These eggs will lose water in quantities ranging from a few % to more than 20% of initial mass depending on variables such as the thermal conductivity of the soil, the heat production of the eggs and the G_{H_2O} of the eggshell (Ackerman & Seagrave, 1987).

3) Rigid-shelled eggs with permeabilities or water vapour conductances intermediate between these two extremes, e.g. *T. triunguis*, are likely to be deposited and incubated in the transitional zone where water potentials vary from -50 to $-40,000$ kPa or more. The transitional zone needed by these eggs will be wide or variable in depth or both. Eggs may experience different hydric microclimates depending on their position in the nest. As a result, the water exchange of these eggs may vary with time and position in the nest. However, the water exchange is unlikely to be as extreme as the water uptake of eggs in the humid zone or the water loss of eggs in the dry zone. Female *Trionyx muticus* nest in the highest points they can find on sandy river bars and they appear to follow these points as the bars migrate along the river in time (Plummer, 1976). These locations may provide a suitable transitional zone for the incubation of the eggs. The distribution of water in soils may also provide insight into crocodilian nesting patterns. The crocodilians can be classified according to their nest type (Campbell, 1972; Greer, 1970) as either mound (Webb *et al.*, 1983; Lutz & Dunbar-Cooper, 1984; Waitkuwait, 1985; Ouboter & Nanhoe, 1987; Hall & Johnson, 1987) or hole (Pooley & Gans, 1976; Whitaker & Whitaker, 1984; Waitkuwait, 1985; Hutton, 1987) nesters. Mound nests may be constructed where the water table is close to the surface as in swampy areas, to avoid placing the eggs in waterlogged soils. Hole nests may be constructed where the water table is sufficiently far from the surface to provide the zonal structure described above. Thus nesting habits among crocodilians are likely to be influenced by the habitat in which they reproduce.

In this chapter several hypotheses that describe 1) the hydric environments available in the soil column, 2) the association between vertebrate eggs with different shell types incubating in soil and the available hydric environments and 3) the water exchange of the eggs are presented. The hypotheses can be tested in the field using straight forward measurements. The collect of data suitable to test the hypotheses will provide new insight into an area of vertebrate development about which we know little. These data will enable us to develop the hypotheses presented here or to replace them with better ones.

Acknowledgements

My work was made possible by the co-operation and assistance of Dr Robert Horton (Department of Agronomy, Iowa State University) who taught me the rudiments of soil physics and allowed me the run of his laboratory, and Dr Richard Seagrave (Department of Chemical Engineering, Iowa State University) who taught me the rudiments of heat and mass transfer and listened critically. Many of the ideas were developed and discussed in the company of Dr Amos Ar, Dr Razi Dmi'el (Department of Zoology, Tel Aviv University) and Dr Adah Leshem (Department of Zoology, Iowa State University). Dr Amos Ar and Ms Ann Belinsky critically reviewed the paper. Any bias expressed in this article and all errors are my own. My work has been supported in part by Israel–US Binational Science Foundation Grants 81–2407 and 84–00357.

References

Ackerman, R. A. (1977). The respiratory gas exchange of sea turtle nests (*Chelonia, Caretta*). *Respiration Physiology*, 31, 19–38.

— (1987). The incubation environment of reptile eggs. *Physiologist*, 30, 137.

Ackerman, R. A., Dmi'el, R. & Ar, A. (1985b). Energy and water vapor exchange by parchment-shelled reptile eggs. *Physiological Zoology*, 58, 129–37.

Ackerman, R. A. & Seagrave, R. C. (1987). Modeling heat and mass exchange of buried avian eggs. *Journal of Experimental Zoology*, Supplement 1, 87–97.

Ackerman, R. A., Seagrave, R. C., Dmi'el, R. & Ar, A. (1985a). Water and heat exchange between parchment-shelled reptile eggs and their surroundings. *Copeia*, 1985, 703–11.

Alho, C. J. R. & Padua, L. F. M. (1982). Reproductive parameters and nesting behavior of the amazon turtle *Podocnemis expansa* (Testudinata: Pelomedusidae) in Brazil. *Canadian Journal of Zoology*, 60, 97–103.

Bagnold, R. A. (1941). *The Physics of Blown Sand and Desert Dunes*. London: Methuen and Co.

Bird, R. B., Stewart, W. E. & Lightfoot, E. N. (1960). *Transport Phenomena*. New York: John Wiley & Sons.

Booth, D. T. & Seymour, R. S. (1987). Effect of eggshell thinning on water vapor conductance of Malleefowl eggs. *Condor*, 89, 453–9.

Bruce, R. R. & Luxmoore, R. J. (1986). Water Retention: Field Methods. In *Methods of Soil Analysis*, ed. A. Klute, pp. 663–86. Madison, Wisconsin: Soil Science Society of America.

Campbell, G. S. (1985). *Soil Physics with Basic*. New York: Elsevier Scientific Publishing.

Campbell, G. S. & Gee, G. W. (1986). Water Potential: Miscellaneous Methods. In *Methods of Soil Analysis*, ed. A. Klute, pp. 619–34. Madison, Wisconsin: Soil Science Society of America.

Campbell, H. W. (1972). Ecological or phylogenetic interpretations of crocodilian nesting habits. *Nature, London*, 238, 404–5.

Cassell, D. K. & Klute, A. (1986). Water Potential: Tensiometry. In *Methods of Soil Analysis*, ed. A. Klute, pp. 563–97. Madison, Wisconsin: Soil Science Society of America.

Cassell, D. K. & Nielsen, D. R. (1986). Field Capacity and Available Water Capacity. In *Methods of Soil Analysis*, ed. A. Klute, pp. 901–26. Madison, Wisconsin: Soil Science Society of America.

Chorley, R. J. (1978). The hillslope hydrologic cycle. In *Hillslope Hydrology*, ed. M. J. Kirkby, pp. 1–42. New York: John Wiley & Sons.

Corey, A. T. & Klute, A. (1985). Application of the potential concept to soil water equilibrium and transport. *Soil Science Society of America Journal*, 49, 3–11.

Deitz, D. C. & Jackson, R. (1979). Use of American alligator nests by nesting turtles. *Journal of Herpetology*, 13, 510–12.

Ewert, M. A. (1985). Embryology of turtles. In *Biology of the Reptilia*, vol. 14 Development A, eds C. Gans, F. Billet & P. F. A. Maderson, pp. 75–268. New York: John Wiley & Sons.

Gardner, W. H. (1986). Water Content. In *Methods of Soil Analysis*, ed. A. Klute, pp. 493–544. Madison, Wisconsin: Soil Science Society of America.

Greer, A. E. (1970). Evolutionary and systematic significance of crocodilian nesting habits. *Nature, London*, 227, 523–4.

Gutzke, W. H. N. (1984). Modification of the hydric environment by eggs of snapping turtles (*Chelydra serpentina*). *Canadian Journal of Zoology*, 62, 2401–3.

Gutzke, W. H. N. & Packard, G. C. (1987). Influence of the hydric and thermal environments on eggs and hatchlings of bull snakes *Pituophis melanoleucus*. *Physiological Zoology*, 60, 9–17.

Gutzke, W. H. N., Packard, G. C., Packard, M. J. & Boardman, T. J. (1987). Influence of the hydric and thermal environments on eggs and hatchlings of painted turtles (*Chrysemys picta*). *Herpetologica*, 43, 393–404.

Hall, P. H. & Johnson, D. R. (1987). Nesting biology of *Crocodylus novaeguineae* in Lake Murray district, Papua New Guinea. *Herpetologica*, 43, 249–58.

Harrison, K. E., Bentley, T. B., Lutz, P. L. & Marszalek, D. S. (1978). Water and gas diffusion in the American Crocodile egg. *American Zoologist*, 18, 637.

Hillel, D. (1971). *Soil and Water Physical Principles and Processes*. New York: Academic Press.

— (1980a). *Fundamentals of Soil Physics*. New York: Academic Press.

— (1980b). *Applications of Soil Physics*. New York: Academic Press.

Hillel, D. & van Bavel, C. H. M. (1976). Simulation of profile water storage as related to soil hydraulic properties. *Soil Science Society of America Journal*, 40, 807–15.

Hutton, J. M. (1987). Incubation temperature, sex ratios and sex determination in a population of Nile crocodiles (*Crocodylus niloticus*). *Journal of Zoology, London*, 211, 143–55.

Kam, Y. C. (1988). Environmental influences on the water exchange and growth of turtle eggs and embryos Ms Thesis. Iowa State University: Ames, Iowa.

Keulen, V. & Hillel, D. (1974). A simulation study of the drying front phenomenon. *Soil Science*, 118, 270–3.

Kirkby, M. A. (1978). *Hillslope Hydrology*. New York: John Wiley & Sons.

Klute, A. (1986). Water Retention: Laboratory Methods. In *Methods of Soil Analysis*, ed. A. Klute, pp. 635–62. Madison, Wisconsin: Soil Science Society of America.

Klute, A. & Dirksen, C. (1986). Hydraulic Conductivity and Diffusivity: Laboratory Methods. In *Methods of Soil Analysis*, ed. A. Klute, pp. 687–734. Madison, Wisconsin: Soil Science Society of America.

Koorevaar, P., Menelik, G. & Dirksen, C. (1983). *Elements of Soil Physics*. New York: Elsevier.

Legler, J. M. (1954). Nesting habits of the western painted turtle, (*Chrysemys picta bellii* (Gray)). *Herpetologica*, 10, 137–44.

Leshem, A. (1989). The effect of temperature and soil water content on embryonic development of the Nile soft-shelled turtle (*Trionyx triunguis*). PhD Dissertation. Tel Aviv University, Ramat Aviv, Tel Aviv, Israel.

Leshem, A., Ackerman, R. A., Ar, A. & Dmi'el, R. (1986). Water exchange of *Trionyx triunguis* eggs during incubation. *Physiologist*, 29, 177.

Leshem, A. & Dmi'el, R. (1986). Water loss from *Trionyx triunguis* eggs incubating in natural nests. *Herpetological Journal*, 1, 115–17.

Lutz, P. L. & Dunbar-Cooper, A. (1984). The nest environment of the American Crocodile (*Crocodylus acutus*). *Copeia*, 1984, 153–61.

McBride, J. F. & Horton, R. (1985). An empirical function to describe measured water distribution from horizontal infiltration experiments. *Water Resources Research*, 21, 1539–44.

Miller, K., Packard, G. C. & Packard, M. J. (1987).

Hydric conditions during incubation influence locomotor performance of hatchling snapping turtles. *Journal of Experimental Biology*, 127, 401–12.

Morris, K. A., Packard, G. C., Boardman, T. J., Paukstis, G. L. & Packard, M. J. (1983). Effect of the hydric environment on growth of embryonic snapping turtle (*Chelydra serpentina*). *Herpetologica*, 39, 272–85.

Mualem, Y. (1986). Hydraulic Conductivity of Unsaturated Soils: Prediction and Formulas. In *Methods of Soil Analysis*, ed. A. Klute, pp. 799–824. Madison, Wisconsin: Soil Science Society of America.

Muth, A. (1980). Physiological ecology of desert iguana (*Dipsosaurus dorsalis*) eggs: Temperature and water relations. *Ecology*, 61, 1335–43.

— (1981). Water relations of desert iguana (*Dipsosaurus dorsalis*) eggs. *Physological Zoology*, 54, 441–51.

Nobel, P. S. (1983). *An Introduction to Biophysical Plant Physiology and Ecology*. San Francisco: W. H. Freeman.

Ouboter, P. E. & Nanhoe, L. M. R. (1987). Notes on nesting and parental care in *Caiman crocodilus crocodilus* in northern Suriname and an analysis of crocodilian nesting habitats. *Amphibia–Reptilia*, 8, 331–48.

Packard, G. C. & Packard, M. J. (1988). The physiological ecology of reptilian eggs and embryos. In *Biology of the Reptilia*, vol. 16 Ecology B, eds C. Gans & R. B. Huey, pp. 523–606. New York: Alan R. Liss, Inc.

Packard, G. C., Packard, M. J. & Boardman, T. J. (1981a). Patterns and possible significance of water exchange by flexible-shelled eggs of painted turtles (*Chrysemys picta*). *Physiological Zoology*, 54, 165–78.

Packard, G. C., Packard, M. J., Boardman, T. J. & Ashen, M. D. (1981b). Possible adaptive value of water exchanges in flexible-shelled eggs of turtle. *Science*, 213, 471–3.

Packard, G. C., Packard, M. J., Boardman, T. J., Morris, K. A. & Shuman, R. D. (1983). Influence of water exchange by flexible-shelled eggs of painted turtles *Chrysemys picta* on metabolism and growth of embryos. *Physiological Zoology*, 56, 217–30.

Packard, G. C., Packard, M. J., Miller, K. & Boardman, T. J. (1987). Influence of moisture, temperature and substrate on snapping turtle eggs and embryos. *Ecology*, 68, 983–93.

Packard, G. C., Paukstis, G. L., Boardman, T. J. & Gutzke, W. H. N. (1985). Daily and seasonal variation in hydric conditions and temperatures inside nests of common snapping turtles (*Chelydra serpentina*). *Canadian Journal of Zoology*, 63, 2422–9.

Packard, G. C., Taigen, T. L., Boardman, T. J., Packard, M. J. & Tracy, C. R. (1979a). Changes in mass of softshell turtle (*Trionyx spiniferus*) eggs. *Herpetologica*, 35, 78–86.

Packard, G. C., Taigen, T. L., Packard, M. J. & Boardman, T. J. (1980). Water relations of pliable-shelled eggs of common snapping turtles (*Chelydra serpentina*). *Canadian Journal of Zoology*, **58**, 1404–11.

Packard, G. C., Taigen, T. L., Packard, M. J. & Shuman, R. D. (1979*b*). Water vapor conductance of Testudinian and Crocodilian eggs (Class Reptilia). *Respiration Physiology*, **38**, 1–10.

Plummer, M. V. (1976). Some aspects of nesting success in the turtle, *Trionyx muticus*. *Herpetologica*, **32**, 353–9.

Pooley, A. C. & Gans, C. (1976). The Nile Crocodile. *Scientific American*, **234**, 114–24:

Rao, R. J. & Singh, L. A. K. (1985). Notes on comparative body size, reproductive effort and areas of management priority for three species of *Kachuga* (Reptilia, Chelonia). *Journal of the Bombay Natural History Society*, **84**, 55–65.

Ratterman, R. J. & Ackerman, R. A. (1989). The water exchange and hydric microclimate of painted turtle (*Chrysemys picta*) eggs incubating in field nests. *Physiological Zoology*, **62**, 1059–79.

Rawlins, S. L. & Campbell, G. S. (1986). Water Potential: Thermocouple Psychrometry. In *Methods of Soil Analysis*, ed. A. Klute, pp. 597–618. Madison, Wisconsin: Soil Science Society of America.

Ruibal, R., Tevis, L., Jr & Roig, V. (1969). The terrestrial ecology of the spadefoot toad *Scaphiopus hammondii*. *Copeia*, **1969**, 571–84.

Seagrave, R. C. (1971). *Biomedical Applications of Heat and Mass Transport*. Ames, Iowa: Iowa State Univ. Press.

Seely, M. K. & Mitchell, D. (1987). Is the subsurface environment of the Namib Desert dunes a thermal haven for chthonic beetles? *South African Journal of Zoology*, **22**, 57–61.

Seymour, R. S., Vleck, D., Vleck, C. M. & Booth, D. T. (1987). Water relations of buried eggs of mound building birds. *Journal of Comparative Physiology*, **157B**, 413–22.

Silberstein, D. (1988). The physical condition prevailing in nests of the loggerhead sea turtle *Caretta caretta*. MS Dissertation, Tel Aviv University: Tel Aviv, Israel.

Snell, H. L. & Tracy, C. R. (1985). Behavioral and morphological adaptations by Galapagoes land iguanas (*Conolophis subcristatus*) to water and energy requirements of eggs and neonates. *American Zoologist*, **25**, 1009–18.

Spotila, J. R., Weinheimer, C. J. & Paganelli, C. V. (1981). Shell resistance and evaporative water loss from bird eggs: Effects of wind speed and egg size. *Physiological Zoology*, **54**, 195–202.

Thompson, M. B. (1987). Water exchange in reptilian eggs. *Physiological Zoology*, **60**, 1–8.

Tracy, C. R. (1980). Water relations of parchment-shelled lizard (*Sceloporus undulatus*) eggs. *Copeia*, **1980**, 478–82.

— (1982). Biophysical modelling in reptilian physiology and ecology. In *Biology of the Reptilia*, vol. 12, eds C. Gans & R. Huey, pp. 275–321. New York: Academic Press.

Tracy, C. R., Packard, G. C. & Packard, M. J. (1978). Water relations of chelonian eggs. *Physiological Zoology*, **51**, 378–87.

Tracy, C. R., & Sotherland, P. R. (1979). Boundary layers of bird eggs: do they every constitute a significant barrier to water loss? *Physiological Zoology*, **52**, 63–6.

van Genuchten, M. Th. (1980). A closed-form equation for predicting the hydraulic conductivity of unsaturated soils. *Soil Science Society of America Journal*, **44**, 892–8.

Waitkuwait, W. E. (1985). Investigations of the breeding biology of the west-African slender-snouted crocodile *Crocodylus cataphractus* Cuvier, 1824. *Amphibia–Reptilia*, **6**, 387–99.

Webb, G. J. W., Sack, G. C., Buckworth, R. & Manolis, S. C. (1983). An examination of *Crocodylus porosus* nests in two Northern Australian freshwater swamps, with analysis of embryo mortality. *Australian Wildlife Research*, **10**, 571–605.

Werner, D. I. (1988). The effect of varying water potential on body weight, yolk and fat bodies in neonate green iguanas. *Copeia*, **1988**, 406–11.

Whitaker, R. & Whitaker, Z. (1984). Reproductive biology of the mugger (*Crocodylus palustris*). *Journal of the Bombay Natural History Society*, **81**, 297–316.

13

Physiological and ecological importance of water to embryos of oviparous reptiles

GARY C. PACKARD

Introduction

Research on the water relations of reptilian embryos apparently originated with the descriptions by Dendy (1899), Brimley (1903) and Coker (1910) of swelling over the course of incubation by the flexible-shelled eggs of several chelonians and lepidosaurians (a term used here to refer to Squamata plus Sphenodontia; Benton, 1985). However, most of the current interest in this subject derives from the landmark studies by Cunningham and his associates (Cunningham & Hurwitz, 1936; Cunningham & Huene, 1938; Cunningham, Woodward & Pridgen, 1939) and by Fitch & Fitch (1967). These investigations were the first to demonstrate that reptilian eggs and embryos are affected profoundly by the availability of water in their environment, and they revealed also that the water relations of reptilian eggs differ in important ways from those of avian eggs. In retrospect, however, the experiments suffered from the absence at the time of a generally accepted method for quantifying the availability of water in substrates. Cunningham *et al.* used a qualitative approach and did not report the water content of substrates used to incubate eggs in their studies, while Fitch & Fitch expressed the water content of substrates gravimetrically. Neither of these methods is based upon thermodynamic principles. Thus, the studies have limited value in elucidating either the patterns or the mechanisms of water exchange by incubating eggs of reptiles (Packard & Packard, 1988a).

The current era of research on water relations of reptilian eggs was made possible by developments in the field of soil physics, where the con-cept of water potential emerged during the 1960s (Taylor & Ashcroft, 1972) and where psychrometric methods for measuring water potential were developed during the 1970s (Savage & Cass, 1984; Rawlins & Campbell, 1986). Water in substrates tends to flow as a liquid from areas of high potential to areas of low potential (Corey & Klute, 1985; Ackerman, Chapter 12), so measurements of water potential are essential in investigations on the water relations of reptilian eggs. However, water also diffuses as a vapour through air spaces in the soil from areas of high partial pressure to areas of low partial pressure (Taylor & Ashcroft, 1972) and vapour pressure gradients often arise secondary to gradients in temperature (Rose, 1968a,b). Thus, careful measurements of temperature are also needed in studies on the water relations of reptilian eggs.

Most of the quantitative information from contemporary studies on the water relations of reptilian eggs comes from experiments performed in the research laboratory, and relatively little information comes from investigations conducted in nature. Nevertheless, numerous predictions concerning the responses of eggs in nature have emerged from the laboratory studies, and some of these predictions have been tested and confirmed in the field (Hotaling, Wilhoft & McDowell, 1985; Snell & Tracy, 1985; Leshem & Dmi'el, 1986). Thus, much of the information gained from carefully designed laboratory experiments can probably be extrapolated to nature. Field research still is needed, however, to assure that the capacity for performance by eggs and embryos in the laboratory actually is expressed under natural conditions (Pough, 1989).

Water relations have been studied most

extensively for eggs and embryos of turtles, relatively few investigations have focused on the eggs of lepidosaurians (Packard & Packard, 1988a), and virtually nothing is known of the water relations of crocodilian eggs (Grigg, 1987). Several studies appearing between 1977 and 1983 suggested that the physiology of embryonic lepidosaurians differs fundamentally from that of embryonic chelonians, because moisture had no apparent effect on length of incubation by embryonic lizards (Muth, 1980) or on size of hatchling lizards (Tracy, 1980). However, these findings may have resulted from weaknesses in experimental design instead of from important differences between embryonic lepidosaurians and embryonic chelonians. For instance, the metabolic response by embryonic painted turtles (*Chrysemys picta*) to different hydric environments escaped detection in an initial study (Packard *et al.*, 1983) because the investigators were unable to block their experiment by clutch (Packard & Packard, 1986). Unassessed variation among clutches in the initial investigation consequently contributed to experimental 'noise' and prevented treatment effects from being detected. Thus, many of the earlier reports that embryonic lepidosaurians are unaffected by the availability of water may have resulted from the inability of the workers to identify (and compensate for) variation among clutches.

Work completed more recently on lepidosaurians indicates that the embryos of all reptiles actually share a common set of physiological responses to variation in the amount of water available inside their eggs (Black *et al.*, 1984; Tracy & Snell, 1985; Gutzke & Packard, 1987; Plummer & Snell, 1988; Phillips *et al.*, 1990; Vleck, Chapter 15). These common patterns of response by embryonic reptiles to variation in hydric conditions will be emphasised here.

Factors affecting water exchange

Many factors affect the exchange of water between a reptilian egg and its environment (Packard & Packard, 1988a) but size of the egg and physical characteristics of the eggshell are among the most important. Large eggs have a greater surface area than small eggs across which exchanges can occur (Iverson & Ewert, Chapter 7). Moreover, porosity of the eggshell presumably governs the movements of both

liquid (Thompson, 1987) and vapour (Packard *et al.*, 1979; Feder, Satel & Gibbs, 1982; Tracy, 1982; Lillywhite & Ackerman, 1984; Ackerman, Dmi'el & Ar, 1985a; Deeming & Thompson, Chapter 17) between the interior of the egg and its environment, and rigidity of the shell determines the extent to which the egg can swell to accommodate the uptake of water. For example, the rigid eggshells of all crocodilians, many turtles, and some lizards have relatively low permeabilities to water (Packard *et al.*, 1979; Dunson & Bramham, 1981; Dunson, 1982; Tracy, 1982), whereas the flexible eggshells of many turtles and most lepidosaurians are highly permeable to water (Packard *et al.*, 1979; Lillywhite & Ackerman, 1984; Ackerman *et al.*, 1985a). Thus, water passes more readily through flexible eggshells than through rigid ones (Packard, Packard & Boardman, 1982b; Deeming & Thompson, Chapter 17). Also, fibres of the shell membrane in flexible eggs may stretch or distend to accommodate uptake of water from the environment (Packard *et al.*, 1982a), while those in the shell membrane of rigid eggs usually are prevented from distending by the interlocking of shell units in the calcareous layer (Packard & Packard, 1979; Packard, Hirsch & Iverson, 1984a; Packard & DeMarco, Chapter 5). Differences in size of eggs, and in characteristics of eggshells, probably underlie most of the differences among species in patterns of water exchange by eggs with the environment (Vleck, Chapter 15).

Another important factor affecting the water exchanges of reptilian eggs is water potential of the substrate on which the eggs rest or in which they are buried (Packard, Packard & Boardman, 1980; Gutzke & Packard, 1987; Gutzke *et al.*, 1987; Packard & Packard, 1987; Packard *et al.*, 1987; Plummer & Snell, 1988; Ackerman, Chapter 12). Eggs resting on substrates at high water potentials generally experience a net absorption of water over the course of incubation and consequently increase in mass (Fig. 13.1). By contrast, eggs resting on substrates at low water potentials usually lose water to their surroundings and therefore decline in mass over the course of development (Fig. 13.1). Responses of rigid-shelled eggs are fairly similar to those of flexible-shelled eggs, except that patterns of net water exchange are muted in the former (Packard *et al.*, 1981b, 1982b; Thompson, 1983). Thus, embryos developing on wet substrates have access to larger reservoirs

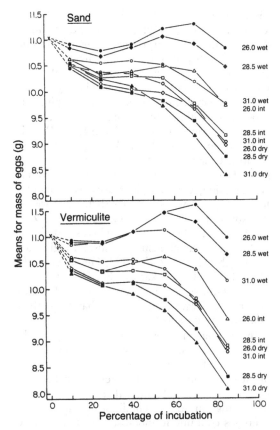

Fig. 13.1. Temporal changes in mass of flexible-shelled eggs of common snapping turtles (*Chelydra serpentina*) incubating on substrates of sand and vermiculite at different temperatures and moisture regimes. Data are expressed as means that have been adjusted by analysis of covariance to compensate for variation in size of eggs at the outset of study. Water potentials at the start of the investigation were −150 kPa, −550 kPa, and −950 kPa for wet, intermediate, and dry substrates, respectively. Eggs had approximately half of their surface in contact with the substrate and half in contact with air trapped inside the covered containers used for incubation. Note the similarity in patterns of change in mass for eggs resting on the two different media. (From Packard *et al.*, 1987.)

of water inside their eggs than do embryos developing on dry media (Morris *et al.*, 1983; Packard *et al.*, 1983) but the differences are more pronounced for eggs with flexible shells.

Temperature also has a profound effect on the water exchanges of reptilian eggs (Gutzke & Packard, 1987; Gutzke *et al.*, 1987; Packard *et al.*, 1987) and seemingly exerts its influence

primarily by altering rates of diffusion of vapour across eggshells (Packard & Packard, 1988*a*). Eggs exposed to relatively low ambient temperatures generally enjoy more positive water balance than eggs incubating at higher temperatures, even when substrates are at the same water potential (Fig. 13.1). Unfortunately, effects on embryos of thermally induced variation in water exchange are difficult to establish with certainty, because effects stemming from changes in the availability of water cannot be distinguished readily from those elicited by direct actions of temperature itself on physiological processes. On the other hand, some of the putative effects of temperature on reptilian embryos (Deeming & Ferguson, Chapter 10) may actually result from differences in the amount of water available to embryos developing in different thermal conditions.

A fourth important factor affecting water exchange is the proportion of the eggshell contacting the substrate (Packard *et al.*, 1980; Tracy & Snell, 1985). Eggs having a large proportion of their shell in contact with the substrate generally enjoy more positive water exchanges than eggs having smaller fractions of their shell in contact with the medium (Fig. 13.2). The degree of contact between an egg and the substrate presumably exerts its effect on the pattern of net water exchange by altering the relative importance of movements across the eggshell by liquid and vapour (Packard & Packard, 1988*a*).

Two attempts have been made to incorporate the aforementioned variables into mechanistic models that characterise the water exchanges experienced by reptilian eggs (Tracy, Packard & Packard, 1978; Ackerman *et al.*, 1985*b*). Both of the models yield insights into the physical processes affecting water exchange, but both suffer from flaws that limit their current utility. For example, the model by Tracy *et al.* (1978) does not address explicitly the exchange of energy between an egg and its environment, so the importance of thermally induced movements of vapour is underestimated. Likewise, the model formulated by Ackerman *et al.* (1985*b*) assumes explicitly that water passes across eggshells only in the form of vapour (*i.e.* no liquid), and that the low hydraulic conductivity of dry soils is unlikely ever to be limiting to water exchanges by eggs. Experimental data suggest that these assumptions are probably incorrect (Gutzke, 1984; Thompson, 1987). The shortcomings in the

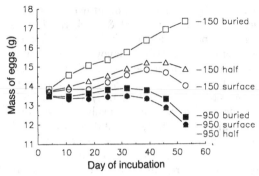

Fig. 13.2. Temporal changes in mass of flexible-shelled eggs of common snapping turtles incubating at 26 °C on substrates of wet (-150 kPa) and dry (-950 kPa) vermiculite. Some of the eggs from each of the two clutches used in this investigation were placed on top of the vermiculite; some were half-buried in the substrates; and some were fully buried in the media. An analysis of covariance with repeated measures confirmed that the degree of contact between eggs and their substrate influenced the pattern of change in mass for eggs on wet substrates but not for those on dry vermiculite ($F_{14,189} = 14.79$, $P < 0.001$, for the three-way interaction between time $\times$ water potential $\times$ treatment); the initial mass of the eggs was the covariate. (G. C. Packard & M. J. Packard, unpublished observations.)

scope or assumptions of the two models may underlie the lack of agreement between predictions from the models and empirical data (Packard *et al.*, 1981*a*, 1987) and the models need to be revised accordingly.

Correlates of water exchange

One of the more obvious results of incubating reptilian eggs in dry environments is a reduction in hatching success, e.g. snapping turtles (*Chelydra serpentina*) in Fig. 13.3. This result is more pronounced for eggs with flexible shells than for those with rigid shells (Packard *et al.*, 1981*b*; Thompson, 1983), and it is more pronounced for small eggs than for large ones (Gutzke & Packard, 1985; Packard, Packard & Birchard, 1989). Eggs with rigid shells tend not to lose large amounts of water even to relatively dry environments (Dunson & Bramham, 1981; Packard *et al.*, 1981*b*; Dunson, 1982; Thompson, 1983), so the water reserves of embryos seemingly are protected in eggs such as these. Hydric conditions during incubation consequently may have relatively little impact on developing embryos in rigid-shelled eggs. Like-

wise, embryos in small eggs have access to smaller reserves of water in the yolk and albumen than do embryos in large eggs. Thus, the loss of a given quantity of water is more likely to bring embryos in small eggs to the limit of their tolerance than is true for embryos in large eggs.

The metabolism of embryos developing on substrates within the range of tolerance increases during incubation (Lynn & von Brand, 1945; Ackerman, 1981; Black *et al.*, 1984; Vleck & Hoyt, Chapter 18). For species with flexible-shelled eggs, metabolism, measured as oxygen consumption, generally increases more rapidly for embryos on wet substrates than for those on dry substrates (Gettinger, Paukstis & Gutzke, 1984; Vleck, Chapter 15). Embryos sustaining high rates of metabolism also sustain high rates of growth (Morris *et al.*, 1983; Gettinger *et al.*, 1984; Packard & Packard, 1986, 1989*a*; Janzen *et al.*, 1990), so that embryos on wet substrates are much larger just before hatching than embryos on dry substrates.

Incubating in moist conditions apparently represents a 'safe harbour' for developing embryos (Shine, 1978), because embryos in moist environments tend to remain in the egg longer before hatching than do embryos incubating in drier conditions (Fig. 13.3). Embryos continue to mobilise yolk and to grow during this interval (Morris *et al.*, 1983; Packard & Packard, 1986, 1989*a*), so this extended period of incubation contributes (together with the higher growth rate) to the larger size of animals hatching in moist environments (Fig. 13.3). The differences in rates of growth and duration of incubation result in the larger hatchlings from wet environments having smaller masses of residual yolk to sustain them through the post-hatching period (e.g., Gutzke & Packard, 1987; Gutzke *et al.*, 1987; Packard *et al.*, 1987, 1988; Phillips *et al.*, 1990; D. Vleck, Chapter 15). These generalisations seemingly apply to embryos developing inside rigid-shelled eggs as well as to those in flexible-shelled eggs (Grigg, 1987; Manolis, Webb & Dempsey, 1987), albeit the responses of these embryos are minimised by the aforementioned constraints on water exchange imposed by the eggshell (Packard *et al.*, 1981*b*; Thompson, 1983).

Preliminary experiments reported from my laboratory raised the possibility that sexual differentiation in embryonic painted turtles (*Chrysemys picta*) is also influenced by hydration of the incubation substrate (Gutzke & Paukstis,

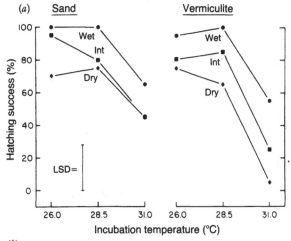

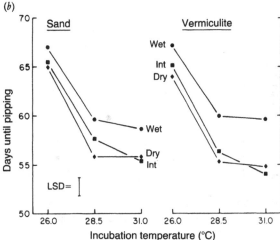

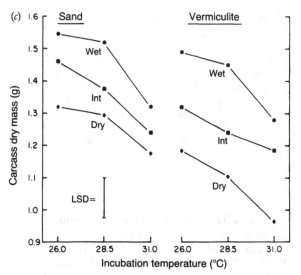

Fig. 13.3. Means for (*a*) hatchability, (*b*) length of incubation to pipping and (*c*) dry mass of carcasses

1983; Paukstis, Gutzke & Packard, 1984). These findings have clear and important bearing on efforts to identify the cellular mechanism(s) underlying the process of sexual differentiation in reptiles (Deeming & Ferguson, 1988, 1989; Crews, Wibbels & Gutzke, 1989; Georges, 1989). However, recent attempts to repeat the work on painted turtles failed to confirm the earlier findings (Packard *et al.*, 1989; Packard, Packard & Benigan, unpublished observations). Thus, whereas moisture in the nest chamber influences a number of physiological processes in embryonic painted turtles (Packard *et al.*, 1983, 1989; Gutzke & Packard, 1985; Packard & Packard, 1986; Gutzke *et al.*, 1987), the process of sexual differentiation may not be among them.

A basis for altered rates of metabolism and growth

Effects of the hydric environment on the physiology of reptilian embryos are probably mediated by the availability of water inside eggs. However, this presumption admittedly is an inference based on correlative data and not on an explicit test of any hypothesis (Packard & Packard, 1988*b*).

Assuming that availability of water is the important variable, how can the amount of water contained by an egg intervene in the physiology of a developing embryo to modulate processes like metabolism and growth? One possible explanation emerges from a consideration of the patterns of increase in concentration of urea in fluid compartments inside eggs incubating in wet and dry environments (Packard & Packard, 1984). Proteins are catabolised by embryos developing in both wet and dry settings, and urea is formed as the primary nitrogenous metabolite (Packard & Packard, 1984, 1987). More urea is formed by embryos in wet environments than by embryos in drier settings (Packard, Packard & Boardman, 1984*a*), simply by virtue of the higher rates of metabolism charac-

(without residual yolk) for hatchlings from flexible-shelled eggs of snapping turtles incubated on substrates of sand and vermiculite at different temperatures and moisture regimens described in Fig. 13.1. Pairs of means differing by the least significant difference (LSD) are statistically different at alpha = 0.05. (From Packard *et al.*, 1987.)

terising embryos in wet surroundings (Gettinger *et al.*, 1984). The concentration of urea in fluid compartments of eggs nonetheless increases faster in dry environments than in wet ones, because more water is available inside eggs in wet environments to disperse the accumulating urea (Packard & Packard, 1989*b*). Consequently, animals hatching from eggs in dry environments have much higher concentrations of urea in their blood than do hatchlings emerging from eggs in wetter settings (Fig. 13.4).

Urea inhibits the activity of mammalian enzymes at concentrations similar to those characterising body fluids of embryonic turtles (*Chelydra serpentina* and *Trionyx spiniferus*) developing in dry environments (Packard & Packard, 1989*b*, 1990). This observation led to an hypothesis which proposes that metabolism and growth are modulated by differential increases in the concentration of urea inside eggs. This hypothesis has been tested explicitly by injecting turtle eggs with different amounts of urea, thereby inducing different levels of uremia in the developing embryos. Although the injection procedure elicited a range of uremias that spanned the range observed in turtles hatching in wet and dry environments, no effect of the injection procedure was detected on the size of hatchlings or on the size of their residual yolk (Packard & Packard, 1989*b*, 1990). Although this 'urea hypothesis' has not been tested in lepidosaurians or crocodilians, these results for turtles suggest that the hypothesis probably should be rejected in its current form.

Several alternative hypotheses emerge from the correlations between osmotic activity of body fluids and moisture regimen of the environment where eggs are incubated (Fig. 13.4). One of the more attractive of these hypotheses states that effects of water on metabolism and growth of reptilian embryos are mediated by changes in the circulation and attendant changes in transport of oxygen. The hypothesis is based largely on similarities between embryonic reptiles developing in wet and dry settings and anuran amphibians subjected to different degrees of desiccation.

For example, embryonic turtles developing in dry environments contain proportionately less body water, and have a higher osmotic activity for their blood (Fig. 13.4), than those developing in wetter settings (Packard *et al.*, 1983, 1988). Similar findings have been reported for anurans (e.g. *Xenopus laevis*) subjected to desiccating

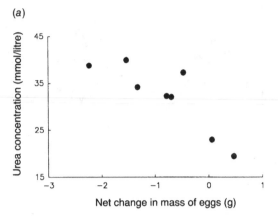

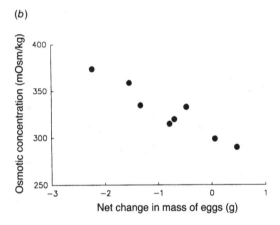

Fig. 13.4. Means for (*a*) concentration of urea in, and (*b*) osmotic activity of, blood plasma of hatchling snapping turtles in relation to corresponding means for net change in mass of their flexible-shelled eggs over the course of incubation. Eggs losing the most mass during incubation were exposed continuously to a dry substrate; eggs gaining the most mass were exposed continuously to a wet substrate; and eggs with intermediate values for change in mass were exposed to wet and dry conditions for different amounts of time. (Data from Packard & Packard, 1989*b*.)

conditions (Hillman, 1984; Hillman, Zygmunt & Baustian, 1987). However, dehydrated anurans also suffer from impaired metabolism (Hillman, 1978, 1987), owing apparently to reductions in cardiac output stemming from reductions in blood volume and increases in blood viscosity

(Hillman, 1984, 1987; Hillman *et al.*, 1985, 1987).

The parallels between reptilian embryos in dry environments and desiccated anurans raise the possibility that embryonic reptiles developing in dry environments experience reductions in blood volume and increases in blood viscosity that lead, in turn, to reductions in perfusion of tissues. These reductions in perfusion may be augmented by reductions also in cardiac performance resulting from elevation of plasma electrolytes (Hillman, 1984).

By the current hypothesis, reductions in the perfusion of tissues lead to reductions in metabolism of those tissues, in much the same way that reductions in perfusion of mammalian liver elicit reductions in oxygen consumption, even at high pO_2 (Edelstone, Paulone & Holzman, 1984). The reductions in metabolism of the embryos lead, in turn, to reductions in the rate of mobilisation of nutrients from the yolk and to reductions in the rate of embryonic growth. Consequently, hatching occurs at a smaller body size than characterises embryos in wet environments, but hatchlings in dry conditions contain larger masses of residual yolk.

This hypothesis is particularly attractive because it also provides an explanation for differences in survival by embryos in wet and dry environments and for differences in length of incubation. For instance, mortality of *X. laevis* at the limit of desiccation has been linked to circulatory collapse (Hillman, 1978). Mortality of embryonic reptiles developing in dry settings may therefore result from a failure of the circulatory system to supply tissues with sufficient oxygen to maintain a minimal level of metabolism. Likewise, hatching of fish (e.g. *Fundulus heteroclitus*) (DiMichele & Powers, 1984), amphibians (Petranka, Just & Crawford, 1982; Bradford & Seymour, 1988) and birds (Visschedijk, 1968) is stimulated by reductions in the availability of oxygen. Embryonic reptiles that hatch the earliest may also be the ones that suffer from impaired circulation, so the hatching of these animals may be linked to the supply of oxygen to their tissues.

Testing this hypothesis by direct manipulation will be difficult, because blood volume of reptilian embryos probably cannot be altered experimentally without altering other variables as well. However, the hypothesis rests on certain assumptions and leads to certain predictions. If experiments fail to support necessary assumptions or critical predictions, doubt is cast on the hypothesis itself. For example, a tacit assumption of the hypothesis is that embryonic reptiles are metabolic conformers (Hochachka, 1988), so one 'test' of the hypothesis would be to determine whether or not reductions in delivery of oxygen are accompanied by gradual, but consistent, decreases in oxygen consumption by embryos. If embryos actually regulate oxygen consumption over a range in rates of oxygen delivery, the hypothesis for control of metabolism and growth would have to be altered or rejected.

Ecological consequences of induced variation

The ecological implications for variation in hydration of the nest environment, and for patterns of net water exchange by reptilian eggs, are manifold. First, embryos developing in relatively dry settings are less likely than those in wetter environments to survive to hatching (Packard *et al.*, 1980, 1987, 1989; Gutzke & Packard, 1985). Thus, the moisture regimen may be a prime determinant of whether embryos complete development successfully. Second, embryos incubating in relatively wet environments grow faster, and incubate longer before hatching, than do embryos in drier surroundings (Gutzke & Packard, 1987; Packard *et al.*, 1987). While the longer period of incubation may cause embryos to be exposed for an extended period to potential predators of nests, the hazards of living outside the nest are probably greater than those associated with remaining in the cavity (see Gibbons & Nelson, 1978; Shine, 1978).

Rapid growth and prolonged incubation by embryos developing in moist environments combine to produce relatively large hatchlings. Is attaining a large size before hatching generally advantageous? The answer to this question depends on your perspective. From the point of view of a hatchling, the answer probably is 'yes'. Large hatchlings of turtles and lizards survive better during the neonatal period than do small hatchlings in the same cohort (Ferguson &. Bohlen, 1978; Fox, 1978; Swingland & Coe, 1979; Ferguson, Brown & DeMarco, 1982; Ferguson & Fox, 1984). Large size at hatching seems to allow higher survival in encounters with predators (Barrows & Schwarz, 1895;

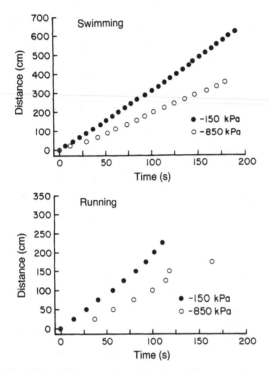

Fig. 13.5. Distances moved by hatchling snapping turtles in relation to time spent swimming in water or running on land. Eggs were incubated on wet (−150 kPa) and dry (−850 kPa) substrates of vermiculite at 29 °C. Thus, turtles hatching on wet substrates were large, whereas those hatching on dry media were small (see Fig. 13.3). The slopes of the plots represent speeds for locomotion. (Redrawn from Miller et al., 1987.)

Ferguson & Fox, 1984), perhaps because large hatchlings are better than small hatchlings at locomotion (Fig. 13.5). Large animals also are superior to small animals in foraging (Froese & Burghardt, 1974; Ferguson & Fox, 1984).

One problem with most of the preceding studies of survival by neonatal animals is that the large size of the successful hatchlings is not known to have resulted from their eggs being incubated in relatively moist environments. The large size of these animals conceivably resulted from their developing in large eggs containing large nutrient reserves (Packard et al., 1987). Large hatchlings from eggs incubated in wet environments are not necessarily equivalent to the large hatchlings emerging from large eggs, so size of eggs may have been a confounding factor in these investigations.

An elegant study that addresses this point directly has been performed using eggs of the lizard Sceloporus virgatus (Vleck, 1988). Eggs were incubated on both wet and dry substrates in the laboratory, and hatchlings were marked for individual identification and then released into the wild. Regular censuses revealed that large hatchlings emerging from eggs incubated on wet substrates enjoyed higher survival over the period of observation than did smaller hatchlings coming from eggs incubated on dry media. Thus, the larger size attained by animals incubating in wet environments conferred a clear benefit during the neonatal period (Vleck, 1988). This finding supports the general proposition that hydration of the nest environment has ecological importance beyond the apparent effect on hatching success.

From the perspective of an embryo, however, circumstances may place an advantage on growing slowly and hatching at a small size. For example, consider an egg that is incubating in a dry environment, where water is not readily available for use by the enclosed embryo. Sustaining high rates of metabolism and growth may result in exhaustion of water reserves before the embryo has reached the point in its development where it can hatch successfully. Under this circumstance, the strategy for survival is to reduce the rates of metabolism and growth, and thereby reduce the demand for water. Although the embryo eventually may hatch at a size that places it at a disadvantage in competition with other hatchlings in the cohort, the point is that the embryo does hatch.

Thus, the ability to respond to reductions in the availability of water with reductions in the demand for water clearly enhances prospects for offspring to survive the prenatal period. In this light, modulation of metabolism by availability of water is a beneficial feature. A reasonable question to ask next is whether the plasticity in developmental response to moisture in the nest environment represents an adaptation of embryogenesis, or whether it merely is a conserved property of embryonic development in reptiles generally (Smith-Gill, 1983). In either case, however, plasticity has a genetic basis and therefore may be subject to evolutionary change in its own right (Bradshaw, 1965; Caswell, 1983; Smith-Gill, 1983; Schlichting, 1986; Stearns, 1989; West-Eberhard, 1989).

Natural selection may also play an indirect role affecting the attributes of hatchlings. The fitness of any particular female depends on the differential survival and reproduction by her offspring. Accordingly, her fitness depends ultimately on her ability to site her nest in a location that will maximise survival by embryos to hatching *and* survival by hatchlings through the hazardous neonatal period. This would be enhanced by locating the nest in a moist substrate and by thereby exploiting developmental plasticity of embryos to the fullest. The female consequently may be under selection for her ability to locate her nest in a substrate that is appropriately moist at the outset of incubation and that is likely to remain moist for the duration of development. Data for the snake *Opheodrys aestivus* illustrate this particularly important point, because females of this species actually choose to oviposit on relatively moist soils (Fig. 13.6).

Gravid reptiles preparing to oviposit in nature do not locate their nests randomly (Ehrenfeld, 1979; Burger & Zappalorti, 1986; Schwarzkopf & Brooks, 1987; Plummer, 1990). Females tend to be most active after a rainfall (Plummer, 1976; Stamps, 1976), and they will refuse to deposit their eggs in dry substrates (Stamps, 1976). The specific cues used by these animals in selecting a nest site are unknown, but moisture probably is a factor of ultimate importance.

Implications for future research

Although research on the water relations of reptilian eggs and embryos is only in its infancy,

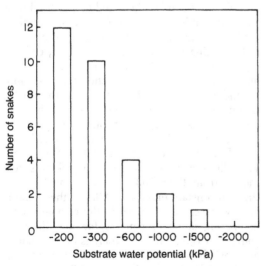

Fig. 13.6. Number of female green snakes (*Opheodrys aestivus*) ovipositing on substrates at different water potentials. (From Plummer & Snell, 1988.)

work performed to date already has important implications for investigations in at least two other areas of current interest to ecological physiologists.

First, investigators concerned with the evolution of life-histories in reptiles usually identify two components of parental investment (Congdon, 1989). One of these is the energy invested by a female in embryogenesis (i.e. the energy committed to producing viable offspring), and the second is the energy invested by the female in parental care (i.e. the energy committed to reproduction in excess of requirements for embryogenesis). In oviparous species that lack maternal care (brooding), these contributions usually are assessed simply by measuring the lipid and non-lipid fractions of unincubated eggs and of newly emerged young (Congdon, Tinkle & Rosen, 1983) and by assessing that part of the original energy reserve consumed by developing embryos and that part carried forward to sustain animals after hatching.

This approach to studying parental investment may be misleading, because the two components are treated like they are constants that are characteristic of each species. In reality, the components of parental investment in oviparous reptiles vary reciprocally in response to the incubation conditions encountered by embryos during incubation (Vleck, Chapter 15). For example, eggs of snapping turtles from north-central Nebraska contain approximately 0.50 g of neutral lipid at oviposition (Janzen *et al.*, 1990). When an egg is incubated on a wet substrate (−150 kPa) at 28.5 °C, approximately 0.27 g of this lipid is consumed by the developing embryo, leaving only 0.23 g to support the hatchling during the postnatal period (Packard *et al.*, 1988). By contrast, when an egg is incubated on a dry substrate (−950 kPa) at the same temperature, only 0.21 g of lipid is consumed by the embryo and nearly 0.29 g remains to support the hatchling (Packard *et al.*, 1988).

If estimates of the components of parental investment were based on the preceding example for snapping turtles, very different conclusions probably would be drawn. Approximately 54% of parental investment goes to support embryogenesis in eggs incubated on wet substrates, whereas only 42% goes to support embryogenesis on dry media. This difference between estimates is similar to those that are thought to distinguish different species of turtles (Congdon *et al.*, 1983), yet was elicited simply by varying the conditions of moisture during incubation.

A second area of contemporary research seeks to demonstrate a genetic basis for individual variation in phenotypic characters that are likely to affect survival (and therefore fitness) of neonatal animals (Feder *et al.*, 1987). Reptiles have become 'model organisms' for much of this work, because of the relative ease of gathering information from large numbers of families with large numbers of siblings (Arnold, 1981; van Berkum & Tsuji, 1987; Garland, 1988). Although these investigations have demonstrated a genetic basis for many components of behavior and physiology, the studies also may have had the unintended (and undesirable) effect of diverting attention from what probably is the most important source of phenotypic variation in neonatal reptiles, namely, environmentally induced variation (see also Deeming & Ferguson, Chapter 10).

Processes such as metabolism and growth in embryonic reptiles are likely to be influenced to varying degrees by factors in the environment. This generalisation applies both to intact embryos and to their component tissues (Smith-Gill, 1983). For purposes of illustration, the potential impact of the environment on various developmental processes can be depicted as a simple, linear function of an environmental factor, E, using the equation

$$K = b + mE \qquad (13.1)$$

where K is the observed rate for the process of interest, and b and m are genetically determined constants (Smith-Gill, 1983). In those instances where m approaches zero, the rate of the developmental process is determined only by the value of the constant b. The environmental factor has no influence on the process, and the rate therefore is highly canalised. In those instances where m is large, however, the rate of the developmental process is extraordinarily sensitive to the value assumed by the environmental factor, and development consequently is quite plastic.

The experimental protocol used in recent investigations of the genetic basis for components of physiological performance in neonatal reptiles minimised the potential impact of the environment on developmental processes and thereby constrained phenotypic variation in neonatal animals. For example, eggs of an oviparous lizard were incubated to hatching at a single temperature and (inadequately charac-

terised) moisture regimen (van Berkum & Tsuji, 1987), while gravid females of a viviparous gartersnake were maintained at the same temperature until parturition (Garland, 1988). The potential effects of temperature and moisture consequently were minimised in the study of lizards, while the influence of temperature (and moisture) on developmental processes was muted in the study of gartersnakes. The value for 'E' in equation 13.1 therefore was forced to become a constant in both of these investigations.

When sprint speed was subsequently examined in neonatal animals, systematic variation in performance was limited to that resulting from genetic differences among individual animals in the values assumed by b and m in equation 13.1. Although this experimental design certainly facilitated the detection of added components of statistical variance in sprint speed attributable to parentage, the procedure inevitably failed to assess the role and importance of environmental factors in eliciting phenotypic variation in neonatal animals. Indeed, investigations such as these may have led other workers to conclude erroneously that most of the variation in phenotypic traits important to survival by animals in the wild results from genetic variation in developmentally canalised characters (Bradshaw, 1965; Caswell, 1983; Schlichting, 1986; Stearns, 1989; West-Eberhard, 1989).

This point can be illustrated in a general way by re-examining data for body size of snapping turtles hatching in different hydric environments. In the original study (Packard & Packard, 1988*b*), eggs from four females were shifted between wet and dry environments at the end of the first and second trimesters of incubation, so that eight treatment groups were distinguished by the environmental conditions encountered by eggs at different times during development (e.g. Wet/Wet/Wet, Wet/Wet/Dry, Wet/Dry/Wet, etc.). Hatchlings were killed, residual yolk was removed from their bodies, and carcasses were dried to constant mass. Variability in the data for carcass mass was assessed in the re-analysis by comparing sums of squares for the different sources of variation in an analysis of covariance and by then evaluating the components of variance (Snedecor & Cochran, 1980).

The results of the two types of analyses are concordant in that approximately 70% of the variability in body size was induced by the

Table 13.1. *Summary of preliminary analysis of covariance on data for dry mass of carcasses (without residual yolks) from snapping turtles hatching in different hydric environments.*

Source of variation	Sum of squares	Percentage of variation	df	Mean square
Clutch	0.22931	7.4	3	0.07644
Trimester 1 (T1)	0.29174	9.4	1	0.29174
Trimester 2 (T2)	1.11405	36.0	1	1.11405
Trimester 3 (T3)	0.69787	22.5	1	0.69787
T1×T2	0.04415	1.4	1	0.04415
T1×T3	0.01083	0.3	1	0.01083
T2×T3	0.02545	0.8	1	0.02545
T1×T2×T3	0.00144	0.0	1	0.00144
Clutch ×T1	0.03559	1.1	3	0.01186
Clutch ×T2	0.01822	0.6	3	0.00607
Clutch ×T3	0.00396	0.1	3	0.00132
Clutch ×T1×T2	0.01137	0.4	3	0.00379
Clutch ×T1×T3	0.00079	0.0	3	0.00026
Clutch ×T2×T3	0.01066	0.3	3	0.00355
Clutch ×T1×T2×T3	0.00785	0.3	3	0.00262
Covariate: Mass 0	0.04480	1.4	1	0.04480
Error	0.54919	17.7	90	0.00610
Total	3.09727	100.0	122	

Table 13.2. *Components of variance in analysis of covariance on data for dry mass of carcasses (without residual yolks) from snapping turtles hatching in different hydric environments. (The several treatment effects (i.e. T1, T2, T3, and their first and second order interactions from Table 13.1) were pooled in order to simplify this analysis. Moreover, none of the interactions involving clutch was a significant source of variation in the preliminary ACOVA (Table 13.1), so the sums of squares and degrees of freedom for these interactions were pooled with those for the error term.)*

Source of variation	df	Mean squares	Expected mean squares	Estimate of variance	Percentage of total variance
Treatment	7	0.31222	$\sigma^2 + nb\sigma_A^2$	0.02016	71
Clutch	3	0.07644	$\sigma^2 + na\sigma_B^2$	0.00233	8
Mass 0	1	0.04480	$\sigma^2 + nab\sigma_{Cov}^2$	0.00032	1
Error	111	0.00574	σ^2	0.00574	20

environment and only 10% was elicited by differences among clutches (Tables 13.1 & 13.2). Variation in body size of hatchlings that is attributed to 'clutch' is a composite of effects, and includes effects resulting from differences in the quality of eggs laid by different females (Kaplan, 1987; Ford & Siegel, 1989), heritable differences among offspring of different parents, and experimental error resulting from minor differences among boxes used to incubate eggs. The contribution of experimental error is small in experiments such as this, and generally is submerged by other sources of variation (Packard, Packard & Boardman, unpublished). The rela-

tive contributions of maternal factors and genetics cannot be assessed but together, these factors contributed no more than 10% of the variability in the current data for body size.

On the other hand, exposing eggs to wet and dry environments at different times in incubation was a major source of variability in the data (~70%). Turtles hatching in wet environments were substantially larger than those hatching in dry conditions (Table 13.3), and the pattern of variation induced by the environment was essentially the same for offspring from all four clutches (Tables 13.1 & 13.3). The significant main effect for moisture regimes in the analysis

Table 13.3. *Means (adjusted by analysis of covariance) for mass (g) of dried carcasses from snapping turtles hatching in different hydric environments. Sample sizes are given parenthetically.*

Treatment	Clutch 1	Clutch 2	Clutch 3	Clutch 4
Dry/Dry/Dry	1.127 (4)	1.247 (4)	1.258 (3)	1.184 (4)
Wet/Dry/Dry	1.327 (4)	1.386 (4)	1.459 (4)	1.247 (4)
Dry/Dry/Wet	1.345 (4)	1.430 (4)	1.477 (4)	1.336 (4)
Dry/Wet/Dry	1.446 (4)	1.461 (4)	1.563 (3)	1.357 (3)
Wet/Dry/Wet	1.491 (4)	1.521 (4)	1.667 (4)	1.416 (4)
Wet/Wet/Dry	1.513 (4)	1.501 (4)	1.689 (3)	1.470 (4)
Dry/Wet/Wet	1.583 (4)	1.630 (4)	1.689 (4)	1.520 (4)
Wet/Wet/Wet	1.613 (4)	1.651 (4)	1.781 (4)	1.516 (3)

of covariance indicates that the growth response of embryonic snapping turtles is highly plastic, whereas the non-significant interactions involving clutch raise the possibility that sensitivity of the growth response to environmental moisture may not vary genetically in this species (Schlichting, 1986).

The preceding example reveals quite clearly that developmental plasticity has considerable potential for introducing phenotypic variation into populations of oviparous reptiles. One of the goals of future research should be to identify the causes of phenotypic variation in neonatal animals in nature, and to assess the ecological and evolutionary importance of the environmentally induced component of variation.

Acknowledgements

I thank D. C. Deeming and M. W. J. Ferguson for inviting me to participate in the international conference on Physical Influences on Embryonic Development in Birds and Reptiles, and K. Miller, M. J. Packard, and F. H. Pough for their constructive criticisms of drafts of this paper. A copy of work in press was kindly provided to me by M. V. Plummer. Preparation of the manuscript was supported, in part, by an operating grant from the U.S. National Science Foundation (DCB 88–16123). My travel to the conference was supported by Colorado State University through funding from the Vice President for Research, the Dean of the College of Natural Sciences, and the Chair of the Department of Biology.

References

Ackerman, R. A. (1981). Oxygen consumption by sea turtle (*Chelonia, Caretta*) eggs during development. *Physiological Zoology*, **54**, 316–24.

Ackerman, R. A., Dmi'el, R. & Ar, A. (1985*a*). Energy and water vapor exchange by parchment-shelled reptile eggs. *Physiological Zoology*, **58**, 129–37.

Ackerman, R. A., Seagrave, R. C., Dmi'el, R. & Ar, A. (1985*b*). Water and heat exchange between parchment-shelled reptile eggs and their surroundings. *Copeia*, **1985**, 703–11.

Arnold, S. J. (1981). Behavioral variation in natural populations. I. Phenotypic, genetic and environmental correlations between chemoreceptive responses to prey in the garter snake, *Thamnophis elegans*. *Evolution*, **35**, 489–509.

Barrows, W. B. & Schwarz, E. A. (1895). The common crow of the United States. U.S. Department of Agriculture Division of Ornithology and Mammalogy Bulletin no. 6. Government Printing Office, Washington [D.C.].

Benton, M. J. (1985). Classification and phylogeny of the diapsid reptiles. *Zoological Journal of the Linnean Society*, **84**, 97–164.

Black, C. P., Birchard, G. F., Schuett, G. W. & Black, V. D. (1984). Influence of incubation water content on oxygen uptake in embryos of the Burmese python (*Python molurus bioittatus* [sic.]). In *Respiration and Metabolism of Embryonic Vertebrates*, ed. R. S. Seymour, pp. 137–45. Dordrecht, The Netherlands: Dr W. Junk.

Bradford, D. F. & Seymour, R. S. (1988). Influence of environmental P_{O2} on embryonic oxygen consumption, rate of development, and hatching in the frog *Pseudophryne bibroni*. *Physiological Zoology*, 61, 475–82.

Bradshaw, A. D. (1965). Evolutionary significance of phenotypic plasticity in plants. *Advances in Genetics*, 13, 115–55.

Brimley, C. S. (1903). Notes on the reproduction of certain reptiles. *American Naturalist*, 37, 261–6.

Burger, J. & Zappalorti, R. T. (1986). Nest site selection by pine snakes, *Pituophis melanoleucus*, in the New Jersey Pine Barrens. *Copeia*, 1986, 116–21.

Caswell, H. (1983). Phenotypic plasticity in life-history traits: demographic effects and evolutionary consequences. *American Zoologist*, 23, 35–46.

Coker, R. E. (1910). Diversity in the scutes of Chelonia. *Journal of Morphology*, 21, 1–75.

Congdon, J. D. (1989). Proximate and evolutionary constraints on energy relations of reptiles. *Physiological Zoology*, 62, 356–73.

Congdon, J. D., Tinkle, D. W. & Rosen, P. C. (1983). Egg components and utilization during development in aquatic turtles. *Copeia*, 1983, 264–8.

Corey, A. T. & Klute, A. (1985). Application of the potential concept to soil water equilibrium and transport. *Soil Science Society of America Journal*, 49, 3–11.

Crews, D., Wibbels, T. & Gutzke, W. H. N. (1989). Action of sex steroid hormones on temperature-induced sex determination in the snapping turtle (*Chelydra serpentina*). *General and Comparative Endocrinology*, 76, 159–66.

Cunningham, B. & Huene, E. (1938). Further studies on water absorption by reptiles eggs. *American Naturalist*, 72, 380–5.

Cunningham, B. & Hurwitz, A. P. (1936). Water absorption by reptile eggs during incubation. *American Naturalist*, 70, 590–5.

Cunningham, B., Woodward, M. W. & Pridgen, J. (1939). Further studies on incubation of turtle (*Malaclemys centrata* Lat.) eggs. *American Naturalist*, 73, 285–8.

Deeming, D. C. & Ferguson, M. W. J. (1988). Environmental regulation of sex determination in reptiles. *Philosophical Transactions of the Royal Society of London B*, 322, 19–39.

— (1989). The mechanism of temperature dependent sex determination in crocodilians: a hypothesis. *American Zoologist*, 29, 973–85.

Dendy, A. (1899). Outlines of the development of the tuatara, *Sphenodon* (*Hatteria*) *punctatus*. *Quarterly Journal of Microscopical Science*, 42(NS), 1–87.

DiMichele, L. & Powers, D. A. (1984). The relationship between oxygen consumption rate and hatching in *Fundulus heteroclitus*. *Physiological Zoology*, 57, 46–51.

Dunson, W. A. (1982). Low water vapor conductance of hard-shelled eggs of the gecko lizards *Hemidactylus* and *Lepidodactylus*. *Journal of Experimental Zoology*, 219, 377–9.

Dunson, W. A. & Bramham, C. R. (1981). Evaporative water loss and oxygen consumption of three small lizards from the Florida keys: *Sphaerodactylus cinereus*, *S. notatus*, and *Anolis sagrei*. *Physiological Zoology*, 54, 253–9.

Edelstone, D. I., Paulone, M. E. & Holzman, I. R. (1984). Hepatic oxygenation during arterial hypoxemia in neonatal lambs. *American Journal of Obstetrics and Gynecology*, 150, 513–8.

Ehrenfeld, D. W. (1979). Behavior associated with nesting. In *Turtles: Perspectives and Research*, eds M. Harless & H. Morlock, pp. 417–34. New York: John Wiley & Sons.

Feder, M. E., Bennett, A. F., Burggren, W. W. & Huey, R. B. (1987). *New Directions in Ecological Physiology*. Cambridge: Cambridge University Press.

Feder, M. E., Satel, S. L. & Gibbs, A. G. (1982). Resistance of the shell membrane and mineral layer to diffusion of oxygen and water in flexible-shelled eggs of the snapping turtle (*Chelydra serpentina*). *Respiration Physiology*, 49, 279–91.

Ferguson, G. W. & Bohlen, C. H. (1978). Demographic analysis: a tool for the study of natural selection of behavioral traits. In *Behavior and Neurology of Lizards*, eds N. Greenberg & P. D. MacLean, pp. 227–43. Rockville, Maryland: U.S. Department of Health, Education, and Welfare, NIMH.

Ferguson, G. W., Brown, K. L. & DeMarco, V. G. (1982). Selective basis for the evolution of variable egg and hatchling size in some iguanid lizards. *Herpetologica*, 38, 178–88.

Ferguson, G. W. & Fox, S. F. (1984). Annual variation of survival advantage of large juvenile side-blotched lizards, *Uta stansburiana*: its causes and evolutionary significance. *Evolution*, 38, 342–9.

Fitch, H. S. & Fitch, A. V. (1967). Preliminary experiments on physical tolerances of the eggs of lizards and snakes. *Ecology*, 48, 160–5.

Ford, N. B. & Seigel, R. A. (1989). Phenotypic plasticity in reproductive traits: evidence from a viviparous snake. *Ecology*, 70, 1768–74.

Fox, S. F. (1978). Natural selection on behavioral phenotypes of the lizard *Uta stansburiana*. *Ecology*, 59, 834–47.

Froese, A. P. & Burghardt, G. M. (1974). Food competition in captive juvenile snapping turtles, *Chelydra serpentina*. *Animal Behaviour*, 22, 735–40.

Garland, T., Jr (1988). Genetic basis of activity metabolism. I. Inheritance of speed, stamina, and antipredator displays in the garter snake *Thamnophis sirtalis*. *Evolution*, **42**, 335–50.

Georges, A. (1989). Female turtles from hot nests: is it duration of incubation or proportion of development at high temperatures that matters? *Oecologia*, **81**, 323–8.

Gettinger, R. D., Paukstis, G. L. & Gutzke, W. H. N. (1984). Influence of hydric environment on oxygen consumption by embryonic turtles *Chelydra serpentina* and *Trionyx spiniferus*. *Physiological Zoology*, **57**, 468–73.

Gibbons, J. W. & Nelson, D. H. (1978). The evolutionary significance of delayed emergence from the nest by hatchling turtles. *Evolution*, **32**, 297–303.

Grigg, G. C. (1987). Water relations of crocodilian eggs: management considerations. In *Wildlife Management: Crocodiles and Alligators*, eds G. J. W. Webb, S. C. Manolis & P. J. Whitehead, pp. 499–502. Chipping Norton, New South Wales: Surrey Beatty & Sons.

Gutzke, W. H. N. (1984). Modification of the hydric environment by eggs of snapping turtles (*Chelydra serpentina*). *Canadian Journal of Zoology*, **62**, 2401–3.

Gutzke, W. H. N. & Packard, G. C. (1985). Hatching success in relation to egg size in painted turtles (*Chrysemys picta*). *Canadian Journal of Zoology*, **63**, 67–70.

— (1987). Influence of the hydric and thermal environments on eggs and hatchlings of bull snakes *Pituophis melanoleucus*. *Physiological Zoology*, **60**, 9–17.

Gutzke, W. H. N., Packard, G. C., Packard, M. J. & Boardman, T. J. (1987). Influence of the hydric and thermal environments on eggs and hatchlings of painted turtles (*Chrysemys picta*). *Herpetologica*, **43**, 393–404.

Gutzke, W. H. N. & Paukstis, G. L. (1983). Influence of the hydric environment on sexual differentiation of turtles. *Journal of Experimental Zoology*, **226**, 467–9.

Hillman, S. S. (1978). The roles of oxygen delivery and electrolyte levels in the dehydrational death of *Xenopus laevis*. *Journal of Comparative Physiology*, **128**, 169–75.

— (1984). Inotropic influence of dehydration and hyperosmolal solutions on amphibian cardiac muscle. *Journal of Comparative Physiology*, **154B**, 325–8.

— (1987). Dehydrational effects on cardiovascular and metabolic capacity in two amphibians. *Physiological Zoology*, **60**, 608–13.

Hillman, S. S., Withers, P. C., Hedrick, M. S. & Kimmel, P. B. (1985). The effects of erythrocythemia on blood viscosity, maximal systemic oxygen transport capacity and maximal rates of oxygen consumption in an amphibian. *Journal of Comparative Physiology*, **155B**, 577–81.

Hillman, S. S., Zygmunt, A. & Baustian, M. (1987). Transcapillary fluid forces during dehydration in two amphibians. *Physiological Zoology*, **60**, 339–45.

Hochachka, P. W. (1988). Metabolic suppression and oxygen availability. *Canadian Journal of Zoology*, **66**, 152–8.

Hotaling, E. C., Wilhoft, D. C. & McDowell, S. B. (1985). Egg position and weight of hatchling snapping turtles, *Chelydra serpentina*, in natural nests. *Journal of Herpetology*, **19**, 534–6.

Janzen, F. J., Packard, G. C., Packard, M. J., Boardman, T. J. & zumBrunnen, J. R. (1990). Mobilization of lipid and protein by embryonic snapping turtles in wet and dry environments. *Journal of Experimental Zoology*, **255**, 155–62.

Kaplan, R. H. (1987). Developmental plasticity and maternal effects of reproductive characteristics in the frog, *Bombina orientalis*. *Oecologia*, **71**, 273–9.

Leshem, A. & Dmi'el, R. (1986). Water loss from *Trionyx triunguis* eggs incubating in natural nests. *Herpetological Journal*, **1**, 115–7.

Lillywhite, H. B. & Ackerman, R. A. (1984). Hydrostatic pressure, shell compliance and permeability to water vapor in flexible-shelled eggs of the colubrid snake *Elaphe obsoleta*. In *Respiration and Metabolism of Embryonic Vertebrates*, ed. R. S. Seymour, pp. 121–35. Dordrecht, The Netherlands: Dr W. Junk.

Lynn, W. G. & von Brand, T. (1945). Studies on the oxygen consumption and water metabolism of turtle embryos. *Biological Bulletin (Woods Hole, Massachusetts)*, **88**, 112–25.

Manolis, S. C., Webb, G. J. W. & Dempsey, K. E. (1987). Crocodile egg chemistry. In *Wildlife Management: Crocodiles and Alligators*, eds G. J. W. Webb, S. C. Manolis & P. J. Whitehead, pp. 445–72. Chipping Norton, New South Wales: Surrey Beatty & Sons.

Miller, K., Packard, G. C. & Packard, M. J. (1987). Hydric conditions during incubation influence locomotor performance of hatchling snapping turtles. *Journal of Experimental Biology*, **127**, 401–12.

Morris, K. A., Packard, G. C., Boardman, T. J., Paukstis, G. L. & Packard, M. J. (1983). Effect of the hydric environment on growth of embryonic snapping turtles (*Chelydra serpentina*). *Herpetologica*, **39**, 272–85.

Muth, A. (1980). Physiological ecology of desert iguana (*Dipsosaurus dorsalis*) eggs: temperature and water relations. *Ecology*, **61**, 1335–43.

Packard, G. C. & Packard, M. J. (1984). Coupling of physiology of embryonic turtles to the hydric environment. In *Respiration and Metabolism of Embryonic Vertebrates*, ed. R. S. Seymour, pp. 99–119. Dordrecht, The Netherlands: Dr W. Junk.

— (1987). Water relations and nitrogen excretion in embryos of the oviparous snake *Coluber constrictor*. *Copeia*, **1987**, 395–406.

— (1988a). The physiological ecology of reptilian eggs and embryos. In *Biology of the Reptilia*, vol. 16, eds C. Gans & R. B. Huey, pp. 523–605. New York: Alan R. Liss.

— (1988b). Water relations of embryonic snapping turtles (*Chelydra serpentina*) exposed to wet or dry environments at different times in incubation. *Physiological Zoology*, **61**, 95–106.

— (1989b). Control of metabolism and growth in embryonic turtles: a test of the urea hypothesis. *Journal of Experimental Biology*, **147**, 203–16.

— (1990). Growth of embryonic softshell turtles is unaffected by uremia. *Canadian Journal of Zoology*, **68**, 841–4.

Packard, G. C., Packard, M. J. & Birchard, G. F. (1989). Sexual differentiation and hatching success by painted turtles incubating in different thermal and hydric environments. *Herpetologica*, **45**, 385–92.

Packard, G. C., Packard, M. J. & Boardman, T. J. (1981a). Patterns and possible significance of water exchange by flexible-shelled eggs of painted turtles (*Chrysemys picta*). *Physiological Zoology*, **54**, 165–78.

— (1984a). Influence of hydration of the environment on the pattern of nitrogen excretion by embryonic snapping turtles (*Chelydra serpentina*). *Journal of Experimental Biology*, **108**, 195–204.

Packard, G. C., Packard, M. J., Boardman, T. J., Morris, K. A. & Shuman, R. D. (1983). Influence of water exchanges by flexible-shelled eggs of painted turtles *Chrysemys picta* on metabolism and growth of embryos. *Physiological Zoology*, **56**, 217–30.

Packard, G. C., Packard, M. J., Miller, K. & Boardman, T. J. (1987). Influence of moisture, temperature, and substrate on snapping turtle eggs and embryos. *Ecology*, **68**, 983–93.

— (1988). Effects of temperature and moisture during incubation on carcass composition of hatchling snapping turtles (*Chelydra serpentina*). *Journal of Comparative Physiology* **158B**, 117–25.

Packard, G. C., Taigen, T. L., Packard, M. J. & Boardman, T. J. (1981b). Changes in mass of eggs of softshell turtles (*Trionyx spiniferus*) incubated under hydric conditions simulating those of natural nests. *Journal of Zoology*, **193**, 81–90.

Packard, G. C., Taigen, T. L., Packard, M. J. & Shuman, R. D. (1979). Water-vapor conductance of testudinian and crocodilian eggs (class Reptilia). *Respiration Physiology*, **38**, 1–10.

Packard, M. J., Burns, L. K., Hirsch, K. F. & Packard, G. C. (1982a). Structure of shells of eggs of *Callisaurus draconoides* (Reptilia, Squamata, Iguanidae). *Zoological Journal of the Linnean Society*, **75**, 297–316.

Packard, M. J., Hirsch, K. F. & Iverson, J. B. (1984b). Structure of shells from eggs of kinosternid turtles. *Journal of Morphology*, **181**, 9–20.

Packard, M. J. & Packard, G. C. (1979). Structure of the shell and tertiary membranes of eggs of softshell turtles (*Trionyx spiniferus*). *Journal of Morphology*, **159**, 131–43.

— (1986). Effect of water balance on growth and calcium mobilization of embryonic painted turtles (*Chrysemys picta*). *Physiological Zoology*, **59**, 398–405.

— (1989a). Environmental modulation of calcium and phosphorus metabolism in embryonic snapping turtles (*Chelydra serpentina*). *Journal of Comparative Physiology*, **159B**, 501–8.

Packard, M. J., Packard, G. C. & Boardman, T. J. (1980). Water balance of the eggs of a desert lizard (*Callisaurus draconoides*). *Canadian Journal of Zoology*, **58**, 2051–8.

— (1982b). Structure of eggshells and water relations of reptilian eggs. *Herpetologica*, **38**, 136–55.

Paukstis, G. L., Gutzke, W. H. N. & Packard, G. C. (1984). Effects of substrate water potential and fluctuating temperatures on sex ratios of hatchling painted turtles (*Chrysemys picta*). *Canadian Journal of Zoology*, **62**, 1491–4.

Petranka, J. W., Just, J. J. & Crawford, E. C. (1982). Hatching of amphibian embryos: the physiological trigger. *Science*, **217**, 257–9.

Phillips, J. A., Garel, A., Packard, G. C. & Packard, M. J. (1990). Influence of moisture and temperature on eggs and embryos of green iguanas (*Iguana iguana*). *Herpetologica*, **46**, 238–45.

Plummer, M. V. (1976). Some aspects of nesting success in the turtle, *Trionyx muticus*. *Herpetologica*, **32**, 353–9.

— (1990). Nesting movements, nesting behavior, and nest sites of green snakes (*Opheodrys aestivus*) revealed by radiotelemetry. *Herpetologica*, **46**, 190–5.

Plummer, M. V. & Snell, H. L. (1988). Nest site selection and water relations of eggs in the snake, *Opheodrys aestivus*. *Copeia*, **1988**, 58–64.

Pough, F. H. (1989). Organismal performance and Darwinian fitness: approaches and interpretations. *Physiological Zoology*, **62**, 199–236.

Rawlins, S. L. & Campbell, G. S. (1986). Water potential: thermocouple psychrometry. In *Methods of Soil Analysis. Part 1. Physical and Mineralogical Methods*, 2nd edn, ed. A. Klute, pp. 597–618. Madison, Wisconsin: American Society for Agronomy and Soil Science Society of America.

Rose, C. W. (1968a). Water transport in soil with a daily temperature wave. I. Theory and experiment. *Australian Journal of Soil Research*, **6**, 31–44.

— (1968b). Water transport in soil with a daily temperature wave. II. Analysis. *Australian Journal of Soil Research*, **6**, 45–57.

Savage, M. J. & Cass, A. (1984). Measurement of water potential using *in situ* thermocouple hygrometers. *Advances in Agronomy*, **37**, 73–126.

Schlichting C. D. (1986). The evolution of

phenotypic plasticity in plants. *Annual Review of Ecology and Systematics*, 17, 667-93.

Schwarzkopf, L. & Brooks, R. J. (1987). Nest-site selection and offspring sex ratio in painted turtles, *Chrysemys picta*. *Copeia*, 1987, 53–61.

Shine, R. (1978). Propagule size and parental care: the 'safe harbor' hypothesis. *Journal of Theoretical Biology*, 75, 417–24.

Smith-Gill, S. J. (1983). Developmental plasticity: developmental conversion *versus* phenotypic modulation. *American Zoologist*, 23, 47–55.

Snedecor, G. W. & Cochran, W. G. (1980). *Statistical Methods*, 7th edition. Ames: Iowa State University Press.

Snell, H. L. & Tracy, C. R. (1985). Behavioral and morphological adaptations by Galapagos land iguanas (*Conolophus subcristatus*) to water and energy requirements of eggs and neonates. *American Zoologist*, 25, 1009–18.

Stamps, J. A. (1976). Egg retention, rainfall and egg laying in a tropical lizard *Anolis aeneus*. *Copeia*, 1976, 759–64.

Stearns, S. C. (1989). The evolutionary significance of phenotypic plasticity. *Bioscience*, 39, 436–45.

Swingland, I. R. & Coe, M. J. (1979). The natural regulation of giant tortoise populations on Aldabra Atoll: recruitment. *Philosophical Transactions of the Royal Society of London B*, 286, 177–88.

Taylor, S. A. & Ashcroft, G. L. (1972). *Physical Edaphology: The Physics of Irrigated and Nonirrigated Soils*. San Francisco: W. H. Freeman.

Thompson, M. B. (1983). *The physiology and ecology of the eggs of the pleurodiran tortoise Emydura macquarii (Gray), 1831*. PhD thesis, University of Adelaide, South Australia.

— (1987). Water exchange in reptilian eggs. *Physiological Zoology*, 60, 1–8.

Tracy, C. R. (1980). Water relations of parchment-shelled lizard (*Sceloporus undulatus*) eggs. *Copeia*, 1980, 478–82.

— (1982). Biophysical modeling in reptilian physiology and ecology. In *Biology of the Reptilia*, vol. 12, eds C. Gans & F. H. Pough, pp. 274–321. London: Academic Press.

Tracy, C. R., Packard, G. C. & Packard, M. J. (1978). Water relations of chelonian eggs. *Physiological Zoology*, 51, 378–87.

Tracy, C. R. & Snell, H. L. (1985). Interrelations among water and energy relations of reptilian eggs, embryos, and hatchlings. *American Zoologist*, 25, 999–1008.

van Berkum, F. H. & Tsuji, J. S. (1987). Inter-familiar differences in sprint speed of hatchling *Sceloporus occidentalis* (Reptilia: Iguanidae). *Journal of Zoology*, 212, 511–9.

Visschedijk, A. H. J. (1968). The air space and embryonic respiration. 3. The balance between oxygen and carbon dioxide in the air space of the incubating chicken egg and its role in stimulating pipping. *British Poultry Science*, 9, 197–210.

Vleck, D. (1988). Embryo water economy, egg size and hatchling viability in the lizard *Sceloporus virgatus*. *American Zoologist*, 28, 87A.

West-Eberhard, M. J. (1989). Phenotypic plasticity and the origins of diversity. *Annual Review of Ecology and Systematics*, 20, 249–78.

14

Roles of water in avian eggs

AMOS AR

Introduction

This chapter is an account of some thoughts and ideas that have been crystallised during the many years of working the field of avian egg physiology. It cannot and will not cover all aspects of the immense quantity of published work. My choice of references is much biased by their relevance to my way of viewing the issue of embryonic development in a cleidoic system – a problem nature has solved while we are still pondering how.

Water loss from avian eggs

Homeothermy and water loss from eggs

Birds have developed an active and intensive lifestyle, which requires homeothermy. Simultaneously, by maintaining a high incubation temperature, their embryos have evolved to develop fast in comparison with reptiles of the same size (Rahn & Ar, 1974; Blueweiss *et al.*, 1978; Deeming & Ferguson, Chapter 10).

Thermoregulation in adult birds is associated with the presence of insulating plumage, which developed to replace reptilian scales. Regarding egg thermoregulation, most birds lay eggs in the still air which acts as an insulating medium in the nest, while many reptiles lay eggs in a soil substrate. This difference is not trivial: soil, and especially wet soil, both conducts heat well and has a high heat storage capacity (Table 14.1) (Hillel, 1978). Hence, eggs in direct contact with the soil exchange heat easily and cannot become much warmer than their surroundings, even if they were actively incubated. On the other hand, still air is a good insulator which prevents fast cooling and permits maintenance of high incubation temperatures (Ackerman & Seagrave, 1984, 1987; Sotherland, Spotila & Paganelli, 1987).

Around this essential feature of high incubation temperature, birds have developed their mode of breeding which includes, in contrast to most reptiles, a series of behavioural, anatomical and physiological adaptations. In general, clutches of reptile eggs are relatively large, and left to the mercy of the environment, whereas birds have evolved towards producing limited numbers of eggs which are totally dependent on parental care for their development (Ar & Yom-Tov, 1978).

Since this care includes regulation of incubation temperature (Huggins, 1941; Drent, 1975) and exposure to the air, avian embryos face a danger shared by all creatures which have evolved to live on land, namely desiccation. The temperature difference between the egg and its environment, as is the case with any water-containing object, enhances this tendency to lose water in the form of vapour into the unsaturated medium. This is true even when the cooler ambient air is saturated with water vapour, as the higher temperature of the egg dictates a higher water vapour pressure within it. Thus, water vapour lost down this pressure gradient will condense in the saturated environment. Data analysis of water loss from eggs of different species during natural incubation to their environments at different ambient temperatures is presented in Fig. 14.1. From the known ambient water vapour pressure, the water vapour pressure at saturation of the same ambient temperature, and the water vapour pressure of the egg contents at egg temperature, the fraction

Table 14.1. *Approximate thermal properties of air, soil components, and egg contents (after Seagrave, 1971; Hillel, 1978; Koorevaar et al., 1983; Breidis & Seagrave, 1984).*

	Heat	
	Capacity J 10^6/(m^3 °C)	Conductance Wm/(m^2 °C)
Substance		
Air	0.001	0.025
Water	4.2	0.6
Soil minerals	2.0	3–9
Organic matter	2.5	0.25
Egg content	2.5–4.5	0.25–0.45

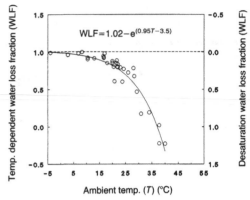

Fig. 14.1. Analysis of diffusive water loss during natural incubation in 33 bird species (Ar, unpublished compilation) as a function of average ambient temperature. The solid line represents the fraction of the total loss that is attributed to the saturation pressure difference between the egg and ambient air due to temperature difference (left ordinate). The right ordinate represents the complementary fraction due to saturation deficit of ambient air.

of the water lost from the egg which can be attributed to the saturation pressure differences due to temperature differences only, are calculated (Fig. 14.1). It is clear that the complementary fraction of water loss (Fig. 14.1) is due to the degree of desaturation of the ambient air. In cold environments up to about 25 °C most of the water loss from eggs is due to a temperature difference between egg and environment (Fig. 14.1). Above this temperature, the saturation pressure deficit plays a more and more important role in water loss from eggs. At high ambient temperatures, eggs may lose water only if the atmosphere is dry. The presence of the nest which may have an intermediate vapour pressure (Rahn, Ackerman & Paganelli, 1977) cannot change this picture in a steady state situation.

The role of the eggshell in water loss

Except for cases of cryptobiosis, life, and certainly active life, cannot proceed without appropriate water concentration present in the cells. It is not surprising, therefore, to discover that birds had to find means (evolutionary speaking) of solving the conflict of keeping eggs warm without desiccating their contents. Birds have thus employed a rigid eggshell, emulating some of their reptilian ancestors (Packard, Tracy & Roth, 1977; Silyn-Roberts & Sharp, 1989; Packard & DeMarco, Chapter 5). The reptilian hard eggshell originally may have developed as an answer to predation pressure, as a protection from environmental contaminants (Packard & Packard, 1980), as a source of calcium for the developing embryo (Bustard et al., 1969), as a

way of providing aeration spaces to individual embryos within a clutch in the nest or any other logical explanation. It is, however, unrealistic to attribute development of the rigid eggshell to one single factor. This may also be deduced by examining the particular physiological properties and ecological features of reptilian eggs which have either lost or have never possessed hard shells (Packard et al., 1977; Ackerman et al., 1985a).

However, for birds, it seems that the rigid eggshell provides, above all, a solution to the water budget of the developing embryo (Ar & Rahn, 1980; Ackerman, Dmi'el & Ar, 1985b). Eggshells cannot be impermeable to water vapour since the embryo must exchange respiratory gases with the environment (Needham, 1932; Wangensteen, Wilson & Rahn, 1970/71). The same gas paths that serve respiration, namely the shell pores, are the avenues through which the egg loses water as vapour to the environment (Ar & Rahn, 1985). This mode of water loss is easy to measure and detect gravimetrically. It is first mentioned in the writings of Hippocrates (according to Needham, 1932) and some 250 years ago, de Reaumur (1738, cited by Landauer, 1967), proposed coating eggs with molten fat in order to prevent this water loss. The efficiency of the shell in reducing water loss can be assessed by comparing it to water loss in adult birds or from 'soft' shelled reptilian eggs. For example, the rate of water efflux of a fowl (*Gallus gallus*) hen per unit mass is 10–15 times higher than the same rate from her eggs (calculated from Booth, 1987). The rate of water loss from a 'soft' shelled reptilian egg the size of a fowl egg in comparable conditions is 10 times that of a chicken egg (Ackerman *et al.*, 1985*b*). The shell itself is impermeable to gases and water vapour. It is the pores in the shell at density of about one per mm^2 that determine the resistance to water loss. They cover an area which is about 0.2% of the total in small thin shells, to about 10% of the total in very large thick shells (Ar & Rahn, 1985).

Diffusive and pipping water loss

In addition to this water loss during incubation, a further phase exists, where after external pipping, more water is lost through the cracks in the shell during hatching (and from the external surface of the wet hatchling after hatching). This additional water loss may amount to 20–40% of the total loss (Ar & Rahn, 1980; Whittow, 1982; Sotherland & Rahn, 1987).

The exact shape and dimensions of the shell pores of various birds vary among species (Tyler, 1956; Board, Tullett & Perrott, 1977; Rahn, Paganelli & Ar, 1987; Toien *et al.*, 1987; Board & Sparks, Chapter 6). Consequently, although there is no doubt that the main mechanism for egg water loss is by diffusion of vapour through the pores, different models have been proposed to explain this diffusive process (Ar *et al.*, 1980*a*; Paganelli, Ar & Rahn, 1981; Simkiss, 1986; Rahn *et al.*, 1987; Sibly & Simkiss, 1987; Toien *et al.*, 1988). The existence of a unidirectional diffusive flux of water vapour across the shell complicates the exchange of the respiratory gases since this diffusion is non-equimolar. As a result, a small static pressure difference in the order of 1mm H_2O (egg content side positive) is established across the shell (Romijn & Roos, 1938; Ar *et al.*, 1980*a*; Paganelli, Chapter 16) such that O_2 and CO_2 must diffuse up and down this pressure gradient in the pores respectively. For water vapour, a somewhat facilitated loss is expected (Paganelli, Ar & Rahn, 1987). The pressure difference may be important in preventing the penetration of spores of contaminating bacteria through the shell pores.

Changes in water loss

The daily rate of diffusive water loss from incubated eggs may be considered a 'constant', varying at random around a daily mean, or it may show a definite trend. The variation around the mean is due in part to the methods of the measurement, but in many cases it is real and stems from variations in egg temperature and nest humidity (Woodall & Parry, 1982). Both of these parameters are influenced by the incubation behaviour of the parent bird and climatic changes around the nest (e.g. temperature, humidity, winds, radiation, rain) which influence the microclimate in the nest, and thus water loss even at a constant water vapour conductance of the eggshell (Carey, 1980, 1986; Woodall & Parry, 1982; Vleck *et al.*, 1983).

There are a few cases in which it has been shown that water loss decreases during incubation (Davis, Platter-Reiger & Ackerman, 1984; Kern, 1986) and an increase in water loss during incubation is noted in many species (Kendeigh, 1940; Swart, Rahn & de Cock, 1987). This is due to an equivalent increase in egg temperature as a result of an increase in nest attentiveness (Spellerberg, 1969; Drent, 1970; Handrich, 1989*a*) and/or to the increase in the metabolic heat production of the growing embryo (Prinzinger, Maisch & Hund, 1979; Grant *et al.*, 1982; Haftorn & Reinertsen, 1982; Ackerman & Seagrave, 1984). Changes in egg temperature at a constant incubation temperature in still air as incubation proceeds, calculated from embryonic heat production, evaporative water cooling and the cooling rate constant of a fowl egg during incubation are shown in Fig. 14.2. The resulting relative changes in daily water loss rate are also shown. Daily water loss increases more than 25% during incubation. Prinzinger *et al.* (1979) compare overall incubation water loss of fertilised and non fertilised eggs of different species, under similar conditions of incubation. The average increase in water loss of fertile eggs over infertile ones, was 31%. This increase is not, however, observed during artificial incubation in forced draught incubators (and will be explained below). In the absence of pronounced changes in parental behaviour on the nest, changes in water loss from the developing egg may indicate that embryonic metabolic heat production is increasing egg temperature, and hence nest air is essentially still air. In these cases, a nest *ventilation* concept, proposed for

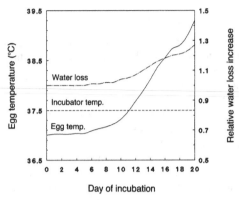

Fig. 14.2. Influence of egg heat production during incubation of fowl egg temperature and egg water loss at a constant incubator temperature and humidity. Values were calculated using published data on oxygen consumption (Romanoff, 1967), egg cooling rate constant (Turner, 1985) at 37.5 °C and 50% relative humidity in a still air incubator.

explaining vapour clearance from nest to ambient air (Rahn *et al.*, 1976), should be replaced at least in part by a nest to ambient vapour *diffusion* concept (Rahn, 1984; Vleck *et al.*, 1983). In some cases the increased rate of water loss is due to changes in water vapour conductance of the eggshell (Carey, 1983) which may increase as a result of outer surface erosion, mainly of the cuticle (Peebles & Brake, 1986; Deeming, 1987; Handrich, 1989*a*) or erosion of the inner surface of the eggshell as calcium is taken up by the growing embryo (Kreitzer, 1972; Booth & Seymour, 1987).

Direct application of the brood patch to the eggshell surface may reduce the contact of the egg surface with the nest atmosphere, and therefore reduce water evaporation from the egg at a constant egg temperature (Rahn, 1984; Yom-Tov, Wilson & Ar, 1986). However, by applying the brood patch, egg temperature may rise, and although the evaporating area would be reduced, the pressure gradient may increase to compensate for the reduced area. A detailed account of these possibilities in penguins is given by Handrich (1989*b*).

The physiological roles of water

Interactions between water and the eggshell

The nature of the shell pores is such that they offer a significant gas phase diffusive resistance to gases and vapour (Paganelli, 1980). As a result, the gas spaces of the fibrous shell membranes between the egg contents and the shell, which are formed by evaporation at the beginning of incubation as the membranes gradually lose liquid water (Lomholt, 1976; Tullett & Board, 1976), must be practically saturated by water vapour. The question is, at what temperature? Is it the temperature corresponding to the evaporating surface of the egg, or to that of the inner surface of the shell? (Simkiss, 1974; Meir & Ar, 1990). The corresponding water potential of the content under the gas space must also be taken into account (Ar *et al.*, 1974; Seymour & Piiper, 1988). Taigen *et al.* (1978) show that water potential of the egg contents significantly influences water loss, but this is, at least, theoretically impossible: the calculated reduction of water vapour from that of pure water due to egg content water potential is negligible. Simkiss (1967) discusses the possibility that secretion of acid or CO_2 is responsible for the resorption of calcium from the shell to build the skeleton of the embryo. By contrast, Ancel, Girard & Ar (unpublished) propose that it is the interaction of metabolic CO_2 and condensed water on the inner surface of the shell which form carbonic acid that dissolves the calcium needed by the embryo, and that this process is enhanced by incubation temperature fluctuations.

The extra-embryonic water reserve of the embryo towards the end of incubation is composed mainly of allantoic and amniotic fluids (Romanoff, 1967; Hoyt, 1979). During internal pipping the walls of these reservoirs are punctured by the embryo and some of the liquid runs along the edges of the air cell. It has been shown that wet eggshells are weaker than dry ones (Tyler & Geake, 1964). Wetting the inner surface of the eggshell may help in the external pipping process, which usually takes place at the edge of the air cell (Bond, Board & Scott, 1988). Egg batches that become dehydrated during incubation possess larger numbers of 'trapped' hatchlings than controls (Snyder & Birchard, 1982; Meir & Ar, 1987). Hence the right amount of fluids may be indispensable to the mechanism of hatching itself and may help explain the apparent relatively fixed 3:1 distribution between diffusive and hatching water loss (Ar & Rahn, 1980).

Water balance of embryo and egg

In spite of an overall normal water loss of about 20–22% of the initial egg mass, the water con-

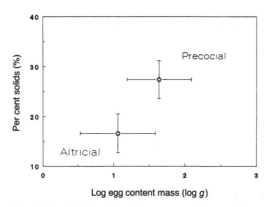

Fig. 14.3. Relationship between per cent solids in egg content and log egg mass. Data comes essentially from Sotherland & Rahn (1987), and is grouped according to the degree of maturity at hatching. Water concentration in egg content is (100 − % solids).

centration of the hatchling (including the spare yolk) is similar to the initial concentration in the fresh egg. Except for rare cases, altricial egg contents and hatchlings both contain about 79–88% water and precocial ones 69–76% (Fig. 14.3), and intermediate forms have values in-between (Ar & Rahn, 1980; Sotherland & Rahn, 1987). This observation can be accounted for quantitatively, if the mass balance takes into account the fact that the embryonic metabolism produces water in proportion to the utilisation of solids in the egg (Ar & Rahn, 1980; Vleck & Vleck, 1987).

Although the 'how' of the water balance is by and large accounted for, we still do not understand this biological 'constant hydration concept' which seems to be unique to birds (Ar & Rahn, 1980). It is perhaps caused by the cleidoic nature of avian eggs, which does not permit a 2-way exchange of water in contrast to some reptiles (Packard, Chapter 13). This feature dictates the finite initial amount of water that has to be loaded into the egg, taking into account metabolic water production and the dilution of the contents as solids are utilised. Thus, a final water concentration in the hatchling which is compatible with life is achieved. In fact, knowing the overall water loss, the gas exchange of the embryo and the water fraction of the hatchling, one can reconstruct the minimal initial water fraction of a fresh egg.

Whatever the avenue for water loss, it can rarely exceed 30% of fresh egg mass (Ar & Rahn, 1980; Sotherland & Rahn, 1987) without causing harmful effects: embryonic mortality towards the end of incubation, when the cumu-

lative effects of water loss are manifested (Tullett & Burton, 1982; Carey, 1986; Meir & Ar, 1987). However, Simkiss (1980) has manipulated eggs by withdrawing 11% of their original mass in liquids from the developing embryos and the eggs still hatch. Snyder & Birchard (1982) show that water loss early during incubation is more harmful than later on. It is interesting to note that too little water loss is also detrimental to the embryonic development. Bird eggs that lose less than 8–10% of their initial mass during incubation rarely survive to hatching (Robertson, 1961; Landauer, 1967; Lundy, 1969; Hulet, Christensen & Bagley, 1987). Since these embryos die regardless of whether the cause for low water loss is low shell conductance or high ambient humidity it can be deduced that it is mainly water and not only the high resistance to oxygen flow which is the cause of reduced hatchability (Meir & Ar, 1986, 1987).

The cause and effect relationship between overall incubation water loss and successful hatching is statistically significant, but with large variation. Some reports indicate that the permitted variability in water loss is larger than is proposed in earlier works, attributing certain compensatory capacities to the embryo (Ar & Rahn, 1980; Simkiss, 1980; Woodall & Parry, 1982; Carey, 1983; Visschedijk, Tazawa & Piiper 1985; Davis, Shen & Ackerman, 1988), but in order to ascertain optimal water loss statistically and beyond doubt, experiments with thousands of eggs are needed. To date this has been possible for domestic species only (Robertson, 1961; Laughlin & Lundy, 1976; Christensen & McCorkle, 1982; Hulet et al., 1987; Meir & Ar, 1987). The establishment of such optima for other species is based mainly on the information coming from domestic species and on the assumption that water loss is relatively invariable among species (Ar et al., 1974; Burnham, 1983; Kuehler & Witman, 1988). Within the water loss limits for embryonic survival, hatchling mass is influenced by incubation water loss and can be predicted (Tullett & Burton, 1982). Eggs that suffer from excess water or excess water loss, produce large and small hatchlings respectively. In spite of the change in wet mass, the dry mass of the hatchling seems to be unaffected (Tullett & Burton, 1982; Davis & Ackerman, 1987). Although it is clear that avian embryos are not capable of regulating their water loss, they may still compensate for suboptimal water losses to some extent by regulating growth and by shifting

Table 14.2. *A comparison of Q_{10} values for water activity and metabolic processes (After Schmidt-Nielson, 1979; Tracy et al., 1980).*

Temp. range °C	Q_{10} for:		Q_{10} ratio
	Vapour pressure	Metabolic rate	
0–10	2.01	2.22	1.1
10–20	1.90	2.10	1.1
20–30	1.82	2.00	1.1
30–40	1.74	1.91	1.1

water among compartments. For instance, over hydrated embryos store water in their skin and muscles (Davis *et al.*, 1988) and dehydrated embryos store water in the liver (Davis & Ackerman, 1987) and utilise mainly allantoic liquid (Hoyt, 1979).

Regulation of water loss

Can an incubating bird regulate the water loss of its egg? The circumstantial evidence that water loss seems to be regulated has led some authors to assume that such a behaviour does exist (Rahn *et al.*, 1976, 1977). Rahn *et al.* (1976) suggest that the brood patch may be the humidity sensing organ. Although some birds cover their nests with wet material or soil when leaving, and others wet the eggs in the nest upon return (Howell, 1979; Davis *et al.*, 1984), so far no correlation has been found between this behaviour and ambient air or nest humidity (Grant, 1982; Walsberg, 1983; Rahn, 1984). This contrasts sharply with the distinct behavioural patterns used by different birds to thermoregulate their eggs (White & Kinney, 1974; Drent 1975). It appears that, given a certain shell conductance, regulation of egg water loss in nature is determined by the inherent behaviour of birds in terms of nesting season, choice of microclimate and nest structure, which keeps nest humidity relatively constant (Rahn, Chapter 21). By regulating egg temperature, the head pressure for water and thus the water vapour pressure difference between the egg and the nest, is also determined. This pressure difference increases with temperature in a predictable manner (Tracy, Welch & Porter, 1980) and has essentially the same 'Q_{10}' as metabolism in general (Schmidt-Nielsen, 1979) and that of the fowl embryo (Tazawa *et al.*, 1989). The increase in water vapour head pressure is directly propor-

tional to the average growth rate of the avian embryo (Table 14.2), or inversely proportional to their incubation length (Rahn & Ar, 1980; Ar & Rahn, 1985). This shows that for a given average humidity in the nest, the rate of diffusive water loss from an egg is matched to the rate of embryonic growth independently of egg temperature. This is especially so since egg temperature rarely falls below the 'physiological zero temperature' (Haftorn, 1988) and that the total diffusive water loss is matched to the total incubation duration in such a way that no further active behavioural signal is needed.

Water loss is, however, regulated by selective adaptation to special conditions such as 'wet' and 'dry' environments, and altitude. In these cases, the main mechanism of adaptation is change in eggshell conductance (Birchard & Kilgore, 1980; Davis *et al.*, 1984; Ar & Rahn, 1985; Carey *et al.*, 1987; Arad, Gavrieli-Levin & Marder, 1988; Monge, Leon-Velarde & de la Torre, 1988).

A glance into the water compartments of the egg

Overall water concentration is only a very crude indication of the water relations within the egg. Water compartments of the egg change in temporal and spatial ways within the egg as the embryo grows and develops. Unfortunately, relatively little is known about water and solute distribution in avian eggs except for the fowl embryo (Romanoff, 1967; Hoyt, 1979; Davis *et al.*, 1988; Deeming, 1989, Chapter 19). It is, however, reasonable to assume that other species differ only in minor details. Most of the water of the egg (75–78%) is initially in the albumen which declines continuously during incubation as a result of water loss to the ambient air and of water movements into other compartments. As the developing embryo transports sodium ions into the space between it and the yolk, water is drawn from the albumen to form the sub-embryonic fluid (Romanoff, 1967). This reaches a peak volume of about 32–40% of the total water contained in the egg by the end of the first third of incubation and declines in volume thereafter. It is worthwhle noting that even for this process, proper incubation behaviour is vital: the absence of egg turning during this stage of incubation alters the rate of formation, and the final volume, of the sub-embryonic fluid and may result in embryonic death (Deeming, Rowlett & Simkiss, 1987; Wittmann & Kaltner, 1988; Deeming, 1989, Chapter 19). The sub-embryonic fluid disappears at about the end of two thirds of incubation. By then the water has moved into two new compartments: the contents of the amniotic and the allantoic sacs, which peak in volume at this time. The allantois holds 18–20%, and the amniotic sac, 10–12% of the total water in the egg (Romanoff, 1967). At about halfway through incubation, the entire embryo is only about 10% of its final size and, therefore, contains (including the blood compartment), some 13–14% of the total water in the egg. As the embryo itself grows in volume, the total amount of the water in the embryo increases at the expense of all the other compartments, and in the last third of incubation, it draws water mainly from the allantoic and amniotic sacs. Metabolic water production also adds water to the developing embryo: at the end of incubation, metabolic water, resulting mainly from the intensive lipid catabolism in the egg (Murray, 1925) accounts for about 8–13% of the total water present in the hatchling. Water movements within the egg are either passive or active and are accompanied by ionic and osmotic changes (Romanoff, 1967; Simkiss, 1980) and future research may reveal the details and their relation to the nitrogen metabolism of the egg (Packard & Packard, 1986).

Although water content of the embryo increases as it grows, its water fraction declines gradually from about the middle of incubation as a linear function of time from around 95% to a final value which depends on the mode of postnatal development (Romanoff, 1967; Pettit, Whittow & Grant, 1984; Ar & Raviv, unpublished). Since water concentration in the embryo itself declines during development, it may be used as an index of development under different experimental conditions. Moreover, water concentration at hatching time may indicate quantitatively the degree of neonatal maturity. Such a trend was depicted for the different maturity groups of birds by Sotherland & Rahn (1987).

Lung respiration and water loss

The transition from the embryonic mode of respiration, from using the chorio-allantoic membrane as the gas exchange organ to lung respiration, is slow in birds. Unlike mammals, where the lungs fill up immediately at birth, lungs and air sacs become gas-filled only gradually in birds during the paranatal period (Green & Vince, 1970; Vince & Tolhurst, 1975; Duncker, 1978; Seymour, 1984). Unfortunately, nothing comparable is known about the lung filling process in reptiles. The mechanism of breathing in birds requires an exact timing in the aerodynamic valving of the respiratory system to direct the gas flow through the parabronchi of the lungs in the typical avian unidirectional pattern (Piiper & Scheid, 1973). This might well be the reason for the pronounced isometrical morphology and histology which the avian respiratory system exhibits between young and adults (Duncker, 1972, 1978), again in contrast to mammals where maturation of the respiratory pattern is gradual (Boyden, 1975). Only the existence of an efficient lung respiratory system permits the embryo to shut down its chorio-allantoic circulation completely and to rely entirely on oxygen supply from the lungs, and only then it can hatch (Ar et al., 1980b). This means that the total respiratory system must be inflated before hatching can take place, and that a certain volume for the inflated lungs should be provided within the egg. The only volume available under the rigid eggshell for the inflation of the respiratory system is the air cell, which has a volume corresponding to the water escaping by evaporation from the egg (Ar & Rahn, 1980). Too little water loss may prevent the full inflation of the respiratory system in the paranatal stage and interfere with normal onset of respiration (Rahn et al., 1976; Davis et al., 1988). Based on Duncker's isometric concept, and using the analysis of Lasiewski & Calder (1971), it is possible to calculate the air volume of the expandable parts of the respiratory system, namely the air sacs (excluding the air in the rigid lungs and trachea). This volume varies from about 21% of total body volume in hatchlings of small species (<10 g) to about 8% in ratites. Since body mass of hatchlings is about two-thirds that of the fresh egg mass (Heinroth, 1922), the corresponding gas volumes that have to be provided for expansion vary between 14% to 5% of the egg volume respectively. It seems therefore that in most bird species, a diffusive water loss of the magnitude close to the observed one is needed for the appropriate transition to lung respiration, although El-Ibiary, Shaffner & Godfrey (1966) can not prove direct involvement of the air cell at the onset on lung ventilation.

The hydric and thermal relations of eggs

Heat and water fluxes

For a certain rate of water loss it is possible to calculate the rate of heat loss from eggs in the form of latent heat of evaporation. Under a given incubation temperature in steady state, fresh egg temperature must therefore be below incubation temperature. As the embryo develops, however, its metabolic heat production increases and so does egg temperature (Fig. 14.2). At about half-way through incubation, egg temperature surpasses that of incubation temperature (Martin & Insko, 1935; Khaskim, 1961; Ackerman & Seagrave, 1984). At the beginning of incubation, egg temperature is about 0.5 °C below incubation temperature, and in the final stages of incubation, it may reach 2 °C above. Evaporation of water may well play a role in cooling eggs below a critical developmental value, especially in species with large eggs, where heat dissipation is relatively limited because of an unfavourable surface area to volume ratio. This may explain the fact that even large species require a water loss which is not less than that of small species and that the potential to lose water, namely the water vapour conductance, increases more than in proportion to the surface area of the egg and the rate of metabolic heat production (Ar & Rahn, 1978; Ar & Rahn, 1985; Meir & Ar, 1990).

The model presented in Fig. 14.2 is consistent with still air incubation, which is probably the case in most natural nests. In artificial forced draught incubation, it is expected that the differences between egg and incubation temperatures will be smaller or negligible. The reason is obvious from the data presented by Tracy & Sotherland (1979) and Sotherland et al. (1987) showing that boundary layers around an egg are decreased and thermal conductance is increased about 6-fold at air velocity of $4 m s^{-1}$, a typical value for artificial incubation. This increase should decrease the difference between egg and ambient temperatures to a sixth of that of still air. The direction of heat and water movements, and the resulting changes in egg temperature are depicted schematically in Fig. 14.4. This scheme does not reveal which of the two fluxes, heat or water vapour, limits the other: is evaporation limited by the resistance to the heat flux towards the evaporating surface needed for maintaining it, or does the saturation caused by the shell (and boundary layer) resistance to water efflux limit the heat inflow? Not surprisingly, calculations show that when water vapour flux is 'translated' into flux of latent heat of evaporation, and when an appropriate temperature gradient is chosen, both resistances are of the same order of magnitude. Metabolic heat production or external heating cannot change this picture once a new steady state is established. However, if the shell material itself has a high thermal conductance, as can be judged from its composition and thickness (Sotherland et al., 1987), then again it is the molecular traffic of water vapour through the shell pores which is limited by the unique shell structure.

Heat capacity and water content

In their extensive review, Sotherland & Rahn (1987) use a model in which water and solid concentrations vary as a function of egg content mass between species. If we go one step further and strip the egg not only from its shell but also from its albumen, which may be considered as the main water reservoir of the egg, we are left with the ovum (the yolk or the true egg). According to Ar & Yom-Tov (1978), the log scale relationship between yolk mass and egg mass of precocial eggs shows a significant increase in yolk proportion with egg mass, while altricial species do not show this trend. Yolk fractions of precocial eggs extrapolated to the size of small altricial eggs are similar to the latter. The data presented by Sotherland & Rahn (1987) show that the yolk size dependence between the groups is the same and essentially linear to egg mass, but yolk fraction of precocial eggs is 1.4–1.5 times that of altricial eggs. No special tendency of eggs of altricial or precocial birds to vary with bird mass is found (Rahn, Paganelli & Ar, 1975). A direct regression of log

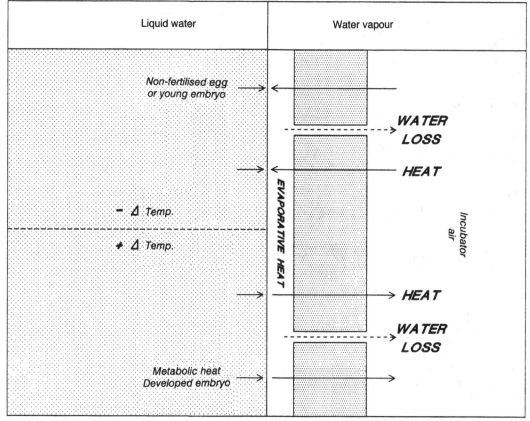

Fig. 14.4. Schematic representation of heat fluxes across the shell of young and fully developed eggs, and the consequences for egg to incubator temperature difference. Water loss represents flux of heat in the form of latent heat of evaporation.

yolk mass on female bird body mass for 102 species has the form of:

$$Yolk(g) = 0.05 \ Bird \ Mass(g)^{0.836 \pm 0.035 SE} \quad (14.1)$$
$$(\bar{X}=1.91; \ r^2=0.85) \ (Ar, \ unpublished)$$

By comparing this equation to the one obtained by Rahn *et al.* (1975) for the relation of egg mass to bird body mass, with a much higher intercept and a bird mass power of 0.770 only, it is obvious that yolk fraction increases (and albumen and water fractions decrease) as egg size increases.

The end result of all these tendencies is that altricial birds which are mostly small, tend to 'wrap' the ovum with a thicker water layer than do large precocial birds (Fig. 14.5). The thickness of the albumen in the left column in Fig. 14.5 is drawn to scale to represent its relative magnitude per unit of yolk mass. The decrease in egg mass to bird mass ratio (specific egg mass) from about 18% in a 10 g passerine bird to

about 4% in the ostrich (solid line) and the percent solids of the same eggs are shown in Fig. 14.6. If the solid and yolk concentrations were the same in all avian eggs, e.g. about 25% solids and 75% water, then the curve would have changed according to the dashed line in Fig. 14.6, demonstrating that small eggs in the range of 10 g and below are almost twice the size compared with the hypothetical situation of constant 75% water fraction. Increasing the water fraction from 70 to 80% for one unit solid mass means a corresponding 1.5 times increase in egg mass (Fig. 14.7). An appropriate cooling time constant can be calculated for eggs of different masses and water fractions (Turner, 1985), and assumes that heat capacity of solids is negligible compared with water. Therefore, all heat capacity is assigned to the water content of the egg. The result of this calculation is also included in Fig. 14.7 (dashed line; right ordinate). There is also a 1.3 increase in cooling time for eggs con-

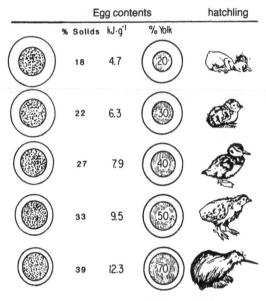

After Sotherland & Rahn, 1987

Fig. 14.5. Two alternative views of yolk and solids and consequently water in avian eggs, in different size and maturity groups. The column on the right-hand side takes egg mass as reference for yolk content (Sotherland & Rahn, 1987). The left-hand column takes yolk as a reference for egg content mass. See text for details.

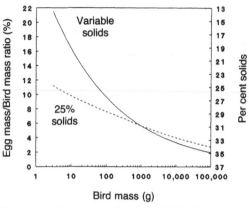

Fig. 14.6. Egg mass relative to bird body mass as a function of bird body mass. *Solid line:* real situation of eggs with variable fractions of water and solids. *Dashed line:* eggs with fixed proportions of solids and water at the level of 25 to 75 respectively.

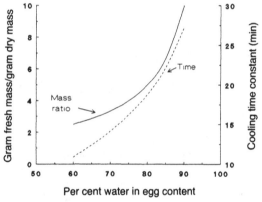

Fig. 14.7. Fresh egg mass per unit dry mass as a function of the water fraction in egg content (*solid line*) and the corresponding calculated cooling time constants (*dashed line*).

taining 80% water per unit solid mass instead of 70%. Thus, 'diluting' the solids of the egg with more water typically grants small altricial birds some 30% more off-nest time for a given degree of permitted cooling (Sidis & Ar, in preparation). The lyrebird (*Menura novaehollandiae*) with its single oversized 60 g egg that contains 80% water and is incubated in a cold climate (Lill, 1986), may represent an example of an egg with extended heat capacity.

Small birds have relatively high metabolic rates (Lasiewski & Dawson, 1967) and thus require longer and more frequent foraging times during incubation (Skutch, 1962; Vleck, 1981). The relatively longer cooling time of their water-loaded eggs may be considered as an adaptation which is compatible with their needs.

Concluding remarks

I suggest that the various aspects of water relations of avian eggs are important in several different ways for normal embryonic development. The rigid eggshell of birds is the most dominant feature in egg-water relationships. It permits conservation of the right amounts of water for development by resisting diffusive loss, yet does not restrict heat exchange. At the same time, it creates the problems of shell cracking and respiratory system expansion during pipping. These are presumably resolved by the right amount of diffusive water loss during incubation to make room for expansion of the respiratory system and the right amount of liquid water for shell wetting and weakening during pipping. Shell and water interact to supply dissolved calcium for the developing embryo. Humidity regulation in nests is only indirect, mainly through egg temperature regulation. The rate of water loss is automatically linked to growth rate by the same effects of temperature. The fraction of water in the embryo at late stages of incubation may be used as an index of its maturity. High water vapour shell conductance of large eggs may help in preventing overheating due to metabolism. The presence of large amounts of water in eggs of small species increases heat capacity of the eggs and helps buffer heat losses during the frequent off-nest periods typical of such species. The 'constant hydration concept' may well be an outcome of water balance and not a determinant of it.

Acknowledgements

I gratefully acknowledge the assistance of Ann Belinsky in preparing this manuscript.

References

Ackerman, R. A., Dmi'el, R. & Ar, A. (1985*b*). Energy and water vapor exchange by parchment-shelled reptile eggs. *Physiological Zoology*, **58**, 129–37.

Ackerman, R. A. & Seagrave, R. C. (1984). Parent–egg interactions: egg temperature and water loss. In *Seabird Energetics*, eds G. C. Whittow & H. Rahn, pp. 73–88. New York: Plenum Publishing Co.

— (1987). Modelling heat and mass exchange of buried avian eggs. *Journal of Experimental Zoology*, Supplement 1, 87–97.

Ackerman, R. A., Seagrave, R. C., Dmi'el, R. & Ar, A. (1985*a*). Water and heat exchange between parchment-shelled reptile eggs and their surroundings. *Copeia*, **1985**, 703–11.

Ar, A., Paganelli, C. V., Farhi, L. E. & Rahn, H. (1980*a*). Convective gas flow caused by diffusion explains positive hydrostatic pressure in hen's eggs. *Physiologist*, **23**, 178.

Ar, A., Paganelli, C. V., Reeves, R. B., Greene, D. G. & Rahn, H. (1974). The avian egg: water vapor conductance, shell thickness, and functional pore area. *Condor*, **76**, 153–8.

Ar, A. & Rahn, H. (1978). Interdependence of gas conductance, incubation length, and weight of the avian egg. In *Respiratory Function in Birds, Adult and Embryonic*, ed. J. Piiper, pp. 227–36. Berlin: Springer-Verlag.

— (1980). Water in the avian egg: overall budget of incubation. *American Zoologist*, **20**, 373–84.

— (1985). Pores in avian eggshells: gas conductance, gas exchange and embryonic growth rate. *Respiration Physiology*, **61**, 1–20.

Ar, A., Visschedijk, A. H. J., Rahn, H. & Piiper, J. (1980*b*). Carbon dioxide in the chick embryo towards end of development: effects of He and F_6 in breathing mixture. *Respiration Physiology*, **40**, 293–307.

Ar, A. & Yom-Tov, Y. (1978). The evolution of parental care in birds. *Evolution*, **32**, 655–69.

Arad, Z., Gavrieli-Levin, I. & Marder, J. (1988). Adaptation of the pigeon egg to incubation in dry hot environments. *Physiological Zoology*, **61**, 293–300.

Birchard, G. F. & Kilgore, D. L. Jr (1980). Conductance of water vapor in eggs of burrowing and nonburrowing birds: implications for embryonic gas exchange. *Physiological Zoology*, **53**, 284–92.

Blueweiss, L., Fox, H., Kudzma, V., Nakashima, D., Peters, R. & Sams, S. (1978). Relationships between body size and some life history parameters. *Oecologica*, **37**, 257–72.

Board, R. G., Tullett, S. G. & Perrott, H. R. (1977). An arbitrary classification of the pore systems in avian eggshells. *Journal of Zoology, London*, **182**, 251–65.

Bond, G. M., Board, R. G. & Scott, V. D. (1988). An account of the hatching strategies of birds. *Biological Review*, **63**, 395–415.

Booth, D. T. (1987). Water flux in malleefowl, *Leipoa ocellata* Gould (Megapodiidae). *Australian Journal of Zoology*, **35**, 147–59.

Booth, D. T. & Seymour, R. S. (1987). Effect of eggshell thinning on water vapor conductance of malleefowl eggs. *Condor*, **89**, 453–9.

Boyden, E. A. (1975). Development of the human lung. In *Practice of Pediatrics*, vol. 4, ed. V. C. Kelley pp. 1–17. Hagerstown: Harper & Row.

Briedis, D. & Seagrave, R. C. (1984). Energy transformation and entropy production in living systems. *Journal of Theoretical Biology*, **110**, 173–93.

Burnham, W. (1983). Artificial incubation of falcon eggs. *Journal of Wildlife Management*, **47**, 158–68.

Bustard, H. R., Simkiss, K., Jenkins, N. J. & Taylor, J. H. (1969). Some analyses of artificially incubated eggs and hatchlings of Green and Loggerhead sea turtles. *Journal of Zoology, London*, **158**, 311–15.

Carey, C. (1980). The ecology of avian incubation. *Bioscience*, **30**, 819–24.

— (1983). Structure and function of avian eggs. In *Current Ornithology*, vol. 1, ed. R. F. Johnston, pp. 69–103. New York: Plenum Press.

— (1986). Tolerance of variation in eggshell conductance, water loss, and water content by red-winged blackbird embryos. *Physiological Zoology*, **59**, 109–22.

Carey, C., Leon-Velarde, F., Castro, G. & Monge, C. C. (1987). Shell conductance, daily water loss, and water content of Andean gull and Puna ibis eggs.

Journal of Experimental Zoology, Supplement 1, 247–52.

Christensen, V. L. & McCorkle, F. M. (1982). Turkey egg weight losses and embryonic mortality during incubation. *Poultry Science*, 61, 1209–13.

Davis, T. A. & Ackerman, R. A. (1987). Effects of increased water loss on growth and water content of the chick embryo. *Journal of Experimental Zoology*, Supplement 1, 357–64.

Davis, T. A., Platter-Reiger, M. F. & Ackerman, R. A. (1984). Incubation water loss by pied-billed grebe eggs: adaptation to a hot, wet nest. *Physiological Zoology*, 57, 384–91.

Davis, T. A., Shen, S. S. & Ackerman, R. A. (1988). Embryonic osmoregulation: consequences of high and low water loss during incubation of the chicken egg. *Journal of Experimental Zoology*, 245, 144–56.

Deeming, D. C. (1987). Effect of cuticle removal on the water vapour conductance of egg shells of several species of domestic bird. *British Poultry Science*, 28, 231–7.

— (1989). Importance of sub-embryonic fluid and albumen in the embryo's response to turning of the egg during incubation. *British Poultry Science*, 30, 591–606.

Deeming, D. C., Rowlett, K. & Simkiss, K. (1987). Physical influences on embryo development. *Journal of Experimental Zoology*, Supplement 1, 341–5.

Drent, R. (1970). Functional aspects of incubation in the Herring Gull. *Behaviour*, Supplement 17, 1–132.

— (1975). Incubation. In *Avian Biology*, vol. 5, eds D. S. Farner, J. R. King & K. C. Parker, pp. 333–420. New York: Academic Press.

Duncker, H.-R. (1972). Die Festlegung des Bauplanes der Vogellunge beim Embryo und das Problem ihres postnatalen Wachstums. *Verh. Anat. Ges.*, 66, 273–77.

— (1978). Development of the avian respiratory and circulatory systems. In *Respiratory Function in Birds, Adult and Embryonic*, ed. J. Piiper, pp. 260–73. Berlin: Springer-Verlag.

El-Ibiary, H. M., Shaffner, C. S. & Godfrey, E. F. (1966). Pulmonary ventilation in a population of hatching chick embryos. *British Poultry Science*, 7, 165–76.

Grant, G. S. (1982). *Avian Incubation: Egg Temperature, Nest Humidity, and Behavioral Thermoregulation in a Hot Environment*. Washington, D.C.: The American Ornithologist's Union.

Grant, G. S., Pettit, T. N., Rahn, H., Whittow, G. C. & Paganelli, C. V. (1982). Water loss from Laysan and Black-footed albatross eggs. *Physiological Zoology*, 55, 405–14.

Green, J. & Vince, M. (1970). Investigation of the lung flotation technique for determining pulmonary respiration in the Japanese quail. *British Poultry Science*, 11, 403–6.

Haftorn, S. & Reinertsen, R. E. (1982). Regulation of body temperature and heat transfer to eggs during incubation. *Ornis Scandinavia*, 13, 1–10.

Haftorn, S. (1988). Incubating female passerines do not let the egg temperature fall below the 'physiological zero temperature' during their absences from the nest. *Ornis Scandinavia*, 19, 97–110.

Handrich, Y. (1989a). Incubation water loss in King Penguin egg. I. Change in egg and brood patch parameters. *Physiological Zoology*, 62, 96–118.

Handrich, Y. (1989b). Incubation water loss in king penguin egg. II. Does the brood patch interfere with eggshell conductance. *Physiological Zoology*, 62, 119–32.

Heinroth, O. (1922). Die Beziehungen zwischen Vogelgewicht, Eigewicht, Gelegegewicht und Brutdauer. *Journal für Ornithologie*, 70, 172–285.

Hillel, D. (1978). *Soil and Water*. New York: Academic Press.

Howell, T. R. (1979). Breeding biology of the Egyptian plover, *Pluvianus aegyptius*. Berkeley: University of California Press.

Hoyt, D. F. (1979). Osmoregulation by avian embryos: the allantois functions like a toad's bladder. *Physiological Zoology*, 52, 354–62.

Huggins, R. A. (1941). Egg temperatures of wild birds under natural conditions. *Ecology*, 22, 148–57.

Hulet, R. M., Christensen, V. L. & Bagley, L. G. (1987). Controlled weight loss during incubation of turkey eggs. *Poultry Science*, 66, 428–32.

Kendeigh, S. C. (1940). Factors affecting length of incubation. *Auk*, 57, 499–513.

Kern, M. D. (1986). Changes in water-vapor conductance of common canary eggs during the incubation period. *Condor*, 88, 390–3.

Khaskim, V. V. (1961). Heat exchange in birds' eggs on incubation. *Biofizika*, 6, 91–9.

Koorevaar, P., Menelik, G. & Dirksen, C. (1983). *Elements of Soil Physics*. Amsterdam: Elsevier.

Kreitzer, J. F. (1972). The effect of embryonic development on the thickness of the egg shells of coturnix quail. *Poultry Science*, 51, 1764–5.

Kuehler, C. M. & Witman, P. N. (1988). Artificial incubation of California condor *Gymnogyps californianus* eggs removed from the wild. *Zoo Biology*, 7, 123–32.

Landauer, W. (1967). *The Hatchability of Chicken Eggs as Influenced by Environment and Heredity*. 2nd edn, Storrs: Storrs Agricultural Experimental Station University of Connecticut.

Lasiewski, R. C. & Calder, W. A. Jr (1971). A preliminary allometric analysis of respiratory variables in resting birds. *Respiration Physiology*, 11, 152–66.

Lasiewski, R. C. & Dawson, W. R. (1967). A reexamination of the relation between standard metabolic rate and body weight in birds. *Condor*, **69**, 13–23.

Laughlin, K. F. & Lundy, H. (1976). The influence of sample size on the choice of method and interpretation of incubation experiments. *British Poultry Science*, **17**, 53–7.

Lill, A. (1986). Time-energy budgets during reproduction and the evolution of single parenting in the superb lyrebird. *Australian Journal of Zoology*, **34**, 351–71.

Lomholt, J. P. (1976). Relationship of weight loss to ambient humidity of birds eggs during incubation. *Journal of Comparative Physiology*, **105**, 189–96.

Lundy, H. (1969). A review of the effects of temperature, turning, and gaseous environment in the incubator on the hatchability of the hen's egg. In *The Fertility and Hatchability of the Hen's Egg*, eds T. C. Carter & B. M. Freeman, pp. 143–76. Edinburgh: Oliver & Boyd.

Martin, J. H. & Insko, W. M. (1935). Incubation experiments with turkey eggs. *Kentucky Agricultural Experimental Station Bulletin* 359.

Meir, M. & Ar, A. (1986). Further increasing hatchability of turkey eggs and poult quality by correcting water loss and incubator oxygen pressure. *Proceedings of the 7th European Poultry Conference*, II, 749–53.

— (1987). Improving turkey poult quality by correcting incubator humidity to match eggshell conductance. *British Poultry Science*, **28**, 337–42.

— (1990). Gas pressures in the air cell of the ostrich egg prior to pipping as related to oxygen consumption, eggshell gas conductance and egg temperature. *Condor*, **92**, 556–63.

Monge, C. C., Leon-Velarde, F. & de la Torre, G. G. (1988). Laying eggs at high altitude. *News in Physiological Sciences*, **3**, 69–71.

Murray, H. A. Jr (1925). Physiological Ontogeny. A. Chicken embryos. II. Catabolism. Chemical changes in fertile eggs during incubation. Selection of standard conditions. *Journal of General Physiology*, **9**, 1–37.

Needham, J. (1932). *Chemical Embryology*. Cambridge: Cambridge University Press.

Packard, G. C. & Packard, M. J. (1980). Evolution of the cleidoic egg among reptilian antecedents of birds. *American Zoologist*, **20**, 351–62.

— (1986). Nitrogen excretion by embryos of a gallinaceous bird and a reconsideration of the evolutionary origin of uricotely. *Canadian Journal of Zoology*, **64**, 691–3.

Packard, G. C., Tracy, C. R. & Roth, J. J. (1977). The physological ecology of reptilian eggs and embryos, and the evolution of viviparity within the class Reptilia. *Biological Reviews*, **52**, 71–105.

Paganelli, C. V. (1980). The physics of gas exchange across the avian eggshell. *American Zoologist*, **20**, 329–38.

Paganelli, C. V., Ar, A. & Rahn, H. (1981). What is the effective Po_2 in a gas-phase diffusion system. *Respiration Physiology*, **45**, 9–11.

Paganelli, C. V., Ar, A. & Rahn, H. (1987). Diffusion-induced convective gas flow through the pores of the eggshell. *Journal of Experimental Zoology*, Supplement, **1**, 173–80.

Peebles, E. D. & Brake, J. (1986). The role of the cuticle in water vapor conductance by the eggshell of broiler breeders. *Poultry Science*, **65**, 1034–9.

Pettit, T. N., Whittow, G. C. & Grant, G. S. (1984). Calorific content and energetic budget of tropical seabird eggs. In *Seabird Energetics*, ed. G. C. Whittow & H. Rahn, pp. 113–37. New York: Plenum Publishing Co.

Piiper, J. & Scheid, P. (1973). Gas exchange in avian lungs: models and experimental evidence. In *Comparative Physiology*, eds. L. Bolis, K. Schmidt-Nielsen & S. H. P. Maddrell, pp. 161–85. Amsterdam: North-Holland Publishing Co.

Prinzinger, R., Maisch, H. & Hund, K. (1979). Untersuchungen zum Gasstoffwechsel des Vogelembryos: I. Stoffwechselbedingter Gewichtsverlust, Gewichtskorrelation, tagliche Steigerungsrate und relative Gesamtenergieproduktion. *Zoologische Jahrbuch Physiologise*, **83**, 180–91.

Rahn, H. (1984). Factors controlling the rate of incubation water loss in bird eggs. In *Respiration and Metabolism of Embryonic Vertebrates*, ed. R. S. Seymour, pp. 271–88. Dordrecht: Dr. W. Junk.

Rahn, H., Ackerman, R. A. & Paganelli, C. V. (1977). Humidity in the avian nest and egg water loss during incubation. *Physiological Zoology*, **50**, 269–83.

Rahn, H. & Ar, A. (1974). The avian egg: incubation time and water loss. *Condor*, **76**, 147–52.

— (1980). Gas exchange of the avian egg: time, structure, and function. *American Zoologist*, **20**, 477–84.

Rahn, H., Paganelli, C. V. & Ar, A. (1975). Relation of avian egg weight to body weight. *Auk*, **92**, 750–65.

— (1987). Pores and gas exchange of avian eggs: a review. *Journal of Experimental Zoology*, Supplement 1, 165–72.

Rahn, H., Paganelli, C. V., Nisbet, I. C. T. & Whittow, G. C. (1976). Regulation of incubation water loss in eggs of seven species of terns. *Physiological Zoology*, **49**, 245–59.

Robertson, I. S. (1961). Studies on the effect of humidity on the hatchability of hens eggs I. The determination of optimum humidity for incubation. *Journal of Agricultural Science*, **57**, 185–95.

Romanoff, A. L. (1967). *Biochemistry of the Avian Embryo*. New York: Wiley-Interscience Publishers.

Romijn, C. & Roos, J. (1938). The air space of the hen's egg and its changes during the period of incubation. *Journal of Physiology*, **94**, 365–79.

Schmidt-Nielsen, K. (1979). *Animal Physiology: Adap-*

tation and Environment. 2nd edn, Cambridge: Cambridge University Press.

Seagrave, R. C. (1971). *Biomedical Applications of Heat and Mass Transfer.* Ames, Iowa: The Iowa State University Press.

Seymour, R. S. (1984). Patterns of lung aeration in the perinatal period of domestic fowl and Brush Turkey. In *Respiration and Metabolism of Embryonic Vertebrates*, ed. R. S. Seymour, pp. 319–32. Dordrecht: Dr. W. Junk.

Seymour, R. S. & Piiper, J. (1988). Aeration of the shell membranes of avian eggs. *Respiration Physiology*, **71**, 101–16.

Sibly, R. M. & Simkiss, K. (1987). Gas diffusion through non-tubular pores. *Journal of Experimental Zoology*, Supplement 1, 187–91.

Silyn-Roberts, H. & Sharp, R. M. (1989). The similarity of preferred orientation development of eggshell calcite of the dinosaurs and birds. *Proceedings of the Royal Society of London, B*, **235**, 347–63.

Simkiss, K. (1967). *Calcium in Reproductive Physiology.* London: Chapman and Hall.

— (1974). The air space of an egg: an embryonic 'cold nose'. *Journal of Zoology, London*, **173**, 225–32.

— (1980). Eggshell porosity and the water metabolism of the chick embryo. *Journal of Zoology, London*, **192**, 1–8.

— (1986). Eggshell conductance – Fick's or Stefan's Law. *Respiration Physiology*, **65**, 213–22.

Skutch, A. F. (1962). The constancy of incubation. *Wilson Bulletin*, **74**, 115–52.

Snyder, G. K. & Birchard, G. F. (1982). Water loss and survival in embryos of the domestic chicken. *Journal of Experimental Zoology*, **219**, 115–17.

Sotherland, P. R. & Rahn, H. (1987). On the composition of bird eggs. *Condor*, **89**, 48–65.

Sotherland, P. R., Spotila, J. R. & Paganelli, C. V. (1987). Avian eggs: barriers to the exchange of heat and mass. *Journal of Experimental Zoology*, Supplement 1, 81–6.

Spellerberg, I. F. (1969). Incubation temperatures and thermoregulation in the McCormick Skua. *Condor*, **71**, 59–67.

Swart, D., Rahn, H. & de Kock, J. (1987). Nest microclimate and incubation water loss of eggs of the African Ostrich (*Struthio camelus* var. *domesticus*). *Journal of Experimental Zoology*, Supplement 1, 239–46.

Taigen, T. L., Packard, G. C., Sotherland, P. R. & Hanka, L. R. (1978). Influence of solute concentration in albumen on water loss from avian eggs. *Auk*, **95**, 422–44.

Tazawa, H., Okuda, A., Nakazawa, S. & Whittow, G. C. (1989). Metabolic responses of chicken embryos to graded, prolonged alterations in ambient temperature. *Comparative Biochemistry and Physiology*, **92A**, 613–7.

Toien, O., Paganelli, C. V., Rahn, H. & Johnson, R. R. (1987). Influence of eggshell pore shape on gas diffusion. *Journal of Experimental Zoology*, Supplement 1, 181–6.

— (1988). Diffusive resistance of avian eggshell pores. *Respiration Physiology*, **74**, 345–54.

Tracy, C. R. & Sotherland, P. R. (1979). Boundary layers of bird eggs: do they ever constitue a significant barrier to water loss. *Physiological Zoology*, **52**, 63–6.

Tracy, C. R., Welch, W. R. & Porter, W. P. (1980). *Properties of Air. A Manual for Use in Biophysical Ecology.* 3rd edn, Madison: The University of Wisconsin.

Tullett, S. G. & Board, R. G. (1976). Oxygen flux across the integument of the avian egg during incubation. *British Poultry Science*, **17**, 441–50.

Tullett, S. G. & Burton, F. G. (1982). Factors affecting the weight and water status of the chick at hatch. *British Poultry Science*, **23**, 361–9.

Turner, J. S. (1985). Cooling rate and size of birds' eggs – a natural isomorphic body. *Journal of Thermal Biology*, **10**, 101–4.

Tyler, C. (1956). Studies on egg shells. VII. – Some aspects of structure as shown by plastic models. *Journal of Food Science and Agriculture*, **7**, 483–93.

Tyler, C. & Geake, F. H. (1964). The testing of methods for crushing eggshells, based on paired readings. from individual eggs and the measurement of some effects of various treatments. *British Poultry Science*, **5**, 29–35.

Vince, M. A. & Tolhurst, B. E. (1975). The establishment of lung ventilation in the avian embryo: the rate at which lungs become aerated. *Comparative Biochemistry and Physiology*, **52A**, 331–7.

Visschedijk, A. H. J., Tazawa, H. & Piiper, J. (1985). Variability of shell conductance and gas exchange of chicken eggs. *Respiration Physiology*, **59**, 339–45.

Vleck, C. M. (1981). Hummingbird incubation: female attentiveness and egg temperature. *Oecologica*, **51**, 199–205.

Vleck, C. M. & Vleck, D. (1987). Metabolism and energetics of avian embryos. *Journal of Experimental Zoology*, Supplement 1, 111–25.

Vleck, C. M., Vleck, D., Rahn, H. & Paganelli, C. V. (1983). Nest microclimate, water-vapor conductance, and water loss in heron and tern eggs. *The Auk*, **100**, 76–83.

Walsberg, G. E. (1983). A test for regulation of nest humidity in two bird species. *Physiological Zoology*, **56**, 231–5.

Wangensteen, O. D., Wilson, D. & Rahn, H. (1970/71). Diffusion of gases across the shell of the hen's egg. *Respiration Physiology*, **11**, 16–30.

White, F. N. & Kinney, J. L. (1974). Avian incubation. *Science*, **186**, 107–15.

Whittow, G. C. (1982). Physiological ecology of incubation in tropical seabirds. In *Tropical Seabird Biology. Studies in Avian Biology 8*, ed. R. W.

Schreiber, pp. 47–72. Lawrence, Kansas: The Cooper Ornithological Society, Allen Press.

Wittmann, J. & Kaltner, H. (1988). Formation and changes of the subembryonic liquid from turned, unturned, and cultured Japanese quail eggs. *Biotechnology and Applied Biochemistry*, **10**, 338–45.

Woodall, P. F. & Parry, D. F. (1982). Water loss during incubation in red bishop (*Euplectes orix*) eggs. *South African Journal of Zoology*, **17**, 75–8.

Yom-Tov, Y., Wilson, R. & Ar, A. (1986). Water loss from Jackass Penguin *Spheniscus demersus* eggs during natural incubation. *Ibis*, **128**, 1–8.

15

Water economy and solute regulation of reptilian and avian embryos

DAVID VLECK

Introduction

The eggs of birds and oviparous reptiles exchange water and respiratory gases with the environment in which they are incubated, primarily via diffusion across the eggshell. Depending on the incubation environment, such eggs may experience either net gains or net losses of water during incubation (Packard, Chapter 13; Ar, Chapter 14). Successful hatching requires that the developing embryo be capable of dealing with the resulting variation in egg water content.

Most bird eggs are incubated in environments where potential gradients favour the continual loss of water throughout incubation. Surfaces on which they lie are usually relatively dry, and parental brooding, plus embryonic metabolism, keep their temperature, and hence water activity, well above that of their environment. Some of the loss is replaced by metabolic water derived from the oxidation of organic molecules within the egg, but water content of all avian eggs declines during incubation, usually by 10–20% of the initial mass of the egg (Ar, Chapter 14). Avian embryos must therefore contend with a gradual dehydration of the egg in which they are developing. Within oviparous reptiles, water budgets are much more variable. The shells of reptile eggs constitute much less of a barrier to water exchange, and reptile eggs are usually laid in close contact with soil of varying moisture content. Depending on soil moisture, reptile eggs may experience either net gain or net loss of water during incubation (Packard, Chapter 13). Embryos face a range of conditions ranging from extreme dehydration of the egg contents to impressive hyperhydration.

Variation in egg water content will affect the volume, surface area, osmotic concentration, and specific solute concentrations of the fluid contained therein. In this chapter I review the factors that affect egg water content, the magnitude of variation in water content, and its developmental consequences. The focus is on the ability of embryos to compensate for variations in water content, and on the ontogenic trade-offs that result. Recent work in my laboratory has concentrated on eggs, embryos, and hatchlings of two species of iguanid lizards, *Urosaurus ornatus* and *Sceloporus virgatus*. Unless otherwise specified, this is the source of data presented below on those species.

Changes in water content: controlling variables

The rate of water exchange between an egg and its incubation environment depends on the water potential gradient (the driving force) and the resistance to water movement imposed by the eggshell and its associated boundary conditions (Ackerman, Dmi'el & Ar, 1985a; Rahn, 1984; Packard & Packard, 1988; Paganelli, Chapter 16; Deeming & Thompson, Chapter 17). The physiological importance of water exchange depends not on rate of exchange, but on the magnitude of that exchange relative to the size of water and solute compartments within the egg. This means that the physiological impact of water exchange during incubation can be modified by adjusting: 1) the incubation environment, or the water potential inside the egg, 2) the shell structure, or 3) water content of the egg at laying. Only one of these, the water potential inside the egg, can be manipulated by the developing embryo. All three can be modified by the maternal parent.

Incubation environment

Females may choose oviposition sites and hence incubation environment with respect to the water potential gradient between an egg and its environment. In many reptiles, this apparently involves choosing a time and place for oviposition that increases the likelihood for successful reproduction. The green snake *Opheodrys aestivus*, when given a choice of oviposition sites that vary in water potential, shows a strong preference for moister sites (Plummer & Snell, 1988). *Anolis* lizards may retain eggs during dry periods, delaying oviposition until moist nesting sites are available (Stamps, 1976). Galapagos land iguanas (*Conolophus subcristatus*) lay their eggs just prior to seasonal rains, so that most of the incubation period coincides with the periods when soil water potentials are highest (Snell & Tracy, 1985). *S. virgatus* is a lizard of the arid mountains of the southwestern USA and northwestern Mexico. In this region, rainfall rarely occurs in late spring and early summer, and soils can be extremely dry. This seasonal drought is broken by late summer rains, which occur reliably, but vary in onset from year to year. Oviposition by *S. virgatus* females always follows the onset of the summer rains, when soil moisture content and water potential rise dramatically. In all of these species, eggs incubated in moist soils take up more water from the soil and produce larger hatchlings than do eggs incubated in drier soils (Snell & Tracy, 1985; Plummer & Snell, 1988). In *S. virgatus*, larger hatchlings have higher post-hatching survivorship than smaller hatchlings, so this timing of oviposition has adaptive significance. The effectiveness of such a choice process is limited, however, by temporal variability of the incubation environment. In many regions, the variability of water potential in incubation sites is likely to be high on the time-scale over which incubation occurs.

The only evidence that birds might actively regulate the water potential around their eggs comes from the observation that Egyptian Plovers (*Pluvianus aegyptius*) cover their eggs with sand and make repeated trips between river water and the nest site during the hottest hours of the day, using water carried in the feathers to moisten the nest area on each trip (Howell,

1979). It seems likely, however, that this is primarily a thermoregulatory response, and that its effects on water exchange are incidental to the necessity to keep eggs from overheating. For some bird species, there is evidence that they do not adjust behaviour to regulate egg water loss in response to changes in nest humidity (Grant *et al.*, 1982; Walsberg, 1983, 1985).

For most reptiles, oviposition marks the end of parental care, and eggs are subject to environmental variability. There are exceptions, and, in some cases, parental behaviour may serve primarily to modify water exchange between eggs and environment after oviposition. Lizards, turtles, and crocodiles have been reported to urinate on the nest site at the time of laying (Bellairs, 1969; Ferguson, 1985; Shine, 1988), but such behaviour is not known to be sustained through incubation. The Malayan pit viper (*Calloselasma rhodostoma*) remains coiled over its eggs after they are laid. Changes in humidity affect the behaviour of a brooding female *C. rhodostoma*, who covers her clutch more completely when ambient humidity decreases but the effects of this behaviour on the water budget of the eggs have not been quantified (York & Burghardt, 1988). Female prairie skinks, *Eumeces septentrionalis*, brood their eggs, and the presence of a brooding female reduces egg mortality (Somma & Fawcett, 1989). The skinks increased behaviours that Somma & Fawcett (1989) characterise as 'water conserving' (for eggs) in drier soils, but they present no data to support this assertion.

Shell permeability

A second maternal option is to modify the permeability of the eggshell to water. Shell structure must satisfy a variety of sometimes conflicting requirements. Shells protect eggs from mechanical damage (Tyler, 1969), aid in denying microorganisms access to egg contents (Board & Fuller, 1974; Packard & Packard, 1980), provide a reservoir of mineral nutrients (Johnston & Comar, 1955; Packard & Packard, 1984*a*), and act as a barrier to exchange of water and respiratory gases. In birds, there is considerable evidence from intra- and interspecific comparisons that there are inverse relationships between conductance of the eggshell to water vapour (G_{H_2O}) and water vapour concentration in the nest (Lomholt, 1976; Vleck, Hoyt & Vleck, 1979; Birchard & Kilgore, 1980; Arad &

Marder, 1982; Vleck *et al.*, 1983; Seymour *et al.*, 1987) or vapour mobility (which increases with altitude as atmospheric pressure declines) in the incubation environment (Wangensteen *et al.*, 1974; Packard, Sotherland & Packard, 1977; Rahn *et al.*, 1977; Carey *et al.*, 1982, 1983; Leon-Valarde *et al.*, 1984). Such inverse relationships tend to minimise variation in water loss between environments.

Control of G_{H_2O} is, however, inadequate to eliminate variation in water loss within a population of birds. There is substantial variance in shell conductance within almost all bird populations (Hoyt, 1979), and variation within a population can be greater than variation between means of populations from very different environments (Bucher & Barnhart, 1984). It has been estimated that in a population of bird eggs that loses, on average, 16% of initial egg mass by evaporation of water across the shell during incubation, about 1/3 of the eggs lose water equal to less than 12% or more than 20% of their initial mass (Hoyt, 1979).

From an evolutionary viewpoint, the high variance in G_{H_2O} between eggs within bird populations makes an interesting contrast with the suggestion that differences of similar or smaller magnitude between populations are adaptations to adjust rates of water loss. Vleck *et al.* (1983) have measured G_{H_2O} and nest humidities for seven species of birds that build stick nests in trees and three species that nest on the ground. All ten species nest in the same local area. The estimated gradient driving water vapour diffusion across the egg shell averages 35% lower in ground nests than in tree nests, due to higher humidity in ground nests. Egg G_{H_2O}, expressed as a percentage of the value expected based on egg size and incubation period, averages 37% higher in the ground-nesters than in the tree-nesters. The relative increase in mean G_{H_2O} matches the increase in nest humidity for ground nesting species, so average water loss during incubation does not differ between nest types. The variability in egg G_{H_2O} is, however, substantial within each species. The average coefficient of variation (= standard deviation/mean) for G_{H_2O} within a species is 27%, close to the average reported for a much larger sample of bird species by Hoyt (1979). This means that about 25% of the eggs from nests of each type have G_{H_2O} values closer to the mean for nests of the other type than to the mean for their own nest type. The basis for

this intra-population variance in G_{H_2O} is poorly known. Its existence has led to two sets of opinions about the adaptive significance of shell conductance, with studies using between population comparisons suggesting that selection on G_{H_2O} has been important (see above), and studies using within species comparisons casting doubt on that importance (Simkiss 1980a,b; Walsberg, 1985; Carey, 1986).

If populations respond readily to directional selection on G_{H_2O}, why doesn't stabilising selection reduce the variation in G_{H_2O} once the population mean approximates the optimum? Possible explanations include any or all of the following: 1) Optimal values within a habitat or nest-type vary on a time-scale too short for evolutionary change in G_{H_2O} to match them precisely. If this is the case, G_{H_2O} will be a heritable character, and the intrapopulation variance in G_{H_2O} should be proportional to the variance in the incubation environment on an appropriate time-scale. 2) Intrapopulation variance in G_{H_2O} represents an equilibrium between selection for an optimal G_{H_2O} and variability produced by mutation. Again, G_{H_2O} will show reasonable heritability, but the variance in G_{H_2O} need not be proportional to environmental variance. 3) Intrapopulation variance in G_{H_2O} is not due to genetic variance, but depends on maternal condition, or is a physical consequence of the possibly chaotic dynamics of pore formation during crystallisation of eggshells. The latter could account for within-clutch variation in G_{H_2O}, but not between-clutch variation. In either case, G_{H_2O} would have a low heritability. Distinguishing between these alternatives would be a constructive step towards resolving the selective importance of variation in G_{H_2O} between populations.

The shells of reptile eggs vary in structure much more than those of avian eggs (Packard, Packard & Boardman, 1982a,b; Packard & DeMarco, Chapter 5). Oviparous snakes and most lizards lay eggs with a compliant, fibrous and non-calcareous shell. These parchment-like shells are highly permeable to water, having G_{H_2O} that exceed those of bird eggs of comparable size by about two orders of magnitude (Ackerman *et al.*, 1985a; Deeming & Thompson, Chapter 17). They lose or gain water very rapidly, depending on the water potential gradient across the shell. Water gain can be substantial, because of the compliance of the shell. Crocodilians (which are more closely related to birds than to the other 'reptiles'), a few lizards, and some turtles lay rigid-shelled, non-compliant eggs. They have G_{H_2O} that are from 0.5 to 5 times the values expected for avian eggs of the same size (Packard *et al.*, 1979; Lutz *et al.*, 1980; Dunson, 1982; Deeming & Thompson, Chapter 17), but because they are incubated in very humid environments, net water exchange is usually small. Other turtles produce pliable-shelled eggs, that are intermediate between these two groups, and which have G_{H_2O} values that are 50–70 times those of bird eggs of the same size (Packard *et al.*, 1979; Packard, Packard & Boardman, 1981; Deeming & Thompson, Chapter 17). At present, we know too little about oviposition sites in most reptiles to judge whether any correlation exists between eggshell permeability and oviposition environment.

Egg water content

A final route to adjusting the physiological impact of water exchange between egg and environment is to modify the size of the water pool in the egg at laying. This may be done by adjusting egg size, relative water content (% water), or both. Changes in egg size affect the importance of water exchange independently from changes in relative water content. The rate of water exchange increases with egg surface area, whereas the size of the affected water pool increases with egg volume. As egg size increases, volume increases more rapidly than surface area, so large eggs are less sensitive to a fixed change in environmental conditions or shell permeability than are small eggs (Ackerman *et al.*, 1985b; Tracy & Snell, 1985). The snake *Coluber constrictor* and the lizard *U. ornatus* both lay flexible-shelled eggs. *C. constrictor* eggs have an initial mass of about 6.1 g, and, when incubated at a soil water potential of −150 kPa, increase in mass by about 3.6 g (60%) over the first 28 days of incubation (Packard & Packard, 1987). *U. ornatus* eggs are relatively small, about 200 mg, yet when incubated in slightly drier soil (−240 kPa), increase in mass by about 270 mg, or 135% of initial mass, during the same time interval. Reptiles that lay relatively large eggs may be capable of successful reproduction in drier environments than species that produce smaller eggs (Tracy & Snell, 1985; Packard, Packard & Gutzke, 1985a).

There is considerable variation in relative water content of eggs at the time of laying. In birds, relative water content, expressed as a percentage of the mass of the contents of the egg excluding the shell, ranges from 86.6% in the magpie *Pica pica* (Carey, Rahn & Parisi, 1980) to 61% in the kiwi *Apteryx australis* (Calder, Parr & Karl, 1978). The avian eggs with the lowest percentages of water at the time of laying, kiwis and megapodes (Vleck, Vleck & Seymour, 1984), are incubated under exceptionally humid conditions (Seymour *et al.*, 1987; Calder, personal communication) but most of the variation between species is associated with differences between developmental modes. Water makes up an average of 84.3% of the contents of eggs of altricial birds, and 74.7% for precocial species (Carey *et al.*, 1980). Eggs of altricial species appear to have higher relative water contents, not because they lose relatively large amounts of water during incubation, but, because embryos hatch at an earlier stage of development when tissue water content is higher, experience higher rates of post-hatching water loss, and cannot leave the nest to obtain water elsewhere after hatching (Ricklefs, 1977; Ar & Yom-Tov, 1978; Vleck & Vleck, 1987).

Measurements of the water content of reptile eggs must be viewed with more caution than those for bird eggs. First, water content of reptile eggs changes with age and stage of development, even prior to laying. Eggs surgically removed from the oviducts of 29 species of lizards and snakes ranged from 39% to 66%, averaging only 54%, with a standard deviation of 7.5% (Ballinger & Clark, 1973; Vitt, 1978; Packard *et al.*, 1985b). Two species included in Vitt's (1978) sample were *S. virgatus* and *U. ornatus*, with water contents of 60% and 50–55%, respectively. Eggs of these species naturally oviposited in my laboratory have significantly higher percentages of water, averaging 67.0% and 66.5% respectively immediately following oviposition. Use of oviductal eggs is likely to lead to underestimates of egg water content and clutch mass at the time of oviposition. Secondly, because of the small size and high permeability of many reptile eggs, water content changes quickly when they are moved from one environment to another, so there is some uncertainty about the initial mass of eggs that are not weighed promptly after laying.

Those cautions aside, it is clear that reptile eggs are laid with a relative water content much lower than most bird eggs. Those data for lizard eggs summarised above, however, include no eggs larger than about 3 g. Eggs of some larger lizards contain much higher proportions of albumen (Tracy & Snell, 1985), and presumably, have higher relative water contents. Eggs from snakes *C. constrictor* and *Pituophis melanoleucus* that voluntarily oviposit in the lab average 59% and 74% water respectively (Packard & Packard, 1987; Gutzke & Packard, 1987). Four species of turtles, none of which lay rigid shelled eggs, produce eggs that contain between 68.9% and 72.6% water (Ricklefs & Burger, 1977; Packard & Packard, 1980; Morris *et al.*, 1983; Wilhoft, 1986). The low initial water content of reptile eggs is correlated with the much higher water potential of their nest sites, compared with those of birds.

Changes in either egg size or relative water content obviously affect aspects of an animal's biology other than egg hydration, and egg structure is likely to be a compromise between possibly conflicting requirements. The relatively low water content of reptile eggs minimises clutch mass, and permits carrying many eggs in large clutches (Tracy, 1980, 1982). Changes in the amount of non-water material in the egg alters the resources available to the embryo and affects maternal material budgets. Changes in egg size can constrain clutch size or alter the burden on a gravid female.

Changes in egg water content during incubation

All bird eggs lose water during incubation (Ar, Chapter 14). Data on rates of water loss during incubation from eggs of 81 species of birds indicate that the fraction of the initial mass of the egg lost by evaporation across the eggshell during incubation ranges from 10% to 23%, with a mean of 15% (S.D. = 2.5%) (Ar & Rahn, 1980). Even in the brush turkey *Alectura lathami*, a megapode, eggs lose about 9.5% of their initial mass by evaporation of water during incubation (Seymour *et al.*, 1987; Booth & Thompson, Chapter 20). The range of fractional water losses that avian embryos can tolerate varies between species. In the domestic fowl (*Gallus gallus*) tolerance to variation in incubation humidity and fractional water loss varies between strains, but hatchability of eggs is usually highest at fractional water losses ranging from about 7% to 12% and declines rapidly at more extreme values (Lundy, 1969), though some chicks can hatch from eggs that lose

26–30% of their initial mass (Snyder & Birchard, 1982). Red-winged blackbird (*Agelaius phoeniceus*) embryos, whose eggshell G_{H_2O} are altered experimentally to produce fractional water losses ranging from 7.4% to 33.0% of initial egg mass, all hatch successfully. The maximum fractional loss of an egg that can hatch is about 43% (Carey, 1986).

Reptile eggs may either lose or gain water during incubation, depending on incubation conditions. Some water uptake appears to be obligate for the smaller flexible-shelled eggs. The available data on the percentage change in egg mass over the course of incubation are shown in Table 15.1; virtually all of the change in mass is due to water exchange. None of these studies was designed to determine tolerance limits, so the results represent minimum ranges over which hatching can occur. Where authors did not provide data for individual eggs, mean values for groups are cited, so the actual range of variation usually exceeds the tabled value. Only four studies, those on the crocodilian *Crocodylus acutus* (Lutz & Dunbar-Cooper, 1984), and the turtles *Chelydra serpentina*, *Trionyx triunguis* and *Chrysemys picta* (Packard *et al.*, 1982a,b; Leshem & Dmi'el, 1986; Ratterman & Ackerman, 1989) are based on field data. It is apparent that relative water exchange spans a relatively small range in species that lay relatively impermeable (especially rigid-shelled) eggs with a relatively high water content (crocodilians and turtles). Field studies on *C. serpentina* and *C. picta* report higher maximum water uptake than laboratory studies on the same species. Laboratory incubation arrangements may not explore the range of hydric environments typical of natural soils (Ratterman & Ackerman, 1989). Among lizards and snakes that lay flexible-shelled eggs, the highest range of variation occurs in the smallest eggs, e.g. *U. ornatus* at 206 mg and *S. virgatus* at 382 mg, but only the largest eggs, e.g. *C. subcristatus* at 50 g (Tracy & Snell, 1985) and *P. melanoleucus* at 24 g (Gutzke & Packard, 1987), are known to be able to tolerate decreases in egg mass below that at laying.

It is instructive to examine the way that the observed ranges in percentage water loss (Fig. 15.1) affect the hydration of the egg contents by the end of incubation. The differences in water exchange exhibited by the red-winged blackbird *A. phoeniceus* and the lizard *S. virgatus* nearly span the range known to occur. The exceptionally complete data for both permit calculation of

Table 15.1. *Range of variation in relative mass of reptile eggs near the end of incubation.*

Species	% of initial mass of egg		Reference
	Maximum	Minimum	
Crocodilians			
Crocodylus acutus	−7.4	−30	Lutz & Dunbar-Cooper (1984)
Crocodylus novaeguineae	+25	−25	Bustard (1971) in Ferguson (1985)
Alligator mississippiensis	−2	−8	Tracy & Snell (1985)
Turtles			
Trionyx triunguis	0	−14.2	Leshem & Dmi'el (1986)
Trionyx spiniferus	0	−7	Packard *et al.* (1981)
Emydoidea blandingii	−3	−15	Packard *et al.* (1982*b*)
Caretta caretta	−8	−19	Tracy & Snell (1985)
Chrysemys picta	+4	−16	Packard *et al.* (1983)
Chrysemys picta	+73	−22	Ratterman & Ackerman (1989)
Chelydra serpentina	+19	−19	Packard *et al.* (1980*a*)
Chelydra serpentina	+43	−12	Packard *et al.* (1982*a*)
Terrapene ornata	+7	−17	Packard, Packard & Gutzke (1985*a*)
Snakes			
Coluber constrictor	+50	0	Packard & Packard (1987)
Pituophis melanoleucus	+49	−12	Gutzke & Packard (1987)
Elaphe obsoletus	+111	+51	Deeming (1989*b*)
Lizards			
Conolophus subcristatus	+47	−21	Tracy & Snell (1985)
Crotaphytus collaris	+80		Tracy & Snell (1985)
Sceloporus undulatus	+175	+100	Tracy (1980)
Sceloporus virgatus	+349	+11	This study
Callisaurus draconoides	+115	+55	Packard, Packard & Boardman (1980*b*)
Amphibolorus barbatus	+37		Packard *et al.* (1985*b*)
Urosaurus ornatus	+366	+27	This study
Eublepharis macularius	+82		Deeming (1989*b*)

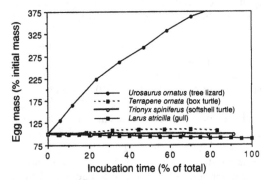

Fig. 15.1. Mass of eggs, as a percentage of their mass when laid, as a function of time during incubation, expressed as a percentage of the total incubation period. (Data for *Urosaurus ornatus* and *Larus atricilla* are Vleck (unpublished); turtle data are from Gettinger, Paukstis & Gutzke, 1984 and Packard, Packard & Gutzke, 1985*a*.)

metabolic water production and the approximate amounts of dry mass in the egg contents just prior to hatching. Red-winged blackbirds lay eggs that weigh about 4.5 g. The egg contents weigh 4.3 g, and are 86% water and 4.7% lipid, with a total energy content of 18.72 kJ (Carey *et al.*, 1980). Prior to hatching, a bird embryo from an egg with an energy content of 18.72 kJ uses about 6.7 kJ of that energy to support development (Vleck & Vleck, 1987) and most of that energy is derived from oxidation of lipids (Romanoff, 1967). *S. virgatus* lays eggs that average about 380 mg, which are 67% water. Over the course of incubation, eggs incubated in moist soils (−100 kPa) use about 82 cm³ of oxygen, equivalent to 1.6 kJ of energy. Embryos in eggs incubated on drier soil (−450 kPa) use only about 67 cm³ of oxygen, or 1.3 kJ of energy. Hatchlings from dry soils are smaller, but retain more residual yolk at the time of hatching than hatchlings from wetter soils (Fig. 15.2).

Oxidation of lipids yields about 39.3 kJ/g, and results in the production of about 1.07 g water per 1 g lipid oxidised (Schmidt-Nielsen, 1983). If we assume for simplicity that lipid is the only substrate oxidised, these values permit calcula-

Table 15.2. *Net water exchange during incubation: effects on egg hydration in the striped plateau lizard,* Sceloporus virgatus, *and the Red-winged Blackbird,* Agelaius phoenixeus

Species	*Sceloporus virgatus*[a]		*Agelaius phoenixeus*	
Water exchange regimen	High gain	Low gain	Low loss	High loss
% change in egg mass during incubation	+240	+45	−7.4	−33
Water in contents near hatching (g)	1.06	0.30	3.6	2.4
Dry mass of contents near hatching (g)[b]	0.111	0.107	0.43	0.43
% water in egg contents near hatching	90.5	73.4	89.2	84.8
Hydration of contents (dry mass/water mass)	0.104	0.362	0.121	0.179

[a]*Sceloporus* eggs were incubated at a soil water potential of −100 kPa (high gain) or −450 kPa (low gain). Neither value is an extreme for the species. See text for sources of *Agelaius* data.
[b]Dry mass of *Sceloporus* egg contents differs between groups because of differences in initial mass of eggs and in metabolism during incubation. I assumed metabolism of *Agelaius* eggs did not differ between groups.

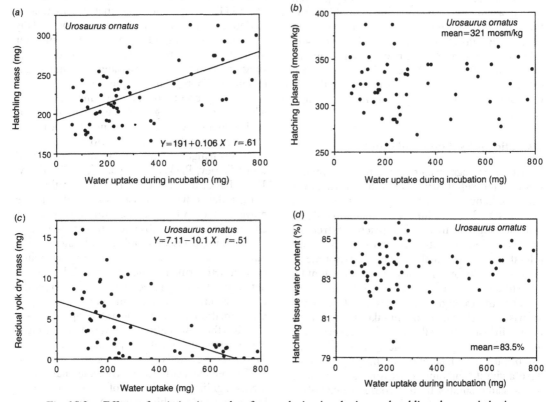

Fig. 15.2.   Effects of variation in uptake of water during incubation on hatchling characteristics in the lizard *Urosaurus ornatus*. (*a*) Hatchling mass increases with water uptake. (*b*) Plasma osmotic concentration is independent of uptake. (*c*) Dry mass of residual yolk decreases with uptake. (*d*) Hatchling tissue water content (%) is independent of water uptake.

tion of the effects of the range of water loss described above (Carey, 1986; Table 15.1) on the hydration of the egg contents just prior to hatching. Non-lipid substances are certainly also oxidised, but the degree to which that occurs (Romanoff & Romanoff, 1949; Seymour *et al.*, 1987) has no significant impact on the results.

The results of this analysis (Table 15.2) may initially seem surprising. First, it is apparent that expressing water content as a percentage can be very deceptive. Dry mass is a small fraction of total mass in any event and so very large changes in water content (1.5 to 3.5-fold) alter the percentage water in the egg contents relatively little (4 to 17%). A much better index to hydration state is the ratio of dry mass to water mass. Second, lizard eggs span a large range in hydration state that completely includes the range for all bird eggs, with the possible exception of megapodes and kiwis. *S. virgatus* can routinely hatch from eggs that are either relatively drier or relatively wetter than bird eggs, even when variability in rates of water loss from bird eggs has been experimentally increased. The modest water uptake by lizard eggs from moderately dry soils is inadequate to dilute the contents of those eggs to the levels that occur in avian eggs, even though avian eggs lose water throughout incubation. In wet soils, however, small lizard eggs can absorb enough water to achieve hyperhydration relative to an avian egg. It may be that embryos in small, parchment-shelled reptile eggs can overcome a greater range of water stress than is normally faced by embryos of any other terrestrial tetrapods.

Osmotic and solute regulation

Variation in egg water content will impose osmotic and solute stress on embryos whenever water exchange between egg and environment is not matched by solute exchange. Available evidence suggests that such a match is extremely unlikely. Consequently, embryos must either tolerate variability in their solute environment, or compensate for variability in water content by adjusting size and/or distribution of solute pools.

For both bird and reptile eggs, most exchange of mass with the environment takes place via gas phase diffusion (Needham, 1931; Ackerman *et al.*, 1985*a*,*b*; Packard & Packard, 1988). Few osmotically or metabolically important solutes have a significant vapour pressure so exchange of inorganic solutes by this route is negligible. One possible exception is ammonia (Clark, 1953), which is produced in modest quantities in both avian and reptilian eggs (Clark, 1953; Clark & Fischer, 1957; Clark, Sisken & Shannon, 1957; Romanoff, 1967; Packard & Packard, 1984*c*, 1986*b*). Organic molecules, of course, can be oxidised to carbon dioxide and water. The carbon dioxide can leave the egg via diffusion across the shell, whereas the metabolic water is added to the pool of water within the egg. Aerobic metabolism, therefore, results in a net loss of solutes and a net gain of water.

Liquid phase exchange between egg and environment can also occur, occasionally in avian eggs, and more frequently in reptilian eggs (Thompson, 1987). Such exchange normally occurs via diffusion rather than bulk flow, because of the resistance imposed by the tiny cross-section of passages through the shell. If avian eggs are left in contact with liquid water, liquid water and aqueous phase solutes may cross the shell (Sotherland *et al.*, 1984), though this rarely happens in nature. Grebe (*Podiceps nigricollis*) eggs that are often in contact with liquid water during incubation are apparently protected against liquid phase exchanges by the structure of their shell and cuticle (Sotherland *et al.*, 1984). Megapode eggs that are incubated buried in soil or organic detritus are potentially in contact with liquid in pore spaces in their incubation medium, but we have no data that suggest liquid phase exchange is important (Booth & Thompson, Chapter 20).

In the present context, the mechanism by which water exchange between eggs and their environment occurs is important only because it may affect solute movement. Some models of such exchange include movement of liquid water (Tracy, Packard & Packard, 1978; Tracy, 1982; Packard & Packard, 1988), but others do not. Models relying exclusively on vapour phase movement of water can account very well for changes in egg mass observed during incubation in several species of reptiles, and suggest that liquid phase movement of water may be unimportant (Ackerman *et al.*, 1985*b*). The match between their predictions and empirical measurements is dependent, however, on *post hoc* estimation of a value for water vapour pressure in the incubation environment. Tracer studies demonstrate that aqueous phase solutes can cross the shells of turtle eggs (Thompson, 1987), but no evidence is available that documents significant exchange of solute molecules between reptile or bird eggs and their environment during natural incubation by avenues other than gaseous diffusion. It is worth noting, however, that for reptile eggs, there are also no data (i.e. careful and complete mineral ion budgets for eggs at the beginning and end of incubation) that permit unqualified rejection of the hypothesis that solutes, as well as water, cross the eggshell via liquid phase diffusion during post-oviposition incubation.

Solute exchange does not covary on a 1:1 basis with water exchange and so variation in water exchange exposes embryos in avian and reptilian eggs to changes in solute concentrations. The extent of possible changes in water content are so large (Table 15.2) that no adult vertebrate could tolerate them. The ability of embryos to hatch successfully over such a range of conditions indicates either exceptional tolerance to variation in tissue hydration, or the ability to regulate their hydration by adjusting the distribution of water and solutes within the shell. In general, we know much more about water economy of eggs and its variation than about how the developing embryo within the egg deals with that variation. In both birds and reptiles, however, it is apparent that the most important part of the solution involves regulation of the distribution of water and solutes inside the egg.

The contents of an egg are far from homogeneous and their compartmentalisation increases as development proceeds. Egg yolk is rich in lipids and consequently has a relatively low water content, 43 to 66% in birds (Ricklefs, 1977). Yolk contains many of the nutrients and most of the energy required for development. The layer of albumen surrounding the yolk contains variable amounts of protein, but is 85% to 90% water in bird eggs, and perhaps up to 98% water in reptile eggs (Tracy & Snell, 1985). Embryos begin to modify the distribution of water very early in development. In bird eggs, as the germinal disc expands, it translocates water from the albumen to form a layer of sub-embryonic fluid between the developing embryo and the yolk (New, 1956). The volume of the sub-embryonic fluid drops as development proceeds, and the amniotic membrane expands to enclose the embryo in the fluid filled amniotic space. The allantois arises as an outpocketing of the posterior digestive tract, which receives fluid from the developing kidneys via the ureters. Deeming (1989a) provides data illustrating temporal changes in fluid volume in each of these compartments in fowl eggs. Embryology of reptiles is qualitatively similar (Gans, Billett & Maderson, 1985; Gans & Billett, 1985), although in reptiles, the allantoic fluid can be relatively gelatinous, appearing similar to the albumen of bird eggs (Moffat, 1985).

Active exchange of water and solutes occurs between all of the fluid compartments within the egg. Measurements of sodium ion concentrations in albumen and sub-embryonic fluid suggest formation of that fluid results from sodium transport across the yolk sac membrane from the albumen to the sub-embryonic space, followed by water movement in the same direction down the resulting osmotic gradient. The sub-embryonic fluid has higher sodium concentrations and lower potassum concentrations than yolk or albumen, and plasma has higher sodium and lower potassium concentrations than does sub-embryonic fluid (Howard, 1957). The existence early in incubation of a sodium-dependent potential difference across the yolk sac membrane, with the albumen side negative, supports Howard's hypothesis (Deeming, Rowlett & Simkiss, 1987).

The allantoic membrane actively transports sodium against both electrical and concentration gradients to produce resorption of water from the allantois late in incubation (Stewart & Terepka, 1969). Simkiss (1980b) reports tracer studies demonstrating substantial net flux of water from albumen to amniotic fluid in chicken embryos, and collected data from other sources which demonstrate that concentration of mineral ions differs significantly between compartments. The allantois, because it can contain a large volume of fluid and is normally hyposmotic compared to the embryos (Stewart & Terepka, 1969) is an obvious candidate for an osmoregulatory organ, and it has proved to be exactly that.

Hoyt (1979) has investigated the effect of variation in water loss from eggs of domestic fowl, and shows that embryos osmoregulate by adjusting water and solute exchange with the allantois. He varied the water loss experienced by fowl eggs by incubating them at different humidities, and sampled eggs after 17 days of incubation, when water loss ranged from 4% to 16% of initial mass (Fig. 15.3). Wet and dry mass of yolk and amniotic fluid are unaffected by the treatment (Fig. 15.3). Embryo dry mass increases with water loss, but embryo wet mass and the percentage of embryonic tissue that was water decreases (Hoyt, 1979). The absolute amount of water contained in embryo tissue decreases slightly (less than 5%) as water loss increases, but the amount of water contained in the allantois decreases by more than 20% in the most dehydrated eggs (Fig. 15.3). This extent of water reabsorption can occur only because sodium is actively transported out of the allantois, and stored, at least in part, in the amniotic

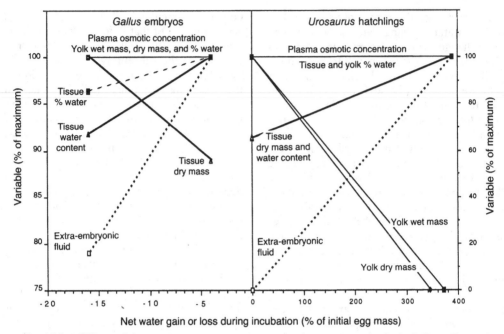

Fig. 15.3. Effects of changes in egg water economy on embryo traits in the fowl, *Gallus gallus*, and hatchling traits in the lizard, *Urosaurus ornatus*. Each value is expressed as a percentage of the maximum value measured for that trait, and water status of the egg is expressed as the % of gain or loss in egg mass since laying. (Data on *Gallus* are calculated from Hoyt, 1979.)

fluid. Although sodium concentration in the plasma increase slightly with water loss, plasma osmotic concentration do not change (Hoyt, 1979).

Davis, Shen & Ackerman (1988) have extended Hoyt's (1979) results by studying a greater range of both water losses and developmental stages in domestic fowl eggs. Embryos in eggs that experience high water loss during incubation have elevated plasma osmolality and electrolyte concentrations, but only after day 17–19 of incubation. Prior to that, plasma osmolality and plasma volume are apparently regulated by reabsorption of water from the allantois, with regulation failing only after allantoic fluid is depleted. They also measured sodium content of different fluid compartments within the egg and confirm the observation that sodium is removed from the allantois. They can not, however, account for all of the sodium apparently removed from the allantois, and suggest that coprecipitation of sodium and urate may remove some sodium from solution in eggs that experience high water loss.

Transport and hormonal studies have supported the hypothesis that solute-linked water movement across the allantoic is the principal method of water regulation within the egg. Electrical potential difference and current across the allantois after day 14 of incubation are due almost completely to sodium reabsorption (Graves, Dunn & Brown, 1986), although the allantoic membrane also reabsorbs calcium from the allantoic fluid (Graves, Helms & Martin, 1984). Allantoic fluid balance appears to be regulated by the same systems that function in adults. Hypophysectomy of developing embryos disrupts fluid balance and results in decreased water reabsorption from the allantois (Murphy, Brown & Brown, 1982). Prolactin stimulates kidney sodium transport and arginine vasopressin (an antidiuretic hormone) reduces sodium concentration and facilitates water reabsorption from the allantois of hypophysectomised embryos (Doneen & Smith, 1982). Removal of sub-embryonic fluid from a domestic fowl egg after 10 days of incubation results in a significant deficit in allantoic, but not amniotic, fluid volume (Deeming, 1989a). The allantois appears to be the fluid compartment most sensitive to changes in fluid volume within the egg.

Regardless of the water status of the egg, most of the allantoic fluid is normally reabsorbed by the avian embryo prior to hatching (Simkiss,

1980*b*) and hatchling size depends on the water economy of the egg (Tullett & Burton, 1982). High water loss produces smaller hatchlings, but the difference in hatchling size is due to a difference in water content. Dry mass of hatchling tissue does not vary with water loss from the egg (Simkiss, 1980*a*; Snyder & Birchard, 1982; Tullett & Burton, 1982; Burton & Tullett, 1985; Davis *et al.*, 1988), so tissue growth is apparently unaffected, though tissue hydration is. In the fowl, the fraction of hatchling tissue made up by water decreases from about 84% at the lowest shell G_{H_2O} and water loss, to about 79% water at the highest (Burton & Tullett, 1985). That 5% difference corresponds to a change in tissue hydration, defined as dry mass–water content, from 0.19 to 0.27. An equivalent change in the water content of a salt solution would inrease the concentration of the solution by about 40%. We do not, as yet, know whether this difference in hydration reflects changes in cell volume, in the size of the extracellular water pool, or both. Davis & Ackerman (1987) compare a variety of tissues of hatchlings from high and low water loss eggs, and conclude the difference in hatchling mass between high water loss and low water loss eggs results primarily from a lower relative water content in leg muscle of hatchlings emerging from drier eggs. In a subsequent study, they conclude that most of the difference in mass between hatchlings from high and low water loss eggs reflects differences in skin hydration (Davis *et al.*, 1988). Unlike hatchling tissue, the relative water content of residual yolk does not vary with water loss during incubation, perhaps because a certain percentage of water is necessary for retraction of the yolk sac into the body prior to hatching (Tullett & Burton, 1982).

Bird embryos, then, deal with changes in water economy by differentially adjusting the volume and composition of compartments within the egg. Water loss is restricted primarily to the allantois, and water content of other compartments, especially the yolk, is defended. Embryos osmoregulate, but tissue hydration is not maintained constant. Tissue growth, as indexed by dry mass, is maintained independent of water loss from the egg, except at very low water loss, where metabolism is probably limited by respiratory gas exchange, not water regulation (Tullett & Deeming, 1982; Burton & Tullett, 1985).

Reptile embryos also regulate the distribution

of water and solutes inside the egg to hatch successfully over a wide range of egg water contents, but the outcome of that regulation differs in some significant ways from the fowl. In most of the species which have been studied, the size of the hatchling varies depending on the water budget of the egg from which it hatched, but hatchling tissue water content is independent of water status of the egg (Packard & Packard, 1984*b*; 1988). The water uptake (lizard eggs gain, rather than lose, water during incubation) by eggs of the lizards *S. virgatus* and *U. ornatus* incubated over a range of soil water potentials has been measured. The osmotic status, hydration state, and body size of the hatchlings from those eggs have then been examined (Fig. 15.2). In both species, plasma osmotic concentration and tissue water content (or hydration) are independent of water uptake during incubation, even though egg water content prior to hatching varies over a 4-fold range. Embryos achieve constant tissue hydration in spite of widely differing water availability by converting less yolk into tissue when water is short. Energy metabolism during incubation (measured as oxygen consumption; Table 15.2) and body size at hatching (measured as wet mass, dry mass, or linear dimensions) increases as water uptake increased. Residual yolk, on the other hand, decreases as water uptake increases. The largest hatchlings from the wettest soils occasionally have no residual yolk at the time of hatching. Tiny hatchlings from dry soils sometimes do not incorporate any residual yolk in their body, because they fail to reabsorb the relatively large amount of residual yolk, leaving it behind in the shell on hatching. Yolk hydration was more variable than hatchling hydration, but does not change systematically with water uptake.

The similarities and differences in the responses of bird and lizard embryos to variation in water availability within the egg are summarised in Fig. 15.3. Embryos in both groups osmoregulate. Water is differentially partitioned between compartments, so that as water availability changes, yolk and tissue hydration are maintained at the expense of other extraembryonic compartments. Differences between the groups are equally apparent. Lizard embryos sacrifice tissue growth when water is short, while bird embryos alter tissue hydration, and may actually have higher dry masses when water is short. As discussed above, however, respiratory gas exchange may limit growth by bird embryos at shell conductances low enough to result in very low water losses. Although the amount of residual yolk varies with water availability in lizards, its composition, as in birds, does not. Compared with birds, lizards can develop over a much wider range of hydration of the egg.

The data summarised here answer some questions, but raise a host of others. We know very little about water and solute regulation in avian embryos prior to hatching, and virtually nothing about the topic in reptile embryos. At present, we must base generalisations about whole orders of vertebrates on data from only a handful of species. We can identify patterns of regulation, but still do not know which steps in these regulatory systems are limiting, leading to embryonic death when water loss is high. Future investigation could profitably focus on providing comparative and systems level data that may enable us to address these questions.

Acknowledgements

Support for the writing of this chapter and some of the research described therein was provided in part by NSF grant BSR 8202880.

References

Ackerman, R. A., Dmi'el, R. & Ar, A. (1985a). Energy and water vapor exchange by parchment-shelled reptile eggs. *Physiological Zoology*, **58**, 129–37.

Ackerman, R. A., Seagrave, R. C., Dmi'el, R. & Ar, A. (1985b). Water and heat exchange between parchment-shelled reptile eggs and their surroundings. *Copeia*, **1985**, 703–11.

Ar, A. & Rahn, H. (1980). Water in the avian egg; Overall budget of incubation. *American Zoologist*, **20**, 373–84.

Ar, A. & Yom-Tov, Y. (1978). The evolution of parental care in birds. *Evolution*, **32**, 655–69.

Arad, Z. & Marder, J. (1982). Egg-shell water vapour conductance of the domestic fowl: comparison between two breeds and their crosses. *British Poultry Science*, **23**, 325–8.

Ballinger, R. E. & Clark, D. R., Jr (1973). Energy content of lizard eggs and the measurement of reproductive effort. *Journal of Herpetology*, **7**, 129–32.

Bellairs, A. (1969). *The Life of Reptiles*, vol. II. London: Weidenfeld and Nicolson.

Birchard, G. F. & Kilgore, D. L., Jr (1980). Conductance of water vapor in eggs of burrowing and non-burrowing birds: Implications for embryonic gas exchange. *Physiological Zoology*, **53**, 284–92.

Board, R. G. & Fuller, R. (1974). Non-specific antimicrobial defenses of the avian egg, embryo, and neonate. *Biological Review*, **49**, 15–49.

Bucher, T. L. & Barnhart, M. C. (1984). Varied egg gas conductance, air cell gas tensions and development in *Agapornis roseicollis*. *Respiration Physiology*, **55**, 277–89.

Burton, F. G. & Tullett, S. G. (1985). The effects of egg weight and shell porosity on the growth and water balance of the chicken embryo. *Comparative Biochemistry and Physiology*, **81A**, 377–85.

Bustard, H. R. (1971). Temperature and water tolerances of incubating crocodile eggs. *British Journal of Herpetology*, **4**, 121–3.

Calder, W. A., Parr, C. R. & Karl, D. P. (1978). Energy content of eggs of the brown kiwi *Apteryx australis*: an extreme in avian evolution. *Comparative Biochemistry and Physiology*, **60A**, 177–9.

Carey, C. (1986). Tolerance of variation in eggshell conductance, water loss, and water content by redwinged blackbird embryos. *Physiologial Zoology*, **59**, 109–22.

Carey, C., Garber, S. D., Thompson, E. L. & James, F. C. (1983). Avian reproduction over an altitudinal gradient. II. Physical characteristics and water loss of eggs. *Physiological Zoology*, **56**, 340–52.

Carey, C., Rahn, H. & Parisi, P. (1980). Calories, water, lipid and yolk in avian eggs. *Condor*, **82**, 335–43.

Carey, C., Thompson, E. L., Vleck, C. M. & James, F. C. (1982). Avian reproduction over an altitudinal gradient: Incubation period, hatchling mass, and embryonic oxygen consumption. *Auk*, **99**, 710–18.

Clark, H. (1953). Metabolism of the black snake embryo I. Nitrogen excretion. *Journal of Experimental Biology*, **30**, 492–501.

Clark, H. & Fischer, D. (1957). A reconsideration of nitrogen excretion by the chick embryo. *Journal of Experimental Zoology*, **136**, 1–15.

Clark, H., Sisken, B. & Shannon, J. E. (1957). Excretion of nitrogen by the alligator embryo. *Journal of Cell and Comparative Physiology*, **50**, 129–34.

Davis, T. A. & Ackerman, R. A. (1987). Effects of increased water loss on growth and water content of the chick embryo. *Journal of Experimental Zoology*, Supplement 1, 357–64.

Davis, T. A., Shen, S. S. & Ackerman, R. A. (1988). Embryonic osmoregulation: Consequences of high and low water loss during incubation of the chicken egg. *Journal of Experimental Zoology*, **245**, 144–56.

Deeming, D. C. (1989a). Characteristics of unturned eggs: Critical period, retarded embryonic growth, and poor albumen utilisation. *British Poultry Science*, **30**, 239–49.

— (1989b). An examination of the residues present in

the eggs of squamate reptiles at hatching. *Herpetological Journal*, 1, 381–5.

Deeming, D. C., Rowlett, K. & Simkiss, K. (1987). Physical influences on embryo development. *Journal of Experimental Zoology*, Supplement 1, 341–5.

Doneen, B. A. & Smith, T. E. (1982). Ontogeny of endocrine control of osmoregulation in chick embryo II. Actions of prolactin, arginine vasopressin, and aldosterone. *General and Comparative Endocrinology*, 48, 310–18.

Dunson, W. A. (1982). Low water vapour conductance of hard-shelled eggs of the gecko lizards *Hemidactylus* and *Lepidodactylus*. *Journal of Experimental Zoology*, 219, 377–9.

Ferguson, M. W. J. (1985). Reproductive biology and embryology of the crocodiles. In *Biology of the Reptilia*, vol. 14, Development A, eds C. Gans, F. Billett & P. A. Maderson, pp. 328–491. New York: John Wiley & Sons.

Gans, C., F. Billett & P. A. Maderson (1985) (eds). *Biology of the Reptilia*, vol. 14, Development A. New York: John Wiley & Sons.

Gans, C. & F. Billett (1985) (eds). *Biology of the Reptilia*, vol. 15, Development B. New York: John Wiley & Sons.

Gettinger, R. D., Paukstis, G. L. & Gutzke, W. H. N. (1984) Influence of the hydric environment on oxygen consumption by embryonic turtles *Chelydra serpentina* and *Trionyx spiniferus*. *Physiological Zoology*, 57, 468–73.

Grant, G. S., Pettit, T.N., Rahn, H., Whittow, G. C. & Paganelli, C. V. (1982). Water loss from Laysan and Black-footed Albatross eggs. *Physiological Zoology*, 55, 405–14.

Graves, J. S., Dunn, B. E. & Brown, S. C. (1986). Embryonic chick allantois: functional isolation and development of sodium transport. *American Journal of Physiology*, 251, C787–94.

Graves, J. S., Helms, E. L. K., III & Martin, H. F., III (1984). Development of calcium reabsorption by the allantoic epithelium in chick embryos grown in shell-less culture. *Developmental Biology*, 101, 522–6.

Gutzke, W. H. N. & Packard, G. C. (1987). Influence of the hydric and thermal environments on eggs and hatchlings of bull snakes *Pituophis melanoleucus*. *Physological Zoology*, 60, 9–17.

Howell, T. R. (1979). Breeding biology of the Egyptian plover, *Pluvianus aegyptius*. *University of California, Publications in Zoology*, 113, 1–76.

Hoyt, D. F. (1979). Osmoregulation by avian embryos: the allantois functions like a toad's bladder. *Physiological Zoology*, 52, 354–62.

Johnston, P. M. & Comar, C. L. (1955). Distribution and contributions of calcium from the albumen, yolk, and shell in developing chick embryo. *American Journal of Physiology*, 183, 365–70.

Leon-Velarde, F., Whittembury, J., Carey, C. &

Monge, C. (1984). Permeability of eggshells of native chickens in the Peruvian Andes. In *Respiration and Metabolism of Embryonic Vertebrates*, ed. R. S. Seymour, pp. 245–57. Dordrecht: Dr W. Junk.

Leshem, A. & Dmi'el, R. (1986). Water loss from *Trionyx triunguis* eggs incubating in natural nests. *Herpetological Journal*, 1, 115–17.

Lomholt, J. P. (1976). Relationship of weight loss to ambient humidity of birds eggs during incubation. *Journal of Comparative Physiology*, B105, 189–96.

Lundy, H. (1969). A review of the effects of temperature, humidity, turning and gaseous environment in the incubator on the hatchability of the hen's egg. In *The Fertility and Hatchability of the Hen's Egg*, eds T. C. Carter & B. M. Freeman, pp. 143–76. Edinburgh: Oliver & Boyd.

Lutz, P. L., Bentley, T. B., Harrison, K. E. & Marszalek, D. S. (1980). Oxygen and water vapour conductance in the shell and shell membrane of the American crocodile egg. *Comparative Biochemistry and Physiology*, 66A, 335–8.

Lutz, P. L. & Dunbar-Cooper, A. (1984). The nest environment of the American crocodile (*Crocodylus acutus*), *Copeia*, 1984, 153–61.

Moffat, L. A. (1985). Embryonic development and aspects of reproductive biology in the tuatara, *Sphenodon punctatus*. In *Biology of the Reptilia*, vol. 14, Development A, eds. C. Gans, F. Billett & P. A. Maderson, pp. 493–521. New York: John Wiley & Sons.

Morris, K. A., Packard, G. C., Boardman, T. J., Paukstis, G. L. & Packard, M. J. (1983). Effect of the hydric environment on growth of embryonic snapping turtles (*Chelydra serpentina*). *Herpetologica*, 39, 272–85.

Murphy, M. J., Brown, S. C. & Brown, P. S. (1982). Hydromineral balance in the chick embryo: effects of hypophysectomy. *Journal of Experimental Zoology*, 220, 321–30.

Needham, J. (1931). *Chemical Embryology*. Cambridge: Cambridge University Press.

New, D. A. T. (1956). The formation of the sub-blastodermic fluid in the hen's egg. *Journal of Embryology and Experimental Morphology*, 32, 365–74.

Packard, G. C. & Packard, M. J. (1980). Evolution of the cleidoic egg among reptilian antecedents of birds. *American Zoologist*, 20, 351–62.

— (1984c). Nitrogen excretion by embryos of an altricial bird, the redwing blackbird (*Agelaius phoeniceus*). *Journal of Experimental Zoology*, 231, 363–6.

— (1984b). Coupling of physiology of embryonic turtles to the hydric environment. In *Respiration and Metabolism of Embryonic Vertebrates*, ed. R. S. Seymour, pp. 99–119. Dordrecht: Dr W. Junk.

— (1986b). Nitrogen excretion by embryos of a gallinaceous bird and a reconsideration of the evolutionary origin of uricotely. *Canadian Journal of Zoology*, 64, 691–3

— (1987). Water relations and nitrogen excretion in embryos of the oviparous snake *Coluber constrictor*. *Copeia*, **1987**, 395–406.

— (1988). The physiological ecology of reptilian eggs and embryos. In *Biology of the Reptilia*, vol. 16, eds C. Gans & R. B. Huey, pp. 523–605. New York: Liss.

Packard, G. C., Packard, M. J. & Boardman, T. J. (1981). Patterns and possible significance of water exchange by flexible-shelled eggs of painted turtles (*Chrysemys picta*). *Physiological Zoology*, **54**, 165–78.

— (1982*b*). An experimental analysis of the water relations of eggs of Blanding's turtles (*Emydoidea blandingii*). *Zoological Journal of the Linnean Society*, **75**, 23–34.

Packard, G. C., Packard, M. J., Boardman, T. J., Morris, K. A. & Shuman, R. D. (1983). Influence of water exchanges by flexible-shelled eggs of painted turtles *Chrysemys picta* on metabolism and growth of embryos. *Physiological Zoology*, **56**, 217–30.

Packard, G. C., Packard, M. J. & Gutzke, W. H. N. (1985*a*). Influence of hydration of the environment on eggs and embryos of the terrestrial turtle *Terrapene ornata*. *Physiological Zoology*, **58**, 564–75.

Packard, G. C., Sotherland, P. R. & Packard, M. J. (1977). Adaptive reduction in permeability of avian eggshells to water vapour at high altitudes. *Nature, London*, **266**, 255–6.

Packard, G. C., Taigen, T. L., Packard, M. J. & Boardman, T. J. (1980*a*). Water relations of pliable-shelled eggs of common snapping turtles (*Chelydra serpentina*). *Canadian Journal of Zoology*, **58**, 1404–11.

— (1981). Changes in mass of eggs of softshell turtles (*Trionyx spiniferus*) incubated under hydric conditions simulating those of natural nests. *Journal of Zoology, London*, **193**, 81–90.

Packard, G. C., Taigen, T. L., Packard, M. J. & Shuman, R. D. (1979). Water-vapor conductance of testudinian and crocodilian eggs (Class *Reptilia*). *Respiration Physiology*, **38**, 1–10.

Packard, M. J. & Packard, G. C. (1984*a*). Comparative aspects of calcium metabolism in embryonic reptiles and birds. In *Respiration and Metabolism of Embryonic Vertebrates*, ed. R. S. Seymour, pp. 155–79. Dordrecht: Dr W. Junk.

— (1986*a*). Effect of water balance on growth and calcium mobilization of embryonic painted turtles (*Chrysemys picta*). *Physiological Zoology*, **59**, 398–405.

Packard, M. J., Packard, G. C. & Boardman, T. J. (1980*b*). Water balance of the eggs of a desert lizard (*Callisaurus draconoides*). *Canadian Journal of Zoology*, **58**, 2051–8.

— (1982*a*). Structure of eggshells and water relations of reptilian eggs. *Herpetologica*, **38**, 136–55.

Packard, M. J., Packard, G. C., Miller, J. D., Jones,

M. E. & Gutzke, W. H. N. (1985*b*). Calcium mobilization, water balance, and growth in embryos of the agamid lizard *Amphibolorus barbatus*. *Journal of Experimental Zoology*, **235**, 349–57.

Plummer, M. V. & Snell, H. L. (1988). Nest site selection and water relations of eggs in the snake, *Opheodrys aestivus*. *Copeia*, **1988**, 58–64.

Rahn, H. (1984). Factors controlling the rate of incubation water loss in bird eggs. In *Respiration and Metabolism of Embryonic Vertebrates*, ed. R. S. Seymour, pp. 271–88. Dordrecht: Dr W. Junk.

Rahn, H., Carey, C., Balmas, K., Bhatia, B. & Paganelli, C. V. (1977). Reduction of pore area of the avian eggshell as an adaptation to altitude. *Proceedings of the National Academy of Sciences*, **74**, 3095–8.

Ratterman, R. J. & Ackerman, R. A. (1989). The water exchange and hydric microclimate of painted turtle (*Chrysemys picta*) eggs incubating in field nests. *Physiological Zoology*, **62**, 1059–79.

Ricklefs, R. E. (1977). Composition of eggs of several bird species. *Auk*, **94**, 350–6.

Ricklefs, R. E. & Burger, J. (1977). Composition of eggs of the diamondback terrapin. *American Midland Naturalist*, **97**, 232–5.

Romanoff, A. L. (1967). *Biochemistry of the Avian Embryos*. New York: Wiley & Sons.

Romanoff, A. L. & Romanoff, A. J. (1949). *The Avian Egg*. New York: Wiley & Sons.

Schmidt-Nielsen, K. (1983). *Animal Physiology: Adaptation and Environment*, 3rd edn. New York: Cambridge University Press.

Seymour, R. S., Vleck, D., Vleck, C. M. & Booth, D. T. (1987). Water relations of buried eggs of mound building birds. *Journal of Comparative Physiology*, **B157**, 413–22.

Shine, R. (1988). Parental care in reptiles. In *Biology of the Reptilia*, vol. 16, eds C. Gans & R. B. Huey, pp. 275–329. New York: Alan R. Liss, Inc.

Simkiss, K. (1980*a*). Eggshell porosity and the water metabolism of the chick embryo. *Journal of Zoology, London*, **192**, 1–8.

— (1980*b*). Water and ionic fluxes inside the egg. *American Zoologist*, **20**, 385–93.

Snell, H. L. & Tracy, C. R. (1985). Behavioral and morphological adaptations by Galapagos land iguanas (*Conolophus subcristatus*) to water and energy requirements of eggs and neonates. *American Zoologist*, **25**, 1009–18.

Snyder, G. K. & Birchard, G. F. (1982). Water loss and survival in embryos of the domestic chicken. *Journal of Experimental Zoology*, **219**, 115–17.

Somma, L. A. & Fawcett, J. D. (1989). Brooding behaviour of the prairie skink, *Eumeces septentrionalis*, and its relationship to the hydric environment of the nest. *Zoological Journal of the Linnean Society*, **95**, 245–56.

Sotherland, P. R., Ashen, M. D., Shuman, R. D. & Tracy, C. R. (1984). The water balance of bird eggs incubated in water. *Physiological Zoology*, **57**, 338–48.

Stamps, J. A. (1976). Egg retention, rainfall, and egg laying in a tropical lizard *Anolis aeneus*. *Copeia*, **1976**, 759–64.

Stewart, M. E. & Terepka, A. R. (1969). Transport functions of the chick chorio-allantoic membrane. *Experimental Cell Research*, **58**, 93–106.

Thompson, M. B. (1987). Water exchange in reptilian eggs. *Physiological Zoology*, **60**, 1–8.

Tracy, C. R. (1980). Water relations of parchment-shelled lizard (*Sceloporus undulatus*) eggs. *Copeia*, **1980**, 478–82.

— (1982). Biophysical modeling in reptilian physiology and ecology. In *Biology of the Reptilia*, vol. 12, eds C. Gans & F. H. Pough, pp. 275–321. New York: Academic Press.

Tracy, C. R., Packard, G. C. & Packard, M. J. (1978). Water relations of chelonian eggs. *Physiological Zoology*, **51**, 378–87.

Tracy, C. R. & Snell, H. L. (1985). Interrelations among water and energy relations of reptilian eggs, embryos, and hatchlings. *American Zoologist*, **25**, 999–1008.

Tullett, S. G. & Burton, F. G. (1982). Factors affecting the weight and water status of the chick at hatch. *British Poultry Science*, **23**, 361–9.

Tullett, S. G. & Deeming, D. C. (1982). The relationship between eggshell porosity and oxygen consumption of the embryo in the domestic fowl. *Comparative Biochemistry and Physiology*, **72A**, 529–33.

Tyler, C. (1969). The snapping strength of eggshells of various orders of birds. *Journal of Zoology, London*, **159**, 65–77.

Vitt, L. J. (1978). Caloric content of lizard and snake (*Reptilia*) eggs and bodies and the conversion of weight to caloric data. *Journal of Herpetology*, **12**, 65–72.

Vleck, C. M., Hoyt, D. F. & Vleck, D. (1979). Metabolism of avian embryos: patterns in altricial and precocial birds. *Physiological Zoology*, **52**, 363–77.

Vleck, C. M. & Vleck, D. (1987). Metabolism and energetics of avian embryos. *Journal of Experimental Zoology*, Supplement 1, 111–25.

Vleck, C. M., Vleck, D., Rahn, H. & Paganelli, C. V. (1983). Nest microclimate, water-vapor conductance, and water loss in heron and tern eggs. *Auk*, **100**, 76–83.

Vleck, D., Vleck, C. M. & Seymour, R. S. (1984). Energetics of embryonic development in the megapode birds, mallee fowl *Leipoa ocellata* and brush turkey *Alectura lathami*. *Physiological Zoology*, **57**, 444–56.

Walsberg, G. E. (1983). A test for regulation of nest humidity in two bird species. *Physiological Zoology*, **56**, 231–5.

— (1985). A test for regulation of egg dehydration by control of shell conductance in mourning doves. *Physiological Zoology*, **58**, 473–7.

Wangensteen, O. D., Rahn, H., Burton, R. R. & Smith, A. H. (1974). Respiratory gas exchange of high altitude adapted chick embryos. *Respiration Physiology*, **21**, 61–70.

Wilhoft, D. C. (1986). Eggs and hatchling components of the snapping turtle (*Chelydra serpentina*). *Comparative Biochemistry and Physiology*, **84A**, 483–6.

York, D. S. & Burghardt, G. M. (1988). Brooding in the Malayan pit viper, *Calloselasma rhodostoma*: Temperature, relative humidity, and defensive behaviour. *Herpetological Journal*, **1**, 210–14.

16

The avian eggshell as a mediating barrier: respiratory gas fluxes and pressures during development

CHARLES V. PAGANELLI

Introduction

The eggshell, underlying shell membranes and chorio-allantois serve the avian embryo as a gas exchange organ, and the air spaces contained within the shell membranes are functionally analogous to the alveolar spaces of mammalian lungs (Rahn & Paganelli, 1985). However, gas exchange through the pores in the shell occurs primarily by diffusion, while alveolar ventilation depends in large part on convective gas flow. This chapter examines the applicable laws of diffusion which govern gas exchange in the avian egg and the consequences of these laws for the developing embryo. Recent additions to our knowledge of the gas phase pathway for oxygen (O_2), carbon dioxide (CO_2) and water vapour exchange are reviewed together with the fluxes of these gases in this pathway and the partial pressures in the air space which drive the fluxes. The regional differences in gas tensions in the air space and convective gas flow induced by non-equimolar diffusion are also considered. Cardiogenic pressure pulses which are transmitted through the pores are also described.

The laws of diffusion in the gas phase

The seemingly impervious shells of bird eggs have been known to contain pores for more than 140 years, according to reports cited by Romanoff (1943). Through the pores and the underlying shell membranes must pass O_2, CO_2 and water vapour as these gases enter or exit the embryonic circulation and tissues during development. The pores in the shell, and the gas-filled spaces between the fibres of the outer and inner shell membranes, provide the pathway

for gas exchange (and hence the resistance) between the chorio-allantoic capillary blood and the ambient atmosphere. This pathway is schematically portrayed for the fowl (*Gallus gallus*) egg in Fig. 16.1 together with a scanning electron micrograph of the shell membranes and capillaries of the chorio-allantois. Oxygen molecules entering the egg from the atmosphere must first traverse the pores in the shell, then the gas-filled interstices of the fibrous shell membranes, and finally the so-called 'film' which coats the fibres of the inner shell membrane and apparently forms a continuous layer on the outward-facing surface of the chorio-allantoic membrane (Wangensteen, Wilson & Rahn, 1970/71; Wangensteen & Weibel, 1982). From this point, O_2 enters the capillaries of the chorio-allantois and is circulated to the embryonic tissues.

It has been appreciated for nearly 100 years that respiratory gas exchange in avian eggs occurs by diffusion through gas-filled pores in the shell (Romanoff & Romanoff, 1949). However, the first measurements of the appropriate diffusing capacity or diffusive conductance of the shell were performed only 20 years ago by Wangensteen *et al.* (1970/71) and Kutchai & Steen (1971). In these investigations, convective gas flow through the shell is avoided and the results are expressed in terms of a *conductance* which explicitly contains the driving partial pressure difference. Early values of egg shell porosity, determined from water loss (Murray, 1925; Pringle & Barott, 1937; Tyler, 1945) or the use of convective gas flow driven by hydrostatic pressure difference (Romijn, 1950; Romanoff, 1943) have previously been compared with diffusive conductance measurements

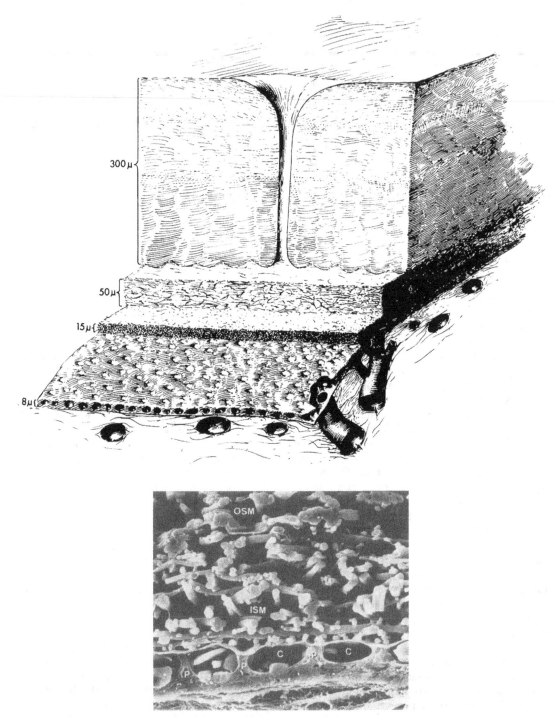

Fig. 16.1. (*Upper*) A sketch of the gas exchange pathway in the fowl egg, drawn to scale. A single pore is shown, with underlying outer shell membrane (OSM, 50 μm), inner shell membrane (ISM, 15 μm) and the chorio-allantoic capillaries (8 μm). (Reprinted from Wangensteen, Wilson & Rahn, 1970/71 with permission.); (*lower*) scanning electron micrograph of OSM and ISM fibres in the fowl egg, showing the chorio-allantoic capillaries (C) and pillars (P) at 3100 × magnification. (Reprinted with permission from Wangensteen & Weibel, 1982.)

(Paganelli, 1980). Many reviews of eggshell porosity and its regulation, the microscopic structure and possible origin of pores, and the respiratory function of pores have appeared in the literature in the last 15 years (Freeman & Vince, 1974; Tullett, 1975, 1984; Board, Tullett & Perrott, 1977; Rahn & Ar, 1980; Board, 1982; Ar & Rahn, 1985; Rahn & Paganelli, 1990; Board & Sparks, Chapter 6). As well as the present volume, the symposium volumes edited by Piiper (1978), Seymour (1984), Metcalfe, Stock & Ingermann (1987) Tullett (1991) and the Symposium on the Physiology of the Avian Egg published in 1980 as volume 20, number 2 of the *American Zoologist*, are of particular interest to students of incubation biology and physiology.

Conductance of the eggshell may be expressed simply in a form analogous to Ohm's law:

$$\dot{M}_x = G_x \Delta P_x \qquad (16.1)$$

where $\dot{M}_x$ is the net flux of gas x (cm³ STPD s⁻¹), G_x is the diffusive conductance or diffusing capacity of the shell to gas x in air (cm³ STPD s⁻¹ torr⁻¹ or other units of quantity time⁻¹ per partial pressure difference), and ΔP_x is the steady-state partial pressure difference of gas x across the pores (torr or other units of pressure or concentration difference). For example, G may be expressed in units of mmol d⁻¹ kPa⁻¹. Multiplying G (cm³ d⁻¹ torr⁻¹) by 0.3349 converts it to G (mmol d⁻¹ kPa⁻¹). Conductance in a diffusive system, as defined above, is directly proportional to the binary diffusion coefficient D_x(cm² s⁻¹) of gas x in air:

$$G_x = kD_x \qquad (16.2)$$

where k is a constant of proportionality.

In order to define G in terms of dimensions, such as pore radius and length, Fick's first law in the steady-state is used:

$$\dot{M}_x = (A_p/L) (D_x/RT) \Delta P_x \qquad (16.3)$$

where R is the gas constant, 2.785 cm³ STPD torr K⁻¹ (cm³ STPD)⁻¹, T is absolute temperature (K), A_p is total effective cross-sectional pore area, (cm²), and L is pore length (cm), usually assumed equal to shell thickness. Thus

$$G_x = (A_p/L) (D_x/RT) \qquad (16.4)$$

and k (equation 16.2) is equal to (A_p/LRT).
The corresponding Fick's law resistance (R_F) is:

$$R_F = L/(A_p D_x) \qquad (16.5)$$

(Note that the RT term has been dropped in equation 16.5. This has the effect of converting resistance units from s torr cm⁻³ to s cm⁻³.)

Use of equations 16.3 and 16.4 implies several simplifying assumptions (Simkiss, 1986; Toien *et al.*, 1987; Sibly & Simkiss, 1987). First, the direction of diffusion is assumed to be perpendicular to the area through which diffusion occurs. Second, end or aperture resistances, caused by boundary layers at the entrance and exit of the pores, are assumed to be negligible. This assumption is equivalent to saying that the partial pressure difference occurs over a distance L, the pore length, and no gradients exist in the gas spaces immediately adjacent to the ends of the pores. Third, convective gas flow through the pores is assumed to play an insignificant role in gas flux. These first two assumptions will be discussed immediately below and the third will be dealt with in a later section.

Although pores are generally cone-, or trumpet-shaped, the error introduced by assuming lines of flux which are perpendicular to the plane of the shell is quite small: less than 1% for pores in the eggshell of the fowl (Toien *et al.*, 1987). Mathematical expressions given by Toien *et al.* (1987) and Sibley & Simkiss (1987) may be used to calculate resistance in conical pores whose sides are sharply tapered.

Aperture or boundary layer resistances in porous media are well-known phenomena in plant physiology; the treatment presented here is derived largely from a monograph on stomatal physiology (Meidner & Mansfield, 1968). Diffusive gas flow *in still air* away from an aperture follows Stefan's law and is proportional to the *diameter* of the aperture, rather than its cross-sectional area. Stefan's law resistance (R_S) to diffusion in the boundary layer in units of s cm⁻³ is:

$$R_S = 1/(4rD_x) \qquad (16.6)$$

where r is the pore radius. The total, diffusive resistance (R_{tot}) through N pores in parallel is then the sum of Fick's law resistance (that offered by the pores themselves) plus the aperture or boundary layer resistances from Stefan's law at the two ends:

$$R_{tot} = L/(A_p D_x) + 2/(4rD_x N) \qquad (16.7)$$

(Here, r is assumed to be the same at both ends and the pore is uniform in radius; thus $N\pi r^2 = A_p$).

From equation 16.7, R_{tot} will be predominantly in the pores themselves when $L/A_p \gg 1/2rN$ or $L/\pi r^2 \gg 1/2r$. This occurs typically when pores are 20 to 30 times longer than their radius (Rahn, Paganelli & Ar, 1987). For example, in a pore of length 0.3 mm and radius circa 0.01 mm, the respective terms in equation 16.7 are 955 and 50; aperture resistance is 5% of the total.

Experimental data which illustrate beautifully the relative contributions of R_F and R_S to egg water loss have been obtained by Simkiss (1986), who has measured water vapour conductance (G_{H_2O}) in intact fowl eggs, and again in the same eggs in which a hole of 1.44 mm² average cross-sectional area is drilled. The effective pore area of the control eggs is 1.96 mm² as calculated from Ar & Rahn (1985). Thus, the control G_{H_2O} should have increased by a factor of 1.73 [(1.96 + 1.44)/1.96] on the basis of equation 16.3. However, G_{H_2O} increased only 16% on average. The relevant resistances in the control and drilled eggs in units of s cm⁻³ are shown in Fig. 16.2. The disc in the centre represents the effective pore area in the control eggs, divided among 10⁴ pores. The Fick's law resistance is circa 30 times the aperture resistance and clearly will be the major factor controlling diffusive water loss. If the number of the pores of the same size ($r = 7.9$ μm), and hence A_p, is doubled R_{tot} is reduced from 6.7 to 3.4 (Fig. 16.2). On the other hand, R_F for a single hole of area 1.44 mm² ($r = 0.68$ mm), placed in parallel with the existing pores, is only 23% of R_{tot}. The aperture resistance now becomes quantitatively the more important of the two. R_{tot} in this instance decreases only from 6.7 to 5.7 or 15% (Fig. 16.2). The calculated decrease in resistance is in very good agreement with the measured increase in conductance shown in Fig. 16.2.

From their measurements on pore casts, pictured in the publications of Tyler (1962, 1964, 1965) and Tyler & Simkiss (1959), Toien et al. (1988) have calculated the resistances offered by 321 individual pores from 23 species. The average total aperture resistance is less than 6.2% of total pore resistance. Eggshell pores with their long, narrow dimensions would seem, therefore, designed to prevent boundary layer or aperture effects; in other words, to prevent shell conductance from fluctuating with wind conditions around the nest. In support of this concept, the theoretical analysis of Tracy & Sotherland (1979) and the experimental results of Spotila,

Weinheimer & Paganelli (1981) both demonstrate that shell conductance is practically invariant with convective air flow around the egg.

Regional differences in air space gas tensions

Regional differences in shell conductance and pore density have been established in 8 species of bird (Rokitka & Rahn, 1987; Seymour & Visschedijk, 1988; Booth, Paganelli & Rahn, 1987; Handrich, 1989). In general, the shell at the blunt pole over the air cell has a higher regional conductance and pore density than the rest of the eggshell. Since gas tensions have always been sampled in the air cell, usually at the blunt pole, the question arises as to the gas tensions in the air space over the remaining larger portion of egg. Regional gas tensions are determined by the ratio of diffusive conductance to chorio-allantoic blood flow, or the G/Q ratio (Rahn & Paganelli, 1985; Paganelli et al., 1988). This is illustrated in Fig. 16.3: the curved lines in the graphs show all the simultaneous O_2 and CO_2 tensions that can theoretically exist in the air space of an egg for given ambient O_2 and CO_2 tensions and blood O_2 and CO_2 contents in the chorio-allantoic artery. Each point along these lines is the result of a given G/Q ratio which varies from zero, when there is only blood flow and no open pores, to infinity, when there are only open pores and no blood flow. The zero point is off the graphs at the far left of the G/Q line and is not shown. The infinity point is at the right of the curve, designated by the point P_E or P_I. The circles show the airspace composition and exchange ratio of either areas 1 and 2 or 1, 2 and 3 of the eggs (see inserts in Fig. 16.3 (b) and (d)). Area 2, the chorio-allantoic region, has a G/Q ratio of 1.29×10^{-3} and area 1, the air space 1.52×10^{-3} torr⁻¹ (Paganelli et al., 1988).

The gas tensions of the chorio-allantoic region can be calculated (Paganelli et al., 1988; Seymour & Visschedijk, 1988), but have also been directly measured by Booth, Paganelli & Rahn (1987) as shown in Fig. 16.3. Table 16.1 shows the average values of these three studies for the chorio-allantoic region and air cell region and the calculated value for the whole egg. This is a general overview of regional gas tension differences in the hen egg and how they are influenced by the regional G/Q ratio (Table 16.1; Fig. 16.3). Further details of the gas

Table 16.1. *Average regional values of oxygen (P_{O_2}) and carbon dioxide (P_{CO_2}) tensions in torr and gas exchange ratio, R, for the chorio-allantoic region, the air cell and the whole egg. For details see text.*

	Chorio-allantois	Air cell	Whole egg
P_{O_2}	100	109	103
P_{CO_2}	41	37	40
R	0.68	0.79	0.71

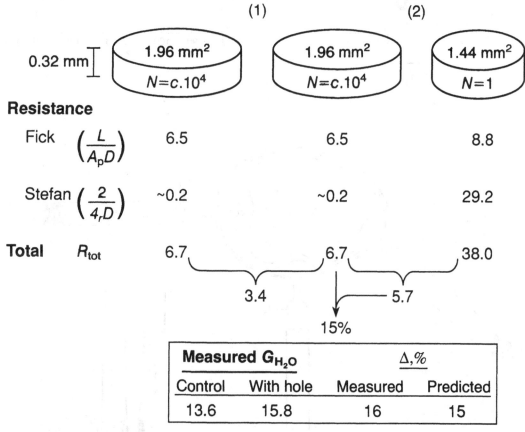

Fig. 16.2. An illustration of the application of Fick's and Stefan's laws to calculation of resistance to water vapour diffusion through the pores of the fowl eggshell. Resistance to diffusion in units of s cm^{-3}; N = number of pores; L = length of diffusion path (shell thickness) = 0.32 mm; A_p = effective total cross-sectional pore area; D = diffusivity of water vapour in air = 0.253 cm^2 s^{-1}; r = pore radius; G_{H_2O} = water vapour conductance in mg d^{-1} torr^{-1}. The figure illustrates the effect on resistance of adding to the already-existing pores in the eggshell either (1) 10^4 pores whose total cross-sectional area just equals that of the original pores (r = 7.9 µm); or (2) a single hole whose cross-sectional area is 1.44 mm^2 (r = 0.677 mm). Total resistance (R_{tot}) is reduced by 1/2 in case (1), but only by 15% in case (2). Measured average values of G_{H_2O} in control eggs and those in which a single hole of 1.44 mm^2 has been drilled clearly confirm the calculated decrease in R_{tot} in case (2). (Data on measured G_{H_2O}O taken from Simkiss, 1986.)

(a)

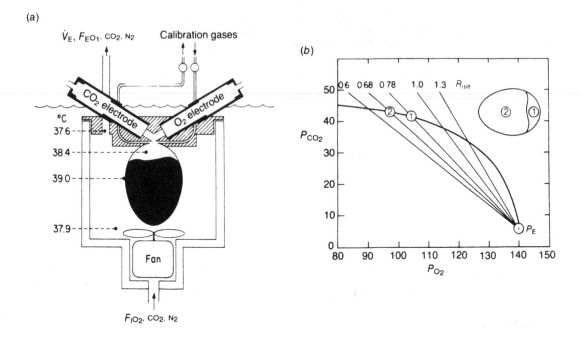

(b)

(c)

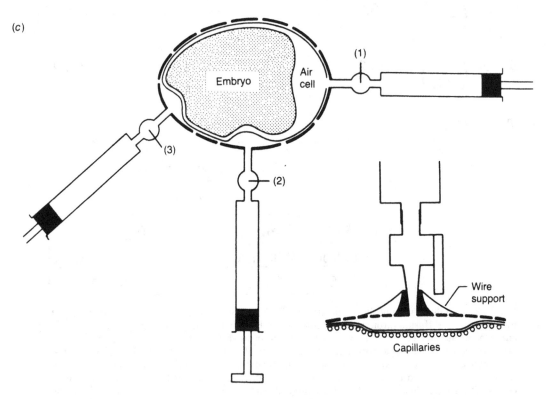

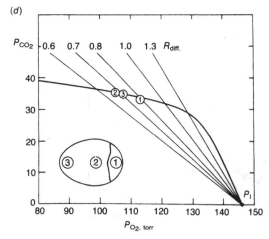

(d)

Fig. 16.3. Regional differences in air space gas tensions in the fowl egg.
(a) Technique of measuring O_2 and CO_2 tensions in the air cell of a fowl egg, using implanted electrodes. Simultaneously, metabolic rate for the whole egg is measured from flow rate ($\dot{V}_E$) and changes in composition of the gas stream ventilating the chamber. F_E is fraction of a particular gas in the flow stream leaving the chamber.
(b) Results obtained using apparatus in (a). O_2–CO_2 diagram showing gas tensions in regions ① (air cell) and ② (chorio-allantois) and the inspired gas mixture (P_E). The straight lines are lines of equal R_{diff} or diffusive gas exchange ratio. See text for details. (Fig. 16.3 (a) and (b) reproduced with permission from Paganelli et al., 1988.)
(c) Technique of measuring gas tensions at three locations. Syringes are fixed to the shell at three positions: air cell end or blunt pole (1), waist (2) and sharp pole (3). Sufficient time is allowed for partial pressure equilibrium between the syringes and the underlying gas spaces, the syringes are removed, and the contents analysed (Booth, Paganelli & Rahn, 1987).
(d) Results obtained using technique shown in (c). The description in (b) applies here as well. The inspired gas mixture is designed P_I. (Booth, Paganelli & Rahn, 1987.)

exchange between the embryonic blood and the ambient atmosphere, including the blood gas tensions in the chorio-allantoic artery and vein, are discussed by Olszowka, Tazawa & Rahn (1988).

The pathway for gas exchange

Pores as respiratory units

The number of pores per egg (N), daily water loss ($\dot{M}_{H2O}$) and daily O_2 consumption ($\dot{M}_{O2}$) at the pre-internal pipping (PIP) stage of incubation are all closely similar functions of egg mass (Fig. 16.4). The data in Fig. 16.4 include eggs ranging from 1 to 1400 g in mass, and encompass 77 species for $\dot{M}_{O2}$, 138 species for $\dot{M}_{H2O}$, and 161 species for pore number. The almost identical slopes of the 3 regressions lead to the following conclusions: both daily water loss and PIP O_2 consumption are proportional to the number of pores per egg, and therefore, the average fluxes of O_2 and water vapour through each pore must be invariant with egg mass (Ar & Rahn, 1985; Rahn & Paganelli, 1990). For example, an egg of a mass of 50 g will have mean fluxes per pore of 49 and 68 $\mu g\ d^{-1}$ for water vapour and oxygen, respectively; the corresponding CO_2 flux is 50 $\mu l\ d^{-1}$ assuming a respiratory quotient of 0.73 (Visschedijk, 1968; Paganelli et al., 1988).

In addition to the relations between O_2 and water fluxes and pore number, Ar & Rahn (1985) show that measured water vapour conductance of the egg shell (131 species) is directly proportional to measured pore number. It follows that conductance per pore is on average independent of egg mass (and by implication shell thickness). The value for the average pore conductance to water vapour is 1.51 $\mu g\ d^{-1}$ $torr^{-1}$ (0.63 $\mu mol\ d^{-1}\ kPa^{-1}$). For pore conductance to remain invariant with increasing shell thickness, Fick's law requires that the ratio of effective cross-sectional area for a single pore to its length remain constant. This remarkable conclusion is illustrated in Fig. 16.5, in which the ratio of measured water vapour conductance to pore number (G/N) is plotted as a function of egg mass for the 131 species mentioned above. The graph illustrates the relative constancy of G/N over an almost 1000-fold change in egg mass. The average shell thickness or pore length for selected egg masses, calculated from the allometric equation given by Ar & Rahn (1985), is also shown in Fig. 16.5. The graph also shows individual pores depicted as cylinders of increasing radius, whose ratio of cross-sectional area to length is 0.68 μm, a value calculated from Fick's law.

As noted earlier, pore conductances for 23 species of avian eggs have also been determined by morphometric analysis of pore casts (Toien et al., 1987, 1988). These results confirm that pore conductance is relatively independent of egg mass, but the average value is 5.4 $\mu g\ d^{-1}\ torr^{-1}$, (2.3 $\mu mol\ d^{-1}\ kPa^{-1}$), 3.6 times the functionally-derived constant. The difference may be caused

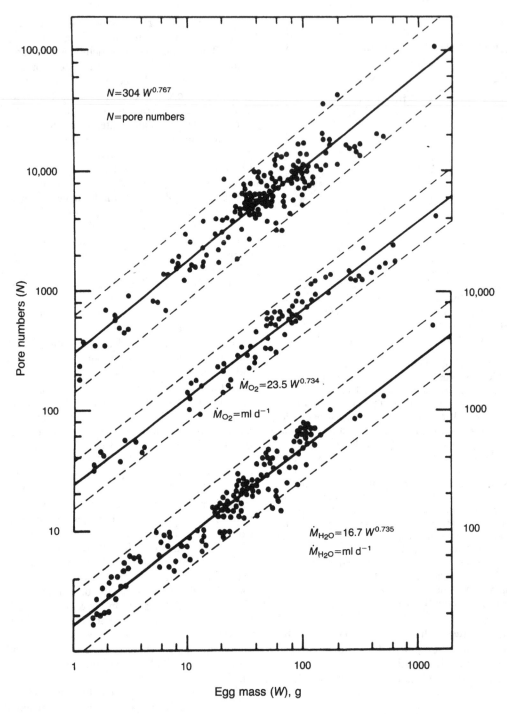

Fig. 16.4. Pore numbers (N), pre-internal pipping oxygen consumption ($\dot{M}_{O_2}$), and daily water loss ($\dot{M}_{H_2O}$), all plotted versus fresh egg mass (W); data from Ar & Rahn (1985). See text for discussion. (Symbol $\dot{M} = \dot{V}$.)

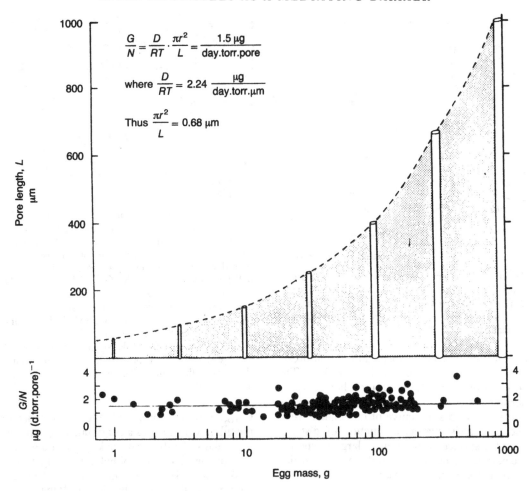

Fig. 16.5. *Lower panel:* Water vapour conductance per pore versus egg mass. The average value of 1.5 µg d^{-1} torr^{-1} pore^{-1} is indicated by the solid line. *Upper panel:* Pore length (shell thickness) thickness) versus egg mass. The dashed line is drawn through points whose value of $\pi r^2/L$ is 0.68 µm. See text for details. (Taken with permission from Rahn & Paganelli, 1990.)

by pore plugs (Becking, 1975), material which partially blocks the entrance to pores (Tullett, Lutz & Board, 1975; Board, Tullett & Perrott, 1977; Board, 1982; Tullett, 1984). This material affects water vapour conductance values in some species of domestic bird (Christensen & Bagley, 1984; Peebles & Brake, 1986; Deeming, 1987).

Given an average water vapour conductance of 1.51 µg d^{-1} torr^{-1} at 25 °C (or 1.92 µg d^{-1} torr^{-1} at 37 °C) O$_2$ and CO$_2$ conductances can be calculated from the proportionality between G and D (equation 16.2) (Paganelli, Ackerman & Rahn, 1978): G_{O_2} and G_{CO_2} are 1.64 and 1.28 µl d^{-1} torr^{-1} (0.55 and 0.43 µmol d^{-1}kPa^{-1}),

respectively. Using equation 16.1 with the above values of $\dot{M}$ and G for O$_2$ and CO$_2$ yields estimates of the partial pressure differences ($\triangle P$) across the shell, for these two gases. From these differences, in turn, and a knowledge of atmospheric or nest air composition, mean gas space O$_2$ and CO$_2$ partial pressures can be predicted (Ar & Rahn, 1985; Rahn & Paganelli, 1990). The results of such calculations are shown in Fig. 16.6 as *predicted* air cell gas tensions; for comparison, average *measured* O$_2$ and CO$_2$ tensions in the air cells of 25 species are also given. The agreement between measured and predicted values is quite satisfactory. The concept of the pore as a respiratory unit, with the

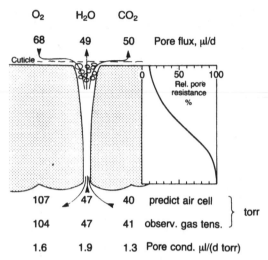

Fig. 16.6. The pore as a respiratory unit, drawn from average pore dimensions determined by Toien *et al.* (1988). Predicted and observed air cell gas tensions agree quite well. See text for description of calculations.

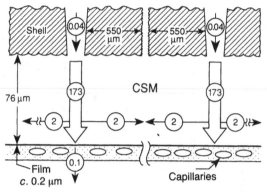

Fig. 16.7. A schematic representation of the relative O_2 conductances of pores, compound shell membrane (CSM) and the 'film' in the fowl egg. The values in circles were calculated from the data of Visschedijk, Girard & Ar (1988), using as unity their $G_{lat\,O_2}$ of 0.024 mmol d^{-1} kPa^{-1}, the O_2 conductance in the plane of the CSM, for a CSM area of *circa* 1 mm^2. The drawing is not to scale; the distance between pores is compressed relative to CSM thickness, and the 'film' thickness is exaggerated. The circled value of 0.1 is the estimated relative conductance of the 'film' itself, and does not include other conductance elements in the chorio-allantoic membrane. See text for further details.

relevant fluxes, conductances and partial pressures is, therefore, summarised in Fig. 16.6. The pore is drawn from average morphometrically-derived dimensions (Toien *et al.*, 1988), and to the right appears the curve of cumulative, relative pore resistance as a function of distance from the surface of the shell. A large fraction of the pore resistance resides in the narrow portion of the pore, as noted previously.

Lateral diffusion in the shell membranes

The gas exchange ratio, shell conductance and air space partial pressures of O_2 and CO_2 change with location on the eggshell. Are the observed local differences in air space O_2 and CO_2 partial pressures compatible with the permeability properties of the compound shell membrane? Visschedijk, Girard & Ar (1988) have answered this question affirmatively with ingeniously-devised experiments, in which they measure the lateral conductance (G_{lat}) of the fowl egg compound shell membrane to hydrogen gas. A value for $G_{lat\,O_2}$ is calculated in the plane of the membrane. The value of 0.024 mmol d^{-1} kPa^{-1} (0.072 ml d^{-1} $torr^{-1}$) for 1 mm^2 area of membrane (the area served by 1 pore) is taken as unity in Fig. 16.7 which is a schematic representation of the pathways for gas diffusion through the shell, compound membrane and 'film'. The perpendicular conductance ($G_{per\,O_2}$) is 173 times

$G_{lat\,O_2}$, but does not include the conductance of the 'film', which forms a continuous, amorphous layer about 0.1–0.2 µm thick on the outer surface of the chorionic epithelium (Wangensteen & Weibel, 1982). The 'film' is distinct from the concept of the inner barrier of Piiper *et al.* (1980), but forms one of the components of the inner barrier. The relative conductance of the 'film' alone, estimated from the work of Kayar *et al.* (1981), is considered to be 0.1 (Fig. 16.7), and other conductance elements in the chorio-allantoic membrane are not included. Resistance in the gas pathway is predominantly in the pores themselves (Fig. 16.7); the shell membrane external to the 'film' increases the total perpendicular resistance by at most 0.02%. From their measurements, Visschedijk *et al.* (1988) conclude that lateral diffusion will have a negligible effect on regional differences in air-space P_{O_2} caused by G/Q differences.

The diffusion path for O_2, CO_2 and water vapour

Is the diffusion path across the pores and internal and external shell membranes the same for O_2, CO_2, and water vapour? Paganelli *et al.*

(1978), on the basis of experimental data propose that these three gases have virtually the
same diffusion path. Therefore, if the conductance of one gas is known, that for the other
gases can be calculated. This model has
generally been accepted, but there are new indications that the diffusive pathway for water
vapour may be shorter than that for CO_2 and O_2.

One can compare, for example, the *measured*
O_2 conductance (calculated from measured
values of total oxygen consumption and air cell
P_{O_2}) with the *derived* O_2 conductance, which is
predicted from measured water vapour conductance assuming both gases have the same diffusion path. Such comparative data are available
from published reports of 20 species and show
on average a *derived* G_{O_2} which is 20% greater
than the *measured* G_{O_2}. However, many of these
comparisons are based on studies from two different laboratories, while in other studies the
measurements were not performed on the same
egg. A carefully executed study of O_2 consumption, air cell O_2 tension and water vapour conductance in the same eggs would resolve this
problem.

In this connection, one must also consider the
source of water vapour. Is it located in the
chorio-allantoic capillaries, which are the source
of CO_2 and the sink for O_2, or is it located closer
to the inner aperture of the pores? A continuous
liquid water column must exist along the fibres
of the inner and outer shell membranes, as can
be surmised from the transport of calcium from
the mammilary knobs of the shell, to which the
outer membrane is attached, to the chorio-allantois. In such a case a liquid water phase is available at the outer surface of the outer shell
membrane to serve as a source for water vapour
(Fig. 16.8).

Kayar *et al.* (1981) have used fowl eggs to
measure the O_2 permeability of the shell and two
shell membranes separately in order to partition
the resistance to diffusion imposed by each of
these layers, and the changes in permeability
during development. During the last week of
incubation the conductances (converted from
their permeability measurements) using 80 cm²
for shell surface area are as follows: shell +
outer and inner membranes (including 'film') =
10.8; shell + outer membrane only = 12.3; and
shell only = 13.1 cm³ STPD d⁻¹ torr⁻¹. Their
observations show that the *total* compound shell
membrane offers about 18% of the total
resistance of O_2 diffusion. From the data of

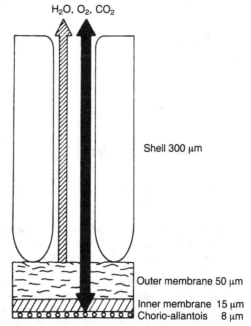

H₂O, O_2, CO_2

Shell 300 μm

Outer membrane 50 μm

Inner membrane 15 μm
Chorio-allantois 8 μm

Fig. 16.8. An hypothesis to explain the possible difference in diffusion paths for water vapour, O_2 and
CO_2 in the egg. The fibres of the outer membrane
contain water and potentially serve as a source of
water vapour in its diffusion from the egg interior to
the atmosphere; thus the pores in the shell constitute
the resistance to diffusion. On the other hand, CO_2
must originate in the chorio-allantoic capillaries, diffuse through the 'film', inner membrane, outer membrane and pores in the shell before reaching the
atmosphere. For O_2 molecules, this pathway is
reversed.

Visschedijk *et al.* (1988), it can be seen that
practically all of this 18% must reside in the
'film' itself, since the shell membrane external to
the 'film' contributes a negligible amount of the
overall O_2 resistance. Thus, the total resistance
of the compound shell membrane could constitute an important factor in differentiating the
O_2, CO_2 and water vapour diffusion pathways, if
water vapour can bypass the membranes (Fig.
16.8).

Hydrostatic pressure in the air space of the fowl egg

Gas exchange through eggshell pores has been
treated so far as a purely diffusive process; any
possible influence of convective gas flow has
been neglected, as noted earlier among the
assumptions implied in using Fick's law.
However, Romijn & Roos (1938) show evidence

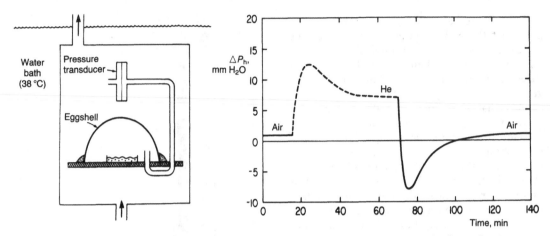

Fig. 16.9. *(Left)* Chamber containing a hemisphere cut from an eggshell and glued to a Lucite plate. The chamber is submerged in a water bath and ventilated with air (N_2) or He. A pressure transducer is connected to the gas space within the shell; the disc under the shell contains liquid water.
(Right) Hydrostatic pressure $\triangle P_h$ in mm H_2O versus time in minutes, recorded across the fowl eggshell preparation shown on the left. The solid line represents pressures measured while the chamber is flushed with dry air. The dashed line represents a change to dry He. At about 70 min, dry air replaces He. (Redrawn from Paganelli, Ar & Rahn, 1987.)

of a higher than atmospheric pressure in the air cell on the order of 1–2 mm H_2O (10–20 Pa). This observation has been confirmed and extended by Paganelli, Ar & Rahn (1987), who measure stable pressures of *circa* 1 mm H_2O (10 Pa) in air and 8 mm H_2O (80 Pa) in helium in an *in vitro* preparation of eggshell. This experimental arrangement and the effect of replacing air flowing into the chamber surrounding the eggshell with helium are shown in Fig. 16.9. There is a transient peak in the pressure, which results from helium diffusing into the shell faster than air molecules ($N_2 + O_2$) can escape. A new steady-state pressure in helium is established over the next hour, at which time helium has replaced all the air under the shell.

A schematic diagram, simplified to show only two gases (water vapour and nitrogen) is given in Fig. 16.10 as an explanation of the steady-state phenomenon. The space within the shell is filled with N_2 and water vapour; the shell is surrounded by dry N_2 only (Fig. 16.10). Since water vapour occupies some of the gas space within the shell, there is a $\triangle P_{N_2}$ across the shell of *circa* 50 torr, and a corresponding flux of N_2 into the shell (solid arrow to right in Fig. 16.10). In the steady-state, the influx of N_2 must be balanced by an outward flux of N_2 driven by convection (dashed arrow to far right in Fig. 16.10). Hence, a positive pressure must exist in the air cell.

Water vapour will also be entrained by the convective flow and carried out, as indicated by the dashed arrow to the left in Fig. 16.10. In their analysis of the convective flows, Paganelli *et al.* (1987) conclude that the convective component is responsible for only about 4% of the total water vapour flux through the eggshell. Hence, the assumption about the negligible influence of convection on gas exchange across the shell seems well justified.

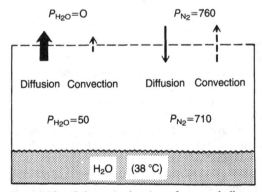

Fig. 16.10. Schematic drawing of an eggshell containing saturated water vapour at 38 °C, surrounded by dry N_2 at 760 torr. The total pressure within the shell is taken as very close to 760 torr. The pores in the shell are represented by the dashed line at the top of the figure. For details, see text. (Reprinted with permission from Paganelli, Ar & Rahn, 1987.)

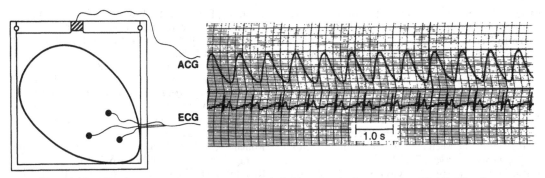

Fig. 16.11. Cardiogenic pressure waves recorded from the fowl egg at *circa* 25 °C. The figure shows simultaneous recording of the acousto-cardiogram (ACG) from the output of the microphone, indicated here as the hatched area at the top of the chamber, and the electrocardiogram (ECG) recorded from leads implanted through the shell. There is perfect synchrony between the two tracings. The heart rate is about 80 min^{-1} at 25 °C.

Cardiogenic pressure oscillations in the air space

With every contraction of the heart, a tissue pressure pulse is generated and transmitted to the air spaces of the egg. There it is converted into an air pressure pulse and propagated through the pores to the ambient atmosphere where it can be sensed, at some distance from the egg, by a small condenser microphone (Fig. 16.11). ECG leads attached to the egg show that the externally-recorded pressure pulses are synchronous with electrical activity associated with heart contractions (Fig. 16.11).

These pressure pulses cannot be detected by even a sensitive pressure transducer capable of resolving fractions of a millimetre of H_2O but can easily be sensed by a microphone. The recorded pressure pulses have been called the acousto-cardiogram or ACG (Rahn, Poturalski & Paganelli, 1989) in analogy to the electro-cardiogram (ECG) and the ballistocardiogram (BCG), measured simultaneously in eggs by Tazawa, Suzuki & Musashi (1989). The ACG provides a simple, non-invasive method for measuring avian embryonic heart rate, which at 37 °C is about 250 beats/min in the fowl and does not change significantly during the last half of incubation. The pressure waves do not measurably influence gas exchange in the fowl egg, even though they must be associated with some convective gas flow through the pores (Paganelli & Rahn, 1989).

References

Ar, A. & Rahn, H. (1985). Pores in avian eggshells: gas conductance, gas exchange and embryonic growth rate. *Respiration Physiology*, **61**, 1–20.

Becking, J. H. (1975). The ultrastructure of the avian eggshell. *Ibis*, **117**, 143–51.

Board, R. G. (1982). Properties of avian egg shells and their adaptive value. *Biological Reviews*, **57**, 1–28.

Board, R. G., Tullett, S. G. & Perrott, H. R. (1977). An arbitrary classification of the pore systems in avian eggshells. *Journal of Zoology, London*, **182**, 251–65.

Booth, D. T., Paganelli, C. V. & Rahn, H. (1987). Regional variation in gas conductance/perfusion ratio in air space gas tensions of the chicken egg. *Physiologist*, **30**, 232.

Christensen, V. L. & Bagley, R. A. (1984). Vital gas exchange and hatchability of turkey eggs at high altitude. *Poultry Science*, **63**, 1350–6.

Deeming, D. C. (1987). Effect of cuticle removal on the water vapour conductance of egg shells of several species of domestic bird. *British Poultry Science*, **28**, 231–7.

Freeman, B. M. & Vince, M. A. (1974). *Development of the Avian Embryo*. London: Chapman & Hall.

Handrich, Y. (1989). Incubation water loss in King Penguin egg. I. Change in egg and brood pouch parameters. *Physiological Zoology*, **62**, 96–118.

Kayar, S. R., Snyder, G. K., Birchard, G. F. & Black, C. P. (1981). Oxygen permeability of the shell and membranes of chicken eggs during development. *Respiration Physiology*, **46**, 209–21.

Kutchai, H. & Steen, J. B. (1971). Permeability of the shell and shell membranes of hen's eggs during development. *Respiration Physiology*, **11**, 265–78.

Meidner, H. & Mansfield, T. A. (1968). *Physiology of Stomata*. New York: McGraw-Hill.

Metcalfe, J., Stock, M. K. & Ingermann, R. L. (1987). *Development of the Avian Embryo*. New York: Alan R. Liss, Inc.

Murray, H. A., Jr (1925). Physiological ontogeny. A. Chicken embryos. II. Catabolism. Chemical changes in fertile eggs during incubation. Selection of standard conditions. *Journal of General Physiology*, 9, 1–37.

Olszowka, A. J., Tazawa, H. & Rahn, H. (1988). A blood-gas nomogram of the chick fetus: blood flow distribution between the chorioallantois and fetus. *Respiration Physiology*, 71, 315–30.

Paganelli, C. V. (1980). The physics of gas exchange across the avian eggshell. *American Zoologist*, 20, 329–38.

Paganelli, C. V., Ackerman, R. A. & Rahn, H. (1978). The avian egg: *in vivo* conductances to oxygen, carbon dioxide and water vapor in late development. In *Respiratory Function in Birds, Adult and Embryonic*, ed. J. Piiper, pp. 212–18. Berlin: Springer-Verlag.

Paganelli, C. V., Ar, A. & Rahn, H. (1987). Diffusion-induced convective gas flow through the pores of the eggshell. *Journal of Experimental Zoology*, Supplement 1, 173–80.

Paganelli, C. V. & Rahn, H. (1989). Do cardiogenic acoustic pressure waves affect gas exchange across the hen's eggshell? *Proceedings of the International Union of Physiological Sciences*, 17, 293.

Paganelli, C. V., Sotherland, P. R., Olszowka, A. J. & Rahn, H. (1988). Regional differences in diffusive conductance/perfusion ratio in the shell of the hen's egg. *Respiration Physiology*, 71, 45–55.

Peebles, E. D. & Brake, J. (1986). The role of the cuticle in water vapor conductance by the eggshell of broiler breeders. *Poultry Science*, 65, 1034–9.

Piiper, J. (1978). *Respiratory Function in Birds, Adult and Embryonic*. Berlin: Springer-Verlag.

Piiper, J., Tazawa, H., Ar, A. & Rahn, H. (1980). Analysis of chorioallantoic gas exchange in the chick embryo. *Respiration Physiology*, 39, 273–84.

Pringle, E. M. & Barott, H. G. (1937). Loss of weight of hen's eggs during incubation under different conditions of humidity and temperature. *Poultry Science*, 16, 49–52.

Rahn, H. & Ar, A. (1980). Gas exchange of the avian egg: time, structure and function. *American Zoologist*, 20, 477–84.

Rahn, H. & Paganelli, C. V. (1985). Transport by gas-phase diffusion: lessons learned from the hen's egg. *Clinical Physiology*, 5, Supplement 3, 1–7.

— (1990). Minireview. Gas fluxes in avian eggs: Driving forces and the pathway for exchange. *Comparative Biochemistry and Physiology*, 95A, 1–15.

Rahn, H., Paganelli, C. V. & Ar, A. (1987). Pores and gas exchange of avian eggs: a review. *Journal of Experimental Zoology*, Supplement 1, 165–72.

Rahn, H., Poturalski, S. A. & Paganelli, C. V. (1989). The acousto-cardiogram (ACG) of the hen's egg. *Proceedings of the International Union of Physiological Sciences*, 17, 293.

Rokitka, M. A. & Rahn, H. (1987). Regional differences in shell conductance and pore density of avian eggs. *Respiration Physiology*, 68, 371–6.

Romanoff, A. L. (1943). Study of various factors affecting permeability of birds' eggshell. *Food Research*, 8, 212–23.

Romanoff, A. L. & Romanoff, A. J. (1949). *The Avian Egg*. New York: John Wiley & Sons.

Romijn, C. (1950). Foetal respiration in the hen. Gas diffusion through the egg shell. *Poultry Science*, 29, 42–51.

Romijn, C. & Roos, J. (1938). The air space of the hen's egg and its changes during the period of incubation. *Journal of Physiology*, 94, 365–79.

Seymour, R. S. (1984). *Respiration and Metabolism of Embryonic Vertebrates*. Dordrecht: Dr W. Junk.

Seymour, R. S. & Visschedijk, A. H. J. (1988). Effects of variation in total and regional shell conductance on air cell gas tensions and regional gas exchange in chicken eggs. *Journal of Comparative Physiology*, 158B, 229–36.

Sibly, R. M. & Simkiss, K. (1987). Gas diffusion through non-tubular pores. *Journal of Experimental Zoology*, Supplement 1, 187–91.

Simkiss, K. (1986). Eggshell conductance – Fick's or Stefan's law? *Respiration Physiology*, 65, 213–22.

Spotila, J. R., Weinheimer, C. J. & Paganelli, C. V. (1981). Shell resistance and evaporative water loss from bird eggs: effects of wind speed and egg size. *Physiological Zoology*, 54, 195–202.

Tazawa, H., Suzuki, Y. & Musashi, H. (1989). Simultaneous acquisition of ECG, BCG, and blood pressure from chick embryos in the egg. *Journal of Applied Physiology*, 67, 478–83.

Tøien, Ø., Paganelli, C. V., Rahn, H. & Johnson, R. R. (1987). Influence of eggshell pore shape on gas diffusion. *Journal of Experimental Zoology*, Supplement 1, 181–6.

— (1988). Diffusive resistance of avian eggshell pores. *Respiration Physiology*, 74, 345–54.

Tracy, C. R. & Sotherland, P. R. (1979). Boundary layers of bird eggs: do they ever constitute a significant barrier to water loss? *Physiological Zoology*, 52, 63–6.

Tullett, S. G. (1975). Regulation of avian eggshell porosity. *Journal of Zoology, London*, 177, 339–48.

— (1984). The porosity of avian eggshells. *Comparative Biochemistry and Physiology*, 78A(1), 5–13.

— (1991). *Avian Incubation*. London: Butterworths.

Tullett, S. G., Lutz, P. L. & Board, R. G. (1975). The fine structure of the pores in the shell of the hen's egg. *British Poultry Science*, 16, 93–5.

Tyler, C. (1945). The porosity of egg shells, and the influence of different levels of dietary calcium upon porosity. *Journal of Agricultural Science*, **35**, 168–76.

— (1962). Some chemical, physical and structural characteristics of eggshells. In *Recent Advances in Food Science, Symposium on Food Sciences 1960*, vol. 1, eds J. Hawthorne & M. M. Leitch, pp. 122–28. London: Butterworth.

— (1964). A study of the eggshells of the Anatidae. *Proceedings of the Zoological Society (London)*, **142**, 547–83.

— (1965). A study of the eggshells of the Sphenisci-formes. *Journal of Zoology (London)*, **147**, 1–19.

Tyler, C. & Simkiss, K. (1959). A study of the egg-shells of ratite birds. *Proceedings of the Zoological Society (London)*, **133**, 201–43.

Visschedijk, A. H. J. (1968). The air space and embryonic respiration: the pattern of gaseous exchange in the fertile egg during the closing stages of incubation. *British Poultry Science*, **9**, 173–84.

Visschedijk, A. H. J., Girard, H. & Ar, A. (1988). Gas diffusion in the shell membranes of the hen's egg: lateral diffusion *in situ*. *Journal of Comparative Physiology*, **158B**, 567–74.

Wangensteen, D. & Weibel, E. R. (1982). Morpho-metric evaluation of chorioallantoic oxygen trans-port in the chick embryo. *Respiration Physiology*, **47**, 1–20.

Wangensteen, O. D., Wilson, D. & Rahn, H. (1970/71). Diffusion of gases across the shell of the hen's egg. *Respiration Physiology*, **11**, 16–30.

17

Gas exchange across reptilian eggshells

DENIS C. DEEMING AND MICHAEL B. THOMPSON

Introduction

In contrast to the open nests of birds, nests of reptiles are generally located in the soil or in decaying vegetation where, effectively separated from the atmospheric gaseous environment, eggs may be exposed to relatively low partial pressures of oxygen and correspondingly high partial pressures of carbon dioxide (Ackerman, 1977; Seymour & Ackerman, 1980; Ferguson, 1985; Packard & Packard, 1988). The deviation from atmospheric oxygen and carbon dioxide tensions may be greater during the later stages of incubation due to a high rate of respiration of the embryos (Booth & Thompson, Chapter 20). In addition, the humidity of these nests is high. Consequently, problems of desiccation are reduced thereby allowing eggs to have high conductances to respiratory gases which result in smaller gradients of respiratory gases across the eggshell. These presumably help to maintain the internal gaseous environment within physiological limits even in nests with low oxygen and high carbon dioxide tensions (Booth & Thompson, Chapter 20). In this chapter, we discuss conductance of different types of reptilian eggshell to water and to respiratory gases, making comparisons, where possible, with avian eggs. As eggshell structure is highly variable (Packard & DeMarco, Chapter 5) conductance and allometric relationships are described separately for each major type of shell.

The eggshell as a mediating barrier

Water vapour

In avian eggs, water vapour, oxygen and carbon dioxide all diffuse across the eggshell through the same pathway – discrete pores in the calcareous eggshell; the relative conductance of the shell to these gases is related to their different rates of diffusion (Paganelli, Ackerman & Rahn, 1978). It has been assumed that water vapour conductance (G_{H_2O}) can be used to calculate oxygen and carbon dioxide conductances (G_{O_2} and G_{CO_2} respectively) of the shell, although recent analysis suggests that this may not always be totally true (Paganelli, Chapter 16). Water vapour conductance can be determined by measuring the loss of water from eggs under conditions of known humidity (Ar et al., 1974) but direct measurement of oxygen and carbon dioxide conductances is a more complex procedure. Consequently, for reptiles (and most birds) G_{O_2} and G_{CO_2} have usually been calculated from measures of G_{H_2O}.

Water vapour conductance values of the eggshells of all reptiles, with the exception of geckos with rigid-shelled eggs, are high relative to values calculated for avian eggs of equivalent mass (Table 17.1). Squamate eggs with parchment-like shells have water vapour conductances up to two orders of magnitude higher than is predicted for avian eggs of similar size (Table 17.1). However, as initial egg mass increases the difference between the observed and predicted values for G_{H_2O} decreases (Table 17.1). The reason for this convergence with increasing sizes is not known. One factor may be egg size: larger eggs have to be stronger, and therefore, thicker eggshells than smaller eggs to provide sufficient strength to hold the greater volume of egg contents. Surface specific shell conductances of squamate eggs are variable but tend to be lower in larger eggs also (Table 17.1). Values have been calculated for pliable- , or rigid-shelled eggs using an estimate of egg sur-

Table 17.1. *Water vapour conductance values* ($G_{H_2O}(A)$, *mg* H_2O, *day*$^{-1}$, *kPa*$^{-1}$) *for various species of reptile compared with the water vapour conductance values* ($G_{H_2O}(P)$) *predicted from avian eggs using the equation* $G_{H_2O} = 3.240 \cdot Mass^{0.780}$ *(Ar et al., 1974). The surface area specific conductance of squamate eggshells* (K, *mg* H_2O, *day*$^{-1}$, *kPa*$^{-1}$, *cm*$^{-2}$) *was also calculated.*

Species	Shell type	Egg mass (g)	$G_{H_2O}(A)$	$G_{H_2O}(P)$	A/P	K*	Ref.†
Squamata							
Scincella laterale	Parchment	0.11	205.95	0.58	355.57	183.6	12
Anolis sagrei	Parchment	0.18	219.02	0.85	257.53	141.0	12
Anolis carolinensis	Parchment	0.23	264.74	1.03	257.11	144.9	12
Eumeces egregius	Parchment	0.24	420.17	1.06	394.74	223.5	12
Eremias guttalata	Parchment	0.32	483.93	1.33	363.86	170.7	1
Sceloporus garmani	Parchment	0.35	515.82	1.43	361.06	213.8	12
Sceloporus woodi	Parchment	0.39	463.54	1.55	298.20	178.8	12
Holbrookia maculata	Parchment	0.45	905.13	1.74	520.79	317.6	12
Eumeces inexpectatus	Parchment	0.49	827.57	1.86	445.56	274.5	12
Sceloporus undulatus	Parchment	0.53	546.41	1.97	276.72	172.0	12
Lacerta sicula	Parchment	0.61	877.29	2.20	389.77	286.6	1
Eumeces septentrionalis	Parchment	0.70	770.80	2.45	314.21	201.9	12
Cnemidophorus sexlineatus	Parchment	0.79	959.98	2.70	356.10	232.1	12
Sceloporus virgatus	Parchment	0.97	1081.65	3.16	341.87	228.3	12
Eumeces laticeps	Parchment	1.06	1076.65	3.39	317.53	214.3	12
Ophisaurus ventralis	Parchment	1.06	1076.65	3.39	317.55	214.3	12
Diadophis punctatus	Parchment	1.12	1074.53	3.54	303.59	206.2	12
Chamaeleo chamaeleon	Parchment	1.14	1245.60	3.59	346.96	174.7	1
Eumeces obsoletus	Parchment	1.70	1501.14	4.90	306.29	218.5	12
Lacerta lepida	Parchment	2.24	2291.20	6.08	376.84	292.6	1
Agama stellio	Parchment	2.28	1307.07	6.16	212.19	159.4	1
Eublepharis macularius	Parchment	3.77	2303.91	9.12	252.62	201.4	1
Natrix tessellata	Parchment	6.83	2062.86	14.50	142.27	169.9	1
Echis colorata	Parchment	10.22	3420.23	19.86	172.21	161.4	1
Elaphe obsoleta	Parchment	11.00	5364.15	21.03	255.07	226.8	12
Vipera xanthina palaestinae	Parchment	14.32	3014.21	25.84	116.65	116.8	1
Uromastix aegyptius	Parchment	15.83	4703.76	27.94	168.35	161.7	1
Spalerosophis diadema	Parchment	23.44	6650.32	37.95	175.24	133.8	1
Python molurus	Parchment	230.00	8300.00	225.29	36.84	46.9	2
Sphaerodactylus sp.	Rigid	0.15	0.150	0.738	0.20	0.1	3
Lepidodactylus lugrubris	Rigid	0.23	0.458	1.030	0.44	0.3	4
Hemidactylus garnoti	Rigid	0.34	0.593	1.397	0.42	0.3	4
Chelonia							
Chrysemys picta	Pliable	5.70	832.00	12.59	66.08	54.4	5
Chelydra serpentina	Pliable	9.96	1065.00	19.46	54.72	48.1	6
Chelydra serpentina	Pliable	11.40	1867.75	21.63	86.35	77.1	7
Caretta caretta	Pliable	37.00	2175.29	43.62	49.87	41.2	8
Trionyx spiniferus	Rigid	10.15	108.00	19.75	5.47	4.8	6
Emydura macquarii	Rigid	10.42	168.77	20.16	8.37	7.4	9
Crocodylia							
Alligator mississippiensis	Rigid	72.27	387.00	91.32	4.24	4.7	6
Crocodylus porosus (inf)‡	Rigid	90.00	82.51	108.37	0.76	0.9	10
Crocodylus porosus (fert)‡	Rigid	90.00	375.05	108.37	3.46	3.9	10
Crocodylus acutus	Rigid	91.30	157.52	109.59	1.43	1.6	11

*In all species except for those from Ackerman, Dmi'el & Ar (1985) an approximate value for surface area was calculated using the relationship $A=4.835 M^{0.662}$ derived from data for avian eggs (Paganelli, Olszowka & Ar, 1974). This is a good approximation for surface area when compared with calculated values (Ackerman *et al.* 1985). $K=G/s.a.$

face area calculated using an equation for avian eggs (Hoyt, 1976) as such data is unavailable for many reptilian eggs. In avian eggs, surface specific shell conductance increases with increasing egg size (Ar *et al.*, 1974; Paganelli *et al.*, 1974). The conductance of more species of parchment-shelled eggs to represent a complete range of egg sizes needs to be measured to help elucidate the area specific conductance versus egg size, or other allometric, relationships in these reptilian eggs. The relationship must concern shell structure and parallel studies of structure and conductance would prove very useful in such analysis.

Greatest resistance to gaseous diffusion is associated with the mineral layer of the eggshell (Feder, Satel & Gibbs, 1982). However, parchment-like shells have very little calcareous material on their outer surface (Packard, Packard & Boardman, 1982; Packard & DeMarco, Chapter 5) and this is reflected in the high values for G_{H_2O}. Although the fibrous shell membrane may not restrict diffusion of water vapour to any great extent, measures of the rate of water loss from model agar eggs show that the eggshell exerts a significant resistance to water loss from real eggs (Lillywhite & Ackerman, 1984; Ackerman, Dmi'el & Ar, 1985). In addition, a boundary layer of stagnant air surrounds the eggshell but its conductance to water vapour is very high. In avian eggs, conductance of the boundary layer is very high compared with that of the eggshell and so does not significantly affect water vapour loss across the eggshell (Tracy & Sotherland, 1979; Spotila, Weinmiller & Paganelli, 1981). However, in parchment-shelled eggs, because of the very high eggshell conductance, the boundary layer may contribute a significant proportion of the total resistance to water vapour exchange of the egg (the boundary layer conductance is 119% that of the shell in one species) (Ackerman *et al.*, 1985). However, as the water vapour conductance of the soil surrounding the egg is much greater than that of the parchment-like eggshell, the conductance of that part of the shell in contact with the soil is larger than that of the shell surface exposed to

the gas phase, which effectively destroys the effects of the boundary layer on water exchange across the shell (Ackerman *et al.*, 1985).

Conductances of pliable-shelled eggs of chelonians fall between those of parchment-shelled eggs of squamates and rigid-shelled avian eggs of similar size (Table 17.1). This almost certainly reflects the thicker, more organised layer of calcium carbonate in the shell, which, in pliable-shelled eggs, is comprised of loosely fitting units of aragonite (Packard *et al.*, 1982; Packard & DeMarco, Chapter 5). For example, eggs of *Chelydra serpentina* have shells with conductances 55–86 times higher than that predicted for equivalent avian eggs (Packard *et al.*, 1979; Feder *et al.*, 1982). By contrast, water vapour conductance of rigid-shelled eggs of chelonians and crocodilians is of the same order of magnitude to that predicted for avian eggs (Table 17.1). In rigid eggshells the individual units of calcium carbonate crystals which make up the mineral layer of the shell are tightly bound to form a continuous layer perforated with discrete pores (Packard & Packard, 1979; Ferguson, 1982; Packard & DeMarco, Chapter 5).

The rigid-shelled eggs of gekkonid lizards have water vapour conductances less than half that for hypothetical avian eggs of equivalent size (Table 17.1) (Dunson & Bramham, 1981; Dunson, 1982). This may be related to the solid layer of calcite in these shells or the presence of a fibrous layer on the outer surface of the eggshell (Packard *et al.*, 1982; Deeming, 1988; Packard & DeMarco, Chapter 5). Pores have been observed in the calcareous layer of some gecko eggshells but they are few in number and have an unusual structure, described as a series of labyrinth channels (Packard & Hirsch, 1989). Pore structure and the pattern of gas exchange in rigid eggshells of geckos need more detailed study.

The relationship between G_{H_2O} and egg mass has been described for 11 species of squamate reptile as:

$$G_{H_2O} = 981.8 \cdot \text{Mass}^{0.51} \qquad (17.1)$$
$$\text{(Ackerman } et\ al., 1985)$$

Notes to Table 17.1 (contd.)

†Where: inf=infertile eggs, and fert=fertile eggs.

‡References: 1 – Ackerman, Dmi'el & Ar (1985), 2 – Black *et al.* (1984), 3 – Dunson & Bramham (1981), 4 – Dunson (1982), 5 – Packard, Packard & Boardman (1981), 6 – Packard *et al.* (1979), 7 – Feder, Satel & Gibbs (1982), 8 – Ackerman (1980), 9 – Thompson (1985), 10 – Grigg & Beard (1985), 11 – Lutz *et al.* (1980), 12 – Somma (1989, unpublished observations).

This relationship changes, however, when all data for parchment-shelled eggs shown in Table 17.1 are analysed:

$$G_{H_2O} = 943.4. \text{Mass}^{0.54} \quad (r^2 = 91.3\%) \quad (17.2)$$

Unlike in birds (Ar et al., 1974), there is not any clear relationship between water vapour conductance and egg mass in non-squamate reptiles, or within all reptiles, so far examined (Table 17.1). This could reflect the lack of data over a broad range of egg sizes, or more likely indicates a significant difference between the patterns of gas exchange of avian and reptilian eggshells.

Length of incubation is related to egg mass and G_{H_2O} in birds (Rahn & Ar, 1974) but no such pattern can be seen in those reptiles examined to date. Eggs of different mass can have similar incubation periods: for example, incubation periods of *Eublepharis macularius* (3 g initial mass), *Alligator mississippiensis* (72 g) and *Python molurus* (230 g) are all around 60 days (Black et al., 1984; Deeming & Ferguson, 1989; Deeming, 1989). In addition, incubation temperature significantly affects the duration of the incubation period in reptiles (Deeming & Ferguson, present volume), so eggs of similar mass (or of one species) can have different incubation periods.

The G_{H_2O} of eggs of *C. serpentina* is affected by the water content of the eggshell. Moist eggshells have higher values for G_{H_2O} than dry eggshells (Feder et al., 1982). This is explained by either bulk flow of water through the shell or, the different hydration states of the shell mechanically altering the geometry of the pores (Feder et al., 1982) by increasing the spaces between individual shell units (Packard, 1980). These results warrant further study of the water exchange across the reptilian eggshell.

Egg fertility affects water vapour conductance in eggs of *Crocodylus porosus* (Grigg & Beard, 1985), with fertile eggs having much higher values for G_{H_2O} than infertile eggs. Although there is no clear hypothesis to explain this difference (Grigg & Beard, 1985), it is worthy of further study. The embryo clearly influences water exchange in parchment-shelled eggs; often the first sign of embryo mortality is often collapse of the egg due to loss of water. Study of the possiblity of embryonic mediation of shell conductance may help explain this area of reptilian egg physiology.

Oxygen and carbon dioxide

In avian eggs, G_{H_2O} is related to G_{O_2} and G_{CO_2} in the following way (Paganelli et al., 1978):

$$G_{O_2} = 0.83. G_{H_2O} \quad (17.2),$$

and

$$G_{CO_2} = 0.64. G_{H_2O} \quad (17.3).$$

These ratios reflect the binary diffusivity ratios of these gases (Paganelli et al., 1978). In eggs of the sea turtle, *Caretta caretta*, G_{H_2O} calculated from the G_{O_2} (255.0 cm³ day⁻¹ kPa⁻¹) measured by Ackerman & Prange (1972) was 307.3 mg H_2O day⁻¹ kPa⁻¹ but the observed G_{H_2O} for this species is 2175.3 mg H_2O day⁻¹ kPa⁻¹ (Ackerman, 1980). Similarly the predicted G_{H_2O} for eggs of *Emydura macquarii* with a G_{O_2} of 22.5 cm³ O_2 day⁻¹ kPa⁻¹ (Thompson, 1985) is 27.1 mg H_2O day⁻¹ kPa⁻¹ but the empirical value is 168.8 (Thompson, 1985). Data on the carbon dioxide conductance (G_{CO_2}) of reptilian eggshells is limited to that for *E. macquarii* (Thompson, 1985). On day 1 of incubation the mean G_{CO_2} is 5.7 cm³ CO_2 day⁻¹ kPa⁻¹ but this increases up to a value around 37.5 by day 25 of incubation. The ratio of G_{CO_2} to G_{O_2} (maximum values) is 1.8 which indicates that 94% of the gaseous diffusion of these gases across the eggshell occurs through air filled pores (Thompson, 1985). These, and other data for eggs of *Crocodylus johnstoni* which show that the actual G_{CO_2} of the eggshell is very much lower than predicted from water vapour conductances (Whitehead, 1987), suggest that water and respiratory gas exchange are not coupled in the same way as the diffusive exchange of these gases across the avian eggshell (Paganelli et al., 1978; Paganelli, Chapter 16). Clearly, we need more data on the patterns of gas diffusion across the reptilian eggshell before we can confirm this suggestion.

Diffusion of oxygen across eggshells of *C. serpentina* and some lizards is affected by the hydration of the shell with moist shells having a lower oxygen conductance than dry shells (Feder et al., 1982; DeMarco, unpublished observations). In *P. molorus*, G_{O_2} is correlated with the water potential of the incubation substrate. On wet substrates (−80 kPa) G_{O_2} is 33.8 cm³ O_2 day⁻¹ kPa⁻¹ but as water potential drops G_{O_2} increases; on a dry substrate (−360 kPa) G_{O_2} is 54.0 cm³ O_2 day⁻¹ kPa⁻¹ (Black et al., 1984). If the eggshell contains liquid water (e.g. in an egg on a wet substrate) perhaps the func-

Table 17.2. *Oxygen conductance values, determined empirically* (G_{O_2}(A)), *water vapour conductance values* (G_{H_2O}(A), mg H_2O day^{-1} kPa^{-1}) *for various species of reptile compared with the oxygen conductance values* (G_{O_2}(P)) *predicted from avian eggs using the equation* $G_{O_2} = 0.83\ G_{H_2O}$ *(Paganelli, Ackerman & Rahn, 1978).*

Species	Shell type	$G_{O_2}(A)$	$G_{H_2O}(A)$	$G_{O_2}(P)$	Ref.†
Crocodylus johnstoni	Rigid	25.58	87.29*	72.45	1
Emydura macquarii	Rigid	22.50	168.77	140.07	2
Caretta caretta	Pliable	255.03	2175.29	1805.49	3,4
Python molurus (−80 kPa)‡	Parchment	33.75	8300.00	6889.00	5
Python molurus (−360 kPa)‡	Parchment	54.00	8300.00	6889.00	5

*Calculated from egg mass 68.2 g (Whitehead, 1987) using avian egg as predictor (Ar *et al.* 1974).
†Substrate potential
‡References: 1 – Whitehead (1987), 2 – Thompson (1985), 3 – Ackerman (1980), 4 – Prange & Ackerman (1974), 5 – Black *et al.* (1984).

tional pore area is reduced to the extent of reducing conductance. Generally, the oxygen conductance of reptilian eggshells is lower than that predicted for avian eggshells (where liquid water is absent from the pores for most of incubation) or from that predicted from water vapour conductance values (Table 17.2).

Eggshell conductance to oxygen has been measured throughout incubation in eggs of *C. johnstoni* and *E. maquarii*. In both cases G_{O_2} is very low at the start of incubation and increases during the early part of incubation in concert with the increase in area of opaque regions of the eggshell (Thompson, 1985; Whitehead, 1987). In *C. johnstoni* the maximum value for G_{O_2} is reached after 65% of the incubation period once the eggshell has become completely opaque (Whitehead, 1987). Water contents of eggshells of crocodilians (8–21%; Manolis, Webb & Dempsey, 1987) and chelonians (22–27%; Thompson, 1985) are higher than those in the fowl (2.5%; Romanoff & Romanoff, 1949). In reptilian eggs, opaque regions are indicative of normal development (Ewert, 1979, 1985; Blanck & Sawyer, 1981; Ferguson, 1982, 1985; Webb, Buckworth & Manolis, 1983*a*; Webb *et al.*, 1983*b*, 1987*a,b*; Thompson, 1985; Chan, 1989) and their development is initiated by the embryo (Ferguson, 1982; Thompson, 1985). Opaque regions do not form in infertile eggs and do not expand in eggs once the embryo dies but neither does the opaque band become translucent again (Ferguson, 1982, 1985). Eggshell from the opaque region has a lower water content (15% compared with 21% in the translucent region) and a higher oxygen conductance

than translucent shell (Thompson, 1985; Whitehead, 1987), reducing the possibility of potentially damaging hypoxic conditions from developing around the embryo during development (Thompson, 1985; Whitehead, 1987). As incubation proceeds the opaque band increases in size in tandem with expansion of the chorioallantoic membrane and the diminishing volume of albumen (Ferguson, 1982, 1985; Webb *et al.*, 1987*a,b*).

When oviposited, pores through avian and reptilian eggshells are fluid-filled. The rate of diffusion of oxygen through water is very much slower than through air (Wangensteen, Wilson & Rahn, 1970/71) and this is reflected in fresh eggs having very low shell conductances to oxygen. In fowl eggs, oxygen conductance of the shell increases rapidly on day 5 of incubation from an initially low level to a level which can support the oxygen requirements of the embryo up to the end of incubation (Kutchai & Steen, 1971; Lomholt, 1976; Tullett & Board, 1976; Kayer *et al.*, 1981; Tranter, Sparks & Board, 1983). The increase in conductance is associated with dehydration of the eggshell membranes and pores by evaporation and formation of sub-embryonic fluid (Tullett & Board, 1976; Seymour & Piiper, 1988) and this probably occurs over the whole egg surface simultaneously. Disintegration of the limiting layer of the inner shell membrane has also been implicated in the increase in G_{O_2} (Tranter *et al.*, 1983).

In reptiles, development of opaque regions on the shell suggests regional drying of the eggshell and membranes. The reason for the difference between avian and reptilian eggs in the pattern of eggshell dehydration is not known, but may involve specialisation of different regions of the shell of reptilian eggs for different functions (Thompson, 1985). Presumably, exchange of respiratory gases occurs predominantly across opaque shell and water exchange across translucent shell. This may facilitate water uptake. As most avian eggs must lose water during incubation (Ar, Chapter 14), no region for specialised water uptake is required, and drying of all pores may even facilitate water loss.

Acknowledgements

We thank Carol Vleck and Ralph Ackerman for reviewing a previous draft of this chapter. DCD was funded by The University of Manchester Research Support Fund. MBT was previously at the Department of Zoology, The University of Florida, Gainesville during preparation of this chapter. We thank Louis Somma and Vince DeMarco for details of their unpublished data.

References

Ackerman, R. A. (1977). The respiratory gas exchange of sea turtle nests (*Chelonia, Caretta*). *Respiration Physiology*, **31**, 19–38.

— (1980). Physiological and ecological aspects of gas exchange by sea turtle eggs. *American Zoologist*, **20**, 575–83.

Ackerman, R. A., Dmi'el, R. & Ar, A. (1985). Energy and water vapor exchange by parchment-shelled reptile eggs. *Physiological Zoology*, **58**, 129–37.

Ackerman, R. A. & Prange, H. D. (1972). Oxygen diffusion across a sea turtle (*Chelonia mydas*) egg shell. *Comparative Biochemistry and Physiology*, **43A**, 905–9.

Ar, A., Paganelli, C. V., Reeves, R. B., Greene, D. G. & Rahn, H. (1974). The avian egg: Water vapor conductance, shell thickness and functional pore area. *Condor*, **76**, 153–8.

Black, C. P., Birchard, G. F., Schuett, G. W. & Black, V. D. (1984). Influence of incubation water content on oxygen uptake in embryos of the Burmese python (*Python molurus bivittatus*). In *Respiration and Metabolism of Embryonic Vertebrates*, ed. R. S. Seymour, pp. 137–45. Dordrecht: Dr W. Junk.

Blanck, C. E. & Sawyer, R. H. (1981). Hatchery practices in relation to early embryology of the loggerhead sea turtle, *Caretta caretta* (Linne). *Journal of Experimental Marine Biology and Ecology*, **49**, 163–77.

Chan, E.-H. (1989). White spot development, incubation and hatching success of leatherback turtle (*Dermochelys coriacea*) eggs from Rantau Abang, Malaysia. *Copeia*, **1989**, 42–7.

Deeming, D. C. (1988). Eggshell structure of lizards of two sub-families of the Gekkonidae. *Herpetological Journal*, **1**, 230–4.

— (1989). The residues in the eggs of squamate reptiles at hatching. *Herpetological Journal*, **1**, 381–5.

Deeming D. C. & Ferguson, M. W. J. (1989). Effects of incubation temperature on growth and development of embryos of *Alligator mississippiensis*. *Journal of Comparative Physiology*, **159B**, 183–93.

Dunson, W. A. (1982). Low water vapor conductance of hard-shelled eggs of the gecko lizards *Hemidactylus* and *Lepidodactylus*. *Journal of Experimental Zoology*, **219**, 377–9.

Dunson, W. A. & Bramham, C. R. (1981). Evaporative water loss and oxygen consumption of three small lizards from the florida keys: *Sphaerodactylus cinereus*, *S. notatus*, and *Anolis sagrei*. *Physiological Zoology*, **54**, 253–9.

Ewert, M. A. (1979). The embryo and its egg: development and natural history. In *Turtles: Perspectives and Research*, eds M. Harless & H. Morlock, pp. 333–413. New York: Wiley-Interscience.

— (1985). Embryology of turtles. In *Biology of the Reptilia*, vol. 14, Development A, ed. C. Gans, F. Billet & P. F. A. Maderson, pp. 75–267. New York: Wiley.

Feder, M. E., Satel, S. L. & Gibbs, A. G. (1982). Resistance of the shell membrane and mineral layer to diffusion of oxygen and water in flexible-shelled eggs of the snapping turtle (*Chelydra serpentina*). *Respiration Physiology*, **49**, 279–91.

Ferguson, M. W. J. (1982). The structure and composition of the eggshell and embryonic membranes of *Alligator mississippiensis*. *Transactions of the Zoological Society of London*, **36**, 99–152.

— (1985). The reproductive biology and embryology of crocodilians. In *Biology of the Reptilia*, vol. 14, Development A, eds C. Gans, F. Billet & P. F. A. Maderson, pp. 329–491. New York: Wiley.

Grigg, G. C. & Beard, L. (1985). Water loss and gain by eggs of *Crocodylus porosus*, related to incubation age and fertility. In *Biology of Australasian Frogs and Reptiles*, ed. G. Grigg, R. Shine & H. Ehmann, pp. 353–59. Sydney: Surrey Beatty Pty Ltd.

Hoyt, D. F. (1976). The effect of egg shape on the surface-volume relationships of birds' eggs. *Condor*, **78**, 343–9.

Kayar, S. R., Snyder, G. K., Birchard, G. F. & Black, C. P. (1981). Oxygen permeability of the shell and membranes of chicken eggs during development. *Respiration Physiology*, **46**, 209–21.

Kutchai, H. & Steen, J. B. (1971). Permeability of the shell and shell membranes of hens' egg during development. *Respiration Physiology*, **11**, 265–78.

Lillywhite, H. B. & Ackerman, R. A. (1984). Hydrostatic pressure, shell compliance and permeability to water vapor in flexible-shelled eggs of the colubrid snake *Elaphe obsoleta*. In *Respiration and Metabolism of Enbryonic Vertebrates*, ed. R. S. Seymour, pp. 121–35. Dordrecht: Dr W. Junk.

Lomholt, J. P. (1976). The development of the oxygen permeability of the avian egg shell and its membranes during incubation. *Journal of Experimental Zoology*, **198**, 177–84.

Lutz, P. L., Bentley, T. B., Harrison, K. E. & Marszalek, D. S. (1980). Oxygen and water vapour conductance in the shell and shell membrane of the american crocodile egg. *Comparative Biochemistry and Physiology*, **66A**, 335–8.

Manolis, S. C., Webb, G. J. W. & Dempsey, K. E. (1987). Crocodile egg chemistry. In *Wildlife Management: Crocodiles and Alligators*, eds G. J. W. Webb, S. C. Manolis & P. J. Whitehead, pp. 445–72. Sydney: Surrey Beatty Pty Ltd.

Packard, G. C. & Packard, M. J. (1988). The physiological ecology of reptilian eggs and embryos. In *Biology of the Reptilia*, vol. 16, Ecology B, eds C. Gans & R. B. Huey, pp. 523–605. New York: Alan R. Liss Inc.

Packard, G. C., Packard, M. J. & Boardman, T. J. (1981). Patterns and possible significance of water exchange by flexible-shelled eggs of painted turtles (*Chrysemys picta*). *Physiological Zoology*, **54**, 165–78.

Packard, G. C., Taigen, T. L., Packard, M. J. & Shuman, R. D. (1979). Water-vapor conductance of testudinian and crocodilian eggs (Class Reptilia). *Respiration Physiology*, **38**, 1–10.

Packard, M. J. (1980). Ultrastructural morphology of the shell and shell membrane of eggs of common snapping turtles (*Chelydra serpentina*). *Journal of Morphology*, **165**, 187–204.

Packard, M. J. & Hirsch, K. (1989). Structure of shells from eggs of the geckos *Gekko gecko* and *Phelsuma madagascarensis*. *Canadian Journal of Zoology*, **67**, 746–58.

Packard, M. J. & Packard, G. C. (1979). Structure of the shell and tertiary membranes of eggs of softshell turtles (*Trionyx spiniferus*). *Journal of Morphology*, **159**, 131–44.

Packard, M. J., Packard, G. C. & Boardman, T. J. (1982). Structure of eggshells and water relations of reptilian eggs. *Herpetologica*, **38**, 136–55.

Paganelli, C. V. Ackerman, R. A. & Rahn, H. (1978). The avian egg: In vivo conductances to oxygen, carbon dioxide and water vapor in late development. In *Respiratory Function in Birds, Adult and Embryonic*, ed. J. Piiper, pp. 212–18. Berlin: Springer-Verlag.

Paganelli, C. V., Olszowka, A. & Ar, A. (1974). The avian egg: Surface area, volume and density. *Condor*, **76**, 319–25.

Prange, H. D. & Ackerman, R. A. (1974). Oxygen consumption and mechanism of gas exchange of green turtle (*Chelonia mydas*) eggs and hatchlings. *Copeia*, **1974**, 758–63.

Rahn, H. & Ar, A. (1974). The avian egg: Incubation time and water loss. *Condor*, **76**, 147–52.

Rahn, H., Paganelli, C. V. & Ar, A. (1974). The avian egg: Air cell gas tensions, metabolism and incubation time. *Respiration Physiology*, **22**, 297–309.

Romanoff, A. L. & Romanoff, A. J. (1949). *The Avian Egg*. New York: Wiley.

Seymour, R. S. & Ackerman, R. A. (1980). Adaptations to underground nesting in birds and reptiles. *American Zoologist*, **20**, 437–47.

Seymour, R. S. & Piiper, J. (1988). Aeration of the shell membranes of avian eggs. *Respiration Physiology*, **71**, 101–16.

Somma, L. A. (1989). The water vapor conductance of squamate reptilian eggs: the influence of scaling on nesting ecology. *First World Congress of Herpetology* Abstract (p. 278), University of Kent, Canterbury.

Spotila, J. R., Weinmiller, C. J. & Paganelli, C. V. (1981). Shell resistance and evaporative water loss from bird eggs: Effects of wind speed and egg size. *Physiological Zoology*, **54**, 195–202.

Thompson, M. B. (1985). Functional significance of the opaque white patch in eggs of *Emydura macquarii*. In *Biology of Australasian Frogs and Reptiles*, eds G. Grigg, R. Shine & H. Ehmann, pp. 387–85. Sydney: Surrey Beatty Pty Ltd.

Tracy, C. R. & Sotherland, P. R. (1979). Boundary layers of bird eggs: do they ever constitute a significant barrier to water loss. *Physiological Zoology*, **52**, 63–6.

Tranter, H. S., Sparks, N. H. C. & Board, R. G. (1983). Changes in structure of the limiting membrane and in oxygen permeability of the chicken egg integument during incubation. *British Poultry Science*, **24**, 537–47.

Tullett, S. G. & Board, R. G. (1976). Oxygen flux across the integument of the avian egg during incubation. *British Poultry Science*, **17**, 441–50.

Wangensteen, O. D., Wilson, D. & Rahn, H. (1970/71). Diffusion of gases across the shell of the hen's egg. *Respiration Physiology*, **11**, 16–30.

Webb, G. J. W., Buckworth, R. & Manolis, S. C. (1983*a*). *Crocodylus johnstoni* in the McKinlay River, N.T. VI. Nesting Biology. *Australian Wildlife Research*, **10**, 607–37.

Webb, G. J. W., Buckworth, R., Sack, G. C. & Manolis, S. C. (1983*b*). An interim method for estimating the age of *Crocodylus porosus* embryos. *Australian Wildlife Research*, **10**, 563–70.

Webb, G. J. W., Manolis, S. C., Dempsey, K. E. & Whitehead, P. J. (1987*b*). Crocodilian eggs: A functional overview. In *Wildlife Management: Crocodiles and Alligators*, eds G. J. W. Webb, S. C. Manolis & P. J. Whitehead, pp. 417–22. Sydney: Surrey Beatty Pty Ltd.

Webb, G. J. W., Manolis, S. C., Whitehead, P. J. & Dempsey, K. E. (1987a). The possible relationship between embryo orientation, opaque banding and the dehydration of albumen in crocodile eggs. *Copeia*, **1987**, 252–7.

Whitehead, P. J. (1987). Respiration of *Crocodylus johnstoni* embryos. In *Wildlife Management: Crocodiles and Alligators*, eds G. J. W. Webb, S. C. Manolis & P. J. Whitehead, pp. 473–97. Sydney: Surrey Beatty Pty Ltd.

18

Metabolism and energetics of reptilian and avian embryos

CAROL MASTERS VLECK
AND DONALD F. HOYT

Introduction

Eggs have served as useful model systems for the study of vertebrate growth and energetics of development because they represent a relatively closed system in terms of energy flow. Measurements of growth and energy utilisation can be made easily because, unlike the situation in mammals, the developing embryos are separate from the mother. In addition, metabolic rates of eggs are easy to interpret without the confounding effects of activity, uncontrolled feeding and general stress that are inherent, for instance, in studies of anuran larval development (Feder, 1981). Several reviews are available on the energetics of avian embryos (C. H. Vleck, Vleck & Hoyt, 1980; Bucher & Bartholomew, 1984; Ar et al., 1987; Bucher, 1987; Hoyt, 1987; Vleck & Vleck, 1987). Less information, however, is available for reptile eggs and there are no reviews dealing specifically with reptilian embryonic growth and energetics. Several authors have compared the energetics of development in birds and reptiles (Dmi'el, 1970; Ricklefs & Cullen, 1973; Ackerman, 1980, 1981; Gettinger, Paukstis & Gutzke, 1984) but no recent comparative analysis has been done, despite the growing interest in reptilian egg biology. In this chapter we will discuss the developmental patterns that are evident in bird and reptile eggs. These patterns are compared and an attempt is made to account for them. The problems inherent in making such comparisons between the Classes Reptilia and Aves are discussed. This chapter is limited to oviparous reptiles and we do not discuss energetics and growth of embryos of viviparous species.

A comparison of reptilian and avian egg biology

One of the major features that distinguishes birds and reptiles from lower vertebrates is the development of a macrolecithal, amniotic egg. The shell offers protection from the environment and the yolk furnishes nutrients for growth and maintenance. Avian and reptilian eggs do differ in important aspects, however. The calcified shell of bird eggs, although it can vary markedly in microscopic structure (Board, 1982; Board & Sparks, Chapter 6), serves in all birds as an important barrier to limit water loss as well as a pathway for the diffusion of respiratory gases (Paganelli, Chapter 16). In reptiles, on the other hand, calcification of the eggshell varies markedly between different groups from the thick, calcareous shells of eggs in the order Crocodilia and some turtles (order Chelonia) and a few lizards (order Squamata) to the flexible, parchment-like eggshells found in most lizards and snakes, and many turtles (Packard & DeMarco, Chapter 5). Consequently the direction and extent of water movement across the shell varies tremendously between species of reptiles. Another major difference in the basic egg biology between the two groups has to do with the placement and care of the eggs. Nearly all bird eggs are laid in the open and all bird eggs are tended by adult birds, primarily to control their temperature. By contrast, most reptile eggs are deposited in sheltered sites beneath rocks or shrubbery or buried in soil or decaying vegetation. Most are given no post-ovipositional attention by parents. Consequently, ambient conditions around reptile eggs are potentially more variable, and gas concentrations may be more extreme than those found in most incubated bird nests. Those features of bird and reptile eggs that complicate a comparative analysis of growth, metabolism, and energetics of development are discussed first.

Incubation temperatures are low and variable in reptiles

Avian development takes place within fairly narrow thermal limits that are determined by the interaction of the incubating adults and the nest environment (White & Kinney, 1974; Drent, 1975; Vleck, 1981; Deeming & Ferguson, Chapter 10). Mean egg temperature in different species of birds varies only from about 32–37 °C although temperatures of the eggs when they are not attended can be more extreme for short periods of time (Drent, 1975; Webb, 1987; Rahn, Chapter 21). The mean egg temperature reported for penguins (Sphenesciformes) is slightly less than 30 °C (Webb, 1987). Incubation temperatures for reptile eggs are almost always lower (average 26–32 °C) and vary widely within and between species (Packard & Packard, 1988; Deeming & Ferguson, Chapter 10). Within a species, incubation temperatures may vary on both a daily and a seasonal basis. For instance, egg temperatures in shallow nests of the turtle, *Emydura macquarii*, in South Australia, can vary up to 12 °C in one day, and are almost 6 °C higher at the top of the nest chamber than at the bottom. The mean minimum temperature in the nest, 16.9 °C, is 10 °C lower than the mean maximum temperature (Thompson, 1988). The mean temperature in American crocodile (*Crocodylus acutus*) nests is 30.9 °C in June and rises to 34.3 °C in August when the eggs hatch (Lutz & Dunbar-Cooper, 1984) although in most crocodile nests, the daily fluctuations are seldom more than 2–4 °C (Packard & Packard, 1988).

In general, reptilian development can proceed over a wide range of temperatures (Packard, Tracy & Roth, 1977), in contrast to the situation in birds, where temperatures must be maintained within just a few degrees for normal development (Lundy, 1969; Drent, 1975; Webb, 1987). For instance, the temperature tolerance for normal development in the lizard, *Dipsosauris Aorsalis*, ranges from 28–38 °C (Muth, 1980) and, in *Sceloporus undulatus*, from about 25–35 °C (Sexton & Marion, 1974).

Temperature will effect the rate of development and incubation period (Deeming & Ferguson, Chapter 10). Zarrow & Pomerat (1937) found about a ten-fold increase in rates of metabolism with temperatures increasing from 12 °C to 30 °C in eggs of the smooth green snake (*Liopeltis vernalis*).

In addition to these well-understood effects of temperature on rate constants, there may be other effects of temperature on energy use patterns. Gutzke *et al.* (1987) examine the effects of incubation at 22, 27 and 32 °C on development of the turtle, *Chrysemys picta*, and report that hatchlings developing at the lower temperature are larger, but have less residual yolk at hatching, and mobilise more non-lipids from their yolks than do hatchlings from eggs incubated at the higher temperatures. Other effects of temperature on growth patterns have been demonstrated in embryos of the snake *Pituophis melanoleucus* (Gutzke & Packard, 1987) and in *Alligator mississippiensis* (Deeming & Ferguson, 1989). Interestingly, in some reptiles (two turtles and *Sphenodon punctatus*, but not *Crocodylus johnstoni*) there is no difference in the total energy used during development at different temperatures (Booth & Thompson, Chapter 20). Apparently this compensation occurs because at higher incubation temperatures, the increased rate of metabolism is maintained for less time, whereas at lower temperatures, the rate of metabolism is less, but incubation period is longer. In two bird species examined, *Leipoa ocellata* and *Gallus gallus*, total energy used increases at lower incubation temperatures (Booth & Thompson, Chapter 20). Comparisons of temperature-dependent processes in birds and reptiles are difficult, given the differences in incubation temperatures that occur between the groups, and the ill-defined effects of temperature on developmental processes. This is particularly true for eggs developing under natural conditions where temperatures tend to fluctuate. Even measurements on metabolism and incubation period made under constant, controlled conditions are difficult to compare between species when the measurements were not made at the same temperature.

Incubation periods are long and variable in reptiles

Incubation periods (the stage after the egg has been laid and before it hatches) are generally longer in reptiles than in birds and vary more widely between species. In reptiles the length of the incubation period depends upon the extent of embryonic development at oviposition that may be extensive in many squamates (see later). In addition, many turtle embryos exhibit various types of developmental arrest either early in development or before hatching (Ewert, Chapter 11). Consequently incubation periods can range from 12–14 days in some lizards such as *Sceloporus aeneus* (Guillette & Gongora, 1986) to more than a year in some turtles (Goode & Russell, 1968; Congdon, Gibbons & Greene, 1983). In birds, incubation periods increase with increasing mass of the egg in a predictable way (Rahn & Ar, 1974), although some birds, notably some sea birds and parrots, have incubation periods that may be more than twice those predicted for eggs of their size (Whittow, 1980; Bucher, 1983). Energy costs of development in these birds with long incubation periods are appreciably higher than for birds with shorter incubation periods (Ackerman *et al.*, 1980; Vleck & Kenagy, 1980; C. Vleck *et al.*, 1980; Vleck, Vleck & Seymour, 1984; Bucher & Bartholomew, 1984; Hoyt, 1987).

Variation in incubation period in reptile eggs is associated with variation in egg temperature (Deeming & Ferguson, Chapter 10). For example, eggs of *Sceloporus undulatus* take more than 60 days to hatch at 25 °C and only 31 days at 35 °C (Sexton & Marion, 1974) and *E. macquarii* eggs laid early in the season in cooler nests, take about 50% longer to hatch than those laid later in the season (Thompson, 1988). In avian eggs, there is some variation in incubation period associated with incubation temperature (Barott, 1937; Boersma, 1982; Deeming & Ferguson, Chapter 10) and there may also be some variance in incubation periods associated with the synchronisation of hatching of a clutch of eggs laid on different days (D. Vleck, Vleck & Hoyt, 1980). The range of incubation periods for a given species, however, is much less for birds than for reptiles because the range of temperatures over which avian development proceeds is much less.

Temperature cannot account for all the variation in reptilian incubation periods. For instance, Lynn & von Brand (1945) found a 20% variance in incubation periods in turtle eggs from the same clutch kept under constant lab conditions. Eggs of *C. acutus* and of the sea turtle *Chelonia mydas*, are of similar mass and incubated at similar temperatures, but the *C. acutus* eggs hatch in 90 days compared to 58 days for the *C. mydas* (Lutz & Dunbar-Cooper, 1984). For some turtles the latitude of origin of the eggs affects incubation period (Deeming & Ferguson, Chapter 10).

Some of the variation in incubation periods in reptiles is related to the extent of development within the oviduct before the egg is laid (Shine, 1983). In birds, all embryos are at about the same stage of development at the time of oviposition because all eggs are in the oviduct for about the same length of time after fertilisation (van Tienhoven, 1983). By contrast, reptiles, especially squamates, often exhibit egg retention which results in short incubation periods such as the 12–14 day incubation period of *S. aeneus* mentioned above (Guillette & Gongora, 1986). In some cases, eggs retained in the oviduct for long periods almost cease development, possibly because of anaerobic conditions (Lynn & von Brand, 1945) and the resulting oxygen debt may explain the relatively high rates of oxygen consumption in newly-laid eggs of the snake *Coluber constrictor* (Clark, 1953a). In other species, development must proceed in the oviduct. Embryos of the snake, *Echis colorata*, are already 88 mm long within hours of laying and Dmi'el (1970) suggests that the eggs of Viperidae are laid when the embryos are at an advanced developmental stage to avoid prolonging incubation in the arid desert habitat in which they are laid. If appreciable development does occur before the eggs are laid, then estimates of the cost of embryonic development based on total oxygen consumption of the eggs (see later) will underestimate the actual total cost of embryonic development.

Energy costs of development usually increase with lower intrinsic rates of development, i.e. long incubation periods (see later), and in birds some attention has been paid to possible evolutionary causes of slow, and energetically costly rates of development which may be associated with feeding habits, body temperature, phylogenetic constraints, or a tropical habitat (Lack, 1968; Whittow, 1980; Boersma, 1982; Ricklefs, 1984). In reptiles, the confounding effects of developmental temperature, variable oviductal development, latitude, and the paucity of data on energy costs of development, in general, make such an analysis premature.

Interaction of hydric environment and energy
use in reptile eggs

In bird eggs, there is no net uptake of water and
most bird eggs lose about 12–15% of their
initial mass as water vapour (Ar & Rahn, 1980;
Ar, Chapter 14). The net loss of water and dry
matter from the egg during development just
about balance such that the fraction of water in
the newly laid egg and in the hatchling are
almost the same (Ar & Rahn, 1980). Variation in
water budgets of individual eggs may influence
energy utilisation (Vleck, Chapter 15), but the
effects are small compared to those found in
reptile eggs.

In reptile eggs, there can be a net uptake or a
net loss of water depending upon the species
and the environment around the egg and the
hydration of the egg and hatchling can be very
different (Vleck, Chapter 15). In reptile eggs
with parchment-like shells, the amount of water
uptake by the egg can have pronounced effects
on metabolic processes. In several species, there
is an increase in the length of the incubation
periods, growth rate, and hatchling size, and a
decrease in yolk reserves, in eggs maintained in
relatively wetter substrates compared to eggs in
drier substrates (Packard & Packard, 1984,
1988; Packard, Chapter 13). Oxygen consumption of eggs of the lizard, *Sceloporus virgatus* is
higher when the eggs are maintained on a wetter
substrate (−100 kPa) than on a drier substrate
(−450 kPa) (Fig. 18.1). On the wetter substrate,
the hatchlings are larger than those incubated on
the drier substrate (403 mg vs. 335 mg). The
total oxygen consumed during incubation is
about 20% greater than that consumed during
development on the drier substrate and residual
yolk is about 85% less. (D. Vleck, personal communication). There is little effect of different
hydric environments on the developing embryos
in rigid-shelled eggs (Packard *et al.*, 1981; Gettinger *et al.*, 1984) and metabolic effects are
intermediate in eggs with shells of intermediate
calcification (Packard, Packard & Boardman,
1982). The mechanisms accounting for the
effect of water on metabolic processes are not
clear (Packard & Packard, 1984).

Reptile eggs in nature can be exposed to soil
water potentials in the range of those known to
have metabolic effects (Packard *et al.*, 1985) and
eggs of a particular species can experience wide

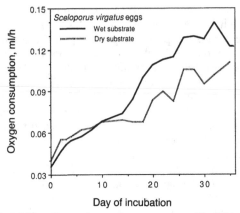

Fig. 18.1.   Rate of oxygen consumption of individual
egg of the lizard, *Sceloporus virgatus*, throughout
incubation when incubated at 30 °C and maintained
in a relatively wet substrate (−100 kPa soil water
potential) and in a relatively dry substrate (−450 kPa
soil water potential). Data are from 3 to 16 eggs
measured each day (D. Vleck, unpublished data).

ranges in net water exchange (Lutz & Dunbar-
Cooper, 1984) which then can affect hatchability
(Leshem & Dmi'el, 1986). Others, however,
have questioned the biological significance of
these effects of hydric condition in the field
(Hotaling, Wilhoft & McDowell, 1985; Plummer & Snell, 1988). A recent study of flexible-
shelled eggs of *C. picta* incubated in natural
nests, indicates that whereas initial egg mass and
incubation period affect hatchling mass, water
exchange of the eggs does not (Ratterman &
Ackerman, 1989). Although the effects of water
exchange in the field on energy budgets for
reptile eggs are not clear, extreme hydric conditions, (causing death by drowning or desiccation) are obviously important to the animal's
biology.

The range in net water movement from eggs
of different species of reptiles is reflected in the
wide range in fractional water contents of newly
laid eggs. The water content of eggs of turtles
Malaclemys temminki and *C. serpentina* that can
take up water from their environment is about
68–72% (Cunningham & Hurwitz, 1936; Ricklefs & Burger, 1977; Wilhoft, 1986). The fraction of water is lower in lizard eggs than in other
reptiles because lizard eggs tend to have very
little albumen at laying (Badham, 1971; Tracy &
Snell, 1985) and it ranges from 40% to 68% in
those species investigated (Ballinger & Clark,
1973; Hadley & Christie, 1974; Vitt, 1978).

Most lizard eggs probably must take up more water from their environment in order to hatch. In fact, in some species, the eggs can take up water equal to three or more times their initial egg mass, depending upon the substrate water potential (Tracy, 1980; D. Vleck, personal communication). Water comprises 39 to 74% of initial egg mass in five species of snakes (Vitt, 1978; Gutzke & Packard, 1987). In eggs of *P. melanoleucus*, water comprises 74% of the initial egg mass, similar to that of precocial bird eggs. The snake eggs, however, can tolerate either a net increase or a decrease in water content, depending upon external conditions, without affecting hatching success (Gutzke & Packard, 1987), whereas hatchability of most bird eggs decreases if 10–20% of the initial mass is not lost as water during development (Ar & Rahn, 1980).

An additional consequence of variable water uptake in reptile eggs has to do with the practical question of what value to use for egg size or mass in allometric analyses. In bird eggs, the initial, or fresh egg mass has served as the scaling factor in most allometric analyses comparing eggs of different sizes, although egg energy content may be a more useful scaling factor (Ar *et al.*, 1987; Vleck & Vleck, 1987). Egg mass is easily measured and widely available for many avian species, and when fresh egg mass is not known, it can be estimated from the egg dimensions or calculated by refilling the egg air cell with water and then weighing the eggs (Hoyt, 1979). In reptile eggs, the mass of the egg may change rapidly after laying and egg dimensions are not constant if the eggshell is flexible. There is no practical method to estimate the size of these eggs at laying if it has not been measured but it is possible for rigid crocodilian and turtle eggs (Deeming & Ferguson, 1990; Iverson & Ewert, Chapter 7). Egg energy content would be more useful than egg mass as a basis for comparing eggs of different species, but this number is relatively difficult to measure, and is not yet available for most species. Even though the energy density (kJ per gram dry mass) is relatively constant between reptilian eggs (Booth & Thompson, Chapter 20), one still must know the dry mass of freshly laid eggs for each species to convert mass values to energy content.

Metabolism

How metabolic patterns are measured

The principal macronutrients in eggs are lipids and proteins which are used in the biosynthesis of the tissues of the developing embryo. In birds, lipids are the primary source of energy for biochemical work (Romanoff, 1967). In at least some reptile eggs, relatively large amounts of protein are consumed in addition to lipids (Lynn & von Brand, 1945; Clark, 1953b; Wilhoft, 1986; Gutzke & Packard, 1987). It is unknown how general the use of protein consumption for energy is by reptile embryos, as are the reasons why reptile embryos would use some proteins for fuel rather than relying on more energy-dense lipids.

The total amount of energy catabolised during embryonic development (energy cost of development) can be calculated from the difference between the total energy contained in eggs at the time of laying and the energy content of hatchlings (including residual yolk and any extra-embryonic membranes). Such methods have been used for both birds (Ricklefs, 1977; Vleck *et al.*, 1984; Ar *et al.*, 1987) and reptiles (Lynn & von Brand, 1945; Wilhoft, 1986). Some workers have measured the decrease in lipids during development in reptile eggs (Congdon & Tinkle, 1982; Congdon *et al.*, 1983; Noble, Chapter 2), but this value will underestimate energy cost of development if significant protein is also catabolised.

Metabolic activity within an egg can also be measured as rates of oxygen consumption, carbon dioxide production or heat production. These measurements can be repeated throughout the incubation period to provide a picture of the ontogeny of energy metabolism throughout development. Integrating metabolic rate measurements over the period of incubation provides a measure of energy cost of development that agrees well with the method described above (Vleck *et al.*, 1984). Numerous workers have measured rates of gas exchange in both bird and reptile eggs (Vleck & Vleck, 1987; Packard & Packard, 1988).

For buried eggs, an indication of embryonic metabolism can also be gained by measuring gas concentrations in the soil around the eggs (Prange & Ackerman, 1974; Seymour & Ackerman, 1980; Lutz & Dunbar-Cooper, 1984).

This can provide an estimate of metabolism of a whole clutch of eggs *in situ*. Interpretation of data using this method is complicated, however, if there are significant or changing rates of respiration of soil organisms or changes in the soil diffusion coefficients such as might occur with changes in water content (Seymour, Vleck & Vleck, 1986). Patterns of nitrogen metabolism as a function of incubation time have been investigated in *C. constrictor* embryos (Clark, 1953*b*), but these can not be used to estimate total energy use.

Patterns in the ontogeny of metabolism

In birds, the ontogeny of metabolism varies with developmental mode (Vleck, Hoyt & Vleck, 1979). Nice (1962) classifies developmental modes of bird into four major groups: precocial, semiprecocial, semialtricial and altricial. The major distinctions between the developmental modes deal with the chicks' relative independence from the parent birds. Precocial hatchlings can thermoregulate at, or soon after, hatching, are mobile, and can feed themselves. Altricial hatchlings are completely dependent on their parents for nourishment and thermoregulation. There is a continuum in hatchling types between extremely precocial and extremely altricial. The patterns of increase in metabolism with time can be seen in different sized eggs by normalising the data. This is done by expressing metabolic rate as a fraction of the rate just prior to internal pipping (pre-IP rate) and time as a fraction of the total incubation patterns (Fig. 18.2). In altricial bird species, embryonic metabolism increases nearly exponentially throughout the entire incubation period, whereas in precocial species, embryonic metabolism increases nearly exponentially only until about 80% of the way through incubation; after which it may stabilise, increase only slowly, or even decline (C. Vleck *et al.*, 1980). These patterns are similar to those found in embryonic growth and are thought to be related to growth patterns (see later).

In reptiles, there are no differences between species in the developmental mode since all reptile hatchlings are independent of parental care and are therefore precocial by avian standards. There are, however, different patterns in the ontogeny of metabolism. The metabolic patterns exhibited by turtle eggs are similar to those of precocial avian eggs (Fig.

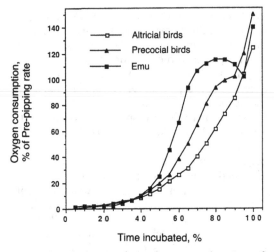

Fig. 18.2. Oxygen consumption as a function of time in egg of precocial species, altricial species and one precocial ratite, the emu, *Dromicaius novaehollandiae* to illustrate the range of patterns in ontogeny of metabolism exhibited by birds. Oxygen consumption is expressed as a percentage of the pre-internal pipping rate and time as a percentage of the incubation period. Curves connect mean values of metabolic rate taken over 5% increments of the incubation period. (Adapted from C. Vleck *et al.*, 1980.)

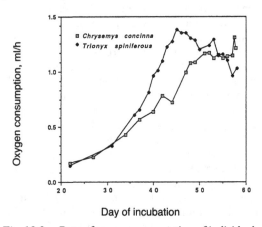

Fig. 18.3. Rate of oxygen consumption of individual turtle eggs measured at 29 °C. The value for *Chrysemys concinna* (open boxes) and *Trionyx spiniferus* (closed diamonds) are daily means based on 5–15 eggs measured each day (Hoyt & Albers, unpublished data).

18.3). The patterns are more or less sigmoidal with an initial exponential increase in oxygen consumption ($\dot{V}_{O_2}$) followed, either by a peak and a subsequent drop, by a plateau, or by a decrease in the rate of increase (Lynn & von Brand, 1945; Ackerman, 1980; Gettinger *et al.*,

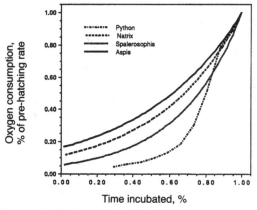

Fig. 18.4. Oxygen consumption, as a percentage of the rate just prior to hatching, plotted as a percentage of the incubation period elapsed, for four species of snakes, measured at 30 ± 1 °C. Data are from Black *et al.*, (1984) for *Python molurus* and from Dmi'el (1970) for two colubrids, *Natrix tessellata* and *Spalerosophis cliffordi* and the viperid, *Aspis cerastes*.

1984; Webb, Choquenot & Whitehead, 1986; Lesham, 1989; Ewert, Chapter 11 Hoyt & Albers, unpublished data). The pattern of increase of V_{O_2} in eggs of one species of lizard, *S. virgatus*, is also sigmoidal (Fig. 18.1). Metabolic rate in developing crocodilian eggs is very similar to the pattern shown in Fig. 18.2 by emu (*Dromicaius novaehollandiae*) eggs (Whitehead, 1987; Thompson, 1989). A peak or a plateau is also seen in the metabolic rate of tuatara eggs, *Sphenodon punctatus* (Thompson, 1989; Booth & Thompson, Chapter 20). In snake embryos, however, the pattern of increase is more similar to that found in altricial bird species (Fig. 18.4). Increases in metabolic rate with time in five species of snake embryos are exponential although there may be an inflection point in the curves a day or two prior to hatching (Dmi'el, 1970). Oxygen consumption of *Python molurus* eggs increases nearly exponentially until about the last 10% of incubation at which time the rate of increase slows (Black *et al.*, 1984). On the other hand, carbon dioxide production of *C. constrictor* embryos maintained at 24 °C is sigmoidal in pattern, as is embryonic growth, and rates of oxygen consumption also appear to begin to level off late in incubation (Clark, 1953*a*).

Growth

Data on embryonic growth throughout incubation are more difficult to obtain than those on

metabolism because the process of making the measurement terminates development. Thus, large numbers of eggs are needed in order to define patterns in growth. The best data sets on growth for birds are from domestic species, for which large numbers of eggs can be obtained from commercial sources (Romanoff, 1960). For non-domestic species, growth data should actually be easier to obtain for reptiles than for birds because large numbers of eggs often can be obtained from a single female. In addition, reptile eggs from one individual are all available at the same time and at the same stage of development, because the whole clutch is laid at once rather than at the rate of one egg per day as in birds.

In birds, patterns in growth vary with developmental mode in the same way that ontogeny of metabolism does. In altricial birds, embryo mass and growth rate (grams of new tissue added per day) increase continuously throughout incubation. In precocial species, embryos approach hatchling mass as early as 80% of the way through incubation, and growth rate declines, sometimes radically, toward the end of incubation (C. Vleck *et al.*, 1980). In all growing embryos, the increase in mass with time can be described by a decaying exponential equation, but the two groups can be distinguished by the position of the inflection point of their growth curves, the time when growth rate begins to decline. In precocial species, growth rate declines and the curve inflects before hatching, whereas in altricial species, the estimated inflection point does not occur until after hatching (Laird, 1966; C. Vleck *et al.*, 1980). The similarities between growth and metabolic rate have led to the suggestion that the two are functonally related (Hoyt, Vleck & Vleck, 1978; C. Vleck *et al.*, 1980; Vleck & Vleck, 1987; Hoyt, 1987). Developing embryos must expend energy to both maintain existing tissue and to synthesise new tissue. As long as growth rate increases, both maintenance and synthesis costs will increase and thus total metabolic rate must also increase. In precocial species, when growth rate declines, so will growth costs, and even though maintenance costs do not decrease, the declining growth costs can produce a plateau or even a decrease in the total metabolic rate (Fig. 18.5).

In reptiles, the increase in metabolic rate parallels increase in body mass, presumably for the same reason as in birds. There are similarities in the logistic patterns of metabolism and

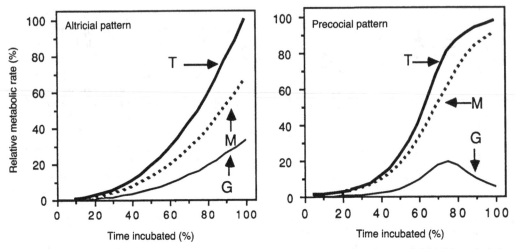

Fig. 18.5. Comparison of embryonic metabolic rate in an altrical and precocial bird based on a cost of growth and maintenance model developed by Hoyt (1987). The fractions of metabolism devoted to growth (*G*) and maintenance (*M*) add together to equal the total (or observed) metabolic rate (*T*). The amount of energy devoted to growth (area under *G*) is similar in both, but the precocial species spend more on maintenance than the altrical species (area under *M*) (adapted from Hoyt, 1987).

growth in *C. constrictor* embryos and growth of other snake embryos is exponential, as is the increase in metabolism (Clark, 1953*a*; Dmi'el, 1970). Growth data for *C. mydas* and *Caretta caretta* are best fit by a logistic equation, as are data on oxygen consumption (Ackerman, 1980, 1981). Lynn & von Brand (1945) do not report masses for the turtle species they studied, but they do report that carapace length increases in an approximately logistic fashion. Dry mass of *C. serpentina* embryos appears to increase in a logistic fashion (Morris *et al.*, 1983) as does rate of oxygen consumption (Gettinger *et al.*, 1984). In *C. picta* embryos, growth rate decreases at the end of incubation when eggs are kept on a moist substrate (although the growth appears to be more nearly exponential on a drier substrate) (Packard & Packard, 1986) and metabolism in the congener, *Chrysemys concinna*, shows a plateau at the end of incubation (Fig. 18.3). Increase in mass of lizard embryos may be either sigmoidal, *Amphibolurus barbatus* or exponential, *Iguana iguana* (Ricklefs & Cullen, 1973; Packard *et al.*, 1985), but metabolic data are not available for these species. Increase in mass of *A. mississippiensis* embryos is logistic in form: there is a distinct plateau in embryo length as well as several other embryonic measurements well before hatching (Deeming & Ferguson, 1989), at the same time the metabolic rate is reaching a peak and then declining (Thompson, 1989).

The reason why growth rates in altrical bird species and many snakes do not decrease appreciably before hatching, as they do in precocial bird species, some lizards and turtles, is not clear. In birds, the altrical condition is evolutionarily derived from the precocial condition (Vleck & Vleck, 1987) and is exhibited by many of those groups generally thought to be the most advanced, e.g. the order Passeriformes (O'Connor, 1984; Sibley, Ahlquist & Monroe, 1988). It is associated with eggs of relatively low energy density and high water content, and with chicks that are relatively immature. The result is a transferral, in time, of energy costs to the parent(s) who spend(s) less energy at the time of laying each egg, and more energy later in care of the hatchlings. Within the class Reptilia, the snakes are probably also the most evolutionarily recent group (Romer, 1966), and they also show the presumably derived pattern in embryonic growth. Whether there is any adaptive significance to this pattern of embryonic growth to life history patterns in snakes remains to be determined.

For some species, it is probably adaptive to complete growth some time before the end of incubation. The time interval at the end of incubation, then, can be shortened or lengthened depending on external conditions. This is true of species that must have synchronous hatching to insure post-hatching survival. For example, eggs of the rhea *Rhea americana* can hatch in as little as 27 days, if stimulated by the presence of older eggs, even though the normal incubation period is 36–37 days (Bruning, 1974). Patterns in oxygen consumption suggest the growth of the rhea embryo is essentially complete by this time, allowing hatching to occur early if need be to synchronise hatching in the clutch (D. Vleck et al., 1980; Cannon, Carpenter & Ackerman, 1986). There may be benefits to synchronous hatching for many reptiles as well (Webb et al., 1986; Thompson, 1989). For many turtles, lizards, and all crocodilians that nest in subterranean nests or mounds of earth, the hatchlings may have to cooperate in digging out of the nest (Carr & Hirth, 1961; Prange & Ackerman, 1974). Even if all the eggs are laid in the nest at the same time, there is probably some individual variation in developmental rates, particularly in shallow nests where temperatures within the clutch vary from place to place (Thompson, 1989), that could result in different hatching times for the eggs in the clutch if this event was not synchronised.

Energetics

Eggs are laid with a pre-determined amount of energy that must be sufficient for development, hatching and any post hatching activities that occur prior to feeding if the embryo is to survive and join the breeding population. In birds and reptiles, much of the emphasis on reproductive energetics has been concerned with energy costs to the parent (maternal investment) and how these vary with clutch size, egg size, season, etc. (Congdon, Dunham & Tinkle, 1982; Walsberg, 1983). Relatively little attention has been given to energy use by the developing embryo. Here we deal specifically only with the energetics of embryonic development in birds and oviparous reptiles, i.e. what happens to the energy that is originally deposited within the egg.

Patterns in energetics of avian embryonic development

Bird eggs have provided a fruitful source of information about patterns in embryonic energetics because: (1) they vary in size over more than three orders of magnitude from the 0.5 g eggs of hummingbirds to the 1.4 kg eggs of the ostrich, *Struthio camelus*; 2) incubation periods vary from as little as 11 to more than 80 days; and 3) the condition of the hatchling varies

from extremely altricial to extremely precocial. Analyses of the functional relationships between egg size and energy content, incubation period, relative maturity at hatching, as well as hatchling mass and energy costs of development have been an important goal of many avian egg biologists. The major results to emerge from these comparative analyses are reviewed briefly here.

Much of the interspecific variation in incubation period, oxygen consumption, hatchling size, total energy costs of development, etc. is associated with differences in egg size and the most commonly used measure of this is initial egg mass (Rahn, Paganelli & Ar, 1974; Hoyt et al., 1978; D. Vleck et al., 1980; Hoyt & Rahn, 1980; Rahn & Ar, 1980; Bucher, 1983; Hoyt, 1987). In these analyses, there has been a dichotomy drawn between altricial and precocial species. When comparing eggs of the same mass, eggs of precocial species have longer incubation periods, higher energy costs of development, produce larger dry mass of hatchling tissue with more residual yolk, but have similar rates of oxygen consumption just prior to hatching (Vleck & Vleck, 1987). The difference in the ontogeny of oxygen consumption and growth described above can account for some of these differences. Precocial species show a plateau or even a decline in oxygen consumption associated with a decrease in growth rate towards the end of incubation whereas altricial species do not (Fig. 18.5). The result of this pattern is that precocial species incur higher total energy costs because the embryo is larger for a greater part of the incubation period than is an altricial embryo; the precocial embryo's total maintenance costs to support the larger body size are greater than those of an altricial embryo (C. Vleck et al., 1980; Hoyt, 1987). The same argument applies to eggs with unusually long incubation periods. The high maintenance costs greatly increase the overall cost of development compared to eggs with shorter incubation periods and lower costs of development (Ackerman et al., 1980; C. Vleck et al., 1980; Pettit & Whittow, 1984; Bucher & Bartholomew, 1984; Hoyt, 1987).

Eggs and hatchlings of altricial species have a lower energy density and higher water content than eggs and hatchlings of precocial species (Ar & Yom-Tov, 1978; Ar & Rahn, 1980; Carey, Rahn & Parisi, 1980). Water makes up 83–84% of egg contents in altricial eggs and hatchlings and 72–75% of contents in precocial species. Energy density is about 4.8 kJ/g contents in altricial species and about 8.0 kJ/g in precocial species (Carey *et al.*, 1980). The result of this fundamental difference is that when similar allometric analyses are carried out using egg energy content rather than egg mass as the scaling factor, the differences between altricial and precocial species virtually disappear (Ar *et al.*, 1987; Vleck & Vleck, 1987). This means that from the viewpoint of how energy in the egg is used, altricial and precocial species do not differ; in both, the energy cost of development per energy incorporated into the hatchling is about the same (about 15 kJ/g yolk-free, dry hatchling mass). It appears that before hatching, all avian embryos expend about the same percentage of the energy stored in the egg at laying (30–35%); and the hatchling, plus residual yolk, represents about 57% of the initial energy content (Ar *et al.*, 1987; Vleck & Vleck, 1987). The rest remains as extra-embryonic membranes. Eggs with long incubation periods fit within the same conceptual framework. These eggs must have unusually high energy densities at laying to offset the high energy costs of development. For example, the bird with the longest incubation period (80 days) is the kiwi (*Apteryx australis*) which also has the highest egg energy density (12.8 kJ/g egg contents) of any bird (Calder, Parr & Karl, 1978).

Energetics of reptilian embryonic development

There is now sufficient information about energy use and related variables in reptile eggs of varying sizes to do similar allometric analyses. Data on eggs and energy use for 19 species of reptiles measured at nearly the same temperature are used in the following analysis (Table 18.1). Egg energy content is available for a few species of reptiles (Lynn & von Brand, 1945; Derickson, 1976; Ballinger & Clark, 1973; Tinkle & Hadley, 1975; Ricklefs & Burger, 1977; Vitt, 1978; Congdon & Tinkle, 1982), but egg energy content is not available for enough of the species in Table 18.1 to make it possible to base the allometric analysis on egg energy content rather than egg mass. This exercise needs to be done when such data become available.

The following caveats concerning the data set must be stated. Only data taken from eggs incubated under constant conditions at 29–31 °C have been used so that assumptions about Q_{10} effects on metabolism, incubation period,

energy use, or hatchling size do not have to be made. Even within this narrow temperature range, however, there are still, undoubtedly, some effects of temperature on developmental processes. Wet egg mass at laying is used as the scaling factor, even though water content, energy content and water uptake during incubation vary between the species. Data from two studies on *Trionyx spiniferus* have been included because the eggs in the two studies vary slightly in mass. Data from two populations of *C. mydas* are included because Ackerman (1980) states they give distinctly different results in his laboratory. The pre-hatch $\dot{V}_{O_2}$ measurements are taken as the highest rate of oxygen consumption reported for the eggs and are not necessarily the rate measured right before hatching. Hatchling masses are reported as wet, yolk-free masses. Dmi'el (1970) does not state whether his hatchling masses include residual yolk or not, but we follow Ackerman's (1980) assumption that they are yolk-free. The hatching data for *P. molurus* probably does include yolk (Black *et al.*, 1984), but we have no information on residual yolk in snake hatchlings in order to correct it, and thus it represents an overestimate of the true hatchling mass. Gettinger *et al.* (1984) report only dry masses for hatchlings and these have been corrected to wet masses based on a water content of 82.1% measured by Hoyt & Albers (unpublished data) in one of the same species (*T. spiniferus*). For those studies in which water content of the incubation medium is varied (Gettinger *et al.*, 1984; Black *et al.*, 1984; D. Vleck, unpublished data), only data from eggs kept in the wettest environment are used. To convert oxygen consumption to energy use we have followed Gettinger *et al.* (1984) in using an energy equivalence of 19.7 kJ litre^{-1} of oxygen consumed.

For this set of reptile species, incubation period, pre-hatch $\dot{V}_{O_2}$, yolk-free hatchling mass and energy used during development scale as power functions of egg mass at laying (Fig. 18.6). Equations and statistics for the least-square regressions are given in Table 18.2.

Comparisons of embryonic energetics in reptiles and birds

We can now compare these scaling variables in reptiles with those previously reported for altricial and precocial birds (Vleck & Vleck, 1987; Hoyt, 1987) using analysis of covariance. For all

Table 18.1. *Variables affecting the energetics of embryonic development in reptiles at incubation temperatures of 30±1 °C. Egg mass is initial egg mass. Pre-hatch $\dot{V}_{O_2}$ is the highest rate of oxygen consumption measured before the eggs hatched. Hatchlings masses are wet masses, without residual yolk. Energy used is the total total energy expended by the embryo during incubation.*

Species	Egg mass (g)	Inc. period (d)	Pre-hatch $\dot{V}_{O_2}$ (ml/d)	Yolk-free hatchling mass (g)	Energy used (kJ)	Ref.
Order Chelonia						
Sternotherus minor	5.6	86	12	3.6	9.94	1
Trionyx spiniferus	9.1	58	25.6	6	14.02	2
Trionyx spiniferus	9.61	57	31.2	7	13	3
Emydura macquarii	10.4	48.4	30.7	5.1	11.3	4,5
Chrysemys concinna	12.2	59	30.5	6.1	11.68	2
Chelydra serpentina	12.75	58	29.4	6.7	14	3
Carettochelys insculpta	33.7	69	65.8	19.76	38.6	6
Caretta caretta	42.6	50	108.1	18.09	38.2	7
Chelonia mydas	48.2	65	90.2	19.9	61.9	7
Chelonia mydas	61.6	63	137	30.84	53.96	7
Dermochelys coriacea	85.5	57	252	45.69	92.49	8
Order Squamata						
Sceloporus virgatus	0.38	35.7	3.35	0.38	1.61	9
Natrix tessellata	7	37	34.3	5.11	10.44	10
Echis colorata	10	43	24.7	6.22	8.77	10
Cerastes cerastes	11	62	30.7	6.47	17.04	10
Vipera xanthina	14	41	35	10.68	11.39	10
Spalerosophis cliffordi	22	60	67	16.33	26.16	10
Python molurus	220	68	360	116	97.6	11
Order Crocodilia						
Alligator mississippiensis	65.6	68	228	44	104	5,8
Crocodylus johnstoni[1]	74.8	91	171.4	37.5	118.9	12
Crocodylus porosus	112.1	94	303.8	—	205.7	12,13

[1]Data for *Crocodylus johnstoni* are reported as the mean of values measured at 29 ° and 31 °C.
1. Ewert, unpublished data; 2. Hoyt & Albers, unpublished data; 3. Gettinger *et al.* (1984); 4. Thompson (1987); 5. Thompson (1989); 6. Webb, Choquenot & Whitehead (1986); 7. Ackerman (1980); 8. Thompson, unpublished data; 9. D. Vleck, personal communication; 10. Dmi'el (1970); 11. Black *et al.* (1984); 12. White-head (1987); 13. Webb *et al.* (1987).

four set of regression equations, the scaling factors for reptile eggs differ either in slope or intercept from those of the bird eggs (Table 18.2). In most cases the slopes of the equations are not all the same and thus comparisons of the values for different species depend upon egg size. We have used the allometric equations for reptiles (Table 18.2) and birds (Vleck & Vleck, 1987; Hoyt, 1987) to compare predicted values for a typical 20 g reptile egg with predicted values for a 20 g altricial or precocial bird egg (Table 18.3).

Over the normal range of egg sizes, incubation periods are much longer in reptile eggs than in bird eggs of the same size and rates of metab-olism are much lower (Fig. 18.6). As with birds, the scatter about the regression line for reptiles is considerable and in fact, within the Order Chelonia, there appears to be no clear relationship between incubation period and egg size (Ewert, 1985). The important point is, however, that incubation periods are much longer in reptiles than in birds. The predicted incubation period for a 20 g reptile egg is more than twice that of the bird eggs, whereas the rate of metabolism just prior to hatching is about one-fourth the rate predicted for the bird eggs (Table 18.3).

The scaling of hatchling mass with egg mass does not differ as much between birds and reptiles as does the scaling of incubation period

Table 18.2. *Regression equations and statistics for variables expressed as a function of fresh egg mass for reptilian embryonic development at 30±1 °C. Data from Table 18.1. Tests of hypotheses compare reptile scaling factors to those for altricial and precocial birds. Asterisks denote statistically significant differences (P<0.05).*

Y variable	a	b±95% c.i.	SE	r^2	Tests of Hypotheses (ANCOVA)	
					Ho: Same as altricial birds	Ho: Same as precocial birds
Incubation Period (d)	40.91	0.12±.07	0.09	0.38	slope: F=11.93 (df=1.64)* intercept:	F=1.28 (df=1.75) F=123 (df=1.76)*
Pre-hatch $\dot{V}_{O_2}$ (ml/d)	4.68	0.82±.05	0.11	0.95	slope: F=5.52 (df=1.38)* intercept:	F=3.59 (df=1.52) F=248 (df=1.53)*
Yolk-free wet hatchling mass (g)	0.80	0.90±.06	0.07	0.98	slope: F=7.75 (df=1.26)*	F=4.44 (df=1.35)*
Energy cost of development, (kJ)	2.18	0.81±.11	0.15	0.91	slope: F=4.38 (df=1.34)* intercept:	F=1.77 (df=1.54) F=67.86 (df=1.55)*

All regression equations are of the form $Y=a$ (egg mass)[b]. All statistical conclusions apply to the log-transformed data. Data for altricial and precocial bird eggs are from Vleck & Vleck (1987). The hatchling masses for the birds were reported as dry masses and these were converted to wet masses for comparison with the reptile data by assuming a water content of altricial hatchlings of 83% and a water content of precocial hatchlings of 74%. The regression equations for the altricial and precocial birds are statistically different from each other, except in the case of prehatching $\dot{V}_{O_2}$ in which the regression equations are not statisically different.

Table 18.3. *Values for variable affecting the energetics of ambryonic energetics calculated from allometric equations based on egg mass. Values are for a 20 g egg of a reptile incubated at 30 °C, and an altricial bird species and a precocial bird species, both incubated at about 37 °C. Values in parenthesis are the percentage the reptilian value represents of the avian values.*

Variable	Reptile	Altricial Bird		Precocial Bird	
[1]Incubation Period (d)	58.3	22.2	(263)	26.1	(223)
[2]Pre-hatching V_{O_2} (ml/d)	55.1	212.2	(26)	212.2	(26)
[1]Yolk-free, wet hatchling mass (g)	11.8	11.3	(104)	10.5	(112)
[1]Energy cost of development (kJ)	24.8	27.0	(92)	47.0	(53)

[1]Allometric equations for altricial and precocial birds are from Vleck & Vleck (1987). Wet hatchling masses are calculated from dry hatchling masses assuming water content is 83% for altricial hatchlings and 74% for precocial hatchlings (Ar & Rahn, 1980).
[2]Allometric equation for birds is from Hoyt (1987).

and rates of metabolism, although the slope for the reptile line is significantly less than that for the birds (Tables 18.2 & 18.3 & Fig. 18.6). In birds, the slope of the regression relating the logarithm of dry, hatchling mass to logarithm of egg mass is not significantly different from one (Vleck & Vleck, 1987), indicating that hatchling mass is nearly a constant fraction of egg mass. In reptiles, the slope (or exponent) is significantly less than one (Table 18.2); the wet hatchling mass as a fraction of initial egg mass decreases with egg size, at least in this data set. In birds, wet hatchling mass, not including residual yolk, averages 57% (S.D. = 6.2, N = 19) of fresh egg mass (Vleck et al., 1980b). For the reptile species reported here, hatchling mass averages 62.2% (S.D. = 15.7) of fresh egg mass, but the inter-specific variation is greater than that found in birds (range was 41.3% to 98.4%). The relationship between hatchling mass and fresh egg mass differs between squamates, crocodilians, rigid-shelled, and pliable-shelled turtles (Deeming & Ferguson, 1991). This greater variation is to be expected given the differences in water content of reptile eggs when they are laid. Reptile eggs with low initial water content, such as those of most lizards, must absorb water from their environment, and the wet hatchling mass can actually end up larger than the egg mass at oviposition (Tracy & Snell, 1985).

Given the generally lower water content of reptile eggs compared to bird eggs at laying, one would predict that wet hatchling mass expressed as a fraction of initial egg mass would be larger in reptiles than in birds. Within turtles, the fraction that hatchling mass (including residual yolk) represents of initial egg mass in rigid-shelled

species (~60−75%) is less than that within flexible shelled turtles (68–80%), presumably because flexible-shelled eggs take up more water from their environment during development than do rigid-shelled eggs (Ewert, 1985). In other reptile groups, hatchling mass comprises a relatively constant fraction of initial egg mass, but the fraction differs in different taxonomic groups (Ewert, personal communication). In order to compare birds and reptiles, this calculation should be based on either dry masses or energy contents of eggs and hatchlings rather than on wet masses.

The total energy used during development is somewhat lower for reptiles than for birds with eggs larger than about 5 g (Tables 18.2 & 18.3 & Fig. 18.6). Only for the very small eggs of the lizard, *S. virgatus*, does the energy cost of development lie above that predicted for birds and this comparison may be inappropriate because there are no data available for energetics of bird eggs as small as 0.4 g. In the mid-range of egg sizes (10–100 g), for which the allometric equations are probably most reliable, energy costs of development for reptile eggs are 41 to 59% of those for precocial bird eggs and 84 to 96% of those for altricial bird eggs. These figures may seem surprisingly similar in view of the much lower rates of metabolism in the reptile eggs compared to bird eggs. The extremely long incubation periods of the reptile eggs, however, mean that the total energy costs of development are more similar to those of birds than the instantaneous rates of metabolism would suggest.

The energy efficiency of development can be assessed by comparing the total energy

(a)

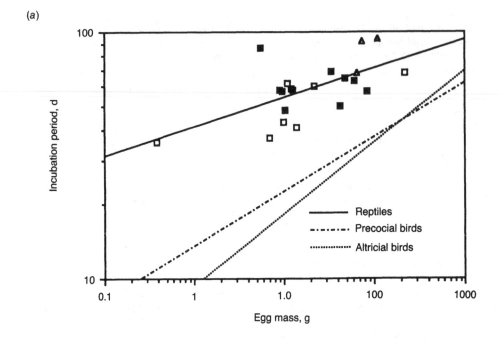

(b)

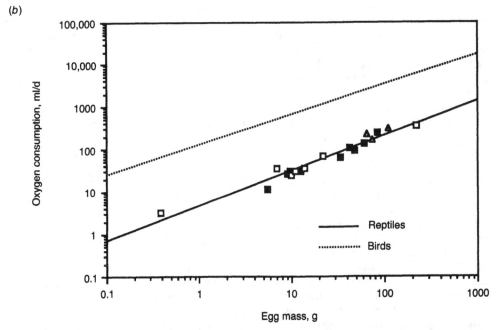

Fig. 18.6.   Increase in (a) length of incubation period, (b) embryonic oxygen consumption just prior to hatching, (c) yolk-free, wet hatchling mass and (d) total energy cost of embryonic development as a function of initial egg mass in birds and lizards. The lines for altricial birds (dotted lines) and precocial birds (dashed lines) are based on least-squares regression equations from data in Vleck & Vleck (1987). The lines for reptiles (solid line) are based on least-squares regression of data in Table 18.1 and the points for reptiles are indicated on the figure. Individual values for chelonian species are plotted with closed squares, for squamates with open squares, and for crocodilians with open triangles. Regression statistics and tests of hypotheses comparing the reptile allometric equations to those of birds are in Table 18.2.

(c)

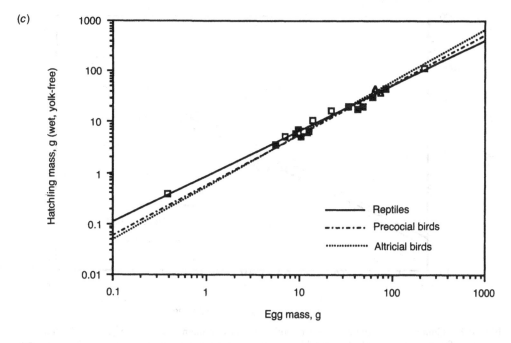

(d)

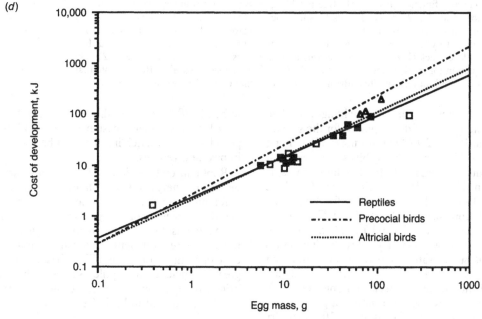

Fig. 18.6 contd.

expended to produce a hatchling of a given size (Fig. 18.7). Over the range of hatchling sizes that have been studied, the line for reptiles lies below those for birds indicating that the energy used to produce the reptile hatchlings is less than that to produce the bird hatchlings. The regression equation describing the relationship for reptiles is not significantly different from that for altricial species, but it differs significantly in intercept (but not slope) from that describing precocial bird species (F = 50.69, df = 1.33). The slope of the relationship for reptiles is less than one (0.856±0.13 c.i.) suggesting that the energy cost to produce a gram of reptile hatchl-

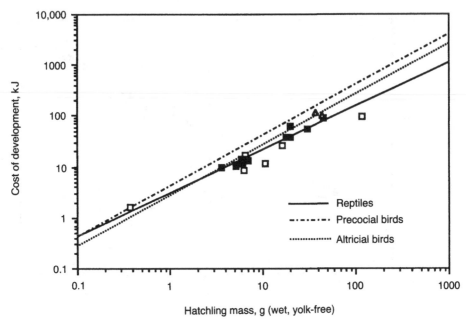

Fig. 18.7. Comparison of the efficiency of embryonic development (energy used during development as a function of wet, yolk-free hatchling mass produced) in reptiles (solid line), altricial birds (dotted line) and precocial birds (dashed line). Individual values for chelonian species are plotted with closed squares, for squamates with open squares, and for crocodilians with open triangles. Lines fit by least squares regression to data for reptiles (see Table 18.1) and that for birds (from Vleck & Vleck (1987) modified to provide wet masses as described in Table 18.2). The regression equation for reptiles (log $Y = 0.462 + 0.86 \log X$) is not significantly different from that for altricial birds, but it differs significantly in intercept from that describing precocial birds.

ing may decrease with wet hatchling size. The mean energy cost of development for the species in Table 18.1 is 2.17 kJ/g wet mass of hatchling produced (S.D. = 0.76). In birds the mean cost to produce a gram of dry hatchling mass is between 15 and 16 kJ/g dry mass (Vleck & Vleck, 1987). The same value for birds calculated from data in Ar *et al.* (1987) yields a mean value between 13 and 14 kJ/g dry mass of embryo. If the water content of reptile hatchlings is about 80% (water content varies from about 77–83%; Ewert, 1979; Hoyt & Albers, unpublished data; D. Vleck, unpublished data), then the energy used to produce a gram of dry hatching mass in reptiles is about 10.8 kJ/g dry mass. This is 70–80% of the value for birds, similar to the values Ackerman (1981) and Gettinger *et al.* (1984) calculated when comparing reptilian and avian costs of development. The energy density of reptile hatchlings may be less than that of avian hatchlings (Booth & Thompson, Chapter 20), but little data are available to test this prediction and the value for energy density of reptile hatchlings would have

to be 20–30% less than that of avian hatchlings in order for the energy cost expressed as kJ used per kJ stored in hatchling tissue to be the same in bird and reptiles.

Reptilian embryonic development appears to be more energy efficient than that of avian embryonic development, although an analysis based on dry mass or energy content of the hatchling needs to be done to indicate just how much more efficient reptilian embryonic development actually is. The basis for the greater efficiency in reptiles seems to be due, primarily, to lower maintenance metabolism in reptiles compared to birds (see later).

Residual yolk

The residual yolk, which is not metabolised, but remains at hatching, is not included in our analysis of energetic efficiency of development. Residual yolk represents biomass and energy that has not been reorganised during development, but simply transported from the maternal parent to her offspring. Thus it represents an energy cost to the parent, but the hatchling has expended essentially no energy for its synthesis or maintenance. Residual yolk does presumably represent an important energy store for the hatchling, however, and possibly source of macromolecules for building. Residual yolk is probably most important for those hatchlings that must engage in energy-expensive posthatching activities before they can begin feeding and are not fed by their parents after hatching. This is particularly true of some megapode birds that must emerge from a mound of soil or vegetation (Vleck *et al.*, 1984) and many reptiles, particularly marine turtles that are buried at considerable depth (Seymour & Ackerman, 1980; Dial, 1987; Werner 1988). It is also important to those reptile embryos that delay hatching (Ewert, Chapter 11).

In birds, residual yolk represents a greater fraction of the wet mass of the hatchling in precocial than in altricial species. Residual yolk averages 14–15% of total hatchling mass in precocial birds, but only 8–10% in altricial birds (Romanoff, 1944; Schmekel, 1960; Ar *et al.*, 1987). Both the residual yolk and the hatchling of altricial species have a higher water content and lower energy density than that in precocial species (Ar *et al.*, 1987); and the energy in the yolk is about 22% of the total energy in an altricial hatchling and about 26% of that in a precocial hatchling (calculated from data in Ar *et al.*, 1987).

There are few data available for residual yolk in reptile hatchlings. The fraction that residual yolk comprises of the total hatchling mass range from 2 to 28% in turtles (Wilhoft, 1986; Hoyt & Albers, unpublished data); from about 1 to 12% in lizards (Werner, 1988; D. Vleck, personal communication); is about 11% in *P. melanoleucus* (Gutzke & Packard, 1987) and ranges from about 5 to 30% in *A. mississippiensis* (Deeming & Ferguson, 1989). It is tempting to try to generalise about the adaptive significance of residual yolk in different species, e.g. calculate the time that metabolism could be sustained in the hatchling using residual yolk as fuel (Werner, 1988) and conclude that some species are more or less dependent on this energy source. It must be kept in mind, however, that the amount of residual yolk within a species varies with the temperature and hydric environment to which the egg is exposed (Gutzke & Packard, 1987; Gutzke *et al.*, 1987; Manolis, Webb & Dempsey, 1987; Werner, 1988; Deem-

ing & Ferguson, 1989; D. Vleck, personal communication), and whether any developmental arrest takes place during incubation (Ewert, Chapter 11). It may also vary with the time of year and maternal energy reserves at laying that affect egg size and composition (Congdon *et al.*, 1983; Tracy & Snell, 1985). The presence of fat bodies as well as residual yolk in some reptile hatchlings, (Snell & Tracy, 1985; Gutzke & Packard, 1987; Werner, 1988) complicates comparisons with bird hatchlings, only a few of which seem to have stores of body fat at hatching.

Comparisons of embryonic energetics to adult energetics

For reptiles, several authors have pointed out the similarities of embryonic oxygen consumption just prior to hatching and those measured in hatchlings (Lynn & von Brand, 1945; Clark, 1953*a*; Prange & Ackerman, 1974). In fact, the embryonic metabolic rates may be even slightly higher than those of the hatchlings (Prange & Ackerman, 1974). By contrast, bird embryos show a marked increase in metabolism after pipping of the eggshell (Fig. 18.2) and metabolic rates of precocial chicks increase to 80–90% of the value of an adult bird of equivalent size within a few days of hatching (Hoyt & Rahn, 1980). This increase in metabolic rate in birds has been attributed to the advent of convective ventilation and the removal of the shell and membranes as a diffusion barrier (Paganelli & Rahn, 1984), the metabolic costs of thermoregulation, and increased costs for activity during and after hatching (Vleck *et al.*, 1984). In reptiles, embryonic metabolic rates are already very similar to those of adults of the same size and at the same temperature, (see later) so it is not surprising that there is no marked increase after hatching as there is in birds.

Metabolic rates of embryos can be compared to those of adult animals of the same mass using the available allometric equations. The metabolic rates of avian embryos are significantly lower than basal metabolic rates of adult birds of the same mass (Hoyt *et al.*, 1978; Vleck *et al.*, 1980*b*). Hoyt (1987) reports a relationship between fresh egg mass (in g) and $\dot{V}_{O_2}$ (in cm^{3-1} day) of avian embryos just before the start of the hatching process. This equation can be expressed in terms of yolk-free hatchling mass (HM in g) as:

$$\dot{V}_{O_2} = 35.4 \text{ HM}^{0.736} \qquad (18.1)$$

by assuming yolk-free hatchlings average 57% of fresh egg mass (Vleck et al., 1980b). For adult passerine birds:

$$\dot{V}_{O_2} = 137 \text{ AM}^{0.724} \qquad (18.2)$$

and for adult non-passerine birds:

$$\dot{V}_{O_2} = 84 \text{ AM}^{0.723} \qquad (18.3)$$

where AM is adult mass in g. These equations (Lasiewski & Dawson, 1967) have been adjusted to a body temperature of 37 °C assuming a Q_{10} of 2.5 and a mean adult body temperature of 40 °C (Calder & King, 1974). The basal metabolic rates of the passerine adults is nearly 200% greater than that of the metabolic rates of the embryos of the same size and the basal metabolic rate of non-passerine adults is more than 100% greater than that of embryos of the same size. The low metabolic rate of the avian embryos is even more striking when we recall that these are growing embryos and presumably are consuming some energy to support growth whereas the adult birds are not growing.

A similar analysis can be done with the reptile data. From the data in Table 18.1, an equation relating pre-hatching $\dot{V}_{O_2}$ (in cm³/day) of reptile embryos at 30 °C to their hatchling mass (HM in g) is:

$$\dot{V}_{O_2} = 5.92 \text{ HM}^{0.90} \quad (r^2 = .96; \text{ SEE} = 0.10)$$
$$(18.4)$$

Bennett & Dawson (1976) provide a similar equation giving $\dot{V}_{O_2}$ of adult reptiles at 30 °C:

$$\dot{V}_{O_2} = 6.67 \text{ AM}^{0.77} \qquad (18.5)$$

where AM is adult body mass (in g). Reptile embryos and adults of the same mass have similar metabolic rates, despite the fact that embryos are probably growing faster than are the adults. For organisms from 1 to 20 g, the predicted values for reptile embryos are about 90% to 130% of those predicted for adult reptiles of the same mass. The similarities in metabolic rates of adult and near-term embryos suggests that the energetic requirements for existence (maintenance metabolism) of adult and embryonic vertebrates are not necessarily very different in ectothermic animals.

The lower metabolic rate of reptilian embryos (equation 18.4) compared to avian embryos (equation 18.1) is, in part, due to the lower temperature of incubation (30 °C for the reptiles vs. 37 °C for the birds). If temperature were the only factor, however, we would have to assume Q_{10} values of from about five to more than 12 (for organisms between 1 and 100 g in body mass) to account for the differences in rate between avian and reptilian embryos. This is clearly unreasonable, and it seems likely that maintenance metabolism, corrected for temperature, of bird embryos is higher than that of reptile embryos, even though metabolic rate in the avian embryo is lower than that of the endothermic adult bird at the same temperature.

This is also supported by a comparison of the pre-hatching $\dot{V}_{O_2}$ in birds (equation 18.1) with those of adult reptiles of the same mass at the same temperature. The metabolic rate of adult lizards at 37 °C is given by:

$$\dot{V}_{O_2} = 10.2 \text{ AM}^{0.82} \qquad (18.6)$$

where AM is adult body mass (in g) (Bennett & Dawson, 1976). In avian embryos, about 80% of energy metabolism is used for maintenance and the rest for growth, although this value appears to vary with egg size and developmental mode (Hoyt, 1987). Even if we decrease the metabolic rates of avian embryos (equation 18.1) by 20% to account for the metabolic costs of growth, the avian embryos appear to have a metabolic rate more than double that of a similar sized, adult reptile at the same temperature. That is, the maintenance metabolism of avian embryos, even when they are not supporting thermoregulation, is still greater than that of reptiles. If maintenance metabolism (analogous to basal metabolic rate) is relatively higher in birds than in reptiles throughout ontogeny, then this would account for the higher total energy costs and lower efficiencies of embryonic development in birds compared to reptiles. The only way that the efficiency of development in birds can approach that of reptiles is to shorten the incubation period so that the maintenance costs are paid over a relatively short time.

It is possible that oviparity is consistent with avian endothermy only because birds have decreased the length of the incubation period, thus minimising the impact of high maintenance costs on the limited energy reserves in an egg. This decrease, of course, follows in part from the rise in incubation temperature which is relatively high in all birds (Drent, 1975) compared to living reptiles. Egg temperature is controlled at levels higher than mean ambient temperatures by the adults in virtually all bird species, usually through direct incubation behaviour, and no bird eggs are known to be able to develop when simply laid in the open and abandoned, with the possible exception of some megapode species of the lowland tropics (Meyer, 1930). In those species of birds that do have exceptionally long incubation periods and must therefore produce very energy rich eggs, it is associated with laying only one or a few eggs (Calder *et al.*, 1978; Whittow, 1980), or laying the eggs at very long intervals (Vleck *et al.*, 1984).

The inherently higher basal metabolic rates of birds compared to reptiles, seems to represent a phylogenetic characteristic (*sensu* Huey, 1987) that is present during ontogeny even before the ability to thermoregulate develops. It is probably part of a larger package of cellular characteristics that homeothermic birds (and possibly mammals) acquired during the evolution of endothermy.

Acknowledgements

The preparation of this manuscript was largely financed by the senior author. Additional support was provided by NSF grant BSR 8202880 to D. Vleck, a Research Scholarship and Creative Activity Award grant to Don Hoyt from the California State Polytechnic University, Pomona, California, and the grandparents of C.V.'s children whose care of their grandchildren provided her with time to complete the manuscript. We gratefully acknowledge the use of unpublished data of Drs Michael Ewert, Michael B. Thompson, and David Vleck. We thank D. Vleck and M. Ewert for their comments.

References

Ackerman, R. A. (1980). Oxygen consumption by sea turtle (*Chelonia, Caretta*) eggs during development. *Physiological Zoology*, 54, 316–24.

— (1981). Growth and gas exchange of embryonic sea turtles (*Chelonia, Caretta*). *Copeia*, 1981, 757–65.

Ackerman, R. A., Whittow, G. C., Paganelli, C. V. & Pettit, T. N. (1980). Oxygen consumption, gas exchange, and growth of embryonic wedge-tailed shearwaters (*Puffinus pacificus chlororhyncus*). *Physiological Zoology*, 53, 210–21.

Ar, A., Arieli, B., Belinsky, A. & Yom-Tov, Y. (1987). Energy in avian eggs and hatchlings: utilization and transfer. *Journal of Experimental Zoology*, Supplement 1, 151–64.

Ar, A. & Rahn, H. (1980). Water in the avian egg; Overall budget of incubation. *American Zoologist*, 20, 373–84.

Ar, A. & Yom-Tov, Y. (1978). The evolution of parental care in birds. *Evolution*, 32, 655–69.

Badham, J. A. (1971). Albumen formation in eggs of the agamid *Amphibolurus barbatus barbatus*. *Copeia*, 1971, 543–5.

Ballinger, R. E. & Clark, D. R., Jr (1973). Energy content of lizard eggs and the measurement of reproductive effort. *Journal of Herpetology*, 7, 129–32.

Barott, H. G. (1937). *Effect of Temperature, Humidity, and Other Factors on Hatch of Hens' Eggs and on Energy Metabolism of Chick Embryos*. United States Department of Agriculture Technical Bulletin No. 553. March, 1937.

Bennett, A. F. & Dawson, W. R. (1976). Metabolism. In *Biology of the Reptilia, vol. 5*, eds. C. Gans & W. R. Dawson, pp. 127–223. New York: Academic Press.

Black, C. P., Birchard, G. F., Schuett, G. W. & Black, V. D. (1984). Influence of incubation water content on oxygen uptake in embryos of the Burmese python (*Python molurus biovittatus*). In *Respiration and Metabolism of Embryonic Vertebrates*, ed. R. S. Seymour, pp. 137–45. Dordrecht: Dr W. Junk Publishers.

Board, R. G. (1982). Properties of avian egg shells and their adaptive value. *Biological Review*, 57, 1–28.

Boersma, P. D. (1982). Why some birds take so long to hatch. *American Naturalist*, 120, 733–50.

Bruning, D. F. (1974). Social structure and reproductive behavior in the greater rhea. *Living Bird*, 1974, 251–94.

Bucher, T. L. (1983). Parrot eggs, embryos, and nestlings: patterns and energetics of growth and development. *Physiological Zoology*, 56, 465–83.

— (1987). Patterns in the mass-independent energetics of avian development. *Journal of Experimental Zoology*, Supplement 1, 139–50.

Bucher, T. L. & Bartholomew, G. A. (1984). Analysis of variation in gas exchange, growth patterns, and energy utilization in a parrot and other avian embryos. In *Respiration and Metabolism of Embryonic Vertebrates*, ed. R. S. Seymour, pp. 359–72. Dordrecht: Dr W. Junk Publishers.

Calder, W. A. & King, J. R. (1974). Thermal and caloric relations of birds. In *Avian Biology, vol. 4*, eds D. S. Farner, J. R. King & K. C. Parkes, pp. 259–413. New York: Academic Press.

Calder, W. A., Parr, C. R. & Karl, D. P. (1978). Energy content of eggs of the brown kiwi *Apteryx australis*: an extreme in avian evolution. *Comparative Biochemistry and Physiology*, 60A, 177–9.

Cannon, M. E., Carpenter, R. E. & Ackerman, R. A.

(1986). Synchronous hatching and oxygen consumption of Darwin's rhea eggs (*Pterocnemia pennata*). *Physiological Zoology*, **59**, 95–108.

Carey, C., Rahn, H. & Parisi, P. (1980). Calories, water, lipid and yolk in avian eggs. *Condor*, **82**, 335–43.

Carr, A. & Hirth, H. (1961). Social facilitation in green turtle siblings. *Animal Behavior*, **9**, 68–70.

Clark, H. (1953b). Metabolism of the black snake embryo I. Nitrogen excretion. *Journal of Experimental Biology*, **30**, 492–501.

— (1953a). Metabolism of the black snake embryo II. Respiratory exchange. *Journal of Experimental Biology*, **30**, 502–5.

Congdon, J. D., Dunham, A. E. & Tinkle, D. W. (1982). Energy budgets and life histories of reptiles. In *Biology of the Reptilia, vol. 13*, eds C. Gans & F. H. Pough, pp. 233–71. New York: Academic Press.

Congdon, J. D., Gibbons, J. W. & Greene, J. L. (1983). Parental investment in the chicken turtle (*Deirochelys reticularia*). *Ecology*, **64**, 419–25.

Congdon, J. D. & Tinkle, D. W. (1982). Reproductive energetics of the painted turtle (*Chrysemys picta*). *Herpetologica*, **38**, 228–37.

Congdon, J. D., Tinkle, D. W. & Rosen, P. C. (1983). Egg components and utilization during development in aquatic turtles. *Copeia*, **1983**, 264–8.

Cunningham, B. & Hurwitz, A. P. (1936). Water absorption by reptile eggs during incubation. *American Naturalist*, **70**, 590–5.

Deeming, D. C. & Ferguson M. W. J. (1989). Effects of incubation temperature on growth and development of embryos of *Alligator mississippiensis*. *Journal of Comparative Physiology*, **B159**, 183–93.

— (1990a). Methods for the determination of the physical characteristics of eggs of *Alligator mississippiensis*: A comparison with other crocodilian and avian eggs. *Herpetological Journal*, **1**, 456–62.

— (1991). Incubation and embryonic development in reptiles and birds. In *Avian Incubation*, ed. S. J. Tullett. pp. 3–37. London: Butterworths.

Derickson, W. K. (1976). Ecological and physiological aspects of reproductive strategies in two lizards. *Ecology*, **57**, 445–58.

Dial, B. E. (1987). Energetics and performance during nest emergence and the hatchling frenzy in loggerhead sea turtles (*Caretta caretta*). *Herpetologica*, **43**, 307–15.

Dmi'el, R. (1970). Growth and metabolism in snake embryos. *Journal of Embryology and Experimental Morphology*, **23**, 761–72.

Drent, R. (1975). Incubation. In *Avian Biology, vol. 5*, eds D. S. Farner, J. R. King & K. C. Parkes, pp. 333–420. New York: Academic Press.

Ewert, M. A. (1979). The embryo and its egg: Development and natural history. In *Turtles: Perspectives and Research*, eds M. Harless & H. Morlock, pp. 333–413. New York: Wiley.

— (1985). Embryology of turtles. In *Biology of the Reptilia, vol. 14*, eds C. Gans, F. Billett & P. F. A. Maderson, pp. 76–267. New York: Wiley.

Feder, M. H. (1981). Effect of body size, trophic state, time of day, and experimental stress on oxygen consumption of anuran larvae: an experimental assessment and evaluation of the literature. *Comparative Biochemistry and Physiology*, **70A**, 497–508.

Gettinger, R. D., Paukstis, G. L. & Gutzke, W. H. N. (1984). Influence of hydric environment on oxygen consumption by embryonic turtles *Chelydra serpentina* and *Trionyx spiniferus*. *Physiological Zoology*, **57**, 468–73.

Goode, J. & Russell, J. (1968). Incubation of eggs of three species of chelid tortoises, and notes on their embryological development. *Australian Journal of Zoology*, **16**, 749–61.

Guillette, L. J., Jr & Gongora, G. L. (1986). Notes on oviposition and nesting in the high elevation lizard, *Sceloporus aeneus*. *Copeia*, **1986**, 232–3.

Gutzke, W. H. N. & Packard, G. C. (1987). Influence of the hydric and thermal environments on eggs and hatchlings of bull snakes *Pituophis melanoleucus*. *Physiological Zoology*, **60**, 9–17.

Gutzke, W. H. N., Packard, G. C., Packard, M. J. & Boardman, T. J. (1987). Influence of the hydric and thermal environments on eggs and hatchlings of painted turtles (*Chrysemys picta*). *Herpetologica*, **43**, 393–404.

Hadley, N. F. & Christie, W. W. (1974). The lipid composition and triglyceride structure of eggs and fat bodies of the lizard *Sceloporus jarrovi*. *Comparative Biochemistry and Physiology*, **48B**, 275–84.

Hotaling, E., Wilhoft, D. C. & McDowell, S. B. (1985). Egg position and weight of hatchling snapping turtles, *Chelydra serpentina*, in natural nests. *Journal of Herpetology*, **19**, 534–6.

Hoyt, D. F. (1979). Practical methods of estimating volume and fresh weight of bird eggs. *Auk*, **96**, 73–7.

— (1987). A new model of avian embryonic metabolism. *Journal of Experimental Zoology*, Supplement **1**, 127–38.

Hoyt, D. F. & Rahn, H. (1980). Respiration of avian embryos – a comparative analysis. *Respiration Physiology*, **39**, 255–64.

Hoyt, D. F., Vleck, D. & Vleck, C. M. (1978). Metabolism of avian embryos: ontogeny and temperature effects in the ostrich. *Condor*, **80**, 265–71.

Huey, R. B. (1987). Phylogeny, history, and the comparative methods. In *New Directions in Ecological Physiology*, eds M. F. Feder, A. F. Bennett, W. W. Burggren & R. B. Huey, pp. 76–98. Cambridge: Cambridge University Press.

Lack, D. (1968). *Ecological Adaptations for Breeding in Birds*. London: Methuen.

Laird, A. K. (1966). Dynamics of embryonic growth. *Growth*, **30**, 263–75.

Lasiewski, R. C. & Dawson, W. R. (1967). A re-examination of the relationship between standard metabolic rate and body weight in birds. *Condor*, **69**, 13–23.

Leshem, A. (1989). The effect of temperature and soil water content on embryonic development of Nile soft-shelled turtles (*Trionyx triunguis*). Unpublished PhD Dissertation. Tel Aviv University, Tel Aviv, Israel.

Leshem, A. & Dmi'el R. (1986). Water loss from *Trionyx triunguis* eggs incubating in natural nests. *Herpetological Journal*, **1**, 115–17.

Lundy, H. (1969). A review of the effects of temperature, humidity, turning and gaseous environment in the incubator on the hatchability of the hen's egg. In *The Fertility and Hatchability of the Hen's Egg*, eds T. C. Carter & B. M. Freeman, pp. 143–76. Edinburgh: Oliver & Boyd.

Lutz, P. L. & Dunbar-Cooper (1984). The nest environment of the American crocodile (*Crocodylus acutus*). *Copeia*, **1984**, 153–61.

Lynn, W. G. & von Brand, T. (1945). Studies on the oxygen consumption and water metabolism of turtle embryos. *Biological Bulletin*, **88**, 112–25.

Manolis, S. C., Webb, G. J. W. & Dempsey, K. E. (1987). Crocodile egg chemistry. In *Wildlife Management: Crocodiles and Alligators*, eds G. J. W. Webb, S. C. Manolis & P. J. Whitehead, pp. 445–72. Chipping Norton, Australia: Surrey Beatty Pty.

Meyer, O. (1930). Untersuchungen an den Eiern von *Megapodius eremita*. *Ornithologische Monatsberichte*, **38**, 1–5.

Morris, K. A., Packard, G. C., Boardman, T. J., Paukstis, G. L. & Packard, M. J. (1983). Effect of the hydric environment on growth of embryonic snapping turtles (*Chelydra serpentina*). *Herpetologica*, **39**, 272–85.

Muth, A. (1980). Physiological ecology of desert iguana (*Dipsosaurus dorsalis*) eggs: Temperature and water relations. *Ecology*, **61**, 1335–43.

Nice, N. N. (1962). Development of behavior in precocial birds. *Transactions of the Linnean Society, New York*, **8**, 1–211.

O'Connor, R. J. (1984). *The Growth and Development of Birds*. New York: Wiley & Sons.

Packard, G. C. & Packard, M. J. (1984). Coupling of physiology of embryonic turtles to the hydric environment. In *Respiration and Metabolism of Embryonic Vertebrates*, ed. R. S. Seymour, pp. 99–119. Dordrecht: Dr W. Junk Publishers.

— (1988).The physiological ecology of reptilian eggs and embryos. In *Biology of the Reptilia, vol. 16*, eds. C. Gans and R. B. Huey, pp. 523–605. New York: Alan R. Liss, Inc.

Packard, G. C., Taigen, T. L., Packard, M. J. & Boardman, T. J. (1981). Changes in mass of eggs of softshell turtles (*Trionyx spiniferus*) incubated under hydric conditions simulating those of natural nests. *Journal of Zoology, London*, **193**, 81–90.

Packard, G. C., Tracy, C. R. & Roth, J. J. (1977). The physiological ecology of reptilian eggs and embryos and the evolution of viviparity within the class Reptilia. *Biological Reviews*, **52**, 71–105.

Packard, M. J. & Packard, G. C. (1986). Effect of water balance on growth and calcium mobilization of embryonic painted turtles (*Chrysemys picta*). *Physological Zoology*, **59**, 398–405.

Packard, M. J., Packard, G. C. & Boardman, T. J. (1982). Structure of eggshells and water relations of reptilian eggs. *Herpetologica*, **38**, 136–55.

Packard, M. J., Packard, G. C., Miller, J. D., Jones, M. E. & Gutzke, W. H. N. (1985). Calcium mobilization, water balance, and growth in embryos of the agamid lizard *Amphibolurus barbatus*. *Journal of Experimental Zoology*, **235**, 349–57.

Paganelli, C. V. & Rahn, H. (1984). Adult and embryonic metabolism in birds and the role of shell conductance. In *Respiration and Metabolism of Embryonic Vertebrates*, ed. R. S. Seymour, pp. 193–204. Dordrecht: Dr W. Junk Publishers.

Pettit, T. N. & Whittow, G. C. (1984). Caloric content and energetic budget of tropical seabird eggs. In *Seabird Energetics*, eds G. C. Whittow & H. Rahn, pp. 113–37. New York: Plenum Press.

Plummer, M. V. & Snell, H. L. (1988). Nest site selection and water relations of eggs in the snake, *Opheodrys aestivus*. *Copeia*, **1988**, 58–64.

Prange, H. D. & Ackerman, R. A. (1974). Oxygen consumption and mechanisms of gas exchange of green turtle (*Chelonia mydas*) eggs and hatchlings. *Copeia*, **1974**, 758–63.

Rahn, H. & Ar, A. (1974). The avian egg: incubation time and water loss. *Condor*, **76**, 147–52.

— (1980). Gas exchange of the avian egg: time, structure, and function. *American Zoologist*, **20**, 477–84.

Rahn, H., Paganelli, C. V. & Ar, A. (1974). The avian egg: air-cell gas tension, metabolism and incubation time. *Respiration Physiology*, **22**, 297–309.

Ratterman, R. J. & Ackerman, R. A. (1989). The water exchange and hydric microclimate of painted turtle (*Chrysemys picta*) eggs incubating in field nests. *Physiological Zoology*, **62**, 1059–79.

Ricklefs, R. E. (1977). Composition of eggs of several bird species. *Auk*, **94**, 350–6.

— (1984). Prolonged incubation in pelagic seabirds: a comment on Boersma's paper. *American Naturalist*, **123**, 710–20.

Ricklefs, R. E. & Burger, J. (1977). Composition of eggs of the diamondback terrapin. *American Midland Naturalist*, **97**, 232–5.

Ricklefs, R. E. & Cullen, J. (1973). Embryonic growth of the green *Iguana iguana*. *Copeia*, **1973**, 296–305.

Romanoff, A. L. (1944). Avian spare yolk and its assimilation. *Auk*, **61**, 235–41.

— (1960). *The Avian Embryo*. New York: Macmillan Co.

— (1967). *Biochemistry of the Avian Embryo*. New York: Wiley & Sons.

Romer, A. S. (1966). *Vertebrate Paleontology*, 3rd edn. Chicago: University of Chicago Press.

Schmekel, L. (1960). Datum über des Gewicht des Vogeldottersackes vom Schlüpftag vis zum Schwinden. *Revue Suisse Zoologie*, **68**, 103–10.

Sexton, O. J. & Marion, K. R. (1974). Duration of incubation of *Sceloporus undulatus* eggs at constant temperature. *Physiological Zoology*, **47**, 91–8.

Seymour, R. S. & Ackerman, R. A. (1980). Adaptations to underground nesting in birds and reptiles. *American Zoologist*, **20**, 437–47.

Seymour, R. S., Vleck, D. & Vleck, C. M. (1986). Gas exchange in the incubation mounds of megapode birds. *Journal of Comparative Physiology*, **B156**, 773–82.

Shine, R. (1983). Reptilian reproductive modes: the oviparity-viviparity continuum. *Herpetologica*, **39**, 1–8.

Sibley, C. G., Ahlquist, J. E. & Monroe, B. L., Jr (1988). A classification of the living birds of the world based on DNA-DNA hybridization studies. *Auk*, **105**, 409–23.

Snell, H. L. & Tracy, C. R. (1985). Behavioral and morphological adaptations by Galapagos land iguanas (*Conolophus subcristatus*) to water and energy requirements of eggs and neonates. *American Zoologist*, **25**, 1009–18.

Thompson, M. B. (1987). Water exchange in reptilian eggs. *Physiological Zoology*, **60**, 1–8.

— (1988). Nest temperatures in the pleurodiran turtle, *Emydura macquarii*. *Copeia*, **1988**, 996–1000.

— (1989). Patterns of metabolism in embryonic reptiles. *Respiration Physiology*, **76**, 243–56.

Tinkle, D. W. & Hadley, N. F. (1975). Lizard reproductive effort: caloric estimates and comments on its evolution. *Ecology*, **56**, 427–34.

Tracy, C. R. (1980). Water relations of parchment-shelled lizard (*Sceloporus undulatus*) eggs. *Copeia*, **1980**, 478–82.

Tracy, C. R. & Snell, H. L. (1985). Interrelations among water and energy relations of reptilian eggs, embryos, and hatchlings. *American Zoologist*, **25**, 999–1008.

van Tienhoven, A. (1983). *Reproductive Physiology of Vertebrates*, 2nd edn, Ithaca, New York: Cornell University Press.

Vitt, L. J. (1978). Caloric content of lizard and snake (Reptilia) eggs and bodies and the conversion of weight to caloric data. *Journal of Herpetology*, **12**, 65–72.

Vleck, C. M. (1981). Hummingbird incubation: female attentiveness and egg temperature. *Oecologia*, **51**, 199–205.

Vleck, C. M., Hoyt, D. F. & Vleck, D. (1979). Metabolism of avian embryos: patterns in altricial and precocial birds. *Physiological Zoology*, **52**, 363–77.

Vleck, C. M., & Kenagy, G. J. (1980). Embryonic metabolism of the fork-tailed storm-petrel: physiological patterns during prolonged and interrupted incubation. *Physiological Zoology*, **53**, 32–42.

Vleck, C. M. & Vleck, D. (1987). Metabolism and energetics of avian embryos. *Journal of Experimental Zoology*, Supplement 1, 111–25.

Vleck, C. M., Vleck, D. & Hoyt, D. F. (1980). Patterns of metabolism and growth in avian embryos. *American Zoologist*, **20**, 405–16.

Vleck, D., Vleck, C. M. & Hoyt, D. F. (1980). Metabolism of avian embryos: Ontogeny of oxygen consumption in the rhea and emu. *Physiological Zoology*, **53**, 125–35.

Vleck, D., Vleck, C. M. & Seymour, R. S. (1984). Energetics of embryonic development in the megapode birds, mallee fowl *Leipoa ocellata* and brush turkey *Alectura lathami*. *Physiological Zoology*, **57**, 444–56.

Walsberg, G. E. (1983). Avian ecological energetics. In *Avian Biology, vol. 7*, eds D. S. Farner, J. R. King & K. C. Parkes, pp. 161–220. New York: Academic Press.

Webb, D. R. (1987). Thermal tolerance of avian embryos: a review. *Condor*, **89**, 874–98.

Webb, G. J. W., Beal, A. M., Manolis, S. C. & Dempsey, K. E. (1987). The effects of incubation temperature on sex determination and embryonic development rate in *Crocodylus johnstoni* and *C. porosus*. In *Wildlife Management: Crocodiles and Alligators*, eds G. J. W. Webb, S. C. Manolis & P. J. Whitehead, pp. 507–31. Chipping Norton, Australia: Surrey Beatty Pty.

Webb, G. J. W., Choquenot, D. & Whitehead, P. J. (1986). Nests, eggs, and embryonic development of *Carettochelys insculpta* (Chelonia: Carettochelidae) from Northern Australia. *Journal of Zoology, London*, **B1**, 521–50.

Werner, D. I. (1988). The effect of varying water potential on body weight, yolk and fat bodies in neonate green iguanas. *Copeia*, **1988**, 406–11.

White, F. N. & Kinney, J. L. (1974). Avian incubation. *Science*, **186**, 107–12.

Whitehead, P. J. (1987). Respiration of *Crocodylus johnstoni* embryos. In *Wildlife Management: Crocodiles and Alligators*, ed. G. J. W. Webb, S. C. Manolis & P. J. Whitehead, pp. 473–97. Chipping Norton, Australia: Surrey Beatty Pty.

Whittow, G. C. (1980). Physiological and ecological correlated of prolonged incubation in sea birds. *American Zoologist*, **20**, 427–36.

Wilhoft, D. C. (1986). Eggs and hatchling components of the snapping turtle (*Chelydra serpentina*). *Comparative Biochemistry and Physiology*, **84A**, 483–6.

Zarrow, M. X. & Pomerat, C. M. (1937). Respiration of the egg and young of the smooth green snake, *Liopeltis vernalis* (Harlan). *Growth*, **1**, 103–10.

19

Reasons for the dichotomy in egg turning in birds and reptiles

DENIS C. DEEMING

Introduction

The need for egg turning during incubation is one of the most dramatic differences between the incubation requirements of birds and reptiles. Almost all avian eggs need to be turned throughout much of incubation (Poulsen, 1953; Drent, 1975) and although there are many studies of the egg turning behaviour in incubating birds (Drent, 1975), these are not discussed here. By stark contrast, reptilian embryos are usually killed by turning during incubation (Ferguson, 1985). Egg turning in birds is thought to prevent deleterious adhesions between the embryo and the shell membranes (New, 1957) but in reptiles such adhesion is normal (Ewert, 1985; Ferguson, 1985). This simplistic view of the phenomenon of egg turning is widely accepted yet if the requirement and effects of turning are examined further the situation is not so clear cut. In this chapter, the effects of egg turning on avian and reptilian development are described and some suggestions are made for the physiological basis of turning in birds. The possible evolutionary relationships between egg turning behaviour in birds and its absence in reptiles, are discussed.

The effects of egg turning on hatchability in birds and reptiles

Avian eggs

The majority of studies of egg turning on avian embryos are concerned with effects on hatchability in poultry (Landauer, 1967; Lundy, 1969). Eycleshymer (1907) first showed that turning rate affected hatchability, although by modern standards the results were poor; 58% of fertile eggs of the fowl (*Gallus gallus*) hatched after being turned five times a day compared with 45% of eggs turned twice a day and only 15% hatchability of unturned eggs. This link between hatchability and turning has been repeatedly confirmed (e.g. Byerly & Olsen, 1936, 1937; Olsen & Byerly, 1936; Funk & Forward, 1952, 1953, 1960; Robertson, 1961*a,b*). These reports have been extensively reviewed by Landauer (1967) and Lundy (1969), since when there have only been two reports of the effects of turning on hatchability (Tullett & Deeming, 1987; Wilson & Wilmering, 1988). In view of these reviews, the effects of turning on hatchability and embryonic mortality are only summarised here.

In the fowl two peaks of mortality are observed in turned eggs, one in the first three days, the other in the last three days of incubation (Payne, 1919). These peaks occur, but are greater, in unturned eggs (Robertson, 1961*a*) with the highest mortality occurring at the end of incubation (Insko & Martin, 1933; Robertson, 1961*b*; Männel & Woelke, 1965). The magnitude of the mortality peaks is inversely related to turning rate (Insko & Martin, 1933; Robertson, 1961*b*; Männel & Woelke, 1965). A turning rate of 96 times a day produced a maximum hatch (Robertson, 1961*b*) but as this was not significantly different from the hatchability of eggs turned 24 times a day, modern incubators have built-in mechanisms which automatically turn the eggs once an hour.

The timing and number of days that eggs are turned during incubation is important. Better hatchability can be achieved by turning fowl eggs for longer periods (in days) of the 21-day incubation period of the fowl (Gamero, 1962). Turning eggs during the first week of incubation produces a good hatchability (Card, 1926; Weinmiller, 1930; New, 1957) yet, by contrast, restricting turning to the third week of incubation has no effect on hatchability (Byerly & Olsen, 1936). Cessation of turning after 10 days of incubation significantly decreases hatchability but after 13 days turning does not affect hatchability (Wilson & Wilmering, 1988). Turning twice a day, only from day 4 to day 7 of incubation, produces hatchabilities similar to those for eggs turned throughout incubation but this contrasts with eggs turned only between days 8 to 11 which hatched as poorly as unturned eggs (New, 1957). This critical period for egg turning (New, 1957), from 3–7 days in the fowl, has been confirmed for both hatchability and embryonic growth (Deeming, Rowlett & Simkiss, 1987; Deeming, 1989a). There is no evidence, however, to support the suggestion that turning in the second week of incubation is important in reducing embryonic mortality at the end of incubation (Kaltofen, 1960).

The cause of embryonic mortality in unturned eggs is not clear, although some authors (Dareste, 1891; Eycleshymer, 1907; New, 1957) consider that failure to turn eggs induces a deleterious adhesion of the extra-embryonic membranes to the shell membranes. The yolk sac, with the sub-embryonic fluid on top of the denser yolk, acts in a turning couple (New, 1957) which produces a tendency of the yolk sac to return to its original orientation after rotation. Such movement is reduced by failure to turn eggs (New, 1957). The hypothesis that turning is required to prevent premature adhesion of the chorion to the inner shell membrane has long been accepted (Robertson, 1961b; Freeman & Vince, 1974; Tazawa, 1980; Wilson & Wilmering, 1988). Malpositioning of embryos late in incubation is also considered to be associated with membrane adhesion (Robertson, 1961b) but it is unclear how adhesion kills the embryos which survive beyond the first few days of incubation (Deeming, 1989a). There are alternative suggestions to explain egg turning. These include the idea that turning improves the physiological environment of the egg (Robertson, 1961b) and turning may destroy 'unstirred layers' which develop within the egg (Deeming et al., 1987). Alternatively, Gamero (1962) suggests that turning stimulates muscular contractions of the embryo which generate metabolic

heat and affect the rate of water utilisation but the reasoning behind this idea is not clear. An early explanation (readily accepted by many) for egg turning (de Reamur, 1751) is that it destroys the temperature gradients that develop within the egg during incubation (Turner, Chapter 9) but as Drent (1975) points out this cannot apply in modern, force-draught incubators in which turning is still required for maximum hatchability despite the uniform and constant incubation temperature.

Reptile eggs

Reptile eggs rarely change position after oviposition (Ewert, 1985) and artificial movement of eggs is generally considered to be deleterious (Bustard, 1972; Raj, 1976; Ferguson, 1985). The effects of turning have only been examined in a few species of reptile, however, and the effects are dependent on the species concerned. Generally, eggs from squamate reptiles (such as the lizards *Calotes versicolor* and *Eublepharis macularius* and the snakes *Elaphe guttata*, and *Python molurus*) are unaffected by turning (Pandha & Thapliyal, 1967; Marcellini & Davies, 1982) though eggs of the lizard *Dipsosaurus dorsalis* appear to be susceptible to movement (Muth, 1980). By contrast, eggs of many turtles and crocodilians are sensitive to turning, particularly after 12 hours of incubation (around the time when the opaque spot appears in the initially translucent eggshell) and before 20 days of incubation (Mitsukuri, 1891; Joanen & McNease, 1977, 1979; Ewert, 1979, 1985; Limpus, Baker & Miller, 1979; Farmenter, 1980; Blanck & Sawyer, 1981; Chan, Salleh & Liew, 1985; Ferguson, 1985; Chan, 1989). Turning during the first few hours of incubation or later than 20 days has little effect (Pooley, 1969; Chan et al., 1985; Ferguson, 1985). Eggs of the turtle *Chelydra serpentina* inverted after 40 days of incubation simply pip upside down (Ewert, 1985). Eggs of the alligator *Alligator mississippiensis* that have been accidentally set upside down can be gently turned over on day 10 of incubation without any adverse effects on embryonic viability (Deeming & Ferguson, 1991). The eggs of some turtles (*Carettochelys insculpta*, *C. serpentina*, *Chrysemys picta*, *Chrysemys scripta* and *Malaclemmys terrapin*) appear to be unaffected by turning (Drajeske, 1974; Ewert, 1979, 1985; Feldman, 1983; Webb, Choquenot & Whitehead, 1986). Other data for *Geochelone carbonaria* are equivocal (Marcellini & Davies, 1982). Interestingly, regular turning of some *C. scripta* eggs shortens their incubation period. Under the same incubation conditions, turned eggs hatched 4 days earlier than unturned eggs: 101.3 days compared with 105.5 days (Drajeske, 1974).

The deleterious effects of turning on embryonic viability are considered to be twofold. Adhesion of the embryo to the inner shell membrane is an important part of development in some reptiles (Blanck & Sawyer, 1981; Ferguson, 1985) for it allows the shell immediately above the embryo to be dehydrated thereby increasing the oxygen conductance of the eggshell and reducing the risk of hypoxia (Thompson, 1985; Whitehead, 1987; Deeming & Thompson, Chapter 17). In inverted reptile eggs, the embryo is trapped under the yolk because its adhesion to the inner shell membrane prevents its return to the top of the egg. The large bulk of the yolk prevents normal development and disrupts growth of the extra-embryonic membranes (Ferguson, 1985). In addition, pathological shortening of the embryonic axis has been observed leading to malformations such as anophthalmia and spina bifida (Ewert, 1985). In some eggs of *C. serpentina*, however, which appear to be less susceptible to movement, the embryo is able to return to the top of the egg after inversion (Ewert, 1985).

A second effect of turning is the tearing of the extra-embryonic membranes after rotation of the egg (Ewert, 1979). Embryonic mortality is associated with shearing forces generated when the egg is rotated (Webb *et al.*, 1987*a,b*). The yolk sac acts in a turning couple in reptile eggs but the rotation of the yolk sac within the egg rips the vitelline and extra-embryonic membranes from the inner shell membrane (Ferguson, 1985; Thompson, 1985; Webb *et al.*, 1987*a,b*). The allantois is particularly sensitive to these shearing forces which effectively destroy it leading to the demise of the embryo (Ferguson, 1982). Prior to adhesion of the embryo and allantois, the yolk is free to rotate and presumably after day 20 of incubation the extra-embryonic membranes are sufficiently developed, and the extent of their adhesion to the shell membrane so great, that they are unaffected by shearing forces.

Exceptional birds and reptiles

A few birds do not turn their eggs. The palm swift (*Cypsiurus parvus*) sticks its eggs to the nest which it has built on the underside of palm leaves (Lack, 1956). Perhaps movement of the leaves by the wind replaces the need for the adult to turn the eggs. The Egyptian plover (*Pluvianus aegyptius*) buries its eggs under a thin layer of sand and sits on the nest throughout the heat of the desert day (Howell, 1979). This prevents loss of water vapour from the nest but presumably it also prevents the adult from turning the egg during the day. Why these eggs have a relatively long incubation period has not been adequately explained (Howell, 1979). The eggs of kiwis from New Zealand (genus *Apteryx*) are disproportionately large (400 g) and comprise 20% of the mass of the adult bird (Calder & Rowe, 1977; Calder, 1979; Rowe, 1980). These large eggs are incubated in a small burrow in which movement of the bird is restricted; as a consequence the eggs are not turned during the long incubation period (Rowe, 1978).

Birds of the Megapodiidae have an unusual reproductive biology (Frith, 1956; Booth & Thompson, Chapter 20). These birds build large mounds of sand, soil or compost in which they bury and abandon their eggs; a behaviour similar to that of many reptiles (Seymour & Ackerman, 1980; Booth & Thompson, Chapter 20). The eggs are not turned during the unusually long incubation period, the adults are absent at hatching and the hatchlings are very precocious. At present, the effects of egg turning have not been examined in megapode birds.

Although reptilian eggs are usually abandoned to the elements some reptiles do exhibit maternal care of the eggs or hatchlings. In its simplest form, as seen in crocodilians and some cobras (e.g. *Ophiphagus hannah*), this involves the female guarding the nest (Ferguson, 1985; Dowling, 1986). Other snakes (genera *Farancia* and *Trimeresurus*) guard their eggs by coiling around them during incubation (Dowling, 1986) but distinct brooding behaviour is only seen in pythons (Pythonidae) from temperate climates which have evolved mechanisms for warming the eggs during incubation. *P. molurus* and *Morilea spilotes* exhibit an advanced form of brooding behaviour. During cold periods, the female snake generates heat by a form of shivering thermogenesis: via rhythmic contractions of the body musculature a constant egg temperature, well above ambient, can be maintained (Hutchison, Dowling & Vinegar, 1966; Vinegar, Hutchison & Dowling, 1970; Vinegar, 1973; Van Mierop & Barnard, 1976a,b, 1978; Slip & Shine, 1988).

The most complex behaviour associated with egg brooding in reptiles is observed in skinks of the genus *Eumeces* (Fitch, 1954). These lizards construct burrows or hollows under logs in which they lay their eggs (Vitt & Cooper, 1985). The female seals herself in the burrow and tends the eggs during incubation, rolling them around and up-ending them (Evans, 1959). Survival of the eggs appears not to be correlated with this movement. If the nest is disturbed the female will retrieve the eggs and rebuild the nest (Evans, 1959; Vitt & Cooper, 1985). Brooding females will even move whole nests if the temperature or humidity regimen of the site becomes unsuitable (Vitt & Cooper, 1986). The enclosed nest chamber is considered to increase nest humidity as well as hiding the female during incubation (Vitt & Cooper, 1986). Like crocodilians, the female also aids in the hatching process (Evans, 1959) though females sometimes eat eggs and hatchlings (Vitt & Cooper, 1985, 1986). The role of this egg brooding behaviour is unclear. Untended eggs are prone to infection and have poor hatchability (Fitch, 1954) so tending eggs may reduce infection. Evans (1959) attributes egg brooding to the same reason that birds turn their eggs (though not suggesting why this was so!). A more recent study, however, suggests that the female tending the eggs is associated with an optimal water exchange of the eggs which thereby maximises hatchability (Somma & Fawcett, 1989).

Physiological effects of egg turning during incubation on embryonic development in birds and reptiles

Failure to turn eggs during incubation clearly reduces hatchability in birds but until recently the mechanism was poorly understood. Recent data on the physiological effects of egg turning in the fowl (Tazawa, 1980; Tullett & Deeming, 1987; Deeming *et al.*, 1987; Deeming, 1989a,b,c) have, however, shown that many aspects of development in unturned eggs are different to those in normal, turned eggs and that the membrane adhesion hypothesis is inadequate to explain why egg turning is necessary during avian development.

Development in unturned fowl eggs

The effects of turning on the orientation of the embryo in relation to the extra-embryonic components of the egg can be described using whole eggs fixed in formaldehyde. By day 12 of incubation the fowl embryo has assumed a specific and predictable position shown in Fig. 19.1 (Romanoff, 1960; Deeming, unpublished observations). The allantois has completed its growth: the inner allantoic membrane covers the amnion and the yolk sac. The outer allantoic membrane, having combined with the chorion to produce the chorio-allantoic membrane, lines the inner surface of the inner shell membrane and has enclosed the albumen in the albumen sac. Albumen is surrounded by chorionic ectoderm except at two places: the remnant of the vitelline membrane separates the albumen from the yolk and mesoderm cells in the sero-amniotic connection separate the albumen from the amniotic fluid (Hirota, 1894; Romanoff, 1960). The embryo lies within the amniotic sac with the sero-amniotic connection close to the nape of its neck (Fig. 19.1). If the egg lies with its long axis parallel to the horizontal plane the amniotic sac lies towards the top of the egg close to the air space. The albumen sac lies towards the sharp pole of the egg but the neck of the albumen sac (effectively a tube) extends up to the sero-amniotic connection (Fig. 19.1). The yolk sac occupies the space behind and to one side of the amniotic sac. The allantois occupies the remainder of the volume of the egg (Fig. 19.1).

The relationships between the embryo and the contents in unturned eggs are different (Tullett & Deeming, 1987; Deeming, unpublished observations). An extreme example of an unturned egg (after 12 days of incubation) is shown in Fig. 19.1. Growth of the allantois is retarded and the chorio-allantois only lines the top half of the egg (Randles & Romanoff, 1950; Tyrell *et al.*, 1954; Tullett & Deeming, 1987). Turning of unturned eggs after day 12 of incubation only stimulates limited growth of the chorio-allantoic membrane (Tullett & Deeming, 1987). The amniotic sac lies at the top of the egg away from the air space and nestles in a hollow on top of the yolk sac (Fig. 19.1). The albumen lies at the bottom of the egg below the yolk sac, and due to the poor growth of the chorio-allantoic membrane, is not enclosed within an albumen sac (Fig. 19.1) (Randles & Romanoff, 1950; Tazawa, 1980; Tullett & Deeming, 1987;

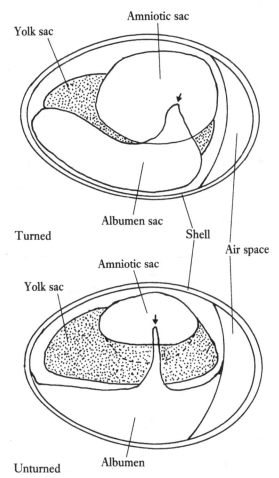

during incubation. During the first half of incubation the amount of sub-embryonic fluid is reduced in unturned eggs (Fig. 19.2) (Deeming *et al.*, 1987; Deeming, 1989*b*). The osmotic pressure and ionic composition of sub-embryonic fluid are unaffected by turning (Deeming *et al.*, 1987) indicating that although less fluid is formed in unturned eggs, it is produced by the same mechanism in turned and

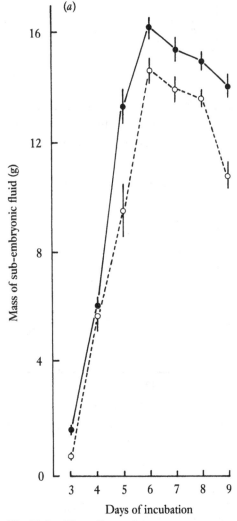

Fig. 19.1. The effects of turning on the orientation of the amniotic, albumen and yolk sacs in the fowl (*Gallus gallus*) egg by day 12 of incubation. In the turned egg (*upper*), the amniotic sac lies to one side of the yolk sac. The neck of the albumen sac extends up to the sero-amniotic connection (marked by arrow) and is not obstructed. In an unturned egg (*lower*), the amniotic sac lies in a hollow in the yolk sac. The albumen sac is incomplete with albumen lying below the yolk sac and the neck of the albumen is precluded by the yolk sac.

Fig. 19.2. The effects of egg turning on the fluid compartments of fowl eggs and the growth of the embryo during development. (*a*) sub-embryonic fluid from days 3 to 9 of incubation. (*b*)–(*e*) fluid compartments were determined during the second week of incubation; (*b*) allantoic fluid, (*c*) amniotic fluid, (*d*) albumen and (*e*) yolk sac. (*f*) embryonic growth during the second half of incubation. Values are means, with standard error bars, for turned (●) and unturned eggs (○). Redrawn from Deeming (1989*a,b*).

Deeming, unpublished observations). The neck of the albumen sac leading up to the sero-amniotic connection is obstructed by the yolk sac (Fig. 19.1). Many unturned eggs do not assume this configuration and their actual state of development lies somewhere between the turned and extreme unturned relationships described here (Deeming, unpublished observations).

Failure to turn eggs affects the mass, and the patterns of change, of the extra-embryonic fluids

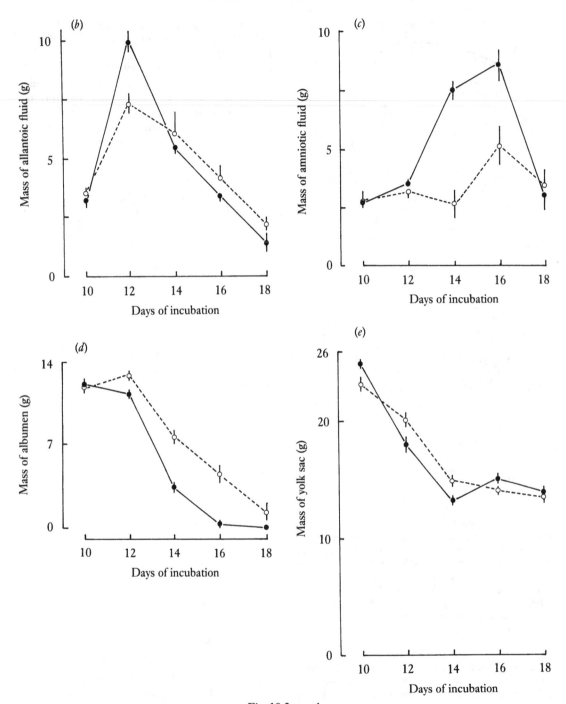

Fig. 19.2 contd.

unturned eggs (Deeming, 1989b). Failure to turn eggs does reduce the solid content, yolk protein and water content of the fluid (Howard, 1957; Elias, 1964; Deeming et al., 1987; Deeming, 1989b).

In the second half of incubation, turning affects the mass of allantoic fluid: on day 12 of incubation there is significantly less allantoic fluid in unturned eggs but on days 14, 16 and 18 there is more fluid in these eggs (Fig. 19.2)

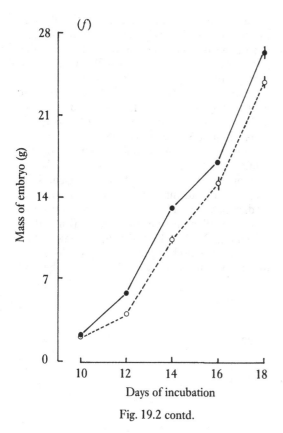

Fig. 19.2 contd.

ment of protein into the amniotic fluid is retarded (Fig. 19.3), resulting in a retarded increase in the specific gravity and refractive index of amniotic fluid which neither attains the maximum values observed in turned eggs nor declines later in incubation (Randles & Romanoff, 1950). The pH of amniotic fluid in turned eggs decreases rapidly after day 12 increasing again thereafter, but in unturned eggs these changes are much slower (Randles & Romanoff, 1950).

The alterations in the pattern of changes of amniotic fluid associated with turning are reflected in the utilisation of albumen by embryos (Deeming, 1989a). The mass of albumen is greater in unturned eggs on day 14 of incubation and thereafter (Fig. 19.2) and unhatched, dead embryos characteristically have large amounts of residual albumen (Tazawa, 1980; Tullett & Deeming, 1987; Deeming et al., 1987; Deeming, 1989a,b). The mass of the yolk is not dramatically affected by failure to turn eggs but yolk is utilised relatively, if not significantly, slower in unturned eggs (Fig. 19.2) (Deeming, 1989a).

The poor utilisation of albumen (and yolk) appears to retard embryonic growth during the second half of incubation. The first report of the

(Deeming, 1989a). The changes in specific gravity, refractive index and pH of allantoic fluid in turned eggs, that are observed during the middle of incubation, are slightly retarded in unturned eggs although the magnitude of the changes are unaffected by turning (Randles & Romanoff, 1950).

Failure to turn eggs drastically reduces the amount of amniotic fluid and prevents the sharp increase in the mass of this fluid normally observed after day 12 of incubation. Although the mass of amniotic fluid in unturned eggs increases later in incubation, the increase is much less than control eggs (Fig. 19.2) (Randles & Romanoff, 1949; Deeming, 1989a). In turned eggs, both the specific gravity and refractive index of amniotic fluid increase rapidly after day 11 of incubation up to a maximum on day 15 after which they decline (Randles & Romanoff, 1950). These changes presumably reflect the movement of albumen into the amniotic fluid (Hirota, 1894; Hanan, 1927; Witschi, 1949; Marshall & Deutsch, 1950; Carincci & Manzoli-Guidotti, 1968; Foote, 1969; Oegema & Jourdian, 1974). In unturned eggs, the move-

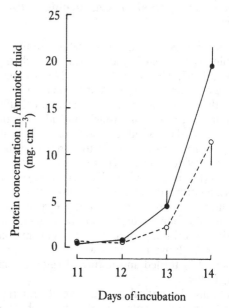

Fig. 19.3. The concentration of protein in the amniotic fluid of fowl embryos in turned (●) and unturned eggs (○) as assessed using a colorimetric method using Coomassie Blue (Pierce & Warriner Ltd). Values are means with standard error bars.

effects of egg turning on embryonic growth rates (Clark, 1933) showed that turning affected embryonic mass during the first half of incubation but not in the second half of incubation. All data collected subsequently have contradicted these results. Embryo mass in the first half of incubation is unaffected by turning (Deeming *et al.*, 1987; Deeming, 1989*a,b*) but failure to turn eggs significantly slows the rate of embryonic growth in the second half of incubation (Fig. 19.2). Embryos in unturned eggs are significantly smaller on all days during the second half of incubation so far examined (Tazawa, 1980; Tullett & Deeming, 1987; Deeming *et al.*, 1987; Deeming, 1989*a,b,c*, unpublished observations). The incubation period of embryos in unturned eggs is extended by 7 hours and only those which utilise all of their albumen are able to hatch (Tullett & Deeming, 1987). One characteristic effect of failure to turn eggs is the wide variation in embryo mass observed in unhatched, unturned eggs together with a wide variation in the amount of residual albumen: the amount of albumen in eggs is inversely correlated with embryo mass after day 16 of incubation (Tullett & Deeming, 1987; Deeming, 1989*a*).

The critical period for egg turning in the fowl

For the domestic fowl, the critical period for egg turning is between 3 and 7 days of incubation (New, 1957; Deeming *et al.*, 1987; Deeming, 1989*a,b*). Turning throughout incubation produces the largest embryos by day 16 of incubation but turning regimens which include turning from 3–7 days produce embryos which are heavier than those exposed to regimens which involve not turning the eggs during the critical period (Deeming *et al.*, 1987; Deeming, 1989*a*). Turning from day 3 to day 7 also affects the amount of amniotic fluid in the egg and the extent of albumen utilisation. In eggs turned during the critical period, the masses of amniotic fluid and albumen are between the extremes observed in turned and unturned eggs (Deeming, 1989*a*).

During the critical period several important developmental events occur. The embryo is undergoing differentiation and growth but there is no evidence at present to suggest that turning during the critical period directly affects changes

in embryonic differentiation later in incubation. Formation of sub-embryonic fluid is maximal during this period and turning clearly affects this process (Deeming *et al.*, 1987). In addition, turning affects the normally rapid expansion of the extra-embryonic membranes. Growth of the chorio-allantoic membrane is retarded by day 10 of incubation in unturned eggs (Tyrell *et al.*, 1954; Tullett & Deeming, 1987).

Turning also has significant effects on the growth of the area vasculosa of the yolk sac membrane. The size of the area vasculosa is unaffected by turning on day 3 of incubation but by days 5, 6, 7 and 8 the area vasculosa is significantly smaller in unturned eggs (Fig. 19.4) (Deeming, 1989*c*). In addition, turning during the critical period improves the growth rate of the area vasculosa by day 7 of incubation beyond

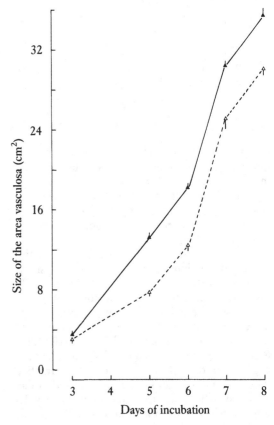

Fig. 19.4. Expansion of the area vasculosa in turned and unturned fowl eggs from day 3 to day 8 of incubation (no data were recorded for day 4). Data for turned (▲) and unturned eggs (△) are means with standard error bars. Data from Deeming (1989*c*).

that in unturned eggs but not to the extent observed in turned eggs (Deeming, 1989c).

Development in turned alligator eggs

Regular turning of alligator (*A. mississippiensis*) eggs, through ± 60° once an hour between days 10 and 45 of incubation, does not affect the development of embryos (Deeming & Ferguson, 1991). Compared with the complete mortality expected from previous results (Ferguson, 1982, 1985) the mortality of alligator embryos turned from days 10 to 45 of incubation is relatively low: 19 out of 25 embryos (76%) in turned eggs survive to day 45 and mortality is not associated with the onset of turning (Deeming & Ferguson, 1991). Mortality is zero in unturned eggs. Turning affects neither the growth rate of alligator embryos, nor their utilisation of albumen or yolk, nor formation of the extra-embryonic fluids (Deeming & Ferguson, 1991).

The physiological effects of turning in fowl eggs

The membrane adhesion hypothesis can not explain the perturbed distribution of fluids and the reduced growth rates of embryos in unturned eggs. Egg turning must affect some physiological processes during development. Turning may destroy concentration gradients in the egg that may develop during incubation. Such 'unstirred layers' may reduce the rate of formation of sub-embryonic fluid, and incorporation of yolk into this fluid, in unturned eggs (Deeming *et al.*, 1987; Deeming, 1989b). In addition, normal respiration produces hypoxic conditions adjacent to the embryo and sets up a gradient in oxygen tensions (P_{O_2}) in the fluids between the embryo and the shell (Lomholt, 1984). If the top halves of unturned eggs are lacquered there is high embryonic mortality but

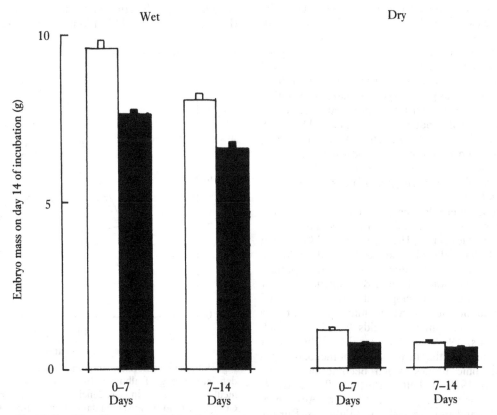

Fig. 19.5. The effects of incubating fowl eggs in an atmosphere of high oxygen tension (304 torr) for the first or second weeks of incubation on the embryos' response to turning. Values are means for wet and dried embryo mass (g) with standard error bars.

lacquering the lower halves of the shells has no effect (Lomholt, 1984). Lacquering the top halves of turned eggshells has no effect on the survival of 4 day chick embryos (Lomholt, 1984). Rotation of the egg, together with the turning couple of the yolk sac, may serve to move the embryo to regions of the egg which have a high P_{O_2}.

The retarded growth of the chorio-allantoic membrane by day 10 of incubation in unturned eggs (Tyrrell et al., 1954; Tullett & Deeming, 1987) probably affects respiratory gas exchange of embryos (Tullett & Deeming, 1987), formation of the albumen sac (Randles & Romanoff, 1950; Tullett & Deeming, 1987) and absorption of calcium from the eggshell (Simkiss, 1980). By day 16 of incubation the blood gas characteristics of chick embryos are affected by turning (Tazawa, 1980). Both P_{O_2} and the haematocrit of the blood are significantly reduced in arterialised blood of embryos in unturned eggs but P_{CO_2} is unaffected (Tazawa, 1980). The poor transfer of oxygen is considered to be due to the presence of residual albumen (Tazawa, 1980); diffusion of oxygen through fluids is very much slower than that of carbon dioxide (Wangensteen, Wilson & Rahn, 1970/71). The retarded growth of embryos in unturned eggs, therefore, may be due to poor oxygen exchange in a similar way to growth retardation observed in eggs with low eggshell conductances (Tullett & Deeming, 1982; Burton & Tullett, 1983). However, incubation of turned and unturned eggs in an atmosphere of oxygen at 304 torr (40%) does not abolish retarded growth rates in unturned eggs (Fig. 19.5).

The water relations of the embryo are important in determining embryonic mass in both birds and reptiles (Hoyt, 1979; Simkiss, 1980; Tullett & Burton, 1982; Packard & Packard, 1988). In the avian egg, albumen acts as an important reservoir of water (Palmer & Guillette, Chapter 3) and during development this is transferred first to the sub-embryonic fluid, and then to the allantoic and amniotic fluids (van Deth, 1962; Simkiss, 1980; Ar, Chapter 14). When water becomes limiting near the end of incubation it is embryonic tissues which become dehydrated (Hoyt, 1979; Tullett & Burton, 1982; Davis & Ackerman, 1987; Davis, Shen & Ackerman, 1988) and not the yolk sac (Tullett & Burton, 1982). The amount of sub-embryonic fluid on day 7 of incubation is reduced in unturned eggs and this leads to a reduction in the amount of

allantoic fluid on day 14 (Deeming et al., 1987; Deeming, 1989b). Therefore, the reduction in embryo mass observed in unturned eggs during the second half of incubation may be associated with an imbalance of the water relations of the embryo. Removal of sub-embryonic fluid from turned eggs on day 7 of incubation reduces the amount of allantoic fluid by day 12 of incubation and thereafter (Deeming et al., 1987; Deeming, 1989b) but does not affect embryo mass until after day 16 (Deeming, 1989b). In addition, this treatment does not significantly affect the amount of amniotic fluid in the egg (Fig. 19.6) and the effects of failure to turn eggs are not well simulated in eggs deficient in sub-embryonic fluid (Deeming, 1989b).

Albumen is an important source of nutrition for the developing embryo (Palmer & Guillette, Chapter 3) and residual albumen is a common characteristic of unturned eggs. If 5 g of albumen are removed from turned eggs on day 3 of incubation, embryo mass is subsequently affected but only after day 12 of incubation (Deeming, 1989b). This reduction in embryo mass is not, however, a simple loss of albumen from the gut of the embryo (Deeming, 1989b).

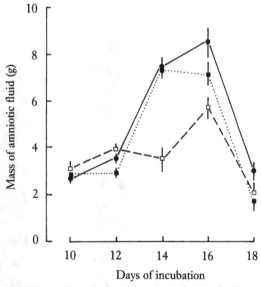

Fig. 19.6. The amount of amniotic fluid in fowl eggs deficient in 5 g of albumen (removed on day 3 of incubation) (□) and in eggs deficient in 3 cm³ of sub-embryonic fluid (removed on day 7 of incubation) (■) on selected days during the second half of incubation. Other data are controls from Fig. 19.2. Values are means with standard error bars. Redrawn using data from Deeming (1989b).

Both removal of albumen and failure to turn eggs have similar effects on the amounts of extra-embryonic fluids. In particular, there is a significant reduction in the mass of amniotic fluid on days 14 and 16 of incubation in both albumen deficient-turned and unturned eggs (Fig. 19.6) (Deeming, 1989*a,b*). Removal of albumen from unturned eggs simply increases the reduction in embryo mass (Deeming, 1989*b*). Generally, the effects of failure to turn eggs during the second half of incubation are simulated well by turned eggs deficient in albumen (Deeming, 1989*b*). Although albumen is very important in the response of the embryo to turning, it cannot explain all of the observations. Unlike those embryos which hatch from unturned eggs (Tullett & Deeming, 1987) the incubation periods of control and albumen deficient eggs are the same (Deeming, 1989*b*).

Events during the critical period are important in explaining the physiological basis of egg turning. Expansion of the extra-embryonic membranes is affected by turning (Tyrrell *et al.*, 1954; Tullett & Deeming, 1987; Deeming, 1989*c*) and this may affect physiological processes. The size of the area vasculosa may affect utilisation of yolk components and may determine the rate of development later in incubation. The effects of an artificial reduction in the size of the area vasculosa have been studied by reducing the incubation temperature which reduces the growth rate of the area vasculosa (Grodzinski, 1934; Deeming, 1989*c*). Incubation of turned eggs at 36 °C during the critical period for turning causes a significant reduction in the size of the area vasculosa by day 7 (Deeming, 1989*c*). If other eggs are returned to 38 °C on day 7 of incubation then, by day 14 of incubation there is more albumen, less amniotic fluid and embryo mass is reduced compared with controls (Deeming, 1989*c*). If unturned eggs are exposed to the same treatment then the retardation in growth of both the area vasculosa and in embryonic growth at day 14 are exacerbated (Deeming, 1989*c*). By contrast, constant incubation at 36 °C reduces the rate of growth of the area vasculosa and the embryo but without diminishing the effects of turning (Deeming, 1989*c*). An increase in incubation temperature (to 39.8 °C) during the critical period increases the rate of expansion of the area vasculosa in unturned eggs but not in turned eggs (Deeming, unpublished observations). The growth of the area vasculosa would appear, therefore, to be maximal in turned eggs. The high incubation temperature is deleterious to normal development (Deeming & Ferguson, Chapter 10), however, and its effects on embryonic growth in unturned eggs during the second half of incubation have not yet been ascertained with any certainty.

The physiological basis of egg turning in birds

To date, the evidence for the fowl embryo suggests that turning during the critical period is important in stimulating the growth of the area vasculosa. Turning may affect these changes by altering the blood pressure within the vascular network (Deeming, 1989*c*). When the egg is rotated, the area vasculosa, which formerly rested at the top of the egg, is shifted to the side of the egg. Gravity may cause blood to accumulate in the side of the area vasculosa at the bottom of the egg. To accommodate this unequal distribution of blood the vessels at the opposite side of the vascular network may collapse in a similar way to the blood circulation in the necks of giraffes (Seymour & Johansen, 1987). The localised increase in blood pressure in the area vasculosa may induce differentiation of peripheral blood vessels (Thomas, 1894; Hughes, 1935). With time, the area vasculosa moves back to the top of the egg. When the egg is turned again then it is likely that another region of the area vasculosa will be exposed to the higher than normal blood pressure. Growth of the area vasculosa continues in unturned eggs through differentiation of blood vessels associated with cardiac blood pressure but the rate of expansion is slower than in turned eggs (Deeming, 1989*c*). Such effects probably also apply to the chorio-allantoic membrane and help to explain the reduction in the area of the chorio-allantois by day 10 (Tullett & Deeming, 1987). Turning may also stimulate the heart directly. The heart rate of embryos late in incubation is accelerated by physical movement (Vince, Clarke & Reader, 1979) and this may also form a part of the physiological response of the embryo to turning earlier in development.

The size of the area vasculosa is likely to have significant effects later in development. Growth of the embryo and the extra-embryonic membranes are dependent on the nutrients supplied by the activity of the yolk sac membrane. If the size of the area vasculosa is reduced then the

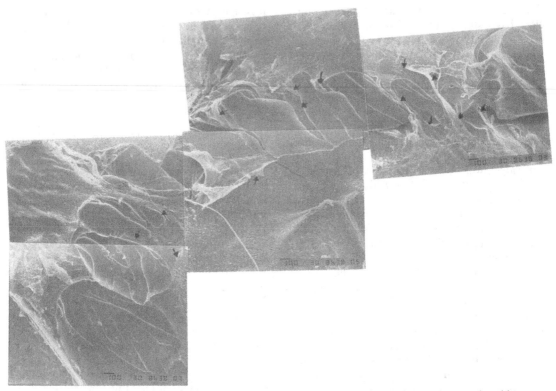

Fig. 19.7. Scanning electron micrograph of the sero-amniotic connection from a thirteen day old chicken embryo. The specimen shows the surface of the amniotic sac containing fluid rich in albumen; the whole egg had been fixed in formaldehyde prior to dissection. Note the crescent shape and the solidified albumen protruding through holes in the membrane (arrowed).

area for secretion of lytic enzymes into the yolk, and absorption of the breakdown products, is presumably restricted (Deeming, 1989c). In addition, normal movement of the area vasculosa relative to the yolk sac contents may expose the area vasculosa to the much more concentrated yolk instead of the more dilute sub-embryonic fluid. The rate of transfer of yolk into the embryonic circulation will be retarded in unturned eggs, which, after day 7, may begin to restrict the growth rate of the embryo and its extra-embryonic membranes (Deeming, 1989c). This initial retardation in the rate of development becomes aggravated by a reduction in the growth rate of the chorio-allantoic membrane which in turn affects formation of the albumen sac.

It is the utilisation of albumen by the embryo which is most affected by turning. Failure to turn eggs prevents formation of the albumen sac by day 12 of incubation and produces the abnormal orientation of the embryo in the amniotic sac relative to the yolk and albumen sacs; these act together to restrict the transfer of albumen into the amniotic fluid in unturned eggs. Turning eggs ensures the correct orientation of egg components in order to maximise the movement of albumen through the sero-amniotic connection (Fig. 19.7) (Hirota, 1894). It is likely that the sero-amniotic connection is functional in unturned eggs but albumen is prevented from entering the amniotic sac by the incorrect orientation of the albumen sac relative to the yolk sac and sero-amniotic connection. The embryo in the unturned egg is, therefore, denied a valuable source of nutrition and its rate of growth is reduced. The period of avian incubation is restricted and if the embryo is unable to utilise the egg's complement of albumen (and yolk) within this time, it will die. The embryos which hatch from unturned eggs do so only seven hours later than the controls (Tullett & Deeming, 1987) but it is unclear what attribute these embryos have that allows them to develop relatively normally under abnormal incubation conditions. Perhaps these embryos are those which have a near normal rate of transfer of albumen during incubation for normal growth and development.

There are major differences between the pattern of albumen utilisation in avian and reptilian eggs. Albumen is absent in many squamate eggs (Tracy & Snell, 1985) and in turtles and crocodilians, it forms much less of the initial mass of the egg (Ewert, 1979; Manolis, Webb & Dempsey, 1987; Burley *et al.*, 1987). In addition, the amount of protein in reptilian albumen is lower than that in avian eggs: only 5% of albumen is protein in eggs of *Crocodylus porosus* compared with 12% in the fowl (Burley *et al.*, 1987). Alligator embryos do not utilise albumen in the same way as in the fowl; although the albumen undergoes dehydration to form sub-embryonic fluid (Webb *et al.*, 1987*a,b*; Deeming & Ferguson, 1989), there is no movement of protein into the amniotic fluid during the middle of incubation (Deeming & Ferguson, 1991). Instead albumen appears to be taken up directly up by the chorio-allantoic membrane (Ferguson, 1985) but this process has yet to be described fully. These differences may help explain why turning is normally absent during incubation of reptile eggs and why artificial turning has no effect upon growth and development in alligators at least.

In summary, the effects of egg turning in birds are multifactorial but the primary function of turning appears to be to maximise the efficiency of the process of albumen utilisation by the embryo. Turning during the critical period ensures that growth of the extra-embryonic membranes is maximal: the area vasculosa is important in uptake of yolk and the chorio-allantois is important in surrounding the albumen and as a respiratory membrane. Turning also ensures that there is maximal formation of sub-embryonic fluid, which reduces the volume of albumen and affects the mass of allantoic fluid later in development. Egg turning also allows the embryo to be in the correct position for the transfer of albumen via the sero-amniotic connection. Failure to turn eggs during the first half of incubation disrupts the normal patterns of development which ensure the full utilisation of albumen which, in turn, leads to a significant retardation in embryonic growth and prevents many embryos from hatching.

Possible reasons for evolution of egg turning during avian incubation

Egg turning is necessary during avian development because of two important aspects of the reproductive strategy of birds: 1) There is a large investment of energy (in the form of protein) incorporated into the albumen by the female bird often resulting in a small clutch. 2) The incubation period of avian eggs is short and the incubation temperature is high: there is a rapid rate of embryonic development. In reptiles, the energy invested in albumen is low (it has a high water content), the incubation period is long and the rate of embryonic development is, compared with birds, slow. In *A. mississippiensis*, for example, failure to absorb all of the albumen in the egg would not impose a great nutritional deficiency on the embryo. This, together with the long incubation period, perhaps allows the embryo to absorb the albumen directly via the chorio-allantois (Ferguson, 1982, 1985; Deeming & Ferguson, 1991). In the fowl egg, the effects of incomplete albumen utilisation are potentially highly deleterious to the nutrition of the embryo. The direct transfer of albumen to the amniotic fluid, embryonic gut and beyond to the yolk sac may have evolved in birds to ensure complete utilisation of this valuable source of nutrition during the short incubation period. Transfer of albumen into the amniotic fluid has the added advantage of allowing any unutilised albumen to be transferred to the yolk sac, where it can be absorbed after hatching.

Avian eggs do vary, however, in their albumen content: altricial species invest a lot of albumen in their eggs (68–86% of egg contents) but precocial species have a much higher yolk content (43–76% albumen content) (Carey, Rahn & Parisi, 1980). Such differences may also be reflected in the turning requirements of different species. The hypothesis above predicts that eggs from altricial species may need a higher rate of, or more vigorous, turning during development than eggs from precocial species or those eggs with a low albumen content such as kiwi (Calder, 1979). Preliminary work on parrot (altricial) and guinea fowl (precocial) eggs suggests that the former do require more frequent turning which involves moving the sharp pole of the egg upwards, for normal development and maximum hatchability (Robert Harvey, unpublished data).

Finally, whilst albumen is a rich source of nutrients, it is possible that albumen may act as a carrier for growth promoting chemicals: hormones such as thyroxine, or growth factors such as epidermal growth factor (Mesiano, Browne & Thorburn, 1985; Palmer & Guillette, Chapter 3) which affect the rate of embryonic growth may be present in the albumen. The failure of the embryo to utilise all of its albumen may, therefore, restrict exposure of the embryo to factors vital for normal embryonic growth and development.

Acknowledgements

Some of the work reported here was carried out during the tenure of a University of Reading Studentship, other data were collected whilst the author was funded by the University of Manchester Research Support Funds. Both are gratefully acknowledged here. The author is grateful to Steve Tullett and David Booth for reviewing this chapter. Many thanks to Steve Tullett, Mark Ferguson, Karen Rowlett, Kenneth Simkiss, Robert Harvey and Dave Unwin for the many hours of discussion which helped formulate the ideas presented here.

References

Blanck, C. E. & Sawyer, R. H. (1981). Hatchery practices in relation to early embryology of the loggerhead sea turtle, *Caretta caretta* (Linne). *Journal of Experimental Marine Biology and Ecology*, **49**, 163–77.

Burley, R. W., Back, J. F., Wellington, J. E. & Grigg, G. C. (1987). Proteins of the albumen and vitelline membrane of eggs of the estuarine crocodile, *Croodylus porosus*. *Comparative Biochemistry and Physiology*, **88B**, 863–7.

Burton, F. G. & Tullett, S. G. (1983). A comparison of the effects of eggshell porosity on the respiration and growth of the domestic fowl, duck and turkey embryos. *Comparative Biochemistry and Physiology*, **75A**, 167–74.

Bustard, H. R. (1972). *Sea Turtles: Their Natural History and Conservation*. London: Collins.

Byerly, T. C. & Olsen, M. W. (1936). Certain factors affecting the incidence of malpositions among embryos of the domestic fowl. *Poultry Science*, **15**, 163–8.

Byerly, T. C. & Olsen, M. W. (1937). Egg turning, pipping position and malpositions. *Poultry Science*, **16**, 371–7.

Calder III, W. A. (1979). The Kiwi and egg design: Evolution as a package deal. *Bioscience*, **29**, 461–7.

Calder III, W. A. & Rowe, B. E. (1977). Body mass changes and energetics of the kiwis's egg cycle. *Notornis*, **24**, 129–35.

Card, L. E. (1926). Incubator eggs turned first six days hatch well. *Report of the Illinois Agricultural Experimental Station*, **39**, 78.

Carey, C., Rahn, H. & Parisi, P. (1980). Calories, water, lipid and yolk in avian eggs. *Condor*, **82**, 335–43.

Carinci, P. & Manzoli-Guidotti, L. (1968). Albumen absorption during chick embryogenesis. *Journal of Embryology and Experimental Morphology*, **20**, 107–18.

Chan, E. H. (1989). White spot development, incubation and hatching success of leatherback turtle (*Dermochelys coriacea*) eggs from Rantau Abang, Malaysia. *Copeia*, **1989**, 42–7.

Chan, E. H., Salleh, H. U. & Liew, H. C. (1985). Effects of handling on hatchability of eggs of the leatherback turtle, *Dermochelys coriacea* (L.). *Pertanika*, **8**, 265–71.

Clark, T. B. (1933). Effect of multiple turning upon the growth of chick embryos. *Poultry Science*, **12**, 279–81.

Dareste, C. (1891). *Recherche Sur La Production Artificielle Des Monstruostites Ou Essais De Teratogenie Experimentale*, 2nd edn, Paris.

Davis, T. A. & Ackerman, R. A. (1987). Effects of increased water loss on growth and water content of the chick embryo. *Journal of Experimental Zoology*, Supplement 1, 357–64.

Davis, T. A., Shen, S. S. & Ackerman, R. A. (1988). Embryonic osmoregulation: consequences of high and low water loss during incubation of the chicken egg. *Journal of Experimental Zoology*, **245**, 144–56.

Deeming, D. C. (1989a). Characteristics of unturned eggs: critical period, retarded embryonic growth and poor albumen utilisation. *British Poultry Science*, **30**, 239–49.

— (1989b). Importance of sub-embryonic fluid and albumen in the embryo's response to turning of the egg during incubation. *British Poultry Science*, **30**, 591–606.

— (1989c). Failure to turn eggs during incubation: Development of the area vasculosa and embryonic growth. *The Journal of Morphology*, **201**, 179–86.

Deeming, D. C. & Ferguson, M. W. J. (1989). Effect of incubation temperature on embryonic growth and development of embryos of *Alligator mississippiensis*. *Journal of Comparative Physiology*, **159B**, 183–93.

Deeming, D. C. & Ferguson, M. W. J. (1991). Effect of egg turning on the embryonic development of *Alligator mississippiensis* A comparison with egg turning in birds. *Acta Zoologica*, (in press).

Deeming, D. C., Rowlett, K. & Simkiss, K. (1987). Physical influences on embryo development.

Journal of Experimental Zoology, Supplement **1**, 341–5.

van Deth, J. H. M. G. (1962). Changes in the water, sodium and potassium distribution in the duck's egg during incubation. *Australian Journal of Experimental Biology and Medical Science*, **40**, 173–89.

Dowling, H. G. (1986). Brooding behaviour in snakes. In *The Encyclopaedia of Reptiles and Amphibians*, eds T. A. Halliday & K. Adler, p. 116. London: George Allen & Unwin.

Drajeske, P. (1974). Movement sensitivity in incubating turtle eggs. *Bulletin of the Chicago Herpetological Society*, **9**, 2–5.

Drent, R. (1975). Incubation. In *Avian Biology*, vol. 5, eds D. S. Farner & J. R. King, pp. 333–420. New York: Academic Press.

Elias, S. (1964). The sub-embryonic liquid in the hen's egg. *Review of Roumaine Embryology and Cytology*, **1**, 165–92.

Evans, L. T. (1959). A motion picture study of maternal behaviour of the lizard, *Eumeces obseletus* Baird and Girard. *Copeia*, **1959**, 103–10.

Ewert, M. A. (1979). The embryo and its egg: development and natural history. In *Turtles: Perspectives and Research*, eds M. Harless & H. Morlock, pp. 333–413. New York: Wiley-Interscience.

— (1985). Embryology of turtles. In *Biology of the Reptilia*, vol. 14, Development A, eds C. Gans, F. Billet & P. F. A. Maderson, pp. 75–267. New York: Wiley.

Eycleshymer, A. C. (1907). Some observations and experiments on the natural and artificial incubation of the egg of the common fowl. *Biological Bulletin*, **12**, 360–74.

Feldman, M. L. (1983). Effects of rotation on the viability of turtle eggs. *Herpetological Review*, **14**, 76–7.

Ferguson, M. W. J. (1982). The structure and composition of the eggshell and embryonic membranes of *Alligator mississippiensis*. *Transactions of the Zoological Society of London*, **36**, 99–152.

— (1985). The reproductive biology and embryology of crocodilians. In *Biology of the Reptilia*, vol. 14, Development A, eds C. Gans, F. Billet & P. F. A. Maderson, pp. 329–491. New York: Wiley.

Fitch, H. S. (1954). Life history and ecology of the five-lined skink, *Eumeces fasciatus*. *University of Kansas, Publications from the Museum of Natural History*, **8**, 1–156.

Foote, F. M. (1969). Translocation of albumen in eggs and embryos of the quail, chick and duck. *Poultry Science*, **48**, 304–6.

Frith, H. J. (1956). Breeding habits in the family Megapodidae. *Ibis*, **98**, 620–40.

Freeman, B. M. & Vince, M. A. (1974). *Development of the Avian Embryo*. London: Chapman & Hall.

Funk, F. M. & Forward, J. F. (1952). Effect of multiple plane turning of eggs during incubation on hatchability. *Report Bulletin of the Missouri Agricultural Experiment Station*, **502**.

— (1953). The effect of angle of turning of eggs during incubation on hatchability. *Report Bulletin of the Missouri Agricultural Experimental Station*, **599**.

— (1960). The relation of angle of turning and position of the egg to hatchability of chicken eggs. *Poultry Science*, **39**, 784–5.

Gamero, A. M. (1962). Frequency of egg turning during incubation in an incubator with forced air movement and uniform temperature. *Rev. Fac. Agron. nac. La Plata*, **38**, 21–31 (*Animal Breeding Abstracts*, **33**, 310).

Grodzinski, Z. (1934). Z badan nad rozrostem Area vasculosa u kurczecia. – Zur kenntnis der wachstumvorgange der area vasculosa beim huhnchem. *Bulletin of the International Academie Polonaise et des lettres, Classe Sciences Mathematique Naturelles, Serie B(II)*, **2**, 415–27.

Hanan, E. B. (1927). Absorption of vital dyes by the fetal membranes of the chick. I. Vital staining of the chick embryo by injections of trypan blue into the air chamber. *American Journal of Anatomy*, **38**, 423–50.

Hirota, S. (1894). On the sero-amniotic connection and the foetal membranes in the chick. *Journal of the College of Science, University of Tokyo*, **6**, 337–97.

Howard, E. (1957). Ontogenetic changes in the freezing point and sodium and potassium content of the sub-germinal fluid and blood plasma of the chick embryo. *Journal of Cellular and Comparative Physiology*, **50**, 451–70.

Howell, T. R. (1979). Breeding biology of the Egyptian plover *Pluvianus aegyptius*. *University of California at Berkley Publications in Zoology*, **113**.

Hoyt, D. F. (1979). Osmoregulation by avian embryos: The allantois functions like a toad's bladder. *Physiological Zoology*, **52**, 354–62.

Hughes, A. F. W. (1935). Studies on the area vasculosa of the embryo chick. I. The first differentiation of the vitelline artery. *Journal of Anatomy*, **70**, 76–122.

Hutchison, V. H., Dowling, H. G. & Vinegar, A. (1966). Thermoregulation in a brooding female Indian python, *Python molurus bivittatus*. *Science, New York*, **151**, 694–6.

Insko, W. M. & Martin, J. H. (1933). Effect of frequent turning on hatchability and distribution of embryo mortality. *Poultry Science*, **12**, 282–6.

Joanen, T. & McNease, L. (1977). Artificial incubation of alligator eggs and post hatching culture in controlled environmental chambers. *Proceedings of the Eighth Annual Meeting of the World Mariculture Society*, pp. 483–90, ed. J. W. Avault Jr. Louisiana: Louisiana State University.

— (1979). Culture of the American alligator *Alligator mississippiensis*. *International Zoo Yearbook*, **19**, 61–6.

Kaltofen, R. S. (1960). De invloed van de keep-frequentie op de embryonensterfte in verschillende fasen van het broed process III. *Veeteelt-en Zuivelbertichten*, **3**, 96–102 (*Animal Breeding Abstracts*, **30**, 112).

Lack, D. (1956). A review of the genera and nesting habits of swifts. *Auk*, **73**, 1–32.

Landauer, W. (1967). *The Hatchability of Chicken Eggs as Influenced by Environment and Heredity*, Monograph 1 (revised). Storrs Connecticut USA, Storrs Agricultural Experimental Station.

Limpus, C. J., Baker, V. & Miller, J. D. (1979). Movement induced mortality of loggerhead eggs. *Herpetologica*, **35**, 335–8.

Lomholt, J. P. (1984). A preliminary study of local oxygen tensions inside bird eggs and gas exchange during early stages of embryonic development. In *Respiration and Metabolism of Embryonic Vertebrates*, ed. R. S. Seymour, pp. 289–98, Dordrecht: Dr W. Junk.

Lundy, H. (1969). A review of the effects of temperature, humidity, turning and gaseous environment in the incubator on the hatchability of the hen's egg. In *The Fertility and Hatchability of the Hen's Egg*, eds T. C. Carter & B. M. Freeman, pp. 143–76. Edinburgh: Oliver & Boyd.

Männel, K. & Woelke, M. E. (1965). Response of the chicken embryo to different turning frequencies. *South African Journal of Agriculture*, **8**, 143–6.

Manolis, S. C., Webb, G. J. W. & Dempsey, K. E. (1987). Crocodile egg chemistry. In *Wildlife Management: Crocodiles and Alligators*, eds G. J. W. Webb, S. C. Manolis & P. J. Whitehead, pp. 445–72. Sydney: Surrey Beatty Pty Ltd.

Marcellini, D. L. & Davies, S. W. (1982). Effects of handling on reptile egg hatching. *Herpetological Review*, **13**, 43–4.

Marshall, M. E. & Deutsch, H. F. (1950). Some protein changes in the developing chick embryo. *Journal of Biological Chemistry*, **185**, 155–61.

Mesiano, S., Browne, C. A. & Thorburn, G. D. (1985). Detection of endogenous epidermal growth factor-like activity in the developing chick embryo. *Developmental Biology*, **110**, 23–8.

Mitsukuri, K. (1891). On the foetal membranes of Chelonia. *Journal of the College of Science, University of Tokyo*, **4**, 1–53.

Muth, A. (1980). Physiological ecology of desert iguana (*Dipsosaurus dorsalis*) eggs: temperature and water relations. *Ecology*, **61**, 1335–43.

New, D. A. T. (1957). A critical period for the turning of hens' eggs. *Journal of Embryology and Experimental Morphology*, **5**, 293–9.

Oegema, T. R. & Jourdian, G. W. (1974). Metabolism of ovomucoid by the developing chick embryo. *Journal of Experimental Zoology*, **189**, 147–62.

Olsen, M. W. & Byerly, T. C. (1936). Multiple turn-ing and orienting eggs during incubation as they affect hatchability. *Poultry Science*, **15**, 88–95.

Packard, G. C. & Packard, M. J. (1988). The physiological ecology of reptilian eggs and embryos. In *Biology of the Reptilia*, vol. 16, Ecology B, eds C. Gans & R. B. Huey, pp. 523–605. New York: Alan R. Liss Inc.

Pandha, S. K. & Thapliyal, J. P. (1967). Egg laying and development in the garden lizard, *Calotes versicolor*. *Copeia*, **1967**, 121–5.

Parmenter, C. J. (1980). Incubation of the eggs of the green sea turtle, *Chelonia mydas*, in Torres Strait, Australia: the effect of movement on hatchability. *Australian Wildlife Research*, **7**, 487–91.

Payne, L. F. (1919). Distribution of mortality during the period of incubation. *Journal of the American Association for the Inst. and Investment in Poultry Husbandry*, **6**, 9–12.

Pooley, A. C. (1969). Preliminary studies on the breeding of the nile crocodile *Crocodylus niloticus*, in Zululand. *Lammergeyer*, **3**, 22–44.

Poulsen, H. (1953). A study of incubation responses and some other behaviour patterns in birds. *Vidensk. Meddr. dansk. naturh. Foren.*, **115**, 1–131.

Raj, U. (1976). Incubation and hatching success in artificially incubated eggs of the hawksbill turtle, *Eretmochelys imbricata*. *Journal of experimental marine Biology and Ecology*, **22**, 91–9.

Randles, C. A. & Romanoff, A. L. (1949). Maldevelopment of the avian embryo as influenced by some environmental factors. *Poultry Science*, **28**, 780–1.

— (1950). Some physical aspects of the amnion and allantois of the developing chick embryo. *Journal of Experimental Zoology*, **114**, 87–101.

de Reamur, R. A. F. (1751). *Konst om Tamme-vogelen van Allerhandesoort in alle Jaartyden Uittebroeijen en Opiebrengen zo door 't Middle van Mest als van 't Gewone Vuur*, vol. I, 1–384, s'Gravenhage: Pieter de Hondt.

Robertson, I. S. (1961a). The influence of turning on the hatchability of hens' eggs II. The effect of turning frequency on the pattern of mortality, the incidence of malpositions, malformations and dead embryos with no somatic abnormality. *Journal of Agricultural Science, Cambridge*, **57**, 57–69.

— (1961b). The influence of turning on the hatchability of hens' eggs I. The effect of rate of turning on hatchability. *Journal of Agricultural Science, Cambridge*, **57**, 49–56.

Romanoff, A. L. (1960). *The Avian Embryo*. New York: MacMillan.

Rowe, B. E. (1978). Incubation temperatures of the North Island Brown kiwi (*Apteryx australis mantelli*). *Notornis*, **25**, 213–17.

— (1980). The kiwi, *Apteryx australis*, an extreme in egg size. *International Zoo Yearbook*, **20**, 196–7.

Seymour, R. S. & Ackerman, R. A. (1980). Adaptations to underground nesting in birds and reptiles. *American Zoologist*, **20**, 437–47.

Seymour, R. S. & Johansen, K. (1987). Blood flow uphill and downhill: does a siphon facilitate circulation above the heart. *Comparative Biochemistry and Physiology*, **88A**, 167–70.

Simkiss, K. (1980). Water and ionic fluxes inside the egg. *American Zoologist*, **20**, 385–93.

Slip, D. J. & Shine, R. (1988). Reptilian endothermy: a field study of thermoregulation by brooding diamond pythons. *Journal of Zoology, London*, **216**, 367–78.

Somma, L. A. & Fawcett, J. D. (1989). Brooding behaviour of the prairie skink, *Eumeces septentrionalis*, and the relationship to the hydric environment of the nest. *Zoological Journal of the Linnean Society*, **95**, 245–56.

Tazawa, H. (1980). Adverse effect of failure to turn the avian egg on the embryo oxygen exchange. *Respiration Physiology*, **41**, 137–42.

Thomas, R. (1894). *Intersuchungen uber die Histogense und Histomechanic des Gefassystems*. Stuttgart: Enke.

Thompson, M. B. (1985). Functional significance of the opaque white patch in eggs of *Emydura macquarii*. In *Biology of Australasian Frogs and Reptiles*, eds G. Grigg, R. Shine & H. Ehmann, pp. 387–85. Sydney: Surrey Beatty Pty Ltd.

Tracy, C. R. & Snell, H. L. (1985). Interrelations among water and energy relations of reptilian eggs, embryos and hatchlings. *American Zoologist*, **25**, 999–1008.

Tullett, S. G. & Burton, F. G. (1982). Factors affecting the weight and water status of the chick at hatch. *British Poultry Science*, **23**, 361–9.

Tullett, S. G. & Deeming, D. C. (1982). The relationship between eggshell porosity and oxygen consumption of the embryo in the domestic fowl. *Comparative Biochemistry and Physiology*, **72A**, 529–33.

— (1987). Failure to turn eggs during incubation: Effects on embryo weight, development of the chorioallantois and absorption of albumen. *British Poultry Science*, **28**, 239–49.

Tyrell, D. A. J., Tamm, I., Forsman, O. C. & Horsfall, F. L. (1954). A new count of allantoic cells of the 10-day chick embryo. *Proceedings of the Society for experimental Biology and Medicine*, **86**, 594–8.

Van Mierop, L. H. S. & Barnard, S. M. (1976a). Thermoregulation in a brooding female *Python molurus bivittatus* (Serpentes: Boidae). *Copeia*, 1976, 398–401.

— (1976b). Observations on the reproduction of *Python molurus bivittatus*. *Journal of Herpetology*, **10**, 333–40.

— (1978). Further observations on thermoregulation in the brooding female *Python molurus bivittatus* (Serpentes: Boidae). *Copeia*, 1978, 615–21.

Vince, M. A., Clarke, J. V. & Reader, M. R. (1979). Heart rate response to egg rotation in the domestic fowl embryo. *British Poultry Science*, **20**, 247–54.

Vinegar, A. (1973). The effects of temperature on the growth and development of embryos of the Indian python, *Python molorus bivittatus* (Reptilia: Serpentes: Boidae). *Copeia*, 1973, 171–3.

Vinegar, A., Hutchison, V. H. & Dowling, H. G. (1970). Metabolism, energetics and thermoregulation during brooding of snakes of the genus *Python* (Reptilia, Boidae). *Zoologica*, **55**, 19–38.

Vitt, L. J. & Cooper, W. E. (1985). The relationship between reproduction and lipid cycling in the skink *Eumeces laticeps* with comments on brooding ecology. *Herpetologica*, **41**, 419–32.

Vitt, L. J. & Cooper, W. E. (1986). Skink reproduction and sexual dimorphism: *Eumeces fasciatus* in the southeastern United States, with notes on *Eumeces inexpectatus*. *Journal of Herpetology*, **20**, 65–76.

Wangensteen, O. D., Wilson, D. & Rahn, H. (1970/71). Diffusion of gases across the shell of the hen's egg. *Respiration Physiology*, **11**, 16–30.

Webb, G. J. W., Choquenot, D. & Whitehead, P. J. (1986). Nests, eggs and embryonic development of *Carettochelys insculpta* (Chelonia: Carettochelidae) from Northern Australia. *Journal of Zoology, London*, **1B**, 521–50.

Webb, G. J. W., Manolis, S. C., Dempsey, K. E. & Whitehead, P. J. (1987b). Crocodilian eggs: A functional overview. In *Wildlife Management: Crocodiles and Alligators*, G. J. W. Webb, S. C. Manolis & P. J. Whitehead (eds), pp. 417–22. Sydney: Surrey Beatty Pty Ltd.

Webb, G. J. W., Manolis, S. C., Whitehead, P. J. & Dempsey, K. E. (1987a). The possible relationship between embryo orientation, opaque banding and the dehydration of albumen in crocodile eggs. *Copeia*, 1987, 252–7.

Weinmiller, L. (1930). Incubation experiments. *Proceedings of the 4th World's Poultry Congress, London*, 21–9.

Whitehead, P. J. (1987). Respiration of *Crocodylus johnstoni* embryos. In *Wildlife Management: Crocodiles and Alligators*, eds G. J. W. Webb, S. C. Manolis & P. J. Whitehead, pp. 473–97. Sydney: Surrey Beatty Pty Ltd.

Wilson, H. R. & Wilmering, R. F. (1988). Hatchability as affected by egg turning in high density plastic egg flats during the last half of incubation. *Poultry Science*, **67**, 685–8.

Witschi, E. (1949). Utilisation of the egg albumen by the avian fetus. *Contributions to Ornithological Biology*, Wissenesch, pp. 111–22. Heidelburg: Carl Winter.

20

A comparison of reptilian eggs with those of megapode birds

DAVID T. BOOTH
AND MICHAEL B. THOMPSON

Introduction

Most reptiles that lay eggs bury them in soil or, in the case of certain crocodilians, in mounds of decaying vegetation. The gaseous environment that these buried eggs experience can be very different from those of the atmosphere (Seymour & Ackerman, 1980). By contrast, avian eggs are usually exposed to more or less atmospheric gaseous conditions (Walsberg, 1980). An exception is the Megapodiidae, a family of galliform birds confined to the Australasian region, which bury their eggs in soil or mounds of decaying vegetation in a manner similar to that of reptiles (Frith, 1956a). In this chapter we discuss the similarities and differences in incubation biology of buried eggs of reptiles and megapodes and compare their gaseous and thermal environment to that of open nesting birds. All avian eggs have hard calcareous shells, so our comparisons tend to focus on rigid-shelled reptilian eggs. Most of the information for megapode eggs comes from just two species, the malleefowl (*Leipoa ocellata*) and Australian brush turkey (*Alectura lathami*). Similarly, information on rigid-shelled reptilian eggs relies heavily on the crocodilians *Crocodylus johnsoni*, *Crocodylus porosus*, and *Alligator mississippiensis*.

Physical characteristics of eggs

Egg and clutch size

Eggs of megapodes range in mass from 80 g in the Polynesian scrubfowl *Megapodius pritchardii* to 220 g in the maleo bird *Macrocephalon maleo* (Schönwetter, 1960–1983). Reptiles that lay

rigid-shelled eggs are restricted to two sub-families of geckos and possibly the Dibamidae amongst the Squamata, some chelonians and all crocodilians (Packard & Packard, 1988). Gecko eggs are very small, e.g. 0.15 g in *Sphaerodactylus* (Dunson & Bramham, 1981), whereas those of rigid-shelled Chelonia range from 2.2 g in *Sternotherus odoratus* to 110 g in *Geochelonia elephantopus* (Ewert, 1979, 1985). Crocodilian eggs range from a mean of 52 g in *Alligator sinensis* to 113 g in *Crocodylus siamensis* (Ferguson, 1985). Thus, although there is some overlap, megapode eggs are generally larger than reptile eggs.

Both *L. ocellata* and *A. lathami* lay their eggs one at a time at 3- to 10-day intervals throughout their 4–8 month breeding seasons so that mounds may contain up to 15 eggs simultaneously (Seymour, Vleck & Vleck, 1986). The total number of eggs laid in a single mound over the entire breeding season varies from 3–33 in *L. ocellata* (Booth, 1987a) and 3–58 in *A. lathami* (Jones, 1988b). However, several *A. lathami* females may lay eggs in the same mound (Jones, 1988b). Hence, at any time after the beginning of the breeding season mounds may contain eggs at several developmental stages. Little is known about clutch size in other megapode species. A captive Wattled brush turkey (*Aepypodius arfakianus*) laid 20 eggs at 2- to 11-day intervals in a single breeding season (Kloska & Nicolai, 1988). Other species also probably lay single eggs at periods of several days apart. By contrast, reptiles generally lay the whole clutch at the same time so that, at any time after oviposition, all embryos are at similar developmental stages and hatch more or less simultaneously. This behaviour has important

Table 20.1. *Shell thickness of three megapode species in unincubated and hatched eggs. Each egg was sampled at least ten times from differing areas of the shell.*

Species	Incubation state	Sample size	Thickness (μm)	% Thinning
Leipoa ocellata	unincubated	8	279	21
	hatched	8	221	
Alectura lathami	unincubated	1	341	12
	hatched	5	305	
Macrocephalon maleo	unincubated	3	349	14
	hatched	2	301	

consequences for both respiratory gas tensions within the micro-environment of the nest and in the hatching-emergence process. Reptilian clutch size varies greatly from species to species, and ranges from one in some lizards (Fitch, 1970) to over 100 in some sea turtles (Bustard, 1972).

Reptiles generally have longer incubation periods compared to equivalent-sized avian eggs. The lower incubation temperature of reptile eggs is partly responsible for longer incubation times (Vleck & Hoyt, Chapter 18). However, among birds, megapodes have notably lengthy incubation periods. Average incubation periods for *L. ocellata* and *A. lathami* are 60 days and 49 days respectively (Vleck, Vleck & Seymour, 1984). For *L. ocellata*, successful incubation periods range from 50 to 90 days (Booth 1987*a*), a range similar to that of crocodilian and sea turtle eggs.

The megapode eggshell is on average 31% thinner than that of similar sized eggs of other galliform birds (Booth, 1988*a*), but is similar in structure to other avian eggshell (Board *et al.*, 1982). The increased fragility resulting from the thin eggshell is not a liability because, once the eggs are laid, they are not subject to mechanical damage from the adult bird during brooding as may occur in open nests. The thin shell is associated with greater than predicted water vapour conductance (G_{H_2O}) which facilitates water loss in the high humidity nesting environment (Seymour & Rahn, 1978; Seymour & Ackerman, 1980). In addition, large increases in G_{H_2O} occur towards the end of incubation (Booth & Seymour, 1987). During incubation, avian embryos utilise calcium from the inside of the eggshell for skeletal development. This process causes eggshell to thin by 4–8% in most species, but has little effect on pore structure because the calcium is removed from the tips of mammillary knobs (Booth, 1989). Pores through megapode eggshell consist of a large diameter outer orifice which rapidly tapers to a narrow diameter canal at the inside surface (Fig. 20.1). The major resistance to gas diffusion occurs in this narrow region. The relatively large degree of shell thinning (12–21%) in megapode eggs (Table 20.1) removes a large proportion of the narrow, high resistance region of the pores, causing shell conductance to increase greatly. External pore orifices are partly occluded with inorganic material which may function as a barrier to micro-organisms (Board *et al.*, 1982).

Compared to avian eggs, reptile eggs exhibit an enormous diversity in eggshell structure (Packard & Hirsch, 1986; Packard & DeMarco, Chapter 5). Most consist of a multi-layered, fibrous shell membrane overlain by a crystalline material, the thickness of which varies considerably. The precise structure determines the gas exchange properties of the shell. In general, the thicker the crystalline layer, the lower the gas conductance of the eggshell. Water vapour conductance of parchment-shelled reptilian eggs may be two orders of magnitude higher than for avian eggs of similar size (Deeming & Thompson, Chapter 17). Conductance is lower in rigid-shelled eggs compared to pliable- and parchment-shelled eggs, but, with the exception of some geckos, never as low as in bird eggs (Deeming & Thompson, Chapter 17). Progression amongst reptiles from rigid-shelled to parchment-shelled eggs is a result of thinner calcareous components to the shell from a typical avian type to an eggshell with no crystalline calcium in the shell at all (Packard, Packard & Boardman, 1982). Unlike avian eggs, reptilian eggs may sustain a net water gain during incubation, and higher shell conductances facilitate this absorption (Packard, Chapter 13).

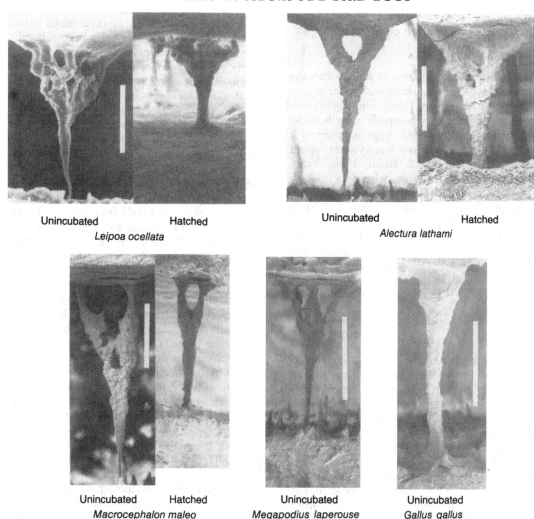

Fig. 20.1. Scanning electron micrographs of pore casts taken from the eggshells of four species of megapodes, and for comparison, a species of open-nesting galliform bird, the fowl *Gallus gallus*. Scale bars are 100 μm.

Egg contents

All megapode eggs so far examined have a relatively large yolk fraction (50–67%; Vleck *et al.*, 1984) compared to most avian species (15–40%; Sotherland & Rahn, 1987). In fact the only species with a larger yolk is the kiwi *Apteryx australis* (Calder, Parr & Karl, 1978). The large yolk fraction results in a relatively large energy content and low water content (Sotherland & Rahn, 1987). Large energy content is needed to maintain the embryo throughout a long incubation period and may also be necessary for the production of an extremely precocial hatchling (Vleck *et al.*, 1984). Megapodes have a similar hatching to

fresh egg mass ratio as other avian species. Like most reptiles, however, megapode hatchlings are completely independent of their parents (Nice, 1962).

Reptilian eggs generally have larger yolk volumes than birds. Values range from 32–55% in chelonians (Ewert, 1979) and 72–99% in squamates (Tracy & Snell, 1985). The albumen fraction is an important reservoir of water in crocodilian and chelonian eggs and many of these eggs do not require water absorption during development. Such eggs have been termed endohydric eggs to distinguish them from softer-shelled ectohydric eggs that must absorb water for successful hatching to occur (Tracy & Snell, 1985).

The gaseous incubation environment

Water vapour

The underground environment is almost always saturated or close to saturated with water vapour (Seymour & Ackerman, 1980). This has important consequences for megapode eggs because optimal hatchability in bird eggs is achieved if about 15% of initial mass is lost as water vapour during incubation (Ar & Rahn, 1980; Rahn, 1984). In contrast to birds, where there is no bulk flow of liquid water, reptilian eggs can exchange water as vapour and as liquid (Thompson, 1987). The rate of water loss ($\dot{M}_{H_2O}$) in avian eggs is determined by three factors: 1) eggshell water vapour conductance (G_{H_2O}); 2) the water vapour pressure inside the egg ($P_{e_{H_2O}}$); and 3) the water vapour pressure in the nest ($P_{n_{H_2O}}$) according to the relationship:

$$\dot{M}_{H_2O} = G_{H_2O} \times (P_{e_{H_2O}} - P_{n_{H_2O}})$$

(Ar et al., 1974)   (20.1)

G_{H_2O} is determined by the number and structure of pores in the eggshell which do not change significantly during incubation in most birds. Water vapour pressure inside an egg is not detectably different from saturated (Taigen et al., 1978) and $P_{e_{H_2O}}$ is determined by egg temperature, which for most birds is in the range 34–38 °C. Thus, $P_{e_{H_2O}}$ ranges between 40 and 50 torr. $P_{n_{H_2O}}$ is determined by factors such as parental brooding behaviour, nest location, nest structure, and atmospheric humidity. Empirical measurements of $P_{n_{H_2O}}$ in open nesting birds range between 8–26 torr (Walsberg, 1980; Rahn, 1984). Hence, in most cases there is a water vapour difference across the eggshell of between 20 and 25 torr. The high humidity of the underground environment greatly reduces this difference and has led to the conclusion that water loss from megapode eggs would be negligible (Seymour & Rahn, 1978; Seymour & Ackerman, 1980).

Relative humidity in *A. lathami* mounds is always above 90% and is probably in excess of 99% most of the time (corresponding to absolute humidities of 35–45 torr). In *L. ocellata* mounds relative humidity is close to 100% at the beginning of the breeding season but decreases to about 80% (corresponding to absolute humidities of 28–36 torr) towards the end (Seymour et al., 1987). Despite these high humidities, eggs of *A. lathami* lose 9.5% and

those of *L. ocellata* 12% of their initial mass during incubation. How can this be explained? Two phenomena appear to be responsible. Firstly, routine measurement of G_{H_2O} by placing eggs in desiccators may underestimate the functional G_{H_2O} in eggs incubated under high humidity conditions, especially when relative humidity exceeds 80% (Seymour et al., 1987). In high humidity liquid water may be drawn out along the pore wall by capillary action and thus reduce functional pore length (Seymour et al., 1987). Secondly, large increases in the rate of mass loss ($\dot{M}_{tot}$) occur during incubation (Fig. 20.2). $\dot{M}_{tot}$ increases 6-fold from 200 mg d^{-1} to 1200 mg d^{-1} in *L. ocellata* and 95-fold from 20 mg d^{-1} to 1900 mg d^{-1} in *A. lathami* during artificial incubation. Three factors cause the increase in $\dot{M}_{tot}$: 1) mass loss due to respiratory gas exchange; 2) an increase in egg temperature as the embryo develops; and 3) an alteration of pore structure which causes G_{H_2O} to increase.

Not all of the mass loss in *L. ocellata* eggs is due to loss of water vapour. In this species, the respiratory gas exchange ratio averages 0.80 over the last half of incubation (Booth, 1988b). When this ratio is 0.727, the mass of oxygen consumed equals the mass of carbon dioxide produced; larger ratios, such as those that occur in *L. ocellata* eggs, result in a net loss of mass. On the day of hatch the high exchange ratio accounts for 24% (300 mg d^{-1}) of $\dot{M}_{tot}$ in this species. However, the rate of oxygen consumption ($\dot{V}_{O_2}$) rises significantly only during the last half of incubation, so mass loss due to high exchange ratio accounts for only 1.7% of the initial egg mass (Booth, 1988b).

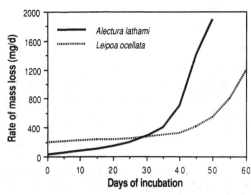

Fig. 20.2. Rates of mass loss of eggs of the megapodes *A. lathami* and *L. ocellata* incubated artificially in mound material. (Data from Seymour et al., 1987.)

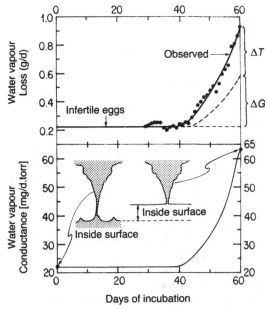

Fig. 20.3. Overview of factors affecting $\dot{M}_{H_2O}$ of artificially incubated *L. ocellata* eggs. Upper drawing documents the increase in $\dot{M}_{H_2O}$ during incubation. $\triangle T$ indicates the contribution of the increase in egg temperature to the increase in $\dot{M}_{H_2O}$, and $\triangle G$ the contribution of the increase in G_{H_2O}. Lower drawing indicates the effect of shell thinning on pore structure and G_{H_2O}. Note that shell thinning causes the loss of the narrow section of the pore canal, the region of greatest diffusive resistance.

In *L. ocellata* eggs, $\dot{M}_{H_2O}$ increases from 200 mg d^{-1} at the beginning of incubation to 900 mg d^{-1} at the end (Fig. 20.3). If we assume a similarly high exchange ratio for *A. lathami* eggs, their $\dot{M}_{H_2O}$ increases from 20 mg d^{-1} to 1600 mg d^{-1}. These increases can be explained by increases in egg temperature and G_{H_2O}. Metabolic heat production increases during development resulting in rises of 2–4 °C in egg temperature above the surrounding mound material (Seymour *et al.*, 1987; Booth, 1987*b*). $P_{e_{H_2O}}$ is temperature-dependent and the partial pressure gradient for water across the eggshell increases with egg temperature facilitating water loss. Increased egg temperature is particularly important in mounds like those of *A. lathami* that have a humidity close to saturation. In this species, the rise in egg temperature causes the partial pressure gradient to increase 40-fold from 0.1 torr at the beginning of incubation to 4 torr at the end (Seymour *et al.*, 1987). In the drier mounds of *L. ocellata*, the increase in egg

temperature is less important but it still causes a 1.5-fold increase in the trans-shell water vapour difference (Seymour *et al.*, 1987).

A three-fold increase in G_{H_2O} during incubation occurs in eggs of *L. ocellata* (Booth & Seymour, 1987). The unusual pore structure and large degree of shell thinning during incubation appear to be a general property of megapode eggs (Fig. 20.1, Table 20.1), thus large increases in $\dot{M}_{H_2O}$ as incubation progresses may be a general characteristic of megapode incubation. These events for eggs of *L. ocellata* are summarised in Fig. 20.3. In *A. lathami* eggs, a 40-fold increase in the water vapour gradient across the eggshell combined with a two-fold increase in G_{H_2O} due to eggshell thinning accounts for the 80-fold increase in $\dot{M}_{H_2O}$ observed during artificial incubation (Seymour *et al.*, 1987).

Responses of reptilian eggs to hydric environment during incubation is quite different from that of birds. The important ecological parameter is how much water exchange occurs, and what the influence of this exchange is on hatching success and hatchling vigour. Unlike birds, reptiles do not characteristically lose a fixed percentage of initial mass for optimal hatching success. Some species with parchment-shelled eggs such as the tuatara (*Sphenodon punctatus*) produce hatchlings that are heavier than the fresh egg contents (Thompson, 1990; Deeming & Ferguson, 1991). Incubation of rigid-shelled chelonian and crocodilian eggs may be equally successful in eggs that gain or lose water during development (Bustard, 1971; Packard *et al.*, 1981*a*; Vleck, Chapter 15).

There have been many studies on the water relations of reptilian eggs in the laboratory (Packard & Packard, 1988; Packard, Chapter 13) although this has included no crocodilian species, and only two turtles with rigid-shelled eggs (Packard *et al.*, 1979*a*; Thompson, 1983). However, it is difficult to place these data into a realistic natural framework because of the lack of data on water potentials in natural nests. Attempts to quantify the hydric environment using relative humidities (Chabreck, 1975) or per cent water contents (Lutz & Dunbar-Cooper, 1984) are difficult to translate to meaningful physiological values. For example, most nests will have relative humidities in excess of 99%. A relatively wet water potential of −100 kPa equates to a relative humidity in excess of

99.9%, yet a relative dry water potential of −1000 kPa still has a relative humidity of 99.3% (Tracy, Packard & Packard 1978).

The lack of field data reflects, in part, the difficulty in measuring natural water potentials (Ackerman, Chapter 12). This results from: 1) the complex (and still not clearly understood) interactions occurring in the soil among changing thermal and hydric conditions and the physical and chemical composition of different soils; and 2) lack of convenient measuring probes. No one probe type covers the whole range of water potentials that reptilian eggs may experience. Tensiometers are useful only in the wet range (down to about −100 kPa) and thermocouple psychrometer in the medium range (about −100−−5000 kPa) (Ackerman, Chapter 12). A few studies have measured water potentials in the field using thermocouple psychrometry (Leshem & Dmi'el, 1986; Packard et al., 1985; Thompson et al., 1990). These studies indicate that nest moisture may fluctuate widely diurnally and vary considerably seasonally. Incubation in different water potentials affects size and vigour of hatchling of eggs of some parchment-shelled squamates and pliable-shelled turtles (Packard et al., 1980, 1981b; Packard, Packard & Boardman, 1984; Miller, Packard & Packard, 1987), but may have little influence on embryos from rigid-shelled eggs. Water flux does not influence hatchling size in the laboratory in hard-shelled Trionyx spiniferus eggs (Packard et al., 1979, 1981a), but data on more species are needed before generalisations can be made.

The one area in which the significance of nest hydration has received attention relates to egg death from flooding. Excess water can be a significant source of mortality in crocodilians and turtles which generally nest close to water (Ragotzkie, 1959; Plummer, 1976; Magnusson, 1982; Webb et al., 1983a,b). The actual cause of death is rarely known, but most likely results from asphyxiation due to reduced oxygen conductance through the eggshell because the pores become occluded with water. Desiccation has also been suggested as a cause of egg mortality in some studies, but other causes, such as overheating, often cannot be precluded as the cause of death. However, good correlations between elevated egg mortality and dry conditions were clearly shown in nature for the soft-shelled eggs of the Galapagos land iguana Conolophus subcristatus (Snell & Tracy, 1985) and the tuatara (Thompson et al., 1990), and the rigid-shelled eggs of the turtle Trionyx triunguis (Leshem & Dmi'el, 1986).

Reptilian eggs exchange water with their nest, but the form of this exchange is still a matter of debate. Unlike avian eggs in which all water exchange occurs via water vapour, reptile eggs can also exchange liquid water (Thompson, 1987). On theoretical grounds, vapour may account for all movement in some cases (Ackerman, Dmi'el & Ar, 1985). Whether liquid water is available (e.g. as condensation on eggs) in natural nests, and how important liquid exchange may be is still being investigated. Many reptile eggs take up water from the nesting environment, and they must increase in volume in order to achieve this. Thus, rigid-shelled eggs that take up water must crack in order to accommodate the increased volume and this has been observed in both the laboratory and field in both turtles and crocodilians (McIlhenny, 1935; Plummer, 1976; Thompson, 1983; Grigg & Beard, 1985). As long as the underlying shell membranes remain intact, shell cracking probably has little influence on hatching success. However, excessive water absorption may result in premature hatching and/or incomplete internalisation of yolk at hatching, at least in A. mississippiensis (Thompson, unpublished data).

Eggs of many reptiles, particularly those with rigid shells, will hatch successfully after sustaining a net water loss. However, eggs of these same species will absorb water during incubation in suitable conditions, even though laboratory studies show that hatchlings gain no advantage from doing so. Indeed, in some laboratory experiments, gaining too much water causes death. Why then, do eggs absorb water they do not need? In some circumstances water absorption may be beyond an embryo's control: if the water potential of the nest becomes far greater than that of the egg due to water inundation, then water absorption by the egg is unavoidable. On the other hand, there is some evidence to suggest that the embryo can influence water exchange under certain conditions (Grigg & Beard, 1985), so why persist in absorbing excess water? One possibility is that water absorption is a 'buffer against the risk of desiccation' later in incubation (Fitch & Fitch, 1967; Bustard, 1971) when the nesting environment may dry considerably.

Oxygen and carbon dioxide

Few measurements of oxygen (P_{O_2}) and carbon dioxide (P_{CO_2}) tensions have been made in open avian nests, but the available data indicate that tensions are not greatly different from the surrounding atmosphere (Walsberg, 1980). By contrast, the P_{O_2} and P_{CO_2} in the underground nest diverge significantly from atmospheric conditions (Fig. 20.4). Three factors interact to determine the respiratory gas tensions experienced by eggs in underground nests: 1) diffusion resistance of the nest substrate; 2) $\dot{V}_{O_2}$ of nest substrate; and 3) $\dot{V}_{O_2}$ of eggs in the nest. Movement of gases through the nest substrate

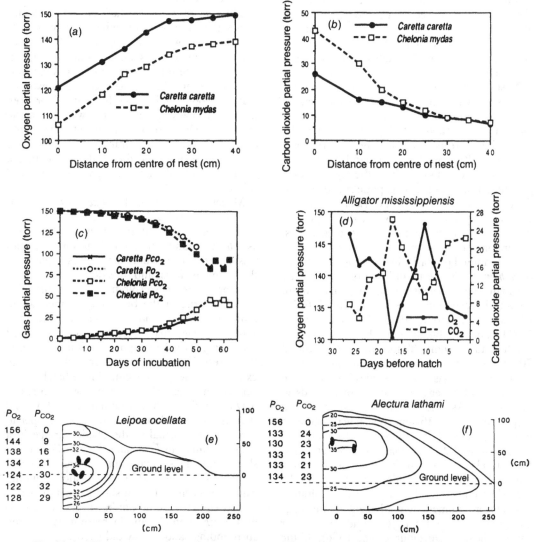

Fig. 20.4. The gaseous and thermal environment of mounds in two mound building megapodes, and the gaseous environment for three species of reptile. Data for *L. ocellata* from Seymour & Ackerman (1980), for *A. lathami* from Seymour (1985), for *C. caretta* and *C. mydas* from Ackerman (1977), and for *A. mississippiensis* from Thompson (unpublished). (*a*), (*b*) Spatial variation in gas tensions at day 50 of incubation. Nests are approximately 30 cm in diameter. (*c*) Temporal variation of gas tensions in centre of nest. (*d*) Temporal variation in nests over the last half of incubation. (*e*), (*f*) Typical profiles in mounds containing eggs. Isotherms in °C.

occurs predominantly by diffusion (Ackerman, 1977; Seymour, 1985; Seymour et al., 1986) and it offers a substantial resistance. Diffusion resistance varies inversely with the gas filled porosity of the substrate, and gas filled porosity varies with the substrate type, degree of compaction, and water content (Seymour, 1985; Seymour et al., 1986). Both A. lathami and L. ocellata adults regularly open and re-work their mounds while eggs are being incubated, which allows periodic convective gas exchange within the mound. It also keeps the mound material friable and loose which increases the gaseous porosity and thus facilitates diffusive gas exchange.

Nest material may harbour many micro-organisms which use aerobic metabolism. In nests that contain large volumes of organic material (L. ocellata and A. lathami) microbial respiration is responsible for the majority of a nest's total $\dot{V}_{O_2}$. On the other hand, nests constructed entirely of sand where the organic component is low (e.g. the megapode M. maleo and the turtles Caretta caretta and Chelonia mydas) the eggs are responsible for most of the $\dot{V}_{O_2}$ of the nest. Thus, in mound building megapodes, respiratory gas tensions experienced by the eggs are determined primarily by the mound itself. In A. lathami when the mound is first constructed the high water content results in a high rate of microbial respiration, and a high diffusive resistance (Seymour, 1985; Seymour et al., 1986). The combination of these two factors result in a P_{O_2} around 80 torr and a P_{CO_2} around 70 torr within the mound (Seymour et al., 1986). Such extreme gas tensions may be lethal to embryos and no eggs are found in mounds at this stage. As the mound dries, microbial respiration and diffusion resistance decrease producing mound P_{O_2} to around 132 torr and P_{CO_2} to around 23 torr. Gas tensions remain at these levels for the remainder of the breeding season (Seymour et al., 1986). Just before hatching, when embryonic $\dot{V}_{O_2}$ is greatest, respiratory gas tensions immediately adjacent to the eggshell are 127 torr for oxygen and 28 torr for carbon dioxide (Seymour et al., 1986). If the eggshells of A. lathami had gas conductances expected of open nesting birds the internal gas tensions may be intolerably extreme (P_{O_2} = 63 torr, P_{CO_2} = 95 torr). However, the high initial conductance combined with the increase in conductances due to shell thinning result in internal gas tensions of 112 torr for O_2 and 43 torr for

CO_2 at the end of incubation. Calculations indicate almost identical gas tensions occur inside naturally incubating eggs of L. ocellata immediately prior to hatching (Seymour et al., 1986). Thus, despite the unusual gas tensions inside the nesting mounds of megapodes, pre-pipping internal gas tensions are statistically indistinguishable from the mean of 23 open nesting species, P_{O_2} = 104 torr, P_{CO_2} = 41 torr (Paganelli & Rahn, 1984).

The increase in eggshell gas conductance during the latter part of incubation in megapode eggs satisfies two conflicting needs. If conductance remained at its initial value throughout incubation, the internal respiratory gas tensions at the end of incubation may become so extreme that development is severely retarded or even fatal. Conversely, if conductance was fixed at the pre-hatch value throughout incubation there is a danger of severe dehydration. Increasing the conductance towards the end of incubation when metabolic demands are greatest is an ideal solution.

Megapodes, such as M. maleo, that lay their eggs singly in volcanically heated sands may not be exposed to respiratory gas tensions as severe as those found in mound nesting species (assuming that the sands themselves contain little organic material and thus do not support significant microbial respiration). Certainly, this would be true at the beginning of incubation. However, by the end of incubation the sand above the egg may have become compacted decreasing its gaseous porosity. Thus, the diffusive resistance may have increased to such an extent that the gas tensions immediately adjacent to the egg, when metabolism is greatest, may be similar to those found in mound nesting species. Pores within M. maleo eggshell taper to a very narrow canal at the inside surface in a manner similar to L. ocellata (Fig. 20.1), and the shell undergoes significant thinning during incubation (Table 20.1). Such shell thinning would result in increased shell gas conductance so that embryos would not be adversely affected by a decrease in P_{O_2} and increase in P_{CO_2} in the surrounding sand if they did occur.

In vegetation mound nests of A. mississippiensis P_{CO_2} is elevated to levels similar to those in megapodes (Fig. 20.4). Some of this results from respiration of mound material, although the contribution of the eggs at the peak of their respiration is sufficient to significantly influence gas tensions in the nest chamber. The difference between *A. mississippiensis* and megapodes is that the whole clutch is at essentially the same developmental stage in alligators whereas in megapodes all stages are present at one time due to sequential oviposition. Thus, the contribution of eggs to deviations in gas tensions from atmospheric late in incubation is higher in alligators than megapodes.

Gas tensions measured in nests of sea turtles buried in relatively sterile sand show deviations from atmospheric for both P_{O_2} and P_{CO_2} (Ackerman, 1977; Seymour & Ackerman, 1980; Packard & Packard, 1988) similar to those measured in mound nests of alligators (Fig. 20.4) and hole nests of *C. johnstoni* (Whitehead, 1987). It has been suggested that the nest's ability to supply oxygen and lose carbon dioxide late in incubation, when embryos are respiring at their peak rate, determines P_{O_2} and P_{CO_2} and constrains clutch size so that eggs are kept within physiological limits (Seymour & Ackerman, 1980). This, of course, implies that females are able to select nest sites, and/or construct nests, that fulfil these requirements. Circumstantial evidence for this is found in the extreme selectivity of sea turtles for nesting beaches (Seymour & Ackerman, 1980).

Nests of sea turtles are constructed in a stereotyped manner, are generally deeper than other reptilian nests, and are located in relatively homogeneous, sterile substrates. Consequently most clutches within a species and beach are likely to experience similar physical environments, and presumably selection operates on these populations to optimise success in those conditions. This almost certainly has occurred in *C. mydas* and *C. caretta* where highest hatching success occurs at gas tensions similar to those in natural nests (Ackerman, 1981*a*). However, most reptilian nests are not as uniform as those of sea turtles. There are a range of nest materials and nest sizes used by crocodilians (Lutz & Dunbar-Cooper, 1984; Whitehead, 1987) and their eggs show great variation in gas conductance (Whitehead, 1987). These factors are reflected in large variations in gas tensions in the nest (Whitehead, 1987) and consequently large variation in gas tensions experienced by the embryos. Gas tensions through incubation do not deviate from atmospheric in *Crocodylus acutus* and *C. johnstoni* as much as they do in sea turtles. The matching of optimal hatching success with gaseous environments has been examined only in sea turtles. However, prelimi-

nary data indicates that CO_2 tensions above atmospheric may speed development in *A. mississippiensis* (Thompson, Guillette & DeMarco, unpublished data), and presumably high carbon dioxide and low oxygen tensions cause premature hatching in *C. johnstoni* (Whitehead, 1987). Additionally, at least one species of turtle (*Chrysemys nelsoni*) as well as other species actively selects alligator nest mounds in which to construct its own nest (Jackson, 1988): perhaps favourable gas tensions or temperatures are responsible for this behaviour.

At their maximum deviation from atmospheric, gas tensions in sea turtle nests are close to those in the air cell of avian eggs. This led to the suggestion that such tensions serve an important physiological function, such as the initiation of shell pipping (Ackerman, 1977). However, in smaller reptile clutches, the nest gas tensions are somewhat closer to atmospheric than those in sea turtles and alligators (Thompson, 1981). Reptilian eggshells have high conductances so the gas tensions experienced by embryos in small clutches is probably much closer to atmospheric than in air cells of avian eggs. Perhaps the similarity of gas tensions in avian egg air cells, turtle and alligator nests indicates some maximum acceptable value, rather than a necessary physiological trigger for hatching (Seymour & Ackerman, 1980).

Like birds, crocodilian and chelonian embryos remove calcium carbonate from the eggshell during development. The influence of this on shell conductance has not been investigated. However, progressive differential drying of the eggshell during incubation, indicated by opaque regions on the shell variably called banding in crocodilians (Ferguson, 1985), chalking and white spot or patch in turtles (Ewert, 1985; Thompson, 1985; Chan, 1989), results in increasing conductance throughout incubation (Thompson, 1985; Whitehead, 1987). The increase in shell conductance closely matches the respiratory requirements of the egg (Thompson, 1985; Whitehead, 1987).

In addition to differential drying of the shell, eggs of *A. mississippiensis* may undergo extrinsic shell degradation (Ferguson, 1981). The exact cause of this is still being investigated and its necessity to successful incubation is a matter of debate (Grigg & Beard, 1985). However, when it occurs it surely increases conductance of the calcareous eggshell weakens the shell facilitating embryonic escape. However, the contribution of the calcareous shell to the overall shell/shell membrane diffusion barrier is small, so degradation of the calcareous component will have little influence on total conductance (Whitehead, 1987). Eggs of *A. mississippiensis* in which the shell surface has been sterilised still produce viable hatchlings able to break free of the shell in the laboratory (Guillette, unpublished data), so shell degradation resulting from the actions of extrinsic micro-organisms may not be necessary in nature. Although extrinsic degradation occurs in nature in *C. porosus*, laboratory data cast doubt on the general necessity for extrinsic degradation in this species, also (Grigg & Beard, 1985; Grigg, 1987).

Temperature and energetics

Temperature

Incubation in reptiles is generally 5–8 °C lower than birds and this has important consequences for both embryonic growth rate and oxygen consumption (Vleck & Hoyt, Chapter 18). The thermal environment of reptilian eggs is often variable; diurnal and longer-term fluctuations in temperature are common, especially in shallow nests (Thompson *et al.*, 1990) but also in mound nests (Chabreck, 1973). By contrast, most avian eggs are incubated at relatively stable temperatures due to the almost constant attention of the incubating parents (Haftorn, 1983; White & Kinney, 1974; Turner, Chapter 9), but eggs can be exposed to short-term drops in temperature when the parent leaves the nest to forage. Consequently, reptilian eggs are more tolerant to a range of incubation temperatures than avian eggs (Deeming & Ferguson, Chapter 10).

In a series of studies Frith (1955, 1956b, 1957) has demonstrated that adult malleefowl are capable of detecting mound temperature with their mouth and have a complex set of mound tending behaviours which serve to regulate mound temperature within narrow limits. The preferred incubation temperature is reported to be around 34 °C. In a different study area, Booth (1987b) also finds the preferred mound temperature to be 34 °C, but, contrary to Frith's findings, temperatures from 27–38 °C are measured, and individual mound temperatures fluctuate markedly over the course of incubation (Fig. 20.5).

Incubation temperatures for *A. lathami* eggs

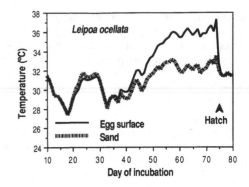

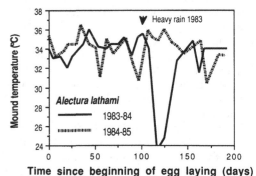

Fig. 20.5. Temperature of the eggshell surface and adjacent sand of a developing *L. ocellata* egg during natural incubation, and mean temperatures of two *A. lathami* mounds throughout the breeding season. (Data for *L. ocellata* from Booth, 1987b, and *A. lathami* from Jones, 1988a, and unpublished observations.)

may range from 30 °C to 38°C (Fleay, 1937; Frith, 1956a), with means being reported as 31.4 °C (Seymour *et al.*, 1986) and 33.3 °C (Jones, 1988a). The temperature within any mound is remarkably constant after the onset of laying (Seymour, 1985; Jones, 1988a; Fig. 20.5), the stability being attributed partly to the mound tending behaviour of the male bird (Fleay, 1937; Seymour, 1985; Jones, 1988a) and partly to the thermal inertia and self regulating thermal properties of the mound (Seymour, 1985). Weather conditions can have a marked effect on mound temperature. Severe rain storms may decrease mound temperature by 8–12 °C for extended periods of time (Fig. 20.5) and result in increased incubation mortality (Jones, 1988b).

Incubation temperatures of other megapodes are not well characterised, but the historical data has been summarised by Frith (1956a): 31–38 °C for *Megapodius eremita*, 35–39 °C for *Megapodius freycinet*, 36 °C for *Megapodius pritch-*

ardi, and 38 °C for *Aepypodius arfakianus*. More recently, *M. pritchardi* eggs were found incubating at 33–35 °C (Todd, 1983), *M. eremita* eggs at 33 °C (Roper, 1983), and *M. maleo* eggs at 31–35 °C (Dekker, 1988). All these temperatures are within the range reported for other avian species (Drent, 1975; Webb, 1987).

Many avian eggs undergo significant increases in egg temperature as incubation proceeds. However, the developing embryo experiences a relatively uniform thermal environment because it floats at the top of the egg where it is in close contact with the brood patch of the incubating adult (Booth, 1987*b*). For example, eggs of the herring gull (*Larus argentatus*) experience a 7 °C increase in mean egg temperature from the beginning to the end of incubation but the embryo increases by only 1.4 °C (Drent, 1970). Megapode eggs are surrounded by a thermally homogeneous environment and the increased heat production by the developing embryo always results in an increase in egg temperature above that of the nesting substrate as incubation progresses. In artificially incubated eggs of *L. ocellata* and *A. lathami*, shell surface temperature rises 2–2.5 °C above incubation substrate during incubation (Booth, 1987*b*; Seymour *et al.*, 1987) and by 3.5 °C in a naturally incubated *L. ocellata* egg (Fig. 20.5). Widely fluctuating nest temperatures, and an increase in egg temperature of several degrees above ambient towards the end of incubation indicate that embryos of *L. ocellata* can survive prolonged exposure to temperatures between 28 °C and 40 °C, a much wider range than for other avian species (Webb, 1987). Generally, avian eggs are more tolerant to short-term (< 1 day) large decreases (> 10 °C) in temperature than to long-term (> 1 day) small decreases (< 10 °C) in temperature (Booth, 1987*b*). Such tolerance to fluctuating temperatures may be an adaptation to mound nesting. Obviously it is advantageous for embryos to continue to develop during periods when the parent birds cannot maintain the preferred incubation temperature due to adverse weather conditions.

Incubation temperature has important consequences for hatching success and incubation period. Artificial incubation of eggs of *L. ocellata* is successful over the range 32–38 °C. Hatching success is greatest at 34 °C (80%), the preferred incubation temperature, and decreases to 22% at 32 °C and 38% at 38 °C (Booth 1987*b*). Although eggs fail to develop at 30 °C in the laboratory, two naturally incubated eggs hatched at an average temperature of 30.7 °C which included 20 days below 30 °C (Fig. 20.5). The natural incubation period in *L. ocellata* is quite variable as it depends on incubation temperature (Frith, 1956*b*; Booth, 1987*b*). Like reptilian embryos, embryos of *L. ocellata* develop slower at lower temperatures, and thus lower temperatures result in longer incubation periods (Deeming & Ferguson, Chapter 10). At the preferred incubation temperature of 34 °C, incubation takes about 60 days, but natural variation in nest temperatures results in hatching in as little as 50 days (Frith, 1962) or as long as 91 days (Bellchambers, 1917). Variable incubation periods in non-megapode avian species such as petrels (Boersma, 1982), are due to varying periods of parental absence when eggs are not incubated and embryonic development is temporarily arrested. By contrast, variation in megapode egg incubation period is caused by continuous exposure to higher or lower temperatures where embryonic development continues, but at different rates.

The range of temperatures over which reptilian eggs develop and the variation that can be sustained by a given species, is generally much greater than that for avian eggs. With the exception of pythons, reptiles do not regulate their egg temperatures and egg manipulation during incubation is rare (Shine, 1988; Somma & Fawcett, 1989; Deeming, Chapter 19). Consequently, nest temperatures are influenced by climatic and physical attributes of the environment. Presumably, selection has moulded the female's behaviour so that she chooses the most appropriate nest location to optimise incubation success. The optimum constant temperature ranges determined in the laboratory vary from 18–22 °C for the tuatara (Thompson, 1990) to 28–38 °C for the iguanid lizard *Dipsosaurus dorsalis* (Muth, 1980). Many species incubate optimally around 30 °C.

Most reptilian eggs are deposited in the ground and left to incubate unattended, so the depth of the nest is an important parameter when considering the thermal regime of incubation. Below about 50 cm there is essentially no diurnal temperature fluctuation (Packard & Packard, 1988). Generally, larger reptiles construct deeper nests and probably all the sea turtles nest deeper than the zone of diurnal temperature variation. Some smaller reptiles, e.g. the lizard *Conolophus pallidus* (Christian &

Tracy, 1982), construct very deep nesting burrows relative to their body size and presumably this is, at least partly, in response to thermal cues. However, most reptiles construct nests where the eggs are influenced by diurnal temperature variations. Such fluctuations can cause considerable thermal gradients within a nest (Thompson, 1988) which may influence development time, hatching synchrony (Thompson, 1989), and sex determination (Bull, 1985). Mound nest temperatures vary between 27 and 33 °C in *C. porosus* (Webb, Messel & Magnusson, 1977) and 24 and 33 °C in *A. mississippiensis* (Chabreck, 1973; Thompson, unpublished data). Other factors which may significantly influence the thermal environment of the nest are shade, e.g. in tuataras (Thompson *et al.*, 1990), aspect, e.g. *Chrysemys picta* (Schwarzkopf & Brooks, 1987) and moisture (Ferguson & Joanen, 1982, 1983).

An important aspect of the thermal environment of incubation in many reptiles (including all the crocodilians and the majority of turtles) is that sex determination is related to incubation temperature (Bull, 1980; Deeming & Ferguson, 1988, Chapter 10). Despite an extensive literature on the occurrence of temperature-dependent sex determination in the *Reptilia*, there is little data on the influence of fluctuating temperatures on sex determination (Bull, 1985), or on the selection of nest sites by females (Schwarzkopf & Brooks, 1987).

Energetics

Interspecific analysis of incubation time and total energetic cost of development in both avian and reptilian eggs indicates that longer incubation times incur a greater energetic cost in eggs of similar mass incubated at the same temperature (Ackerman *et al.*, 1980; Vleck & Kenagy, 1980; Ackerman, 1981b; Whitehead, 1987; Hoyt, 1987; Vleck & Hoyt, Chapter 18). It may be concluded, therefore, that incubation of eggs of the same species but at lower temperature would lead to greater energy expenditure due to the consequential increase in incubation time.

In *L. ocellata*, incubation at temperatures below 36 °C extends incubation and increases maximum $\dot{V}_{O_2}$ (Fig. 20.6), resulting in greater total energy expenditure at low incubation temperatures (Fig. 20.7). The increased $\dot{V}_{O_2}$ at low temperature is unexpected and difficult to

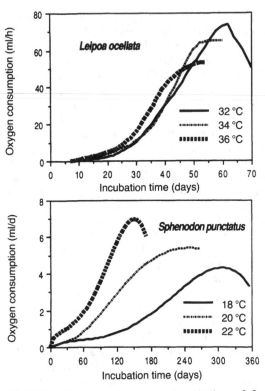

Fig. 20.6. Patterns of oxygen consumption of *L. ocellata* (Booth, 1987b) and *S. punctatus* (Thompson, unpublished observations) eggs at various incubation temperatures.

explain. Metabolism of a developing embryo at any point in time may be conceptually divided into two components: 1) energy expenditure due to tissue synthesis; and 2) energy expenditure due to tissue maintenance (Vleck, Vleck & Hoyt, 1980; Hoyt, 1987). Cost of tissue synthesis is proportional to growth rate. Maintenance is dependent on the amount of tissue present and should be proportional to the 0.72 power of embryo mass. Using this model, smaller absolute rates of energy expenditure at lower temperature are expected because at any equivalent developmental stage, both tissue synthesis (due to slower growth rates) and tissue maintenance costs (due to a Q_{10} effect on metabolic processes) are reduced. If this model is applicable for eggs of *L. ocellata*, a possible conclusion is that the energetic cost of synthesising tissue is itself a function of temperature, i.e. it requires more energy to synthesise the same amount of tissue at lower temperatures (Booth, 1987b). A second possible explanation for the elevated oxygen consumption at low incubation temperatures could be that the embryo is elevating its metabolic rate in a primordial attempt to thermoregulate. Embryos incubated at 34 °C show a homeothermic response to chilling 3 days before hatch, but such responses are absent earlier in incubation (Booth, 1987*b*). However, the elevation in $\dot{V}_{O_2}$ in eggs incubated at lower

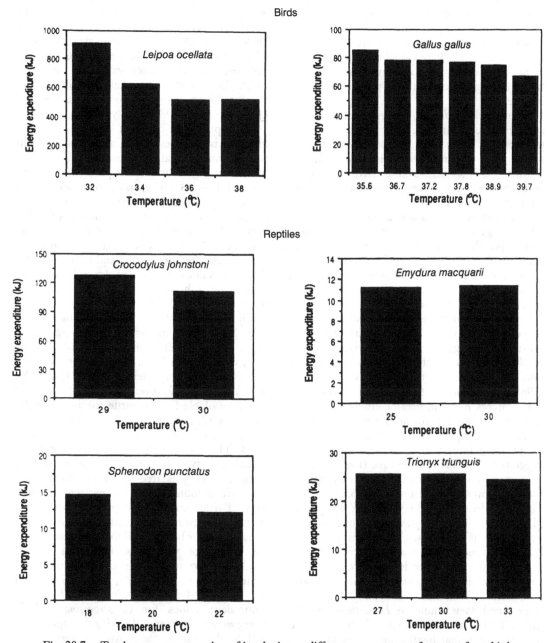

Fig. 20.7. Total energy consumption of incubation at different temperatures for eggs of two bird and four reptile species. Oxygen consumption data was converted to joules using the conversion factor 19.64 J/ml O_2 (Vleck *et al.* 1984). (Data for *L. ocellata* from Booth, 1987*b*, *G. gallus* from Barott, 1937, *C. johnstoni* from Whitehead, 1987, *E. macquarii* from Thompson, 1983, *S. punctatus* from Thompson, unpublished observations, and *T. triunguis* from Leshem, Ar, Dmi'el & Ackerman, unpublished observations.)

temperatures occurs several weeks before hatching so it is unlikely that thermoregulation is the true explanation. A third possible explanation is that at cooler incubation temperatures more yolk is converted to tissue, hence chicks hatching at lower temperatures have similar masses to chicks at higher temperatures, but have smaller residual yolks and larger yolk-free hatchling masses (C. M. Vleck, personal communication). Data on residual yolk mass have not been collected in megapodes, so this hypothesis remains untested.

When fowl (*Gallus gallus*) eggs are incubated at different temperatures, those incubated at low temperature have an extended incubation period, but maximum $\dot{V}_{O_2}$ is also reduced (Barott, 1937) so that, over a narrow range of temperatures, total energy expenditure is the same (Fig. 20.7). However, at the lowest possible incubation temperature, a slight increase in total energy consumption occurs, and, at the highest possible incubation temperature, a slight decrease is observed (Fig. 20.7). Thus, the data from *L. ocellata* and *G. gallus* suggest that in avian eggs there is an increased energetic cost to development at low temperatures.

Comparison of the energetic cost of incubation of *C. mydas* and *C. caretta* eggs at 30 °C with those of avian eggs of similar size incubated at 36 °C has shown that the cost of development is lower in these turtles than in birds. Ackerman (1981*b*) concludes that lower incubation temperatures result in greater efficiency during incubation. However, further investigation comparing the energetic cost of incubation at different temperatures in four reptilian species suggests that this is not the case (Fig. 20.7). In all four species incubation is longer but peak $\dot{V}_{O_2}$ lower at lower temperatures (e.g. *S. punctatus*; Fig. 20.6). In three of the species, *E. macquarii*, *T. triunguis*, and *S. punctatus*, total energy expenditure is similar at all temperatures (Fig. 20.7) and much lower than for avian eggs of equivalent size. However, in *C. johnstoni*, significantly more energy is consumed at the lower temperature (Whitehead, 1987; Fig. 20.7), the same trend as is found in *L. ocellata* and *G. gallus*. Thus, intraspecific data do not support the hypothesis that incubation is less costly at lower temperature. Although the data are too few to be conclusive, it is tempting to speculate that developmental energy expenditure is independent of incubation temperature in non-crocodilian reptiles, but dependent on temperature in crocodilians and birds. If this hypothesis is correct, the mechanism behind the differences needs to be explained and clearly, this area warrants further investigation.

If temperature *per se* cannot explain why bird eggs consume more energy during development than equivalent-sized reptile eggs, is there a phylogenetic difference in tissue synthesis cost? The energy density of avian egg contents varies with hatchling type when expressed on a wet weight basis, but is remarkably uniform when expressed on a dry weight basis, averaging 29.2 kJ g^{-1} (Ar *et al.*, 1987). Analysis of energy density of reptile eggs on a wet basis is difficult to interpret because eggs may absorb water from the environment after oviposition. Using dry weights as a basis avoids this problem, and data from 24 lizard, 4 snake, and 3 turtle species are remarkably uniform averaging 26.9 kJ g^{-1} (Table 20.2). However, the majority of lizard data, and all of the snake data comes from one study (Vitt, 1978) where the data are reported in terms of ash-free dry weight, and not dry weight. Thus the values in terms of dry weight will be slightly lower than indicated in Table 20.2. The mean value of 26.9 kJ g^{-1} for reptiles is significantly lower ($P < 0.005$) than that of birds (29.2 kJ g^{-1}) so it would appear that reptile eggs have a lower energy density compared to avian eggs on a dry weight basis. The avian yolk-free hatchling has an average energy density of 25.1 kJ g^{-1} (dry wt) (Ar *et al.*, 1987). Equivalent data for reptiles is only available for two turtle species, *Chelydra serpentina* (21.4 kJ g^{-1}; Wilhoft, 1986) and *T. triunguis* (23.5 kJ g^{-1}; Leshem *et al.*, unpublished data). These data suggest that yolk-free turtle hatchlings have lower energy densities (dry wt) compared to yolk-free avian hatchlings, the same trend found in the comparisons of fresh eggs. Turtles have a high proportion of bone tissue, however, and equivalent data for squamates may be required for a more appropriate comparison with birds. The mean cost for producing a gram of dry hatchling tissue for 34 avian species is 13.9 kJ g^{-1}, range 9.7–21.9 kJ g^{-1} (calculated from data in Ar *et al.*, 1987). Equivalent values for *C. serpentina* and *T. triunguis* are 15.8 and 9.9 kJ g^{-1} respectively, within the range for birds. If the energy expenditure per gram of dry hatchling tissue is similar for reptiles and birds as the data suggest, but avian hatchlings have higher energy densities, then it follows that the amount of energy expended per unit of energy in the yolk-free

Table 20.2. *Energy content of various reptile eggs. Energy content is expressed on a dry weight basis and is for the egg contents excluding shell and membranes unless otherwise stated.*

Species	Energy density (kJ g^{-1})	Reference
Lizards		
Anolis carolinensis	25.8	Ballinger & Clark, 1973
Callisaurus draconoides	27.0	Vitt, 1978[a]
Cnemidophorus gularis	26.6	Ballinger & Clark, 1973
Cnemidophorus sexlineatus	26.3	Ballinger & Clark, 1973
Cnemidophorus tigris	28.1	Vitt, 1978[a]
Coleonyx variegatus	26.5	Vitt, 1978[a]
Cophosaurus texanus	28.0	Vitt, 1978[a]
Cophosaurus texanus	25.7	Ballinger & Clark, 1973
Crotaphytus collaris	27.6	Vitt, 1978[a]
Eumeces fasciatus	25.5	Ballinger & Clark, 1973
Gambelia wislizeni	27.9	Vitt, 1978[a]
Holbookia maculata	27.1	Vitt, 1978[a]
Lygosoma laterale	25.8	Ballinger & Clark, 1973
Phrynosoma cornutum	26.8	Vitt, 1978[a]
Phrynosoma cornutum	25.2	Ballinger & Clark, 1973
Phrynosoma modestum	27.0	Vitt, 1978[a]
Phrynosoma platyrhinos	26.5	Vitt, 1978[a]
Sceloporus clarki	26.9	Vitt, 1978[a]
Sceloporus magister	26.1	Vitt, 1978[a]
Sceloporus olivaceus	25.8	Ballinger & Clark, 1973
Sceloporus scalaris	26.0	Vitt, 1978[a]
Sceloporus undulatus	27.8	Vitt, 1978[a]
Sceloporus undulatus	25.9	Ballinger & Clark, 1973
Sceloporus virgatus	27.5	Vitt, 1978[a]
Urosaurus graciosus	27.4	Vitt & Oharmt, 1975: Vitt, 1978[a]
Urosaurus ornatus	28.0	Vitt, 1978[a]
Uta stansburiana	27.0	Vitt, 1978[a]
Mean	26.7	
Snakes		
Chilomeniscus cinctus	27.6	Vitt, 1978[a]
Contia tenuis	26.9	Vitt, 1978[a]
Salvadora hexalepis	26.7	Vitt, 1978[a]
Sonora semiannulata	27.7	Vitt, 1978[a]
Mean	27.2	
Turtles		
Chelydra serpentina	28.5	Wilhoft, 1986
Chyrsemys picta	26.4	Congdon & Tinkle, 1982
Trionyx triunguis	28.5	Leshem *et al.*, unpublished
Mean	27.8	
Mean lizards, snakes & turtles	26.9	

[a]Data are expressed in terms of kJ per gram ash-free dry weight, and include shell and membranes.

hatchling, will be lower in birds compared to reptiles. For 34 avian species the mean total production efficiency is 0.57 J J^{-1}, range 0.31–0.91 J J^{-1} (Ar *et al.*, 1987). The equivalent values for *C. serpentina* and *T. triunguis* are 0.76 J J^{-1} and 0.42 J J^{-1} respectively. Clearly, the great variability in the avian data, and scarcity of equivalent reptile data make any confident conclusions about the relative production efficiencies of reptilian and avian eggs premature. A more detailed analysis made on a wet weight basis came to a tentative conclusion that reptilian development is more efficient than avian embryonic development (Vleck & Hoyt, Chapter 18).

Hatching

Megapode eggs are unique among birds in that they do not form a fixed air-cell at the blunt pole of the egg during incubation. Air spaces do form, but they can occur anywhere under the shell (Baltin, 1969; Seymour, 1984; Vleck *et al.*, 1984). The air-cell plays an important role in fowl eggs in the transition from chorio-allantoic to pulmonary respiration (Visschedijk, 1968; Seymour, 1984). Approximately a day before hatching the fowl embryo pierces the chorio-allantois below the air-cell with its beak and begins to ventilate its lungs with gas from the air-cell. Circulation in the remainder of the chorio-allantois remains functional but slowly decreases as the transition to pulmonary respiration is made (Visschedijk, 1968). After external pipping, the chorio-allantoic circulation shuts down completely and respiration occurs entirely via the lungs. The embryo then completes the hatching process without any loss of blood. The lack of a fixed air-cell precludes this process in megapode embryos. Hatching, therefore, is a rapid process, the legs kick and shatter the shell which has become very fragile by shell thinning, and hatching is achieved through further kicking and shoulder movements (Frith, 1962; Baltin, 1969; Seymour, 1984; Vleck *et al.*, 1984). Prior to the shell being broken there is no gas in the lungs (Seymour, 1984), hence the transition from chorio-allantoic to pulmonary respiration is rapid, taking only minutes to a few hours (Seymour, 1984; Vleck *et al.*, 1984). Once the hatchling is free of the shell it must ascend through approximately 60 cm of mound material, a process that takes approximately one day in *L. ocellata* and two days in *A. lathami* (Vleck *et al.*, 1984). This is an energetically demanding process and 8% (*L. ocellata*) and 33% (*A. lathami*) of the total energy consumed during incubation is expended making the escape (Vleck *et al.*, 1984). Routine working of the mound by the adult facilitates the hatchlings' escape by keeping the mound material loose and friable. The cost of nest escape is probably even higher in species such as *M. maleo* where the soil above the egg may become compacted.

The regular formation of an air-cell between the shell membranes is not a feature of reptilian eggs, although it may occur in some species sometimes (Packard & Packard, 1979; Whitehead, 1987). In *C. johnstoni* (Whitehead, 1987) and some chelonians (Ewert, 1985), airspaces can form between the calcareous shell and shell membranes or between the shell membranes and chorio-allantoic membranes but, unlike birds, apparently not between the shell membranes. Rigid-shelled eggs of turtles and crocodilians may absorb water during incubation (Plummer, 1976), so air space formation is probably not necessary, although it may be common during artificial incubation in at least *C. johnstoni* and *A. mississippiensis* (Whitehead, 1987; Deeming & Ferguson, 1989). Like megapodes, the transition to pulmonary respiration in reptiles probably does not occur until the eggshell is pipped, but, unlike megapodes, crocodilians and turtles commonly do not emerge from the eggshell for 12–24 h after pipping, although this period is variable (Gutzke, Paukstis & Packard, 1984). During this time the hatchlings frequently swallow, presumably as they clear fluid from the lungs (Thompson, unpublished data).

In many reptiles, hatching synchrony is probably important in maximising hatchling survival (Carr & Hirth, 1961; Thompson, 1989). In very deep nests exposed to constant thermal regimes, developmental asynchrony probably never occurs, but in shallower nesting species different rates of development can occur within a clutch due to thermal gradients in the nest, especially from top to bottom (Webb *et al.*, 1983b; Thompson, 1988). In species where asynchronous development may occur, there is a period late in incubation where V_{O_2} of eggs falls after reaching a peak. Hatching may occur at any time after the peak, so the peak-pip period varies amongst eggs within a clutch to allow synchronous hatching. A similar argument has been made for ratite birds where hatching synchrony is important (Vleck, Vleck & Hoyt, 1979). This pattern of V_{O_2} occurs in crocodilians (Whitehead, 1987; Thompson, 1989; Vleit & Demarco, unpublished data) and some turtles (Webb, Choquenot & Whitehead, 1986; Thompson, 1989; Ewert, Chapter 11).

Although parental care in reptiles is exceptional, many species of crocodilians excavate the nest to release the young at the time of hatching. Unlike megapodes, crocodiles do not work the nest material after oviposition. Crocodilian mound nesters may lie on, or across, the nest during incubation (Joanen, 1969), thereby packing the nest material, but they then release the hatchlings. Neonate crocodilians vocalise frequently and this alerts the female to the presence of hatchlings in the nest. Presumably, this parental behaviour provides the selective pressure for synchrony of hatching in crocodilians. Similar behaviour is completely unknown in chelonians, although it may occur in some squamates (Somma & Fawçett, 1989). Adult megapodes do not deliberately liberate their young from the nest.

Acknowledgements

Scanning electron microscope work was done by D. T. B. while he was a Visiting Research Fellow in the Department of Physiology at State University of New York, Buffalo and funded by N.I.H. grant RO1 HL33437. Preparation of the manuscript and figures was done with the help of N.S.F. grant DCB 87–18264. We thank R. A. Ackerman and D. N. Jones for providing us with their unpublished data on *T. triunguis* and *A. lathami* respectively.

References

Ackerman, R. A. (1977). The respiratory gas exchange of sea turtle nests (*Chelonia, Caretta*). *Respiration Physiology*, **31**, 19–38.

— (1981*a*). Growth and gas exchange of embryonic sea turtles (*Chelonia, Caretta*). *Copeia*, **1981**, 757–65.

— (1981*b*). Oxygen consumption by sea turtle (*Chelonia, Caretta*) eggs during development. *Physiological Zoology*, **54**, 316–24.

Ackerman, R. A., Dmi'el, R. & Ar, A. (1985). Energy and water vapor exchange by parchment-shelled reptile eggs. *Physiological Zoology*, **58**, 129–37.

Ackerman, R. A., Whittow, G. C., Paganelli, C. V. & Pettit, T. N. (1980). Oxygen consumption, gas exchange, and growth of embryonic Wedge-tailed shearwater (*Puffinus pacificus chlororhynchus*). *Physiological Zoology*, **53**, 210–21.

Ar, A., Arieli, B., Belinsky, A. & Yom-Tov, Y. (1987). Energy in avian eggs and hatchlings: utilization and transfer. *Journal of Experimental Zoology*, Supplement, **1**, 151–64.

Ar, A., Paganelli, C. V., Reeves, R. B., Greene, D. G. & Rahn, H. (1974). The avian egg, water vapor conductance, shell thickness and functional pore area. *Condor*, **76**, 153–8.

Ar, A. & Rahn, H. (1980). Water in the avian egg, overall budget of incubation. *American Zoologist*, **20**, 373–84.

Ballinger, R. E. & Clark, D. R. (1973). Energy content of lizard eggs and the measurement of reproductive effort. *Journal of Herpetology*, **7**, 129–32.

Baltin, S. (1969). Zur biologie und ethologie des Talegalla-Huhns (*Alectura lathami* Gray) unter besonderer berücksichtigung des verhaltens während der brutperiode. *Z. Tierpsychol.*, **26**, 524–72.

Barott, H. G. (1937). Effect of temperature, humidity, and other factors on the hatch of the hen's eggs and on energy metabolism of chick embryos. *Technical Bulletin* No. 553, US Department of Agriculture, Washington, D.C.

Bellchambers, T. D. (1917). Notes on the Malleefowl *Leipoa ocellata rosiae*. *South Australian Ornithologist*, **3**, 78–81.

Board, R. G., Perrott, H. R., Love, G. & Seymour, R. S. (1982). A novel pore system in the eggshells of the Mallee fowl, *Leipoa ocellata*. *Journal of Experimental Zoology*, **220**, 131–4.

Boersma, P. D. (1982). Why some birds take so long to hatch. *American Naturalist*, **120**, 733–50.

Booth, D. T. (1987*a*). Home range and hatching success of Malleefowl, *Leipoa ocellata* Gould (Megapodiidae), in Murray mallee near Renmark, S.A. *Australian Wildlife Research*, **14**, 95–104.

— (1987*b*). Effect of temperature on development of malleefowl *Leipoa ocellata* eggs. *Physiological Zoology*, **60**, 437–45.

— (1988*a*). Shell thickness in megapode eggs. *Megapode Newsletter*, **2**, 13.

— (1988*b*). Respiratory quotient of Malleefowl (*Leipoa ocellata*) eggs late in incubation. *Comparative Biochemistry and Physiology*, **90A**, 445–7.

— (1989). Regional changes in shell thickness, shell conductance, and pore structure during incubation in eggs of the mute swan. *Physiological Zoology*, **62**, 607–20.

Booth, D. T. & Seymour, R. S. (1987). Effect of eggshell thinning on water vapor conductance of malleefowl eggs. *Condor*, **89**, 453–9.

Bull, J. J. (1980). Sex determination in reptiles. *Quarterly Review of Biology*, **55**, 3–21.

— (1985). Sex ratio and nest temperatures in turtles, comparing field and laboratory data. *Ecology*, **66**, 1115–22.

Bustard, H. R. (1971). Temperature and water tolerances of incubating crocodile eggs. *British Journal of Herpetology*, **4**, 198–200.

Bustard, R. (1972). *Sea Turtles. Natural History and Conservation*, p. 220. London: W. Collins Sons & Co Ltd.

Calder, W. A., Parr, C. R. & Karl, D. P. (1978). Energy content of eggs of the Brown kiwi *Apteryx australis*; an extreme in avian evolution. *Comparative Biochemistry and Physiology*, **60A**, 177–9.

Carr, A. & Hirth, H. (1961). Social facilitation in green turtle siblings. *Animal Behavior*, **9**, 68–70.

Chabreck, R. H. (1973). Temperature variation in nests of the American alligator. *Herpetologica*, **29**, 48–51.

— (1975). Moisture variation in nests of the American alligator (*Alligator mississippiensis*). *Herpetologica*, **31**, 385–9.

Chan, E. (1989). White spot development, incubation and hatching success of leatherback turtle (*Dermochelys coriacea*) eggs from Rantau Abang, Malaysia. *Copeia*, **1989**, 42–7.

Christian, K. A. & Tracy, C. R. (1982). Reproductive behavior of Galapagos land iguanas, *Conolophus pallidus*, on Isla Santa Fe, Galapagos. In *Iguanas of the World*, eds G. M. Burghardt & A. S. Rand, pp. 366–79. Park Ridge, New Jersey: Noyes Publications.

Congdon, J. D. & Tinkle, D. W. (1982). Reproductive energetics of the painted turtle (*Chrysemys picta*). *Journal of Herpetology*, **38**, 228–37.

Deeming, D. C. & Ferguson, M. W. J. (1988). Environmental regulation of sex determination in reptiles. *Philosophical Transactions of the Royal Society of London*, **B322**, 19–39.

— (1989). Effects of incubation temperature on growth and development of embryos of *Alligator mississippiensis*. *Journal of Comparative Physiology*, **159B**, 183–93.

— (1991). Incubation and embryonic development in reptiles and birds. In *Avian Incubation*, ed. S. G. Tullett. pp. 3–37 London: Butterworths.

Dekker, R. W. R. J. (1988). Notes on ground temperatures at nesting sites of the Maleo *Macrocephalon maleo* (Megapodiidae). *Emu*, **88**, 124–7.

Drent, R. H. (1970). The Herring gull and its egg. *Behaviour Supplement*, XVII.

— (1975). Incubation. In *Avian Biology*, vol. 5, ed. D. S. Farner & J. R. King, pp. 333–421. New York: Academic Press.

Dunson, W. A. & Bramham, C. R. (1981). Evaporative water loss and oxygen consumption of three small lizards from the Florida Keys, *Sphaerodactylus cinereus*, *S. notatus*, and *Anolis sagrei*. *Physiological Zoology*, **54**, 253–9.

Ewert, M. A. (1979). The embryo and its egg: development and natural history. In *Turtles, Perspectives and Research*, eds M. Harless & H. Morlock, pp. 333–413. New York: John Wiley & Sons.

— (1985). Embryology of turtles. In *Biology of the Reptilia*, vol. 14, Development A, eds C. Gans, F. Billett & P. F. A. Maderson, pp. 75–267. New York: John Wiley & Sons.

Ferguson, M. W. J. (1981). Extrinsic microbial degradation of the alligator eggshell. *Science*, **214**, 1135–7.

— (1985). Reproductive biology and embryology of the crocodilians. In *Biology of the Reptilia*, vol. 14, Development A, ed. C. Gans, F. Billett & P. F. A. Maderson, pp. 329–491. New York: John Wiley & Sons.

Ferguson, M. W. J. & Joanen, T. (1982). Temperature of egg incubation determines sex in

Alligator mississippiensis. *Nature, London*, **296**, 850–3.

— (1983). Temperature-dependent sex determination in *Alligator mississippiensis*. *Journal of Zoology, London*, **200**, 143–77.

Fitch, H. S. (1970). Reproductive cycles in lizards and snakes. *Miscellaneous Publications University of Kansas Museum of Natural History* No. 52, 1–247.

Fitch, H. S. & Fitch, A. V. (1967). Preliminary experiments on physical tolerances of the eggs of lizards and snakes. *Ecology*, **48**, 160–5.

Fleay, D. H. (1937). Nesting habits of the Brush Turkey. *Emu*, **36**, 153–63.

Frith, H. J. (1955). Incubation in the Mallee fowl (*Leipoa ocellata*), Megapodiidae. *Proceedings XI International Ornithological Congress, Basel. Experimentia*, Supplement, **3**, 570–4.

— (1956a). Breeding habits of the family Megapodiidae. *Ibis*, **98**, 620–40.

— (1956b). Temperature regulation in nesting mounds of the Mallee-fowl *Leipoa ocellata* Gould. *C.S.I.R.O. Wildlife Research*, **1**, 79–95.

— (1957). Experiments on the control of temperature in the nesting mounds of Mallee fowl *Leipoa ocellata*. *C.S.I.R.O. Wildlife Research*, **2**, 101–10.

— (1962). *The Mallee-fowl*. Sydney: Angus & Robertson.

Grigg, G. (1987). Water relations of crocodilian eggs, management considerations. In *Wildlife Management: Crocodiles and Alligators*, eds G. J. W. Webb, S. C. Manolis & P. J. Whitehead, pp. 498–502. Chipping Norton, Australia: Surrey Beatty & Sons.

Grigg, G. & Beard, L. (1985). Water loss and gain by eggs of *Crocodylus porosus*, related to incubation age and fertility. In *Biology of Australasian Frogs and Reptiles*, eds G. Grigg, R. Shine & H. Ehman, pp. 353–9. Chipping Norton, Australia: Surrey Beatty & Sons.

Gutzke, W. H. N., Paukstis, G. L. & Packard, G. C. (1984). Pipping versus hatching as indices of time of incubation in reptiles. *Journal of Herpetology*, **18**, 494–6.

Haftorn, S. (1983). Egg temperature during incubation in the Great tit *Parus major* in relation to ambient temperature, time of day, and other factors. *Cinclus*, **6**, 22–38.

Hoyt, D. F. (1987). A new model of avian embryonic metabolism. *Journal of Experimental Zoology*, Supplement, **1**, 127–38.

Jackson, D. R. (1988). Reproductive strategies of sympatric freshwater Emydid turtles in northern Peninsular Florida. *Bulletin Florida State Museum Biological Science*, **33**, 113–58.

Joanen, T. (1969). Nesting ecology of alligators in Louisana. *Proceedings of the Game and Fish Commission Annual Conference*, **23**, 141–51.

Jones, D. N. (1988a). Construction and maintenance of the incubation mounds of the Australian Brush turkey *Alectura lathami. Emu*, **88**, 210–18.

— (1988*b*). Hatching success of the Australian Brush turkey *Alectura lathami* in South-east Queensland. *Emu*, **88**, 260–2.

Kloska, C. & Nicolai, J. (1988). Fortpflanzangsverhalten des Kamn-Talegalla (*Aepypodius arfakianus* Salvad). *Journal fur Ornithologie*, **129**, 185–204.

Leshem, A. & Dmi'el, R. (1986). Water loss from *Trionyx triunguis* eggs incubating in natural nests. *Herpetological Journal*, **1**, 115–17.

Lutz, P. L. & Dunbar-Cooper, A. (1984). The nest environment of the American crocodile (*Crocodylus acutus*). *Copeia*, **1984**, 153–61.

Magnusson, W. E. (1982). Mortality of eggs of the crocodile *Crocodylus porosus* in northern Australia. *Journal of Herpetology*, **16**, 121–30.

McIlhenny, E. A. (1935). *The Alligators Life History*, p. 117. Boston: Christopher Publishing House.

Miller, K., Packard, G. C. & Packard, M. J. (1987). Hydric conditions during incubation influence locomotor performance of hatchling snapping turtles. *Journal of Experimental Biology*, **127**, 401–12.

Muth, A. (1980). Physiological ecology of desert iguana (*Dipsosaurus dorsalis*) eggs, temperature and water relations. *Ecology*, **61**, 1335–43.

Nice, M. M. (1962). Development of behaviour in precocial birds. *Transactions of the Linnaean Society of New York*, **8**, 1–211.

Packard, G. C. & Packard, M. J. (1988). The physiological ecology of reptilian eggs and embryos. In *Biology of the Reptilia*, vol. 16, Ecology B, eds C. Gans & R. B. Huey, pp. 523–605. New York: John Wiley & Sons.

Packard, G. C., Packard, M. J. & Boardman, T. J. (1984). Influence of hydration of the environment on the pattern of nitrogen excretion by embryonic snapping turtles (*Chelydra serpentina*). *Journal of Experimental Biology*, **108**, 195–204.

Packard, G. C., Packard, M. J., Boardman, T. J. & Asheen, M. D. (1981*b*). Possible adaptive value of water exchange in flexible-shelled eggs of turtles. *Science*, **213**, 471–3.

Packard, G. C., Paukstis, G. L., Boardman, T. J. & Gutzke, W. H. N. (1985). Daily and seasonal variation in hydric conditions and temperature inside nests of common snapping turtles (*Chelydra serpentina*). *Canadian Journal of Zoology*, **63**, 2422–9.

Packard, G. C., Taigen, T. L., Boardman, T. J., Packard, M. J. & Tracy, C. R. (1979). Changes in mass of softshell turtle (*Trionyx spiniferus*) eggs incubated on substrates differing in water potential. *Herpetologica*, **35**, 78–86.

Packard, G. C., Taigen, T. L., Packard, M. J. & Boardman, T. J. (1980). Water relations of pliable-shelled eggs of common snapping turtles (*Chelydra serpentina*). *Canadian Journal of Zoology*, **58**, 1404–11.

— (1981*a*). Changes in mass of eggs of softshell turtles (*Trionyx spiniferus*) incubated under hydric conditions simulating those of natural nests. *Journal of Zoology, London*, **193**, 81–90.

Packard, M. J. & Hirsch, K. F. (1986). Scanning electron microscopy of eggshells of contemporary reptiles. *Scanning Electron Microscopy*, **4**, 1581–90.

Packard, M. J. & Packard, G. C. (1979). Structure of the shell and tertiary membranes of eggs of softshell turtles (*Trionyx spiniferus*). *Journal of Morphology*, **159**, 131–43.

Packard, M. J., Packard, G. C. & Boardman, T. J. (1982). Structure of eggshells and water relations of reptilian eggs. *Herpetologica*, **38**, 136–55.

Paganelli, C. V. & Rahn, H. (1984). Adult and embryonic metabolism in birds and the role of shell conductance. In *Respiration and Metabolism of Embryonic Vertebrates*, ed. R. S. Seymour, pp. 193–204. Dordrecht: Dr W. Junk Publishers.

Plummer, M. V. (1976). Some aspects of nesting success in the turtle, *Trionyx muticus. Herpetologica*, **32**, 353–9.

Ragotzkie, R. A. (1959). Mortality of loggerhead turtle eggs from excessive rainfall. *Ecology*, **40**, 303–5.

Rahn, H. (1984). Factors controlling the rate of incubation water loss in birds eggs and the role of shell conductance. In *Respiration and Metabolism of Embryonic Vertebrates*, ed. R. S. Seymour, pp. 271–88. Dordrecht: Dr W. Junk Publishers.

Roper, D. S. (1983). Egg incubation and laying behaviour of the incubator bird *Megapodius freycinet* on Savo. *Ibis*, **125**, 384–9.

Schönwetter, M. (1960–1983). *Handbuch der oologie*, vol. 1–3. Berlin: Akademie Verlag.

Schwarzkopf, L. & Brooks, R. J. (1987). Nest site selection and offspring ratio in painted turtles, *Chrysemys picta. Copeia*, **1987**, 53–61.

Seymour, R. S. (1984). Patterns of lung aeration in the perinatal period of domestic fowl and brush turkey. In *Respiration and Metabolism of Embryonic Vertebrates*, ed. R. S. Seymour, pp. 319–22. Dordrecht: Dr W. Junk Publishers.

— (1985). Physiology of megapode eggs and incubation mounds. *Acta XVIIIth Ornithological Congress*, Moscow, pp. 854–63.

Seymour, R. S. & Ackerman, R. A. (1980). Adaptations to underground nesting in birds and reptiles. *American Zoologist*, **20**, 437–47.

Seymour, R. S. & Rahn, H. (1978). Gas conductance in the eggshell of the mound-building Brush turkey. In *Respiratory Functions in Birds, Adult and Embryonic*, ed. J. Piiper, pp. 343–6. Berlin: Springer-Verlag.

Seymour, R. S., Vleck, D. & Vleck, C. M. (1986). Gas exchange in the incubation mounds of Megapode birds. *Journal of Comparative Physiology*, **156B**, 773–82.

Seymour, R. S., Vleck, D., Vleck, C. M. & Booth, D. T. (1987). Water relations of buried eggs of mound building birds. *Journal of Comparative Physiology*, **157B**, 413–22.

Shine, R. (1988). Parental care in reptiles. In *Biology of the Reptilia*, vol. 16, Ecology B, eds C. Gans & R. B. Huey, pp. 275–329. New York: John Wiley & Sons.

Snell, H. L. & Tracy, C. R. (1985). Behavioral and morphological adaptations by Galapagos land iguanas (*Conolophus subcristatus*) to water and energy requirements of eggs and neonates. *American Zoologist*, 25, 1009–18.

Somma, L. A. & Fawcett, J. D. (1989). Brooding behaviour of the prairie skink, *Eumeces septentrionalis*, and its relationship to the hydric environment of the nest. *Zoological Journal of the Linnaean Society*, 95, 245–56.

Sotherland, P. R. & Rahn, H. (1987). On the composition of bird eggs. *Condor*, 89, 48–65.

Taigen, T. L., Packard, G. C., Sotherland, P. R. & Hanka, L. R. (1978). Influence of solute concentration in albumen on water loss from avian eggs. *Auk*, 95, 422–3.

Thompson, M. B. (1981). Gas tensions in natural nests and eggs of the tortoise *Emydura macquarii*. In *Proceedings of the Melbourne Herpetological Symposium*, eds C. B. Banks & A. A. Martin, pp. 74–7. Parkville: Zoological Board of Victoria.

— (1983). The physiology and ecology of the eggs of the pleurodiran tortoise *Emydura macquarii* (Gray), 1831. PhD Thesis, University of Adelaide, South Australia.

— (1985). Functional significance of the opaque white patch in eggs of *Emydura macquarii*. In *Biology of Australasian Frogs and Reptiles*, eds G. Grigg, R. Shine & H. Ehmann, pp. 387–95. Chipping Norton, Australia: Surrey Beatty & Sons.

— (1987). Water exchange in reptile eggs. *Physiological Zoology*, 60, 1–8.

— (1988). Nest temperatures in the pleurodiran turtle *Emydura macquarii*. *Copeia*, 1988, 998–1002.

— (1989). Patterns of metabolism in embryonic reptiles. *Respiration Physiology*, 76, 243–56.

— (1990). Incubation of eggs of tuatara, *Sphenodon punctatus*. *Journal of Zoology, London*, (in press).

Thompson, M. B., Packard, G. C., Packard, M. J. & Rose, B. (1990). Analysis of the nest environment of tuatara *Sphenodon punctatus*. *Functional Ecology*, (in prep.).

Todd, D. (1983). Pritchard's megapode on Niuafo'os Island, Kingdom of Tonga. *Journal of the World Pheasant Association*, 8, 69–88.

Tracy, C. R., Packard, G. C. & Packard, M. J. (1978). Water relations of chelonian eggs. *Physiological Zoology*, 51, 378–87.

Tracy, C. R. & Snell, H. L. (1985). Interrelationships among water and energy relations of reptilian eggs, embryos, and hatchlings. *American Zoologist*, 25, 999–1088.

Visschedijk, A. H. J. (1968). The air space and embryonic respiration. I. The pattern of gaseous exchange in the fertile egg during the closing stages of incubation. *British Poultry Science*, 9, 173–84.

Vitt, L. J. (1978). Caloric content of lizard and snake (*Reptilia*) eggs and bodies and the conversion of weight to caloric data. *Journal of Herpetology*, 12, 65–72.

Vitt, L. J. & Oharmt, R. D. (1975). Ecology, reproduction, and reproductive effort of the iguarid lizard *Urosaurus graciosus* on the lower colorado river. *Herpetologica*, 31, 56–65.

Vleck, C. M. & Kenagy, G. J. (1980). Embryonic metabolism of the Fork-tailed storm petrel, Physiological patterns during prolonged and interrupted incubation. *Physiological Zoology*, 53, 32–42.

Vleck, C. M., Vleck, D. & Hoyt, D. F. (1979). Metabolism of avian embryos, ontogeny of oxygen consumption in the rhea and emu. *Physiological Zoology*, 53, 125–35.

— (1980). Patterns of metabolism and growth in avian embryos. *American Zoologist*, 20, 405–16. *Zoology*, 53, 125–35.

Vleck, C. M., Vleck, D. & Seymour, R. S. (1984). Energetics of embryonic development in the megapode birds, Mallee fowl *Leipoa ocellata* and Brush turkey *Alectura lathami*. *Physiological Zoology*, 57, 444–56.

Walsberg, G. E. (1980). The gaseous microclimate of the avian nest. *American Zoologist*, 20, 363–72.

Webb, D. R. (1987). Thermal tolerance of avian embryos, A review. *Condor*, 89, 874–98.

Webb, G. J. W., Buckworth, R. & Manolis, S. C. (1983*b*). *Crocodylus johnstoni* in the McKinlay River, N.T. VI. Nesting biology. *Australian Wildlife Research*, 10, 607–37.

Webb, G. J. W., Choquenot, D. & Whitehead, P. J. (1986). Nests, eggs, and embryonic development of *Carettochelys insculpta* (Chelonia, Carettochelidae) from Northern Australia. *Journal of Zoology, London* B1, 521–50.

Webb, G. J. W., Messel, H. & Magnusson, W. (1977). The nesting of *Crocodylus porosus* in Arnhem Land, Northern Australia. *Copeia*, 1977, 238–49.

Webb, G. J. W., Sack, G. C., Buckworth, R. & Manolis, S. C. (1983*a*). An examination of *Crocodylus porosus* nests in two northern Australian freshwater swamps, with an analysis of embryo mortality. *Australian Wildlife Research*, 10, 571–605.

White, F. N. & Kinney, J. L. (1974). Avian incubation. *Science*, 186, 107–15.

Whitehead, P. (1987). Respiration by *Crocodylus johnstoni* embryos. In *Wildlife Management, Crocodiles and Alligators*, eds G. J. W. Webb, S. C. Manolis & P. J. Whitehead, pp. 473–97. Chipping Norton, Australia: Surrey Beatty & Sons.

Wilhoft, D. C. (1986). Eggs and hatchling components of the snapping turtle (*Chelydra serpentina*). *Comparative Biochemistry and Physiology*, 84A, 483–6.

21

Why birds lay eggs

HERMANN RAHN

Introduction

It has long been a puzzle why 'birds constitute the only vertebrate class that is both rich in species number and exclusively oviparous' (Blackburn & Evans, 1986). Viviparity has evolved in all other vertebrate classes and has evolved independently 93 times among the reptiles (Shine, 1985, Chapter 22), the ancestors of birds. The absence of avian viviparity has frequently been explained by citing morphological or physiological factors which presumably are incompatible with live birth. However, as Blackburn & Evans (1986) show, such arguments cannot be defended and thus argue that birds have achieved most of the advantages of viviparity 'by such specialisations as endothermy, egg incubation, nest construction, uricotelism, shell pigmentation, parental care, altricial hatchlings, albumen provision, and calcareous eggshells'. Other interpretations are discussed by Anderson, Stoyan & Ricklefs (1987) and Dunbrack & Ramsay (1989).

Whatever the reasons for maintenance of oviparity, their success is obvious judging from their colonisation in all parts of the world and the large number of species that have evolved, more than double the number for mammals. One can also look at birds as successful competitors with mammals by adapting incubation behaviour, shell structure and shell porosity to various environmental conditions to deliver hatchlings which are behaviourly, functionally and in gross composition similar to mammalian neonates. Some common features are: 1) Avian embryos are maintained at temperature of about 36 °C and mammalian embryos at 37 °C. 2) During late stages of development, the O_2 and CO_2 tensions of the embryonic and extra-embryonic blood vessels in avian embryos attain values similar to those in the mammalian fetus. In both, the acid–base status is essentially that of the adult, characterised by a plasma P_{CO_2} of 30–40 torr and a pH of 7.4 to 7.5, the hallmark of warm-blooded vertebrates. 3) Both birds and mammals deliver neonates which range from altricial species, born naked and blind, to precocial species, born furred or feathered, able to walk and thermoregulate. 4) The relative water content of neonates is inversely proportional to their state of maturity and, in birds, varies from *circa* 85% in altricial species to *circa* 65% in the most precocial species.

Homeostasis, reflecting the constancy of the internal environment and body temperature, is the hallmark of adult warm-blooded animals (Cannon, 1929; Barcroft, 1932). It describes equilibrium rates of numerous physiological and biochemical processes, such as renal and ventilatory functions, which throughout life regulate the plasma bicarbonate and CO_2 tensions to maintain a pH between 7.4 and 7.5. There are renal and other functions which also maintain a constant state of hydration in our tissues. The mammalian fetus develops in this environment and is eventually delivered with adult-like functions as an immature (rat, mouse) or mature (ungulate) neonate. A similar state of physiological development is achieved outside the body in birds and is clearly seen in Fig. 21.1 for there is similarity in the respiratory gas tensions and transport between a precocial fetus (lamb) during the last trimester and that of a precocial chick at the end of incubation (Rahn, Matalon & Sotherland, 1985). Oxygen delivery to the placenta is provided by the uterine artery and has a typical value of 90 torr, similar to the typical value of 100 torr in the air space between the

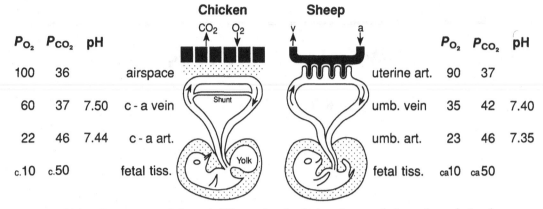

Fig. 21.1. Comparison of O_2 and CO_2 tension in fowl embryo and sheep fetus during late development. Data from Rahn, Matalon & Sotherland (1985) and Olszowka, Tazawa & Rahn (1988); art. = artery, tiss. = tissue, umb. = umbilical, c-a = chorio-allantoic.

shell and the chorio-allantois. While the egg-shell 'placenta' has direct access to the atmosphere, its conductance is fixed throughout the development, while uterine blood flow increases as the metabolic demand increases. It means that shell conductance must be designed to match the metabolic demand, just prior to initiation of lung ventilation, so as to provide a CO_2 tension of *circa* 40 torr to match the plasma bicarbonate level, both values which determine a pH of 7.5, the typical value of adult birds.

Thus, to achieve a state of functional development at the end of incubation, similar to that of the mammalian newborn, has required avian eggs to be incubated at a temperature close to that of the parent, to develop a pore system where gas diffusion is independent of fluctuating boundary layer conditions, and where the metabolism at the end of incubation is matched to the shell conductance. It also requires a finite rate of egg water loss to attain the appropriate state of hydration at hatching by maintenance of an absolute nest humidity matched to the shell conductance and egg temperature, whatever the ambient climate.

Egg temperature

For eggs to be maintained close to body temperature it becomes necessary to develop a mechanism for heat transfer from the incubating parent and the construction of an incubator, the nest, with sufficient insulation from ambient convective currents. In addition, a hard-shelled egg is required which can withstand its rotation, as well as the weight of the parent, yet is fragile enough to be broken by the hatchling. It also requires microscopic organic or inorganic 'plugs' in the pore openings to filter out dust and micro-organisms (Board, 1982; Tullett, 1984) and hydrophobic properties to exclude water or rain from clogging or entering the communication channels.

Eggshells

Shell mass increases 13 times for every ten-fold increase in egg weight:

$$\text{shell mass, g} = 0.055 \, W^{1.11} \quad (21.1)$$

(Rahn & Paganelli, 1989), presumably a common bio-engineering principle, since a similar exponent is found among other invertebrates and vertebrates when exo- or endoskeletal support is regressed against body mass (Anderson, Rahn & Prange, 1979). The breaking point of the shell, expressed as a force, F (kg), yielding the first gross sign of shell deformation is proportional to the square of shell thickness:

$$F = 17.5 \, T^2, \text{(mm)} \quad (21.2).$$

For many species, F is about twice that of the mass of the parent giving a safety factor of 2.0 (Ar, Rahn & Paganelli, 1979). About 97% of the shell is calcium carbonate and has a density of 2.02 g cm^{-3} (Rahn & Paganelli, 1989) a composition and density similar to that of bone of birds and mammals (Blitz & Pellegrino, 1969).

For liquid water to enter the microscopic pores of a fowl egg requires a pressure of about 100 torr, a value that is predicted by Laplace's law for a mean pore diameter of 10 μm (Visschedijk & Rahn, 1983). This is an important property when the bottom of eggs may be in contact with water throughout incubation, such as in the floating nests of the Eared Grebe, *Colymbus nigricollis*, (Sotherland *et al.*, 1984) or other eggs incubated in wet nests.

Heat transfer

'Most birds develop a naked brood patch by shedding feathers on parts of the breast and belly in the beginning of the breeding season. Along with defeathering, the size and number of blood vessels in the cutaneous and subcutaneous tissues increase significantly. It is generally believed that these changes, which occur under hormonal control, enhance the transfer of heat from the incubating bird to the eggs. Several studies have shown that the incubating bird regulates the temperature of the egg by adjusting the nest attentiveness and the "tightness of sit" on the eggs. In addition to this behavioral regulation, there is also evidence that the incubating birds respond physiologically to egg temperatures that deviate from normal. For example, it has been found that egg cooling increases breathing rate, shivering, and oxygen consumption. These studies show that warming of the eggs to normal incubation temperature is facilitated by an increase in heat production, and apparently the heat is distributed to the brood patch by the circulating blood' (Mitgård, Sejrsen & Johansen, 1985). For example, with cooling of an artificial egg the oxygen consumption of the incubating bird can double in the zebra finch (*Poephila guttata*) (Vleck, 1981*a*), triple in the European starling (*Sturnus vulgaris*) (Biebach, 1979) and reach a level of 4.6 times that of the resting oxygen consumption in the Bantam fowl (*Gallus gallus*) when eggs are maintained below 15 °C (Toien, Aulie & Steen, 1986).

Using a ^{133}Xenon washout method Mitgård *et al.* (1985) show that egg cooling results in a vasodilatation of the brood patch vessels, and suggest that cold vasodilatation is an important mechanism providing increased heat transfer whenever egg temperature falls below a normal level. Similar mechanisms have been reported for the circulation of the webbed feet of the arctic fulmar (*Fulmarus glacialis*), and the antarctic fulmar (*Fulmarus glacialoides*), (Johansen & Millard, 1974; Murrish & Guard, 1977; Krog & Toien, 1984) which presumably serve as auxiliary warming devices for eggs incubated in cold

climates. Such a cold-vasodilatation might also occur in the upper portion of the feet of the king and emperor penguin (*Aptenodytes patagonica* and *Aptenodytes forsteri*), which serve as a resting place for the egg under the skin flap (Handrich, 1989).

Egg temperatures

During incubation, egg temperatures are conventionally recorded from thermocouples or thermistors placed in the centre of the egg or on the inner membrane of the air cell where the probe does not interfere with the developing embryo. In Table 21.1 the temperatures of the egg, the brood patch, the mean ambient air and the egg mass are listed for a large number of species extending the list of 27 species reported by Drent (1975), while Webb (1987) provides extensive data on thermal tolerance of avian embryos. The frequency distribution of these egg temperatures, as well as the body temperature of birds reported by McNab (1966), is shown in Fig. 21.2. The mean egg temperature is 35.7 °C, 5 °C below the mean body temperature. Egg temperatures are plotted against an ambient temperature range from −30 to +50 °C in Fig. 21.3. The point at the far left is the mean value and SD (vertical bar) of 42 eggs of the emperor penguin (*A. forsteri*) measured by Prevost (1961) at ambient temperatures ranging from −10 to −27 °C (horizontal bar). These are the only egg temperatures in Table 21.1 which are obtained by plunging a mercury thermometer into the egg centre of the 500 g egg and because of their large mass, these values may be quite accurate. On the other end of the scale are eggs incubated above 38 °C where the isotherm crosses egg

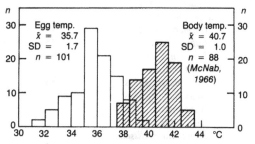

Fig. 21.2. Frequency distribution of central egg temperature compared with frequency distribution of body temperature of birds taken from table of McNab (1966).

Table 21.1. *Egg mass (g) and egg, brood-patch, body and ambient temperatures (°C) for avian species.*

| Order — Species | Egg mass | Temperature | | | | Ref. |
		Egg	Brood patch	Body	Amb.	
Struthioniformes						
Struthio camelus	1440	36.0	—	38.7	—	1
Struthio camelus	—	32.9	37.3	—	—	2
Struthio camelus	1370	34.9	38.0	—	20	40
Sphenisciformes						
Aptenodytes forsteri	425	31.4	36.1	36.6	−16	3
Aptenodytes patagonica	306	32.5	38.0	—	2	4
Aptenodytes patagonica	303	36.5	38.2	—	7	43
Pygoscelis adeliae	124	33.7	—	37.9	—	1
Pygoscelis adeliae	125	35.9	—	—	−4	5
Pygoscelis adeliae	124	35.2	—	—	—	41
Pygoscelis papua	136	36.1[a]	—	—	—	42
Eudyptes chrysocome	92	34.2[a]	—	—	7	42
Spheniscus demersus	103	35.9[a]	—	—	—	42
Procellariiformes						
Puffinus pacificus	60	36.3	37.8	39.5	—	1
Puffinus pacificus	60	35.0	37.6	—	29	6
Diomedea immutabilis	285	36.0	38.7	—	22	7
Diomedea nigripes	305	35.6	38.3	—	22	7
Pterodroma hypoleuca	40	33.8	34.9	—	21[c]	8
Oceanodroma leucorrhoa	10	34.8	37.4	37.9	—	1
Oceanodroma leucorrhoa	10	32.3	36.4	38.9	15[c]	9
Oceanodroma leucorrhoa	10	33.1	36.7	38.1	8[c]	9
Oceanodroma leucorrhoa	10	35.9	—	—	—	10
Oceanites oceanicus	9	36.0	—	38.8	—	1
Oceanites oceanicus	9	35.7	—	—	—	10
Pelecanoides urinatrix	21	35.8	—	38.0	—	1
Macronectes giganteus	250	35.5	—	—	—	10
Podicipediformes						
Podiceps nigricollis						
(wet nest)	21	33.4	—	—	17	11
(dry nest)	21	36.0	—	—	17	11
Pelecaniformes						
Fregata minor	66	36.8	—	—	—	12
Sula sula	58	36.0	—	—	—	13
Sula sula	58	36.0	—	—	—	12
Phaeton rubricauda	71	35.0	—	—	—	12
Phaeton rubricauda	71	36.0	—	—	—	13
Ciconiformes						
Casmerodius albus	49	34.3	—	—	17	14
Bubulcus ibis	28	36.8	—	—	22	14
Eudocimus albus	51	35.5	—	—	20	14
Anseriformes						
Anas platyrhynchos	54	37.0	—	41.2	—	1
Anas platyrhynchos	—	36.3[a]	—	—	—	15
Anas crecca	—	37.9[a]	—	—	—	19
Anas acuta	—	37.9[a]	—	—	—	19
Somateria mollissima	108	33.6	38.5	39.1	3	16
Branta leucopsis	115	33.2[b]	37.7	—	2	16
Branta canadensis	161	34.3[a]	—	—	7	17
Branta canadensis	161	37.9[a]	—	—	7	17

Table 21.1. *contd.*

Order — Species	Egg mass	Temperature				Ref.
		Egg	Brood patch	Body	Amb.	
Galliformes						
Gallus gallus	58	37.5	40.7	41.0	—	1
Phasianus colchicus	30	38.6	—	—	22	18
Gruiformes						
Gallinula chloropus	20	35.0	—	—	—	1
Charadriiformes						
Pluvianus aegyptius						
(wetted nest)	9	38.0	—	—	44	20
Charadrius vociferus	14	38.2	—	—	35	21
Recurvirostra americana	32	37.5	—	—	35	21
Himantopus mexicanus	21	37.7	—	—	35	21
Catharacta skua	91	35.9	—	41.2	—	1
Catharacta skua	—	35.6	39.3	—	—	1
Catharacta skua	—	36.4	39.0	41.4	—	1
Larus argentatus	92	38.3	40.5	41.2	—	1
Larus glaucescens	98	36.0	—	—	11	22
Larus heermanni	53	36.8	—	—	23	23
Larus modestus	52	36.0	—	—	—	33
(early morning)	52	33.0	—	—	6	33
Rissa tridactyla	50	37.4	—	—	11	22
Rissa tridactyla	51	37.4	—	—	3	24
Rissa tridactyla	—	36.4	38.9	—	—	25
Sterna fuscata	36	38.0	39.6	40.5	—	1
Sterna maxima	68	37.8	—	—	24	14
Sterna forsteri	21	38.1	—	—	35	21
Sterna nilotica	28	35.6	—	—	15	26
Gygis alba	21	35.4	—	—	28	27
Gygis alba	23	35.3	36.7	—	20	28
Anous tenuirostris	24	37.4	—	—	28	27
Rhynchops nigra	27	35.1	—	—	15	26
Columbiformes						
Zenaida asiatica	9	39.2	38.5	—	42	29
Zenaida macroura	6	37.0	—	—	16	32
Zenaida macroura	6	39.8	38.6	38.3	45	30
Streptopelia risoria	—	42.5	—	40.0	45	30
Columba livia	—	39.0	—	38.9	45	44
Columba livia	—	36.7	—	40.8	28	44
Strigiformes						
Tyto alba	26	34.2	39.3	40.8	—	1
Speotyto cunicularia	15	35.5	39.8	40.2	—	1
Caprimulgiformes						
Chordeiles acutipennis	6	38.0	—	—	35	21
Apodiformes						
Chaetura pelagica	1.9	35.2	—	—	—	1
Stellula calliope	0.5	34.6	—	—	—	1
Archilochus alexandri	0.5	32.5	—	—	14	31
Calypte anna	0.6	35.8	—	—	9	31
Calypte costae	0.5	35.8	—	—	24	31
Passeriformes						
Empidonax oberholseri	1.6	37.6[d]	42.4	—	13	34
Empidonax difficilis	1.5	34.3	—	—	18	32

Table 21.1. *contd.*

| Order — Species | Egg mass | Temperature | | | | |
		Egg	Brood patch	Body	Amb.	Ref.
Hirundo rustica	1.9	35.3	—	—	—	1
Riparia riparia	1.4	33.7	—	—	20	35
Parus ater	1.1	35.4	—	—	—	1
Parus caeruleus	1.1	35.5	—	—	—	1
Parus montanus	1.3	36.4	—	—	—	36
Regulus regulus	1.3	36.5	—	—	—	37
Troglodytes aedon	1.4	35.1	40.6	41.3	—	1
Dumatella carolinensis	3.8	34.0	—	—	—	1
Toxostoma redivivum	—	38.0	—	—	15	32
Sturnus vulgaris	7.0	35.0	—	—	—	1
Icterus galbula	—	32.6	—	—	19	32
Passer moabiticus	1.8	33.7	—	—	33	38
Carpodacus mexicanus	—	31.8	—	—	17	32
Carpodacus mexicanus	—	35.5	—	—	19	32
Pipilo fuscus	4.3	36.0	—	—	17	32
Pipilo erythrophthalmus	—	37.0	—	—	19	32
Zonotrichia leucophris	2.8	37.3	—	—	8	39
Melospiza melodia	2.6	34.4	—	—	—	1
Poephila guttata	1.0	35.2	—	40.2	—	1
Psaltriparus minimus	0.7	36.0	—	—	21	32

[a]Air cell temperature.
[b]Interpolation between top and bottom egg temperature.
[c]Burrow air temperature.
[d]24 hour average.

1. Drent, 1975; 2. Bertram & Burger, personal communication; 3. Prevost, 1961; 4. Stonehouse, 1960; 5. Rahn & Hammel, 1982; 6. Whittow *et al.*, 1982; 7. Grant *et al.*, 1982b; 8. Grant *et al.*, 1982a; 9. Ricklefs & Rahn, 1979; 10. Williams & Ricklefs, 1984; 11. Sotherland, personal communication; 12. Howell & Bartholomew, 1962; 13. Whittow, 1980; 14. Vleck *et al.*, 1983; 15. Caldwell & Cornwell, 1975; 16. Rahn *et al.*, 1983; 17. Cooper, 1978; 18. Rahn *et al.*, 1977; 19. Afton, 1978; 20. Howell, 1979; 21. Grant, 1982; 22. Morgan *et al.*, 1978; 23. Rahn & Dawson, 1979; 24. Rahn, unpublished observations; 25. Barrett, 1980; 26. Grant *et al.*, 1984; 27. Rahn *et al.*, 1976; 28. Pettit *et al.*, 1981; 29. Russell, 1969; 30. Walsberg & Voss-Roberts, 1983; 31. Vleck, 1981b; 32. Vleck, personal communication; 33. Howell *et al.*, 1974; 34. Morton & Pereyra, 1985; 35. Ellis, 1982; 36. Haftorn, 1979; 37. Haftorn, 1978; 38. Yom-Tov *et al.*, 1978; 39. Zerba & Morton, 1983; 40. Swart *et al.*, 1987; 41. Derksen, 1977; 42. Burger & Williams, 1979; 43. Handrich, 1989; 44. Marder & Gavrieli-Levin, 1986.

temperatures (Fig. 21.3). Note that these eggs are cooler than the air and indicate a reversal of the normal temperature gradient from the body to the egg. Thus, the general ability of birds to maintain their egg temperatures over a very narrow range in all climates has undoubtedly been one of the major reasons for their success in competition with endothermic mammals.

Temperature difference between body and egg

The average values for body, brood patch and central egg temperature for 12 incubating species are shown on the right-hand side of Fig. 21.4 and provide a general overview of the heat gradient from body to the centre of an egg, a difference of about 5 °C. On the left-hand side of Fig. 21.4 a reverse gradient is shown from the egg to the parent. This phenomenon is illustrated by eggs of the white-winged dove (*Zenaida asiatica*) which are about 1 °C warmer than the brood patch temperature at an ambient temperature of 42.5 °C (Russell, 1969). Additional evidence of heat flux reversal has been

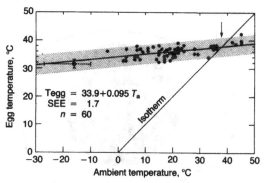

Fig. 21.3. Central egg temperatures as a function of ambient temperature (from Table 21.1). Note arrow where isotherm crosses the egg temperature regression. To the right of this point, the general heat gradient from body to egg is reversed. For details see text.

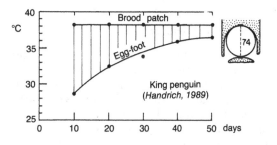

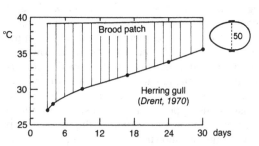

Fig. 21.5. Two examples of how the temperature differences across the egg diminish during incubation. The top lines represent the brood patch temperature which stays remarkably constant. The bottom line represents the bottom of the egg in the herring gull and in the king penguin the temperature between the eggshell and the foot upon which the egg rests. See cross-section, which also indicates the thickness of the egg, mm. Illustration is based on the data presented by the authors.

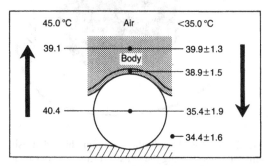

Fig. 21.4. On the right, average body, brood patch and central egg temperatures and SD for 12 incubating species when the ambient temperature is below 35 °C. Large arrow indicates direction of heat flow. On the left, the reversal of heat flow where the ambient temperature was 45 °C. These values are based on three species of the pigeon family, for details see text. The average nest air temperature from Drent (1975).

shown in 3 species of the pigeon family when ambient temperatures average 45 °C, and egg and body temperature average 40.4 °C and 39.1 °C respectively (Walsberg & Voss-Roberts, 1983; Marder & Gavrieli-Levin, 1986). In these situations the parents 'sit tight' on the egg and body cooling is achieved by gular flutter and panting in some species, but in the acclimated pigeon *Columba livia* there is an efficient heat loss mechanism of cutaneous evaporation (Marder & Gavrieli-Levin, 1986), a mechanism not previously recognised in birds.

That large temperature gradients exist within the incubating egg and that these are reduced

during development is well illustrated by the differences that exist between the brood patch temperature, which stays remarkably constant, and the bottom surface of the egg. Data for the herring gull (*Larus argentatus*; Drent, 1970) and king penguin (Handrich, 1989) are shown in Fig. 21.5. The large temperature differences during the early stages are most likely due to inattentiveness, and are eventually reduced as attentiveness increases and embryo metabolism begins to increase during the last half of incubation. The contribution of the embryo's metabolism to central egg temperature is tested by Grant *et al.* (1982a) in two species of albatross (*Diomedea immutabilis* and *Diomedea nigripes*), whose egg temperatures increase from *circa* 34 °C to 36.5 °C and 37.3 °C, respectively after 40 and 50 days of development. At this stage, formalin is injected into the egg to kill the embryo and the following day the egg temperature has reverted to their early stage temperature of *circa* 34 °C (*n* = 8).

Egg vs embryo temperatures

While temperature gradients exist within the egg, in seven species the average value was 0.68 °C cm^{-1} (Rahn, Krog & Mehlum, 1983), what is the temperature to which the embryo is exposed, particularly during the early stage of development when the largest gradients exist? Early in incubation the embryo floats to the top of the egg where it is closest to the brood patch, and considerably warmer than the conventionally measured central temperature (Turner, Chapter 9). To assure such floating behaviour, however, requires that a density gradient in the yolk is established, presumably in the ovarian follicle, and that the future blastoderm is located at the least dense pole. The density difference between the top and bottom of the yolk of fresh fowl eggs is shown in Fig. 21.6, but note, however, that the top of the yolk is not in contact with the shell, being constrained by the strands of the chalazae. Only after two days of incubation have they dissolved so that the area pellucida is now floating in close proximity to the pores of the shell for maximal temperature exposure and access to oxygen.

The importance of temperature

To provide throughout development the warmest temperature for the developing embryo requires many diverse strategies. First are the hormonal changes responsible for the defeathering of the brood-patch area, the increased vascularity and the initiation of a cold-vasodilatation reflex of the brood-patch vessels. In addition, there is construction of the proper nesting site. During development of the ovum within the follicle a density gradient must be established so that the future germinal disc is located at the least dense portion of the yolk. Thus, during the early stages of incubation, when the largest temperature differences exist within the egg, the embryo will be buoyed upwards to be positoned under the brood patch, the warmest area of the egg. It is the cold-vasodilitation reflex and the brooding behaviour which provides the heat input, assisted during the last third of the incubation period by the metabolism of the embryo (Turner, Chapter 9).

Conventionally recorded central egg temperatures provide only a general account of egg or embryo temperature and the technique of

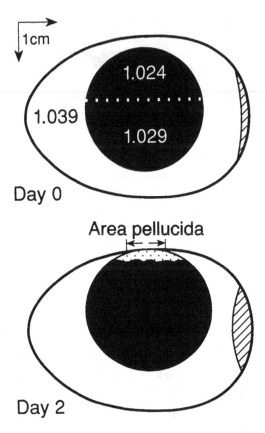

Day 0

Area pellucida

Day 2

Fig. 21.6. *Top:* Longitudinal section through fresh chicken eggs showing the density differences between the top and lower half of the yolk and the albumen. Note that the top of the yolk is still 3 mm from the shell, constrained by the strands of the chalazae. Not until the second day of incubation have the latter between loosened to allow the top of the yolk and the blastoderm to contact the egg shell.

placing a temperature probe may actually kill the embryo. It gives little insight about the temperature gradients within the egg and how they change during development (Turner, 1987). Nor will central egg temperatures taken during the day provide evidence of large temperature changes that occur at night, particularly in desert nesting species (Grant, 1982).

Development of carbon dioxide tensions

The hydrogen ion concentration (pH) to which the mammalian fetus is exposed is controlled within very narrow limits by the arterial carbon dioxide tension (P_{CO_2}), which in turn is regulated through the respiratory centre by the alveolar

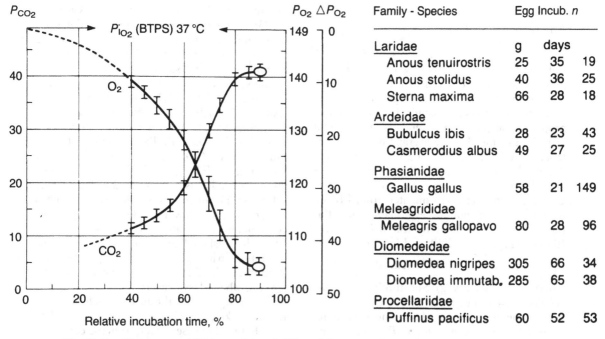

Family - Species		
	g	days	n

Family - Species	g	days	n
Laridae			
Anous tenuirostris	25	35	19
Anous stolidus	40	36	25
Sterna maxima	66	28	18
Ardeidae			
Bubulcus ibis	28	23	43
Casmerodius albus	49	27	25
Phasianidae			
Gallus gallus	58	21	149
Meleagrididae			
Meleagris gallopavo	80	28	96
Diomedeidae			
Diomedea nigripes	305	66	34
Diomedea immutab.	285	65	38
Procellariidae			
Puffinus pacificus	60	52	53

Fig. 21.7. The mean and SE bars of air cell CO_2 and O_2 tension during development measured in ten species, listed in the table, whose egg mass ranged from 21 to 300 g, incubation time from 17 to 65 days and n = number of analyses. At the PIP stage (*circa* 90% of incubation time) the circles represent the mean value of 15 additional species. (From Rahn & Paganelli, 1990.)

ventilation. How is the acid–base balance established in the avian embryo where alveolar ventilation is not initiated prior to internal pipping and where carbon dioxide transport depends upon the CO_2 conductance of the eggshell which is formed within the shell gland before the egg is laid and appears to remain constant during incubation? The relationship between shell conductance, CO_2 production and air cell CO_2 tension is expressed by the diffusive conductance equation, namely,

$$P_{CO_2} = \dot{M}_{CO_2} / G_{CO_2} \qquad (21.3)$$

Where $\dot{M}_{CO_2}$ = CO_2 production, cm³ d⁻¹
  G_{CO_2} = diffusive CO_2 conductance of the shell, cm³ (d torr)⁻¹
and P_{CO_2} = air space CO_2 tension, torr, assuming nest air P_{CO_2} to be equal to zero.
Since G_{CO_2} is constant, the air space P_{CO_2} is directly proportional to the CO_2 production and gradually rises with development as metabolism increases.

The changes in P_{CO_2} during embryonic development are shown together with simultaneous changes in air cell oxygen tensions (P_{O_2}) in Fig. 21.7, remarkably similar among 10 species, whose egg mass varied from 21 to 300 g and incubation times from 17 to 65 days. Furthermore, at the pre-internal pipping stage (PIP), just prior to the introduction of lung ventilation, the average P_{CO_2} value is equal to 41 torr, with a standard error of 1.2 based on a total of 25 species (Rahn & Paganelli, 1989*b*).

To achieve this common CO_2 tension requires that in all these species the shell conductance, established prior to egg laying, is matched to the PIP CO_2 production (equation 21.3). From blood analysis of fowl embryos, it is well established that the plasma bicarbonate concentration, controlled by the renal function, rises in parallel with P_{CO_2} during development keeping the pH within normal limits and establishing a value of *circa* 7.5 at the PIP stage (Tazawa, 1980, 1986). By these means avian embryos are able to establish an acid–base balance which is similar to that of the mammalian neonate as well as that of the adult bird, namely a P_{CO_2} between 30–40 torr and a pH of 7.5 (Tazawa, 1986).

A pore model for CO_2

In view of the pore models presented by Ar & Rahn (1985) and Rahn & Paganelli (1990) the number of pores required to establish a given P_{CO_2} in the air space for any egg mass can be estimated (Paganelli, Chapter 16). The model proposes that pore conductance is independent of egg mass and that the average pore conductance for CO_2 at 37 °C = 1.28 mm³ (d torr)⁻¹. Thus, G_{CO_2} of equation 21.3 = $N \times 0.00128$ cm³ (d torr)⁻¹, where N = number of pores. The CO_2 production can be estimated from PIP O_2 uptake as a function of egg mass when this regression is multiplied by the R.Q. of 0.73 (Rahn & Paganelli, 1990), and is equal to:

$$\dot{M}_{CO_2} = 17.2\ W^{0.734}\ (cm^3\ d^{-1}) \quad (21.4)$$

where W = egg mass (g). Substituting these two relationships into equation 21.3 and rearranging allows one to solve for the number of pores required to establish a given P_{CO_2} in the air space. If a P_{CO_2} of 40 torr is selected then the number of pores is

$$N = 17.2\ W^{0.734} / (40 \times 0.00128)\ (21.5)$$

For example, if $W = 65$ g, $\dot{M}_{CO_2} = 368$ cm³ d⁻¹ and $N = 7200$ and for a 200 g egg 16,400 pores are predicted.

What are the P_{CO_2} values in wet nesters, such as the pied-billed grebe (*Podilymbus podiceps*) and the western grebe (*Aechmophorus occidentalis*), which have about twice the predicted shell conductance and pores, but whose pore conductance is similar to that of other species (Ar & Rahn, 1985)? Assuming these eggs to have a 'normal' rate of metabolism, one would predict air space P_{CO_2} values at the PIP stage of about 20 instead of 40 torr and a 20 torr higher O_2 tension. Since plasma bicarbonate concentrations follow the changes in P_{CO_2} during development, one would predict about half the normal value so that the pH would still be 7.5.

CO_2 tension in nest air

If CO_2 is released from the eggs, what are the concentrations found in nest air? Prior to the 1930s, hatchability of fowl eggs in artificial incubators was generally lower than naturally incubated eggs. This led to inquiries of the CO_2 concentrations and humidity of natural nests and three classical studies by Lamson & Edmond (1914), Burke (1925) and Chattock (1925). These have never been repeated in such detail and provide the best description of the gaseous environment of the fowl nest and serve as models for other species.

Lamson & Edmond (1914) and Burke (1925) have measured the CO_2 concentration of the fowl nest throughout incubation by continuously aspirating nest air from a hollow dummy egg, absorbing the CO_2 with barium hydroxide, and expressing the CO_2 concentration in parts per 10,000 in air. The general agreement between these two studies is remarkable and the average values for 11 clutches of fertile eggs are shown in Fig. 21.8. The lower line gives the CO_2 concentrations of the ambient air, (*circa* 0.03%) while the dashed line above is the CO_2 concentrations when glass or boiled eggs are substituted for the fertile clutch. The rise of CO_2 tension, when glass eggs are introduced, is due to CO_2 release from the brood patch and surrounding skin, and when fertile eggs are used, to the additional release of CO_2 from the eggs. The shape of the nest air curve is of special interest, since it appears to be identical to the CO_2 production of eggs which are shown as circles, superimposed upon the nest air line. If nest air CO_2 concentration is directly proportional to egg CO_2 production, it suggests that the resistance offered by the feathers and nest lining to the escape of CO_2 from the nest air to the ambient atmosphere remains the same

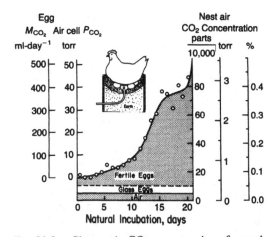

Fig. 21.8. Changes in CO_2 concentration of nest air (solid line) during incubation of fowl eggs, plotted from the data of Lamson & Edwards (1914) and Burke (1925). Circles indicate the CO_2 production of chicken eggs. For details see text.

throughout incubation, whether this transport is diffusive, convective or a combination of both.

The right-hand side of equation 21.3 should correctly be written as $(P_{aCO_2} - P_{nCO_2})$, where a = air space and n = nest air. The data above show, for example, that P_{nCO_2} at the PIP stage is 3 torr and therefore would raise the calculated P_{aCO_2} by 3 torr. Since nest P_{CO_2} is only known for the fowl and depends in other birds on the nest construction, incubation behaviour, clutch size and other factors, it is generally assumed to be zero.

Water budget

The embryos of mammals and birds (Needham, 1963; Romanoff, 1967; Altman & Dittmer, 1974) have relative water contents of *circa* 92 to 95% early in development. During incubation the water content gradually declines and as Needham (1963) states 'A decreasing water content is a universal accompaniment of growth'. At the time of birth or hatching this relative water content is inversely proportional to the degree of maturity of the neonate, and among birds can vary from *circa* 85% in altricial species (Ar *et al.*, 1987) to 68% in the most precocial megapodes (Vleck, Vleck & Seymour, 1984). Among mammals the altricial rat is born with a relative water content of 86.6%, the precocial guinea pig with a water content of 71% (Altman & Dittmer, 1974), both being supported by the maternal environment whose average water content is of the order of 60 to 65%. Avian embryos achieve their different states of hydration at the end of incubation by having their egg contents provisioned by a given amount of water and losing a finite fraction which is common among all the species and appears to be crucial for hatching success.

The relative water content of fresh eggs varies greatly and in 127 species, representing 44 families, ranges from *circa* 85% in the most altricial species to 68% in the megapodes and 60% in *Apteryx*, the most precocial forms (Sotherland & Rahn, 1987). During development eggs lose water first by diffusion of water vapour through the pores of the shell and after external pipping and emergence from the shell by convection. What appears to be most unusual is the fact that the relative water content of the hatchling is essentially the same as that of the fresh egg (Ar & Rahn, 1980; Carey, 1983; Ar *et al.*, 1987). The water of the egg contents (egg mass + shell

mass), as well as the hatchling in a large number of altricial and precocial species, has been measured and the overall mean difference of the relative water content between the hatchling and fresh egg content for 38 species is essentially zero (Ar *et al.*, 1987).

These observations suggest that the loss of water during development is precisely regulated. The total water loss established by Ar *et al.*, (1987) was 28% of the initial egg content and a similar value can be estimated when one compares the differences between hatchling and egg content for 98 species cited in the literature (Rahn & Paganelli, 1991). Accepting an overall value of 28%, the total water loss can then be partitioned between a 15% loss of water by diffusion across the egg shell (Ar & Rahn, 1980; Rahn & Paganelli, 1990) and an additional 13% by convection.

The rate of water loss

As the relative diffusive water loss appears to be similar in most species, 15% (SD = 2.6%) of the initial egg mass (n = 117 species; Rahn & Paganelli, 1990) the daily water loss can be described by a general equation, namely

$$\dot{M}_{H_2O} = 150 \, (W/I) \qquad (21.6)$$

Where $\dot{M}_{H_2O}$ = diffusive daily water loss (mg d^{-1}); W = egg mass (g); I = incubation time (d); and 150 = constant which changes g to mg. This equation predicts the daily water loss for any egg, if the initial egg mass and incubation time are known. In order to understand how this average rate is achieved in various climates where the absolute humidity may vary from 3–4 torr in dry polar regions to *circa* 22 torr in moist, tropical climates and even higher humidities in eggs incubated in wet nests, we turn to the diffusive water loss equation (Paganelli, 1980), where

$$\dot{M}_{H_2O} = G_{H_2O} \, (P_e - P_n) \qquad (21.7)$$

Where G_{H_2O} = water vapour conductance of the shell (mg (d torr)$^{-1}$); P_e = water vapour pressure of the egg (torr): and P_n = water vapour pressure in the nest (torr).

P_e is assumed to be fully saturated and therefore determined by the mean egg temperature, which in turn is regulated by the cold-vasodilation reflex of the brood-patch blood flow (Mitgård *et al.*, 1985). On the other hand, the microclimate of the nest air is maintained by the

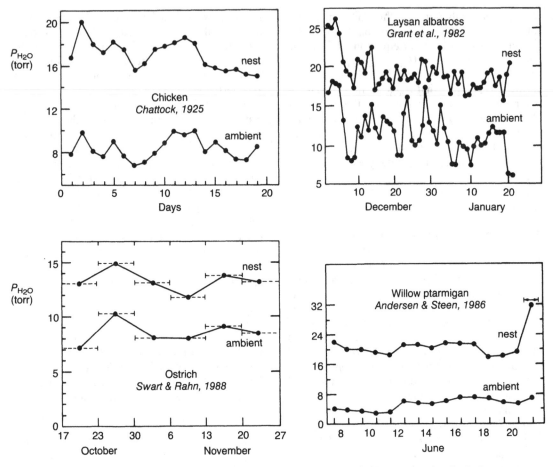

Fig. 21.9. The absolute nest and ambient humidity during incubation in the fowl, Laysan albatross, African ostrich and willow ptarmigan. (Reconstructed from the authors' data.)

overall brooding behaviour of the parent. Such behaviour is presumably a genetic trait which includes the type of nest construction, the plumage which insulates the egg and nest, and the incubation bouts, periodic stand ups, egg turning, etc. The daily water loss cannot accumulate in the nest and so the average overall nest vapour pressure must exceed by some degree the ambient vapour pressure.

Thus, there are two variables, egg temperature and brooding behaviour which determine the difference between the vapour pressure of the egg and the nest $(P_e - P_n)$. These two variables, in turn, must be matched to the shell conductance (or number of invariant pores) so that the average daily water loss is equal to 150 (W/I) in various climates (equation 21.6). The nest humidities throughout incubation are regulated in four species exposed to different

ambient humidities and temperatures (Fig. 21.9). These reveal several generalisations: 1) while the ambient *relative* humidities at nesting sites vary greatly throughout the day (Rahn *et al.*, 1977; Andersen & Steen, 1986) the ambient *absolute* humidities in most nesting environments are relatively constant; 2) each species maintains its own specific absolute nest humidity; and 3) the difference between ambient and nest humidity differs in each species, but is maintained when slight differences in ambient humidity do occur, suggesting that this is achieved by a programmed brooding behaviour.

The differences between nest and ambient humidity have now been established in 20 species (Rahn & Paganelli, 1990) where the ambient humidities varied from 4 torr in dry polar climates to 22 torr in warm moist regions, and include species nesting on the ground, cliffs,

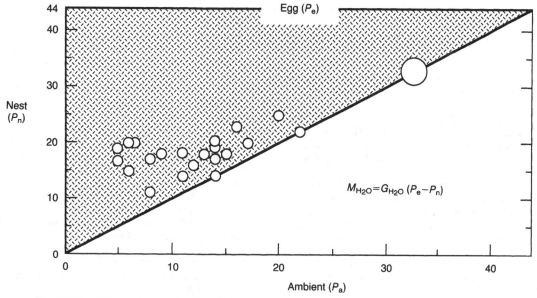

Fig. 21.10. The nest humidity, P_n, of 20 species incubating in various climates where the ambient humidity varied from 4 to 22 torr. For details see text. (From Rahn & Paganelli, 1990.)

trees and burrows. The observed nest humidity is plotted against the ambient humidity in Fig. 21.10 and shows how different species have regulated their brooding behaviour in various climates. For illustrative purposes only, a common vapour pressure in the egg of 44 torr is shown, equivalent to the saturation pressure of 35.7 °C, the mean egg temperature (Fig. 21.2). Thus all nest water vapour pressures are confined to the shaded triangle and $(P_e - P_n)$ must be matched in each species to the appropriate shell conductance to achieve a daily water loss which at the end of incubation will result in a total diffusive loss of about 15% of the initial egg mass. The large circle at the right in Fig. 21.10 is a predicted nest vapour pressure for the pied-billed grebe, based on the egg cup temperatures of their wet nests (Davis, Platter-Rieger & Ackerman, 1984). The water vapour pressure gradient is obviously reduced in these eggs and accounts for their twice normal shell conductance, but normal water loss.

Thus, the rate of diffusive water loss is governed by various factors: the egg temperature and brood patch blood flow which determine P_e, the nest humidity, P_n, which in all climates must exceed the ambient humidity, P_a, by a finite difference, and in turn is determined by the par-

ticular nest construction and brooding behaviour of the parent. Finally, shell conductance must be matched to the $P_e - P_n$ difference so that the daily diffusive water loss times the days of incubation will account for a total diffusive water loss of about 15% of the initial egg mass.

References

Afton, A. D. (1978). Incubation rhythms and egg temperatures of an American Green-winged Teal and a renesting Pintail. *The Prairie Naturalist*, **10**, 115–19.

Altman, P. L. & Dittmer, D. S. (1974). *Biology Data Book, 2nd edn, vol. 3*, p. 1989. Bethesda, Maryland: Federation of American Society of Experimental Medicine.

Andersen, Ø. & Steen, J. B. (1986). Water economy in bird nests. *Journal of Comparative Physiology*, **156B**, 823–8.

Anderson, D. J., Stoyan, N. C. & Ricklefs, R. E. (1987). Why are there no viviparous birds? A comment. *American Naturalist*, **130**, 941–7.

Anderson, J. F., Rahn, H. & Prange, H. D. (1979). Scaling of supportive tissue mass. *Quarterly Review of Biology*, **54**, 139–48.

Ar, A., Arieli, B., Belinski, A. & Yom-Tov, Y. (1987). Energy in avian eggs and hatchlings: utilization and transfer. *Journal of Experimental Zoology*, Supplement 1, 151–64.

Ar, A. & Rahn, H. (1980). Water in the avian egg: overall budget of incubation. *American Zoologist*, **20**, 373–84.

— (1985). Pores in avian eggshells: Gas conductance, gas exchange and embryonic growth rate. *Respiration Physiology*, **61**, 1–20.

Ar, A., Rahn, H. & Paganelli, C. V. (1979). The avian egg: Mass and strength. *Condor*, **81**, 331–7.

Barcroft, J. (1932). La fixité du milieu intérieur est la condition de la vie libre (Claude Bernard). *Biological Reviews*, **7**, 24–51.

Barrett, R. T. (1980). Temperature of Kittiwake, *Rissa tridactyla*, eggs and nest during incubation. *Ornis Scandinavica*, **11**, 50–9.

Biebach, H. (1979). Energetik des Brütens beim Star (*Sturnus vulgaris*). *Journal für Ornithologie*, **120**, 121–38.

Blackburn, D. G. & Evans, H. E. (1986). Why are there no viviparous birds? *American Naturalist*, **128**, 165–90.

Blitz, R. M. & Pellegrino, E. D. (1969). The chemical anatomy of bone. *Journal of Bone and Joint Surgery*, **51A**, 456–66.

Board, R. G. (1982). Properties of avian egg-shells and their adaptive value. *Biological Reviews*, **57**, 1–28.

Burger, A. E. & Williams, A. J. (1979). Egg temperatures of the Rockhopper Penguin and some other penguins. *Auk*, **96**, 100–5.

Burke, E. (1925). A study of incubation. *University of Montana Agriculture Station, Bozeman, Montana, Bulletin* **178**, pp. 1–44.

Caldwell, P. J. & Cornwell, G. W. (1975). Incubation behavior and temperatures of the Mallard Duck. *Auk*, **92**, 706–31.

Cannon, W. B. (1929). Organization for physiological homeostasis. *Physiological Reviews*, **9**, 339–431.

Carey, C. (1983). Structure and function of avian eggs. In *Current Ornithology*, vol. 1, ed. R. F. Johnston, pp. 69–103. New York: Plenum Press.

Chattock, A. P. (1925). On the physics of incubation. *Philosophical Transactions of the Royal Society, Series B*, **213**, 397–450.

Cooper, J. A. (1978). The history and breeding biology of the Canada Geese of Marshy Point, Manitoba. *Wildlife Monographs*, **61**, 1–87.

Davis, T. A., Platter-Rieger, M. F. & Ackerman, R. A. (1984). Increased water loss by Pied-billed Grebe eggs: Adaptation to a hot, wet nest. *Physiological Zoology*, **57**, 384–91.

Derksen, D. V. (1977). A quantitative analysis of the incubation behavior of the Adelie Penguin. *Auk*, **94**, 552–66.

Drent, R. H. (1970). Functional aspects of incubation in the Herring Gull. *Behaviour*, Supplement 17, 1–132.

— (1975). Incubation. In *Avian Biology*, vol. 5, eds D.

S. Farner & J. R. King, pp. 334–420. New York: Academic Press.

Dunbrack, R. L. & Ramsay, M. A. (1989). The evolution of viviparity in amniote vertebrates: Egg retention versus egg size reduction: *American Naturalist*, **133**, 138–48.

Ellis, J. H. (1982). The thermal nest environment and parental behavior of a burrowing bird, the Bank Swallow. *Condor*, **84**, 441–3.

Grant, G. S. (1982). Avian incubation: Egg temperature, nest humidity, and behavioral thermoregulation in a hot environment. *Ornithological Monographs*, No 30. Washington, DC: American Ornithological Union.

Grant, G. S., Pettit, T. N., Rahn, H., Whittow, G. C. & Paganelli, C. V. (1982a). Water loss from Laysan and Black-footed Albatross eggs. *Physiological Zoology*, **55**, 405–14.

Grant, G. S., Pettit, T. N., Rahn, H., Whittow, G. C. & Paganelli, C. V. (1982b). Regulation of water loss from Bonin Petrel (*Pterodroma hypoleuca*) eggs. *Auk*, **99**, 236–42.

Grant, G. S., Paganelli, C. V. & Rahn, H. (1984). Microclimate of Gull-billed Tern and Black Skimmer nests. *Condor*, **86**, 337–8.

Haftorn, S. (1978). Egg laying and regulation of egg temperature during incubation in the Goldcrest *Regulus regulus*. *Ornis Scandinavica*, **9**, 2–21.

— (1979). Incubation and regulation of egg temperature in the Willow Tit *Parus montanus*. *Ornis Scandinavica*, **10**, 220–34.

Handrich, Y. (1989). Incubation water loss in King Penguin egg. I. Change in egg and brood pouch parameters. *Physiological Zoology*, **62**, 96–118.

Howell, T. R. (1979). Breeding biology of the Egyptian Plover, *Pluvianus aegyptius*. *University of California Publications – Zoology*, vol. 113. Los Angeles: University of California Press.

Howell, T. R., Araya, B. & Millie, W. R. (1974). Breeding biology of the Gray Gull, *Larus modestus*. *University of California Publications – Zoology*, vol. 104. Los Angeles: University of California Press.

Howell, T. R. & Bartholomew, G. A. (1962). Temperature regulation in the Red-tailed Tropic Bird and the Red-footed Booby. *Condor*, **64**, 6–18.

Johansen, K. & Millard, R. W. (1974). Cold-induced neurogenic vasodilatation in skin of the Giant Fulmar, *Macronetes giganteus*. *American Journal of Physiology*, **227**, 1232–5.

Krog, J. O. & Tøien, Ø. (1984). Circulatory and metabolic reaction in arctic birds to the immersion of the feet in ice water. In *The Peripheral Circulation*, eds S. Hunyor, J. Ludbrook, J. Shaw & M. McGrath, pp. 157–60. Amsterdam: Elsevier.

Lamson, G. H. & Edmond, H. D. (1914). Carbon dioxide in incubation. *Bulletin Storrs Agricultural Experiment Station*, **76**, 215–58.

Marder, J. & Gavrieli-Levin, I. (1986). Body and egg temperature regulation in incubating pigeons exposed to heat stress: The role of skin evaporation. *Physiological Zoology*, **59**, 532–8.

McNab, B. K. (1966). An analysis of body temperature of birds. *Condor*, **68**, 47–55.

Mitgård, U., Sejrsen, P. & Johansen, K. (1985). Blood flow in the brood patch of Bantam hens: evidence of cold vasodilation. *Journal of Comparative Physiology* **155B**, 703–9.

Morgan, K. R., Paganelli, C. V. & Rahn, H. (1978). Egg weight loss and nest humidity during incubation in two Alaskan gulls. *Condor*, **80**, 272–5.

Morton, M. L. & Pereyra, M. E. (1985). The regulation of egg temperatures and attentiveness patterns in the Dusky Flycatcher (*Empidonax oberholseri*). *Auk*, **102**, 25–37.

Murrish, D.E. & Guard, C. L. (1977). Cardiovascular adaptations of the Giant Petrel, *Macronectes giganteus*, to the antarctic environment. In *Adaptations within the Antarctic Ecosystems*, ed. G. A. Llano, pp. 511–30. Washington, DC: Smithsonian Institution.

Needham, J. (1963). *Chemical Embryology*, vol. 2. New York: Hafner Publishing Company.

Olszowka, A. J., Tazawa, H. & Rahn, H. (1988). A blood-gas nomogram of the chick fetus: blood flow distribution between the chorioallantois and fetus. *Respiration Physiology*, **71**, 315–30.

Paganelli, C. V. (1980). The physics of gas exchange across the avian eggshell. *American Zoologist*, **20**, 329–38.

Pettit, T. N., Grant, G. S., Whittow, G. C., Rahn, H. & Paganelli, C. V. (1981). Respiratory gas exchange and growth of White Tern embryos. *Condor*, **83**, 355–61.

Prevost, J. (1961). *Ecologie du Manchot Empereur*. Paris: Hermann.

Rahn, H., Ackerman, R. A. & Paganelli, C. V. (1977). Humidity in the avian nest and egg water loss during incubation. *Physological Zoology*, **50**, 269–83.

Rahn, H. & Dawson, W. R. (1979). Incubation water loss in eggs of Heermann's and Western Gulls. *Physiological Zoology*, **52**, 451–60.

Rahn, H. & Hammel, H. T. (1982). Incubation water loss, shell conductance, and pore dimensions in Adeliae Penguin eggs. *Polar Biology*, **1**, 91–7.

Rahn, H., Krog, J. & Mehlum, F. (1983). Microclimate of the nest and egg water loss of the Eider *Somateria mollissima* and other water fowl in Spitsbergen. *Polar Research*, **1**, 171–83.

Rahn, H., Matalon, S. & Sotherland, P. R. (1985). Circulatory changes and oxygen delivery in the chick embryo prior to hatching. In *Cardiovascular Shunts*, eds K. Johansen & W. W. Burggren, pp. 199–215. Copenhagen: Munksgaard.

Rahn, H. & Paganelli, C. V. (1989). Shell mass, thickness and density of avian eggs derived from the tables of Schönwetter. *Journal für Ornithologie*, **130**, 59–68.

— (1990). Gas fluxes in avian eggs: Driving forces and the pathway for exchange. *Comparative Biochemistry and Physiology*, **95A**, 1–15.

— (1991). Energy budget and gas exchange of avian eggs. In *Avian Incubation*, ed. S. G. Tullett, pp. 175–93. London: Butterworths.

Rahn, H., Paganelli, C. V., Nisbet, I. C. T. & Whittow, G. C. (1976). Regulation of incubation water loss in seven species of Terns. *Physiological Zoology*, **49**, 245–59.

Ricklefs, R. E. & Rahn, H. (1979). The incubation temperature of Leach's Storm-Petrel. *Auk*, **96**, 625–7.

Romanoff, A. L. (1967). *Biochemistry of the Avian Embryo*. New York: John Wiley & Sons.

Russell, S. M. (1969). Regulation of egg temperatures by incubating White-winged Doves. In *Physiological Systems in Semi-arid Environments*, eds C. C. Hoff & M. L. Riedesel, pp. 107–12. Albuquerque, New Mexico: University of New Mexico Press.

Shine, R. (1985). The evolution of viviparity in reptiles: an ecological analysis. In *Biology of the Reptilia*, vol. 15, eds C. Gans & F. Billett, pp. 605–94. New York: John Wiley & Sons.

Sotherland, P. R., Ashen, M. D., Shuman, R. D. & Tracy, C. R. (1984). The water balance of bird eggs incubated in water. *Physiological Zoology*, **57**, 338–48.

Sotherland, P. R. & Rahn, H. (1987). On the composition of bird eggs. *Condor*, **89**, 48–65.

Stonehouse, B. (1960). The King Penguin of South Georgia. I. Breeding behaviour and development. *Falkland Island Dependency Survey, Scientific Report*, 1–81.

Swart, D., Rahn, H. & de Kock, J. (1987). Nest microclimate and incubation water loss of eggs of the African Ostrich (*Struthio camelus* var. *domesticus*). *Journal of Experimental Zoology*, Supplement 1, 239–46.

Tazawa, H. (1980). Oxygen and CO_2 exchange and acid-base regulation in the avian embryo. *American Zoologist*, **20**, 395–404.

— (1986). Acid-base equilibrium in birds and eggs. In *Acid-base regulation in Animals*, ed. N. Heisler, pp. 203–33. Amsterdam: Elsevier.

Tøien, Ø., Aulie, A. & Steen, J. B. (1986). Thermoregulatory responses to egg cooling in incubating bantam hens. *Journal of Comparative Physiology*, **156B**, 303–7.

Tullett, S. G. (1984). The porosity of avian eggshells. *Comparative Biochemistry and Physiology*, **78A**, 5–13.

Turner, J. S. (1987). Blood circulation and the flow of heat in an incubated egg. *Journal of Experimental Zoology*, Supplement 1, 99–104.

Visschedijk, A. H. J. & Rahn, H. (1983). Replacement of diffusive by convective gas transport in the developing hen's egg. *Respiration Physiology*, **52**, 137–47.

Vleck, C. M. (1981a). Energetic cost of incubation in the Zebra Finch. *Condor*, **83**, 229–37.

— (1981b). Humming bird incubation: Female attentiveness and egg temperature. *Oecologia*, **51**, 199–205.

Vleck, C. M., Vleck, D., Rahn, H. & Paganelli, C. V. (1983). Nest microclimate, water vapor conductance, and water loss in Heron and Tern eggs. *Auk*, **100**, 76–83.

Vleck, D., Vleck, C. M. & Seymour, R. S. (1984). Energetics of embryonic development in the Megapode birds, Mallee Fowl *Leipoa ocellata* and Brush Turkey *Alectura lathami*. *Physiological Zoology*, **57**, 444–56.

Walsberg, G. E. & Voss-Roberts, K. A. (1983). Incubation in desert-nesting doves: mechanism for egg cooling. *Physiological Zoology*, **56**, 88–93.

Webb, D. R. (1987). Thermal tolerance of avian embryos: A review. *Condor*, **89**, 874–98.

Whittow, G. C. (1980). Physiological and ecological correlates of prolonged incubation in sea birds. *American Zoologist*, **20**, 427–36.

Whittow, G. C., Ackerman, R. A., Paganelli, C. V. & Pettit, T. N. (1982). Prepipping water loss from the eggs of the Wedgetailed Shearwater. *Comparative Biochemistry and Physiology*, **72A**, 29–34.

Williams, J. B. & Ricklefs, R. E. (1984). Egg temperature and embryo metabolism in some high-latitude procellariiform birds. *Physiological Zoology*, **57**, 118–27.

Yom-Tov, Y., Ar, A. & Mendelssohn, H. (1978). Incubation behavior of the Dead Sea Sparrow. *Condor*, **80**, 340–3.

Zerba, E. & Morton, M. L. (1983). Dynamics of incubation in Mountain White-crowned Sparrows. *Condor*, **85**, 1–11.

22

Influences of incubation requirements on the evolution of viviparity

RICHARD SHINE

Introduction

Although oviparity characterises most living vertebrate species, including all birds and most reptiles, viviparity has evolved many times in animal phylogeny. Approximately 100 independent evolutionary origins of viviparity (herein defined to mean production of fully formed young rather than laying of eggs) have been identified within squamate reptiles (Blackburn, 1982; Shine, 1985). These multiple evolutionary origins provide excellent opportunities for comparative analyses of the implications of this change in incubation conditions, from the external nest to the female's oviduct, for the physiology of developing embryos. Similarly, studies of embryonic physiology, and especially of the effects of incubation conditions on the fitness of hatchlings, may clarify evolutionary models on the selective forces important in the shift from oviparity to viviparity.

In this chapter, I consider the relationship between physical influences on embryonic development on the one hand, and the evolution of viviparity on the other. Each of these factors could affect the other, so two main questions arise:
1) Have direct effects of incubation conditions on viability of hatchlings played a significant role in promoting or constraining the acquisition of viviparity in various vertebrate lineages, or the probability of subsequent re-evolution of oviparity? This question is the main focus of the present chapter.
2) What adaptations of embryonic and maternal physiology have resulted from changes in incubation conditions brought about by the evolution of prolonged uterine retention of eggs,

ovoviviparity, and placentotrophy? There has been considerable interest in this topic from physiologists, and modifications of maternal and embryonic physiology in viviparous taxa, e.g., shifts in oxygen affinity of fetal haemoglobins (Grigg & Harlow, 1981), are well-documented. Presumably, shifts have also occurred in other variables such as thermal tolerances, ability of embryos to withstand movement, and so forth. Although this topic deserves further research, it will be considered only peripherally in the present chapter.

Shifts in incubation conditions

What is the likely nature of the changes in incubation conditions faced by embryos in a lineage when viviparity evolves? The most obvious changes are modifications of the thermal, hydric and gaseous environments to which the developing embryos are exposed. Also, the embryo *in utero* must be able to tolerate movement during maternal locomotion, whereas this tolerance is not necessary for oviposited eggs of many species (Deeming, Chapter 19). Except for characteristics such as tolerance to movement, it is difficult to generalise about the direction of changes in incubation conditions brought about by the evolution of prolonged uterine retention of eggs. For example, comparisons of nest temperatures and maternal body temperatures suggest that the evolution of viviparity may increase mean incubation temperature quite dramatically in heliothermic alpine lizards, but not in related sympatric species in which females do not bask (Shine, 1983*a*). Thus, it may be difficult to predict the relationship between soil temperature and mean body temperature of an ectotherm in the same habitat. Similar arguments can be raised with respect to the availability of water and oxygen, and the ease of eliminating carbon dioxide. The extent of the difference in incubation conditions between natural nest sites and the maternal oviduct depends upon the specific location of the nest sites, and the physical characteristics of the oviducts of various species. Although it is thus difficult to generalise about shifts in incubation conditions consequent upon the evolution of viviparity, there seems little doubt that such shifts may often be profound.

So far, the changes in *mean* incubation conditions, e.g. mean temperature, mean water potential, mean oxygen tension have been emphasised as the important variables modified by a shift in reproductive mode. However, it may often be true that the *variance* changes more than the mean, and effects on the variance of incubation conditions may be biologically more significant than changes in mean values. Presumably, water potential is generally less variable inside an oviduct than in an external nest (and more uniform among eggs within a single clutch, as well as in a clutch through time). This reduced variance of water potentials in both time and space may be a significant benefit of prolongation of uterine retention of embryos in some environments.

The situation is less clear-cut in the case of the thermal environment. In many squamate species, maternal thermoregulatory behaviour (especially basking, and choice of warm nocturnal refugia) may result in relative thermal constancy for retained eggs. However, the reverse may also be true, where maternal temperatures fall to ambient levels at night but rapidly increase with morning basking. In such a case, the variance in temperatures of retained eggs may be very low over a 12-hour period (at the mother's selected temperature) but very high over a 24-hour period (when overnight minima are incorporated). Hence, the evolution of viviparity in these circumstances may change the embryonic thermal environment from one of gradual cycling to one with a virtual step function, even if 'mean incubation temperature' is unaffected. The nature of likely adaptive modifications to embryonic thermal sensitivity in such a situation is unclear. One possibility is that developmental processes, like other physiological processes within the mother's body (Dawson,

1975) may adapt to the relatively constant selected temperature of the mother. That is, the embryo may lose its ability to develop over a wide range of incubation temperatures, and instead become specialised for developing at a constant high temperature for much of the day. If such shifts in embryonic thermal tolerances occur, they should be evident if one compares the range of temperatures at which successful embryonic development can proceed in closely-related oviparous and viviparous forms. Preliminary data on thermal tolerances of montane scincid lizards provide a possible example of this phenomenon (see later).

Can incubation requirements constrain the evolution of viviparity?

How does the extent and nature of such a shift in incubation conditions affect the probability that viviparity will arise in a lineage? Here, the ways in which this factor might prevent the evolution of viviparity in a taxon in which this change in reproductive modes would otherwise have evolved are discussed. The basic hypothesis is that prolonged uterine retention of eggs may be impossible if too great a difference exists between the physical conditions for incubation in the nest site versus in the oviduct. Embryonic physical requirements could prevent the evolution of viviparity for a variety of reasons. Embryos may be unable to develop under the physical conditions existing *in utero* because 1) embryos have very specific physical requirements; 2) uteri of some taxa may be unable to support prolonged retention of developing embryos; or 3) although embryos can tolerate a wide range of physical conditions, and uteri are potentially capable of sustaining embryos, the conditions *in utero* are simply too different from those in natural nest sites. The first two of these alternative explanations involve biological attributes of a taxon, and are likely to be phylogenetically conservative (e.g. chelonian embryos and squamate embryos might differ consistently in their tolerance to some aspect of incubation conditions). By contrast, the third hypothesis relies upon local environmental factors as selective forces on embryonic physical tolerances, and hence suggests that the resulting constraints may differ even between closely related species nesting in different habitats. Each of these hypotheses are considered in turn.

Lineage-wide effects

Has natural selection for prolonged uterine retention of eggs been opposed by specific embryonic requirements characteristic of particular vertebrate lineages? Hypotheses of this type have rarely been popular in the evolutionary literature, perhaps because optimality arguments have received more attention than has the role of constraints (Gould & Lewontin, 1979). Hence, the general idea that interspecific differences in embryonic requirements may preclude the evolution of viviparity has rarely been tested. One well-known hypothesis of this type is the idea that viviparity has been precluded in some reptilian taxa because of embryonic dependence on calcium in the eggshell (Packard, Tracy & Roth, 1977). Although this hypothesis has attracted strong criticism (Tinkle & Gibbons, 1977), it does generate testable predictions and has stimulated considerable research.

Other hypotheses on physiological obstacles to viviparity also deserve attention. For example, the absence of viviparity in chelonians might be due to their tendency to suspend embryogenesis *in utero* at an early stage of development (Ewert, Chapter 11), or the vulnerability of developing embryos to rotation of the eggs (Limpus, Baker & Miller, 1977; Shine, 1983*a*; Deeming, Chapter 19). If pre-ovipositional embryonic diapause occurs in all turtles, it may be a significant barrier to the evolution of viviparity in this group. Intermediate stages of prolonged embryogenesis *in utero* are difficult to imagine, because any factor (environmental or genetic) promoting prolonged uterine retention of eggs will not affect the amount of embryonic development *in utero*. Intermediate stages towards viviparity could arise only if mutations for both uterine retention *and* cessation of embryonic diapause occurred in the same lineage. Similarly, extreme sensitivity to physical disturbance would preclude prolonged retention of developing embryos, and may in fact have selected for pre-ovipositional embryonic diapause. Other hypotheses on physiological constraints include suggestions that avian viviparity has been precluded by high temperatures and low oxygen tension in the oviduct (Anderson, Stoyan & Ricklefs, 1987) or by respiratory biomechanics (Duncker, 1989).

There have been many additional suggestions as to why birds, chelonians and crocodilians have *not* evolved viviparity (Tinkle & Gibbons, 1977; Packard *et al.*, 1977; Blackburn & Evans, 1986; Dunbrack & Ramsay, 1989; Shine, 1989). These ideas may be broadly divided into proximate and ultimate hypotheses. The former suggest that viviparity has been precluded by physiological or morphological constraints, whereas the latter suggest that particular features of these taxa have selected against any prolongation of uterine development (e.g. impairment of flight; sacrifice in fecundity). Although our ability to provide convincing answers to questions of the general form 'why has some attribute *not* evolved in a particular taxon?' is questionable, such speculation may ultimately provide a useful set of testable hypotheses about the phenomena in question. Nonetheless, the most rigorous tests of such ideas are likely to come from comparisons among closely related species differing in reproductive mode, rather than from broad comparisons among Orders or Classes.

Physical requirements for embryogenesis may affect the evolution of viviparity, and cause differences among lineages in the frequency of evolution of viviparity, even if the lineages are identical in all significant aspects of embryonic and maternal physiology. This apparently paradoxical result derives from the possibility that physiological constraints to prolonged uterine retention of eggs may be particularly intense early in development. If so, any lineage that overcomes such an obstacle, early in its ancestry, may then face only minor physiological barriers to subsequent multiple evolutions of much more prolonged retention of embryos. The squamates may be an example of such a group, so that the high frequency of independent evolutionary origins of viviparity within this lineage may have been possible only because some ancestral squamate overcame initial physiological obstacles to (relatively limited) uterine retention. In keeping with this hypothesis, the duration of retention of the developing embryo *in utero* in oviparous species is phylogenetically conservative among vertebrates. It is thus worth considering in more detail whether basic differences among major lineages in the duration of uterine retention might help to explain why viviparity has arisen so frequently in the Squamata, and not at all in birds, crocodilians or turtles.

Although there are undoubtedly exceptions, the general rule seems to be that embryos of turtles, crocodilians, the tuatara (*Sphenodon punctatus*) and birds usually develop only slightly *in utero* (Bellairs, Chapter 23) with the overwhelming majority of embryonic development occurring outside of the maternal reproductive tract. For example, freshly laid eggs of most species of turtle investigated to date contain embryos at the gastrula stage (Ewert, 1985). Birds and the tuatara are similar (Gilbert, 1971; Moffat, 1985), and crocodilian embryogenesis is only slightly more advanced at oviposition (Ferguson, 1985). Although a similar situation is reported in some squamates (Blanc, 1974), most oviparous lizards and snakes retain developing eggs *in utero* for a high proportion of the total period of embryonic development (Shine, 1983*b*). At oviposition, the embryos are well differentiated with obvious limbs and internal organs, and functional circulatory systems. More prolonged embryonic development occurs in some squamate species, including several with incubation periods so brief (< 48 hours) that in most respects the population is functionally viviparous (Blanchard, 1933; Bustard, 1964).

These differences among major 'oviparous' vertebrate taxa in the duration of uterine retention of eggs (and more importantly, the degree of embryonic development at oviposition) may have major implications for the evolution of viviparity. Relatively little is known about temporal shifts in physical requirements for embryogenesis in reptiles, but such shifts are likely to be common. For example, the influence of incubation temperature on embryonic gender is limited to certain critical periods in development (Bull & Vogt, 1981; Deeming & Ferguson, Chapter 10) and embryonic vulnerability to physical disturbance changes rapidly through incubation (Limpus *et al.*, 1979; Deeming, Chapter 19). The duration of uterine retention of embryos in present-day oviparous species suggests that the evolution of viviparity in some squamate lineages may have involved a modification of incubation conditions relatively early in embryogenesis (e.g. chamaeleons) whereas, in others, the changes may have involved embryos already well along in development (e.g. scincids). If embryonic requirements are particularly narrow and inflexible at one stage of embryogenesis, then this stage may pose a major barrier to the acquisition of viviparity in that lineage. Such may have been the case, for example, with the two major vertebrate lineages that are entirely oviparous: the archosaurs and turtles. Intrinsically different patterns of embryonic

dependence on incubation conditions in ancestral squamate reptiles, or perhaps a fortunate combination of selective forces and incubation environments in early squamates, may have overcome this putative evolutionary obstacle, and allowed the present-day situation of widespread, prolonged embryonic development *in utero* among members of this group.

In such a scenario, the multiple evolutionary origins of viviparity in present-day squamates, and the lack of such transitions among birds, crocodilians and turtles, would ultimately reflect differences between these taxa in the degree of embryogenesis prior to oviposition. To put it simply, the lack of viviparity in birds may be due to massive physiological obstacles to uterine retention of very early embryos, whereas the high frequency of independent origins of viviparity in squamates reflects the relative ease with which uterine retention may proceed once these initial stages have been passed. Most attempts to explain the absence of viviparous birds, turtles and crocodilians begin with the premise that these groups must differ in some important way from the squamates, because of the great difference in the frequency of evolutionary origins of viviparity between squamates and the other lineages. The above hypothesis, however, suggests that this apparently 100-fold difference could actually result from a single evolutionary transition (from almost no uterine embryogenesis to limited uterine embryogenesis) in a single lineage of ancestral squamates. If so, the difference in the frequency of origins of viviparity among major lineages may be ultimately attributable to a single event, and hence might be due entirely to chance. Although this scenario may seem fanciful, it is amenable to testing. It predicts that detailed studies of embryonic incubation requirements will show major shifts during embryogenesis in vulnerability of developing embryos to the kinds of changes in incubation conditions consequent upon prolonged retention of eggs in the maternal oviduct.

Species-specific effects

The preceding discussion concentrated on phylogenetically conservative features that might help to explain broad taxonomic patterns in the distribution of viviparity. The same types of arguments may also be applied at a much finer taxonomic scale, and may even apply to congeneric species. In such cases, the crucial difference between the taxa, constraining the evolution of viviparity in one species but not the other, is likely to result from adaptations of embryonic physiology to specific conditions in local nest-sites (and hence, tolerance of particular incubation conditions). For example, a reptile species inhabiting a desert and one inhabiting a mesic environment may be exposed to very different selective pressures on water utilisation by oviposited eggs. Such differences might constrain the evolution of viviparity in one taxon (because embryos are unable to develop under the physical conditions experienced in the oviduct) while permitting it in the other. For example, eggs of one xeric-adapted species of Australian agamid lizards swell excessively and burst when incubated under moist conditions, whereas eggs of mesic-zone congeners do not (Badham, 1971). Such differences in hydric tolerances between closely related species might influence the probability of successful uterine retention of developing eggs.

The same type of argument could be raised with respect to any physical variable important for embryonic development. For example, although there is strong empirical support for the hypothesis that reptilian viviparity has arisen primarily in response to occupancy of cold environments (Blackburn, 1982; Shine, 1985), many high-elevation and high-latitude reptiles continue to reproduce by egg-laying, e.g. *Scincella ladacensis* of the Himalayas, > 4000 m above sea level (Shine, 1985). Viviparity in such lineages may be precluded by specific adaptations of embryonic physiology. For example, embryos adapted to survive low soil temperatures may be unable to develop successfully at elevated maternal body temperatures. Such hypotheses are testable, through 1) hormonal manipulation to prolong uterine retention of developing embryos of oviparous species (Shine & Guillette, 1988), with monitoring of subsequent embryonic viability; and 2) detailed studies of physical requirements for embryonic development in a range of oviparous forms, especially those with viviparous congeners. An example of the latter approach is given below.

Can incubation requirements facilitate the evolution of viviparity?

So far, only the possible negative effects of the shift from external to uterine incubation on the evolutionary fitness of the reproducing female and her offspring have been considered. It is equally likely that this shift has had positive effects on hatchling survivorship and viability in many lineages, and that such effects have been major selective pressures for the evolution of viviparity. Although some hypotheses for the evolution of reptilian viviparity invoke biological factors as selective forces (e.g. reduction of competition for nest sites, competitive advantages to early hatching, reduction of nest predation and parasitism), most authors have concentrated upon advantages to hatchling survivorship accruing from incubation under physical conditions experienced in the maternal oviduct rather than the external nest (Shine, 1985). A wide variety of physical incubation conditions have been suggested to act in this way, including temperature, moisture, and oxygen tension (Mell, 1929; Neill, 1964; Guillette, 1982). The hypothesis that cold climates somehow stimulate the evolution of squamate viviparity has the strongest empirical support (Shine, 1985), although the nature of the advantage to prolonged uterine retention of embryos in cold climates has received little scrutiny. The conventional idea has been that hatchling survivorship is improved because accelerated development at high maternal body temperatures enables hatching prior to lethal autumn frosts (Shine, 1983a). However, this accelerated development might reduce the effects of any source of mortality on eggs in the nest, not simply cold-induced mortality (Shine, 1989).

Hatchling viability

Until recently, discussions of selective forces for the evolution of reptilian viviparity have focused on factors likely to kill developing embryos, rather than simply reduce their viability. The growing literature on the effects of incubation conditions on hatchling viability (Miller, Packard & Packard, 1987) suggests that major selective advantages could accrue from prolongation of uterine retention of eggs even if survivorship to the point of hatching is unaffected. Research on long-term effects of incubation conditions on individual survivorship, growth and ultimately reproductive success might do much to clarify the adaptive basis of shifts in reproductive mode among squamate reptiles.

Reversibility of the evolution of viviparity

As well as affecting the relative 'costs' and 'benefits' of prolongation of uterine retention, embryonic sensitivity to incubation conditions may influence the probability that viviparity, once having evolved, is able to revert to oviparity. Most well-corroborated phylogenetic hypotheses for reptilian lineages which are reproductively bimodal suggest that the transition has been from oviparity to viviparity rather than the reverse, but several cases are ambiguous. For example, the only oviparous New World viperid snake is *Lachesis muta*, which is unlikely to be close to the basal stock that gave rise to sympatric viviparous species (Shine, 1985). One possible reason for the apparent scarcity (perhaps lack) of re-evolution of oviparity in squamates is that adaptations of embryonic physiology subsequent to the evolution of viviparity make it difficult or impossible for the embryo to tolerate physical incubation conditions outside of the mother's body. The loss of the eggshell is an obvious example of a morphological shift that might preclude the re-evolution of oviparity in a viviparous lineage, especially in xeric environments (Packard, 1966). Such an 'evolutionary ratchet' might also occur for physiological rather than morphological reasons.

Preliminary data for Australian scincid lizards provide a possible example of this phenomenon (Fig. 22.1). Eggs of five oviparous skinks (*Lampropholis delicata, Lampropholis guichenoti, Lampropholis mustelinum, Leiolopisma duperreyi, Nannoscincus maccoyi*), and gravid females of three viviparous skinks (*Eulamprus tympanum, Leiolopisma coventryi, Leiolopisma entrecasteauxii*), were maintained over a range of constant temperatures (10, 15, 25 or 30 °C). Gravid females were collected in high-elevation habitats in southern Australia, with *L. mustelinum* coming from Mount Wilson (60 km west of Sydney, New South Wales) and the other species coming from Coree Flats in the Brindabella Ranges (40 km west of Canberra, A.C.T.). Data represent 185 eggs of oviparous species, and 31 successful births of viviparous species. Little embryonic development occurred at 10 °C (*n* = 23 eggs incubated, and 30 gravid females). Eggs of the oviparous species developed successfully at 15 to 25 °C, and with lower success at 30 °C. By contrast, viviparous species completed gestation only at 30 °C and (infrequently) at 25 °C (Fig.

22.1). Relative hatching success was higher in oviparous species than in viviparous species at 15, 20 and 25 °C (Mann-Whitney 'U' test, $P < 0.05$) but not at 30 °C ($P > 0.13$). In summary, eggs of the oviparous skinks developed successfully over a wide range of ambient temperatures, but embryos of vivparous forms developed successfully only at relatively high temperatures.

The optimal temperature for embryonic survivorship in viviparous forms in this study (30 °C) was close to mean maternal selected temperatures in the field (Shine, 1983*a*). By contrast, eggs of the oviparous species develop in the field at a mean temperature around 18 °C (Shine, 1983*b*). It is particularly interesting to

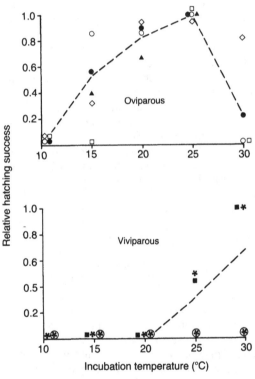

Fig. 22.1. Proportions of embryos of oviparous and viviparous Australian scincid lizards completing development at various constant temperatures in the laboratory. Upper graph shows oviparous species: circle is *Nannoscincus maccoyi*, dot is *Lampropholis mustelinum*, cross is *Lampropholis delicata*, open diamond is *Lampropholis guichenoti*, and open square is *Leiolopisma duperreyi*. Lower graph shows viviparous species: asterisk is *Leiolopisma coventryi*, enclosed asterisk is *Leiolopisma entrecasteauxii*, and square is *Eulamprus tympanum*. Dotted lines connect mean relative hatching success at each temperature. See text for methods and sample sizes.

note that mean body temperatures of most of the oviparous species studied are also close to 30 °C during daylight hours (Shine, 1983a), apparently above the thermal tolerances of the developing eggs (Fig. 22.1). Retention of developing eggs for 40 to 60 days prior to oviposition in these species may be possible only because of regular diurnal fluctuations in maternal body temperatures: cool nights mean that overall mean maternal temperatures average around 25 °C (Shine, 1983a). More importantly in the present context, the embryos of viviparous species appear to have lost the ability to develop at nest temperatures, possibly militating against the re-evolution of oviparity in this lineage.

Sample sizes are low in this comparison, and methodology is a problem because the uterine embryos may have died due to indirect effects of constant temperatures on maternal physiology, rather than directly on thermal tolerances of the embryos. However, the females appeared healthy even months after their embryos had died. Further investigation of the question will be of value, to evaluate whether specific incubation requirements may expedite or constrain the evolution of viviparity, to investigate adaptations of embryonic physiology to prolonged uterine retention, and to determine whether adaptations of embryonic physiology to viviparity may preclude the re-evolution of oviparity.

Need for gradual changes

Laboratory studies that prolong uterine retention of eggs by means of administration of exogenous hormones might also reveal the extent to which facultative retention is possible, and hence provide data to test a cornerstone of the evolutionary–ecological literature on this topic. Most recent discussions in this field have assumed that the transition from 'normal' oviparity to viviparity occurs via a series of intermediate stages of gradual increases in the duration of embryonic retention of embryos. The assumption is crucial, because the conditions likely to favour such intermediate stages may be more restrictive than the conditions likely to favour a viviparous species over an oviparous one (Packard et al., 1977; Shine & Bull, 1979). The reason for this contrast is that many of the 'advantages' of viviparity (e.g. emancipation of the need for suitable nesting sites) accrue only to a fully viviparous animal and not to an oviparous organism with prolonged uterine retention of eggs. If rapid evolutionary shifts in the duration of retention are possible, the significance of this argument is greatly weakened.

Prospects for further research

Studies on the effects of prolonged uterine retention of eggs on embryonic physiology would be of great interest. Such experiments could include hormonal manipulation of gestation periods, and observations on squamate populations with high variance in the degree of uterine retention of eggs, e.g. the scincid lizard *Lerista bougainvillii* in south-eastern Australia (Shine, 1985). By focusing on variation within a single species, and perhaps even within a single population, it should be possible to discern probable consequences of prolongation of uterine retention to 1) embryonic physiology, 2) embryonic survivorship, 3) hatchling viability, and 4) evolutionary fitness of the reproductive female.

A close taxonomic focus may well enable us to discriminate between alternative interpretations. Traditionally, many hypotheses about adaptations of embryonic and maternal systems to the evolution of viviparity have relied upon comparisons between distantly related organisms such as birds and mammals. Such comparisons may be useful as a source of hypotheses, but are unlikely to offer a robust test of the ideas. These broad comparisons introduce many confounding factors which relate to the differing morphologies, physiologies and evolutionary histories of the taxa being compared. Restriction of comparisons to closely related oviparous and viviparous forms makes it more likely that the correct evolutionary pathways may be identified. For example, the hypothesis that strong ontogenetic shifts in oxygen affinities of haemoglobins evolved as adaptations to anoxic nest environments, and hence pre-date the evolution of viviparity (Grigg & Harlow, 1981), may most usefully be tested with data from closely related oviparous and viviparous forms. More generally, we can use these taxonomically focused tests to discriminate between adaptations to viviparity versus preadaptations to viviparity.

Accurate reconstruction of evolutionary pathways is essential for useful analyses of selective forces. For example, a straightforward comparison between a viviparous reptile and a eutherian mammal might lead one to conclude that the current mammalian condition evolved via an intermediate stage involving retention of large yolky eggs within the female's body. It seems likely that no such intermediate ever existed; instead, mammalian viviparity arose by dramatic reduction in offspring size rather than prolongation of uterine retention (Dunbrack & Ramsay, 1989). Hence, the relevant intermediate forms probably resembled present-day monotremes and metatherians rather than viviparous reptiles. This conclusion has a major impact on hypotheses concerning incubation conditions of embryos of early mammals and the transition from oviparity to viviparity in this lineage. The crucial physiological challenges facing such embryos probably involved the dramatic reduction in hatchling size, and a shift in nutritional systems from lecithotrophy to oligotrophy (with lactation) in oviparous lineages. Although the subsequent evolution of viviparity (and then, placentotrophy) may have also modified eutherian embryonic physiology substantially, many of the distinctive physiological characteristics of eutherian embryos may well have evolved in response to egg-size reduction and lactation of neonates, rather than to placentotrophy. Present-day viviparous squamates show a wide range of degrees of maternal–fetal nutrient transfer, and hence provide an ideal opportunity to assess the evolutionary pressures on these alternative means of provisioning the young (Vitt & Blackburn, 1983; Stewart, 1989).

Acknowledgements

My ideas on these topics have benefited from discussions with many colleagues, particularly G. C. Packard, M. J. Packard, L. J. Guillette, M. B. Thompson, C. H. Tyndale-Biscoe, J. Lombardo, M. H. Wake and G. C. Grigg. I thank M. B. Thompson for comments on the MS, and D. T. Anderson and M. Byrne for discussions on reproductive systems of invertebrates. This work was supported by the Australian Research Council.

References

Anderson, D. J., Stoyan, N. C. & Ricklefs, R. D. (1987). Why are there no viviparous birds? A comment. *American Naturalist*, **130**, 941–7.

Badham, J. (1971). *A Comparison of Two Variants of the Bearded Dragon,* Amphibolurus barbatus *(Cuvier)*. PhD thesis, University of Sydney, NSW.

Blackburn, D. G. (1982). Evolutionary origins of viviparity in the Reptilia. I. Sauria. *Amphibia-Reptilia*, 3, 185–205.

Blackburn, D. G. & Evans, H. E. (1986). Why are there no viviparous birds? *American Naturalist*, **128**, 165–90.

Blanc, F. (1974). Table de developpement de *Chamaeleo lateralis* Gray, 1831. *Annales d'Embryologie et Morphologie*, 7, 99–115.

Blanchard, F. N. (1933). Eggs and young of the smooth green snake, *Liopeltis vernalis* (Harlan). *Papers of the Michigan Academy of Science, Arts and Letters*, 17, 493–508.

Bull, J. J. & Vogt, R. C. (1981). Temperature-sensitive periods of sex determination in emydid turtles. *Journal of Experimental Zoology*, **218**, 435–40.

Bustard, H. R. (1964). Reproduction in the Australian rainforest skinks, *Saiphos equalis* and *Sphenomorphus tryoni. Copeia*, 1964, 715–16.

Dawson, W. R. (1975). On the physiological significance of the preferred body temperatures of reptiles. In *Perspectives of Biophysical Ecology*, eds D. M. Gates & R. Schmerl, pp. 443–73. New York: Springer-Verlag.

Dunbrack, R. L. & Ramsay, M. A. (1989). The evolution of viviparity in amniote vertebrates: egg retention versus egg size reduction. *American Naturalist*, **133**, 138–48.

Duncker, H. R. (1989). Structural and functional integration across the reptile–bird transition: locomotor and respiratory systems. In *Complex Organismal Functions: Integration and Evolution in Vertebrates*, eds D. B. Wake and G. Roth, pp. 147–69, Dahlem Konferenzen. Chichester: John Wiley & Sons Ltd (in press).

Ewert, M. A. (1985). Embryology of turtles. In *Biology of the Reptilia*, vol. 14, Development A, eds C. Gans, F. Billett & P. F. Maderson, pp. 75–268. New York: John Wiley & Sons.

Ferguson, M. W. J. (1985). Reproductive biology and embryology of the crocodilians. In *Biology of the Reptilia*, vol. 14, Development A, eds C. Gans, F. Billett & P. F. Maderson, pp. 329–491. New York: John Wiley & Sons.

Gilbert, A. B. (1971). Transport of the egg through the oviduct and oviposition. In *Physiology and Biochemistry of the Domestic Fowl*, vol. 3, eds D. J. Bell & B. M. Freeman, pp. 1345–50. New York: Academic Press.

Gould, S. J. & Lewontin, R. C. (1979). The spandrels of San Marco and the panglossian paradigm: a critique of the adaptationist programme. *Proceedings of the Royal Society London*, **B205**, 581–98.

Grigg, G. C. & Harlow, P. (1981). A fetal-maternal shift of blood oxygen affinity in an Australian viviparous lizard, *Sphenomorphus quoyii* (Reptilia, Scincidae). *Journal of Comparative Physiology*, **142**, 495–9.

Guillette, L. J. Jr (1982). The evolution of viviparity

and placentation in the high elevation Mexican lizard *Sceloporus aeneus*. *Herpetologica*, **38**, 94–103.

Limpus, C. J., Baker, V. & Miller, J. D. (1979). Movement induced mortality of loggerhead eggs. *Herpetologica*, **35**, 335–8.

Mell, R. (1929). *Beitrage zur Fauna Sinica. IV. Grundzuge einer Okologie der chinesischen Reptilien und einer herpetologischen Tiergeographie Chinas*. Walter de Gruyter, Berlin.

Miller, K., Packard, G. C. & Packard, M. J. (1987). Hydric conditions during incubation influence locomotor performance of hatchling snapping turtles. *Journal of Experimental Biology*, **127**, 401–12.

Moffat, L. A. (1985). Embryonic development and aspects of reproductive biology in the tuatara, *Sphenodon punctatus*. In *Biology of the Reptilia*, vol. 14, Development A, eds C. Gans, F. Billett & P. F. Maderson, pp. 493–522. New York: John Wiley & Sons.

Neill, W. T. (1964). Viviparity in snakes: some ecological and zoogeographical considerations. *American Naturalist*, **48**, 35–55.

Packard, G. C. (1966). The influence of ambient temperature and aridity on modes of reproduction and excretion of amniote vertebrates. *American Naturalist*, **100**, 667–682.

Packard, G. C., Tracy, C. R. & Roth, J. J. (1977). The physiological ecology of reptilian eggs and embryos, and the evolution of viviparity within the Class Reptilia. *Biological Reviews*, **52**, 71–105.

Shine, R. (1983*a*). Reptilian viviparity in cold climates: testing the assumptions of an evolutionary hypothesis. *Oecologia*, **57**, 397–405.

— (1983*b*). Reptilian reproductive modes: the oviparity-viviparity continuum. *Herpetologica*, **39**, 1–8.

— (1985). The evolution of viviparity in reptiles: an ecological analysis. In *Biology of the Reptilia*, vol. 15, Development B, eds C. Gans & F. Billett, pp. 605–94. New York: John Wiley & Sons.

— (1989). Ecological influences on the evolution of vertebrate viviparity. In *Complex Organismal Functions: Integration and Evolution in Vertebrates*, eds D. B. Wake & G. Roth. pp. 263–78. Dahlem Konferenzen. Chichester: John Wiley & Sons Ltd (in press).

Shine, R. & Bull, J. J. (1979). The evolution of live-bearing in lizards and snakes. *American Naturalist*, **113**, 905–23.

Shine, R. & Guillette, L. J. Jr (1988). The evolution of viviparity in reptiles: a physiological model and its ecological consequences. *Journal of Theoretical Biology*, **132**, 43–50.

Stewart, J. R. (1989). Facultative placentotrophy and the evolution of squamate placentation: quality of eggs and neonates in *Virginia striatula*. *American Naturalist*, **133**, 111–37.

Tinkle, D. W. & Gibbons, J. W. (1977). The distribution and evolution of viviparity in reptiles. *Miscellaneous Publications of the Museum of Zoology, University of Michigan*, **154**, 1–55.

Vitt, L. J. & Blackburn, D. G. (1983). Reproduction in the lizard *Mabuya heathi* (Scincidae): a commentary on viviparity in New World *Mabuya*. *Canadian Journal of Zoology*, **61**, 2798–806.

23

Overview of early stages of avian and reptilian development

RUTH BELLAIRS

Introduction

Almost all the experimental analyses of the embryology of birds have been carried out on the domestic fowl (*Gallus gallus*), though sufficient work has been published on other domesticated species to indicate that there is little variation in the patterns of development, at least among the carinates. The situation is sadly different for reptiles, where there has been little experimental work, and even morphological studies are still needed for some of the most critical stages of certain groups. Some patterns emerge, but owing to the wide variation among different groups of reptiles we must maintain great caution in making generalisations.

This chapter is concerned principally with the patterns of development which are evident at the earliest stages, i.e. until the beginning of organogenesis. It is during this period that the basic events of embryogenesis take place and the patterns of development that emerge are fundamental for the establishment of the entire embryonic body. Important differences which occur at later stages and play an essential role in the divergence of body form are well covered in a series of reviews in *The Biology of the Reptilia* (Ewert, 1985; Ferguson, 1985; Hubert, 1985a,b; Moffat, 1985) and will not be discussed here.

The ovum and fertilisation

Birds and reptiles have many shared characteristics, not least of which is the possession of the cleidoic egg. The large yolk is formed from lipids and proteins which are transmitted from the maternal circulation through the follicle cells into the ovum (White, Chapter 1). There is

evidence that maternal cytoplasmic materials are also extruded into the ovum from the follicle cells. In the fowl, 'lining bodies', which are complex finger-like processes of follicle cell membranes, are pushed into the oocyte (Fig. 23.1) and appear to be engulfed and broken down (Bellairs, 1965). In certain reptiles, such as the lizards, *Acanthodactylus scutellatus* and *Chalcides ocellatus* (Bou-Resli, 1974; Ibrahim & Wilson, 1989); the turtle, *Pseudemys scripta* (Rahil & Narbitz, 1973) and the snake, *Natrix rhombifera* (Betz, 1963), the cell membrane of the oocyte becomes fused with those of certain follicle cells (Fig. 23.1), so that there is a continuity of the cytoplasm and it seems likely that, as in birds, maternal material passes into the oocyte. The significance of the transfer is not known, although one possibility is that it is associated with the passage of large amounts of RNA. There is evidence that, in most vertebrates, maternal RNA is deposited in the ovum and used in the earliest cleavage stages. In lizards, at least, it is possible that the entire cytoplasmic contents may pass from a follicle cell into the oocyte, so that nuclei as well as cytoplasmic organelles may play an active role in differentiation of the oocyte.

The yolk in a bird egg is covered by the vitelline membrane, a two-layered structure, of which the inner part is laid down in the ovary and the outer part in the oviduct immediately after fertilisation (Bellairs, Harkness & Harkness, 1963). An endogenous lectin which is found in the outer layer of the fowl vitelline membrane (Cook *et al.*, 1985) is also present not only in the outer layer of the vitelline membrane of other domestic species but also in that of crocodilians (unpublished data). The two layers

Follicle : oocyte relationship

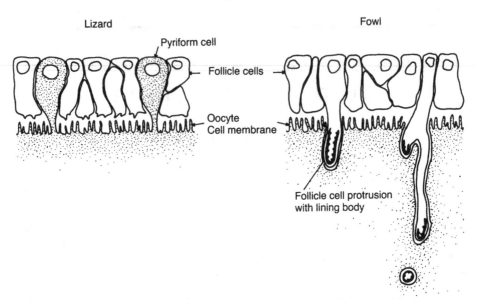

Fig. 23.1.   The relationship between follicle and oocyte in a typical lizard and the fowl. In lizards, the cell membrane of the oocyte becomes fused with that of certain follicle cells (pyriform cells). In the fowl, specialised regions of the follicle cells protrude into the oocyte and appear to become engulfed.

of the fowl vitelline membrane are each composed of fibres, but they differ in their structure and their composition (Bellairs *et al.*, 1963). The fine structure of the vitelline membrane does not appear to have been described yet for any reptile (but see Webb *et al.*, 1987*a*).

In all amniotes, fertilisation takes place high up the female genital tract. A common feature to birds, and many reptiles, is that sperm may be stored and remain viable in the mated female for long periods. This is unusual in mammals though it occurs in certain bats (Wimsatt, Krutsch & Napolitano, 1966). Such a facility has clear advantages for species scattered over large areas or which do not live in social groups. Sperm storage as a normal feature of reproduction has been claimed for many lizards, snakes and turtles (Fox, 1977; Saint-Girons, 1985) and has been cautiously suggested for crocodilians (Ferguson, 1985) and the tuatara (*Sphenodon punctatus*) (Moffat, 1985) but unfortunately, the evidence is not always clear-cut. A delay of several months between mating and egg-laying, which is sometimes considered to be evidence for delayed fertilisation, might equally be due to a delay in development of the fertilised egg, or be the result of parthenogenetic development.

The only satisfactory evidence must be a combination of the following: 1) The presence of sperm in the female tract some time after mating. Such an arrangement has been described for the fowl (Schindler *et al.*, 1967), where the sperm are closely packed in a sac at the junction of the so-called uterus and vagina. These authors suggest that the sperm might be nourished by the secretions from the glands. Similar sacs are present in the oviducts of certain lizards and snakes (Fox, 1956; Cuellar, 1966; Saint-Girons, 1975). Saint-Girons (1975) states that, in various iguanid and agamid lizards, the sperm are generally fixed to the uterine epithelial cells by their acrosomes, and suggests that they are nourished by the uterine wall. Sacs of this type have not been found in crocodilians (Ferguson, 1985). 2) Sperm must be capable of fertilisation after a prolonged interval in the oviduct. The fact that they may still be motile is not adequate evidence and it is necessary to establish that viable embryos can develop from ova fertilised by them. Sperm of the fowl remain viable for at least a week after mating (Lorenz, 1973) but there seems to be no comparable information for reptiles. 3) The possibility of parthenogenetic development must be eliminated by investigating the chromosomal composition of the resulting embryo. True parthenogenetic development is recorded in both birds and reptiles. Olsen (1965), using inbred Beltsville Small White turkeys, has obtained parthenogones which survive to hatching. All are male (the homogametic sex in birds) and are thought to have doubled their chromosomes by the retention of the second polar body. Parthenogenesis occurs in nature in a range of lizards (Darevsky, Kupriyanova & Uzzell, 1985) and the chromosomes appear to be diploid in 15 species and triploid in 13 taxa. These authors suggest that in many of these taxa parthenogenesis is associated with hybridity, which would appear to be opposite to that in turkeys (Olsen, 1965).

Primordial germ cells

Hubert (1985a) reviews the work on the primordial germ cells in reptiles and makes some comparisons with the situation in birds. Primordial germ cells have attracted much attention because of a controversy about their origin: do they arise *de novo* from somatic cells, or are they derived from germ plasm set aside in the fertilised egg? One of the major problems in settling this question is that it has often been difficult to identify the primordial germ cells with complete confidence.

In the fowl embryo germ cells were until recently identifiable only by staining for glycogen, using the technique of Meyer (1958) and could not be recognised clearly at stages earlier than the head process stage (stage 5 of Hamburger & Hamilton, 1951). However, specific carbohydrate antibodies have been found to act as markers (Loveless *et al.*, 1990), and these show that the primordial germ cells are present even in the embryo of the unincubated egg. Meanwhile, experimental work in which blastoderms have been transected has suggested that primordial germ cells originate from the epiblast (early ectoderm) (Ginsberg & Eyal-Giladi, 1987).

There appears to be no evidence as to whether the primordial germ cells of any reptile are stainable with these specific carbohydrate antibodies. Germ cells of both the amphibian *Xenopus laevis* and the laboratory mouse stain with similar carbohydrate antibodies (Heath & Wylie, 1981; Wylie *et al.*, 1986), so it is likely that reptile germ cells will also stain. All data

available for reptiles show that the primordial germ cells are first identifiable at a stage long after we might expect them to arise (reviewed by Hubert, 1985a). This is probably because of a failure to identify them any earlier. Whether they develop in the ectoderm or endoderm, and if they arise at the anterior or posterior end of the blastoderm are still points of contention. These discussions are likely to be irrelevant because the primordial germ cells of reptiles probably resemble those of the fowl and arise throughout the entire area pellucida, subsequently becoming more apparent at one end rather than the other.

Cleavage

In both birds and reptiles, cleavage takes place whilst the embryo is still in the female genital tract and the details are not well understood. Unlike the situation in amphibians, the cleavage furrows are irregular and their pattern does not appear to be related to the future orientation of the embryo. In both birds and reptiles, the early embryo takes the form of a disc on top of the yolk, and in the earliest stages the cells at the periphery are not fully closed from the yolk. Descriptions of the early cleavage stages exist for birds (Bellairs, Lorenz & Dunlap, 1978), turtles (Pasteels, 1937a; Ewert, 1985), the lizard *Lacerta vivipara* (Hubert, 1962, 1985b) and the snake *Vipera aspis* (Hubert & Dufaure, 1968), but nothing seems to be known about the earliest stages of crocodilians or the tuatara.

In both birds and those reptiles which have been studied, the open cells at the periphery of the embryo gradually become completely surrounded by cell membranes. In birds, the periphery of the disc now attaches to the inner surface of the vitelline membrane and the blastoderm expands centrifugally over the yolk. It is not normally attached elsewhere, but is liable to adhere over its entire upper surface if too much fluid evaporates from the egg. This usually leads to developmental anomalies, e.g. the neural tube may fail to roll up. The blastoderm of a snake or lizard adheres so closely over its entire dorsal surface to the vitelline membrane under normal conditions, that it is almost impossible to dissect it free from the membrane. In my opinion, this adhesion may be related to the fact that only a negligible amount of albumen is present in the eggs of lizards and snakes (Saint-Girons, 1985).

In birds, the embryo draws fluid from the albumen through the vitelline membrane and deposits it in a subgerminal cavity (New, 1956). A similar transport of fluid probably occurs in crocodilians (Webb *et al.*, 1987*a*,*b*). The formation of the subgerminal cavity in birds corresponds with the appearance of the area pellucida. No subgerminal cavity is illustrated by Hubert (1962) in his histological study of cleavage stages of *L. vivipara*, but a more clearly defined cavity develops in certain turtles and an area opaca and area pellucida become identifiable. In birds, the formation of the area pellucida is linked to the establishment of the antero-posterior polarity.

Polarity

The establishment of *dorso-ventral polarity* is associated with yolk throughout the animal kingdom, with the exception of those groups which almost or totally lack yolk (e.g. eutherian mammals). The yolk is heavier than the cytoplasm and so sinks to the lower side of the egg whilst the cytoplasm rises to the upper surface. In the large yolked eggs of birds and reptiles, the dorsal side of the blastoderm is thus the furthest away from the yolk, whilst the ventral side rests on the surface of the yolk. It appears likely, therefore, that gravity is responsible for this initial polarity. In birds, the presence of the large amount of albumen ensures that, if the egg is turned, the yolk rotates within the shell and the embryo returns to its dorsal position (New, 1957). Webb *et al.* (1987*b*) suggest that a similar effect occurs with the crocodilian egg. In my opinion, the paucity or absence of albumen in the eggs of lizards and snakes makes this recovery unlikely. Ultimately, the ability of a bird embryo to adjust its position in the egg is reduced as the chorio-allantoic membranes become fixed to the inside of the shell. A similar process takes place in turtles and crocodilians and is indicated by the appearance of 'chalking' on the outer surface of the shell (Ewert, 1985; Ferguson, 1985).

There is some evidence that gravity may be important also in the establishment of the *cranio-caudal polarity*. Towards the end of cleavage, a thicker area, Köller's sickle (Fig. 23.3), becomes apparent at one side of the area pellucida, and this marks the future caudal end, both in birds (Bellairs, 1971) and in reptiles (Pasteels, 1937*a*). If this region is destroyed in the fowl embryo, two similar regions form, one on either side, and

two embryonic axes develop (Spratt & Haas, 1961). If the entire embryo is cut into two or more pieces, two or more embryonic axes are formed and these axes are not necessarily orientated in the same direction (Lutz, 1965). A similar situation exists in *L. vivipara* which if it is cut in two will form twinned axes (Hubert, 1964). It seems, therefore, that the establishment of polarity is a gradual process which may not be completed by the end of cleavage.

In both birds and reptiles, the critical period appears to be passed in the lower part of the oviduct. It has long been recognised that, in birds, there is a relationship between the shape of the laid egg and the orientation of the developing embryo. Thus, the long axis of the embryo usually lies at right angles to the long axis of the egg, with the embryo's future right side being nearer the more pointed end of the egg: this Rule of von Baer (Fig. 23.2) applies in all domestic species of bird investigated (Clavert, 1963; Fargeix, 1963). In general, the adherence to, or deviation from, the rule appears to be consistent for individual females. Clavert (1963) suggests that in reptiles too, the Rule of von Baer generally applies; a similar relationship between embryonic axis and egg axis has been noted for the lizards *Camaesaura anguinea*, *Chamaeleo bitaeniatus*, *Lacerta stirpium* and *Anguis fragilis* (Pasteels, 1937*a*; Clavert & Zahnd, 1955; Ancel & Raynaud, 1959), and for various, though not all, turtles (Ewert, 1985).

The eggs of both birds and reptiles spend a comparatively long time in the uterus and there is evidence that they undergo rotation during this period. This rotation coincides with a tilting of the egg, so that one end of the blastoderm comes to lie at a lower level than the other. The lower end becomes the cranial end of the embryo, the higher end the caudal. Using *A. fragilis*, Raynaud (1962) cut windows in the uterine wall and observed these rotations directly. Pasteels (1970), in reviewing the work of several authors on reptile embryos, concluded 'the orientation of embryos corresponds to a rotation of the egg around its long axis in the uterus . . .'. It was suggested therefore that this rotation was the prime event in the establishment of the cranio-caudal axis, and that it was responsible for the tilting of the egg.

This view has been challenged as a result of experiments on the cleavage stages of the fowl (Kochav & Eyal-Giladi, 1971; Eyal-Giladi & Fabian, 1980). In these experiments an egg was

The rule of von Baer

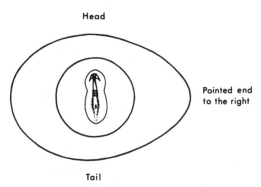

Fig. 23.2. The Rule of von Baer. When a bird's egg is held with its pointed end to the right, the embryo lies at right-angles to the main axis of the egg, with its head furthest from the observer.

removed from the hen shortly after it had entered the uterus and was suspended by the chalazae in a bath of saline. The head developed at the lower end of the tilted blastoderm and the tail at the higher end, but rotation was not essential, although it is the means by which tilting is normally brought about. They concluded that gravity was the prime agent in establishing cranio-caudal polarity, however, although they were unable to explain the phenomenon. Meanwhile, Callebaut (1983) has shown that in the eggs of the quail there are four different types of yolky cytoplasm present at this time, one of which, β-ooplasm, becomes associated with Köller's sickle. It seems likely that with changes in position, this ooplasm always moves to the highest point. This being so, then the implications are that in the absence of gravity, normal development will not occur. This is a proposition which can be put to the test only in a spacecraft, and such an experiment is now within the realms of possibility (Bellairs, 1990).

The rotation of the ovum in the oviduct may subserve some quite different function, such as preventing adhesions between the blastoderm and the uterine wall, a possibility not testable with the experiments of Kochav & Eyal-Giladi (1971). Alternatively it may ensure that the uterine secretions which form the shell are spread evenly over the surface of the shell membranes (Palmer & Guillette, Chapter 3; Packard & DeMarco, Chapter 5; Board & Sparks, Chapter 6).

Gastrulation

Gastrulation is the process by which endoderm and mesoderm are formed and which results in an embryo with several layers. It involves the migration of cells which converge on a specific region of the young embryo. This region may take the form of a blastopore, as is the case in amphibians, or a primitive streak, as is the case in birds.

Some reptiles have both a blastopore, like amphibians, and a primitive streak, like birds. Therefore, is gastrulation in reptiles a half-way house between that in amphibians and birds? In most groups of reptiles (Fig. 23.3) the blastopore arises near the posterior border of the area pellucida. In birds and mammals, the blastopore is replaced by the primitive streak. Blastopores have occasionally been described at the posterior end of the primitive streak of birds (Patterson, 1909) though it is usually considered that these 'blastopores' are the result of damage to the embryo. Certainly, there is normally no true blastopore at the caudal end of the primitive streak in the fowl embryo. However, it is relevant to this discussion that the presumptive areas of the fowl embryo are initially derived from the caudal end of the area pellucida and are then displaced cranialwards by migration (Fig. 23.4). It is this forward movement which causes the primitive streak to appear.

Morphological studies tell us little, other than whether we are dealing with a blastopore and its associated archenteron, or a primitive streak. There are many examples of investigators being led astray by trying to make deductions about cell movements from the morphological appearance of a group of cells. The only reliable method is to label groups of cells and follow their displacements. Studies of this type have been carried out in the chick embryo by several authors (Pasteels, 1937a; Rosenquist, 1966). Although they may differ in the details of their descriptions, there is general agreement on the overall patterns of migration. In reptiles, only one good experimental analysis of gastrulation movements exists: that of Pasteels (1937b, 1956) on the turtle, *Clemmys leprosa*.

An early gastrula of *C. leprosa*, and an early fowl gastrula at a comparable stage are illustrated in Fig. 23.5. In each case there is a *convergence* of cells towards the midline at the caudal end of the area pellucida. Indeed, the early movements are very similar in the two

Lizards
and snakes

Turtles

Crocodilians

Birds

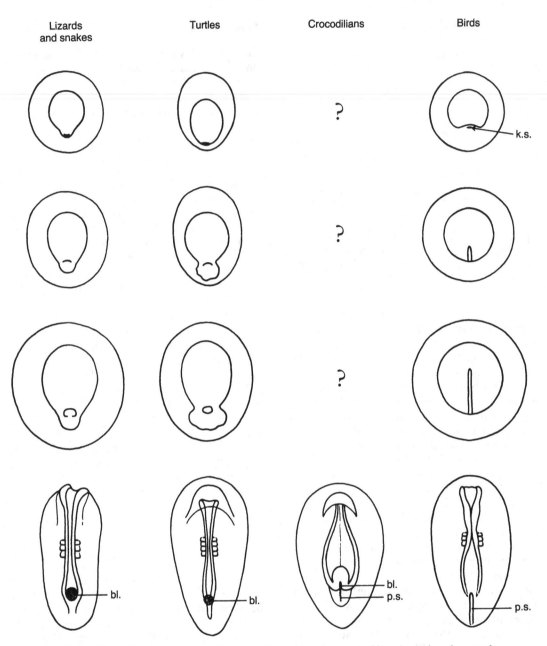

— k.s.

?

?

?

— bl.

— bl.

— bl.
— p.s.

— p.s.

Fig. 23.3.   The table shows comparable stages in the development of lizards, and snakes, turtles, crocodilians and birds. In the earliest stages illustrated (*top row*) each embryo consists of a central region, the area pellucida, and a peripheral region, the area opaca. Köller's sickle (k.s.) marks the future caudal end of the fowl embryo. Gastrulation takes place through a blastopore (bl.) in lizards, snakes and turtles, but through a primitive streak (p.s.) in birds, whilst both a blastopore and a primitive streak are present in crocodilians.

animals but there is, however, a most significant difference. The midline movement from anterior to posterior seen in the turtle is absent in the fowl. The absence of this movement at

this stage may be the main reason for the shift from a blastopore to a primitive streak.

Once the cell movements start in the turtle embryo, then medial, posteriorly directed migra-

Fate maps for mesoderm

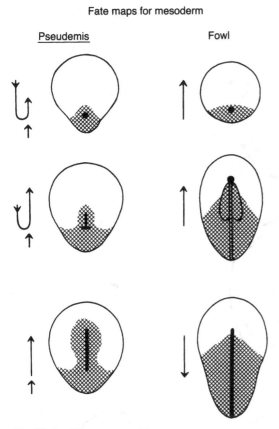

Pseudemis   Fowl

Fig. 23.4. Diagram to illustrate the cranialward migration of the presumptive mesoderm in a turtle, *Pseudemis virginica*, and in the fowl (after Pasteels, 1937*a,b*, 1970). In each case, the mesoderm (hatched) appears first at the caudal end and spreads cranialwards in succeeding stages. In the turtle, there is initially a caudalward migration of cells on the upper surface of the embryo, followed by a forward migration after the cells have invaginated through the blastopore; these movements are indicated by the curved arrows at the left of the diagram. In the fowl, there is no caudalward movement in the early stages and only a cranialward movement is indicated by the arrows. The caudalward movement is however not lost in the fowl but occurs later in development as regression takes place (final diagram). The thick black lines indicate the developing notochord.

tions balance those from the lateral and caudal ends and they all become focused on one region (Fig. 23.5). Inevitably, the cells must invaginate, or in abnormal cases bulge upwards and exogastrulate. At first, the dorsal lip of the blastopore is a crescent-shaped groove, as in amphibians, but gradually, as the lateral and posterior material migrates towards this region, the lateral and ventral lips are formed (as in amphibians). The embryonic body develops anterior to this region.

In birds, it is as if the *absence* of the medial, posteriorly directed movements at this stage leads to an imbalance of cell migration. This results in a dominance of the migratory movements from the posterior end, so that material is pushed along in a trough on the surface instead of invaginating (Fig. 23.5). Cells coming in from the sides exacerbate the situation, so that the dorsal surface of the blastoderm becomes raised up in folds, i.e., the sides of the primitive streak. These can perhaps be regarded as excessively elongated lateral lips of a potential blastopore. Once the cells stop migrating toward the midline, then there is no further migration forward of surface material of the developing primitive streak. After a short pause, material now migrates beneath the surface to form the head process, and at the same time a posteriorly directed movement takes place. This so-called *regression* has been discussed in detail by Bellairs (1971, 1986). It is suggested here that regression is a resurgence of the caudalward movement which was lost at the earlier stage. If this is so, then it would appear that from the evolutionary point of view, it is the change in balance of the migratory movements which has shifted the birds from having a blastopore to having a primitive streak.

Unfortunately, we have no direct evidence as to the gastrulation movements in crocodilians (Fig. 23.5). From the embryological point of view, the possession of both blastopore and primitive streak in this group has been difficult to understand (Bellairs, 1971), since other groups have either one or the other. A possible explanation is that the strength of the posteriorly directed movements has been reduced in these animals in the early stages of gastrulation, though not so greatly as in birds. If the balance of cell migrations has thus been disturbed, it might be expected that the blastopore would be shifted cranially without its being completely abolished (Figs 23.3 & 23.5). In the process, however, the cells caudal to it would become compressed and raised into folds by the migration of additional tissue from the sides of the area pellucida. Thus a short primitive streak would form. It would be especially interesting to have further information on gastrulation in crocodilians in view of their close similarity to birds, both groups being of archosaurian stock.

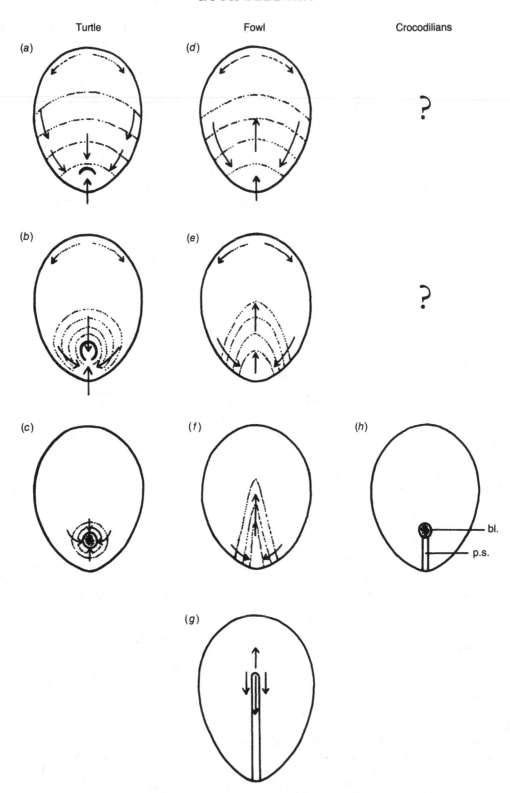

Fig. 23.5. Comparison of the gastrulation movements of the turtle, *Clemmys leprosa* (after Pasteels, 1937a) and the fowl (after Pasteels, 1937b). Dotted lines indicate compression in the upper layer.

An important aspect of gastrulation is the question of how the endoderm forms. Formerly, it was thought that the endoderm developed in birds before the appearance of the primitive streak and was not derived from material that ingressed. The arguments circulated around the question of whether it formed by polyingression or delamination and were based on the study of cell shapes in sectioned material. This misapprehension was corrected by Rosenquist (1966) who traced the passage of labelled cells through the primitive streak and into the endoderm. The difficulty had arisen because at the outset of gastrulation, at the time that the egg is laid, there are already two layers of cells, and the lower one, which is rather yolky, was thought to be the embryonic endoderm. Rosenquist (1966) showed that this layer was the hypoblast, the endoderm of the future yolk sac stalk; this becomes invaded by the true embryonic endoderm at the time when the mesoderm begins to ingress through the primitive streak. There is evidence that a similar situation exists in mammals (Lawson & Pedersen, 1987).

There is no direct evidence as to how the embryonic endoderm forms in reptiles. An extensive literature exists which is based largely on morphological studies; Pasteels (1956), using fixed material, concluded that the endoderm formed in turtles by gastrular invasion and in squamates by delamination. It seems probable, however, that if, and when, some reliable marking experiments are carried out it will be found that all reptiles conform to the situation in fishes, amphibians, birds and mammals and form their endoderm by invagination or ingression at the same time as they form their mesoderm.

The embryonic axis

Even before gastrulation is completed in amniotes, the formation of the embryonic axis is underway. The amniotes, therefore, differ from amphibians where gastrulation is virtually finished before axis formation begins. The main tissues of the embryonic axis, the neural tube, the notochord, paired somites and lateral plate, are laid down in all amniotes in a cranio-caudal direction and the general patterns of formation seem to differ little between birds and reptiles. By the time that the primitive streak is fully formed in birds the presumptive areas of cells are grouped around Hensen's node. Some cells are still on the surface but others have begun to ingress and contribute to the layer of mesoderm. The mesodermal cells immediately on each side of the midline become arranged as a block of tissue called the segmental plate. As the ingression process continues caudalwards in birds, more and more mesoderm from the primitive streak is added to the caudal end of each segmental plate. Meanwhile the somites form pair by pair by separating off from the cranial ends of the two segmental plates. In the fowl, a new pair of somites is formed about every 100 minutes. The result is that the length of the segmental plates remains fairly constant during the time that the somites are forming.

The origin of the segmental plate mesoderm in reptiles is less clear, but since there is no primitive streak (apart from the short one in crocodiles), then it seems likely that it forms from the mesoderm which has invaginated through the blastopore at an earlier stage. The situation is probably not unlike that in amphibians, except that the latter do not develop a segmental plate but form their somites directly from the invaginated mesoderm.

Much attention has been paid to the way in which the segmentation of the somites takes place from the segmental plates. One of the most significant findings was that of Meier (1979) who examined scanning electron micrographs of the segmental plate of the fowl embryo in stereo and was able to recognise a segmental pattern. This consisted of groups of cells, 'somitomeres', each of which corresponded with a future somite. Each somitomere consisted of concentric circular arrangements. Not everyone was convinced by his initial examples, but sufficient evidence has now accumulated for most workers in the field to accept the validity of his conclusions. In the fowl

---

Fig. 23.5 *contd.*

The major difference in the patterns of migration is that the midline movement from anterior to posterior seen in turtles is absent at this stage in the fowl. This absence, which may be important in the evolutionary shift from gastrulation by a blastopore to gastrulation by a primitive streak, does not appear to be total, since a cranio-caudal migration of cells occurs in bird embryos during regression (Fig. 23.4) at the end of gastrulation. The early gastrulation of crocodilians is not known, but this group appears to be intermediate in type, having both blastopore and primitive streak.

and quail, about 11 somitomeres were identified in each segmental plate (Packard & Meier, 1984) and this corresponded with the number of somites which formed if a segmental plate is isolated in organ culture (Packard & Jacobson, 1976).

A similar situation exists in the snapping turtle *Chelydra serpentina*, which appears to be the only reptile whose segmental plates have been studied in this way (Packard & Meier, 1984). These somitomeres closely resemble those found in birds but are fewer in number. There are about seven somitomeres in each segmental plate and this corresponds with the number of somites which develop if the segmental plate is isolated (Packard, 1980).

The number of somites which form in any species is important because they are the first segmental structures to develop in an embryo and therefore set the pattern for all the other segmental structures which follow, e.g. the vertebrae, the segmental nerves and the intersegmental arteries. The number of somites must be at least equal to, and usually more than, the number of vertebrae which will form from them. The precise number of somites which develop is not easy to ascertain because by the time the last ones have segmented along the tail, the first ones which had formed in the trunk have undergone further development and differentiated into dermatome, myotome and sclerotome. Nevertheless, in birds about 50–52 pairs normally develop (Bellairs, 1986) but the numbers do not appear to have not been counted for any reptile.

The mechanism by which the correct number of somites is formed is not fully understood, though a number of theories exist which attempt to explain the process (Bellairs, 1986; Primmett *et al.*, 1988). An important feature is that the somites never extend to the tip of the tail, a small band of segmental plate-like mesoderm remaining unsegmented. This mesoderm may provide a reservoir of material which may enable the final number of somites to be regulated by the embryo during development. In birds at least, somitomeres develop in this terminal mesoderm (Bellairs & Sanders, 1986) although they do not normally complete their segmentation. This appears to be related to two major factors: the cell surface receptors in this region differ from those in the trunk segmental plate (Mills, Lash & Bellairs, 1990); and the segmentation process is interrupted and overcome by a wave of cell death which spreads from the tip towards the

trunk and gradually overtakes the segmenting process (Mills & Bellairs, 1989).

Unfortunately, the situation in the tails of reptile embryos is as yet unknown, but it seems likely that a similar mechanism also exists. The number of vertebrae varies greatly among reptiles (Hoffstetter & Gasc, 1969), and shifts in the balance of forces promoting segmentation, and in those leading to cell death, may have played a role during evolution in regulating the shape and proportions of the vertebral column.

Limbs and limblessness

The prominent position of the limb buds on the body surface and their easy access for experimentation has made them favoured structures for embryologists and the focus of many theories of development. The subject has been reviewed extensively and will not be considered in detail here. Suffice it to point out that the developing limb buds of reptiles (Raynaud, 1985) present many of the same features as those of birds, but that in those reptiles which possess vestigial limbs or no limbs in the adult (e.g. *A. fragilis*) there are significant differences. Thus, although vestigial limb buds form in *A. fragilis* (Raynaud, 1962) the apical ridge is poorly developed and degenerates at an early stage.

Staging

A Normal Table consists of a number of arbitrarily selected stages of development, with a brief description and illustration of the major external features of each stage (see Deeming & Ferguson, Chapter 10). Good Normal Tables are essential for any species on which sustained embryological investigations are carried out because they provide a standardised framework which permits consistency of staging between different authors.

The most widely used Normal Tables for birds is that of Hamburger & Hamilton (1951) and this has been supplemented by Eyal-Giladi & Kochav (1976) who have inserted additional stages for cleavage and early gastrulation. Amongst the most useful for representatives of the reptile groups are the following: lizards; e.g. *Lacerta vivipara* and *Anguis fragilis* (Dufaure & Hubert, 1961; Raynaud, 1961); snakes, e.g. *Vipera berus* (Hubert & Dufaure, 1968); crocodilians: e.g. *Alligator mississippiensis* (Ferguson, 1985); turtles: e.g. *Chelydra serpentina* and several sea turtles (Yntema, 1968; Miller, 1985); and the tuatara, *Sphenodon punctatus* (Moffat, 1985).

It must be emphasised that stages are arbitrarily allocated by the authors and attempts to co-ordinate the numbering between the different groups of reptiles have not been successful. This is not only because the time spent in each stage varies, but also because the external characteristics do not always appear in precisely the same order or combination in each group. Moreover, in the early stages, one of the key features is the number of somites which have formed, and this varies greatly from one group to another. For example, the forelimb buds appear in the fowl embryo when about 28 pairs of somites have developed, and the hind limbs when about 30 have formed, and the forelimbs are always in advance of the hind limbs (Hamburger & Hamilton, 1951). By contrast, the hind limbs of crocodilians appear when about 40 pairs of somites have formed, whilst the forelimb buds do not appear until later (Ferguson, 1985).

Concluding remarks

Our understanding of early development is much greater for birds than for any reptile, due primarily to the accessibility of fowl eggs. Little experimental work has been carried out on the early stages of reptiles and even morphological studies are sparse. Some out-of-date controversies about development in birds have been abandoned as a result of recent work (e.g. those concerned with the establishment of polarity and the origin of the endoderm). Similar controversies still exist for reptiles but will probably be found to be redundant when modern investigations have been carried out. The study of avian development can therefore help to focus future enquiries into reptile development. Conversely, the study of reptilian embryos may provide the most important key to understanding the evolution of the blastopore into the primitive streak. Perhaps the biggest challenge of all is to unravel the mysteries of crocodilian gastrulation.

Acknowledgements

I am most grateful to Mrs R. Cleevely for drawing the figures, and to A. Bellairs, E. J. Sanders and L. Vakaet for helpful discussions. I acknowledge assistance from the British Heart Foundation and the National Kidney Research Fund.

References

Ancel, P. & Raynaud, A. (1959). Observations préliminaire sur l'orientation de l'embryon dans l'oeuf d'Orvet (*Anguis fragilis* L.). *Comptes Rendus de L'Académie des Sciences*, 248, 1086–90.

Bellairs, R. (1965). The relationship between oocyte and follicle in the hen's ovary as shown by electron microscopy. *Journal of Embryology and Experimental Morphology*, 13, 215–33.

— (1971). *Developmental Processes in Higher Vertebrates*. London: Logos Press.

— (1986). The tail bud and cessation of segmentation. In *Somites in Developing Embryos*, eds R. Bellairs, D. A. Ede & J. W. Lash, pp. 161–78, vol. 118 NATO ISI series A. London: Plenum Press.

— (1990). The role of gravity in bird development. *Symposium of the European Space Agency*. In: *Microgravity as a Tool in Developmental Biology* (ed. T. Duc Guyenne) sp-1123, 59–63.

Bellairs, R., Harkness, M. & Harkness, R. D. (1963). The vitelline membrane of the hen's egg: a chemical and electron microscopical study. *Journal of Ultrastructure Research*, 8, 339–59.

Bellairs, R., Lorenz, F. W. & Dunlap, T. (1978). Cleavage in the chick embryo. *Journal of Embryology and Experimental Morphology*, 43, 55–69.

Bellairs, R. & Sanders, E. J. (1986). Somitomeres in the chick tail bud: an SEM study. *Anatomy and Embryology*, 175, 235–40.

Betz, T. W. (1963). The ovarian histology of the diamond-backed watersnake, *Natrix rhombifera* during the reproductive cycle. *Journal of Morphology*, 113, 245–60.

Bou-Resli, M. (1974). Ultrastructural studies on the intercellular bridges between the oocyte and follicle cells in the lizard *Acanthodactylus scutellatus* Hardyi. *Zeitschrift für Anatomie und Entwicklungsgeschichte*, 143, 239–54.

Callebaut, M. (1983). The constituent oocytal layers of the avian germ and the origin of the primordial germ cell yolk. *Archives d'anatomie microscopique et de morphologie expérimentale*, 72, 199–214.

Clavert, J. (1963). Symmetrization of the egg of vertebrates. *Advances in Morphogenesis*, 2, 27–60.

Clavert, J. & Zahnd, J. P. (1955). Sur le determinisme de la symétrie bilatérale et la regle de von Baer chez les reptiles. *Comptes Rendus de la Societé de Biologie*, 149, 1650–1.

Cook, G. M. W., Bellairs, R., Rutherford, N. G., Stafford, C. A. & Alderson, T. (1985). Isolation, characterization and localization of a lectin within the vitelline membrane of the hen's egg. *Journal of Embryology and Experimental Morphology*, 90, 389–407.

Cuellar, O. (1966). Oviducal anatomy and sperm storage in lizards. *Journal of Morphology*, 119, 7–17.

Darevsky, I. S., Kupriyanova, L. A. & Uzzell, T. (1985). Parthenogenesis in reptiles. In *Biology of the Reptilia*, vol. 15, Development B, eds C. Gans & F. Billett, pp. 411–526. New York: John Wiley & Sons.

Dufaure, J. P. & Hubert, J. (1961). Table de développement du lézard vivipare *Lacerta (Zootoca) vivipara* Jacquin. *Archives d'anatomie microscopique et de morphologie expérimentale*, **50**, 309–28.

Ewert, M. A. (1985). Embryology of turtles. In *Biology of the Reptilia*, vol. 14, Development A, eds C. Gans, F. Billett & P. F. A. Maderson, pp. 75–269. New York: John Wiley & Sons.

Eyal-Giladi, H. & Kochav, S. (1976). From cleavage to primitive streak formation: a complementary normal table and a new look at the first stages of the development of the chick. *Developmental Biology*, **49**, 321–37.

Eyal-Giladi, H. & Fabian, B. C. (1980). Axis determination in uterine chick blastodiscs under changing spatial positions during the sensitive period for polarity. *Developmental Biology*, **77**, 228–32.

Fargeix, N. (1963). L'orientation dominante de l'embryon de la Caille domestique (*Coturnix coturnix japonica*) et le règle de von Baer. *Comptes rendus des séances de la Société de Biologie*, **157**, 1431–4.

Ferguson, M. W. J. (1985). Reproductive biology and embryology of the crocodilians. In *Biology of the Reptilia*, vol. 14, Development A, eds C. Gans, F. Billett & P. F. A. Maderson, pp. 329–492. New York: John Wiley & Sons.

Fox, H. (1977). The urinogenital system in reptiles. In *Biology of the Reptilia*, vol. 6, eds C. Gans & T. S. Parsons, pp. 1–158. New York: Academic Press.

Fox, W. (1956). Seminal receptacles in snakes. *Anatomical Record*, **124**, 519–40.

Ginsburg, M. & Eyal-Giladi, H. (1987). Primordial germ cells of the young chick blastoderm originate from the central region of the area pellucida irrespective of the embryo-forming process. *Development*, **101**, 209–19.

Hamburger, V. & Hamilton, H. H. (1951). A series of normal stages in the development of the chick embryo. *Journal of Morphology*, **88**, 49–92.

Heath, J. & Wylie, C. C. (1981). Cell surface molecules of mammalian foetal germ cells. In *Development and Function of Reproductive Organs*, eds A. G. Byskov & H. Peters, pp. 83–92. Amsterdam: Excerpta Medica.

Hoffstetter, R. & Gasc, J. P. (1969). Vertebrae and ribs of modern reptiles. *Biology of the Reptilia*, vol. 1, ed. C. Gans, pp. 201–301. New York: Academic Press.

Hubert, J. (1962). Etude histologique des jeunes stades du développement embryonnaire du lézard vivipere. (*Lacerta vivipara* Jacquin). *Archives d'anatomie microscopique et de morphologie expérimentale*, **51**, 11–26.

— (1964). Essais de fissuration de l'oeuf du lézard vivipara (*Lacerta vivipara* Jacquin). *Comptes rendus des séances de la Société de biologie*, **158**, 523–5.

— (1985*a*). Origin and development of oocytes. In *Biology of the Reptilia*, vol. 14, Development A, eds C. Gans, F. Billett & P. F. A. Maderson, pp. 41–74. New York: John Wiley & Sons.

— (1985*b*). Embryology of the Squamata. In *Biology of the Reptilia*, vol. 15, Development B, eds C. Gans & F. Billett, pp. 1–34. New York: John Wiley & Sons.

Hubert, J. & Dufaure, J. P. (1968). Table de développement de la vipère aspic: *Vipera aspis*. *Bulletin de la Société de Zoologie de France*, **93**, 135–48.

Ibrahim, M. M. & Wilson, I. B. (1989). Light and electron microscope studies on ovarian follicles in the lizard *Chalcides ocellatus*. *Journal of Zoology, London*, **218**, 187–208.

Kochav, S. & Eyal-Giladi, H. (1971). Bilateral symmetry in chick embryo determination by gravity. *Science*, **171**, 1027–9.

Lawson, K. A. & Pedersen, R. A. (1987). Cell fate, morphogenetic movement and population kinetics of embryonic endoderm at the time of germ layer formation in the mouse. *Development*, **101**, 627–52.

Lorenz, F. W. (1973). Influence of age on avian gametes. In *Aging Gametes*, ed. R. J. Blandau, pp. 278–99. Basel: S. Karger.

Loveless, W., Bellairs, R., Thorpe, S. J., Page, M. & Feizi, T. (1990). Developmental patterning of the carbohydrate antigen FC10.2 during early embryogenesis in the chick. *Development* **108**, 97–106.

Lutz, H. (1965). Symétrisation de l'oeuf d'oiseau. *Annals de la Faculté des Sciences, Marseille*, **26**, 71–84.

Meier, S. (1979). Development of the chick embryo mesoblast. Formation of the embryonic axis and establishment of the metameric pattern. *Developmental Biology*, **73**, 25–45.

Meyer, D. B. (1958). The application of histochemical reactions to whole chick embryos. *Anatomical Record*, **130**, 339.

Miller, J. D. (1985). Embryology of marine turtles. In *Biology of the Reptilia*, vol. 14, Development A, eds C. Gans, F. Billett & P. F. A. Maderson, pp. 269–328. New York: John Wiley & Sons.

Mills, C. L. & Bellairs, R. (1989). Mitosis and cell death in the tail of the chick embryo. *Anatomy and Embryology*, **180**, 301–8.

Mills, C. L., Lash, J. W. & Bellairs, R. (1990). Evidence for the involvement of the VLA/integrin family of receptors in the promotion of chick tail segmentation. *Anatomy and Embryology* (in press).

Moffat, L. A. (1985). Embryonic development and

aspects of reproductive biology in the Tuatara, *Sphenodon punctatus*. In *Biology of the Reptilia*, vol. 14, Development A, eds C. Gans, F. Billett & P. F. A. Maderson, pp. 493–522. New York: John Wiley & Sons.

New, D. A. T. (1956). The formation of the sub-blastodermic fluid in hens' eggs. *Journal of Embryology and Experimental Morphology*, 4, 221–7.

— (1957). A critical period for the turning of hens' eggs. *Journal of Embryology and Experimental Morphology*, 5, 293–9.

Olsen, M. W. (1965). Delayed development and atypical cellular organization in blastodiscs of fertilized turkey eggs. *Developmental Biology*, 12, 1–14.

Packard, D. S. (1980). Somite formation in cultured embryos of the snapping turtle, *Chelydra serpentina*. *Journal of Embryology and Experimental Morphology*, 59, 113–19.

Packard, D. S. & Meier, S. (1984). Morphologcal and experimental studies of the somitomeric organisation of the segmental plate in snapping turtle embryos. *Journal of Embryology and Experimental Morphology*, 84, 35–48.

Packard, D. S. & Jacobson, A. G. (1976). The influence of axial structures on chick somite formation. *Developmental Biology*, 53, 36–48.

Pasteels, J. (1937a). Etudes sur la gastrulation des vertebrées meroblastiques. II. Reptiles. *Archives de Biologie*, 48, 105–84.

— (1937b). Etudes sur la gastrulation des vertebrées meroblastiques. III. Oiseaux. *Archives de Biologie*, 48, 381–488.

— (1956). Une table analytique du développement des reptiles. I. Stades de gastrulation chez les Chéloniens et les Lacertiliens. *Annales de la Société Royale Zoologique de Belgique*, 87, 217–41.

— (1970). Développement embryonnaire. In *Traité de Zoologie XIV*, fascicule III ed. P. Grassé, pp. 893–971. Parso, Masson et Cie.

Patterson, J. T. (1909). Gastrulation in the pigeon's egg. A morphological and experimental study. *Journal of Morphology*, 20, 65–124.

Primmett, D. R. N., Norris, W. E., Carlson, G. J., Keynes, R. J., Schlesinger, M. J. & Stern, C. D. (1988). Periodic segmental anomalies induced by heat shock and other treatments in the chick embryo are associated with the cell division cycle. *Development*, 105, 119–30.

Rahil, K. S. & Narbaitz, R. (1973). Ultrastructural studies on follicular cell oocyte relationship in the turtle *Pseudemys scripta*. *Anatomical Record*, 175, 419.

Raynaud, A. (1961). Quelques phases du développement des oeufs chez l'orvet (*Anguis fragilis* L.).

Bulletin biologique de la France et de la Belgique, 65, 365–82.

— (1962). Lés ebauches des membres de l'embryon d'orvet (*Anguis fragilis* L.). *Comptes Rendus de L'Académie des Sciences*, 254, 3449–51.

— (1985). Development of limbs and embryonic limb reduction. In *Biology of the Reptilia*, vol. 15, Development B, eds C. Gans & F. Billett, pp. 59–148. New York: John Wiley & Sons.

Rosenquist, G. C. (1966). A radioautographic study of labeled grafts in the chick blastoderm: development from primitive streak stages to stage 12. *Contributions to Embryology*, 38, 111–21.

Saint-Girons, H. (1975). Sperm survival and transport in the female genital tract of reptiles. In *The Biology of Spermatozoa*, eds E. S. E. Hafez & C. G. Thibault, pp. 106–13. Basel: S. Karger.

— (1985). Comparative data on lepidosaurian reproduction and some time tables. In *Biology of the Reptilia*, vol. 15, Development B, eds C. Gans & F. Billett, pp. 35–58. New York: John Wiley & Sons.

Schindler, H., Ben-David, E., Hurwitz, S. & Kempenich, O. (1967). The relation of spermatozoa to the glandular tissue in the storage sites of the hen's oviduct. *Poultry Science*, 46, 1462–71.

Spratt, N. T. & Haas, H. (1961). Integrative mechanisms in development of the early chick blastoderm. *Journal of Experimental Zoology*, 147, 57–94.

Webb, G. J. W., Manolis, S. C., Whitehead, P. J. & Dempsey, K. E. (1987a). The possible relationship between embryo orientation, opaque banding and the dehydration of albumen in crocodile eggs. *Copeia*, 1987, 252–7.

Webb, G. J. W., Manolis, S. C., Dempsey, K. E. & Whitehead, P. J. (1987b). Crocodilian eggs: a functional overview. In *Wildlife Management: Crocodiles and Alligators*, eds G. J. W. Webb, S. C. Manolis & P. J. Whitehead, pp. 417–22. Sydney: Surrey Beatty & Sons.

Wimsatt, W. A., Krutsch, P. H. & Napolitano, L. (1966). Studies on sperm survival mechanisms in the female reproductive tract of hibernating bats. I Cytology and ultrastructure of intra-uterine spermatozoa in *Myotis lucifugis*. *American Journal of Anatomy*, 119, 25–60.

Wylie, C. C., Scott, D. & Donovan, P. J. (1986). Primordial germ cell migration. In *Developmental Biology a comprehensive synthesis*. vol. 2, *The Cellular Basis of Morphogenesis*, ed. L. W. Browder, pp. 433–48. New York: Plenum Press.

Yntema, C. L. (1968). A series of stages in the embryonic development of *Chelydra serpentina*. *Journal of Morphology*, 125, 219–52.

24

Ions and ion regulating mechanisms in the developing fowl embryo

JAMES I. GILLESPIE, JOHN R. GREENWELL,
ELSPETH RUSSELL AND CLAIRE J. DICKENS

Introduction

This chapter outlines one approach to the study of the basic physiology of the early avian embryo, but, due to limitations of space this has to be a restricted view. No attempt has been made to provide a comprehensive review of all the literature but simply to describe some new results obtained from our laboratory which suggest areas for future study. This physiological approach towards an understanding of embryogenesis assumes that embryonic cells will possess ion transporting mechanisms similar to those which have been observed in adult cells. Our hypothesis suggests that the embryonic cells will utilise the ion regulatory mechanisms to effect growth, promote morphogenic movements and regulate cell proliferation, migration and differentiation (Moolenaar, Tsein & Van der Saag, 1983; Busa, 1986).

Interest in the basic physiology of the early embryo can be traced back for over 60 years. Much of the experimental work on the early embryo deals with amphibian (Barth & Barth, 1974; Turin & Warner, 1980; Gillespie, 1984) and echinoderm blastulae (Rapkine & Prenant, 1925), amphibian gastrula (Gillespie, 1984) and pre-implantation mammalian blastocyst (Cross, 1973; Borland, Biggers & Lechene, 1976, 1977) and has centred on the characterisation of the mechanisms regulating ion and water movements across the developing embryonic epithelia. There is considerable interest in the hydrogen and calcium ion concentrations of the interstitial fluid of blastulae and gastrulae. Direct chemical measurements on this fluid indicate that there is a high interstitial pH and distinct changes in pH and calcium occur during early development. For example, in the amphibian, the pH of the blastocoel fluid is 8.0 to 8.5, considerably more alkaline than the plasma or interstitial fluid of the tadpole (Holtfretter, 1944) while in the echinoderm, immediately prior to gastrulation, the pH of the blastocoel fluid becomes alkaline, increasing from approximately 7.0 to 9.0 (Rapkine & Prenant, 1925). This work, together with observations that the experimental manipulation of the extracellular pH and calcium ion concentrations affects the aggregation and movement of embryonic cells, has led to the simple hypothesis, postulated by Holtfretter (1944), that embryos actively modify their internal environment as a means of influencing morphogenic movements and affecting cell behaviour.

Evidence has accumulated to support this idea (Stableford, 1967; Barth & Barth, 1974; Gillespie, 1984) particularly for the amphibian embryo where there are clear indications that the concentrations of Ca^{2+}, H^+ and Cl^- have a role in influencing the behaviour of mesodermal cells during gastrulation (Kubota & Durston, 1978; Keller et al., 1985). Chemical analysis of aspirated samples of blastocoel fluid from axolotl (Amblystoma mexicanum) embryos show that during gastrulation the calcium concentration is low (0.15 mM) and increases during the early stages of neurulation (Stableford, 1967).

Early research was limited by technical difficulties in making accurate measurements from living embryos. The subsequent developments in electrophysiology, the use of ion sensitive microelectrodes, and optical probe techniques, now allow precise measurements of ion concentrations and ion movements to be made with minimal tissue or cell damage. The application of these techniques in the amphibian embryo has generated some interesting results.

Using pH-sensitive microelectrodes, the pH of the blastocoel fluid is found to be pH 8.4 throughout gastrulation and neurulation (Turin & Warner, 1980; Gillespie, 1984). By contrast, the interstitial Ca^{2+} concentration falls from 1.5 mM to 0.4 mM just prior to and during gastrulation. During neurulation the Ca^{2+} concentration increases and approaches 2 mM (Gillespie, 1984). During these stages, the intersutial chloride concentration is also low, 30 mM, compared to a plasma value of 120 mM in the tadpole. The importance of these ionic conditions is seen when mesodermal cells are removed from the embryo and their behaviour studied *in vitro*. *In vivo* these cells undergo defined shape changes prior to migration across the blastocoel roof but *in vitro* these changes in behaviour are not observed unless the ionic composition of the tissue culture medium has a high pH and low Ca^{2+} and Cl^- concentrations (Kubota & Durston, 1978; Keller *et al.*, 1985). This demonstrates that the external environment can have important effects on the behaviour of these embryonic cells.

It is more than likely that concomitant alterations in the internal environment of the cell will also affect the physiology and behaviour of avian embryonic cells. In the fowl (*Gallus gallus*) embryo, at about the 4 somite stage, the internal environment is being segregated from the sub-embryonic fluid by the formation of the outer embryonic epithelia. At this time the cardio-vascular system has not developed and is thus not yet a factor in the regulation of the embryonic internal environment. After closure of the neural tube, the embryo, particularly in the head region, is separated from its external environment by the surface ectoderm. The materials, solutes and water, necessary for the growth of the embryo at this stage must be carried across this surface ectoderm. The transport characteristics of this simple epithelium must therefore play a key role in the production of the intra-embryonic environment.

In an attempt to characterise the internal environment of the early chick embryo the hydrogen ion concentration (pH) has been measured using conventional electrophysiological techniques and ion-sensitive microelectrodes (Gillespie & McHanwell, 1987). Measurement of pH was chosen for the initial experiments, since the data from echinoderm and amphibian

embryos suggest that H^+ ions play an important role. In embryos aged between 6 and 22 somites, the pH adjacent to the neural tube and between the somites is in the region of 8.0 (Fig. 24.1). This is considerably higher than might be expected from the pH of the plasma of adult

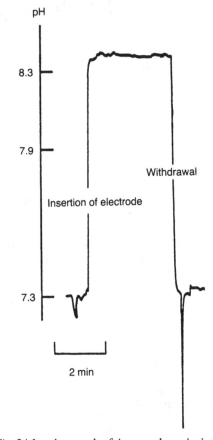

Fig. 24.1. A record of intra-embryonic interstitial pH made from a fowl embryo '*in ovo*' (i.e. still attached to the yolk) using a H^+ sensitive microelectrode. As the electrode passes from the bathing solution and is just touching the ectodermal epithelium there is a small downward deflection representing a touching artefact. Advancing the electrode, moves the tip through the epithelium putting the tip approximately 20 μm into the embryo. The electrode was placed adjacent to the neural tube (50 μm laterally) and between somites 7 and 8 in a 10 somite embryo. The signal recorded (assuming a 5 mV transepithelial potential) corresponds to an interstitial pH of 8.1. On removing the electrode, there is a withdrawal artefact before the signal returns to the original baseline. Similar records were obtained in embryos taken from the yolk and maintained in a small organ bath in balanced salt solution (see Gillespie & McHanwell, 1987).

fowl and much higher than the pH of the sub-embryonic fluid which bathes the embryo *in ovo*.

A second and unexpected observation is that the interstitial pH is different in different regions of the same embryo. Typically, the pH in the head region (mesencephalon) is higher than in the region of the somites and at more caudal regions towards the unsegmented mesoderm (Fig. 24.2). Measurements of interstitial potassium and calcium ions also show rostro-caudal differences but these ion concentration gradients depend on the composition of the bathing solution (Fig. 24.3; Carnan & Gillespie, 1987).

These observations pose three basic questions: 1) How does the embryo maintain such a high interstitial pH? 2) What are the mechanisms underlying the regional variation in pH, K^+ and Ca^{2+}? 3) Does the modified internal environment influence the behaviour of cells and tissues to facilitate the processes of development? Some data is presented here which expands on each of these questions. The background to this approach is large and thus, for reasons of space, much of the experimental detail has been omitted.

Mechanisms underlying the high interstitial pH and ion gradients

A pH difference across the ectodermal epithelium could be maintained by several synergistic mechanisms. For example, in the interstitial spaces the mesenchymal cells may function as a buffering system to modulate the pH. The macromolecular composition of the extracellular environment may have a buffering capacity which produces a basic pH. Alternatively, the ectodermal epithelium may actively, or passively, transport acid equivalents out of the embryo.

Preliminary experiments have investigated whether active transport processes within the ectodermal cells are involved in the maintenance of the pH difference. The application of metabolic inhibitors (cyanide, di-nitro-phenol and iodoacetic acid) abolish the pH gradient across the ectoderm suggesting that the pH gradient is maintained by energy dependent processes (Fig. 24.4). The mechanisms underlying the high interstitial pH have been investigated further by an extensive series of experiments involving substitution of specific ions in the bathing solution. The most important result is

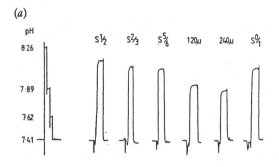

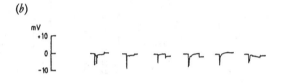

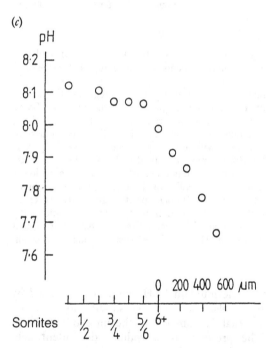

Fig. 24.2. An example of the head to tail difference observed in interstitial pH in an isolated 6-somite fowl embryo maintained in an organ bath in balanced salt solution. The inset shows the original traces from the pH sensitive microelectrode (*a*) and the conventional voltage recording electrode (*b*). Between somites 1 and 2 ($S^{1/2}$) the pH is 8.15 while at a point 240 µm distal to the last somite, in the unsegmented mesoderm, the pH is 7.78. Section (*c*) shows a plot of the complete data for this 6-somite embryo demonstrating the high interstitial pH in the region of the somites and the fall in pH in the region of the unsegmented mesoderm.

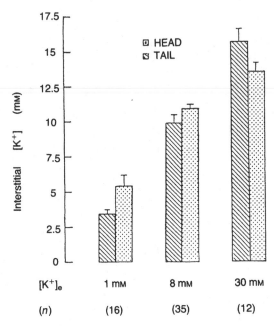

Fig. 24.3. Data from 63 different fowl embryos where measurements of the interstitial potassium concentration were made in the region of the mesencephalon (head) and in the region of the unsegmented mesoderm (tail) in each embryo. Three groups of embryos were used, bathed in balanced salt solutions containing either 1, 8 or 30 mM potassium (numbers within each group are shown in parentheses). In embryos bathed in 1 mM external potassium the concentration in the 'tail' is significantly lower than in the 'head'. On the other hand, in embryos bathed in 30 mM potassium this gradient is reversed and the concentration in the 'tail' is significantly greater than the 'head'. Embryos bathed in 8 mM potassium showed little rostro-caudal potassium gradient.

that the interstitial pH could be influenced by the replacement of the sodium ions in the external medium bathing the embryo (Fig. 24.4). The presence of a sodium dependent acid extrusion mechanism or mechanisms, e.g. Na^+/H^+ exchange or $Na^+ : HCO_3^-$ co-transport on the outer surface of the ectodermal epithelium is, therefore, indicated. Such Na^+-dependent mechanisms do not use energy directly but the Na^+-gradient which drives them is actively maintained via the ATP-dependent $Na^+ : K^+$ pump. A simple combination of active transport and ion exchange, or co-transport, mechanisms could account for the inhibition of the gradient by the metabolic poisons and its Na^+ dependence.

The origin of the head to tail pH gradient

There are several hypotheses which could account for the regional differences in interstitial pH. These include: 1) regional differences in interstitial buffering, either cellular or associated with the macromolecular composition of the extracellular spaces; 2) differences in the passive movements of H^+ ions across the ectoderm; or 3) regional differences in the pH regulating or transporting capacities of the ectodermal epithelium.

The buffering power of the interstitial spaces has been estimated (Gillespie & McHanwell, 1987) and indicates that there are no regional differences in buffering power thus eliminating this as a major contributary mechanism. Measurements of the interstitial potassium concentration in different regions of the same embryo, bathed in a solution containing 1 mM potassium, suggest that there is a rostro-caudal gradient, similar to that of the pH gradient. The interstitial K^+ concentration in the head region is typically 6 mM while in the region of the unsegmented mesoderm it is 3 mM. However, in embryos bathed in a solution containing 30 mM K^+ the rostro-caudal gradient is reversed (Carnan & Gillespie, 1987). The K^+ concentration in the head region is 13 mM while in the unsegmented mesoderm it is 16 mM. Experiments done in solutions containing 8 mM, a concentration close to that of the sub-embryonic fluid (Howard, 1957; Carnan & Gillespie, 1987), therefore, the normal environment for the embryo, show that the rostro-caudal K^+ gradient is all but abolished (Fig. 24.3).

One possible explanation for these observations is that there are regional differences in the ability of certain small ions to move passively across the epithelium of the surface ectoderm. This difference in 'leakage' could be 'tighter' in the head regions than in more caudal regions. This hypothesis has been examined by measuring the rate of change in interstitial K^+, on a change in external K^+ from 8 to 30 mM (Fig. 24.5). Potassium accumulates at a faster rate in the more caudal region of the embryo suggesting that the influx of ions into the embryo is higher in this region. The rate at which potassium moves across the epithelium is not influenced by incubating the embryo in a solution containing 2 mM KCN. Thus, we conclude that ions such as potassium do move passively across the ectodermal epithelium and that there are regional

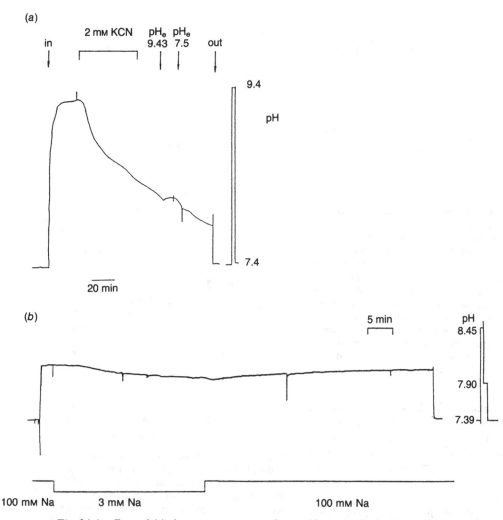

Fig. 24.4.   Record (*a*) shows an experiment from a 12-somite fowl embryo of a trace of interstitial pH. In this experiment, the interstitial pH was 8.6. For the time period indicated, the embryo was exposed to a bathing solution containing 2 mM potassium cyanide (KCN). This resulted in a rapid and irreversible fall in the interstitial pH showing that the high pH was dependent on cellular metabolism. During the fall in interstitial pH induced by KCN, the pH of the bathing solution was increased briefly to 9.43. This resulted in a relatively small change in interstitial pH suggesting that the fall in interstitial pH does not simply result from an increase in the passive leakage of H^+ ions into the embryo from the bathing solution. The insertion and removal of the ion sensitive electrode are indicated by arrows. Calibration of the electrode was done at the end of the experiment. (*b*) An example of an experiment in which sodium in the bathing solution was replaced with an impermeant ion (Tris). During the wash period in low Na solution the interstitial pH fell. On returning Na to the bathing solution, the pH gradually returned to its original value. These experiments suggest that Na^+ ions on the outer surface of the ectodermal epithelium are required for the regulation of the interstitial pH. The electrode was calibrated at the end of the experiment.

differences in epithelial 'tightness'. The rostro-caudal variation in the concentration of K^+ and Ca^{2+} could be accounted for by the 'leakage' characteristics of the ectodermal epithelium. However, *in ovo* the sub-embryonic fluid has a pH of close to 7.4 and the pH of the interstitial fluid is 8 (Howard, 1957; Gillespie & McHanwell, 1987). This high pH in the inter-stitial fluid cannot be simply explained by passive movements.

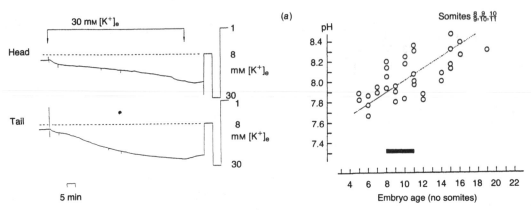

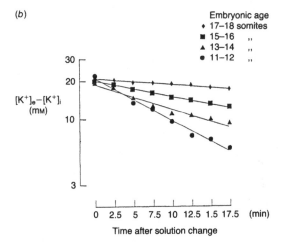

Fig. 24.5. The time-course of changes in interstitial potassium measured simultaneously in the same fowl embryo (15 somites) with two potassium-sensitive microelectrodes. One electrode was inserted into the embryo at the level of the mesencephalon, the other in the region of the unsegmented mesoderm. Where indicated, the bathing solution was changed from one containing 8 mM potassium to one containing 30 mM. This resulted in an increase in interstitial potassium in both regions of the embryo. Note that the rate of increase in potassium concentration and the extent of the change was greater in the more caudal 'tail' region. The calibration of each electrode is shown at the end of each trace.

As the embryo develops, the ectoderm becomes progressively tighter. This process occurs in a rostro-caudal temporal sequence with the more rostral regions becoming tighter first. This tightening appears to continue for some time even at one location in the embryo. There is evidence to show that there is a gradual increase in interstitial pH and a progressive reduction in permeability of the ectodermal epithelium to potassium ions (Fig. 24.6). The ultrastructural basis for these physiological changes has not yet been determined.

So far we have no information on the proposed regional differences in the ion transporting characteristics of the ectodermal epithelium. Such studies are, in part, hindered by the experimental difficulties encountered in measuring the ion transporting mechanisms in single cells or small well defined epithelial areas.

Optical techniques using ion sensitive fluorescent probes have been developed to the point where we are now able to measure the concentrations of ions in single living cells, or even in regions of a cell. Using these techniques, the intracellular concentrations of H^+, Ca^{2+}, Na^+, Mg^{2+} and Cl^-, can be measured and the mechanisms involved in their regulation at the

Fig. 24.6. Section (*a*) shows measurements of interstitial pH made from 33 different fowl embryos. The measurements were all made in the region of the 8th, 9th and 10th somites. (In embryos younger than 8–10 somites, measurements were made in the region of the unsegmented mesoderm at the position where these somites would develop.) Although there is considerable scatter, it is clear that, if we consider the pooled data as representative of what occurs in a single embryo, then the pH in any particular region increases with age. This could suggest that the H^+ pumping capacity of the ectodermal epithelium increases or alternatively that the leakage of H^+ ions into the embryo decreases with age. (*b*) Shows the collected data from 15 embryos arbitrarily divided into four age groups. From recordings of the change in interstitial potassium when the bathing solution was increased from 8 to 30 mM (similar to those shown in Fig. 24.5) the rate of change was calculated from semi-logarithmic plots of the observed changes. The rate of change in potassium concentration was slower in older embryos (17–18 somites) than the rate observed in younger embryos (11–12 somites). These data suggest that the ectodermal epithelium in the region of the mesencephalon gets progressively tighter as an embryo develops.

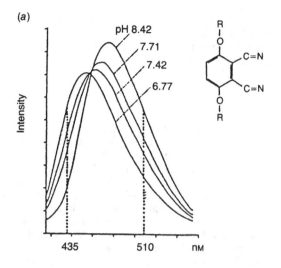

(a)

pH 8.42
7.71
7.42
6.77

Intensity

435   510   ɴᴍ

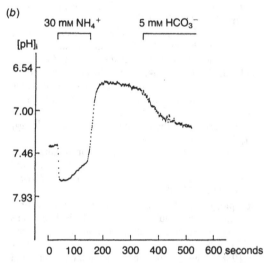

(b)

30 ᴍᴍ NH₄⁺   5 ᴍᴍ HCO₃⁻

[pH]ᵢ

6.54

7.00

7.46

7.93

0   100   200   300   400   500   600 seconds

Fig. 24.7. A record of intracellular pH obtained from a section of neuroectodermal epithelium microdissected from a 10-somite fowl embryo. To measure intracellular pH, the cells were loaded with the pH sensitive fluorochrome di-cyano-hydroquinone (DCH: see Gillespie & Greenwell, 1987, 1988) the structure and emission spectra for which are shown in (a). The principle of this dye and technique is to determine the shift in the emission spectrum which is pH dependent. The record (b) shows the consequences of acid loading the cells with a pulse of ammonium chloride. Recovery from this acid load proceeds in the absence of added HCO₃⁻ in the superfusion solution. Recovery of this type is indicative of a Na⁺/H⁺ exchange mechanism (see Gillespie & Greenwell, 1988). Adding HCO₃ to the superfusion solution accelerated the recovery process. This HCO₃⁻ dependent recovery is also dependent on external Na⁺ (see Gillespie & Greenwell, 1988) and is indicative of a Na⁺ : HCO₃⁻ co-transport mechanism.

cellular level studied (Cobbold & Rink, 1987; Gillespie & Greenwell, 1987, 1988). The particular advantage of these optical probes and techniques is that very small cells can be loaded with a cell permeant ester of the fluorochromes. Once inside the cell, these dye esters are converted by non-specific intracellular esterases into the anionic and active form of the fluorochrome. The electrical charge on the dye molecule makes them cell membrane impermeant and the dye is trapped inside the cell (Cobbold & Rink, 1987). As there is no need to inject the fluorochrome into the cell, or even to mechanically touch the cell, these probes are ideal for studies in embryonic material where small defined pieces of embryo (20–50 cells) can be used and measurements made from single cells. Preliminary experiments to determine the H⁺ handling mechanisms in the ectodermal epithelium reveal that these cells have two mechanisms, a Na⁺/H⁺ exchanger and a Na⁺ : HCO₃⁻ co-transporter (Fig. 24.7; Gillespie & Greenwell, 1988; Gillespie, Greenwell & Russell, 1989).

Variations in the intracellular properties of cells

As the internal environment of the embryo is not the same as that of the adult, is there any evidence for the hypothesis is that this might be important for developmental processes? In the amphibian blastula and gastrula, there is experimental evidence that the environmental conditions are important in facilitating changes in the shape of mesodermal cells and activating their migration. This begs the question whether similar phenomena occur in the fowl embryo. At the time when a high interstitial pH is observed, several developmental processes are under way. The somites have formed and the cells within them are proliferating. The neural crest is migrating from the dorsal surface of the neural tube to different locations within the embryo where the cells differentiate into a variety of different types (LeDourain, 1982). Our work suggests that intracellular pH and calcium may play a role in each of these major events. In this section, experiments with isolated neural crest will be described. Details of experiments on somites are given by Gillespie & Greenwell (1988).

There is some evidence that ions, particularly Ca²⁺, may influence the onset of crest cell migration (Newgreen & Gooday, 1985). Using optical techniques to study Ca^{2+} and H^+ ions in migrating crest cells has shown that the embryonic cells show surprising results. One of the major difficulties in studying crest cells *in vivo* is that there is only a relatively small number of cells and that they are, during their migratory stages, intermingled with and indistinguishable from other cell types. As with the amphibian mesenchymal cells, one approach to answering basic questions about the physiology of these cells is to isolate a reasonably pure population of the cells using tissue culture techniques. When microdissected sections of the neural tube are removed from the embryo and placed in culture, crest cells migrate onto the substrate (Newgreen & Gooday, 1985). Although this approach removes the crest cells from their natural environment it does provide a preparation with which to study the physiology of these cells. With knowledge of cell behaviour in this preparation it should then be possible to predict their behaviour in the embryo.

In culture it is possible to identify cells which have actively migrated from the neural tube and cells which have remained near the explant and are in close contact with other cells. The intracellular concentrations of Ca^{2+} and H^+ in these two functionally different populations of cells have been measured. In the closely packed cells the resting Ca^{2+} concentration is $137 \pm n$M (mean $\pm$ S.D., $n = 19$) and the mean pH is 7.30 ($n = 30$). However, in the actively migrating isolated cells, the Ca^{2+} is significantly lower, 50 ± 5 nM ($n = 15$), and the pH of 7.48 is significantly higher. One interpretation of these data is that the onset of crest cell migration is activated and maintained by a fall in intracellular Ca^{2+} and a rise in intracellular pH. The precise details of how these differences occur and the mechanisms by which they exert their action on the cells are the subject of active investigation.

A general hypothesis to account for the observations in relation to the whole embryo, and in single crest cells, would be that although the intracellular pH is influenced by the extracellular pH, the cells themselves also possess mechanisms to potentiate this effect. The alkaline environment inside the embryo would also decrease cell–cell adhesion (Holtfretter, 1944) and promote the onset of migration. Either the change in intracellular pH *per se* or the activation of other, as yet unknown mechanisms, alter and regulate intracellular Ca^{2+}.

Athough it is only speculation that these changes in intracellular ions affect cell behaviour, preliminary experiments show that changes in intracellular pH and Ca^{2+} can affect crest cell shape and mobility.

References

Barth, L. G. & Barth, L. J. (1974). Ionic regulation of embryonic induction and differentiation in *Rana pipiens*. *Developmental Biology*, **39**, 1–22.

Borland, R. M., Biggers, J. D. & Lechene, C. P. (1976). Kinetic aspects of rabbit blastocoele fluid accumulation. An application of electron probe microanalysis. *Developmental Biology*, **50**, 201–11.

— (1977). Studies on the composition and formation of mouse blastocoele fluid using electron probe microanalysis. *Developmental Biology*, **55**, 1–8.

Busa, W. B. (1986). Mechanisms and consequences of pH-mediated cell regulation. *Annual Review of Physiology*, **48**, 389–402.

Carnan, E. & Gillespie, J. I. (1987). Measurements of interstitial K^+ and Ca^{2+} ion concentrations in the early chick embryo. *Journal of Physiology*, **394**, 82P.

Cobbold, P. H. & Rink, T. J. (1987). Fluorescence and bioluminescence measurement of cytoplasmic free calcium. *Biochemical Journal*, **248**, 313–28.

Cross, M. H. (1973). Active sodium and chloride transport across the rabbit blastocoele wall. *Biology of Reproduction*, **8**, 566–75.

Gillespie, J. I. (1984). The distribution of small ions during the early development of *Xenopus laevis* and *Amblystoma mexicanum* embryos. *Journal of Physiology*, **344**, 359–77.

Gillespie, J. I. & Greenwell, J. R. (1987). Measurement of intracellular pH in single isolated cells using fluorescent probes. *Journal of Physiology*, **391**, 10P.

— (1988). Changes in the intracellular pH and pH regulating mechanisms in somatic cells in the early chick embryo: A study using fluorescent pH-sensitive dye. *Journal of Physiology*, **405**, 385–95.

Gillespie, J. I., Greenwell, J. R. & Russell, E. (1989). pH regulation in the developing chick neuroectodermal eptithelium. *Proceedings of the International Congress of Physiological Sciences. XXXI International Congress, Helsinki*, P4346.

Gillespie, J. I. & McHanwell, S. (1987). Measurement of intra-embryonic pH during the early stages of development of the chick embryo. *Cell and Tissue Research*, **247**, 445–51.

Holtfretter, J. (1944). A study of the mechanisms of gastrulation. II. *Journal of Experimental Zoology*, **95**, 171–212.

Howard, E. (1957). Ontogenic changes in the freezing point and sodium and potassium content of the subgerminal fluid and blood plasma of the chick embryo. *Journal of Cellular and Comparative Physiology*, **50**, 451–70.

Keller, R. E., Danichik, M., Gimlich, R. & Shih, J. (1985). The function and mechanism of convergent extension of *Xenopus laevis*. *Journal of Embryology and Experimental Morphology*, **89**, 185–209.

Kubota, H. & Durston, A. J. (1978). Cinomatographical study of cell migration in the open gastrula of *Amblystoma mexicanum*. *Journal of Embryology and Experimental Morphology*, **44**, 71–80.

LeDourain, N. (1982). *The Neural Crest*. London: Cambridge University Press.

Moolenaar, W. H., Tsein, R. Y. & Van der Saag (1983). Na/H exchange and cytoplasmic pH in the action of growth factors in human fibroblasts. *Nature, London*, **304**, 645–8.

Newgreen, D. F. & Gooday, D. (1985). Control of the onset of migration of neural crest cells in avian embryos. *Cell and Tissue Research*, **239**, 329–36.

Rapkine, L. & Prenant, M. (1925). Reaction du liquide blastocoellen chez le pleuteus d'Oursin dans le premier phase du developpement. *Comptes Rendes d'Academie de Science*, **181**, 1099–101.

Stableford, L. T. (1967). A study of calcium in the early development of the amphibian embryo. *Developmental Biology*, **16**, 303–14.

Turin, L. & Warner, A. E. (1980). Intracellular pH in early *Xenopus* embryos: its effects on current flow between blastomeres. *Journal of Physiology*, **300**, 489–504.

25

Electrochemical processes during embryonic development

PAUL O'SHEA

Introduction

Much has been written and more has been said about putative fundamental concepts or philosophies of biological growth and development. These range from the 'mechanics' of Wilhelm Roux, and the 'determinants' of August Weismann in the latter part of the nineteenth century, to the elegance of simple geometrical transformations as pronounced by D'Arcy Thompson (1917). Contemporary views include the notion that living systems are dynamic, self-evolving structures popularised by Prigogine and Turing (Harrison, 1987). Conceptually attractive as these latter views are, they are not easily accessible experimentally, and it seems that the most dramatic advances have taken place in molecular genetics/biology. This has undeniably permitted us a glimpse of some of the most fundamental processes of morphogenesis (Ingham, 1988). Despite the evident fecundity, however, a true molecular picture of the elemental mechanisms of differentiation is not attainable exclusively from molecular genetics. It is necessary to address the problem from a physical and physiological point of view as well as from these more established genetic-based approaches. This is perhaps exemplified best by the approach of Williams and co-workers (Perry, Wilcock & Williams, 1988) who emphasise, and indeed demonstrate, that morphogenesis is a process of many interconnecting physical, electrical, chemical and genetic processes. Thus, multi-disciplinary approaches are necessary to understand the many interactions which constitute growth and development (O'Shea, 1988a). Here I shall barely mention the role that differential gene activity may play in morphogenesis (which in any case is probably very familiar to most developmental biologists). Instead I will emphasise more physically based mechanisms. These two approaches, however, should not be deemed mutually exclusive or alternative interpretations of morphogenesis; rather they represent complementary views of hierarchical interacting processes.

Whilst no individual would claim total originality, there have been two major initiatives which appear to provide a strong foundation for embarking on physical studies of biological development. The first, is a quantitative assessment of the role mechanical forces and geometry play in eliciting changes in biological form (Odell et al., 1981; Oster, Murray & Harris, 1983). The resultant differential stresses and strains which affect cell and tissue morphology are thought to be under the control of Turing-type reaction-diffusion mechanisms (Harrison, 1987).

The second initiative (which is related) suggests that electrochemical processes exert a profound effect on growth and development (Lund, 1947; Jaffe, 1986). Experimental investigations of such growth-associated electrochemical phenomena have been pioneered by Jaffe (1986) and comprehensively documented (Nuccitelli, 1986; O'Shea, 1988b). Jaffe suggests that living systems may generate electrical and chemical gradients by the asymmetric translocation of ions from one region of a cell to another or from one population of cells to another. Electrochemical fields once established, represent a spatial polarisation of the cell(s) which may prelude other more comprehensive and sophisticated differentiation processes (Nuccitelli, 1986; O'Shea, 1989). Despite the abundant documented affects of long-range electrochemical processes on growth and development, I have some reservations about the ubiquity of their relevance.

In this chapter, a logical sequence of physical events which initially elicit the polarisation of a single cell at the molecular level, and lead to the establishment of a bundle of differentiated cells is described. Much of this is fairly general in that it ought to apply to early development in all eukaryotic cells but the basic mechanisms will be illustrated with examples taken from experimental work carried out with the domestic fowl (*Gallus gallus*) at not later than the 20-somite stage. Some of this experimental work includes a novel technique which involves the non-invasive spatial imaging of electromagnetic fields around the developing fowl egg. My colleagues and I are of the opinion that these fields *reflect* morphogenetic processes. I will describe appropriate mechanisms which would be expected to generate such fields and illustrate their developmental consequences with experimental results obtained from viewing single cells whereby membrane protein targeting seems to be the result. A potentially important 'spin-off' from this work, which may be of economic interest to bird producers, is that we can identify whether or not eggs are fertilised and viable. The analytical technique, outlined below, is completely non-invasive and of short duration, thus we expect in the near future to be able to detect in a commercial environment, those eggs which will hatch and those which will not.

Electrochemical processes and cells

What are cells?

Living cells are not simple 'bags of water full of proteins' or even 'bags of water with a little scaffolding', rather they ought to be conceived as elaborate and highly sophisticated dynamic structures which exhibit long-range order. This long-range organisation does not simply take the form of reticulated organelles or multi-enzyme systems, e.g. so-called 'metabolons' (Batke, 1989) and an extensive cytoskeleton coupled to an extracellular matrix; it is augmented by the more intangible organisation (i.e. biochemically) exhibited by scalar and vector fields (O'Shea, 1988*b*). Whilst several types of field may exist and play important roles in biological organisa-

tion, for the purposes of this chapter I shall concentrate on the role electrochemical processes play in avian development.

The nature of electrochemical fields

An electrochemical field may be established in a cell when ions (i) flow through a medium of appropriate resistance (Fig. 25.1). For these present purposes it is enough to state that such a current would elicit an electrochemical potential difference ($\Delta_{\mu i}$) of i. In other words, both a chemical potential difference ($\Delta_{[i]}$) and an electrical potential difference ($\Delta\psi$) may be generated:

$$\Delta_{\mu i} = \Delta\psi + Z\Delta_{[i]} \qquad (25.1)$$

(where Z is a composite constant). Their respective magnitudes would depend upon the ionic buffering capacity and electrical capacitance over the spatial region that i flows.

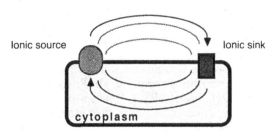

A transcellular electrochemical field

Fig. 25.1. A transcellular electrochemical potential difference is generated by an asymmetric disposition of ionic sources and sinks within the plasma membrane. Ionic current flow through the intra- and extra-cellular resistance generates a potential difference (see equation 25.1).

By way of a summary: the different components of an electrochemical potential difference ($\Delta\mu i$) are suggested to elicit different biological responses. The electrical component for example, could be utilised to drive charged components of a cell to their corresponding pole. The other component, a chemical potential difference exists as a concentration gradient ($Z\Delta_{[i]}$). Thus, from a developmental point of view, even if a chemical morphogen were to be homogeneously distributed about the cell, it may behave quite differently dependent upon its position within the ion field.

How cells generate electrochemical fields

Electrochemical fields may be established simply by the transport of ions along a defined spatial/cellular axis which has an appropriate resistance. In order to establish such a translocation of ions, i.e. from one part of a cell to another or from one group of cells to another, there must be an asymmetric disposition of ion transporting devices in the plasma membrane(s) or the respective cells (Figs 25.1 & 25.2). The question of how this state of affairs comes into existence, however, has rarely been addressed. The problem is both general and important; localisation of membrane components appears to be a ubiquitous property of cells (Schreiner *et al.*, 1977; McClosky & Poo, 1984; Oliver &

Berlin, 1982). In terms of membrane proteins, I have called this process *Domain Condensation* and have shown that it may take place spontaneously, provided certain conditions are met (O'Shea, 1989, 1990). The functional significance of such localised arrangements depends to a large extent upon the particular system in question but, for growth and development, domain condensation of specific ion transporters seems to be a fundamental determinant of (i.e. cause, or in some cases, one must also concede, coincident with) further differentiation.

Developmental consequences of transcellular electrical fields

The vast body of literature detailing most of the observations of transcellular ionic currents in

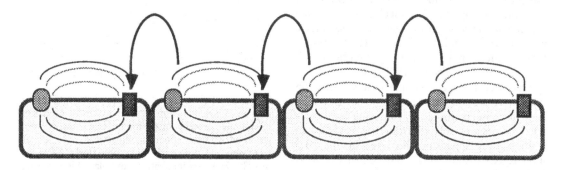

Intercellular interactions leading to further differentiation

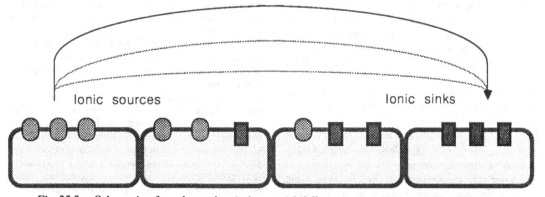

Ionic sources   Ionic sinks

Fig. 25.2. Schematic of an electrochemical potential difference spanning many cells. The upper drawing indicates several cells (such as shown in Fig. 25.1) which are linked together by an ionic current. The ionic sources, however, are not constrained to sinks within their own cells and may interact with other cells. Under some circumstances this may lead to further differentiation as shown in the lower drawing. The result of this interaction generates an electrochemical potential difference over many cells.

single and multicellular systems has been reviewed by Nuccitelli (1986). In fact, such ionic activity has been identified in virtually all eukaryotic systems and as a result Jaffe (1986) has suggested they are of major developmental significance. Despite the apparent ubiquity of such transcellular activity and the great abundance of morphogenetic events to which they are suggested to correlate, Stern (1986) argues that such claims are excessive, although much of what he finds disagreeable is based on some misconceptions of elementary physics and is strongly repudiated by Nuccitelli, Robinson & Jaffe (1987). There remains, therefore, a substantial body of information which implies that long-range electrochemical processes may influence cellular differentiation. As appealing as these hypotheses are, however, there are some grievous quantitative problems. Similarly, the precise molecular mechanisms which accompany or result from these processes have not received anything like as much documentation as the more phenomenological physiological observations (O'Shea, 1988b, 1989).

The effects of transcellular ionic currents on cytodifferentiation during early avian development have been studied by imposing ionic currents on tissue isolated from the early avian embryo using apparatus illustrated in Fig. 25.3 (Poo & Robinson, 1977). The construction is fairly modest and consists of a small rectangular cell (either glass, perspex or teflon) with a base made up of a small slide lined with polylysine to facilitate cell adhesion. A suitable electrolyte surrounds the cells and an ionic current (*circa* 1–2 mA) is passed between two electrodes. The voltage which is imposed across each embryonic cell amounts to a few mvolts. The experiments shown in Fig. 25.4 indicate the profound effect that a significant transcellular voltage may have on the redistribution of a charged membrane protein (in this case acetylcholine binding proteins). The voltage (*circa* 6 mV) is slightly larger than that generated in a few types of cell (Woodruff, Kulp & LaGaccia, 1988) but much larger than typically encountered.

Under *most* circumstances, the observed ionic current densities flowing about individual cells are around 1–10 $\mu A\ cm^{-2}$ (Nuccitelli, 1986) which would be expected to generate a transcellular voltage of around 10 μvolts. This ought not to be discernible from electrical noise at the membrane surface and so cannot be used to electrophorese charged cellular components

about the cell. This type of quantitative analysis seriously questions the ubiquitous role that extracellular ionic currents have been suggested

Monitoring the field-dependent diffusion of membrane proteins

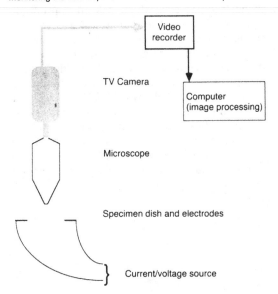

Fig. 25.3. Schematic of the apparatus utilised to impose electric fields on a cell and view the effects on membrane protein disposition. The images were recorded with the apparatus illustrated; muscle cells were removed from early embryonic fowl (*Gallus gallus*) (30–75 hours incubation) and placed in a viewing cell coated with polylysine. A balanced osmotic support medium was provided from a reservoir which was similar in composition to that described by Gillespie & McHanwell (1987) although the Ca^{2+} component was varied from 1 μM to 2 mM as $CaCl_2$, depending upon the desired circumstances. The osmolarity and ionic strength due to the variable $CaCl_2$ was maintained by balancing with KCl. Plasma membrane acetylcholine-binding proteins were visualised by perfusing the vessel (under reflux) with fluorescein-isothiocyanate–acetylcholine and observing the fluorescein fluorescence located on the cell membrane. The microscope was focused onto a specific region of a single embryonic muscle cell. The voltage resulting from current flow between the two platinum electrodes was controlled by a stabilised power supply. Monochrome images were recorded and processed in real-time with the aid of a computer (Intel 80386 cpu) equipped with elementary image-processing software. As a result of the crude image processing, particularly with the amplification routine, some of the resultant images have taken on a degree of pattern that was visible and, therefore, not quantitatively reliable. The results (Figs. 25.4) should be taken as qualitative trends. All experiments were carried out at 24 °C.

to play in morphogenesis (Nuccitelli *et al.*, 1987). There are exceptions to this including one of interpretation, for such a could be voltage summed over many cells would be significant and may play a role in morphogenesis. Other cells produce larger currents including the huge currents associated with primitive streak development in the chick (Stern, 1986). On the other hand, ions flowing through the cytoplasm may generate a more substantial voltage (of the order of millivolts) because the cytoplasmic impedance is much greater than the extracellular aqueous phase (*circa* $> 10^4$ times). It would seem more likely, therefore, that it is worth considering that it is the intracellular, and not the extracellular, component of the ionic current loop which is developmentally important.

With these points in mind it is particularly important to note that the protein redistribution shown in Fig. 25.4 results from purely extracellular electrical activity. This is also the case with all the other work so far published mainly using *Xenopus laevis* as an experimental organism. Thus, the voltage which is generated by a transcellular ion current, does not influence the cellular interior because the resistance of the plasma membrane is sufficiently insulating. With this in mind one should also consider that an intracellular voltage may also have a significant effect on protein targeting (O'Shea, 1989). An intracellular voltage generated by inserting electrodes into the cell, certainly does elicit an electrophoresis of the observed membrane proteins (Fig. 25.4). The current required to generate equivalent voltages in the cytoplasm is much smaller than that necessary in the extracellular aqueous phase.

Electrochemical fields across many cells

Much of what I have discussed about the field-dependent mechanisms of development for single cells is also relevant when such fields exist over several cells (Fig. 25.2). Following on from this, it is not difficult to imagine that a single cell which spontaneously polarises could influence its neighbours. These interactions may then be consolidated by changes in the nature of the protein content of the cell membranes such that a large current loop is established. And, indeed, a macroscopic ionic gradient (i.e. over many cells) has been observed in the developing fowl (Gillespie *et al.*, Chapter 24).

In addition, because of the added complexity of multicellular systems, a number of other developmental mechanisms may come into play. A dramatic example is illustrated by the work of Erickson & Nuccitelli (1984) with embryonic fibroblasts. They show that both cell motility and orientation is strongly influenced by the presence of electric fields of 'physiological magnitudes'. Trinkaus (1985) also emphasises the possible importance of electric fields as a controlling factor of cell migration or cell orientation, the latter of which is clearly demonstrated by McCaig (1989).

Romanoff (1967) draws attention to electrical phenomena and metabolic gradients and the possible roles they may play in avian development. In particular, at a very early stage of development there is a small potential difference amounting to 0.8 mV between the blastoderm and the albumen, after 24 hours incubation it had risen to 7.5 mV, but, after around 80–100 hours, it was found to decline. A stable pH gradient exists in the extracellular space adjacent to the neural tube of the early fowl embryo (Gillespie & McHanwell, 1987; Gillespie *et al.*, Chapter 24). This pH gradient appears to be maintained by active processes, judged by a sensitivity to metabolic poisons (such as KCN), as opposed to a static system whereby the pH gradient would result from a difference in the cell surface charge density. Further examination of some of this work (Gillespie & McHanwell, 1987) indicates that, in addition to demonstrating the presence of a pH gradient, there is information concerning electrical gradients along the rostro-caudal axis in the early fowl embryo. Although, any possible developmental roles of this voltage which amounts to a few mV (up to 5) are not discussed, it is clear that a small electrical potential difference may elicit a profound redistribution of cellular material. Similarly, Deeming, Rowlett & Simkiss (1987) report a voltage of about 7 mV across the yolk sac membrane of embryos of approximately the same age. This voltage, however, appears to be generated by a Na^+ current rather than by a H^+ or OH^- current as reported by Gillespie & McHanwell (1987).

Non-invasive investigations of morphogenesis

Despite the laudable efforts to obtain physiological and other physical measurements *in vitro* during avian development (and, indeed, with

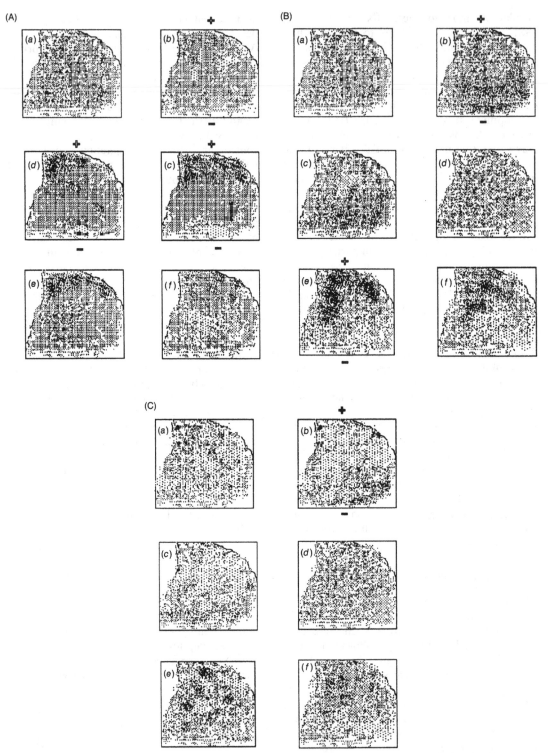

Fig. 25.4. (A) Views of the effects of electric fields on the disposition of acetylcholine binding proteins (ach) in single cells 30 hours old. These images were obtained as a screen-dump of a processed image to a laser printer. The original border of the cell (*a*) was outlined on the screen-

just about any invasive technique in biology) a significant but often unspoken problem lurks: is it reasonable to extrapolate what is observed *in vitro* to the *in vivo* state? The answer, of course, is that, under most circumstances, there is no other choice. By taking as many precautions as possible to mimic *in vivo* conditions, however, it is often a fair assumption. Recently, however, several non-invasive or non-perturbing techniques have been developed which will hopefully permit the *in vitro* to *in vivo* extrapolation to be made with much more confidence as well as

---

Fig. 25.4 *contd*

dump with indian ink for purposes of clarity as there was some movement of the cell envelope (but no cell migration). The cells were supplemented with 2 mM $CaCl_2$.

(*a*) ach disposition with no transcellular potential difference (0 volts).

(*b*) (*a*) is subjected to a transcellular voltage of 6 mvolts for a period of 20 minutes and the ach disposition visualised. The top of the picture was anodic and the bottom cathodic.

(*c*) ach disposition after further 25 minutes at same voltage.

(*d*) ach disposition after further 50 minutes at same voltage after which point the voltage is reset to 0 volts.

(*e*) ach disposition after 20 minutes at 0 volts.

(*f*) ach disposition after further 45 minutes at 0 volts.

Although most of the ach reattains a random disposition once the voltage is set to 0 volts, the original homogeneous disposition (i.e. as in (*a*)) is not reattained. This is thought to reflect domain condensation due to the concentration of ach at the anode. Ach are thought to be negatively charged, consequently their direction of motion (towards the anode) is as one might expect.

(B) Views of the disposition of ach in single cells which are 30 hours old: (*a*)–(*c*) effects of an electric field in a medium with low levels of Ca^{2+}: (*d*)–(*f*) intracellular current sources with medium as in (*a*). Images were obtained in an identical manner to (*a*) but the medium was supplemented with an EGTA-Ca buffer which maintained the Ca^{2+} at 1 μm.

(*a*) ach disposition with no transcellular potential difference (0 volts).

(*b*) ach disposition after 75 minutes at 6 volts. The top of the picture was anodic and the bottom cathodic.

(*c*) ach disposition after 75 minutes at 0 volts after (*b*).

As in (*a*) after the electrical treatment the ach does not reattain its former random disposition. The direction of motion (towards the cathode) of the ach is counterintuitive (on the basis of (*a*)). Under these circumstances, it is expected that electroosmosis is the dominant force. It should be noted that the direction of protein migration is affected by merely changing the ambient ($CaCl_2$).

(*d*) ach disposition with intracellular electrodes (out of view): 0 volts.

(*e*) ach disposition after 75 minutes at 6 mvolts (intracellular potential difference). The top of the picture was anodic and the bottom cathodic.

(*f*) ach disposition after 75 minutes at 0 volts after (*e*).

An intracellular potential difference appears to be as effective at promoting protein migration as the extracellular potential difference shown in (*a*). The current necessary to achieve the same voltage, however, was about 10^{-4} of that in (*a*).

(C) Views of the disposition of ach in single cells: (*a*)–(*c*), effects of an electric field on cells 75 hours old, with medium as in Fig. 25.4: (*d*)–(*f*), effect of $CaCl_2$ on cells 30 hrs old maintained at a transcellular potential of 0 volts.

Images were obtained in an identical manner to 25.4(A) but the medium was supplemented with 2 mM $CaCl_2$.

(*a*) ach disposition with no transcellular potential difference (0 volts).

(*b*) ach disposition after 75 minutes at 6 mvolts.

(*c*) ach disposition after 75 minutes at 0 volts after (*b*).

These observations (*a*)–(*c*) indicate that the rate of motion of the ach is much slower than that shown in (A) but the overall direction of migration appears to be the same.

(*d*) ach disposition with transcellular potential difference maintained at 0 volts.

(*e*) ach disposition after 75 minutes at 0 volts + 2.8 mM $CaCl_2$.

(*f*) ach disposition after 180 minutes after (*e*) with all the added $CaCl_2$ removed.

Patches or clusters of ach are observed to spontaneously appear in the presence of $CaCl_2$, removal appears to promote some reduction in the patch intensity which is taken to reflect repeptisation of the ach clusters.

providing information that could not be realised in any other way. In particular, Nuclear Magnetic Resonance (NMR) Imaging (MRI) and magnetometry, utilising the SuperConducting QUantum Interfering Device (SQUID), appear to offer a novel means of obtaining spatio-temporal information about the organisation and dynamics of structures and physiological processes taking place during biological development. For those readers encountering these techniques as applied to biological developmental for the first time, I recommend reviews of MRI by Lohman & Ratcliffe (1988) and SQUID magnetometry by Swithenby (1988).

The first MRI (sometimes referred to as chemical microscopy) work addressing the problem of embryonic development was carried out with the locust (Lohman & Ratcliffe, 1988). It is possible to discriminate between various types of structure as well as determining the chemical nature of some embryonic environments. As yet, however, there has been no published information regarding similar studies of early development in the bird or reptile.

MRI has been used to record anatomical changes which take place during the later stages (*circa* 11 days) of avian morphogenesis and the 3-D pictures which are obtained are startling for their clarity. From a developmental point of view, however, the value of MRI is far greater than that of a new tool for 'seeing through the shell' and merely obtaining pictures with a· spatial resolution equivalent to that of a poor light microscope. NMR spectroscopy is an essential analytical tool in the chemist's repertoire and in the same manner, it is also possible to identify the nature of chemical species in the imaging mode as well as tool for elucidating gross architecture.

The generally held view that MRI is a non-invasive technique is not strictly true; measurements take place by subjecting the biological specimen to a rather large gradient or static magnetic field strengths. MRI is used routinely as a diagnostic tool in many hospitals throughout the world (i.e. as a so-called 'body scanner') and one would expect that some effort would have gone into identifying any possible clinical effects the technique could have on mature or developing biological tissues. Indeed, there have been several studies undertaken to characterise the effects that magnetic and electromagnetic fields may have on living tissue (Budinger, 1979; Prato

et al., 1987), but, cumulatively, they do not amount to a rigorous and comprehensive study. Similarly the absence of observable effects on more simple biochemical reactions (Moore, Ratcliffe & Williams, 1983) does not permit one to make the same deduction about a very complex and changing system such as tissue development.

During a typical MRI data collection routine, a gradient reversal of a field of 0.5 T m^{-1} with a frequency of 1 ms for example, would elicit a current density of about 2.5 mA m^{-2} around the circumference of a spherical biological object of diameter 1 cm. Current densities of this order appear to be routinely generated by developing systems (Nuccitelli, 1986), it would not be surprising, therefore, that by interfering with these 'natural' currents, normal development could be perturbed. Despite my apparent pessimism, and given the fact that magnetic fields do not seem to have any observable effect on lymphocyte division (Cooke & Morris 1981), amphibian development (Prasad, Wright & Forster, 1982) or of course, during clinical usage, one might presume that magnetic field-induced currents or indeed macroscopic ionic currents generated by the cell have no influence on cell function. It is my belief, however, that there are temporal 'windows' during the developmental process when it is particularly sensitive to the transcellular electrochemical activity and perhaps fortuitously, the fields associated with MRI do not impinge on these windows or if they do, not for long enough. This is consistent with the observations shown in Fig. 25.4 whereby membrane protein redistribution occurs only for embryonic muscle cells which are about 24 hours old but cells which are much older (i.e. *circa* 3 days) do not exhibit protein movement to the same extent. It is possible, therefore, that transcellular ion fields are important during certain phases of early development. Consequently, it is only during these phases that perturbation by extraneous electromagnetic fields may influence development.

SQUID magnetometry

A novel technique to be directed towards studies of biological development is SQUID magnetometry (Swithenby, 1988). Unlike MRI this technique does not impose magnetic fields upon a biological subject. Rather, the extremely weak magnetic fields which result from the electrical activity of cells are measured. Magnetic field measurements take place by simply scanning across the biological sample (or more usually moving the sample across the SQUID detector). This technique, therefore, is truly non-invasive but this is not to say that no problems exist with the SQUID technique. In fact as judged by how well known SQUID magnetometry is, as compared with the MRI technique, illustrates how unfamiliar SQUID magnetometers are, and it will be sometime yet before we see them as expensive boxes in hospitals.

There are two fundamental constraints on the application of the SQUID magnetometer to research in developmental biology. The first relates to the hardware, i.e. the basic measurement of magnetic fields associated with cellular activity. One might anticipate that ionic currents which flow over macroscopic dimensions (i.e. >100 μm), and appear to promote biological morphogenesis, would generate magnetic fields that are in principle within the resolution of the SQUID. Nevertheless, these magnetic fields are very weak and their measurement requires some considerable expertise to operate and maintain the apparatus. A discussion of the spatial and electrical resolution of the SQUID magnetometer is given by Swithenby (1988).

The second problem relates to the interpretation of the magnetic field data. This is the so-called 'inverse problem', whereby the current distribution must be deduced from the magnetic field data using Maxwell's laws. Unfortunately, this is not as straightforward as it may appear for the resultant current distribution depends upon the adoption of a source model. The magnetic profile may not relate to a unique current distribution and, therefore, would profit with supplemental information about the architecture of the developing organism. Thus *in vitro* work may well be indispensable (at least under some circumstances) to interpret the magnetic information.

Non-invasive measurements of electrical activity in the developing fowl embryo

A preliminary study of the magnetic fields produced *in ovo* by the developing fowl embryo (Swithenby, 1988) has been carried out with pick-up coils 24–11 mm in diameter and a scan interval of 5 mm. This limits the spatial resolution of the SQUID so that only the gross features of the current distribution can be resolved. Nevertheless, at least some data are obtained which not only demonstrate that such measurements are possible but also provide some new information. A simple current dipole is identified which increases above noise after about 20 hours incubation (Fig. 25.5). After an incubation of *circa* 70 hours the current dipole

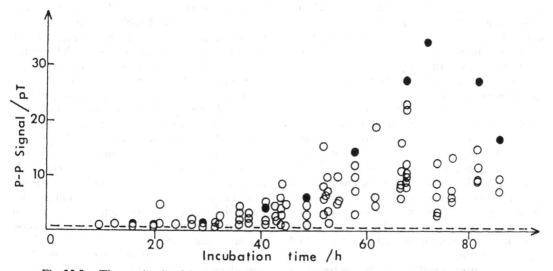

Fig. 25.5. The amplitude of the peak-to-peak (p–p) signal changes with the incubation time (h). The filled symbols refer to signals recorded from one fowl egg. The open signals are indicative of the variations and were recorded with several eggs. The overall trends, however, are consistent with the filled points.

appears to be at a maximum and corresponds to 25 µA cm. After this time, the patterns become highly elaborate and unpredictable and in some cases there is polarity reversal. The ionic current distribution, therefore, must also be presumed, to become more elaborate, possibly reflecting the mounting elaboration of the embryo. The signal is estimated to be oriented approximately parallel to the caudal-capital axis about 7 mm from the centre of the egg. Some eggs have been windowed in order to correlate the respective positions and orientations of the signals and the embryo. These observations tentatively confirm that the current dipole is displaced horizontally about 1 cm to the side of the embryo (Thomas, unpublished observations).

A summary of the present state of knowledge of the patterns of magnetic activity emanating from the egg during incubation is shown in Fig. 25.6. The field contours appear to be dipolar which if correct, permit the calculation of the position and orientation of the equivalent source as well as its strength. Throughout the first three days of incubation the dipole appears to be located towards the centre of the yolk with a strength of about 30 µA cm. This position is highly consistent from egg to egg (± 3 mm). The embryo is usually located on the surface of the yolk, so is not considered to be the most likely source of the field signal. The signal is thought to be of metabolic origin, however, because of its sensitivity to metabolic poisons (such as KCN) and ion transport inhibitors (such as ouabain, valinomycin and detergents). In addition, the signal disappears if the incubation temperature is lowered to around 4–6 °C (Fig. 25.7) but when the egg is permitted to return to the normal incubation temperature, the magnetic signal also returns. Both of these observations indicate that the source of the signal is biological in origin and not due to any contamination from magnetic material. Similarly, control experiments which involve removal of the embryo, the extra-embryonic membranes and the yolk followed by SQUID measurements, indicate that the signal was associated with them and not the remaining shell. In other cases, removal of the embryo but not the associated membranes does not affect the signal for *circa* 20 minutes when it begins to diminish.

I should stress that these results are based on the simplest interpretation of a dipolar current distribution. It is worrying that there is no obvious structure at the centre of the yolk which could be responsible for such electrical activity. In addition, the strength of the dipole appears to be fairly large as compared to other biological current sources. This casts doubt on the simple localised dipole model, consequently other models must be considered. This presents no problem theoretically, as fields produced by a number of dipole sources or by an extended source will smear and look like a simple dipolar form as the separation from the source increases. This is exacerbated by the use of the large coil (23/24 mm) which is about half the diameter of the egg. By utilising a 4 mm coil it should prove possible to increase the spatial resolution and identify any fine structure to the current distribution. Preliminary work with a new coil indicates that a more sophisticated pattern can be resolved. In addition, analytical methods have been developed to identify 3-D current density distributions from external magnetic field data. The future looks promising for SQUID magnetometry as a tool for research in biological development but, of course, its likely that the most spectacular advances will occur in clinical science.

Non-invasive identification of fertilisation and viability

Perhaps the most valuable information that magnetometry could provide the agricultural industry is that it may identify whether eggs have been fertilised and are viable. Without this information, considerable resources are expended on unfertilised eggs which do not ultimately return the investment. Such information about the fate of an egg gleaned from an elementary SQUID magnetometer study would, therefore, have economic implications for the large scale bird producer. There are, however, several hindrances to the installation of a suitable working magnetometer into the egg producing environment. The most significant would be problems of cost i.e. the magnetometer itself is not a cheap instrument and the running costs must include about $100 worth of liquid helium every day or so. Electromagnetic noise due to the environment would seriously compromise measurements, and the machine would fill a respectably sized room so space may also be a consideration. All of these would conspire to interfere with the establishment of a working magnetometer. With the advent of 'room-temperature' superconductors, or, at least,

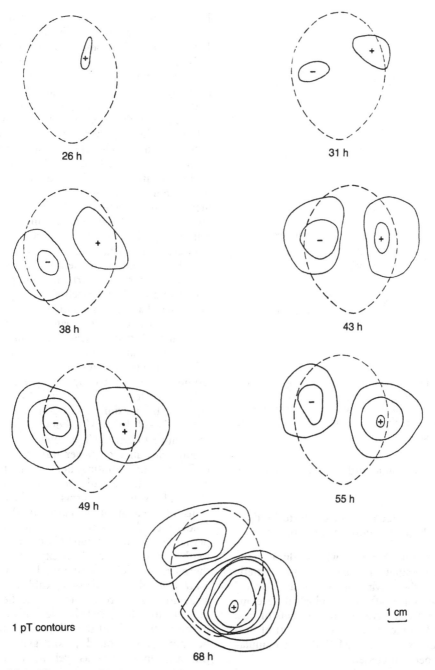

26 h

31 h

38 h

43 h

49 h

55 h

1 pT contours

68 h

1 cm

Fig. 25.6.   Summary of the magnetic signals produced by the fowl egg at various stages of develop-ment as measured by the SQUID magnetometer. The drawings represent the magnetic field contours measured as fowl development progresses. The times in hours refer to the period of incubation.

warmer SQUIDS (Watts, 1989), which will also facilitate a considerable miniaturisation and cheaper computer processing power, these problems will hopefully be solved. In the not too distant future, every egg at various incubation times could be automatically assayed for fertility or viability and those which do not conform could then be recycled. This would lead to a

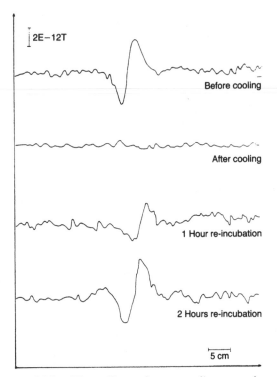

2E−12T

Before cooling

After cooling

1 Hour re-incubation

2 Hours re-incubation

5 cm

Fig. 25.7. The effect of egg cooling on the amplitude of a representative peak-to-peak signal. Similar results to cooling were obtained by treatment of the egg with KCN, valinomycin, ouabain and several detergents.

much more economic bird production and save resources.

Influences of electric and magnetic fields of non-biological origin on development

To complete this chapter, I wish to draw attention to the possibility that electric and magnetic fields from the environment act as external physical influences affecting biological development. To a certain extent, some aspects of this, with regard to the interference (or not) of biological processes during MRI data accumulation, have been discussed.

Given the fundamental nature of electric and magnetic activity, it is not surprising that they may have an influence on living systems. As far as I know, however, there is no published work which reports the effects of subjecting incubating fowl eggs to static or oscillating, electric or magnetic fields. On the other hand, much data is accumulating which indicates the concern and interest that magnetic and electric field effects have on biological systems (Biomagnetism,

1987). Some of their effects have proved to be extremely valuable for clinical use (Brown, 1987) whilst others appear to be much more sinister (Sadgrove, 1989). Unfortunately, there is little or no information available about the pathological effects of magnetic or electric fields, on biological systems at the molecular level, most studies are merely phenomenological. Nevertheless, there is an ominous consistency about such reports which vary from farmers who claim a high rate of miscarriage and stillbirth in animals grazing under or very near to power transmission lines, to an alarming, high incidence of leukaemia in children, from homes clustered around transmission lines and power stations (including those which are non-nuclear). Whilst I have reservations about some of the data relating to the apparent leukaemia clustering, there are certainly enough data to stimulate a more concerted and comprehensive effort than has been attempted so far.

It is possible to speculate how extraneous physical fields may interfere with biological processes. The most obvious source of man-made electromagnetic interference is, of course, the national electricity grid systems prevalent in the industrially developed countries. In the UK for example, the maximum voltage that power lines operate is 400 kV (RMS) between phases (230 kV relative to Earth). The minimum permitted ground clearance of power transmission cables is 7.6 m; the resulting field strength on the ground directly beneath the cables is maximally about 11 kV m^{-1}. In terms of cellular dimensions, this field strength is around 300 mV/cell. Under most circumstances, however, the lowest permitted ground clearance is not reached due to the siting of the electricity 'pylons'. Furthermore, the field decays rapidly as one moves away from the transmission cables and many structures can 'screen' the electric fields. It is unlikely, therefore, that electric fields will be all-pervasive throughout the UK; rather, there will be 'hot spots' around power generating and transmission systems as well as 'cooler' spots around some household electrical appliances; most areas, however, ought to have relatively 'cold' field strengths.

In view of the results of 'physiological' DC fields (Figs 25.4 & 25.5) some 50 times smaller than the environmental fields in close proximity to living systems, some concern ought to be expressed as to their influence on cellular activity. Depending upon the frequency, one might anticipate that electric fields emanating from power cables may have a considerable effect on the disposition of charged species within the cell. In addition, oscillating fields may induce abnormal dipole moments in proteins (soluble or membrane bound) which may then respond anomalously to the 'normal' developmental electric field. The frequency of AC fields is also an important variable and the biological threshold of effect appears to be some five times greater at 50 Hz than at 15–20 Hz.

There is a strong possibility that bio-electric fields (either AC or DC) are fundamental (albeit cryptic) components of the developmental programme. On the other hand, I consider magnetic fields of biological origin unlikely to play any such role, a view which is also taken by Swithenby (1988). Despite the fact that the SQUID magnetometer can certainly measure the magnetic fields emanating from the fowl egg, we believe that they play no role in avian development, not least because the fields are so weak. This is not to say that magnetic fields cannot interfere with morphogenesis. This most probably arises through the induction of ionic currents and thus through the generation of electrochemical fields. The magnitude of the induced currents is fairly large during MRI but even with such large induced currents there seems to be no consistent and dramatic effect on biological systems. There are reasons for this (i.e. the 'temporal window' sensitivity) but one might presume that the much weaker magnetic fields, resulting from power transmission lines, household and industrial electrical appliances, would exert no biological effect. The magnitudes of urban magnetic fields (ignoring those associated with a MRI facility) are of the order of about 0.3 μT to 30 μT (RMS) directly below power lines which reduces to 10 μT at about 20–30 m from a perpendicular ground position. Fields close to electrical appliances are of the order of 10–50 μT but attenuate rapidly with distance from the source. Based on a comparison of their magnitude to MRI fields, these weaker fields might be expected to be inert as far as biological development is concerned. By and large this is certainly the case, although, as I have stressed, there may well be 'developmental windows' wherein they may also be very influential. In other cases, there are a number of observations which are only explainable if some kind of concerted effect takes place and this seems to be the case for both magnetic and electromagnetic fields (Adey, 1988; Frohlich, 1983).

Future prospects

Cellular electrical activity may complement other molecular systems which appear to play a role in morphogenesis. The latter includes the cytoskeleton which has been emphasised (Burn, 1988; Busa, 1988; O'Shea, 1989) as a 3-D structural component under control of second-messenger systems such as the phosphotidyl-inositol cycle and ionic gradients (e.g. Ca^{2+}). I would like to emphasise, however, that these mechanisms merely provide one aspect (albeit a rather fundamental one) of a number of very different sequential, parallel and hierachical interacting processes. A clear understanding of growth and development in highly sophisticated structures such as avian embryos, may only be attained if multi-disciplinary ventures are undertaken. This must include elucidating the nature of multi-layered interactions of biochemical, genetic, physical, and physiological processes.

Acknowledgements

I would like to express thanks to my research colleague, Dr Irene Ridge for valuable discussion, incisive criticism and good company at all stages of this work. I am also grateful to Mr Ian Thomas for providing me with some unpublished data relating to his PhD research project.

References

Adey, R. (1988). Effects of microwaves on cells and molecules. *Nature, London*, 333, 401.

Batke, J. (1989). Remarks on the supramolecular organisation of the glycolytic system *in vivo*. *FEBS Letters*, 12, 13–16.

Brown, M. (1987). Taking the pain out of diagnosis. *The Sunday Times*, 27 March 1988, p. 34.

Biomagnetism (1987). *Proceedings of the 6th International Conference on Biomagnetism*, Tokyo, Japan. Abstracts: E-58, JA-7, JA-1, JA-3 and JA-61.

Budinger, T. F. (1979). Thresholds for physiological effects due to rf and magnetic fields used in NMR imaging. *IEEE Transactions. Nuclear Science*, NS-26, 2821–5.

Burn, P. (1988). Phosphotidylinositol cycle and its possible involvement in the regulation of cytoskeleton–membrane interactions. *Journal of Cell Biochemistry*, 36, 15–24.

Busa, W. B. (1988). Roles for the phosphotidylinositol cycle in early development.

Philosophical Transactions of the Royal Society of London, **B320**, 415–26.

Cooke, P. & Morris, P. G. (1981). The effects of NMR exposure on living organisms. 2 A genetic study of human lymphocytes. *British Journal of Radiology*, **54**, 622–5.

Deeming, D. C., Rowlett, K. & Simkiss, K. (1987). Physical influences on embryo development. *Journal of Experimental Zoology*, Supplement **1**, 341–5.

Erickson, C. A. & Nuccitelli, R. (1984). Embryonic fibroblast motility and orientation can be influenced by physiological electric fields. *Journal of Cell Biology*, **98**, 296–307.

Frohlich, H. (1983). Coherence in biology. In *Coherent excitations in biological systems*, pp. 1–5, eds H. Frohlich & F. Kremer. Heidelberg: Springer-Verlag.

Gillespie, J. I. & McHanwell, S. (1987). Measurement of intra-embryonic pH during the early stages of development in the chick embryo. *Cell and Tissue Research*, **247**, 445–51.

Harrison, L. (1987). What is the status of reaction-diffusion theory 34 years after Turing? *Journal of Theoretical Biology*, **125**, 369–84.

Ingham, P. W. (1988). The molecular genetics of embryonic pattern formation in Drosophila. *Nature, London*, **335**, 25–34.

Jaffe, L. (1986). Ionic currents in development: an overview. In *Ionic currents in development*. R. Nuccitelli (ed.), pp. 351–7. New York: Alan R. Liss Inc.

Lohman, J. A. B. & Ratcliffe, R. G. (1988). Prospects for NMR imaging in the study of biological morphogenesis. *Experientia*, **44**, 666–72.

McCaig, C. (1989). Studies on the mechanism of embryonic nerve reorientation in a small applied electric field. *Journal of Cell Science*, **93**, 723–30.

McClosky, M. & Poo, M.-M. (1984). Protein diffusion in cell membranes: some biological implications. *International Review of Cytology*, **87**, 19–81.

Moore, G. R., Ratcliffe, R. G. & Williams, R. J. P. (1983). NMR and the biochemist. *Essays in Biochemistry*, **19**, 142–95.

Nuccitelli, R. (1986). *Ionic currents in development* (ed.). New York: Alan R. Liss Inc.

Nuccitelli, R., Robinson, K. & Jaffe, L. (1987). On electrical currents in development. *BioEssays*, **5**, 292–4.

— (1988). Ionic currents in morphogenesis. *Experientia*, **44**, 657–66.

Odell, G., Oster, G. F., Burnside, B. & Alberch, P. (1981). The mechanical basis of morphogenesis. *Developmental Biology*, **85**, 456–62.

Oliver, J. M. & Berlin, R. D. (1982). Mechanisms that regulate the structural and functional architecture of cell surfaces. *International Review of Cytology*, **74**, 55–94.

Oster, G. F., Murray, J. D. & Harris, A. (1983). Mechanical aspects of mesenchymal morpho-

genesis. *Journal of Embryology and Experimental Morphology*, **78**, 83–125.

O'Shea, P. (1988*a*). New perspectives in morphogenesis. (ed.) *Experientia*, **44**, No. 8.

— (1988*b*). Physical fields and cellular organisation: field dependent mechanisms of morphogenesis. *Experientia*, **44**, 684–94.

— (1989). Biophysical mechanisms of development. In *Signals in Plant Development*, pp. 25–44. J. Krekule & F. Seidlova, eds. The Netherlands: SPB Academic Publishing.

— (1990). Primary physical events in biological growth and development: self-organisation in biological membranes. *Journal of Cell Biophysics* (in press).

Perry, C. C., Wilcock, J. R. & Williams, R. J. P. (1988). A physico-chemical approach to morphogenesis: the roles of inorganic ions and crystals. *Experientia*, **44**, 638–50.

Poo, M.-M. & Robinson, K. (1977). Electrophoresis of concanavalin A receptors in the embryonic muscle cell membrane. *Nature, London*, **265**, 602–5.

Prasad, N., Wright, D. A. & Forster, J. D. (1982). Effects of nuclear magnetic resonance on early stages of amphibian development. *Magnetic Resonance Imaging*, **1**, 35–7.

Prato, F. S., Ossenkopp, K. P., Kavaliers, M., Sestine, E. & Teskey, G. C. (1987). Attenuation of morphine-induced analgesia in mice by exposure to magnetic resonance imaging: separate effects of the static radiofrequency and time-varying magnetic fields. *Magnetic Resonance Imaging*, **5**, 9–14.

Romanoff, A. L. (1967). *The Biochemistry of the Avian Embryo*. New York: Wiley-Interscience.

Sadgrove, J. (1988). Electricity and cancer. *The Guardian*, 27 July 1988, p. 23.

Schreiner, G. F., Fujiwara, K., Pollard, T. D. & Unanue, E. R. (1977). Redistribution of myosin accompanying capping of surface Ig. *Journal of Experimental Medicine*, **145**, 1393–8.

Stern, C. (1986). Do ionic currents play a role in the control of development. *BioEssays*, **4**, 180–4.

Swithenby, S. (1988). Non-invasive monitoring of ionic current flow dusing development by SQUID magnetometry. *Experientia*, **44**, 673–8.

Thompson, D'Arcy (1917). *On Growth and Form*. 2nd edition (1968). London: Cambridge University Press.

Trinkaus, J. P. (1985). Further thoughts on directional cell movement during morphogenesis. *Journal of Neuroscience Research*, **13**, 1–19.

Watts, S. (1989). Superconductors warm to the beating heart. *New Scientist*, 23 September 1989, p. 33.

Woodruff, R. I., Kulp, J. H. & LaGaccia, E. D. (1988). Electrically mediated protein movement Drosophila follicles. *Roux's Archives of Developmental Biology*, **197**, 231–8.

26

Methods for shell-less and semi-shell-less culture of avian and reptilian embryos

BRUCE E. DUNN

Introduction

Shell-less culture of embryos is a technique which involves culturing embryos with associated yolk and albumen outside of the eggshell and shell membranes. In contrast to shell windowing techniques, shell-less culture allows continuous observation of, and access to developing embryos and extra-embryonic membranes almost to the time of hatching. Injections can be made into, or operations performed on, a specific portion of the embryo, and resultant changes in heart rate, behaviour, regional growth or other parameters observed continuously without having to sacrifice the embryo. Due to lack of adhesion of extra-embryonic membranes to an overlying shell window, multiple injections or operations can be performed on a given embryo in shell-less culture. Additionally, grafts consisting of embryonic, neoplastic or other tissue onto the chorio-allantoic membrane of a single embryo can be followed routinely without having to sacrifice either host or graft tissue. Of particular interest, avian embryos maintained in shell-less culture become calcium-deficient due to the absence of the eggshell (Dunn & Boone, 1977; Tuan, 1980, 1983; Richards, 1982; Ono & Wakasugi, 1984). Thus, shell-less culture provides a unique experimental system for studying the relationships between calcium homeostasis and other metabolic and physiological functions during development (Tuan *et al.*, Chapter 27).

In the past, investigators have cultured avian embryos in a variety of vessels including: glass dishes (Assheton, 1896; Fere, 1900; Loisel, 1900; Paton, 1911; Vogelaar & van den Boogert, 1925; Vollmar, 1935; Schmidt, 1937), ceramic dishes (Corner & Richter, 1973), beakers (Romanoff, 1943; Boone, 1963; Palen & Thorneby, 1976) evaporating dishes (DeGennaro *et al.*, 1980), plastic shells (Quisenberry & Dillon, 1962), petri dishes (Williams & Boone, 1961), sandwich bags (Schlesinger, 1966; Elliott & Bennett, 1971), and in sandwich-type plastic wrap in several configurations (Criley, 1971; Dunn, 1974) with varying degrees of success. Shell-less culture is now an established laboratory technique. Two techniques for shell-less culture are currently in use: 1) the petri dish technique described by Auerbach *et al.* (1974) and 2) the plastic wrap/culture tripod technique described by Dunn & Boone (1976). Survival and growth of embryos in plastic wrap/tripod chambers (Fig. 26.1) are superior to survival and growth in petri dishes (Dunn, Fitzharris & Barnett, 1981*a*). Shell-less culture techniques have been utilised in studies of cornea- and tumour-induced angiogenesis (Vu *et al.*, 1985; Castellot *et al.*, 1986), calcium metabolism in the developing embryo (Dunn & Fitzharris, 1979; Tuan, 1980, 1983; Clark & Dunn, 1986; Ono & Tuan, 1986; Packard & Clark, 1987) and in a variety of other developmental studies (Fitzharris, Markwald & Dunn, 1980; Watanabe & Imura, 1983; Seed & Hauschka, 1984; Lemmon & McCloon, 1986; Gagote, 1986; Thompson, Abercrombie & Wong, 1987).

In contrast to shell-less culture, semi-shell-less culture involves transfer of embryos and egg contents to a 'surrogate' eggshell. This method has the advantages of providing egg shell minerals to embryos, of allowing diffusion of gases through the bottom of the eggshell (corresponding to the location of plastic wrap in shell-less culture methods), of promoting downgrowth of the chorio-allantoic membrane and of allowing albumen uptake by embryos. Semi-shell-less culture allows nutritional and immunological studies involving reconstitution of the yolk and albumen of various species (Rowlett & Simkiss, 1987) and should prove useful in development of transgenic embryos in which injections of nucleic acid into or near the germinal disc of oviductal eggs are necessary (Perry, 1988).

The goals of this chapter are three-fold. First, survival and growth of avian and reptilian embryos in shell-less culture and semi-shell-less culture methods will be reviewed. 'Window' methods, in which a portion of the avian egg shell is removed and covered with a transparent window without removing egg contents from the shell (Rugh, 1962), will not be discussed. Second, physiological studies on embryos in shell-less culture and in semi-shell-less culture which have helped to define limitations of these culture methods will be discussed. Finally, recommendations for improvements in shell-less and semi-shell-less culture techniques to increase embryonic survival and growth will be presented.

Current status of methods for shell-less culture of avian embryos

To facilitate comparison of results between species with different incubation periods, incubation age will be expressed as a percentage of total incubation (% I). Thus, three days of incubation in the fowl (*Gallus gallus*) embryo (21-day incubation period) is equivalent to 14% I. To date, best long-term survival and growth have been obtained when three day (14% I) fowl embryos and egg contents are suspended in hemispherical pouches of sandwich-type plastic wrap using simple tripods (Fig. 26.1) incubated in saturation humidity with 1% CO_2 in a forced draught tissue culture incubator. Culture tripods are covered loosely with the top of a sterile petri dish to prevent contamination and to promote gas exchange (Fig. 26.1). Prior to 14% I, the amnion is incompletely formed (Romanoff, 1960; Freeman & Vince, 1974). As a result, embryos placed in shell-less culture before 14% I are subject to significant desiccation, despite elevated humidity in the culture environment (Dunn *et al.*, 1981*a*; Wittmann, Kugler & Kaltner, 1987). Survival in shell-less culture is diminished in the absence of CO_2 (Auerbach *et*

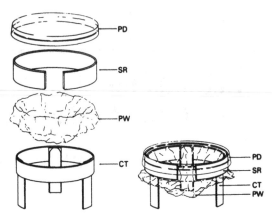

Fig. 26.1. Diagram of a culture tripod constructed of commercial thin wall plastic 'rain drain' pipe. *Left:* 'exploded' view. *Right:* completed chamber. CT = culture tripod consisting of a rim with legs cut from plastic pipe; PW = plastic wrap for supporting embryos and egg contents; SR = split ring cut from plastic pipe. SR serves to hold plastic wrap in place on CT and also to raise the petri dish (PD) cover approximately 2 mm above the upper edge of the CT to allow gas exchange. (From Dunn *et al.*, 1981*a*.)

al., 1974; Slavkin, Slavkin & Bringas, 1980) or when CO_2 incubators which lack forced ventilation are utilised (Dunn, unpublished observation). Under optimal conditions, mean survival of fowl embryos is 18 days (86% I) of total incubation with 50% survival through 18.5 days (88% I) (Dunn & Boone, 1976). To date, avian embryos have not been hatched from shell-less culture chambers. Embryonic growth in shell-less culture is significantly retarded by 13 days (62% I), although embryos become calcium deficient as early as nine days (43% I) of total incubation (Dunn & Boone, 1976). Culture chamber diameter, type of plastic wrap and duration of incubation *in ovo* are some factors which influence survival and growth of embryos in shell-less culture (Dunn *et al.*, 1981*a*). *In ovo*, 75% to 80% of the calcium and 25% to 30% of the magnesium present in the hatchling are resorbed from the eggshell during incubation (Dunn & Boone, 1976; Ono & Wakasugi, 1984). In the absence of the eggshell, embryos in shell-less culture become hypocalcaemic with serum calcium levels of 1.1–1.5 mM (normal approximately 2.5 mM) as early as 10 days (48% I) of total incubation (Burke, Narbaitz & Tolnai, 1979; Graves, Helms & Martin, 1984).

Survival and growth of turkey (*Meleagris gallopavo*) embryos (28-day incubation period) in shell-less culture are similar to survival and growth of fowl embryos when normalised for differences in embryo size and duration of incubation (Richards, 1982; Richards, Rosebrough & Steele, 1984). Specifically, 90% survival through 24 days (86% I) is obtained in plastic wrap pouches suspended in tripods when embryos and egg contents are removed from shells after four days (14% I) of incubation. Growth is significantly retarded by 18 days (64% I). Embryos are hypocalcaemic by 16 days (57% I) and are consistently hypoproteinaemic by 17 days (61% I). Abnormalities in trace mineral metabolism in turkey embryos in shell-less culture have also been identified (Richards, 1984; Richards et al., 1984; Richards & Steele, 1987).

Survival and growth of Japanese quail (*Coturnix coturnix japonica*) embryos (16-day incubation period) in shell-less culture are similar to those of fowl and turkey embryos when normalised for differences in embryo size and duration of incubation (Ono & Wakasugi, 1983, 1984). Mean survival is 13.2 days (83% I) when embryos and egg contents are transferred from shells at 2.5 days (16% I). Survival is 80% through 9.5 days (59% I) and 59% through 14.5 days (91% I). Growth, calcium and magnesium content of quail embryos are retarded by 9–10 days (56–63% I).

Current status of methods for semi-shell-less culture of avian embryos

Ono & Wakasugi (1984) have developed a semi-shell-less culture method in which the contents of Japanese quail eggs are transferred into fowl eggshells after 2.5 days (16% I) of incubation *in ovo*. Due to the much larger size of fowl egg-shells, the quail egg contents flatten out considerably in fowl eggshells, similar to conditions in shell-less culture. Mean survival of quail embryos in semi-shell-less culture is 14.3 days (89% I) and 3/93 (3%) of quail embryos hatched. Quail embryos in semi-shell-less culture show improved growth and mineral content compared to embryos in shell-less culture. Although hatching of quail embryos from surrogate fowl eggshells is possible, the hatchlings in this study are unable to stand, have wet and sticky down and die within 1.5 days post-hatch (Ono & Wakasugi, 1984). Growth and mineral (calcium, magnesium) content of embryos in semi-shell-less culture are intermediate

between those in embryos in intact quail eggs and in shell-less culture.

Rowlett & Simkiss (1987) have developed a similar method for semi-shell-less culture of fowl embryos. After three days (14% I) of incubation *in ovo*, egg contents are transferred to surrogate turkey or fowl eggshells. Embryo growth and calcium uptake in semi-shell-less culture are improved compared with similar parameters in shell-less culture, although embryos in semi-shell-less culture are smaller than those *in ovo*. The chorio-allantoic membrane develops normally in surrogate egg shells, thus promoting resorption of calcium from the shell and allowing formation of a functional albumen sac. Survival of embryos in semi-shell-less culture is approximately 55% to 17 days (81% I). Growth of fowl embryos in semi-shell-less culture is retarded at 16 days (76% I) and thereafter, compared with control embryos in intact eggs. Fowl embryos have hatched from semi-shell-less culture and been reared to laying.

A three-stage method for incubating fertilised oviductal eggs to hatching in semi-shell-less culture has been devised by Perry (1988) and Naito & Perry (1989). After 24 hours of incubation in a glass beaker containing medium designed to simulate plumping fluid, the oviductal egg is transferred sequentially to two different surrogate eggshell configurations. Hatchability of embryos incubated for a total of 22 days (corresponding to one day in the oviduct and 21 days *in ovo*) is low (approximately 7%) but improvements are anticipated. This method allows injection of nucleic acid or other substances of interest into avian eggs at the single cell-stage and holds promise in the development of methods for germ line transformation in poultry.

Shell-less and semi-shell-less culture of reptilian embryos

The use of techniques to culture reptilian embryos is less widespread than for birds (Bellairs, 1971; Ferguson, 1985). Embryos of viviparous lizards (*Lacerta vivipara* and *Anguis fragilis*) have been removed from the female and cultured, with their membranes intact, in petri dishes (Panigel, 1956; Maderson & Bellairs, 1962; Holder & Bellairs, 1962) or glass cups (Raynaud, 1959a,b). Some of these embryos develop through to hatching. A petri dish culture system similar to that of Auerbach *et al.* (1974) has been used for embryos of *Alligator mississippiensis* but with limited success: there is growth up to stage 18 of development (day 28 at 30 °C; 33% I) but there is a high incidence of abnormalities and the embryos do not hatch (Ferguson, 1981, 1985).

A form of semi-shell-less culture system has also been used for *A. mississippiensis* embryos but, instead of transferring the contents of eggs between different eggshells (*cf.* Ono & Wakasugi, 1984; Rowlett & Simkiss, 1987), only the top third of the shell is removed (Ferguson, 1981, 1985). The shell has to be removed within 12 hours of laying (0.6% I at 30 °C), before the embryo has adhered to the eggshell (Deeming, Chapter 19); the viscosity of the albumen and yolk keeps them in the shell and embryonic development can proceed normally to hatching (Ferguson, 1981, 1985).

Physiological studies relating to limitations of shell-less and semi-shell-less culture methods for avian embryos

Retardation of growth of embryos in shell-less culture is due to a combination of factors (Dunn & Boone, 1976; Richards, 1982; Ono & Wakasugi, 1984; Rowlett & Simkiss, 1987). First, eggshell calcium, which contributes 75% to 80% of the calcium in the newly hatched fowl (Romanoff, 1967; Simkiss, 1967) is not available to embryos in shell-less culture. Secondly, the combination of abnormal egg geometry and lack of turning restricts albumen assimilation in shell-less culture (Dunn & Boone, 1976). Third, gas exchange in shell-less culture is presumably altered, as cultured fowl egg contents have approximately 38 cm^2 of surface area unobstructed by plastic wrap which is approximately two-thirds the surface area of a standard 60-gram fowl egg (Wangensteen & Rahn, 1970/1971). Finally, without a shell to restrict water vapour diffusion, water loss during incubation is potentially greater *in vitro* than *in ovo*, possibly contributing to embryonic death. In all shell-less culture methods reported to date, embryos have been cultured in high humidity. Thus, humidity requirements during incubation in shell-less culture and semi-shell-less culture appear to be different from those for intact eggs. Each of these areas will be reviewed in detail below.

Calcium

As noted above, avian embryos in shell-less culture become calcium-deficient and hypocalcaemic as early as 43% I (Dunn & Boone, 1977; Tuan, 1980, 1983; Richards, 1982; Ono & Wakasugi, 1984; Graves *et al.*, 1984). In addition, such embryos become hypomagnesaemic (Burke *et al.*, 1979). Supplementation of calcium by placing eggshell pieces onto the chorio-allantoic membrane, or by injecting calcium into the allantoic sac, results in increased serum and total levels of calcium, but is insufficient *per se* to restore growth to *in ovo* levels (Tuan, 1980, 1983; Dunn, Clark & Scharf, 1987). In fowl embryos in shell-less culture, the specific activity of the renal enzyme 25 hydroxyvitamin D-1 hydroxylase is four- to six-fold higher than is the corresponding enzymatic activity in intact embryos (Turner, Graves & Bell, 1987). The increased specific activity in embryos in shell-less culture is due to an increase in maximal enzyme velocity (V_{max}); the Michaelis constant (K_m) remains unchanged. In addition, the allantoic epithelium of embryos in shell-less culture resorbs calcium from within the allantoic sac against an electrochemical gradient, which is strong evidence for expression of an active calcium pump (Graves *et al.*, 1984). By contrast, the allantoic epithelium *in ovo* fails to resorb calcium from the allantoic sac (Ranly, 1972). Further, based on ultrastructural studies, the parathyroid glands of fowl embryos in shell-less culture appear hyperactive compared to those *in ovo* during the latter portion of incubation (Clark & Dunn, 1986). To my knowledge, the level of circulating parathyroid hormone in avian embryos in shell-less culture has not been reported.

Elevation of the specific activity of renal 25 hydroxyvitamin D-1 hydroxylase, expression of active calcium uptake by the allantoic epithelium and hyperactivity of parathyroid glands all represent compensatory phenomena presumably aimed at counter-acting calcium deficiency in the avian embryo in shell-less culture. By contrast, the chorionic epithelium of shell-less culture embryos fails to transport calcium normally (Tuan, 1980; Dunn, Graves & Fitzharris, 1981*b*), since interaction with either the shell membranes, eggshell calcium or both appear to be required for maximal expression of calcium transport by the chorio-allantoic membrane (Tuan, 1987). Taken together, these studies demonstrate that the avian embryo in shell-less culture provides an interesting model with which to investigate developmental regulation of calcium metabolism (Tuan *et al.*, Chapter 27).

Egg turning and albumen

The physiological role of turning in avian eggs has been extensively studied (Deeming, Chapter 19). In general, turning promotes fluid redistribution among the various compartments within the avian egg during incubation and stimulates growth of extra-embryonic membranes. Specifically, turning promotes formation of sub-embryonic fluid, stimulates growth of the area vasculosa and expansion of the chorio-allantoic membrane promoting formation of the albumen sac. The albumen sac plays an essential role in transfer of albumen through the sero-amniotic connection into the amnion at 12 to 14 days (57–67% I) in the fowl embryo (Randles & Romanoff, 1950; Tullett & Deeming, 1987; Deeming, 1989; Chapter 19). Albumen is an important store of protein and water which is ingested directly from the amnion by the embryo during the latter portion of incubation (Freeman & Vince, 1974). In unturned, intact eggs, hatchability is decreased and embryonic growth is retarded. A pool of unabsorbed albumen is frequently present beneath the chorio-allantoic membrane of unhatched embryos (dead in shell) from unturned, intact eggs (Tazawa, 1980; Tullett & Deeming, 1987).

In plastic wrap/tripod chambers, downgrowth of the chorio-allantoic membrane is frequently incomplete and the albumen sac fails to form completely, leaving unabsorbed albumen on the bottom of culture chambers (Dunn & Boone, 1976, 1978). Thus, growth reduction in cultured embryos probably results, in part, from incomplete resorption of albumen. Fowl, turkey and quail embryos in unturned plastic wrap/tripod chambers show significantly decreased growth from approximately 62% I, the time at which albumen resorption through the sero-amniotic connection begins *in ovo*. Rocking of fowl embryos in shell-less culture chambers from side to side at an angle of approximately 30° promotes normal growth through to day 14 (67% I) compared with *in ovo* controls (Deeming, Rowlett & Simkiss, 1987; Rowlett & Simkiss, 1987). Stimulation of growth of embryos in shell-less culture by rocking may result from disruption of unstirred fluid

layers in the overlying albumen, thus promoting resorption of sodium ions and water to form sub-embryonic fluid (Deeming *et al.*, 1987). Whether rocking promotes uptake of albumen through the sero-amniotic connection in shell-less culture has not been reported. Embryos in shell-less culture have reduced levels of serum protein (Burke *et al.*, 1979; Richards, 1982; Richards *et al.*, 1984) which may result from decreased uptake of albumen protein, or may be a manifestation of growth retardation. Fowl embryos in unturned intact eggs not only demonstrate reduced growth and utilisation of albumen, but are hypoxaemic (Tazawa, 1980), similar to embryos in shell-less culture (Rowlett & Simkiss, 1989). Thus, egg turning may also influence embryonic gas exchange (Deeming, Chapter 19).

Gas exchange

The respiration and acid–base balance in developing avian embryos has been studied extensively (Romijn & Roos, 1938; Wangensteen & Rahn, 1970/1971; Rahn, Paganelli & Ar, 1987). Throughout incubation, shell conductance to gas and water vapour remains constant, whereas embryonic O_2 consumption and CO_2 production increase significantly. As a result, the blood and air space oxygen tension (P_{CO_2}) decrease while the carbon dioxide tension (P_{CO_2}) in these components increases during incubation (Rahn, Chapter 21). At the time of pipping, blood P_{O_2} is approximately 7.6 kPa while P_{CO_2} is 4.3 kPa (Tazawa, Mikami & Yoshimoto, 1971). Blood pH remains stable at approximately 7.4. Bicarbonate levels increase normally during incubation due to resorption of bicarbonate from the eggshell and to metabolic alkalosis (Simkiss, 1980).

The surface area of chorio-allantoic membrane which develops in shell-less culture is approximately one-third less than is the surface area of this membrane *in ovo* (Dunn & Boone, 1976). In many cases, the chorio-allantoic membrane grows down along the plastic wrap in shell-less culture embryos (Dunn & Boone, 1978). Gas exchange through the plastic wrap appears to be important since growth and survival of embryos in shell-less culture are enhanced in plastic wrap with high permeability, and growth is decreased when gas exchange through the plastic wrap is restricted experimentally (Dunn *et al.*, 1981a). Rowlett & Simkiss

(1989) have recently studied respiratory gases and acid–base status in embryos in shell-less culture maintained in ambient air. Embryonic survival in ambient air is significantly less than reported in the presence of 1% CO_2 (Dunn & Boone, 1976), although whether the presence of CO_2 *per se* promotes survival or growth in shell-less culture has not been studied in detail. Fowl embryos in shell-less culture have lower P_{O_2} and lower P_{CO_2} both in venous (oxygenated) and in arterial (deoxygenated) blood from the chorioallantoic membrane compared with embryos *in ovo*. Blood pH of embryos in shell-less culture is stable at approximately 7.4, similar to *in ovo*. A small increase in base excess is observed in embryos in shell-less culture suggesting that renal proton excretion occurs as *in ovo* (Simkiss, 1980). Turkey embryos in shell-less culture have higher haematocrit values than *in ovo* (Richards, 1982). Presumably the elevated haematocrit is an adaptation to increase O_2 carrying capacity of the blood, thereby compensating for the reduced surface area of the chorioallantoic membrane (Richards, 1982).

Taken together, the data above suggest that the avian embryo in shell-less culture, similar to that *in ovo*, is capable of regulating its acid–base and respiratory status. The avian embryo in shell-less culture should provide a unique model by which to study acid–base regulation in the absence of eggshell bicarbonate ion.

Evaporative water loss

During incubation *in ovo*, water is continually lost from eggs. In general, optimal weight loss is approximately 15% (Ar & Rahn, 1980). Weight loss significantly higher or lower than 15% is associated with decreased hatchability. Normal fowl embryos are reportedly capable of some degree of osmoregulatory function, especially during the last third of incubation (Hoyt, 1979; Simkiss, 1980). Embryos in shell-less culture (in ambient air, at 37.5 °C, 80% relative humidity) show a cumulative loss of water which is only slightly greater than that observed in embryos *in ovo* (Rowlett & Simkiss, 1987). No other studies have reported weight loss in shell-less culture egg contents, nor have the effects of weight loss on survival and growth of embryos in shell-less culture been reported.

Miscellaneous physiological abnormalities in avian embryos in shell-less culture

Turkey embryos in long-term shell-less culture exhibit hyperglycaemia in concert with extremely low levels of circulating insulin. In addition, blood glucose levels fail to decrease significantly in response to exogenous avian insulin, thus a type II diabetes-like condition exists (McMurtry *et al.*, 1989). A variety of evidence suggests that insulin and insulin receptors influence differentiation and organogenesis in the fowl embryo *in ovo* (Bassas *et al.*, 1987). Thus hyperglycaemia and hypoinsulinaemia may contribute to retarded growth and developmental abnormalities in avian embryos in shell-less culture (McMurtry *et al.*, 1989).

Turkey embryos in long-term shell-less culture also exhibit evidence of copper deficiency with reduced hepatic copper content (Richards *et al.*, 1984). Copper is distributed abnormally among hepatic cytoplasmic protein fractions in such embryos. Turkey embryos in shell-less culture also exhibit hyperlipidaemia which possibly results from the observed copper deficiency (Richards *et al.*, 1984).

Fowl embryos in shell-less culture reportedly exhibit hypertension and tachycardia, which may be reversed partially by addition of exogenous calcium (Tuan & Nguyen, 1987; Tuan *et al.*, Chapter 27). A development-specific necrosis of liver tissues has been reported in fowl embryos both in shell-less culture and *in ovo*, although the frequency and extent of tissue abnormality are significantly greater in shell-less culture (Ono & Tuan, 1986). However, no such abnormalities were reported in a previous comparison of liver development in fowl embryos in shell-less culture and *in ovo* (Narbaitz, Kacew & Burke, 1980). The explanation for these discrepant results is not known.

Suggestions for improvement

The rate and optimal duration of rocking during incubation of embryos in shell-less culture should be determined. Quantitative comparisons of changes in volume and composition of sub-embryonic fluid, allantoic fluid, amniotic fluid and albumen would be useful in determining how incubation in shell-less culture affects water redistribution in avian egg contents. The volume of amniotic fluid in embryos in shell-less culture appears to be significantly reduced during the last third of incubation (Dunn, unpublished observations). Thus, shell-less culture may provide a useful model for studying the effects of oligohydramnios on embryonic development and growth. Perhaps the nutritional status of shell-less culture embryos can be improved by periodically injecting albumen into the amniotic sac to mimic normal uptake of albumen through the sero-amniotic connection. If successful, this method should prove useful in studies of embryonic nutrition.

Elevation of the P_{O_2} within the incubator during the last third or possibly during the entire period of incubation may stimulate embryonic survival and growth both in shell-less culture and in semi-shell-less culture. Elevated P_{O_2} stimulates embryonic O_2 consumption and growth in ovo, as hypoxaemia is known to limit growth in ovo (Stock & Metcalfe, 1987; Tullett & Burton, 1987). Elevation of CO_2 within the incubator, following CO_2 levels known to occur in the air space in ovo (Wangensteen & Rahn, 1970/1971), may also stimulate survival and growth of embryos in vitro, by facilitating acid–base regulation.

Since embryos in shell-less culture have a large surface area of extra-embryonic membranes exposed directly to the environment, it seems likely that best results in culture will be obtained with elevated ($\geq 80\%$) relative humidity, thus minimising evaporative water loss. Use of low relative humidities would most likely result in increased mortality. However, the effects of varying incubator humidity levels on survival and growth of avian embryos in shell-less culture and in semi-shell-less culture have not been reported.

To improve survival and growth of embryos in shell-less culture, supplementation of calcium (and magnesium) will be necessary to replace those minerals normally resorbed from the egg shell in ovo. Suggested methods for calcium supplementation include: 1) addition of shell pieces with intact membranes onto the chorio-allantoic membrane; 2) injection of calcium into the allantoic fluid; 3) injection of calcium, possibly in a protein-bound form, into the yolk; and 4) slow infusion of calcium into an allantoic vessel using a continuous method (Drachman & Coulombre, 1964). As noted above, calcium supplementation per se is insufficient to promote growth of embryos in shell-less culture to in ovo levels (Tuan, 1983; Dunn et al., 1987). The

effects of calcium supplementation may be most dramatic after other parameters of shell-less culture incubation (incubator gas composition, rocking of culture chambers) have been optimised.

Significant improvements in survival and growth of embryos in shell-less culture (and by analogy, in semi-shell-less culture) will probably require improvements in several parameters. It seems likely that a combination of elevated incubator P_{O_2}, elevated P_{CO_2} (possibly following air space P_{CO_2} levels) and rocking of plastic wrap/tripod shell-less culture chambers in a forced draught tissue culture-type incubator (to provide maximum gas mixing) with high humidity may lead to significant increases in embryonic growth and survival. Injection of concentrated albumen into the amnion may help to overcome lack of albumen uptake by embryos in shell-less culture. Finally, supplementation of embryos with egg shell minerals, in combination with some or all of the suggested improvements above, may help to stimulate survival and growth of embryos in shell-less culture to hatching. Shell-less and semi-shell-less techniques should continue to prove useful in a variety of studies of the physiology of avian embryos.

References

Ar, A. & Rahn, H. (1980). Water in the avian egg: overall budget of incubation. *American Zoologist*, 20, 373–84.

Assheton, R. (1896). An experimental examination into the growth of the blastoderm of the chick. *Proceedings of the Royal Society of London, Series B*, 60, 349–56.

Auerbach, R., Kubai, L., Knighton, D. & Folkman, J. (1974). A simple method for long-term cultivation of chick embryos. *Developmental Biology*, 41, 391–4.

Bassas, L., Lesniak, M. A., Girbau, M. & DePablo, F. (1987). Insulin-related receptors in the early chick embryo: from tissue patterns to possible function. *Journal of Experimental Zoology*, Supplement 1, 299–307.

Bellairs, R. (1971). *Developmental Processes in Higher Vertebrates*. London: Logos Press Ltd.

Boone, M. A. (1963). A method of growing chick embryos in vitro. *Poultry Science*, 42, 916–21.

Burke, B., Narbaitz, R. & Tolnai, S. (1979). Abnormal characteristics of the blood from chick embryos maintained in 'shell-less' culture. *Revue Canadienne de Biologie*, 38, 63–6.

Castellot, J. J., Jr, Kambe, A. M., Dobson, D. E. & Speigelman, S. (1986). Heparin potentiation of

3T3-adipocyte stimulated angiogenesis: mechanisms of action on endothelial cells. *Journal of Cellular Physiology*, **127**, 323–9.

Clark, N. B. & Dunn, B. E. (1986). Calcium regulation in the embryonic chick II. Ultrastructure of the parathyroid gland in shell-less and *in ovo* embryos. *Journal of Morphology*, **190**, 1–8.

Corner, M. A. & Richter, A. P. J. (1973). Extended survival of the chick embryo *in vitro*. *Experientia*, **29**, 467–8.

Criley, B. M. (1971). Sustained observation of chick embryos. *American Biology Teacher*, **33**, 356–7.

Deeming, D. C. (1989). Characteristics of unturned eggs: critical period, retarded embryonic growth and poor albumen utilisation. *British Poultry Science*, **30**, 239–49.

Deeming, D. C., Rowlett, K. & Simkiss, K. (1987). Physical influences on embryo development. *Journal of Experimental Zoology*, Supplement, **1**, 341–5.

DeGennaro, L. D., Packard, D. S., Stach, R. W. & Wagner, B. J. (1980). Growth and differentiation of chicken embryos in simplified shell-less cultures under ordinary conditions of incubation. *Growth*, **44**, 343–54.

Drachman, D. B. & Coulombre, A. J. (1964). Method for continuous infusion of fluids into the chorioallantoic circulation of the chick embryo. *Science*, **138**, 144–5.

Dunn, B. E. (1974). Technique for shell-less culture of the 72-hour avian embryo. *Poultry Science*, **53**, 409–12.

Dunn, B. E. & Boone, M. A. (1976). Growth of the chick embryo *in vitro*. *Poultry Science*, **55**, 1067–71.

— (1977). Growth and mineral content of cultured chick embryos. *Poultry Science*, **56**, 662–72.

— (1978). Photographic study of chick embryo development *in vitro*. *Poultry Science*, **57**, 370–7.

Dunn, B. E., Clark, N. B. & Scharf, K. E. (1987). Effects of calcium supplementation on growth of shell-less culture chick embryos. *Journal of Experimental Zoology*, Supplement, **1**, 33–8.

Dunn, B. E. & Fitzharris, T. P. (1979). Differentiation of the chorionic epithelium of chick embryos maintained in shell-less culture. *Developmental Biology*, **71**, 216–27.

Dunn, B. E., Fitzharris, T. P. & Barnett, B. D. (1981*a*). Effects of varying chamber construction and embryo pre-incubation age on survival and growth of chick embryos in shell-less culture. *Anatomical Record*, **199**, 33–43.

Dunn, B. E., Graves, J. S. & Fitzharris, T. P. (1981*b*). Active calcium transport in the chick chorioallantoic membrane requires interaction with the shell membrane and/or shell calcium. *Developmental Biology*, **88**, 259–68.

Elliott, J. H. & Bennett, J. (1971). Growth of chick embryos in polyethylene bags. *Poultry Science*, **50**, 974–5.

Fere, C. (1900). Remarques sur l'incubation des oeufs de poule prives de leur coquille. *Comptes rendus des Séances. Société de Biologie et de Ses Filiales et Associées (Paris)*, **52**, 601–2.

Ferguson, M. W. J. (1981). Review: The value of the American alligator (*Alligator mississippiensis*) as a model for research in craniofacial development. *Journal of Craniofacial Genetics and Developmental Biology*, **1**, 123–44.

— (1985). Reproductive biology and embryology of the crocodilians. In *Biology of the Reptilia*, vol. 14, Development A, eds C. Gans, F. Billett & P. F. A. Maderson, pp. 329–491. New York: John Wiley & Sons.

Fitzharris, T. P., Markwald, R. R. & Dunn, B. E. (1980). Effect of beta-amino propionitrile fumarate (BAPN) on early heart development. *Journal of Molecular and Cellular Cardiology*, **12**, 553–77.

Freeman, B. M. & Vince, M. A. (1974). *Development of the Avian Embryo. A Behavioral and Physiological Study*. London: Chapman & Hall.

Gagote, S. S. (1986). A modified method of long-term cultivation of shell-less chick embryos *in vitro*. *Journal of Advanced Zoology*, **7**, 118–20.

Graves, J. S., Helms, E. L. & Martin, H. F. (1984). Development of calcium reabsorption by the allantoic epithelium in chick embryos grown in shell-less culture. *Developmental Biology*, **101**, 522–6.

Holder, L. A. & Bellairs, A. d'A. (1962). The use of reptiles in experimental embryology. *British Journal of Herpetology*, **3**, 54–61.

Hoyt, D. F. (1979). Osmoregulation by avian embryos: the allantois functions like a toad's bladder. *Physiological Zoology*, **52**, 354–62.

Lemmon, V. & McLoon, S. C. (1986). The appearance of an L1-like molecule in the chick primary visual pathway. *Journal of Neuroscience*, **6**, 2987–94.

Loisel, G. (1900). Incubation d'oeufs de poule retires de leur coquille. *Comptes Rendus des Séances. Société de Biologie et de Ses Filiales et Associées (Paris)*, **52**, 582–3.

Maderson, P. F. A. & Bellairs, A. d'A. (1962). Culture methods as an aid to experiment on reptile embryos. *Nature, London*, **195**, 401–2.

McMurtry, J. P., Richards, M. P., Rosebrough, R. W. & Steele, N. C. (1989). A diabetic-like condition of turkey embryos maintained in shell-less culture. *Proceedings of the Society for Experimental Biology and Medicine*, **190**, 324–9.

Naito, H. & Perry, M. M. (1989). Development in culture of the chick embryo from cleavage to hatch. *British Poultry Science*, **30**, 265–70.

Narbaitz, R., Kacew, S. & Burke, B. (1980). Ultrastructural and biochemical alterations in the livers from chick embryos maintained in shell-less culture. *Anatomy and Embryology*, **159**, 307–16.

Ono, T. & Tuan, R. S. (1986). Effect of experimentally induced calcium deficiency on development, metabolism, and liver morphogenesis of the chick embryo. *Journal of Embryology and Experimental Morphology*, **92**, 207–22.

Ono, T. & Wakasugi, N. (1983). Development of cultured quail embryos. *Poultry Science*, **62**, 532–6.

— (1984). Mineral content of quail embryos cultured in mineral-rich and mineral-free conditions. *Poultry Science*, **63**, 159–66.

Packard, M. J. & Clark, N. B. (1987). Calcium regulation in the embryonic chick III. Calcium and phosphate in serum and allantoic fluid of normal and shell-less embryos. *Journal of Experimental Zoology*, Supplement 1, 25–32.

Palen, K. & Thorneby, L. (1976). A simple method for cultivating the early chick embryo *in vitro*. *Experientia*, **32**, 267–8.

Panigel, M. (1956). Contribution a l'etude de l'ovoviviparite chez les reptiles: gestation at parturition chez le lezard vivipare *Zootoca vivipara*. *Annals de Science naturelle (Zoologie)*, **11**, 569–668.

Paton, S. (1911). Experiments on developing chickens' eggs. *Journal of Experimental Zoology*, **11**, 469–72.

Perry, M. M. (1988). A complete culture system for the chick embryo. *Nature, London*, **331**, 70–2.

Quisenberry, J. H. & Dillon, E. J. (1962). Growing embryos in plastic shells. *Poultry Science*, **41**, 1675.

Rahn, H., Paganelli, C. V. & Ar, A. (1987). Pores and gas exchange of avian eggs: a review. *Journal of Experimental Zoology*, Supplement, 1, 165–75.

Randles, C. A. & Romanoff, A. L. (1950). Some physical aspects of the amnion and allantois of the developing chick embryo. *Journal of Experimental Zoology*, **114**, 87–101.

Ranly, D. M. (1972). The retention of ^{45}Ca by the chorioallantoic sac of the chick embryo. *Poultry Science*, **51**, 1044–5.

Raynaud, A. (1959a). Une technique permettant d'obtenir le developpement des oeufs d'Orvet (*Anguis fragilis* L.) hors de l'organisme maternel. *Comptes Rendes de Academie de Sciences, Paris*, **249**, 1715–17.

— (1959b). Developpement et croissance des embryons d'Orvet (*Anguis fragilis* L.) dans l'oeuf incube *in vitro*. *Comptes Rendes de Academie de Sciences, Paris*, **249**, 1813–15.

Richards, M. P. (1982). Long-term, shell-less culture of turkey embryos. *Poultry Science*, **61**, 2089–96.

— (1984). Synthesis of a metallothionein-like protein by developing turkey embryos maintained in long-term shell-less culture. *Journal of Pediatric Gastroenterology and Nutrition*, **3**, 128–36.

Richards, M. P., Rosebrough, R. W. & Steele, N. C. (1984). Hepatic zinc, copper and iron of turkey embryos (*Meleagris gallopavo*) maintained in long-term, shell-less culture. *Comparative Biochemistry and Physiology*, **78A**, 525–31.

Richards, M. P. & Steele, N. C. (1987). Trace element metabolism in the developing avian embryo: a review. *Journal of Experimental Zoology*, Supplement 1, 39–52.

Romanoff, A. L. (1943). Cultivation of early chick embryo *in vitro*. *Anatomical Record*, **87**, 365–9.

— (1960). *The Avian Embryo*. New York: Macmillan.

— (1967). *Biochemistry of the Avian Embryo: A Quantitative Analysis of Prenatal Development*. New York: Wiley Interscience.

Romijn, C. & Roos, J. (1938). The air space of the hen's egg and its changes during the period of incubation. *Journal of Physiology*, **94**, 365–79.

Rowlett, K. & Simkiss, K. (1987). Explanted embryo culture: *in vitro* and *in ovo* techniques for domestic fowl. *British Poultry Science*, **28**, 91–101.

Rowlett, K. & Simkiss, K. (1989). Respiratory gases and acid-base balance in shell-less avian embryos. *Journal of Experimental Biology*, **143**, 529–36.

Rugh, R. (1962). *Experimental Embryology*. Minneapolis: Burgess.

Schlesinger, A. B. (1966). Plastic bag culture method for chick embryos. *CUEBS News*, **2**, 10–1.

Schmidt, G. (1937). On the growth stimulating effect of egg white and its importance for embryonic development. *Enzymologia*, **4**, 40–8.

Seed, J. & Hauschka, S. D. (1984). Temporal separation of the migration of distinct myogenic precursor populations into the developing chick wing bud. *Developmental Biology*, **106**, 389–93.

Simkiss, K. (1967). *Calcium in Reproductive Physiology*. London: Chapman & Hall.

— (1980). Water and ionic fluxes inside the egg. *American Zoologist*, **20**, 385–93.

Slavkin, H. C., Slavkin, M. D. & Bringas, P. (1980). Mineralization during long-term cultivation of chick embryos *in vitro*. *Proceedings of the Society for Experimental Biology and Medicine*, **163**, 249–57.

Stock, M. K. & Metcalfe, J. (1987). Modulation of growth and metabolism of the chick embryo by a brief (72-hour) change in oxygen availability. *Journal of Experimental Zoology*, Supplement 1, 351–6.

Tazawa, H. (1980). Adverse effect of failure to turn the avian egg on the embryo oxygen exchange. *Respiration Physiology*, **41**, 137–42.

Tazawa, H., Mikami, F. & Yoshimoto, C. (1971). Effect of reducing the shell area on the respiratory properties of chicken embryo blood. *Respiration Physiology*, **13**, 353–60.

Thompson, R. P., Abercrombie, V. & Wong, M. (1987). Morphogenesis of the truncus arteriosus of the chick embryo heart: movements of auto-radiographic tattoos during septation. *Anatomical Record*, **218**, 434–40.

Tuan, R. S. (1980). Calcium transport and related functions in the chorioallantoic membrane of cultured shell-less chick embryos. *Developmental Biology*, **74**, 196–204.

— (1983). Supplemented egg shell restores calcium transport in chorioallantoic membrane of cultured shell-less embryos. *Journal of Embryology and Experimental Morphology*, **74**, 119–31.

— (1987). Mechanism and regulation of calcium transport by the chick embryonic chorioallantoic membrane. *Journal of Experimental Zoology*, Supplement 1, 1–14.

Tuan, R. S. & Nguyen, H. Q. (1987). Cardiovascular changes in calcium-deficient chick embryos. *Journal of Experimental Medicine*, **165**, 1418–23.

Tullett, S. G. & Burton, F. G. (1987). Effect of two gas mixtures on growth of the domestic fowl embryo from days 14 through 17 of incubation. *Journal of Experimental Zoology*, Supplement 1, 347–50.

Tullett, S. G. & Deeming, D. C. (1987). Failure to turn eggs during incubation: the effects on embryo weight, development of the chorioallantois and absorption of albumen. *British Poultry Science*, **28**, 239–49.

Turner, R. T., Graves, J. S. & Bell, N. H. (1987). Regulation of 25-hydroxyvitamin D_3 metabolism in chick embryo. *American Journal of Physiology*, **252**, E38–43.

Vogelaar, J. P. M. & van den Boogert, J. B. (1925). Development of the egg of *Gallus domesticus in vitro*. *Anatomical Record*, **30**, 385–95.

Vollmar, H. (1935). Eine methode zur beobachtung der Entwicklung des Huhnembryo *in vitro*. *Zeitschrift fur Zellforschung und Mikroskopische Anatomie*, **23**, 566–70.

Vu, M. T., Smith, C. F., Burger, P. C. & Klintworth, G. K. (1985). Methods in Laboratory Investigation. An evaluation of methods to quantitate the chick chorioallantoic membrane assay in angiogenesis. *Laboratory Investigation*, **53**, 499–508.

Wangensteen, O. D. & Rahn, H. (1970/71). Respiratory gas exchange by the avian embryo. *Respiration Physiology*, **11**, 31–45.

Watanabe, K. & Imura, K. (1983). Significance of the egg shell in the development of the chick embryo: a study using shell-less culture. *Zoological Magazine*, **92**, 64–72.

Williams, W. L. & Boone, M. A. (1961). Inhibition of growth of chick embryo in shell-less culture. *Proceedings of the Federation of the American Society for Experimental Biology*, **20**, 419.

Wittmann, J., Kugler, N. & Kaltner, H. (1987). Cultivation of the early quail embryo: induction of embryogenesis under *in vitro* conditions. *Journal of Experimental Zoology*, Supplement 1, 325–8.

27

Experimental studies on cultured, shell-less fowl embryos: calcium transport, skeletal development, and cardio-vascular functions

ROCKY S. TUAN, TAMAO ONO, ROBERT E. AKINS
AND MASAFUMI KOIDE

Introduction

During development, the eggshell supplies the majority of the calcium needed by the fowl (*Gallus gallus*) embryo, a finding drawn from the cumulative work of many chemical embryologists (Simkiss, 1961) and the ^{45}calcium tracer study of Johnston & Comar (1955). This dependence on the eggshell is particularly evident when embryos are placed *ex ovo* in long-term shell-less culture (Dunn & Boone, 1977; Tuan, 1980*a*; Slavkin, Slavkin & Bringas, 1980). These cultures are produced by removing the entire content of a fertilised fowl egg from the eggshell after three days of incubation *in ovo*, and incubating it in a plastic sac suspended within a ringstand (Fig. 27.1) (Dunn, 1974; Dunn & Boone, 1976; Tuan, 1980*a*; Dunn, Fitzharris & Barnett, 1981*a*; Dunn, Chapter 26). These shell-less embryos develop severe systemic calcium deficiency (Tuan, 1980*a*; Watanabe & Imura, 1983; Narbaitz & Jande, 1983; Ono & Tuan, 1986), since their only available calcium source is the egg yolk, which constitutes less than 20% of the total calcium found in a hatchling (Packard & Packard, 1984; Romanoff, 1967). The onset of calcium deficiency in the shell-less embryo roughly coincides with the period when shell calcium mobilisation would normally begin, around incubation days 10–12 (Terepka, Stewart & Merkel, 1969; Crooks & Simkiss, 1975; Tuan & Zrike, 1978). The hypocalcaemic state of the embryo is indicated by the significantly lowered serum calcium values (Fig. 27.2). Interestingly,

the hypocalcaemia induced by nutritional calcium deficiency is accompanied by a concomitant hyperphosphataemic state in the shell-less embryo (Fig. 27.2).

Calcium is a vital requirement of proper embryonic development, and is required for skeletal mineralisation, neuromuscular activities, coagulation, and many other physiological functions. Since the embryo is undergoing rapid organogenesis and growth, perturbations in its calcium homeostasis during development are likely to cause major anomalies in various metabolic and cell-tissue functions. The singular calcium deficiency of the shell-less fowl embryo makes it, therefore, uniquely suited as an experimental model to examine the functional importance of calcium homeostasis in avian development. Work in our laboratory during the last ten years has focused on the application of the shell-less fowl embryo to study three aspects of avian development: calcium transport, skeletal development, and cardio-vascular functions; and much insight has been gained concerning the mechanism and regulation of these functions.

Calcium transport

The developing fowl embryo acquires calcium from two sources. Until about day 10 of incubation, the egg yolk is the only source; thereafter, calcium is mobilised from the eggshell (Johnston & Comar, 1955; Simkiss, 1961; Terepka *et al.*, 1976). Calcium transport from the egg yolk is carried out by the yolk sac epithelium, a highly

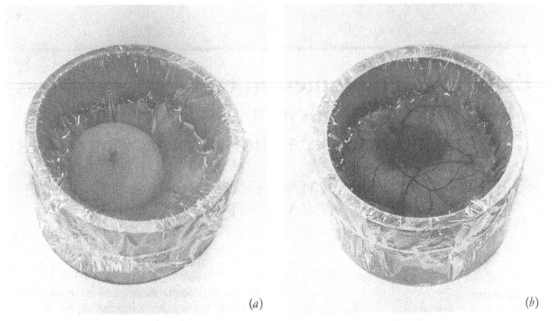

(a)

(b)

Fig. 27.1. Shell-less culture of fowl embryo. Fertilised eggs are first incubated for 3 days *in ovo*, then cracked open, and the contents placed inside a hemispherical pouch made of plastic wrap. The cultures are incubated at 37.5 °C in a humidified, constant air-flow atmosphere. Two shell-less embryos are shown: (*a*) day-3 (beginning of culture) and (*b*) day-10 (total incubation time). By incubation day 10, the chorio-allantoic membrane has spread completely across the surface of the culture.

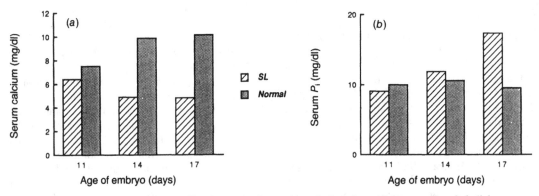

Fig. 27.2. Developmental profile of serum calcium (*a*) and phosphate (*b*) in normal and shell-less (SL) embryos. (Data from Ono & Tuan, 1986.)

vacuolated and vascularised columnar epithelium formed from the combination of endoderm and mesoderm (the splanchnopleure) and is connected to the midgut of the embryo by an open tube, the yolk duct, such that the walls of the yolk sac and walls of the gut are continuous (Romanoff, 1960; Lambson, 1970; Juurlink & Gibson, 1973). Functionally, the circulation within the mesoderm of the splanchnopleure serves to transport nutrients, including calcium, from the yolk to the body of the embryo. Eggshell calcium, on the other hand, is mobilised by the chorio-allantoic membrane (Tuan, 1987), formed by the fusion of the respective mesoderm of the chorionic and allantoic membranes during early to mid stages of embryonic development. The chorio-allantoic membrane consists of three distinct cell layers, termed the ectodermal (chorionic, facing the eggshell and shell membrane), mesodermal, and the endodermal (allantoic, facing the allantoic cavity) layers (Leeson & Leeson, 1963; Coleman & Terepka, 1972a; Narbaitz, 1972). The tissue is highly vascularised, with its main blood vessels embedded in the sparsely populated mesoderm, and a capillary network intercalated within the ectoderm (Narbaitz, 1977).

Mobilisation of yolk calcium

During development of the fowl, yolk calcium begins to be mobilised around day 7–8 of incubation and is accumulated in the embryo (Fig. 27.3) (Romanoff, 1967; Packard & Packard, 1984). The onset of yolk calcium mobilisation corresponds to the beginning of skeletal mineralisation, since centres of ossification, first observed in the head bones of the embryo, also appear at about the same time (Romanoff, 1967). Little is known about the mechanism of calcium transport by the yolk sac, although it is generally assumed that yolk materials are mobilised mostly by endocytosis by the endodermal cells of the yolk sac. Histology of the yolk sac reveals a large number of lipid-containing vacuoles in the endodermal epithelium, representing yolk materials which have been engulfed (Fig. 27.4). The calcium reserve in the egg yolk is presumed to be complexed with various phospholipids and protein, including phosvitin, a negatively charged phosphoprotein (Grizzuti & Perlmann, 1974). Thus, bulk phase endocytosis should suffice for the mobilisation

of yolk calcium, although there have been no reported experimental studies actually substantiating this assumption.

Studies using the shell-less fowl embryo have shed some light on how yolk calcium transport may be regulated (Tuan & Ono, 1986; Ono & Tuan, 1988, 1990a,b). The unavailability of shell calcium means that the shell-less embryo is solely dependent on the yolk reserve, and its ultimate insufficiency as the only supply of calcium leads to systemic hypocalcaemia (Ono & Tuan, 1986). Interestingly, shell-less embryos respond to the administration of 1,25-dihydroxy vitamin D_3, the active metabolite of vitamin D, and an elevation of circulating calcium is observed (Narbaitz, 1979; Tuan & Ono, 1986; Gawande & Tuan, 1988), suggesting possible increased yolk calcium mobilisation. Radiolabelling the yolk calcium compartment with ^{45}Ca shows that the hypercalcaemic effect of vitamin D is, indeed, due to enhanced mobilisation of yolk calcium, and is not derived from resorption of skeletal calcium (Tuan & Ono, 1986). This phenomenon is seen in both normal and shell-less embryos (Narbaitz & Tolnai, 1978; Narbaitz & Fragisco, 1984; Gawande & Tuan, 1988), and is thus likely to be the regulatory mechanism for yolk calcium mobilisation. The hypercalcaemic effect of vitamin D is seen at various stages of development, from incubation day 10 to day 17 (Narbaitz, 1979; Narbaitz & Fragisco, 1984; Tuan & Ono, 1986; Gawande & Tuan, 1988; Clark, Murphy & Lee, 1989), i.e. at the onset of yolk calcium mobilisation as well as during the time of shell calcium usage by the embryo (Tuan & Zrike, 1978). Based on these analyses, it has been postulated that during late development, the yolk calcium reserve is continuously being replenished with calcium derived from the eggshell, which is then subsequently transported in a regulated manner into the embryo (Tuan, 1987). This pathway would account for the age profile of yolk calcium content, which first decreases and then increases during late development (Fig. 27.3).

The idea that calcium transport by the yolk sac is regulated by vitamin D has received further support from our studies (Ono & Tuan, 1988, 1990a,b, 1991). In a number of systems, in particular the intestinal mucosa, vitamin D regulation of calcium transport is mediated by the induced expression of a calcium-binding protein, calbindin (Wasserman & Fullmer, 1983). Calbindin is characterised by its strong

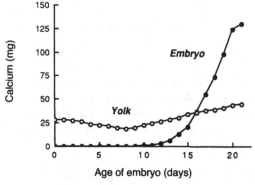

Fig. 27.3. Developmental profile of embryonic calcium accumulation and yolk sac calcium content. (Data from Romanoff, 1967.)

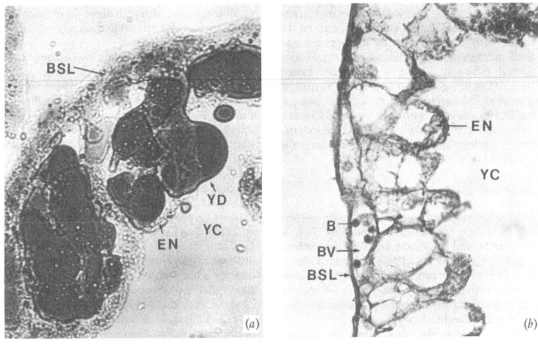

Fig. 27.4. Histology of day-14 fowl embryonic yolk sac. (*a*) Oil red O staining. (*b*) Haematoxylin-eosin staining. Note the highly vacuolated endodermal epithelium, and the lipid-rich nature of these vacuoles (stained with Oil red O) containing yolk droplets. B, blood cell; BV, blood vessel; BSL, basal serosal layer; EN, endodermal cell; YC, yolk cavity; YD, yolk droplets (vacuoles). Magnification: 850 ×.

and specific affinity for calcium ions, its vitamin D dependent expression, and shares structural homologies with calmodulin, with Ca^{2+}-binding, E–F hand domains (Cheung, 1982; Strynadka & James, 1989). The cytosolic location of calbindin suggests that it may act as an intracellular buffer of Ca^{2+}, although its exact mechanism of action in transcellular calcium transport remains unknown (Carafoli, 1987). We have examined the embryonic yolk sac with respect to the presence, regulation of expression, and cellular location of calbindin, using specific antibodies against calbindin kindly supplied by Dr R. H. Wasserman (Cornell University) (Wasserman, Taylor & Fullmer, 1974). Our findings (Ono & Tuan, 1988, 1990*a,b*, 1991) are summarised: 1) Calbindin is expressed in the embryonic yolk sac, detectable in both day 9 and day 14 samples; 2) The calbindin of the embryonic yolk sac (calbindin-YS) resembles that of the adult duodenum in both molecular weight (28,000) and isoelectric point, as well as the presence of E–F hand Ca^{2+}-binding structural domains; 3) Systemic calcium deficiency caused by shell-less culture of fowl embryos

results in enhanced expression of calbindin-YS in the yolk sac during late development; 4) Calbindin-YS expression is inducible by $1,25(OH)_2$ vitamin D_3 treatment of the yolk sac *in vivo* and *in vitro*; and 5) Immunohistochemistry reveals that calbindin-YS is localised exclusively to the non-vacuolar regions of the cytoplasm of the yolk sac endoderm (Fig. 27.5).

Thus the embryonic yolk sac appears to be a genuine target tissue under vitamin D-mediated endocrine control, with the expression of a vitamin D-dependent gene product, calbindin. It is not known, however, how calbindin-YS functions in transport of calcium by the yolk sac. Since the bulk of yolk sac transport is likely to be via endocytosis, digestive dissolution of the yolk material within the endocytic vesicles (or vacuoles) would result in a highly elevated concentration of calcium due to the high content of calcium in the yolk (estimated to be ~35–50 mM). It is conceivable that calbindin-YS may be involved in: 1) safeguarding the cytosol from calcium poisoning due to leakage from the endocytic vesicles, or 2) actually facilitating the transcellular mobilisation of calcium from the

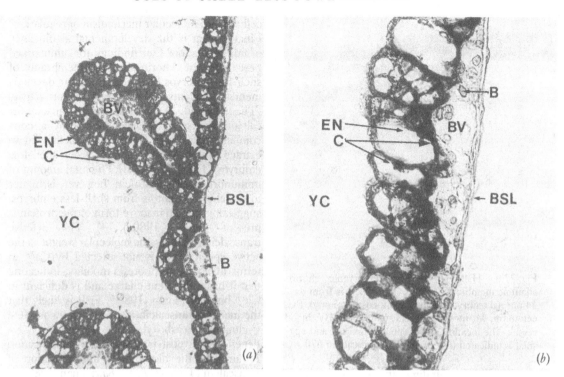

Fig. 27.5. Immunohistochemical localisation of calbindin in day-9 fowl embryonic yolk sac. (*a*) Low magnification showing the general nature of positive calbindin staining along the yolk sac endoderm. (*b*) Higher magnification showing the cytosolic pattern of calbindin immunostaining. Abbreviations used are same as those in Fig. 27.4; C, calbindin localisation. Magnification: (*a*) 175 ×; (*b*) 720 ×.

vesicles to the basoserosal side of the endoderm by serving as a Ca^{2+}-binding 'bucket-brigade' intermediate, possibly also involving the action of a basolateral ATP-driven Ca^{2+} pump. Although additional studies are clearly needed to elucidate the mechanism of yolk sac calcium transport, the information gathered using the shell-less embryo, with its complete dependence on yolk calcium, has provided the direction for future investigations. Finally, it should be pointed out that proper vitamin D metabolism indeed is vital for the development and hatchability of the fowl embryo (Henry & Norman, 1978; Sunde, Turk & DeLuca, 1978; Norman, Leathers & Bishop, 1983; Hart & DeLuca, 1984; Narbaitz, Tsang & Grunder, 1987).

Mobilisation of eggshell calcium

The eggshell represents the primary source for calcium needed by the developing avian embryo and is mobilised by the chorio-allantoic membrane (Tuan, 1987). The calcium-transporting epithelium of the chorio-allantois is the ectoderm (Coleman & Terepka, 1972*b*; Terepka *et al.*, 1976), a columnar epithelium consisting of several cell types, among which are the capillary-covering cells with thin cytoplasmic processes that separate the capillary network from the calcium-rich shell membrane and eggshell (Fig. 27.6). Work from our laboratory has yielded a great deal of information concerning the mechanism for calcium transport by the chorio-allantois (Tuan & Scott, 1977; Tuan, Scott & Cohn, 1978*a,b,c*; Tuan & Zrike, 1978; Tuan, 1979; Tuan, 1980*a,b,c*, 1983, 1984; Tuan & Knowles, 1984; Tuan *et al.*, 1986*a,b,c*; Tuan & Fitzpatrick, 1986; Akins, Love & Tuan, 1988; Akins & Tuan, 1989, 1990). In short, the current working model (Fig. 27.7) entails an adsorptive pinocytosis process, with the involvement of an extrinsic membrane Ca^{2+}-binding protein (CaBP or transcalcin) acting as a Ca^{2+} receptor to sequester extracellular calcium. This is released from the calcite of the eggshell by the action of carbonic anhydrase mediated acidifica-

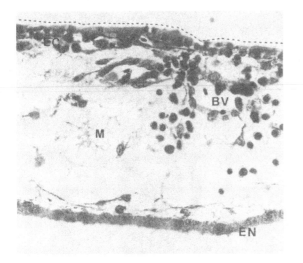

Fig. 27.6. Histology of fowl embryonic chorio-allantoic membrane (CAM). The CAM is from day-14 normal embryo (haematoxylin–eosin staining). EC, ectoderm; M, mesoderm; EN, endoderm; BV, blood vessel. The location of the shell membrane and egg-shell is indicated (*broken line*). Magnification: 670 ×.

tion. Following binding, the Ca^{2+}-transcalcin complex is adsorbed by pinocytosis. The membrane of the endosome contains a plasma membrane derived Ca^{2+}-activated ATPase which is a near neighbour ($\leqslant$ 1.2 nm) of transcalcin (Fig. 27.7). The Ca^{2+}–ATPase acts as a Ca^{2+} pump which is directed lumenally, thereby ensuring a high calcium content within the endosomes. Subsequent internal acidification of the vesicles leads to dissociation of Ca^{2+} from transcalcin, and the calcium-loaded vesicles eventually fuse with the basolateral membrane to release their content to the serosal compartment. The expression of calcium transport activity is development-specific; cellular differentiation and expression of the transport components, transcalcin, Ca^{2+}–ATPase, and carbonic anhydrase, all occur around incubation day 12. This working model is supported by a number of experimental findings (Tuan, 1987).

The calcium-deficient shell-less embryo again provides an interesting model to analyse the regulation of the calcium transport function of the chorio-allantoic membrane. An initial question is directed towards an understanding of the relationship between calcium transport by the chorio-allantoic membrane and the transport substrate, the eggshell. Is the activity of calcium transport dependent on and perhaps regulated by the eggshell? If so, what is the

cellular and molecular mechanism of regulation? Finally, what is the developmental significance of this regulation? Our findings are summarised here: 1) The chorio-allantoic membrane of shell-less embryos does not exhibit the development-specific onset of calcium-transport activity (Tuan, 1980*a*). 2) The diminished capacity for calcium transport is accompanied by a concomitant decrease in Ca^{2+}-binding activity in an extract of the chorio-allantois from shell-less embryos (Tuan, 1980*a*). 3) The total amount of immunoreactive transcalcin, however, is higher in the chorio-allantois from shell-less embryos, suggesting that an inactive form of the protein is present (Tuan, 1980*a*). 4) The inactive transcalcin has the same molecular weight as the active species, but is not affected by Ca^{2+} in terms of its electrophoretic mobility, indicating that it has a different charge and is deficient in Ca^{2+} binding (Tuan, 1980*c*). 5) It is likely that the inactive transcalcin is a result of incomplete γ-glutamyl carboxylation, a vitamin K-dependent post-translational modification required in the biosynthesis and activation of transcalcin (Tuan *et al.*, 1978*a,c*; Tuan, 1980*b,c*). 6) The higher level of total transcalcin protein in the chorio-allantois from shell-less embryos is reflected by a concomitant increase in the level of translatable transcalcin mRNA, suggesting that increased transcalcin gene expression is taking place in the chorio-allantois of the shell-less embryo (Tuan, Fitzpatrick & Mulligan, unpublished observations). 7) These changes with respect to alterations in the level of Ca^{2+} binding activity and transcalcin are partially remedied when eggshell and/or shell membrane are replaced onto the chorio-allantoic membrane of cultured embryos. The regions of the membrane directly underlying and adhering to the exogenous shell/shell membrane exhibit close-to-normal phenotype (Dunn, Graves & Fitzharris, 1981*b*; Tuan, 1983).

These findings strongly indicate that the calcium transport function of the chorio-allantois is closely regulated by the presence of the eggshell, the transport substrate. The regulation is exerted ultimately at a cellular and molecular level, i.e. the gene expression of specific transport components and the accompanied specific cytodifferentiation into a transport epithelium. In this manner, once the chorio-allantoic membrane has grown to a size large enough to surround the embryo completely, and therefore attach to the shell membrane, the ectoderm is

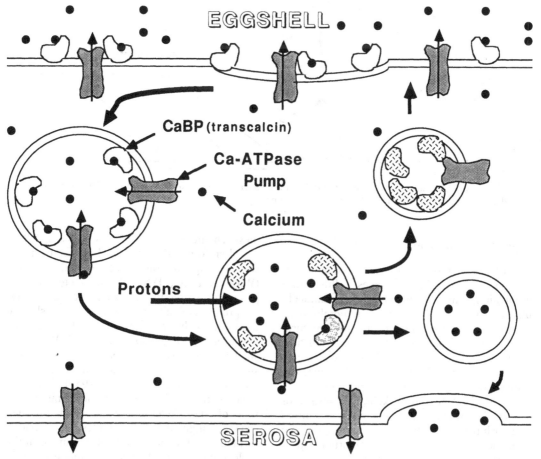

Fig. 27.7. Adsorptive endocytosis mechanism of calcium transport by the chorio-allantoic membrane of the fowl embryo. (For details, see text and Tuan, 1987.)

signalled to undergo differentiation into a transporting epithelium with concomitant gene expression and biosynthesis of functional components. On the other hand, the shell-less embryo lacks its eggshell and is incapable of carrying out similar transport functions. The influence of the eggshell/shell membrane on the function of the chorio-allantoic membrane appears to act in a short-range manner, since in the shell-less embryo only those regions of the membrane which adhere directly to the exogenously added shell membrane may be restored to normal (Dunn *et al.*, 1981*b*; Tuan, 1983). The nature of this influence remains to be elucidated.

For further analysis of the cellular basis of ion transport by the chorio-allantoic membrane, it is obviously advantageous to obtain a pure population of ectodermal cells which may be cultured and manipulated *in vitro*. Again the shell-less embryo may be used for this purpose (Akins & Tuan, 1989, 1990). A ring is placed over the exposed chorio-allantois of the shell-less embryo, a physiological buffer containing dissociative enzymes is added and limited digestion allows cells to be dissociated from the top ectodermal layer of the chorio-allantois (Fig. 27.8). An enzyme-free, EDTA-based dissociation solution (Specialty Media, Inc.) has also been used successfully to isolate ectodermal cells from the chorio-allantois of day 10 embryos: the limited dissociation procedure selectively removes the ectodermal layer (Fig. 27.9). These ectodermal cells survive and grow in culture for periods of up to several weeks, and are currently being used for studies on the cellular mechanism of calcium transport (Akins & Tuan, 1990).

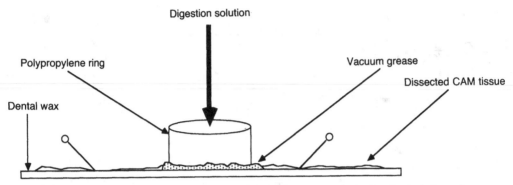

Fig. 27.8. Diagrammatic representation of the dissociative procedure for the isolation of ecto-dermal cells from the chorio-allantois of fowl embryos.

Skeletal development

The development of the embryonic skeleton is a complex process which involves cellular differentiation, synthesis and assembly of extracellular matrix, and its subsequent mineralisation, to give rise to a unique form and structure. Studies in our laboratory have focused on how this process may be regulated, particularly how the synthesis of the organic matrix may be influenced by the inorganic components associated with it. Since the skeleton consists of mineralising tissues, understanding the interrelationship between the organic and inorganic components of the matrix is an important aspect of its biology.

The shell-less embryo again provides a unique system for such studies. By virtue of its singular calcium deficiency, occurring early in the time course of skeletal development, the shell-less embryo represents a system in which mineralisation is severely perturbed. This allows the investigator to examine how organic matrix synthesis, which is the consequence of cell differentiation, may be affected. Our investigations started by analysing the expression of collagen types in the embryonic skeleton. The rationale for this approach is that the transition of collagen types is a hallmark of cellular differentiation during skeletal development (Linsenmayer, Toole & Trelstad, 1973; von der Mark, von der Mark & Gay, 1976*a,b*). For example, endochondral ossification in the long bones of the limb involves a mesenchyme/cartilage/bone sequence, accompanied by the transitional expression of type I/type II/type I collagens. On the other hand, intramembranous ossification of the parietal head bones proceeds directly from

mesenchyme to bone (Bernard & Pease, 1969), and therefore maintains type I collagen expression throughout the morphogenetic process (Wiestner *et al.*, 1981). By characterising the pattern of expression of specific collagen types, the progress of cellular differentiation in skeletal tissues may be precisely delineated and any perturbations may also be detected. When the shell-less embryo is analysed in this manner, significant anomalies in its skeletal development are observed (Tuan & Lynch, 1983). The most obvious anomaly is the severe under-mineralisation of the skeleton, resulting in retarded growth, angulated long bones, and poorly formed vertebrae and digits of the limbs. In the endochondral long bones, a temporal delay in the collagen type II to type I transition is seen, most likely a result of deficient bone formation due to systemic calcium deficiency, as in the case of adult rickets (Barnes *et al.*, 1973). A most intriguing finding is that type II collagen is expressed in the intramembranous parietal bone, the calvarium. Immunohistochemistry revealed that the cells synthesising type II collagen are located medially in a central zone of the bone (Jacenko & Tuan, 1986*a*), and are unrelated to periosteal cells which have been shown to possess chondrogenic potential (Thorogood, 1979). Preliminary analysis (Jacenko & Tuan, 1986*b*) indicates that the calvaria of shell-less embryos also express type X collagen, a short-chained collagen which is generally believed to be highly specific for chondrocytes undergoing hypertrophy and mineralisation (Burgeson, 1988). Our most recent study utilising *in situ* cDNA–mRNA hybridisation confirms that the centrally localised cells of the calvarium are actively

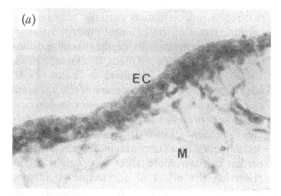

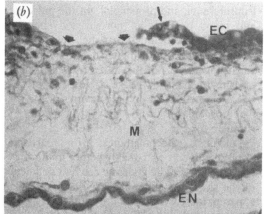

Fig. 27.9. Histology of the chorio-allantois of day 10 fowl embryos before (a) and after (b) dissociation of the ectoderm as described in Fig. 27.8. EC, ectoderm; M, mesoderm; EN, endoderm. Note the selective removal of the ectoderm (large arrows) after dissociation. A segment of ectoderm which is in the process of being lifted off is indicated (small arrow). Magnification: 540 ×.

undergoing gene expression for type II collagen and are positively identified for type II collagen mRNA (McDonald & Tuan, 1989). Furthermore, it is found that cells positive for type II collagen mRNA are actually found in the calvaria of both normal and shell-less embryos, the only difference being a quantitative one. Consequently, it may be surmised that the expression of type II collagen in the calvarium may be regulated at several levels, such that transcription may be constitutive but biosynthesis is 'de-repressed' only under calcium deficiency. It is worthy of further speculation that intramembranous ossification may in fact resemble endochondral ossification, the difference being that the former has a highly abbreviated or transient cartilaginous phase.

(Note: the cartilaginous phase is being equated here with type II collagen gene expression.) The tissue calcium deficiency of the shell-less embryo (Dunn & Boone 1977; Tuan, 1983) has thus somehow created a permissive environment for the overt expression of the chondrogenic phenotype. Further substantiation of the cartilage phenotype in the calvaria of the shell-less embryo is provided by the finding that high-molecular weight, highly sulphated, and chondroitin sulphate rich proteoglycans are also synthesised by these bones. Immunohistochemistry has shown that these are specifically associated with the same cellular population which express type II collagen (Jacenko & Tuan, 1986b; Jacenko, 1988; Tuan, Jacenko & Mattessich, unpublished observations). The properties of these proteoglycan species are characteristic of those found in cartilage (Ruoslahti, 1988). Finally, when the shell-less embryo is supplemented with calcium, either in the form of eggshell pieces or a slurry of $CaCO_3$ (Tuan, 1983), the appearance of the aberrant cartilage phenotype is not observed (Jacenko & Tuan, 1986b; Jacenko, 1988).

How does localised and systemic calcium deficiency give rise to the emergence of an apparently aberrant cartilage phenotype in a tissue which is normally completely osteogenic? Calcium deficiency most likely results in a relatively looser or less structured extracellular matrix (Jacenko, 1988; Tuan, 1989), and thus permits the cell–cell aggregation and interactions thought to be vital for cellular condensation, a prerequisite for chondrogenic differentiation of mesenchymal cells (Ede, 1983). The role of calcium (or the influence of calcium deficiency) in regulating chondrogenesis in the intramembranous bones is therefore not a direct one, but instead depends on its contribution to the formation of the appropriate extracellular matrix, which in turn regulates chondro-/osteogenic differentiation (Jacenko, 1988; Tuan, 1989).

These studies on skeletal development again illustrate the usefulness and uniqueness of the shell-less embryo as an experimental model in which calcium deficiency is made to occur during a time of vital tissue morphogenesis and organogenesis. Concerning the formation of the skeleton, which is quantitatively the most calcium-dependent activity of embryonic development, the shell-less embryo system has thus permitted us to dissect precisely the cytodifferentiation events underlying chondrogenesis and osteogenesis and, in particular, to reveal the pluripotency of skeletal mesenchymal cells. Interestingly, several recent reports using cells isolated from embryonic calvaria have, indeed, confirmed the multiple differentiation potential of the calvarial cells (Aubin *et al.*, 1982; Rifas *et al.*, 1982; Nefussi *et al.*, 1985; Villanueva, Nishimoto & Nimni, 1989).

Cardio-vascular functions

Compared to adult animals, little is known about the regulation of cardio-vascular activities in the developing embryo. Another series of experiments on the shell-less embryo has addressed a separate fundamental question: what is the functional relationship between calcium homeostasis and cardio-vascular functions? One of the motivations for our studies is the suggestion that, in human and animal populations, calcium deficiency may be functionally linked to essential hypertension (McCarron, 1985). These studies are based, however, mostly on either nutritional or epidemiological surveys (McCarron, 1983; Harlan *et al.*, 1984), or animals already predisposed to hypertension (Ayachi, 1979; McCarron *et al.*, 1985). Again, our rationale is that the shell-less embryo, because of its unique calcium deficiency, should serve as a convenient and valid experimental model to analyse this relationship.

Systemic calcium deficiency appears to be related to the pathogenesis of hypertension and tachycardia (Tuan & Nguyen, 1987). The shell-less fowl embryo is clearly tachycardiac and hypertensive, with increases in both systolic and diastolic pressures (Fig. 27.10). Interestingly,

calcium supplementation partially restores the calcium status of the cultured embryo and significantly ameliorates its cardio-vascular functions. Our studies (Koide & Tuan, 1988, 1989*a,b,c,d*, Koide, Miyahara & Tuan, 1990) have focused on the mechanism of the apparent calcium deficiency mediated hypertension and are summarised as follows: 1) The systemic calcium deficiency in the shell-less embryo is not accompanied by a concomitant decrease in myocardial calcium content per wet weight. 2) By observing the effect of adrenergic regulators, including noradrenalin, phentolamine, isoproterenol, and propanolol, on the pulse rate and blood pressure of normal and shell-less embryos, it is concluded that shell-less embryos develop a significantly higher α-adrenergic sensitivity, whereas the β-adrenergic sensitivity remains unchanged. These responses represent the integrated effect on the activities of the cardiac pacemaker, and the myocardial and vascular muscle tissues. 3) The level of total circulating catecholamines, in particular noradrenaline, is remarkably higher in the shell-less embryo. 4) The systemic calcium deficiency of the shell-less embryo is accompanied by an apparent increase in serum sodium (Fig. 27.11), but the circulating blood volume is unchanged. 5) Using erythrocytes, which possess catecholamine-sensitive membrane enzyme systems (Kregenow, 1977), as a convenient cell-type for membrane analysis, it is found that cellular calcium handling in the shell-less embryo is characterised by greater ion permeability. There is also higher calcium pumping activity by the Ca^{2+}–ATPase and sodium–calcium exchange. This altered membrane

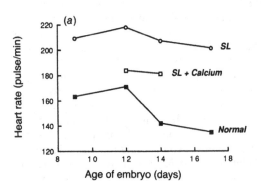
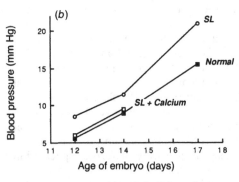

Fig. 27.10. Developmental profile of cardio-vascular activities in normal and shell-less fowl embryos, and shell-less embryos supplemented with calcium (see text). (*a*) Pulse rate; (*b*) mean blood pressure. Data from Tuan & Nguyen (1987).

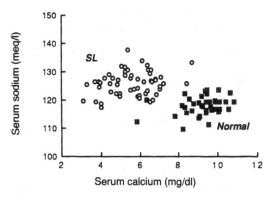

Fig. 27.11. Relationship between serum sodium and calcium levels in normal and shell-less fowl embryos. (Data from Koide & Tuan, 1989a.)

property may account for the hypernatraemia but normal cellular calcium in the shell-less embryo by the involvement of sodium–calcium exchange.

In the shell-less fowl embryo significant alterations have occurred in at least two aspects important for cardio-vascular functions as a result of the chronic, systemic calcium deficiency. 1) The adrenergic system is hypersensitised, particularly via the α-pathway. Coupled with its highly elevated level of catecholamines, the shell-less embryo is likely to be maximally stimulated with respect to both blood pressure and pulse rate. 2) Changes in cellular calcium handling also take place in the shell-less embryo, particularly those involving sodium–calcium exchange and the ATP-driven calcium pump. These changes may serve either to compensate for, or actually to augment, the effects of the systemic calcium deficiency in the shell-less embryo. Therefore, the whole body responses to calcium deficiency represent an integrated, complex combination of these changes. Current studies aim to examine whether these changes observed in the erythrocytes also occur in the smooth muscle and myocardial cells, and how they may be related to the altered cardio-vascular activities of the shell-less embryo.

The shell-less fowl embryo thus provides an excellent, well-defined experimental model for the evaluation of the relationship between calcium homeostasis and cardio-vascular functions. Experimentally, the shell-less embryo is highly amenable to various manipulations, including the measurement of blood pressure, pulse rate, as well as the administration of various pharmacochemical agents. For example, blood pressure

is directly determined by cannulating a derivative of the vitelline or chorio-allantoic artery with a KCl-filled glass microelectrode, and then measuring the hydrostatic pressure needed to maintain the salt concentration gradient at the tip of the microelectrode in dynamic equilibrium (Tuan & Nguyen, 1987). In addition, application of pharmacological agents, e.g. the adrenergic agents, directly onto the chorio-allantoic membrane results in rapid, stable responses (Koide & Tuan, 1989a), suggesting that pinocytic absorption by the ectodermal cells of the chorio-allantois (Dunn & Fitzharris, 1987) is an effective means of drug delivery. In fact, a 'case history' may be worked up for each embryo with respect to its serum parameters, cardio-vascular activities, and pharmacological treatments. Interestingly, the hypertension observed in the shell-less embryo is not as severe as that in spontaneous hypertensive rats or other experimental animals, but is, instead, similar in relative magnitude to human essential hypertension which are without complications of advanced arteriosclerosis or nephropathy. Moreover, unlike some of the other experimental animal models, the hypertension in shell-less embryos does not include a genetic component. Thus, the fowl embryo *in vitro* may represent an unique and useful experimental model system for the study of the relationship between calcium deficiency (perhaps nutritionally based) and hypertension, and may also serve as a potential bioassay system to examine the potency of various pharmacological agents on calcium deficiency related hypertension.

Concluding remarks

This review has highlighted some of the applications of shell-less culture of the fowl embryo for studying a number of physiological and developmental functions. We have primarily taken advantage of the unique calcium deficiency of the animal to examine how calcium homeostasis may be functionally linked to other vital processes of embryonic development, including calcium transport, skeletal development, and cardio-vascular functions. In addition, one should also take into consideration that the fowl embryo is one of the most easy experimental animals to manipulate, and that shell-less culture renders it even more accessible because of the exposure of the entire egg content. Consequently, experimentations that involve surgical manipulations, chorio-allantoic grafts, gross morphology, etc, also find the shell-less embryo an ideal system. Finally, the shell-less culture procedure has recently been successfully adapted to the production of hatchlings (Ono & Wakasugi, 1984; Rowlett & Simkiss, 1987; Perry, 1988), thereby permitting the observation of the long-term effect of experimental biochemical or transgenic manipulations, (Freeman & Messer, 1985; Sang & Perry, 1989) in the adult animal.

Acknowledgements

The authors acknowledge grant support (R.S.T.) from the National Institutes of Health (HD 15822, HD 21355), United States Department of Agriculture (88–37200–3746), and the March of Dimes Birth Defects Foundation (1–1146). Tamao Ono is a Fogarty International Research Fellow of the National Institutes of Health (TWO3964), Robert E. Akins was supported by National Institutes of Health Training Grant (HD 07152), and Masafumi Koide is an International Rotary Scholar. The expert help of Ms Jennie Platt in the preparation of the manuscript is gratefully acknowledged.

References

Akins, R., Love, J. & Tuan, R. (1988). Cellular calcium uptake by chick embryo chorioallantoic membrane. *Journal of Cell Biology*, **107**, 784a.

Akins, R. & Tuan, R. (1989). Molecular components of transepithelial calcium transport in the chick chorioallantoic membrane. *Journal of Cell Biology*, **109**, 302a.

Akins, R. & Tuan, R. (1990). Endosomal compartmentalization during calcium transport by the chick chorioallantoic membrane. *Journal of Cell Biology*, **111**, 191a.

Aubin, J., Heersche, J., Merriless, M. & Sodek, J. (1982). Isolation of bone cell clones with differences in growth, hormone responses and extracellular matrix production. *Journal of Cell Biology*, **92**, 452–61.

Ayachi, S. (1979). Increased dietary calcium lowers blood pressure in the spontaneously hypertensive rat. *Metabolism*, **28**, 1234.

Barnes, M., Constable, B., Morton, L. & Kodicek, E. (1973). Bone collagen metabolism and vitamin D deficiency. *Biochemical Journal*, **132**, 113–15.

Bernard, G. & Pease, D. (1969). An electron microscopic study of initial intramembranous osteogenesis. *American Journal of Anatomy*, **125**, 271–90.

Burgeson, R. (1988). New collagens, new concepts. *Annual Reviews of Cell Biology*, **4**, 551–77.

Carafoli, E. (1987). Intracellular calcium homeostasis. *Annual Review of Biochemistry*, **56**, 395–433.

Cheung, W. Y. (1982). Calmodulin: an overview. *Federal Proceedings*, **41**, 2253–7.

Clark, N., Murphy, M. & Lee, S. (1989). Ontogeny of vitamin D action on the morphology and calcium transport properties of the chick embryonic yolk sac. *Journal of Developmental Physiology*, **11**, 243–51.

Coleman, J. & Terepka, A. (1972*a*). Fine structural changes associated with the onset of calcium, sodium, and water transport by the chicken chorioallantoic membrane. *Journal of Membrane Biology*, **7**, 111–27.

— (1972*b*). Electron probe analysis of the calcium distribution of cells of the embryonic chick chorioallantoic membrane. II. Demonstration of intracellular location during active transcellular transport. *Journal of Histochemistry and Cytochemistry*, **20**, 414–24.

Crooks, J. & Simkiss, K. (1975). Calcium transport by the chick chorioallantois *in vivo*. *Quarterly Journal of Experimental Physiology and Cognitive Medical Science*, **60**, 55–63.

Dunn, B. (1974). Technique for shell-less culture of the 72-hour avian embryo. *Poultry Science*, **53**, 409–12.

Dunn, B. & Boone, M. (1976). Growth of the chick embryo *in vitro*. *Poultry Science*, **56**, 1067–71.

Dunn, B. & Boone, M. (1977). Growth and mineral content of cultured chick embryos. *Poultry Science*, **56**, 662–72.

Dunn, B. & Fitzharris, T. (1987). Endocytosis in the embryonic chick chorionic epithelium. *Journal of Experimental Zoology*, Supplement 1, 75–9.

Dunn, B. E., Fitzharris, T. P. & Barnett, B. D. (1981*a*). Effects of varying chamber construction and embryo pre-incubation age on survival and growth of chick embryos in shell-less culture. *Anatomical Record*, **199**, 33–43.

Dunn, B., Graves, J. & Fitzharris, T. (1981*b*). Active calcium transport in the chick chorioallantoic membrane requires interaction with the shell membrane and/or shell calcium. *Developmental Biology*, **88**, 259–68.

Ede, D. (1983). Cellular condensations and chondrogenesis. In *Cartilage: Development, Differentiation and Growth*. New York: Academic Press.

Freeman, B. & Messer, L. (1985). Genetic manipulation of the domestic fowl – A review. *World Poultry Science Journal*, **41**, 124–32.

Gawande, S. & Tuan, R. (1988). Effect of 1,25(OH)$_2$ vitamin D$_3$ on calcium metabolism and skeletal development in calcium deficient chick embryos. *Journal of Cell Biology*, **107**, 691a.

Grizzuti, K. & Perlmann, G. (1974). Further studies

on the binding of divalent cations to the phosphoglycoprotein phosvitin. *Biochemistry*, **14**, 2171–5.

Harlan, W., Hull, A., Schmouder, R., Landis, J., Thompson, F. & Larkin, F. (1984). Blood pressure and nutrition in adults. *American Journal of Epidemiology*, **120**, 17.

Hart, L. & DeLuca, H. (1984). Hydroxylation of carbon-24 of 25-OH-cholecalciferol is not necessry for normal hatchability in chickens. *Journal of Nutrition*, **114**, 2059–65.

Henry, H. & Norman, A. (1978). Vitamin D: Two dihydroxylated metabolites are required for normal chicken egg hatchability. *Science*, **201**, 835–7.

Jacenko, O. (1988). Calcium and Cell Differentiation During Chick Embryonic Skeletogenesis. PhD Thesis, University of Pennsylvania, Philadelphia.

Jacenko, O. & Tuan, R. (1986a). Calcium deficiency induces expression of cartilage-like phenotype in chick embryonic calvaria. *Developmental Biology*, **115**, 215–32.

— (1986b). Changes in the extracellular matrix of chick embryonic bone during induced calcium deficiency. *Progress in Clinical Related Research*, **217B**, 401–4.

Johnston, P. & Comar, C. (1955). Distribution of calcium from the albumen, yolk and shell to the developing chick embryo. *American Journal of Physiology*, **183**, 365–70.

Juurlink, B. & Gibson, M. (1973). Histogenesis of the yolk sac in the chick. *Canadian Journal of Zoology*, **51**, 509–19.

Koide, M. & Tuan, R. (1988). Characterization of hypertension in calcium deficient chick embryo. *Circulation*, Supplement II, **78**, 443.

— (1989a). Characterization of chick embryonic cardiovascular function: Calcium deficiency, hypertension, and adrenergic regulation. *American Journal of Physiology*, **257**, H1900–9.

— (1989b). Characterization of Ca-deficient hypertension in chick embryos: Adrenergic regulation of cardiovascular function and cellular Ca handling. *Annals New York Academy of Sciences*, (in press).

— (1989c). Effect of adrenergic modifiers on Ca deficiency-related hypertension in chick embryos. *3rd Cardiovascular Pharmacotherapy, International Symposium*, (in press).

— (1989d). Ca handling by erythrocytes (RBC) of Ca-deficient chick embryo. *Journal of Cell Biology*, **109**, 305a.

Koide, M., Miyahara, T. & Tuan, R. (1990). Adrenergic regulation of Ca handling in erthrocytes (RBC) of hypertensive, Ca-deficient chick embryos *Circulation*, **82(III)**, 152.

Kregenow, F. (1977). Avian erythrocytes. In *Membrane Transport in Red Cells*, eds J. Ellory and V. Lew, pp. 383–426. New York: Academic Press.

Lambson, R. (1970). An electron microscopic study of the endodermal cells of the yolk sac of the chick during incubation and after hatching. *American Journal of Anatomy*, **129**, 1–20.

Leeson, T. & Leeson, C. (1963). The chorio-allantois of the chick. Light and electron microscopic observations at various times of incubation. *Journal of Anatomy*, **97**, 585–95.

Linsenmayer, T., Toole, B. & Trelstad, R. (1973). Temporal and spatial transition in collagen types during embryonic chick limb development. *Developmental Biology*, **35**, 232–9.

McCarron, D. (1983). Calcium and magnesium nutrition in human hypertension. *Annals of International Medicine*, **98**, 800–5.

— (1985). Is calcium more important than sodium in the pathogenesis of essential hypertension? *Hypertension (Dallas)*, **7**, 607–27.

McCarron, D., Lucas, P., Schneidman, R. & Drueke, T. (1985). Blood pressure development of the spontaneously hypertensive rat after concurrent manipulations of dietary Ca^{2+} and Na^+. *Journal of Clinical Investigation*, **76**, 1147–54.

McDonald, S. & Tuan, R. (1989). Expression of collagen type transcripts in chick embryonic bone detected by *in situ* cDNA–mRNA hybridization. *Developmental Biology*, **133**, 221–34.

Narbaitz, R. (1972). Cytological and cytochemical study of the chick choironic epithelium. *Review of Canadian Biology*, **31**, 259–67.

— (1977). Structure of the intra-chorionic blood sinus in the chick embryo. *Journal of Anatomy*, **124**, 347–54.

— (1979). Response of shell-less cultured chick embryos to exogenous parathyroid hormone and 1,25-dihydroxycholecalciferol. *General and Comparative Endocrinology*, **37**, 440–2.

Narbaitz, R. & Fragisco, B. (1984). Hypervitaminosis D in the chick embryo: Comparative study on the activity of various vitamin D_3 metabolites. *Calcified Tissue International*, **36**, 392–400.

Narbaitz, R. & Jande, S. (1983). Differentiation of bones and skeletal muscles in chick embryos culured on albumen. *Review of Canadian Biological Experimentation*, **42**, 271–7.

Narbaitz, R. & Tolnai, S. (1978). Effects produced by the administration of high doses of 1,25-dihydroxycholecalciferol to the chick embryo. *Calcified Tissue Research*, **26**, 221–6.

Narbaitz, R., Tsang, C. & Grunder, A. (1987). Effects of vitamin D deficiency in the chick embryo. *Calcified Tissue International*, **40**, 109–13.

Nefussi, J., Boy-Lefevre, M., Boulekbache, H. & Forest, N. (1985). Mineralization *in vitro* of matrix formed by osteoblasts isolated by collagenase digestion. *Differentiation*, **29**, 160–8.

Norman, A., Leathers, V. & Bishop, J. (1983). Normal egg hatchability requires the simultaneous administration to the hen of 1,25-dihydroxycholecalciferol and 24R,25-dihydroxychole-calciferol. *Journal of Nutrition*, 113, 2505–15.

Ono, T. & Tuan, R. (1986). Effect of experimentally induced calcium deficiency on development, metabolism, and liver morphogenesis of the chick embryo. *Journal of Embryology and Experimental Morphology*, 92, 207–22.

— (1988). Presence of vitamin D dependent calcium binding protein in chick embryonic yolk sac. *Journal of Cell Biology*, 107, 691a.

— (1989). Vitamin D regulation of yolk calcium mobilization during chick embryonic development: Expression of vitamin D-dependent Ca^{2+}-binding protein (Calbindin) in the yolk sac. In *Avian Incubation*, ed. S. G. Tullett, p. 325. London: Butterworths.

— (1990a). Double staining of immunoblot using enzyme histochemistry and India ink. *Analytical Biochemistry*, 187, 324–327.

— (1990b). Vitamin D and chick embryonic yolk calcium mobilization: identification and regulation of vitamin D-dependent Ca^{++}-binding protein, calbindin-D_{28K}, in the yolk sac. **Developmental Biology** (in press).

Ono, T. & Wakasugi, N. (1984). Mineral content of quail embryos cultured in mineral-rich and mineral-free conditions. *Poultry Science*, 63, 159–66.

Packard, M. J. & Packard, G. C. (1984). Comparative aspects of calcium metabolism in embryonic reptiles and birds. In *Respiration and Metabolism of Embryonic Vertebrates*, ed. R. S. Seymour, pp. 155–79. Dordrecht: Dr W. Junk.

Perry, M. (1988). A complete culture system for the chick embryo. *Nature, London*, 331, 70–2.

Rifas, L., Uitto, J., Memoli, V., Keuttner, K., Henry, R. & Peck, W. (1982). Selective emergence of differentiated chondrocytes during serum-free culture of cells derived from fetal rat calvaria. *Journal of Cell Biology*, 92, 493–504.

Romanoff, A. L. (1960). *The Avian Embryo*. New York: Macmillan.

— (1967). *Biochemistry of the Avian Embryo. A Quantitative Analysis of Prenatal Development*. New York: John Wiley & Sons.

Rowlett, K. & Simkiss, K. (1987). Explanted embryo culture: *In vitro* and *in ovo* techniques for domestic fowl. *British Poultry Science*, 28, 91–101.

Ruoslahti, E. (1988). Structure and biology of proteoglycans. *Annual Review of Cell Biology*, 4, 229–55.

Sang, H. & Perry, M. (1989). Episomal replication of cloned DNA injected into the fertilized ovum of the hen, *Gallus domesticus*. *Molecular Reproduction and Development*, 1, 98–106.

Simkiss, K. (1961). Calcium metabolism and avian reproduction. *Biological Reviews*, 36, 321–67.

Slavkin, H., Slavkin, M. & Bringas, P., Jr (1980). Mineralization during long-term cultivation of chick embryos *in vitro*. *Proceedings of the Society for*

Experimental Biology and Medicine, 163, 249–57.

Strynadka, N. & James, M. (1989). Crystal structures of the helix–loop–helix calcium-binding proteins. *Annual Review of Biochemistry*, 58, 951–98.

Sunde, M., Turk, C. & DeLuca, H. (1978). The essentiality of vitamin D metabolites for embryonic chick development. *Science*, 200, 1067–9.

Terepka, A., Coleman, J., Armbrecht, H. & Gunther, T. (1976). Transcellular transport of calcium. *Calcium in Biological Systems. Symposia of the Society for Experimental Biology*. Cambridge: Cambridge University Press.

Terepka, A., Stewart, M. & Merkel, N. (1969). Transport functions of the chick chorioallantoic membrane. II. Active calcium transport *in vitro*. *Experimental Cell Research*, 58, 107–17.

Thorogood, P. (1979). In vitro studies on skeletogenic potential of membrane bone periosteal cells. *Journal of Embryology and Experimental Morphology*, 54, 185–207.

Tuan, R. S. (1979). Vitamin K-dependent γ-glutamyl-carboxylase activity in the chick embryonic chorioallantoic membrane. *Journal of Biological Chemistry*, 254, 1356–64.

— (1980a). Calcium transport and related functions in the chorioallantoic membrane of cultured shell-less chick embryos. *Developmental Biology*, 74, 196–204.

— (1980b). Biosynthesis of calcium-binding protein of chick embryonic chorioallantoic membrane: *In vitro* organ culture and cell-free translation. *Cell Calcium*, 1, 411–29.

— (1980c). Requirement of eggshell for expression of vitamin K-dependent calcium-binding protein in the chick embryonic chorioallantoic membrane. In *Vitamin K Metabolism and Vitamin K-Dependent Protein*, ed. J. Suttie, pp. 294–8. Baltimore: University Park Press.

— (1983). Supplemented eggshell restores calcium transport in chorioallantoic membrane of cultured shell-less chick embryos. *Journal of Embryology and Experimental Morphology*, 74, 119–31.

— (1984). Carbonic anhydrase and calcium transport function of the chick embryonic chorioallantoic membrane. *Annals of the New York Academy of Sciences*, 429, 459–72.

— (1987). Mechanism and regulation of calcium transport by the chick embryonic chorioallantoic membrane. *Journal of Experimental Zoology*, Supplement 1, 1–13.

— (1991). Ionic regulation of chondrogenesis. In *Cartilage Vol. IV. Molecular Aspects*, eds B. K. Hall and S. Newman. Telford Press, Caldwell (in press).

Tuan, R., Carson, M., Jozefiak, J., Knowles, K. & Shotwell, B. (1986). Calcium transport function of the chick embryonic chorioallantoic membrane. I. *In vivo* and *in vitro* characterization. *Journal of Cell Science*, 82, 73–84.

— (1986b). Calcium transport function of the chick embryonic chorioallantoic membrane. II. Func-

INDEX

435